AF868531

Materials Interaction with Femtosecond Lasers

Bernd Bauerhenne

Materials Interaction with Femtosecond Lasers

Theory and Ultra-Large-Scale Simulations of Thermal and Nonthermal Pheomena

Springer

Bernd Bauerhenne
Theoretical Physics II
University of Kassel
Kassel, Germany

ISBN 978-3-030-85134-7 ISBN 978-3-030-85135-4 (eBook)
https://doi.org/10.1007/978-3-030-85135-4

© The Editor(s) (if applicable) and The Author(s), under exclusive license to Springer Nature Switzerland AG 2021
This work is subject to copyright. All rights are solely and exclusively licensed by the Publisher, whether the whole or part of the material is concerned, specifically the rights of translation, reprinting, reuse of illustrations, recitation, broadcasting, reproduction on microfilms or in any other physical way, and transmission or information storage and retrieval, electronic adaptation, computer software, or by similar or dissimilar methodology now known or hereafter developed.
The use of general descriptive names, registered names, trademarks, service marks, etc. in this publication does not imply, even in the absence of a specific statement, that such names are exempt from the relevant protective laws and regulations and therefore free for general use.
The publisher, the authors and the editors are safe to assume that the advice and information in this book are believed to be true and accurate at the date of publication. Neither the publisher nor the authors or the editors give a warranty, expressed or implied, with respect to the material contained herein or for any errors or omissions that may have been made. The publisher remains neutral with regard to jurisdictional claims in published maps and institutional affiliations.

This Springer imprint is published by the registered company Springer Nature Switzerland AG
The registered company address is: Gewerbestrasse 11, 6330 Cham, Switzerland

Foreword

Femtosecond laser pulses are extremely short and extremely intense light pulses that were first generated four decades ago and have been further developed and improved since then. Nowadays, femtosecond laser pulses are used in research to, among other things, observe chemical reactions in real time, create nanostructures or manipulate material surfaces. Femtosecond laser pulses are currently being applied in different industries, in archaeology and in medicine. The major advantage of using such light pulses is the possibility to manipulate matter far beyond the thermodynamic limits. Femtosecond laser pulses create an extreme non-equilibrium state in materials, in which the electrons can reach temperatures of tens of thousands of degrees Celsius, while the ions initially remain at near room temperature. In this state, very exciting and novel non-thermal phenomena can be observed, in which collective ionic motions take place that cannot occur under normal conditions, and which are theoretically described by means of precise density functional simulations. However, and due to their complexity, such calculations are only feasible for small systems up to a maximum of around thousand atoms. The above-mentioned laser-induced non-equilibrium state has a very short lifetime, which mainly depends on the so-called electron-phonon interaction, *i.e.*, on the collisions between electrons and ions. During this process, energy is transferred from the electrons to the ions, which leads to the buildup of a common temperature of the ionic lattice and the electrons. This happens on a time scale of a few picoseconds. From then on, the motion of the ions is thermal and can be determined using large-scale classical molecular dynamics simulations based on analytical interatomic potentials. Such molecular dynamics simulations can account for hundreds of millions of atoms. Laser-induced structural effects only become experimentally noticeable on a nano- to microseconds time scale.

The understanding of the transition from non-thermal to thermal thermal ionic motion is of fundamental importance for the description of laser laser processing of materials. Until Dr. Bauerhenne's works, there was no theory that accurately describes this transition. Dr. Bauerhenne was aiming at developing a unified theory for non-thermal and thermal laser-induced structural phenomena with atomic precision able solve this major open problem of Solid State Physics. And I can confirm that he has perfectly succeeded. His work enables the description of ultrafast structural phase transitions in materials with millions of atoms without having to sacrifice the accuracy of density functional theory. Dr. Bauerhenne has not only developed the desired theory but also invented a method for constructing complex interatomic interactions that depend on the electron temperature, in order to describe laser-excited systems. On the basis of these interatomic potentials and the unified theory described before, Bernd Bauerhenne performed large-scale atomistic simulations on various materials. His results reproduce one-to-one the experimental observations based upon sophisticated pump-probe techniques. The present book describes the achievements of Bernd Bauerhenne and provides a solid basis for the understanding of mechanical properties of materials under extreme conditions. The results shown in this book and obtained by B. Bauerhenne during his PhD work has resulted in numerous publications in renowned journals. Many other novel results of his doctoral thesis, also presented here, are the subject of publications that are still in preparation.

I got to know Mr. Bauerhenne as a student in the 3rd semester when he attended my lecture "Theoretical Mechanics". He immediately caught my attention due to his interest as well as his very original solutions to the exercises. Later on he joined my group for performing his Diploma work. Dr. Bauerhenne completed his studies of physics and mathematics simultaneously and with honors. His outstanding performance during his studies also qualified him to participate in the 66th Lindau Nobel Laureate Meeting in 2016, which was dedicated to physics. With the method for generation of high-precision laser-excited potentials, he won the doctoral student prize at European Materials Research Society (EMRS) meeting in Strasbourg in 2017. Dr. Bauerhenne is both an extremely talented and a very hardworking young scientist with a great deal of perseverance. He certainly belongs to the top 10% scientists at his current career stage. His Ph.D. thesis is one of the best ones I have ever supervised or reviewed.

The present book is scientifically rigorous and, at the same time, easy to understand. It is very well written, perfectly structured and its chapters look like an extremely successful lecture notes. In addition, many basic formulas that are in the work are explained in more detail than in most textbooks. The notation is well thought out. Each variable and quantity has its own designation. There are no repetitions of letters, which is remarkable with respect to the large amount of quantities that are defined throughout the book. I am therefore very pleased that Mr. Bauerhenne is making his work available to a wide readership in the form of this extremely valuable book.

Prof. Dr. Martin E. Garcia
Theoretical solid-state and ultrafast physics
University of Kassel
Kassel, Germany

Acknowledgements

At first, I am grateful to Zachary Evenson and his editorial team of Springer Nature for their support and endeavors, which has enabled the publication of this academic work as a monograph. I also thank very much Prof. Dr. Martin E. Garcia from the University of Kassel for supporting me during writing this comprehensive book project, which was originally a dissertation with the title "Unified theory for non-thermal and thermal effects occurring in matter following a femtosecond laser-excitation" at the University of Kassel in the department of natural sciences with the date of the disputation 10.12.2020. I would also like to thank very much Prof. Dr. Felipe Valencia Hernandez from the National University of Columbia for giving me very useful comments and suggestions on this book and for the calculation of the electron-phonon coupling constant in antimony, which was essential for my simulations of laser-excited antimony.

I acknowledge the wonderful cooperation with Dr. Sascha Epp from the Max Planck Institute for Structure and Dynamics of Matter. He performed ultrafast X-ray diffraction experiments on femtosecond laser-excited thin antimony films, with which I compared my theoretical calculations, and he provided me useful comments on this book. I am also grateful to Dr. Dmitry Ivanov for giving me very useful hints and tips regarding ultra large scale MD simulations with classical interatomic potentials to simulate laser-excited matter. I would like to acknowledge Dr. Vladimir Lipp for the useful scientific discussions.

I acknowledge gratefully the financial support by the "Promotionsstipendium des Otto-Braun Fonds" and by the "Abschlussstipendium der Universtät Kassel". It was essential for my research that I was able to perform my large scale calculations on the IT Servicecenter (ITS) University of Kassel, on the Lichtenberg High Performance Computer (HHLR) TU Darmstadt, and on the computing cluster FUCHS in Frankfurt.

About This Book

This book presents a unified view of the response of materials as a result of femtosecond laser excitation, introducing a general theory that captures both ultrashort-time non-thermal and long-time thermal phenomena. It includes a novel method for performing ultra-large-scale molecular dynamics simulations extending into experimental and technological spatial dimensions with ab-initio precision. For this, it introduces a new class of interatomic potentials, constructed from ab-initio data with the help of a self-learning algorithm, and verified by direct comparison with experiments in two different materials—the semiconductor silicon and the semimetal antimony.

In addition to a detailed description of the new concepts introduced, as well as giving a timely review of ultrafast phenomena, the book provides a rigorous introduction to the field of laser–matter interaction and ab-initio description of solids, delivering a complete and self-contained examination of the topic from the very first principles. It explains, step by step from the basic physical principles, the underlying concepts in quantum mechanics, solid-state physics, thermodynamics, statistical mechanics, and electrodynamics, introducing all necessary mathematical theorems as well as their proofs. A collection of appendices provide the reader with an appropriate review of many fundamental mathematical concepts, as well as important analytical and numerical parameters used in the simulations.

Contents

About the Author

Bernd Bauerhenne conducts research in the Solid State and Ultrafast Physics Group at the Institute of Theoretical Physics of the University of Kassel. One focus of his research is the theory of ultrafast phenomena in solids and nanostructures, in particular, the description of ultrafast structural changes induced by an intense femtosecond laser. Among other things, he develops highly accurate interatomic potentials using self-learning algorithms, performs ultra large scale molecular dynamics simulations, and applies electronic temperature-dependent density functional theory. He studied mathematics and physics at the University of Kassel and at the University of Luxemburg with a focus on numerics and dynamical systems in mathematics and theoretical modeling of solids in physics. He received a diploma in mathematics and a diploma in physics and he completed his Ph.D. in theoretical physics at the University of Kassel. He obtained the Otto Braun Fund doctoral fellowship and the University of Kassel final fellowship, won the Ph.D. student award of the Symposion X at the European Materials Research Society meeting in Strasbourg 2017, and had the honor of attending the 66th Lindau Nobel Laureate Meeting, which was dedicated to physics.

Acronyms

Au	Gold
bcc	Body centered cubic
BNNT	Boron nitride nanotube
CHIVES	Code for Highly excIted Valence Electron Systems
DFT	Density functional theory
dia	Diamond-like structure
e^-	Electron
e.g.	Exempli gratia $\widehat{=}$ for example
EPC	Electron-phonon coupling
fcc	Face centered cubic
fs	Femtosecond
FWHM	Full width at half maximum
GGA	Generalized gradient approximation
i.e.	Id est $\widehat{=}$ that is
LDA	Local density approximation
MD	Molecular dynamics
MEAM	Modified Embedded Atom Method
Mo	Molybdenum
MPI	Message passing interface
OpenMP	Open multi-processing
PES	Potential energy surface
ps	Picosecond
RPA	Random-phase approximation
sc	Simple cubic
Sb	Antimony
Si	Silicon
TTM	Two-temperature model
TTM-MD	Two-temperature model combined with molecular dynamics simulations
W	Tungsten

List of Symbols

$\mathbb{1}$	Unity matrix
$\hat{\mathbb{1}}$	Unity operator
a	Lattice parameter
α_{s}	Super diffusion exponent
α_{f}	Fractional diffusion exponent
α_{abs}	Absorption coefficient
$\aleph_2$	Parameter to manipulate the two-body potential in order to correct the melting temperature of the interatomic potential
$\aleph_3$	Parameter to manipulate the three-body potential in order to correct the melting temperature of the interatomic potential
$\arg(\mathbf{r})$	Argument function
$\mathbf{a}_1, \mathbf{a}_2, \mathbf{a}_3$	Lattice vectors in $\mathbb{R}^3$ of the Bravais grid
$\mathbf{A}$	Vector potential in $\mathbb{R}^3$ used in Electrodynamics
$\hat{b}^{\dagger}_{j\mathbf{q}}$	Phonon creation operator that creates a phonon in state $j\mathbf{q}$
$\hat{b}_{j\mathbf{q}}$	Phonon annihilation operator that creates a phonon in state $j\mathbf{q}$
$\mathbf{b}_1, \mathbf{b}_2, \mathbf{b}_3$	Lattice vectors in $\mathbb{R}^3$ of the reciprocal grid
$\mathbf{B}$	Magnetic field vector in $\mathbb{R}^3$
$C_{\mathrm{i}p}$	Specific heat of the ions at constant pressure
C_{e}	Specific heat of the electrons
$C_{\mathrm{e}}^{(\mathbf{i})}$	Local specific heat of the electrons in sub cell $\mathbf{i}$
$\mathbb{c}$	Speed of light in vacuum
$\mathbb{C}$	Set of all complex numbers
$\mathscr{C}_{11}, \mathscr{C}_{12}, \mathscr{C}_{44}$	Independent elastic constants of Si in the bulk diamond-like crystal structure
c_x, c_y, c_z	Cell lengths of the simulation cell in the x, y, and z direction
$c_2^{(q)}$	Coefficients of the two-body potential
$c_3^{(q_1 q_2 q_3)}$	Coefficients of the three-body potential
$c_4^{(q_1 q_2 q_3 q_4 q_5 q_6)}$	Coefficients of the four-body potential
$c_{\rho}^{(q_1 q_2)}$	Coefficients of the embedded atom potential

$\mathbf{C}$	Orthonormal matrix in $\mathbb{R}^{3N_{\text{at}}\times 3N_{\text{at}}}$ that diagonalizes the dynamical matrix $\mathbf{D}$
$\mathbf{C}(\mathbf{q_n})$	Unitary matrix in $\mathbb{C}^{3N_{\text{b}}\times 3N_{\text{b}}}$ that diagonalizes the Fourier transformed dynamical matrix $\mathbf{D}(\mathbf{q_n})$
$C_{jk}(\mathbf{q})$	Matrix element in row j and column k of matrix $\mathbf{C}(\mathbf{q_n})$
$\hat{c}^{\dagger}_{\ell\mathbf{k}}$	Electron creation operator that creates an electron in state $\ell\mathbf{k}$
$\hat{c}_{\ell\mathbf{k}}$	Electron annihilation operator that destroys the electron in state $\ell\mathbf{k}$
d_{film}	Thickness of the film
D	Diffusion coefficient
Δx, Δy, Δz	Cell lengths of the sub cells in the x, y, and z direction
$\mathbf{d}$	Electrical dipole moment vector in $\mathbb{R}^3$
$\mathbf{d}^{(\text{I})}$	Electrical dipole moment vector in $\mathbb{R}^3$ caused by the atoms of set I
$\mathbf{d}^{(\text{I})}_{\text{e}}$	Electrical dipole moment vector in $\mathbb{R}^3$ caused by the electrons of the atoms of set I
$\mathbf{d}^{(\text{I})}_{\text{ions}}$	Electrical dipole moment vector in $\mathbb{R}^3$ caused by the ions of set I
$\mathbf{d}_{\upsilon}$	Relative position vector in $\mathbb{R}^3$ of atoms belonging to basis υ
δ_{ij}	Kronecker delta
$\delta(\mathbf{r})$	Delta function
$\mathfrak{D}(E)$	Electron density of states at energy E
$\mathbf{D}$	Dynamical matrix in $\mathbb{R}^{3N_{\text{at}}\times 3N_{\text{at}}}$
$D_{\mathbf{T_n}\upsilon\alpha,\mathbf{T_{n'}}\upsilon'\alpha'}$	Dynamical matrix element
$\mathbf{D}(\mathbf{q_n})$	Fourier transformed dynamical matrix in $\mathbb{C}^{3N_{\text{b}}\times 3N_{\text{b}}}$ at wave vector $\mathbf{q_n}$
$D_{\upsilon\alpha,\upsilon'\alpha'}(\mathbf{q_n})$	Matrix element of the Fourier transformed dynamical matrix at wave vector $\mathbf{q_n}$
e	Electronic charge
e	Euler's number
$\text{erf}(x)$	Gauss error function
$E^{(s_\ell)}_{\text{err}}(T_{\text{e}})$	Normalized root-mean square deviation in cohesive energies of reference simulation s_ℓ at electronic temperature T_{e}
$E^{(s_\ell)}_{\text{sum}}(T_{\text{e}})$	Sum of the cohesive energies of the ab-initio forces of reference simulation s_ℓ at electronic temperature T_{e}
E	(Internal) Energy
E_{total}	Total energy
E_{phot}	Photon energy
E_{kin}	Kinetic energy of the nuclei
$E^{(\text{T})}_{\text{kin}}$	Thermal kinetic energy of the nuclei of a link cell
$E^{(\mathbf{i})}_{\text{kin}}$	Local kinetic energy of the nuclei in sub cell $\mathbf{i}$
$E_{\text{kin}_{\mathcal{M}_k}}$	Kinetic energy in the phonon modes of set $\mathcal{M}_k$
$E^{(\mathbf{i})}_{\text{kin}_{\mathcal{M}^{(\mathbf{i})}_k}}$	Local kinetic energy in the phonon modes of set $\mathcal{M}^{(\mathbf{i})}_k$ in sub cell $\mathbf{i}$

$E_{\mathrm{L_{tot}}}$	Total energy that is absorbed from the laser
$E_{\mathrm{L}}(t_\ell)$	Total laser energy that is penetrating into the total simulation volume up to time t_ℓ
$\Delta E_{\mathrm{L}}^{(\mathbf{i})}(t_\ell)$	Total laser energy that is penetrating into the sub cell **i** during time step t_ℓ
$\Delta E_{\mathrm{L}}^{(\mathbf{i}_{\mathrm{top}})}(t_\ell)$	Laser energy that is penetrating into the top sub cells $\mathbf{i}_{\mathrm{top}}$ during time step t_ℓ
$E_{\mathrm{L_{abs}}}(t)$	Energy that is deposited by the laser in the electronic system up to time t
$E_{\mathrm{L_{abs}}}^{(\mathbf{i})}(t)$	Energy that is locally deposited by the laser in the electronic system of sub cell **i** up to time t
$\Delta E_{\mathrm{L_{abs}}}(t_\ell)$	Energy that is deposited by the laser in the electronic system during time step Δt_ℓ
$\Delta E_{\mathrm{L_{abs}}}^{(\mathbf{i})}(t_\ell)$	Energy that is locally deposited by the laser in the electronic system in sub cell **i** during time step Δt_ℓ
E_{c}	Cohesive energy
$E_{\mathrm{c}}^{(s_\ell)}(T_{\mathrm{e}}, t_k)$	Cohesive energy in reference simulation s_ℓ at time step t_k and electronic temperature T_{e}
E_{F}	Fermi energy
E_{fusion}	Enthalpy of fusion
E_{i}	Total energy of the ions
$E_{\mathrm{D}}^{(\mathbf{i})}(t_\ell)$	Energy that is transferred from sub cell **i** to the neighboring sub cells up to time t_ℓ due to heat conductivity caused by different electronic temperatures
$E_{\mathrm{D}}(t_\ell)$	Total energy that was transferred in or out of the complete simulation cell due to heat conductivity up to time t_ℓ
$\Delta E_{\mathrm{D}}^{(\mathbf{i})}(t_\ell)$	Total electronic heat diffusion of sub cell **i** during time step t_ℓ
$\Delta E_{\mathrm{D}}^{(\mathbf{i},\mathbf{i}')}(t_\ell)$	Electronic heat diffusion from sub cell $\mathbf{i}'$ to sub cell **i** during time step t_ℓ
E_{e}	Total energy of the electrons
$E_{\mathrm{e}}^{(\mathbf{i})}$	Local internal energy of the electrons of sub cell **i**
$\Delta E_{\mathrm{e}}(t_\ell)$	Total change of the internal energy of the electrons at time step t_ℓ
$\Delta E_{\mathrm{e}}^{(\mathbf{i})}(t_\ell)$	Total change of the local internal energy of the electrons of sub cell **i** at time step t_ℓ
$\dot{E}_{\mathrm{ep}}$	Energy transfer rate between electrons and nuclei due to electron-phonon coupling
$E_{\mathrm{ep}}^{(\mathbf{i})}$	Local energy transfer rate between electrons and nuclei due to electron-phonon coupling in sub cell **i**
$\Delta E_{\mathrm{ep}}(t_\ell)$	Energy that is transferred from the nuclei to the electrons due to electron-phonon coupling at time step t_ℓ
$\Delta E_{\mathrm{ep}}^{(\mathbf{i})}(t_\ell)$	Energy that is locally transferred in sub cell **i** from the nuclei to the electrons due to electron-phonon coupling at time step t_ℓ
E_{gap}	Electronic band gap

E_{nn}	Interaction energy of the nuclei
E_{HK}	Hohenberg-Kohn energy functional
E_{Ha}	Hartree energy functional
E_{KS}	Kohn-Sham energy functional
E_{xc}	Exchange correlation functional
$E_{\text{xc}}^{(\text{LDA})}$	Exchange correlation functional in the local density approximation
$E_{\text{xc}}^{(\text{GGA})}$	Exchange correlation functional in the generalized gradient approximation
$\mathcal{E}_\alpha(\mathbf{r}_1, \ldots, \mathbf{r}_{N_{\text{at}}})$	Potential energy surface of the electrons in state α for nuclei at positions $\mathbf{r}_1, \ldots, \mathbf{r}_{N_{\text{at}}}$
$\mathbf{E}$	Electrical field vector in $\mathbb{R}^3$
$\widehat{\mathbf{e}}_x, \widehat{\mathbf{e}}_y, \widehat{\mathbf{e}}_z$	Unit vectors in Cartesian coordinates x, y, z
$\widehat{\mathbf{e}}_r, \widehat{\mathbf{e}}_\phi, \widehat{\mathbf{e}}_\vartheta$	Unit vectors in spherical coordinates r, ϕ, ϑ
$e^{(i)}$	i-th Column vector in $\mathbb{R}^{3N_{\text{at}}}$ of $\mathbf{C}^{\text{t}}$ describing the nuclei motions of the i-th eigenmode
$e^{(i)}(\mathbf{q_n})$	i-th Column vector in $\mathbb{R}^{3N_{\text{b}}}$ of $\mathbf{C}^{\dagger}(\mathbf{q_n})$ describing the nuclei motions in the basis cell of the i-th eigenmode at wave vector $\mathbf{q_n}$
ε_0	Vacuum permittivity
$\varepsilon_{\text{xc}}^{(\text{hom})}$	Exchange-correlation energy density of the homogeneous electron gas
f_β	Equilibrium occupation number of orbital β
$\mathcal{F}_{n\gamma}$	Force on atom n in direction γ obtained from the interatomic potential
F	Helmholtz free energy
F_{HK}	Hohenberg-Kohn functional describing the kinetic and interaction energy of the electrons
F_{M}	Mermin's Helmholtz free energy functional
F_{KS}	Kohn-Sham Helmholtz free energy functional
$f_{\text{c}}(r)$	Cutoff function
$f_{\text{G}}(x)$	Gaußian probability distribution function
$f_{\text{F}}(E, T_{\text{e}})$	Fermi distribution function
$f_{\text{epsf}}(E, E', \omega)$	Electron-phonon spectral function
$f_{\text{El}}(\omega)$	Eliashberg function
$f_{\text{err}}^{(s_\ell)}(T_{\text{e}})$	Normalized root-mean square deviation in atomic forces of reference simulation s_ℓ at electronic temperature T_{e}
$f_{\text{sum}}^{(s_\ell)}(T_{\text{e}})$	Sum of the squares of the ab-initio forces of reference simulation s_ℓ at electronic temperature T_{e}
$\mathbf{f}$	Force vector in $\mathbb{R}^3$
$\mathbf{f}_i$	Force vector in $\mathbb{R}^3$ on atom
$\mathbf{f}_i^{(s_\ell)}((T_{\text{e}}, t_k))$	Force vector in $\mathbb{R}^3$ on atom i in reference simulation s_ℓ at time step t_k and electronic temperature T_{e}
$\mathbf{f}_i^{(\text{s})}$	Stochastic force vector in $\mathbb{R}^3$ on atom i
$\mathcal{F}$	Vector in $\mathbb{R}^{3N_{\text{at}}}$ of all atomic forces obtained from the interatomic potential

$\mathcal{F}^{(\mathbf{i})}$	Vector in $\mathbb{R}^{3N_{\text{at}}^{(\mathbf{i})}}$ of forces of all atoms located in the sub cell $\mathbf{i}$ obtained from the interatomic potential
$\mathcal{F}_{\text{tot}}$	Vector in $\mathbb{R}^{3N_{\text{at}}}$ of all total atomic forces
$\mathcal{F}_{\text{tot}}^{(\mathbf{i})}$	Vector in $\mathbb{R}^{3N_{\text{at}}^{(\mathbf{i})}}$ of total forces of all atoms located in the sub cell
$\mathfrak{g}_i$	Random number following a normal Gaußian distribution
$\mathbf{G}_{\mathbf{n}}$	Vector in $\mathbb{R}^3$ belonging to the reciprocal grid
G_{ep}	Electron-phonon coupling constant
$G_{\text{ep}}^{(\mathbf{i})}$	Local electron-phonon coupling constant in sub cell $\mathbf{i}$
$G_{\text{ep}_{\mathcal{M}_k}}$	Electron-phonon coupling constant for the phonon modes of set $\mathcal{M}_k$
$G^{(\mathbf{i})}_{\text{ep}_{\mathcal{M}_k^{(\mathbf{i})}}}$	Local electron-phonon coupling constant for the phonon modes of set $\mathcal{M}_k^{(\mathbf{i})}$ in sub cell $\mathbf{i}$
$\mathcal{G}(x)$	Integral of the normal Gaußian distribution
$\mathcal{G}^{-1}(x)$	Inverse function of $\mathcal{G}(x)$
Γ	Gamma function
$\Gamma(i \to j)$	Transition rate from state i to state j
H	Hamilton function
$H^{(\text{harm})}$	Hamilton function in the harmonic approximation
H_{L}	Additional term in the Hamilton function describing the electromagnetic field interaction on a charged particle
$\hat{H}_{\text{L}}$	Operator describing the electromagnetic field interaction on the electrons
$\hat{H}$	Hamilton operator
$\hat{H}^{(\text{harm})}$	Hamilton operator in the harmonic approximation
$\hat{H}_{\text{e}}$	Electronic Hamilton operator
$\hat{H}_i^{(1)}$	One-particle Hamilton operator for electron i
$\hat{H}_i^{(\text{L})}$	Operator describing the laser field interaction on electron i
$\hat{H}_{\text{KS}}$	Kohn-Sham Hamilton operator
$\mathcal{H}$	Set of quantum numbers of the Hamilton operator
$\mathcal{H}_{\mathcal{M}_k}(t_{\ell+1})$	Kinetic energy of the phonon mode set $\mathcal{M}_k$ at time step $t_{\ell+1}$, if the influence of the electron-phonon coupling is neglected at the time step $t_{\ell+1}$
$\hbar$	Reduced Planck's constant
$I_{\text{inc}}(x, y, t)$	Spatial incident laser intensity at position x, y and time t
$I_{\text{inc}_{\text{tot}}}(x, y)$	Total incident laser fluence at position x, y
$I_{\text{L}_{\text{tot}}}(x, y)$	Total absorbed laser fluence at the surface at position x, y
$I_{\text{L}_{\text{tot}}}^{(\text{damage})}$	Damage threshold for the incident laser fluence
$I_{\mathbf{q}}(t)$	Diffraction peak intensity for reciprocal vector $\mathbf{q}$ at time t
$I_{hkl}(t)$	Diffraction peak intensity for the Miller indices (hkl) at time t
$\text{Im}(x)$	Imaginary part of x
i	Imaginary unit
$\mathbf{i}$	Integer vector in $\mathbb{N}_0^3$ labeling the sub cells in the simulation cell

$\mathbf{I}$	Set of all integer vectors $\mathbf{i}$ labeling the sub cells that are located in the simulation cell
$\mathbf{J}$	Charge current density vector in $\mathbb{R}^3$
k_B	Boltzmann constant
K_e	Electronic heat conductivity
$K_\mathrm{e}^{(\mathbf{i})}$	Local electronic heat conductivity of sub cell $\mathbf{i}$
$\kappa^{(4i)}_{q_2 q_4 q_7 q_6 q_9}$	Term used to calculate the four-body potential like a two-body potential
$\mathbf{k}$	Electron wave vector in $\mathbb{R}^3$
$\mathbb{L}$	Lorenz constant
L_min	Minimal periodical length of a nanotube
L_o	Optical penetration depth
L_b	Ballistic range
λ	Wave length
$\mathrm{MSD}(t)$	Mean-square displacements of the atoms at time t
$\mathrm{MSD}_{\mathcal{M}}(t)$	Mean-square displacements of the atoms at time t in direction of the phonon modes of set $\mathcal{M}$
$\mathrm{MSD}_x(t)$	Mean-square displacements of the atoms at time t in the x-direction
$\mathrm{MSD}_y(t)$	Mean-square displacements of the atoms at time t in the y-direction
$\mathrm{MSD}_z(t)$	Mean-square displacements of the atoms at time t in the z-direction
m	Mass
m_k	Mass of atom k
m_υ	Mass of atoms belonging to basis υ
m_e	Electron mass
$M^{(j\mathbf{q})}_{\ell\mathbf{k}+\mathbf{q}+\mathbf{G},\ell'\mathbf{k}}$	Electron-phonon coupling matrix element
$\mathcal{M}_\mathrm{s}(T_\mathrm{e})$	Set of ab-initio reference simulations s_ℓ used for the interatomic potential fit at electronic temperature T_e
$\mathcal{M}_k$	k-th subset containing the indices of the corresponding phonon modes
$\mathcal{M}^{(\mathbf{i})}_k k$	k-th subset containing the indices of the corresponding phonon modes in sub cell $\mathbf{i}$
$\mathcal{M}_{N_4^{(\theta)}}(q_1, q_2, q_3)$	Set that contains the allowed powers (q_4, q_5, q_6) of the bond angles as a function of $N_4^{(\theta)}$ and the distance powers (q_1, q_2, q_3)
μ	Chemical potential
μ_e	Chemical potential of the electrons
μ_0	Vacuum permeability
n	Index of refraction
N_b	Number of atoms in the basis cell
N_e	Total number of electrons
N_at	Total number of atoms
$N_\mathrm{at}^{(\mathbf{i})}$	Number of atoms in sub cell $\mathbf{i}$

$N_{\mathrm{at}}^{(s_\ell)}$	Number of atoms used in reference simulation s_ℓ for fitting
$N_t^{(s_\ell)}$	Number of (time) steps used in reference simulation s_ℓ for fitting
N_{D}	Number of time steps Δt_{D} for solving the diffusion equations that are performed during one time step Δt for solving the ionic equations of motion
N_x, N_y, N_z	Number of sub cells used in the simulation cell in the x, y, and z direction
$N_{\mathcal{M}}$	Total number of different subsets $\mathcal{M}_k$ of phonon modes
$N_2^{(r)}$	Degree of the two-body potential
$N_3^{(r)}$	Radial degree of the three-body potential
$N_3^{(\theta)}$	Angular degree of the three-body potential
$N_4^{(r)}$	Radial degree of the four-body potential
$N_4^{(\theta)}$	Angular degree of the four-body potential
$N_\rho^{(r)}$	Number of measures $\rho_i^{(q_1)}$ of the atomic density used in the embedded atom potential
$N_\rho^{(\rho)}$	Density degree of the embedded atom potential
N_{C}	Total number of coefficients used in the polynomial potential
N_{d}	Total number of data points used for fitting of the interatomic potential
$\mathcal{N}(q_3, q_4)$	Function to numerate the coefficients of the spherical harmonics
$\hat{n}_{\ell\mathbf{k}}$	Electron occupation number operator that counts the number of electrons in state $\ell\mathbf{k}$
$n_{\ell\mathbf{k}}$	Number of electrons in state $\ell\mathbf{k}$
$\hat{n}_{j\mathbf{q}}$	Phonon occupation number operator that counts the number of phonons in state $j\mathbf{q}$
$n_{j\mathbf{q}}$	Number of phonons in state $j\mathbf{q}$
$\hat{n}_{\mathrm{e}}(\mathbf{r})$	Electron density operator
$n_{\mathrm{e}}(\mathbf{r})$	Electron density
$n_{\mathrm{e}_0}(\mathbf{r})$	Ground state electron density
$n_{\mathrm{e}}^{(\mathrm{out},k)}(\mathbf{r})$	Output electron density in mixing step k
$n_{\mathrm{e}}^{(\mathrm{in},k)}(\mathbf{r})$	Input electron density in mixing step k
$\mathbb{N}$	Set of all positive integer numbers
$\mathcal{N}_i$	Set containing the indices of the neighboring atoms of atom i
$\mathbf{n}$	Vector in $\mathbb{Z}^3$ characterizing the position in the Bravais or reciprocal grid
η_{ij}	Measure for the bond order of bond between atom i and j
η_{loc}	Function determining how a bond contributes to the bond order of a given bond
Ω	Number of states
ρ	Charge density
ρ_{at}	Atomic density at ambient conditions
ρ_i	Atomic density surrounding atom i
$\rho_i^{q_1}$	Measure for the atomic density surrounding atom i

ρ_{loc}	Function determining how an atom contributes to the atomic density ρ_i surrounding atom i
$\hat{\rho}$	Density matrix
$\hat{\rho}^{(\text{mc})}$	Density matrix of the microcanonical ensemble
$\hat{\rho}^{(\text{c})}$	Density matrix of the canonical ensemble
$\hat{\rho}_{\text{e}}$	Electronic density matrix
$\hat{\rho}_{\text{e}}^{(1)}$	One-electron reduced density matrix
$\hat{\rho}_{\text{e}}^{(2)}$	Two-electron reduced density matrix
p	Momentum
p	Pressure
$\mathbb{P}$	Probability
$\mathbb{P}(t)$	Total emitted power at time t
$\wp_\ell(x)\ell$	ℓ-th Legendre polynomial
$\mathfrak{p}_i^{(\ell)}(r)$	Radial projector in the pseudopotential
$\hat{\mathbf{p}}$	Momentum vector operator
$\mathbf{p}_{\mathbf{n}\upsilon}$	Momentum vector in $\mathbb{R}^3$ of nucleus at position $\mathbf{n}\upsilon$
$\hat{p}_{\mathbf{n}\upsilon\alpha}$	Operator describing the momentum in direction α of nucleus at position $\mathbf{n}\upsilon$
$p_{\mathbf{n}\upsilon\alpha}$	Momentum in direction α of nucleus at position $\mathbf{n}\upsilon$
$\mathbf{P}$	Vector in $\mathbb{R}^{3N_{\text{at}}}$ of all mass-normalized nuclei momenta
$P_{\upsilon\alpha}(\mathbf{T}_\mathbf{n})$	Mass-normalized momentum in direction α of nucleus at position $\mathbf{n}\upsilon$
$\mathscr{P}$	Vector of all mass-normalized nuclei momenta in direction of the eigenmodes
$\mathfrak{P}(\mathbf{q}_\mathbf{n})$	Vector of all mass-normalized nuclei momenta in reciprocal space at wave vector $\mathbf{q}_\mathbf{n}$
$\mathfrak{P}_{\upsilon\alpha}(\mathbf{q}_\mathbf{n})$	Mass-normalized nuclei momentum in direction α of nuclei of basis υ in reciprocal space at wave vector $\mathbf{q}_\mathbf{n}$
$\hat{\mathfrak{P}}_{\upsilon\alpha}(\mathbf{q}_\mathbf{n})$	Operator describing the mass-normalized nuclei momentum in direction α of nuclei of basis υ in reciprocal space at wave vector $\mathbf{q}_\mathbf{n}$
$\mathcal{P}(\mathbf{q}_\mathbf{n})$	Vector of all mass-normalized nuclei momenta in direction of the eigenmodes in reciprocal space at wave vector $\mathbf{q}_\mathbf{n}$
$\hat{\mathcal{P}}_j(\mathbf{q}_\mathbf{n})$	Momentum operator of j-th phonon eigenmode in reciprocal space at wave vector $\mathbf{q}_\mathbf{n}$
$\mathbf{P}_{\mathcal{M}}$	Projection matrix that projects onto the directions of the phonon modes in set $\mathcal{M}$
$\mathbf{P}^{(\mathbf{i})}_{\mathcal{M}_k^{(\mathbf{i})}}$	Local projection matrix of sub cell $\mathbf{i}$ that projects onto the directions of the phonon modes in set $\mathcal{M}_k^{(\mathbf{i})}$
π	Circle number
Φ	Interatomic potential
Φ_{loc}	Local energy contribution
Φ_0	Potential energy of an isolated atom
Φ_2	Pair or two-body potential

Φ_3	Three-body potential
$\Phi_3^{(\mathrm{tot})}$	Total three-body term
Φ_4	Four-body potential
$\Phi_4^{(\mathrm{tot})}$	Total four-body term
$\Phi_{2\mathrm{b}}$	Bond-order potential
Φ_ρ	Embedded atom potential
ϕ_s	Scalar potential used in Electrodynamics
Ψ	Wave function
Q	Heat
$\mathfrak{q}$	Electrical charge
qi	i-th generalized coordinate
$\mathbf{q_n}$	Phonon wave vector in $\mathbb{R}^3$
$\mathfrak{R}$	Reflectivity
$r^{(\mathrm{c})}$	(Global) Cutoff radius
$r_\rho^{(\mathrm{c})}$	Cutoff radius of the embedded atom potential
$r_2^{(\mathrm{c})}$	Cutoff radius of the two-body potential
$r_3^{(\mathrm{c})}$	Cutoff radius of the three-body potential
$r_4^{(\mathrm{c})}$	Cutoff radius of the four-body potential
$r_{2\mathrm{b}}^{(\mathrm{c})}$	Cutoff radius of the bond-order potential
r_{ij}	Distance between atom i and j
$\mathbf{r}_{ij}$	Distance vector in $\mathbb{R}^3$ between atom i and j
$\hat{\mathbf{r}}_{ij}$	Normalized distance vector in $\mathbb{R}^3$ between atom i and j
r_s	Wigner-Seitz sphere radius
r_k	Random number following an uniform distribution in the interval $[0, 1]$
$\mathrm{RMSD}_z(t)$	Root mean-square displacement of the atoms at time t in the z-direction
$\mathrm{RMSD}_y(t)$	Root mean-square displacement of the atoms at time t in the y-direction
$\mathrm{RMSD}_{xy}(t)$	Root mean-square displacement of the atoms at time t in the x, y-directions
$\mathrm{RMSD}_{xz}(t)$	Root mean-square displacement of the atoms at time t in the x, z-directions
$\mathbb{R}$	Set of all real numbers
$\mathcal{R}$	Vector in $\mathbb{R}^{3N_\mathrm{at}}$ of all atomic coordinates
$\mathcal{R}^{(\mathbf{i})}$	Vector in $\mathbb{R}^{3N_\mathrm{at}^{(\mathbf{i})}}$ of the coordinates of all atoms located in sub cell $\mathbf{i}$
$\mathbf{r}_i$	Coordinate vector in $\mathbb{R}^3$ of atom i
$\mathbf{r}_i^{(s_\ell)}(T_\mathrm{e}, t_k)$	Coordinate vector in $\mathbb{R}^3$ of atom i in reference simulation s_ℓ at time step t_k and electronic temperature T_e
$\mathbf{r}_i^{(\mathrm{opt})}$	Coordinate vector in $\mathbb{R}^3$ of the equilibrium position of atom i
$\mathbf{R}_i$	Coordinate vector in $\mathbb{R}^3$ of electron i
S	Entropy
S_e	Electronic entropy

$S_e^{(\mathbf{i})}$	Local electronic entropy of sub cell **i**
$\mathcal{S}_{best}$	Subset of efficient polynomial-degree combinations
$SF_{\mathbf{q}}$	Structure factor for wave vector **q**
$\mathbb{S}(\mathcal{V})$	Surface of volume $(\mathcal{V})$
$\mathbf{S}$	Poynting vector in $\mathbb{R}^3$
σ	Spin of the electron
σ_{el}	Electrical conductivity
$\mathfrak{s}$	Permutation
$\mathfrak{S}_{N_e}$	Set of all permutations $\mathfrak{s}$ of the integer set $\{1, \ldots, N_e\}$
t	Time
Δt	Time increment
Δt_D	Time increment to solve the heat diffusion equation
t_r	Retarded time
τ	Full width at half maximum (FWHM) time width
T	Temperature
T_m	Melting temperature
T_e	Electronic temperature
$T_e^{(\mathbf{i})}$	Local electronic temperature of sub cell **i**
$\Delta T_e(t_\ell)$	Change of the electronic temperature at time step t_ℓ
$\Delta T_e^{(\mathbf{i})}(t_\ell)$	Change of the local electronic temperature in sub cell **i** at time step t_ℓ
$\hat{T}_e$	Kinetic energy operator of the electrons
T_i	Ionic temperature
$T_i^{(\mathbf{i})}$	Local ionic temperature in sub cell **i**
$T_{i_{\mathcal{M}_k}}$	Ionic temperature of the phonon modes of set $\mathcal{M}_k$
$T^{(\mathbf{i})}_{i_{\mathcal{M}_k^{(\mathbf{i})}}}$	Local ionic temperature of the phonon modes of set $\mathcal{M}_k^{(\mathbf{i})}$ in sub cell **i**
$\hat{T}_i$	Kinetic energy operator of the nuclei
T_{aux}	Kinetic energy functional of the auxiliary system of non-interacting electrons
$\mathrm{Tr}(\hat{A})$	Trace of the operator $\hat{A}$
$\hat{\mathcal{T}}_{\mathbf{n}}$	Translation operator related to the vector $\mathbf{T_n}$
$\mathfrak{t}_{\mathbf{n}}$	Eigenvalue of the translation operator $\hat{\mathcal{T}}_{\mathbf{n}}$
$\mathbf{T_n}$	Vector in $\mathbb{R}^3$ belonging to the Bravais grid
θ_{ijk}	Angle between the bond from atom i to j and i to k
$\mathfrak{u}_{j\mathbf{k}}(\mathbf{r})$	Periodical function within the Bloch function
u_{em}	Energy density of the electromagnetic fields
u_{mech}	Mechanical energy density
$\mathbf{u}_{\mathbf{n}\upsilon}$	Displacement vector in $\mathbb{R}^3$ of nucleus at position $\mathbf{n}\upsilon$
$u_{\mathbf{n}\upsilon\alpha}$	Displacement in direction α of nucleus at position $\mathbf{n}\upsilon$
$\hat{u}_{\mathbf{n}\upsilon\alpha}$	Operator describing the displacement in direction α of nucleus at position $\mathbf{n}\upsilon$
U	Vector in $\mathbb{R}^{3N_{at}}$ of all mass-normalized nuclei displacements
$\mathbf{u}$	Vector in $\mathbb{R}^{3N_{at}}$ of all nuclei displacements

$U_{\upsilon\alpha}(\mathbf{T_n})$	Mass-normalized displacement in direction α of nucleus at position $\mathbf{n}\upsilon$
$\hat{U}_{\upsilon\alpha}(\mathbf{T_n})$	Operator describing the mass-normalized displacement in direction α of nucleus at position $\mathbf{n}\upsilon$
$\mathscr{U}$	Vector in $\mathbb{R}^{3N_{\mathrm{at}}}$ of all mass-normalized nuclei displacements in direction of the eigenmodes
$\mathfrak{U}(\mathbf{q_n})$	Vector in $\mathbb{R}^{3N_{\mathrm{b}}}$ of all mass-normalized nuclei displacements in reciprocal space at wave vector $\mathbf{q_n}$
$\mathfrak{U}_{\upsilon\alpha}(\mathbf{q_n})$	Mass-normalized nuclei displacement in direction α of nuclei of basis υ in reciprocal space at wave vector $\mathbf{q_n}$
$\hat{\mathfrak{U}}_{\upsilon\alpha}(\mathbf{q_n})$	Operator describing the mass-normalized nuclei displacement in direction α of nuclei of basis υ in reciprocal space at wave vector $\mathbf{q_n}$
$\mathcal{U}(\mathbf{q_n})$	Vector in in $\mathbb{R}^{3N_{\mathrm{b}}}$ of all mass-normalized nuclei displacements in direction of the eigenmodes in reciprocal space at wave vector $\mathbf{q_n}$
$\hat{\mathcal{U}}_j(\mathbf{q_n})$	Displacement operator of j-th phonon eigenmode in reciprocal space at wave vector $\mathbf{q_n}$
$\mathbf{v}_j$	Velocity vector in $\mathbb{R}^3$ of nuclei j
$\mathbf{v}^{(\mathrm{c})}$	Velocity vector in $\mathbb{R}^3$ of the collective motion in the link cell
$\mathbf{v}_i^{(\mathrm{T})}$	Velocity vector in $\mathbb{R}^3$ of the thermal motion of atom i
$\hat{V}_{\mathrm{ext}}$	External potential operator of the electrons
$V_{\mathrm{ext}}(\mathbf{R})$	External potential of the electrons
$V_{\mathrm{xc}}(\mathbf{R})$	Exchange and correlation potential
$V_{\mathrm{xc}}^{(\mathrm{hom})}(\mathbf{R})$	Exchange and correlation potential of the homogeneous electron gas
$V_{\mathrm{eff}}(\mathbf{R})$	Effective potential of the electrons in the Kohn-Sham equations
$\hat{V}_{\mathrm{int}}$	Operator describing the interaction of the electrons with each other
$\hat{V}_{ij}^{(\mathrm{int})}$	Operator that describes the interaction of electrons i and j
$\hat{V}_i^{(\mathrm{ep})}$	Operator describing the interaction of electron i with the phonons
$\hat{V}^{(\mathrm{ep})}$	Operator describing the interaction of the electrons with the phonons
$V_{\ell\mathbf{k},\ell'\mathbf{k}'}^{(\mathrm{ep})}$	Matrix element of the electron-phonon coupling operator
$V^{(\upsilon)}(\mathbf{r})$	Single nuclei potential
$V_{\mathbf{k}}^{(\upsilon)}(\mathbf{r})$	Fourier coefficient at wave vector $\mathbf{k}$ of the single nuclei potential
$V_{\mathrm{loc}}(r)$	Local part of the pseudopotential
$V_{\mathrm{nl}}(\mathbf{r},\mathbf{r}')$	Non-local part of the pseudopotential
$\mathcal{V}$	Volume
$\mathcal{V}_{\mathrm{s}}$	Volume of the simulation cell
$\mathcal{V}_{\mathrm{b}}$	Volume of the basis cell
$\mathcal{V}$	Vector in $\mathbb{R}^{3N_{\mathrm{at}}}$ of all atomic velocities
$\mathcal{V}^{(\mathrm{i})}$	Vector in $\mathbb{R}^{3N_{\mathrm{at}}^{(\mathrm{i})}}$ of velocities of all atoms located in sub cell $\mathbf{i}$

W	Work
$\mathbb{W}_{err}$	Interatomic potential fit error function
$\mathcal{W}(t_{\ell+1})$	Vector in $\mathbb{R}^{3N_{at}}$ that contains the velocities of all nuclei at time step $t_\ell + 1$, if the electron-phonon coupling is neglected at time step $t_\ell + 1$
ω_i	Eigenfrequency of the i-th eigenmode
$\mathbf{w}$	Wrapping vector in $\mathbb{R}^3$ to form the BNNT
$\mathbf{W}$	Diagonal matrix in $\mathbb{R}^{3N_{at}\times 3N_{at}}$ that contains ω_i^2 on the diagonal
$\mathbf{W}(\mathbf{q_n})$	Diagonal matrix in $\mathbb{R}^{3N_b\times 3N_b}$ that contains $\omega_i^2(\mathbf{q_n})$ on the diagonal at wave vector $\mathbf{q_n}$
$\mathbb{w}_f^{(s_\ell)}(T_e)$	Fit weight for the forces of reference simulation s_ℓ at electronic temperature T_e
$\mathbb{w}_E^{(s_\ell)}(T_e)$	Fit weight for the cohesive energies of reference simulation s_ℓ at electronic temperature T_e
ξ	Constant describing the acceleration strength of the ions due to the electron-phonon coupling
$\xi_{\mathcal{M}_k}$	Constant describing the acceleration strength of the ions in direction of the phonon modes of set $\mathcal{M}_k$ due to the electron-phonon coupling
$\xi^{(\mathbf{i})}_{\mathcal{M}_k^{(\mathbf{i})}}$	Local constant describing the acceleration strength of the ions in direction of the phonon modes of set $\mathcal{M}_k^{(\mathbf{i})}$ due to the electron-phonon coupling in sub cell $\mathbf{i}$
ξ_{ij}	Measure for the bond order of bond between atom i and j
ξ_{loc}	Function determining how a bond contributes to the bond order of a given bond
χ_{ij}	Measure for the bond order of bond between atom i and j
χ_{loc}	Function determining how a bond contributes to the bond order of a given bond
$\chi^{(3i)}_{q_1q_3q_4}$	Term used to calculate the three-body potential like a two-body potential
$\chi^{(4i)}_{q_1q_4q_7q_5q_8}$	Term used to calculate the four-body potential like a two-body potential
$Y_{\ell m}(\hat{\mathbf{r}})$	Spherical harmonic function
Z_k	Proton number of atom k
$\mathscr{Z}^{(c)}$	Quantum mechanical partition function of the canonical ensemble
$\mathcal{Z}^c$	Classical partition function of the canonical ensemble
$\mathbb{Z}$	Set of all integer numbers
$\mathbb{z}$	Peierls parameter
ζ	Constant occurring in the classical partition function $\mathcal{Z}^{(c)}$ of the canonical ensemble
∞	Infinity

Chapter 1
Introduction

Atoms, molecules, and condensed matter are formed by the forces governing the interaction between the contained charged particles—the nuclei and the electrons. Electronic excitations determine the electrical, optical, and magnetic properties of matter. The theoretical treatment of the interacting system of electrons and nuclei in condensed matter and molecules is still one of the great challenges in any complete theoretical description, although, in principle, only the well-known Coulomb interaction between charged particles needs to be considered. The main reason for this is that an accurate description of the electronic equations of motion requires a quantum mechanical based modeling, since phenomena that occur on a atomic or subatomic scale can only be explained within a quantum mechanical framework. At the beginning of the twentieth century the development of quantum mechanics and the description of electrons by wave functions allowed to explain, for the fist time, the observed discrete energy levels of the electrons in an atom. By solving the Schrödinger equation one can obtain the wave function and the different energy eigenstates of the system. Furthermore, the formation of covalent bonds can be explained by the overlap of electronic orbitals, which are these space regions with a high probability to find an electron. Since an exact wave functional treatment of the many electrons-nuclei interaction in condensed matter physics is just impossible, simplifying but sufficiently accurate approximations must be developed. A ground breaking step in this direction was the Born-Oppenheimer approximation in 1927 [1], which provides a framework to decouple the degrees of freedom of the electrons and the nuclei, so that one needs to consider much less effective degrees of freedom at one calculation step. Together with the semiclassical approximation, the Born-Oppenheimer separation set the foundation for a classical description for the nuclei, so that only the electrons needed to be treated quantum mechanically. In such a physical picture the nuclei move as classical particles on a potential energy surface (PES) created by the electrons and one can determine, for example, the eigenmodes of the nuclei oscillations, which are called phonons. Practically, in order to perform calculations for molecules or solids, one can use, for instance, tight-binding like the-

© The Author(s), under exclusive license to Springer Nature Switzerland AG 2021

B. Bauerhenne, *Materials Interaction with Femtosecond Lasers*,
https://doi.org/10.1007/978-3-030-85135-4_1

ories, in which the electronic wave function is constructed as a linear combination of atomic one-electron wave functions. For the description of many thermodynamical processes, the electrons can be assumed to be in the ground state. For such electrons, Hohenberg and Kohn [2] showed in 1964 that knowledge of the electron density as a main variable is sufficient to determine all relevant properties and the determination of the complicated many body wave function is not required. This is the foundation of modern density functional theory (DFT). Any observable like the energy is just a functional of the ground state electron density. However, only the formulation of the Kohn-Sham Ansatz [3] in 1965 allowed practical calculations within DFT, since it provides a way to calculate the energy functional as a function of the electron density.

Light-matter interactions provide a powerful tool to investigate and manipulate molecules and solids. A leap forward was the invention of the laser[1] in 1960, which then allowed to generate coherent, directed, and high intense radiation. Starting from the early 1980s, the further development of lasers able to produce pulses shorter than a picosecond, opened new dimensions to study and manipulate matter. Atomic motions in molecules and solids, chemical reactions, the formation or breaking of bonds take place on a sub-picosecond timescale. Thus, deep insights into the processes occurring in matter became possible by utilizing femtosecond laser-pulses. The long sought-for experiment of direct observation of atomic motions in solids is still to be done, although promising progress was made in this direction. Shock waves driven by short laser pulses have been exploited extensively in order to generate and observe matter under extreme conditions of density, temperature, and pressure [4–8]. Pumping a sample with femtosecond laser-pulses and then probing its optical properties allowed to observe, at least indirectly, the influence of the laser-excitation on the atomic dynamics [9–12]. Ultrafast diffraction experiments with electrons or x-rays as probe made it possible to study time-resolved intensities of several Bragg peaks following a femtosecond laser-excitation [13, 14]. Beside this advanced study of matter, femtosecond laser-pulses allow a very precise material processing. In a single process step, a nanometer-scale surface structuring [15–20] is possible.

The modeling of the femtosecond laser-excitation of matter is yet another great challenge in theoretical physics. At the end of the nineteenth century, electromagnetic radiation was introduced in the framework of classical electrodynamics. In this description, Maxwell's equations fully explain the generation and propagation of electric and magnetic fields and the introduction of the Lorenz force describes classically the interaction between radiation and matter. In order to obtain a deeper understanding, the radiation-matter interaction had to be translated to the quantum mechanical picture. For this, an alternative formulation of quantum mechanics—the second quantization—is necessary. Due to the ultrashort interaction time, the laser field has a negligible influence on the nuclei whereas the electrons undergo strong interaction. Very simple test systems allow for detailed calculations of the electron response to the laser field [21]. The laser field excites electron hole pairs and induces a polarization. After the laser excitation the polarization vanishes and the

[1] **L**ight **A**mplification by **S**timulated **E**mission of **R**adiation.

electrons thermalize to a Fermi distribution with a higher temperature T_e within tens of femtoseconds due to the Coulomb interaction. On a picosecond timescale, incoherent electron-phonon collisions induce an energy transfer such that T_e decreases and that the ionic temperature T_i increases until both temperatures are equal. Thus, the laser-generated transition state of hot electrons and cold ions, far away from thermodynamical equilibrium, exists for several picoseconds. The increased electron temperature T_e changes dramatically the interatomic bonding and has therefore a significant impact on the nuclei motions and gives raise to many non-thermal phenomena.

In order to be able to perform calculations for more realistic systems, further approaches were introduced. In 1984, Runge and Gross [22] extended ground state DFT of Hohenberg and Kohn to time-dependent DFT. This theory allows to treat an electronic system in a time-dependent external field by considering the time-dependent electron density instead of the many body wave function. An extension of the Kohn-Sham Ansatz allows to derive the energy functional as a function of the time-dependent electron density. Time-dependent DFT describes properly the excitation of the electrons during the interaction with the laser pulse. However, for times after the excitation, the electrons would stay in a coherent state and would not relax to a thermal distribution with a common T_e due to the inaccurate description of the Coulomb and exchange interactions between the electrons and the limited system sizes that can be treated. Thus, time-dependent DFT can only describe very early stages of the femtosecond laser-excitation. The state with thermalized electrons at increased electron temperature T_e and its influences needs to be described by other approaches. One possible treatment is tight binding theory at increased T_e [23, 24]. A more accurate description is given by T_e-dependent DFT, which was already formulated by Mermin [25] in 1965. For this, he extended ground state DFT using the concepts of thermodynamics and statistical mechanics. The first molecular dynamics (MD) simulations with T_e-dependent DFT was performed by Alavi et al. in 1994 for a dense, hot hydrogen gas [26]. In 1996, Silvestrelli et al. [27] performed MD simulations of the ultrafast melting of silicon (Si) at increased T_e. Many more calculations using T_e-dependent DFT followed [28, 29]. In all of these calculations, T_e is kept constant, since the thermal relaxation of T_e due to the electron-phonon coupling caused by incoherent electron-phonon collisions is neglected. So far, MD simulations using T_e-dependent DFT to treat non-thermal effects and including the electron-phonon coupling were not possible, but the strength of the electron-phonon coupling could be derived from first principles. Allen et al. [30] derived a method to calculate the electron-phonon coupling strength for metals in 1987 using the concepts introduced by Eliashberg in the framework of the theory of superconductivity [31, 32]. Sadasivam et al. [33] extended this approach to semiconductors in 2017.

MD simulations including the thermal effects of a laser excitation—the electron-phonon coupling—were only possible within a classical framework, so far. In 1974, Anisimov et al. [34] introduced a two-temperature model (TTM) by associating a temperature to the nuclei and a separate temperature T_e to the electrons and relating both temperatures using a continuum modeling. In 2003, Ivanov and Zhigilei [35] extended the TTM method and introduced a combined atomistic-continuum modeling, which allows to perform classical MD simulations including the effect of the electron-phonon coupling. Similar to the two-temperature model, the electrons have a individual temperature and are treated in a continuum. The ions are considered by a classical interatomic potential for electrons in the ground state. Due to the usage of a interatomic potential, MD simulations with hundreds of million of atoms are feasible. This allows for atomistic simulations of the surface structuring by femtosecond laser-pulses [20], something which is impossible for first principle simulations that are limited to around thousand atoms. By performing a velocity scaling, Ivanov and Zhigilei were able to take the effect of the electron-phonon coupling into account. This approach named TTM-MD allows for a correct energy conservation, but the non-thermal effects—the dramatic changes in interatomic bonding due to the increased electron temperature T_e—are nevertheless neglected, since the used interatomic potential is constructed for electrons in the ground state. In order to include also the T_e-induced changes on the interatomic bonding, T_e-dependent interatomic potentials were introduced by Khakshouri et al. [36] in 2008. Up to now, T_e-dependent interatomic potentials are available for the metals tungsten [36, 37], gold [38], molybdenum [39] and for the semiconductor silicon [40, 41]. However, the T_e-dependent interatomic potentials for the semiconductor silicon exhibit a very rough if not wrong description of the PES at increased electron temperatures. Beside this, there exists no physical meaningful MD simulation setup including the electron-phonon coupling and using a T_e-dependent interatomic potential. So far, only MD simulations that do not allow for energy conservation [42] or that are based on rather nonphysical assumptions [41] were performed.

In the present work, this challenge was overcome and an ab-initio theory was developed, which leads to a MD simulation method that not only allows for energy conservation but also includes both, the non-thermal effects of the hot electrons on interatomic bonding and the electron-phonon coupling in a, as shown, physical meaningful way. This approach can be used in MD simulations with T_e-dependent DFT or with a T_e-dependent interatomic potential. In addition, a solution was found to the problem of lacking accurate T_e-dependent interatomic potentials by deriving a method to construct T_e-dependent interatomic potentials from T_e-dependent DFT calculations. The obtained T_e-dependent interatomic potentials describe the T_e-dependent DFT results with a high accuracy and can be very efficiently evaluated. Covalent bonding is much more influenced by the presence of hot electrons than metallic bonding. At ambient conditions the equilibrium structure of the semiconductor silicon shows tetragonal bonding geometry caused by covalent bonds. Furthermore, silicon is the most important semiconductor for electronic devices like computers, since high purity silicon monocrystalline wafers can be produced and patterned at the nanoscale. For the latter, femtosecond laser-pulse processing is a promising tool. For example, the formation of laser-induced periodic surface structures (LIPPS) [43] or the production of black silicon surfaces to increase the efficiency of solar cells [44] were reported. However, so far, the influence of non-thermal effects on this processing is unknown. Thus, silicon was chosen as a candidate for applying the previously described approaches to study the influences of a femtosecond laser excitation. An intense femtosecond laser-pulse induces non-thermal melting of silicon, which corresponds to an ultrafast solid-to-liquid transition. An even more interesting non-thermal phenomenon is a laser-induced solid-to-solid transition, since the structure remains in a condensed state. Antimony, arsenic and bismuth, which are located in the same group of the periodic table, crystallize in the A7 structure. Arsenic transforms from the A7 into the simple cubic structure under the influence of pressure [45]. The necessary pressure for this transition is reduced, if arsenic is excited by a femtosecond-laser pulse [46, 47]. Thus, antimony was chosen as an additional candidate for the study, since it is a promising material to observe a laser-induced solid-to-solid transition.

In this book, a T_e-dependent interatomic potential was constructed for silicon (Si) and antimony (Sb). The high accuracy of both interatomic potentials was verified. MD simulations of a femtosecond laser-excitation were performed for Si and Sb using the previously mentioned simulation method, were compared with available experiments using ultrafast x-ray or electron diffraction, and an excellent agreement was found, which also validates the approach. In addition, the influences of non-thermal and thermal effects were studied in both materials occurring after a femtosecond laser-excitation. The present work is the first one to treat non-thermal and thermal laser-induced ultrafast dynamics on the same level of accuracy. This book is organized as follows:

Chapter 2 describes the treatment of solids from first principles. I present the theoretical development of ground state DFT and T_e-dependent DFT. I prove all needed theorems in a mathematical rigorous way, something in which the hitherto literature commonly fails. In addition, I provide state of the art approaches for an efficient implementation as a computer algorithm.

Chapter 3 covers the light matter interaction from first principles. I review the electrodynamics as well as the second quantization description. Again from first principles, I describe the influence of a femtosecond laser-excitation on the electrons in a solid. I furthermore review the derivations of the electron-phonon coupling in the framework of second quantization and present the theories to derive the coupling strength. In this chapter, I review and prove all relevant statements in a mathematical precise way to expose the connections between the different related fields of theoretical physics to the reader. At the end of the chapter, I introduce the T_e-dependent DFT code CHIVES,[2] which was developed in the group of Prof. Dr. M. E. Garcia to study the structural response of matter following a femtosecond laser-excitation.

Chapter 4 deals with the relevant non-thermal phenomena in Si, Sb, and boron nitride nanotubes (BNNTs) from T_e-dependent DFT calculations. At first, I review the basics about performing MD simulations. Then I present T_e-dependent DFT calculations and MD simulations of Si, Sb and BNNTs at increased T_e, indicating the different effects caused by the excited PES in these materials. I also take a glance at BNNTs, since the femtosecond laser-excitation causes THz radiation relevant for realizing THz emitters.

In Chap. 5, I present classical calculations of the laser-excitation, which allow ultra large scale MD simulations. For that, I introduce classical interatomic potentials for

---

[2] Code for **H**ighly exc**I**ted **V**alence **E**lectron **S**ystems.

electrons in the ground state and T_e-dependent interatomic potentials. I also review the available methods for the description of femtosecond laser-excitations within the MD framework. I indicate how ultra large scale MD simulations can be performed using an efficient parallelization on modern super computers. I analyze the accuracy of the two available T_e-dependent interatomic potentials for Si. The chapter concludes presenting an universal interatomic potential parameter fitting program.

In Chap. 6, I describe the MD simulation method and the theory beyond that allows to consider the excited PES and the electron-phonon coupling. I introduce the polynomial T_e-dependent interatomic potential model and present the advanced fitting procedure. At the end of the chapter, I present how the T_e-dependent interatomic potentials can be efficiently evaluated.

In Chap. 7, I apply the previously described approaches to the femtosecond laser-excitation of Si. I present a T_e-dependent interatomic potential $\Phi^{(\mathrm{Si})}(T_e)$ for Si and verify its high accuracy. I also show that such an accuracy cannot be reached by fitting the coefficients of widely used classical interatomic potential models developed for electrons in the ground state. Using $\Phi^{(\mathrm{Si})}(T_e)$, I perform MD simulations of femtosecond laser excited Si and study the influences of the excited PES and of the electron-phonon coupling. I derive the time-dependent Bragg peak intensities and compare the obtained results with available experiments based on ultrafast electron diffraction and show that there is an excellent agreement. $\Phi^{(\mathrm{Si})}(T_e)$ exhibits a melting temperature that is close to the DFT value. However, both deviate from the experimental value. Thus, in the end of the chapter, a modification of the potential parameters of $\Phi^{(\mathrm{Si})}(T_e)$ are presented, so that the experimental melting temperature is reproduced without changing other properties at higher T_e's.

In Chap. 8, I study the femtosecond laser-excitation of Sb with the previously described approaches. I present a T_e-dependent interatomic potential $\Phi^{(\mathrm{Sb})}(T_e)$ for Sb and verify its high accuracy. In order to compare the obtained results with experiments, I further determine the optical properties of Sb from first principles. Then I present the results of my MD simulations of femtosecond laser-excited Sb, compare the time-dependent Bragg peak intensities with results obtained from ultrafast x-ray diffraction experiments and show that there is an excellent agreement at low and moderate intensities. At the end of the chapter, I discuss large scale MD simulations of a femtosecond laser excited Sb film and study the laser-induced A7 to sc transition.

Finally in Chap. 9, I present a general summary and conclusion of the work with the main achievements. I also provide possible directions for future research.

References

1. M. Born, R. Oppenheimer, Annalen der Physik **389**(20), 457 (1927). https://doi.org/10.1002/andp.19273892002. https://onlinelibrary.wiley.com/doi/abs/10.1002/andp.19273892002
2. P. Hohenberg, W. Kohn, Phys. Rev. **136**, B864 (1964). https://doi.org/10.1103/PhysRev.136.B864. https://link.aps.org/doi/10.1103/PhysRev.136.B864
3. W. Kohn, L.J. Sham, Phys. Rev. **140**, A1133 (1965). https://doi.org/10.1103/PhysRev.140.A1133. https://link.aps.org/doi/10.1103/PhysRev.140.A1133

4. F. Cottet, M. Hallouin, J.P. Romain, R. Fabbro, B. Faral, H. Pepin, Applied Physics Letters **47**(7), 678 (1985). https://doi.org/10.1063/1.96055
5. R. Cauble, D.W. Phillion, T.J. Hoover, N.C. Holmes, J.D. Kilkenny, R.W. Lee, Phys. Rev. Lett. **70**, 2102 (1993). https://doi.org/10.1103/PhysRevLett.70.2102. https://link.aps.org/doi/10.1103/PhysRevLett.70.2102
6. T. Löwer, R. Sigel, K. Eidmann, I.B. Földes, S. Hüller, J. Massen, G.D. Tsakiris, S. Witkowski, W. Preuss, H. Nishimura, H. Shiraga, Y. Kato, S. Nakai, T. Endo, Phys. Rev. Lett. **72**, 3186 (1994). https://doi.org/10.1103/PhysRevLett.72.3186. https://link.aps.org/doi/10.1103/PhysRevLett.72.3186
7. M. Koenig, B. Faral, J.M. Boudenne, D. Batani, A. Benuzzi, S. Bossi, C. Rémond, J.P. Perrine, M. Temporal, S. Atzeni, Phys. Rev. Lett. **74**, 2260 (1995). https://doi.org/10.1103/PhysRevLett.74.2260. https://link.aps.org/doi/10.1103/PhysRevLett.74.2260
8. R. Evans, A.D. Badger, F. Falliès, M. Mahdieh, T.A. Hall, P. Audebert, J.P. Geindre, J.C. Gauthier, A. Mysyrowicz, G. Grillon, A. Antonetti, Phys. Rev. Lett. **77**, 3359 (1996). https://doi.org/10.1103/PhysRevLett.77.3359. https://link.aps.org/doi/10.1103/PhysRevLett.77.3359
9. G.C. Cho, W. Kütt, H. Kurz, Phys. Rev. Lett. **65**, 764 (1990). https://doi.org/10.1103/PhysRevLett.65.764. https://link.aps.org/doi/10.1103/PhysRevLett.65.764
10. K.J. Yee, K.G. Lee, E. Oh, D.S. Kim, Y.S. Lim, Phys. Rev. Lett. **88**, 105501 (2002). https://doi.org/10.1103/PhysRevLett.88.105501. https://link.aps.org/doi/10.1103/PhysRevLett.88.105501
11. M. Hase, K. Ushida, M. Kitajima, J. Phys. Soc. Jpn. **84**(2), 024708 (2015). https://doi.org/10.7566/JPSJ.84.024708. http://dx.doi.org/10.7566/JPSJ.84.024708
12. O.V. Misochko, J. Exp. Theor. Phys. **123**(2), 292 (2016). https://doi.org/10.1134/S1063776116070116. http://dx.doi.org/10.1134/S1063776116070116
13. G. Sciaini, M. Harb, S.G. Kruglik, T. Payer, C.T. Hebeisen, F.J.M.z. Heringdorf, M. Yamaguchi, M.H.V. Hoegen, R. Ernstorfer, R.J.D. Miller, Nature **458**, 56 (2009). https://doi.org/10.1038/nature07788
14. R.J.D. Miller, R. Ernstorfer, M. Harb, M. Gao, C.T. Hebeisen, H. Jean-Ruel, C. Lu, G. Moriena, G. Sciaini, Acta Crystallographica Sect. A **66**(2), 137 (2010). https://doi.org/10.1107/S0108767309053926
15. X. Liu, D. Du, G. Mourou, IEEE J. Quantum Electron. **33**(10), 1706 (1997). https://doi.org/10.1109/3.631270. https://doi.org/10.1109/3.631270
16. F. Courvoisier, P.A. Lacourt, M. Jacquot, M.K. Bhuyan, F. L., J.M. Dudley, Opt. Lett. **34**(20), 3163 (2009). https://doi.org/10.1364/OL.34.003163. http://ol.osa.org/abstract.cfm?URI=ol-34-20-3163
17. X.C. Wang, H.Y. Zheng, C.W. Tan, F. Wang, H.Y. Yu, K.L. Pey, Opt. Express **18**(18), 19379 (2010). https://doi.org/10.1364/OE.18.019379. http://www.opticsexpress.org/abstract.cfm?URI=oe-18-18-19379
18. S. Höhm, M. Rohloff, A. Rosenfeld, J. Krüger, J. Bonse, Appl. Phys. A **110**(3), 553 (2013). https://doi.org/10.1007/s00339-012-7184-z. https://doi.org/10.1007/s00339-012-7184-z
19. S. Höhm, M. Herzlieb, A. Rosenfeld, J. Krüger, J. Bonse, Opt. Exp. **23**(1), 61 (2015). https://doi.org/10.1364/OE.23.000061. http://www.opticsexpress.org/abstract.cfm?URI=oe-23-1-61
20. D.S. Ivanov, V.P. Lipp, A. Blumenstein, F. Kleinwort, V.P. Veiko, E. Yakovlev, V. Roddatis, M.E. Garcia, B. Rethfeld, J. Ihlemann, P. Simon, Phys. Rev. Appl. **4**, 064006 (2015). https://doi.org/10.1103/PhysRevApplied.4.064006. https://link.aps.org/doi/10.1103/PhysRevApplied.4.064006
21. A.V. Kuznetsov, Phys. Rev. B **44**, 13381 (1991). https://doi.org/10.1103/PhysRevB.44.13381. https://link.aps.org/doi/10.1103/PhysRevB.44.13381
22. E. Runge, E.K.U. Gross, Phys. Rev. Lett. **52**, 997 (1984). https://doi.org/10.1103/PhysRevLett.52.997. https://link.aps.org/doi/10.1103/PhysRevLett.52.997
23. P. Stampfli, K.H. Bennemann, Phys. Rev. B **42**, 7163 (1990). https://doi.org/10.1103/PhysRevB.42.7163. https://link.aps.org/doi/10.1103/PhysRevB.42.7163
24. P. Stampfli, K.H. Bennemann, Phys. Rev. B **49**, 7299 (1994). https://doi.org/10.1103/PhysRevB.49.7299. https://link.aps.org/doi/10.1103/PhysRevB.49.7299

25. N.D. Mermin, Phys. Rev. **137**, A1441 (1965). https://doi.org/10.1103/PhysRev.137.A1441. https://link.aps.org/doi/10.1103/PhysRev.137.A1441
26. A. Alavi, J. Kohanoff, M. Parrinello, D. Frenkel, Phys. Rev. Lett. **73**, 2599 (1994). https://doi.org/10.1103/PhysRevLett.73.2599. https://link.aps.org/doi/10.1103/PhysRevLett.73.2599
27. P.L. Silvestrelli, A. Alavi, M. Parrinello, D. Frenkel, Phys. Rev. Lett. **77**, 3149 (1996). https://doi.org/10.1103/PhysRevLett.77.3149. https://link.aps.org/doi/10.1103/PhysRevLett.77.3149
28. V. Recoules, J. Clérouin, G. Zérah, P.M. Anglade, S. Mazevet, Phys. Rev. Lett. **96**, 055503 (2006). https://doi.org/10.1103/PhysRevLett.96.055503. http://link.aps.org/doi/10.1103/PhysRevLett.96.055503
29. E.S. Zijlstra, A. Kalitsov, T. Zier, M.E. Garcia, Adv. Mater. **25**(39), 5605 (2013). https://doi.org/10.1002/adma201302559
30. P.B. Allen, Phys. Rev. Lett. **59**, 1460 (1987). https://doi.org/10.1103/PhysRevLett.59.1460. https://link.aps.org/doi/10.1103/PhysRevLett.59.1460
31. J. Bardeen, L.N. Cooper, J.R. Schrieffer, Phys. Rev. **106**, 162 (1957). https://doi.org/10.1103/PhysRev.106.162. https://link.aps.org/doi/10.1103/PhysRev.106.162
32. F. Marsiglio, Eliashberg theory at 60: strong-coupling superconductivity and beyond. Ann. Phys. **417**, 168102 (2020). https://doi.org/10.1016/j.aop.2020.168102. http://www.sciencedirect.com/science/article/pii/S000349162030035X
33. S. Sadasivam, M.K.Y. Chan, P. Darancet, Phys. Rev. Lett. **119**, 136602 (2017). https://doi.org/10.1103/PhysRevLett.119.136602. https://link.aps.org/doi/10.1103/PhysRevLett.119.136602
34. S.I. Anisimov, B.L. Kapeliovich, T.L. Perel'man, JETP **39**(2), 375 (1974). http://www.jetp.ac.ru/cgi-bin/e/index/e/39/2/p375?a=list
35. D.S. Ivanov, L.V. Zhigilei, Phys. Rev. B **68**, 064114 (2003). https://doi.org/10.1103/PhysRevB.68.064114. https://link.aps.org/doi/10.1103/PhysRevB.68.064114
36. S. Khakshouri, D. Alfè, D.M. Duffy, Phys. Rev. B **78**, 224304 (2008). https://doi.org/10.1103/PhysRevB.78.224304. http://link.aps.org/doi/10.1103/PhysRevB.78.224304
37. S.T. Murphy, S.L. Daraszewicz, Y. Giret, M. Watkins, A.L. Shluger, K. Tanimura, D.M. Duffy, Phys. Rev. B **92**, 134110 (2015). https://doi.org/10.1103/PhysRevB.92.134110. http://link.aps.org/doi/10.1103/PhysRevB.92.134110
38. G.E. Norman, S.V. Starikov, V.V. Stegailov, J. Exp. Theor. Phys. **114**(5), 792 (2012). https://doi.org/10.1134/S1063776112040115
39. J.A. Moriarty, R.Q. Hood, L.H. Yang, Phys. Rev. Lett. **108**, 036401 (2012). https://doi.org/10.1103/PhysRevLett.108.036401
40. L. Shokeen, P.K. Schelling, J. Appl. Phys. **109**(7), 073503 (2011). https://doi.org/10.1063/1.3554410
41. R. Darkins, P.W. Ma, S.T. Murphy, D.M. Duffy, Phys. Rev. B **98**, 024304 (2018). https://doi.org/10.1103/PhysRevB.98.024304. https://link.aps.org/doi/10.1103/PhysRevB.98.024304
42. L. Shokeen, P.K. Schelling, Computational Materials Science **67**, 316 (2013). https://doi.org/10.1016/j.commatsci.2012.07.042. http://www.sciencedirect.com/science/article/pii/S0927025612004910
43. J. Reif, O. Varlamova, S. Varlamov, M. Bestehorn, Appl. Phys. A **104**(3), 969 (2011). https://doi.org/10.1007/s00339-011-6472-3. https://doi.org/10.1007/s00339-011-6472-3
44. A.Y. Vorobyev, C. Guo, Appl. Surf. Sci. **257**(16), 7291 (2011). https://doi.org/10.1016/j.apsusc.2011.03.106. http://www.sciencedirect.com/science/article/pii/S0169433211004703
45. T.N. Kolobyanina, S.S. Kabalkina, L.F. Vereshchagin, L.V. Fedina, Zh. Eksp. Teor. Fiz. **55**, 164 (1968). http://www.jetp.ac.ru/cgi-bin/dn/e_031_02_0259.pdf
46. E.S. Zijlstra, N. Huntemann, M.E. Garcia, New J. Phys. **10**(3), 033010 (2008). http://stacks.iop.org/1367-2630/10/i=3/a=033010
47. N. Huntemann, E.S. Zijlstra, M.E. Garcia, Appl. Phys. A **96**(1), 19 (2009). https://doi.org/10.1007/s00339-009-5175-5. http://dx.doi.org/10.1007/s00339-009-5175-5

Chapter 2
Ab-initio Description of Solids

Abstract Solids are formed by atoms, which consist of electrons dwelling in the field of the nucleus composed of protons and, with exceptions, a number of neutrons. The nucleus is of a very small diameter of $\sim 10^{-15}$ m and contains most of the atomic mass. The much lighter electrons are located around the nucleus, extending the total diameter of the atom to $\sim 10^{-10}$ m. In a solid, the electrons are crucial to form bonds between the atoms finally holding the solid spatially confined. Hence, a precise description of the electron interaction is paramount to describe accurately the inter-atomic bonding. In this chapter, the description of the electrons is presented from first principles, which is called ab-initio. Also the conventional treatment of the nuclear motion is described, which can be done classically. All the statements necessary to develop the theory are derived and proved from the physical grounds to provide the reader the necessary deep understanding to the findings in later chapters.

2.1 Quantum Mechanical Description

The fact that electrons can also behave like waves, and therefore observation cannot be explained using a point-like or small-particle description, was already established at the beginning of the twentieth century. Accordingly the wave-particle duality was introduced. To any particle, a wavelength

$$\lambda = \frac{2\pi \hbar}{p} \tag{2.1}$$

can be related to its momentum p using the reduced Planck's constant $\hbar$. In this way, small particles are described by waves. This wave description can explain, for example, why only discrete energy levels are allowed for the electrons in an atom, or why electrons show a diffraction pattern after passing a Young's double slit. Although the idea of a wave-particle duality remains a useful conceptual tool, the precise formulation is done in quantum mechanics, which is based on the following [1]:

© The Author(s), under exclusive license to Springer Nature Switzerland AG 2021

B. Bauerhenne, *Materials Interaction with Femtosecond Lasers*,
https://doi.org/10.1007/978-3-030-85135-4_2

- The state of any system consisting of N particles with coordinates $\mathbf{r}_1, \ldots, \mathbf{r}_N$ is described by a wave function $\Psi(\mathbf{r}_1, \ldots, \mathbf{r}_N, t)$, which is a one dimensional and complex function of all particle coordinates and the time t. The term

$$\int_{\mathcal{V}_1} d^3r_1 \cdots \int_{\mathcal{V}_N} d^3r_N \left|\Psi(\mathbf{r}_1, \ldots, \mathbf{r}_N, t)\right|^2 \tag{2.2}$$

describes the probability to find the particles $1, \ldots, N$ in the volumes $\mathcal{V}_1, \ldots, \mathcal{V}_N$ at time t. The wave function is taken to be normalized, *i.e.*,

$$1 = \int_{\mathbb{R}^3} d^3r_1 \cdots \int_{\mathbb{R}^3} d^3r_N \left|\Psi(\mathbf{r}_1, \ldots, \mathbf{r}_N, t)\right|^2. \tag{2.3}$$

- A scalar product is defined between any two given wave functions Ψ_1 and Ψ_2 by

$$\langle\Psi_1|\Psi_2\rangle = \int_{\mathbb{R}^3} d^3r_1 \cdots \int_{\mathbb{R}^3} d^3r_N\, \Psi_1^*(\mathbf{r}_1, \ldots, \mathbf{r}_N, t)\, \Psi_2(\mathbf{r}_1, \ldots, \mathbf{r}_N, t). \tag{2.4}$$

Here $\Psi^*(\mathbf{r}_1, \ldots, \mathbf{r}_N, t)$ denotes the complex conjugated of $\Psi(\mathbf{r}_1, \ldots, \mathbf{r}_N, t)$. By definition, one has $\langle\Psi_1|\Psi_2\rangle = \langle\Psi_2|\Psi_1\rangle^*$ and $\langle\Psi_1|a_2\,\Psi_2 + a_3\,\Psi_3\rangle = a_2\,\langle\Psi_1|\Psi_2\rangle + a_3\,\langle\Psi_1|\Psi_3\rangle$, where $a_2, a_3 \in \mathbb{C}$ and Ψ_3 is any other wave function.
- Identical particles are indistinguishable: Hence, $|\Psi|^2$ does not change when two particles of the same type are interchanged. If two fermions, for instance electrons, are interchanged, the sign of Ψ changes. If two bosons are interchanged, Ψ keeps unaffected. The requirement of antisymmetry for fermionic wave functions, within the context of electronic orbital theory, leads to the Pauli exclusion principle.
- Electrons have got an intrinsic angular momentum of $\frac{1}{2}\,\hbar$ called spin and denoted by σ. Due to this, any state can be filled with two electrons: One electron with spin up and one electron with spin down.
- A wave function Ψ can be manipulated by an operator $\hat{A}$ to get a different wave function $\hat{A}\,\Psi$. A linear operator $\hat{A}$ obeys $\hat{A}\,(a_1\,\Psi_1 + a_2\,\Psi_2) = a_1\,\hat{A}\,\Psi_1 + a_2\,\hat{A}\,\Psi_2$, where $a_1, a_2 \in \mathbb{C}$ and Ψ_1, Ψ_2 are arbitrary wave functions. The complex conjugated operator $\hat{A}^\dagger$ of $\hat{A}$ is defined by

$$\left\langle\hat{A}^\dagger\,\Psi\middle|\Psi\right\rangle = \left\langle\Psi\middle|\hat{A}\,\Psi\right\rangle. \tag{2.5}$$

A Hermitian operator $\hat{A}$ is linear and obeys $\hat{A}^\dagger = \hat{A}$.
- The definition (2.4) of the scalar product between two wave functions allows to introduce the so called bra-ket notation, which was originally invented by Paul Dirac in 1939 [2]. The state of a system, which is described by the wave function Ψ_2, corresponds to the ket $|\Psi_2\rangle$. All system states belong to a Hilbert space. For a given element $|\Psi_1\rangle$ of the Hilbert space, the corresponding element of the related

dual space is notated by $\langle \Psi_1 |$, is called a bra and is defined from the scalar product $\langle \Psi_1 | \Psi_2 \rangle$ with a ket $| \Psi_2 \rangle$ from the Hilbert space.

- For any operator $\hat{A}$, an eigenket $|\psi_\alpha\rangle$ is defined by $\hat{A}\,|\psi_\alpha\rangle = a_\alpha\,|\psi_\alpha\rangle$, where $a_\alpha \in \mathbb{C}$ is the related eigenvalue. Almost every operator $\hat{A}$ has got a complete set of orthonormal eigenkets $\{|\psi_\alpha\rangle, \alpha \in \mathcal{A}\}$. With the help of a complete set of orthonormal eigenkets $\{|\psi_\alpha\rangle, \alpha \in \mathcal{A}\}$, one can represent any ket $|\Psi\rangle$ as

$$|\Psi\rangle = \sum_{\alpha \in \mathcal{A}} |\psi_\alpha\rangle \, \langle \psi_\alpha | \Psi \rangle. \tag{2.6}$$

Due to this, a complete set of orthonormal kets is characterized by

$$\hat{\mathbb{1}} = \sum_{\alpha \in \mathcal{A}} |\psi_\alpha\rangle\langle\psi_\alpha|. \tag{2.7}$$

Furthermore, one can represented any operator $\hat{B}$ as

$$\hat{B} = \sum_{\alpha, \beta \in \mathcal{A}} |\psi_\alpha\rangle \left\langle \psi_\alpha \middle| \hat{B} \middle| \psi_\beta \right\rangle \langle \psi_\beta |, \tag{2.8}$$

where $\left\langle \psi_\alpha \middle| \hat{B} \middle| \psi_\beta \right\rangle \in \mathbb{C}$ is called the matrix element.

- Any physically measurable quantity A is called observable and is related to an Hermitian operator $\hat{A}$. For example, the particle position $\mathbb{r}$ is related to the operator $\hat{\mathbf{r}} = \mathbf{r}$ acting on the space of wave functions such that $\hat{\mathbf{r}}\,\Psi(\mathbf{r}) = \mathbf{r}\,\Psi(\mathbf{r})$, and the particle momentum $\mathbf{p}$ to the operator $\hat{\mathbf{p}}$ acting in the same space as $\hat{\mathbf{p}}\,\Psi(\mathbf{r}) = -\mathrm{i}\,\hbar\,\nabla_{\mathbf{r}}\,\Psi(\mathbf{r})$. Furthermore, a function $f(\hat{A})$ of an operator $\hat{A}$ is defined by the corresponding MacLaurin series $f(\hat{A}) = \sum_{n=0}^{\infty} f^{(n)}(0)\,\hat{A}^n$ with $\hat{A}^0 = \hat{\mathbb{1}}$. Every Hermitian operator $\hat{A}$ exhibits a complete set $\{\psi_\alpha, \alpha \in \mathcal{A}\}$ of orthonormal eigenfunctions ψ_α with real eigenvalues $a_\alpha \in \mathbb{R}$. $\mathcal{A}$ is called set of quantum numbers. An eigenfunction ψ_α obeys $\hat{A}\,\psi_\alpha = a_\alpha\,\psi_\alpha$, whereas a_α is the corresponding eigenvalue. An eigenvalue is called degenerate if two or more linearly independent eigenfunctions exist that are related to this eigenvalue. Due to the completeness, any function Ψ can be represented as a linear combination of the eigenfunctions using coefficients $c_\alpha \in \mathbb{C}$: $\Psi = \sum_{\alpha \in \mathcal{A}} c_\alpha\,\psi_\alpha$.

- If two operators $\hat{A}$ and $\hat{B}$ commute with each other, *i.e.*,

$$\begin{aligned} \hat{A}\,\hat{B} &= \hat{B}\,\hat{A}, \\ \Leftrightarrow \qquad 0 &= \hat{A}\,\hat{B} - \hat{B}\,\hat{A} =: \left[\hat{A}, \hat{B}\right]_-, \end{aligned} \tag{2.9}$$

then the eigenfunctions of $\hat{A}$ can be constructed to be also eigenfunctions of $\hat{B}$. $[.,\,.]_-$ denotes the commutator.

- If the system is in the state described by the normalized wave function $\Psi(\mathbf{r}_1, \ldots, \mathbf{r}_N, t)$, the expected value of the observable A with related Hermitian operator $\hat{A}$ is given by

$$\langle\Psi|\hat{A}|\Psi\rangle := \langle\Psi|\hat{A}\,\Psi\rangle = \int_{\mathbb{R}^3} d^3r_1 \cdots \int_{\mathbb{R}^3} d^3r_N\, \Psi^*(\mathbf{r}_1, \ldots, \mathbf{r}_N, t)\, \hat{A}\, \Psi(\mathbf{r}_1, \ldots, \mathbf{r}_N, t). \tag{2.10}$$

- In any given experiment, the measured value of any observable A corresponds always to one of the eigenvalues a_α of the related Hermitian operator $\hat{A}$. After the measurement, in which the eigenvalue a_α was measured, the wave function Ψ of the system transforms to the corresponding eigenfunction ψ_α of a_α.
- The total energy of the system is described by the Hamilton operator $\hat{H}$, which is straight forward constructed from the classical Hamilton function H. The time propagation of the wave function Ψ is given by the **Schrödinger equation**

$$\mathrm{i}\,\hbar\,\frac{\partial\Psi}{\partial t} = \hat{H}\,\Psi. \tag{2.11}$$

Here i denotes the imaginary unit. If the Hamilton operator $\hat{H}$ does not depend explicitly on time, *i.e.*, $\frac{\partial\hat{H}}{\partial t} = 0$, then the time propagation is just given by

$$\Psi(\mathbf{r}_1, \ldots, \mathbf{r}_N, t) = \sum_{\alpha\in\mathcal{H}} \mathrm{e}^{-\mathrm{i}\frac{1}{\hbar}E_\alpha\,(t-t_0)}\, c_\alpha\, \psi_\alpha(\mathbf{r}_1, \ldots, \mathbf{r}_N), \tag{2.12}$$

where $\Psi(\mathbf{r}_1, \ldots, \mathbf{r}_N, t_0) = \sum\limits_{\alpha\in\mathcal{H}} c_\alpha\, \psi_\alpha(\mathbf{r}_1, \ldots, \mathbf{r}_N)$ and $\{\psi_\alpha, \alpha \in \mathcal{H}\}$ is the complete set of eigenfunctions ψ_α with eigenvalues E_α of $\hat{H}$, which are determined from the **time-independent Schrödinger equation**

$$\hat{H}\,\psi = E\,\psi. \tag{2.13}$$

The functions $\mathrm{e}^{-\mathrm{i}\frac{1}{\hbar}E_\alpha\,(t-t_0)}\,\psi_\alpha(\mathbf{r}_1, \ldots, \mathbf{r}_N)$ are called the stationary states. The ground state corresponds to the stationary state with minimal energy E_0.

Now we come back to the solid: We consider a system of N_{at} atoms, each with a nucleus of charge $Z_k\,e$ and mass m_k. There are in total N_{e} electrons with charge $-e$ and mass m_{e}. The entire system is electrically neutral, *i.e.*,

$$N_{\mathrm{e}} = \sum_{k=1}^{N_{\mathrm{at}}} Z_k. \tag{2.14}$$

We denote the position of the ith nucleus by $\mathbf{r}_i$ and the position of the jth electron by $\mathbf{R}_j$. The state of this system of interacting nuclei and electrons is described by the wave function

$$\Psi \equiv \Psi\left(\mathbf{r}_1, \ldots, \mathbf{r}_{N_{\mathrm{at}}}, \mathbf{R}_1, \ldots, \mathbf{R}_{N_{\mathrm{e}}}, t\right). \tag{2.15}$$

By taking the kinetic energy of the nuclei and electrons and the Coulomb interaction between the different particles into account, the Hamilton operator is constructed as

$$\begin{aligned}\hat{H} = &-\sum_{i=1}^{N_{\mathrm{e}}} \frac{\hbar^2}{2m_{\mathrm{e}}} \nabla^2_{\mathbf{R}_i} \\ &+\sum_{i=1}^{N_{\mathrm{e}}} \sum_{j=i+1}^{N_{\mathrm{e}}} \frac{e^2}{4\pi\varepsilon_0} \frac{1}{\left|\mathbf{R}_i - \mathbf{R}_j\right|} - \sum_{i=1}^{N_{\mathrm{e}}} \sum_{k=1}^{N_{\mathrm{at}}} \frac{e^2}{4\pi\varepsilon_0} \frac{Z_k}{\left|\mathbf{r}_k - \mathbf{R}_i\right|} + \sum_{k=1}^{N_{\mathrm{at}}} \sum_{\ell=k+1}^{N_{\mathrm{at}}} \frac{e^2}{4\pi\varepsilon_0} \frac{Z_k Z_\ell}{\left|\mathbf{r}_k - \mathbf{r}_\ell\right|} \\ &-\sum_{k=1}^{N_{\mathrm{at}}} \frac{\hbar^2}{2m_k} \nabla^2_{\mathbf{r}_k},\end{aligned} \tag{2.16}$$

if relativistic effects are neglected. ε_0 denotes the vacuum permittivity. Since this Hamilton operator does not depend explicitly on time, finding the solution reduces solving the time-independent Schrödinger equation (2.13), which now reads

$$\hat{H}\,\psi(\mathbf{r}_1, \ldots, \mathbf{r}_{N_{\mathrm{at}}}, \mathbf{R}_1, \ldots, \mathbf{R}_{N_{\mathrm{e}}}) = E\,\psi(\mathbf{r}_1, \ldots, \mathbf{r}_{N_{\mathrm{at}}}, \mathbf{R}_1, \ldots, \mathbf{R}_{N_{\mathrm{e}}}), \tag{2.17}$$

for the ground state and any excited state of the system. However, complete analytical solutions of the Schrödinger equation are available only for the important but limited case of a single hydrogen atom. At least, for systems with one or two atoms, containing a few electrons, it is still possible to find numerical solutions. For heavier atoms with many electrons or a larger number of atoms, the Schrödinger equation cannot be solved. Moreover, the wave function cannot even be stored on the currently available computers, nor on those that may become available in the near future. To illustrate this, we consider the wave function of a simple Si_2 molecule. This wave function depends on the three dimensional coordinates of two nuclei and 28 electrons. Consequently, the wave function has $3 \times 30 = 90$ degrees of freedom. If we use a 10 point grid to describe each of these degrees and store at each point the value of the wave function just in single precision with 4 bytes, storing the complete wave function needs 4×10^{90} bytes in the general case, where all degrees of freedom are coupled. This number is even larger than the total number of protons in the universe, which is approximately 10^{79} and is called the Eddington number [3].

2.2 Born-Oppenheimer Approximation

To decouple the degrees of freedom of the electrons and the nuclei, the so-called **Born-Oppenheimer** or **adiabatic** approximation [4] was developed in 1927. The following derivation is taken from Ref. [5]. The nuclei are much heavier than the electrons, since the proton mass is $\sim$ 1836 times bigger than the electron mass and a nucleus contains several protons, except for the hydrogen atom. In addition, most nuclei contain also neutrons, which are even a little bit heavier than protons. Hence, the heavy nuclei move much slower than the light electrons, so that the dynamic of the nuclei and the electrons can be separated under certain conditions. To do so, the kinetic energy $\hat{T}_\mathrm{i}$ of the nuclei is considered as a perturbation in the Hamiltonian (2.16):

$$\begin{aligned} \hat{H} =& \hat{H}_\mathrm{e} \underbrace{- \sum_{k=1}^{N_\mathrm{at}} \frac{\hbar^2}{2m_k} \nabla^2_{\mathbf{r}_k},}_{=\hat{T}_\mathrm{i}} \\ \hat{H}_\mathrm{e} =& - \sum_{i=1}^{N_\mathrm{e}} \frac{\hbar^2}{2m_\mathrm{e}} \nabla^2_{\mathbf{R}_i} \\ &+ \sum_{i=1}^{N_\mathrm{e}} \sum_{j=i+1}^{N_\mathrm{e}} \frac{e^2}{4\pi\varepsilon_0} \frac{1}{\left|\mathbf{R}_i - \mathbf{R}_j\right|} - \sum_{i=1}^{N_\mathrm{e}} \sum_{k=1}^{N_\mathrm{at}} \frac{e^2}{4\pi\varepsilon_0} \frac{Z_k}{\left|\mathbf{r}_k - \mathbf{R}_i\right|} \\ &+ \sum_{k=1}^{N_\mathrm{at}} \sum_{\ell=k+1}^{N_\mathrm{at}} \frac{e^2}{4\pi\varepsilon_0} \frac{Z_k Z_\ell}{\left|\mathbf{r}_k - \mathbf{r}_\ell\right|}. \end{aligned} \tag{2.18}$$

For each fixed set of $\mathbf{r}_1, \ldots, \mathbf{r}_{N_\mathrm{at}}$ values, $\hat{H}_\mathrm{e}$ can be taken as a Hamiltonian for the electrons in an external potential. Solving the corresponding electronic time-independent Schrödinger equation

$$\boxed{\hat{H}_\mathrm{e}\, \phi_{\alpha,\mathbf{r}_1,\ldots,\mathbf{r}_{N_\mathrm{at}}}(\mathbf{R}_1, \ldots, \mathbf{R}_{N_\mathrm{e}}) = \mathcal{E}_\alpha(\mathbf{r}_1, \ldots, \mathbf{r}_{N_\mathrm{at}})\, \phi_{\alpha,\mathbf{r}_1,\ldots,\mathbf{r}_{N_\mathrm{at}}}(\mathbf{R}_1, \ldots, \mathbf{R}_{N_\mathrm{e}})} \tag{2.19}$$

yields a complete set $\left\{\phi_{\alpha,\mathbf{r}_1,\ldots,\mathbf{r}_{N_\mathrm{at}}}(\mathbf{R}_1, \ldots, \mathbf{R}_{N_\mathrm{e}}), \alpha \in \mathcal{H}\right\}$ of orthonormal eigenfunctions for any configuration $\mathbf{r}_1, \ldots, \mathbf{r}_{N_\mathrm{at}}$ of the nuclei. The coordinates of the nuclei are just parameters of the eigenvalues $\mathcal{E}_\alpha(\mathbf{r}_1, \ldots, \mathbf{r}_{N_\mathrm{at}})$ and eigenfunctions $\phi_{\alpha,\mathbf{r}_1,\ldots,\mathbf{r}_{N_\mathrm{at}}}(\mathbf{R}_1, \ldots, \mathbf{R}_{N_\mathrm{e}})$. Due to the completeness, any solution of the time-independent Schrödinger equation (2.17) of the complete Hamiltonian $\hat{H}$ at any given nuclear configuration can be written as

$$\psi(\mathbf{r}_1, \ldots, \mathbf{r}_{N_\mathrm{at}}, \mathbf{R}_1, \ldots, \mathbf{R}_{N_\mathrm{e}}) = \sum_{\alpha\in\mathcal{H}} \chi_\alpha(\mathbf{r}_1, \ldots, \mathbf{r}_{N_\mathrm{at}})\, \phi_{\alpha,\mathbf{r}_1,\ldots,\mathbf{r}_{N_\mathrm{at}}}(\mathbf{R}_1, \ldots, \mathbf{R}_{N_\mathrm{e}}) \tag{2.20}$$

with coefficients $\chi_\alpha(\mathbf{r}_1, \ldots, \mathbf{r}_{N_\mathrm{at}})$ that depend on the nuclei coordinates. Putting this in (2.17) yields

$$
\begin{aligned}
0 =& \left(\hat{H} - E\right) \psi(\mathbf{r}_1, \ldots, \mathbf{r}_{N_{\mathrm{at}}}, \mathbf{R}_1, \ldots, \mathbf{R}_{N_{\mathrm{e}}}) \\
=& \sum_{\alpha \in \mathcal{H}} \left(\hat{H}_{\mathrm{e}} + \hat{T}_{\mathrm{i}} - E\right) \chi_\alpha(\mathbf{r}_1, \ldots, \mathbf{r}_{N_{\mathrm{at}}}) \phi_{\alpha, \mathbf{r}_1, \ldots, \mathbf{r}_{N_{\mathrm{at}}}}(\mathbf{R}_1, \ldots, \mathbf{R}_{N_{\mathrm{e}}}) \\
=& \sum_{\alpha \in \mathcal{H}} \left(\mathcal{E}_\alpha(\mathbf{r}_1, \ldots, \mathbf{r}_{N_{\mathrm{at}}}) + \hat{T}_{\mathrm{i}} - E\right) \chi_\alpha(\mathbf{r}_1, \ldots, \mathbf{r}_{N_{\mathrm{at}}}) \phi_{\alpha, \mathbf{r}_1, \ldots, \mathbf{r}_{N_{\mathrm{at}}}}(\mathbf{R}_1, \ldots, \mathbf{R}_{N_{\mathrm{e}}}).
\end{aligned} \tag{2.21}
$$

By multiplying with $\phi^*_{\beta, \mathbf{r}_1, \ldots, \mathbf{r}_{N_{\mathrm{at}}}}(\mathbf{R}_1, \ldots, \mathbf{R}_{N_{\mathrm{e}}})$, $\beta \in \mathcal{H}$ from the left, integrating over all electronic coordinates $\mathbf{R}_1, \ldots, \mathbf{R}_{N_{\mathrm{e}}}$ and using that the functions $\left\{\phi_{\alpha, \mathbf{r}_1, \ldots, \mathbf{r}_{N_{\mathrm{at}}}}(\mathbf{R}_1, \ldots, \mathbf{R}_{N_{\mathrm{e}}}), \alpha \in \mathcal{H}\right\}$ are orthonormal, one gets

$$
\begin{aligned}
0 =& \int_{\mathbb{R}^3} d^3 R_1 \cdots \int_{\mathbb{R}^3} d^3 R_{N_{\mathrm{e}}}\, \phi^*_{\beta, \mathbf{r}_1, \ldots, \mathbf{r}_{N_{\mathrm{at}}}}(\mathbf{R}_1, \ldots, \mathbf{R}_{N_{\mathrm{e}}}) \times \\
& \times \sum_{\alpha \in \mathcal{H}} \left(\mathcal{E}_\alpha(\mathbf{r}_1, \ldots, \mathbf{r}_{N_{\mathrm{at}}}) + \hat{T}_{\mathrm{i}} - E\right) \chi_\alpha(\mathbf{r}_1, \ldots, \mathbf{r}_{N_{\mathrm{at}}}) \phi_{\alpha, \mathbf{r}_1, \ldots, \mathbf{r}_{N_{\mathrm{at}}}}(\mathbf{R}_1, \ldots, \mathbf{R}_{N_{\mathrm{e}}}) \\
=& \left(\mathcal{E}_\beta(\mathbf{r}_1, \ldots, \mathbf{r}_{N_{\mathrm{at}}}) - E\right) \chi_\beta(\mathbf{r}_1, \ldots, \mathbf{r}_{N_{\mathrm{at}}}) \\
& + \sum_{\alpha \in \mathcal{H}} \int_{\mathbb{R}^3} d^3 R_1 \cdots \int_{\mathbb{R}^3} d^3 R_{N_{\mathrm{e}}}\, \phi^*_{\beta, \mathbf{r}_1, \ldots, \mathbf{r}_{N_{\mathrm{at}}}}(\mathbf{R}_1, \ldots, \mathbf{R}_{N_{\mathrm{e}}}) \times \\
& \qquad \times \underbrace{\sum_{k=1}^{N_{\mathrm{at}}} \frac{-\hbar^2}{2 m_k} \nabla^2_{\mathbf{r}_k}}_{\hat{T}_{\mathrm{i}}} \chi_\alpha(\mathbf{r}_1, \ldots, \mathbf{r}_{N_{\mathrm{at}}}) \phi_{\alpha, \mathbf{r}_1, \ldots, \mathbf{r}_{N_{\mathrm{at}}}}(\mathbf{R}_1, \ldots, \mathbf{R}_{N_{\mathrm{e}}}).
\end{aligned} \tag{2.22}
$$

Using the product rule

$$
\begin{aligned}
& \nabla^2_{\mathbf{r}_k} \left(\chi_\alpha(\mathbf{r}_1, \ldots, \mathbf{r}_{N_{\mathrm{at}}}) \phi_{\alpha, \mathbf{r}_1, \ldots, \mathbf{r}_{N_{\mathrm{at}}}}(\mathbf{R}_1, \ldots, \mathbf{R}_{N_{\mathrm{e}}})\right) \\
=& \phi_{\alpha, \mathbf{r}_1, \ldots, \mathbf{r}_{N_{\mathrm{at}}}}(\mathbf{R}_1, \ldots, \mathbf{R}_{N_{\mathrm{e}}}) \nabla^2_{\mathbf{r}_k} \chi_\alpha(\mathbf{r}_1, \ldots, \mathbf{r}_{N_{\mathrm{at}}}) \\
& + 2 \nabla_{\mathbf{r}_k} \chi_\alpha(\mathbf{r}_1, \ldots, \mathbf{r}_{N_{\mathrm{at}}}) \cdot \nabla_{\mathbf{r}_k} \phi_{\alpha, \mathbf{r}_1, \ldots, \mathbf{r}_{N_{\mathrm{at}}}}(\mathbf{R}_1, \ldots, \mathbf{R}_{N_{\mathrm{e}}}) \\
& + \chi_\alpha(\mathbf{r}_1, \ldots, \mathbf{r}_{N_{\mathrm{at}}}) \nabla^2_{\mathbf{r}_k} \phi_{\alpha, \mathbf{r}_1, \ldots, \mathbf{r}_{N_{\mathrm{at}}}}(\mathbf{R}_1, \ldots, \mathbf{R}_{N_{\mathrm{e}}}),
\end{aligned} \tag{2.23}
$$

one obtains

$$
\begin{aligned}
0 =& \left(\mathcal{E}_\beta(\mathbf{r}_1, \ldots, \mathbf{r}_{N_{\mathrm{at}}}) - E\right) \chi_\beta(\mathbf{r}_1, \ldots, \mathbf{r}_{N_{\mathrm{at}}}) \\
& - \sum_{\alpha \in \mathcal{H}} \sum_{k=1}^{N_{\mathrm{at}}} \frac{\hbar^2}{2m} \int_{\mathbb{R}^3} d^3 R_1 \cdots \int_{\mathbb{R}^3} d^3 R_{N_{\mathrm{e}}} \times \\
& \times \Big(\phi^*_{\beta, \mathbf{r}_1, \ldots, \mathbf{r}_{N_{\mathrm{at}}}}(\mathbf{R}_1, \ldots, \mathbf{R}_{N_{\mathrm{e}}}) \phi_{\alpha, \mathbf{r}_1, \ldots, \mathbf{r}_{N_{\mathrm{at}}}}(\mathbf{R}_1, \ldots, \mathbf{R}_{N_{\mathrm{e}}}) \nabla^2_{\mathbf{r}_k} \chi_\alpha(\mathbf{r}_1, \ldots, \mathbf{r}_{N_{\mathrm{at}}})
\end{aligned}
$$

$$
\begin{aligned}
&+ 2\, \phi^*_{\beta,\mathbf{r}_1,\ldots,\mathbf{r}_{N_{\mathrm{at}}}}(\mathbf{R}_1, \ldots, \mathbf{R}_{N_{\mathrm{e}}}) \times \\
&\quad \times \nabla_{\mathbf{r}_k} \chi_\alpha(\mathbf{r}_1, \ldots, \mathbf{r}_{N_{\mathrm{at}}}) \cdot \nabla_{\mathbf{r}_k} \phi_{\alpha,\mathbf{r}_1,\ldots,\mathbf{r}_{N_{\mathrm{at}}}}(\mathbf{R}_1, \ldots, \mathbf{R}_{N_{\mathrm{e}}}) \\
&+ \phi^*_{\beta,\mathbf{r}_1,\ldots,\mathbf{r}_{N_{\mathrm{at}}}}(\mathbf{R}_1, \ldots, \mathbf{R}_{N_{\mathrm{e}}})\, \chi_\alpha(\mathbf{r}_1, \ldots, \mathbf{r}_{N_{\mathrm{at}}})\, \nabla^2_{\mathbf{r}_k} \phi_{\alpha,\mathbf{r}_1,\ldots,\mathbf{r}_{N_{\mathrm{at}}}}(\mathbf{R}_1, \ldots, \mathbf{R}_{N_{\mathrm{e}}})\Big) \\
=& \Big(\hat{T}_{\mathrm{i}} + \mathcal{E}_\beta(\mathbf{r}_1, \ldots, \mathbf{r}_{N_{\mathrm{at}}}) - E\Big)\, \chi_\beta(\mathbf{r}_1, \ldots, \mathbf{r}_{N_{\mathrm{at}}}) \\
&- \sum_{\alpha\in\mathcal{H}} \sum_{k=1}^{N_{\mathrm{at}}} \frac{\hbar^2}{2m} \int_{\mathbb{R}^3} d^3 R_1 \cdots \int_{\mathbb{R}^3} d^3 R_{N_{\mathrm{e}}} \times \\
&\times \Big(2\, \phi^*_{\beta,\mathbf{r}_1,\ldots,\mathbf{r}_{N_{\mathrm{at}}}}(\mathbf{R}_1, \ldots, \mathbf{R}_{N_{\mathrm{e}}}) \times \\
&\quad \times \nabla_{\mathbf{r}_k} \chi_\alpha(\mathbf{r}_1, \ldots, \mathbf{r}_{N_{\mathrm{at}}}) \cdot \nabla_{\mathbf{r}_k} \phi_{\alpha,\mathbf{r}_1,\ldots,\mathbf{r}_{N_{\mathrm{at}}}}(\mathbf{R}_1, \ldots, \mathbf{R}_{N_{\mathrm{e}}}) \\
&+ \phi^*_{\beta,\mathbf{r}_1,\ldots,\mathbf{r}_{N_{\mathrm{at}}}}(\mathbf{R}_1, \ldots, \mathbf{R}_{N_{\mathrm{e}}})\, \chi_\alpha(\mathbf{r}_1, \ldots, \mathbf{r}_{N_{\mathrm{at}}})\, \nabla^2_{\mathbf{r}_k} \phi_{\alpha,\mathbf{r}_1,\ldots,\mathbf{r}_{N_{\mathrm{at}}}}(\mathbf{R}_1, \ldots, \mathbf{R}_{N_{\mathrm{e}}})\Big).
\end{aligned}
\tag{2.24}
$$

By defining the nuclear coordinate-dependent operator

$$
\begin{aligned}
&\hat{A}_{\beta\alpha}(\mathbf{r}_1, \ldots, \mathbf{r}_{N_{\mathrm{at}}}) \\
&:= - \sum_{k=1}^{N_{\mathrm{at}}} \frac{\hbar^2}{2m} \int_{\mathbb{R}^3} d^3 R_1 \cdots \int_{\mathbb{R}^3} d^3 R_{N_{\mathrm{e}}} \times \\
&\times \Big(2\, \phi^*_{\beta,\mathbf{r}_1,\ldots,\mathbf{r}_{N_{\mathrm{at}}}}(\mathbf{R}_1, \ldots, \mathbf{R}_{N_{\mathrm{e}}})\, \nabla_{\mathbf{r}_k} \phi_{\alpha,\mathbf{r}_1,\ldots,\mathbf{r}_{N_{\mathrm{at}}}}(\mathbf{R}_1, \ldots, \mathbf{R}_{N_{\mathrm{e}}}) \cdot \nabla_{\mathbf{r}_k} \\
&\quad + \phi^*_{\beta,\mathbf{r}_1,\ldots,\mathbf{r}_{N_{\mathrm{at}}}}(\mathbf{R}_1, \ldots, \mathbf{R}_{N_{\mathrm{e}}})\, \nabla^2_{\mathbf{r}_k} \phi_{\alpha,\mathbf{r}_1,\ldots,\mathbf{r}_{N_{\mathrm{at}}}}(\mathbf{R}_1, \ldots, \mathbf{R}_{N_{\mathrm{e}}})\Big),
\end{aligned}
\tag{2.25}
$$

one gets

$$
\begin{aligned}
0 =& \Big(\hat{T}_{\mathrm{i}} + \mathcal{E}_\beta(\mathbf{r}_1, \ldots, \mathbf{r}_{N_{\mathrm{at}}}) - E\Big)\, \chi_\beta(\mathbf{r}_1, \ldots, \mathbf{r}_{N_{\mathrm{at}}}) \\
&+ \sum_{\alpha\in\mathcal{H}} \hat{A}_{\beta\alpha}(\mathbf{r}_1, \ldots, \mathbf{r}_{N_{\mathrm{at}}})\, \chi_\alpha(\mathbf{r}_1, \ldots, \mathbf{r}_{N_{\mathrm{at}}}).
\end{aligned}
\tag{2.26}
$$

This equation combined with Eq. (2.19) is still exact. The approximation is given by setting all $\hat{A}_{\beta\alpha}(\mathbf{r}_1, \ldots, \mathbf{r}_{N_{\mathrm{at}}})$ to zero, which are the transition matrix elements between the electronic states β and α for given coordinates $\mathbf{r}_1, \ldots, \mathbf{r}_{N_{\mathrm{at}}}$ of the nuclei. This yields

$$\boxed{\left(\hat{T}_{\mathrm{i}} + \mathcal{E}_\beta(\mathbf{r}_1, \ldots, \mathbf{r}_{N_{\mathrm{at}}})\right) \chi_\beta(\mathbf{r}_1, \ldots, \mathbf{r}_{N_{\mathrm{at}}}) = E\, \chi_\beta(\mathbf{r}_1, \ldots, \mathbf{r}_{N_{\mathrm{at}}}),} \tag{2.27}$$

which takes the form of a time-dependent Schrödinger equation for the nuclei in an effective potential given by the electron energy $\mathcal{E}_\beta(\mathbf{r}_1, \ldots, \mathbf{r}_{N_{\mathrm{at}}})$ at every fixed nuclear configuration. This potential is also called potential energy surface (PES). β controls, whether the ground state or an excited state of the electrons is considered. Now we summarize:

Theorem 2.1 (Born-Oppenheimer approximation) *The dynamics of electrons and nuclei is separated, which leads to a two steps solution of the Schrödinger equation:*

1. *For fixed nuclei positions $\mathbf{r}_1, \ldots, \mathbf{r}_{N_{at}}$, the electronic time-independent Schrödinger equation (2.19) is solved to get the electronic wave eigenfunctions $\phi_{\alpha,\mathbf{r}_1,\ldots,\mathbf{r}_{N_{at}}}(\mathbf{R}_1, \ldots, \mathbf{R}_{N_e})$ that only depend on the electronic coordinates $\mathbf{R}_1, \ldots, \mathbf{R}_{N_e}$ and contain the nuclei coordinates as parameters. The obtained electronic energy eigenvalues $\mathcal{E}_\alpha(\mathbf{r}_1, \ldots, \mathbf{r}_{N_{at}})$ depend also on the nuclei positions.*
2. *For the considered state β of the electrons, the nuclei time-independent Schrödinger equation (2.27) is solved to get the nuclei wave eigenfunction $\chi_\beta(\mathbf{r}_1, \ldots, \mathbf{r}_{N_{at}})$. Here $\mathcal{E}_\beta(\mathbf{r}_1, \ldots, \mathbf{r}_{N_{at}})$ forms the effective potential for the nuclei.*

As mentioned previously, the physical ground of this approximation is given by the fact that the electronic dynamics is much faster than the nuclear dynamics. Hence, the electrons adapt their motions immediately to a change in the nuclei configuration. In other words, the nuclei move so slowly that the fast electrons can always reach their equilibrium dynamics. By separating the electronic and nuclei dynamics, only the electronic wave function as a function of the electronic coordinates and the nuclei wave function as a function of the nuclei coordinates must be considered when solving the electronic and nuclear time-independent Schrödinger equation, respectively. In this way, the effective degrees of freedom of the used wave functions are reduced in contrast to the full wave function of the original problem (2.17).

In addition, the slowly moving nuclei exhibit a relatively large momentum p due to their high masses, so that they show a very small wavelength λ corresponding to the wave-particle duality (2.1). Hence, quantum effects can often be neglected for the nuclei and they can be considered as classical particles obeying classical mechanics without the need to solve the nuclei wave function. This can be proven by focusing the expected values of position and momentum and using the Ehrenfest theorem [1].

2.2.1 Nuclei Motion in the Harmonic Approximation in Crystalline Systems

In the ground state, the nuclei oscillate slightly around their equilibrium positions. Therefore, the nuclei motions can be described in the harmonic approximation, which is presented in this section based on Ref. [5]. In many solids, the equilibrium positions correspond to a periodic crystal structure. An ideal crystal structure can be constructed by using a basis cell, which is spanned by three lattice vectors $\mathbf{a}_1, \mathbf{a}_2, \mathbf{a}_3$ and contains N_b nuclei. These nuclei have got the relative positions $\mathbf{d}_\upsilon$, $\upsilon = 1, \ldots, N_\text{b}$ in the basis cell. To construct the ideal crystal structure, this basis cell extends infinitely in all three directions, so that the zero points of the repeated basis cells form a grid with an infinite number of grid points

$$\mathbf{T_n} = n_1\,\mathbf{a}_1 + n_2\,\mathbf{a}_2 + n_3\,\mathbf{a}_3, \tag{2.28}$$

where $\mathbf{n} = (n_1, n_2, n_3)^\text{t}$ and $n_1, n_2, n_3 \in \mathbb{Z}$. This is the so called **Bravais grid**. The equilibrium positions of the nuclei are given as $\mathbf{T_n} + \mathbf{d}_\upsilon$, where $\mathbf{n}$ characterizes in which basis cell the nuclei is localized and $\mathbf{d}_\upsilon$ denotes the relative position of the nuclei in the basis cell. By construction, a translation with any Bravais grid vector $\mathbf{T_n}$ keeps the ideal crystal structure invariant. However, real solids have a finite extension. On the other hand, solids usually are large compared to the atomic distances, so that most of the atoms are localized far away from the surface and, therefore, are not influenced by surface effects. To use a finite number of atoms in the calculations but still maintaining the translation invariance, one uses a simulation cell with a finite number of atoms and periodic boundary conditions. The simulation cell is created by repeating the basis cell N_1 times in $\mathbf{a}_1$, N_2 times in $\mathbf{a}_2$ and N_3 times in $\mathbf{a}_3$ direction with $N_1, N_2, N_3 \in \mathbb{N}$, so that it is spanned by the three vectors $N_1\,\mathbf{a}_1$, $N_2\,\mathbf{a}_2$, $N_3\,\mathbf{a}_3$. In the calculations, one only deals with the $N_\text{at} = N_1\,N_2\,N_3\,N_\text{b}$ atoms contained in the simulation cell.

To also allow nuclei displacements from the equilibrium positions, we introduce nuclei displacement vectors $\mathbf{u}_{\mathbf{n}\upsilon}$ and write the nuclei positions as

$$\mathbf{r}_{\mathbf{n}\upsilon}(t) = \mathbf{T_n} + \mathbf{d}_\upsilon + \mathbf{u}_{\mathbf{n}\upsilon}(t) \tag{2.29}$$

and the nuclei momenta as

$$\mathbf{p}_{\mathbf{n}\upsilon}(t) = m_\upsilon\,\dot{\mathbf{u}}_{\mathbf{n}\upsilon}(t). \tag{2.30}$$

In the Born-Oppenheimer approximation, the nuclei move on the PES generated from the electrons, so that the Hamilton function of the nuclei is given by (see Eq. 2.27)

$$H = \sum_{\mathbf{n}\upsilon} \frac{\mathbf{p}_{\mathbf{n}\upsilon}^2}{2\,m_\upsilon} + \mathcal{E}_\beta\big(\{\mathbf{r}_{\mathbf{n}\upsilon}\}\big), \tag{2.31}$$

where $\mathcal{E}_\beta(\{\mathbf{r}_{\mathbf{n}\upsilon}\})$ is the PES generated by the electrons in the state β. To simplify things, we perform a Taylor series expansion of the PES $\mathcal{E}_\beta(\{\mathbf{r}_{\mathbf{n}\upsilon}\})$ around the nuclei equilibrium positions $\mathbf{T_n} + \mathbf{d}_\upsilon$:

$$\begin{aligned}
\mathcal{E}_\beta(\{\mathbf{r}_{\mathbf{n}\upsilon}\}) = & \underbrace{\mathcal{E}_\beta(\{\mathbf{T_n} + \mathbf{d}_\upsilon\})}_{=0} \\
& + \underbrace{\sum_{\mathbf{n}\upsilon\alpha} \left.\frac{\partial \mathcal{E}_\beta}{\partial r_{\mathbf{n}\upsilon\alpha}}\right|_{\{\mathbf{T_n}+\mathbf{d}_\upsilon\}}}_{=0} u_{\mathbf{n}\upsilon\alpha} \\
& + \frac{1}{2} \sum_{\mathbf{n}\upsilon\alpha} \sum_{\mathbf{n}'\upsilon'\alpha'} \left.\frac{\partial^2 \mathcal{E}_\beta}{\partial r_{\mathbf{n}\upsilon\alpha}\, \partial r_{\mathbf{n}'\upsilon'\alpha'}}\right|_{\{\mathbf{T_n}+\mathbf{d}_\upsilon\}} u_{\mathbf{n}\upsilon\alpha}\, u_{\mathbf{n}'\upsilon'\alpha'} \\
& + \ldots
\end{aligned} \tag{2.32}$$

Here the sum over α and α' accounts for the x-, y-, and z-direction. The first term is just a constant and can be set to zero. The second term is zero, since $\{\mathbf{T_n} + \mathbf{d}_\upsilon\}$ are the equilibrium positions. In the **harmonic approximation**, only the terms up to the second order are taken into account, which will be a good approximation, if the nuclei slightly oscillate around their equilibrium positions. In this way, the Hamilton function in the harmonic approximation is given by

$$\begin{aligned}
H^{(\mathrm{harm})} = & \frac{1}{2} \sum_{\mathbf{n}\upsilon} \frac{\mathbf{p}_{\mathbf{n}\upsilon}^2}{m_\upsilon} \\
& + \frac{1}{2} \sum_{\mathbf{n}\upsilon\alpha} \sum_{\mathbf{n}'\upsilon'\alpha'} \frac{1}{\sqrt{m_\upsilon\, m_{\upsilon'}}} \left.\frac{\partial^2 \mathcal{E}_\beta}{\partial r_{\mathbf{n}\upsilon\alpha}\, \partial r_{\mathbf{n}'\upsilon'\alpha'}}\right|_{\{\mathbf{T_n}+\mathbf{d}_\upsilon\}} \sqrt{m_\upsilon}\, u_{\mathbf{n}\upsilon\alpha} \sqrt{m'_\upsilon}\, u_{\mathbf{n}'\upsilon'\alpha'}.
\end{aligned} \tag{2.33}$$

Definition 2.1 (*Dynamical matrix*) One defines the dynamical matrix $\mathbf{D} \in \mathbb{R}^{3\,N_{\mathrm{at}} \times 3\,N_{\mathrm{at}}}$ by

$$D_{\mathbf{T_n}\upsilon\alpha, \mathbf{T_{n'}}\upsilon'\alpha'} = \frac{1}{\sqrt{m_\upsilon\, m_{\upsilon'}}} \left.\frac{\partial^2 \mathcal{E}_\beta}{\partial r_{\mathbf{n}\upsilon\alpha}\, \partial r_{\mathbf{n}'\upsilon'\alpha'}}\right|_{\{\mathbf{T_n}+\mathbf{d}_\upsilon\}}, \tag{2.34}$$

which is symmetric, by definition.

We define the mass-normalized nuclei momenta

$$P_{\upsilon\alpha}(\mathbf{T_n}) = \frac{p_{\mathbf{n}\upsilon\alpha}}{\sqrt{m_\upsilon}} = \sqrt{m_\upsilon}\, \dot{u}_{\mathbf{n}\upsilon\alpha}, \tag{2.35}$$

which all are written in the $3\,N_{\mathrm{at}}$-dimensional vector $\mathbf{P}$, and the mass-normalized nuclei displacements

$$U_{\upsilon\alpha}(\mathbf{T_n}) = \sqrt{m_\upsilon}\, u_{\mathbf{n}\upsilon\alpha}, \tag{2.36}$$

which all are written in the $3\,N_{\mathrm{at}}$-dimensional vector $\mathbf{U}$. Now the Hamilton function in the harmonic approximation reads

$$H^{(\mathrm{harm})} = \frac{1}{2}\mathbf{P}^{\mathrm{t}} \cdot \mathbf{P} + \frac{1}{2}\mathbf{U}^{\mathrm{t}} \cdot \mathbf{D} \cdot \mathbf{U}. \tag{2.37}$$

This is the Hamilton function of $3\,N_{\mathrm{at}}$ coupled harmonic oscillators. We can decouple these oscillators by performing a linear transformation of the momenta and displacements in the directions of the eigenmodes, which are called **phonons**. For this, we utilize that the dynamical matrix $\mathbf{D}$ is symmetric. Therefore, a orthonormal matrix $\mathbf{C} \in \mathbb{R}^{3\,N_{\mathrm{at}} \times 3\,N_{\mathrm{at}}}$ with $\mathbf{C}^{\mathrm{t}} \cdot \mathbf{C} = \mathbb{1}$ exists that diagonalizes the dynamical matrix

$$\mathbf{C} \cdot \mathbf{D} \cdot \mathbf{C}^{\mathrm{t}} = \mathbf{W} = \begin{bmatrix} \omega_1^2 & 0 & \dots & 0 \\ 0 & \omega_2^2 & 0 & \vdots \\ \vdots & 0 & \ddots & 0 \\ 0 & \dots & 0 & \omega_{3N_{\mathrm{at}}}^2 \end{bmatrix}. \tag{2.38}$$

with real ω_i^2. Moreover, these ω_i^2 appearing on the diagonal of $\mathbf{W}$ are positive, since $\{\mathbf{T_n} + \mathbf{d_n}\}$ are the equilibrium positions of the PES, so that $\mathbf{D}$ is positive definite. Therefore, we can evaluate the square root and obtain the angular eigenfrequencies $\omega_i \in \mathbb{R}$ of the eigenmodes. The $3\,N_{\mathrm{at}}$-dimensional column vectors $\mathbf{e}^{(i)}$ of $\mathbf{C}^{\mathrm{t}}$, *i.e.*,

$$\mathbf{C}^{\mathrm{t}} = \left[\mathbf{e}^{(1)}, \mathbf{e}^{(2)}, \dots, \mathbf{e}^{(3N_{\mathrm{at}})}\right], \tag{2.39}$$

describe the nuclei motions of the different eigenmodes. These vectors are orthonormal and form a complete basis set of $\mathbb{R}^{3N_{\mathrm{at}}}$. We construct the vector

$$\mathscr{U} = \mathbf{C} \cdot \mathbf{U}, \tag{2.40}$$

which contains the mass-normalized displacements in direction of the eigenmodes, and the vector

$$\mathscr{P} = \mathbf{C} \cdot \mathbf{P}, \tag{2.41}$$

which contains the mass-normalized momenta in direction of the eigenmodes. Using these vectors, we can write the Hamilton function in the harmonic approximation as

$$
\begin{aligned}
H^{(\text{harm})} &= \frac{1}{2}\,\mathbf{P}^{\mathrm{t}} \cdot \underbrace{\mathbf{C}^{\mathrm{t}} \cdot \mathbf{C}}_{=\mathbb{1}} \cdot \mathbf{P} + \frac{1}{2}\,\mathbf{U}^{\mathrm{t}} \cdot \underbrace{\mathbf{C}^{\mathrm{t}} \cdot \mathbf{C}}_{=\mathbb{1}} \cdot \mathbf{D} \cdot \underbrace{\mathbf{C}^{\mathrm{t}} \cdot \mathbf{C}}_{=\mathbb{1}} \cdot \mathbf{U} \\
&= \frac{1}{2}\left(\mathbf{C}\cdot\mathbf{P}\right)^{\mathrm{t}} \cdot \mathbf{C}\cdot\mathbf{P} + \frac{1}{2}\left(\mathbf{C}\cdot\mathbf{U}\right)^{\mathrm{t}} \cdot \underbrace{\mathbf{C}\cdot\mathbf{D}\cdot\mathbf{C}^{\mathrm{t}}}_{=\mathbf{W}} \cdot \mathbf{C}\cdot\mathbf{U} \\
&= \frac{1}{2}\,\mathscr{P}^{\mathrm{t}} \cdot \mathscr{P} + \frac{1}{2}\,\mathscr{U}^{\mathrm{t}} \cdot \mathbf{W} \cdot \mathscr{U} \\
&= \frac{1}{2}\sum_{i=1}^{3N_{\mathrm{at}}} \left(\mathscr{P}_i^2 + \omega_i^2\,\mathscr{U}_i^2\right).
\end{aligned} \tag{2.42}
$$

This is the Hamilton function of $3\,N_{\mathrm{at}}$ independent harmonic oscillators. In the classical limit, that we expect to be enough in many cases due to the short de Broglie's wavelengths, the related equations of motion in the harmonic approximation can be obtained from from the second law of Newton:

$$
\begin{aligned}
m_\upsilon\,\ddot{u}_{\mathbf{n}\upsilon\alpha} &= -\frac{\partial \mathcal{E}_\beta^{(\text{harm})}}{\partial \mathbf{r}_{\mathbf{n}\upsilon\alpha}}, \\
\Leftrightarrow \quad m_\upsilon\,\ddot{u}_{\mathbf{n}\upsilon\alpha} &= -\sum_{\mathbf{n}'\upsilon'\alpha'} \left.\frac{\partial^2 \mathcal{E}_\beta}{\partial r_{\mathbf{n}\upsilon\alpha}\,\partial r_{\mathbf{n}'\upsilon'\alpha'}}\right|_{\{\mathbf{T}_\mathbf{n}+\mathbf{d}_\upsilon\}} u_{\mathbf{n}'\upsilon'\alpha'}, \\
\Leftrightarrow \quad \sqrt{m_\upsilon}\,\ddot{u}_{\mathbf{n}\upsilon\alpha} &= -\sum_{\mathbf{n}'\upsilon'\alpha'} \frac{1}{\sqrt{m_\upsilon\,m_{\upsilon'}}} \left.\frac{\partial^2 \mathcal{E}_\beta}{\partial r_{\mathbf{n}\upsilon\alpha}\,\partial r_{\mathbf{n}'\upsilon'\alpha'}}\right|_{\{\mathbf{T}_\mathbf{n}+\mathbf{d}_\upsilon\}} \sqrt{m_{\upsilon'}}\,u_{\mathbf{n}'\upsilon'\alpha'}, \\
\Leftrightarrow \quad \ddot{U}_{\upsilon\alpha}(\mathbf{T}_\mathbf{n}) &= -\sum_{\mathbf{T}_{\mathbf{n}'}}\sum_{\upsilon'\alpha'} D_{\mathbf{T}_\mathbf{n}\upsilon\alpha,\mathbf{T}_{\mathbf{n}'}\upsilon'\alpha'}\,U_{\upsilon'\alpha'}(\mathbf{T}_{\mathbf{n}'}).
\end{aligned} \tag{2.43}
$$

From the above equation, we see that

$$
-\left.\frac{\partial^2 \mathcal{E}_\beta}{\partial r_{\mathbf{n}\upsilon\alpha}\,\partial r_{\mathbf{n}'\upsilon'\alpha'}}\right|_{\{\mathbf{T}_\mathbf{n}+\mathbf{d}_\upsilon\}} u_{\mathbf{n}'\upsilon'\alpha'} = -\frac{1}{\sqrt{m_\upsilon}}\,D_{\mathbf{T}_\mathbf{n}\upsilon\alpha,\mathbf{T}_{\mathbf{n}'}\upsilon'\alpha'}\,U_{\upsilon'\alpha'}(\mathbf{T}_{\mathbf{n}'})
$$

is the force in α-direction on the υth nucleus in the $\mathbf{n}$th basis cell, if the υ'th nucleus in the $\mathbf{n}'$th basis cell is displaced in α'-direction. This allows to derive a further property of the dynamical matrix:

Theorem 2.2 (Translation invariance of the dynamical matrix) *The dynamical matrix* $\mathbf{D}$ *obeys*

$$
D_{\mathbf{T}_\mathbf{n}\upsilon\alpha,\mathbf{T}_{\mathbf{n}'}\upsilon'\alpha'} = D_{\mathbf{T}_\mathbf{n}-\mathbf{T}_{\mathbf{n}'}\upsilon\alpha,\mathbf{0}\upsilon'\alpha'}\,. \tag{2.44}
$$

Proof The υ'th nucleus in the $\mathbf{n}'$th basis cell is displaced in α'-direction by $u_{\mathbf{n}'\upsilon'\alpha'} \neq 0$ and we consider the force in α-direction on the υth nucleus in the $\mathbf{n}$th basis cell.

Because of the translation invariance with respect to vectors of the Bravais-grid we can perform a translation by $-\mathbf{T}_{\mathbf{n}'}$, so that we obtain the same force in α-direction on the υth nucleus in the $(\mathbf{n}-\mathbf{n}')$th basis cell, if the υ'th nucleus in the $\mathbf{0}$th basis cell is displaced in α'-direction by $u_{\mathbf{0}\upsilon'\alpha'} = u_{\mathbf{n}'\upsilon'\alpha'}$:

$$-\left.\frac{\partial^2 \mathcal{E}_\beta}{\partial r_{\mathbf{n}\upsilon\alpha}\,\partial r_{\mathbf{n}'\upsilon'\alpha'}}\right|_{\{\mathbf{T}_\mathbf{n}+\mathbf{d}_\upsilon\}} u_{\mathbf{n}'\upsilon'\alpha'} = -\left.\frac{\partial^2 \mathcal{E}_\beta}{\partial r_{\mathbf{n}-\mathbf{n}'\upsilon\alpha}\,\partial r_{\mathbf{0}\upsilon'\alpha'}}\right|_{\{\mathbf{T}_\mathbf{n}+\mathbf{d}_\upsilon\}} \overbrace{u_{\mathbf{0}\upsilon'\alpha'}}^{=u_{\mathbf{n}'\upsilon'\alpha'}},$$

$$\Leftrightarrow \quad \frac{1}{\sqrt{m_\upsilon\, m_{\upsilon'}}}\left.\frac{\partial^2 \mathcal{E}_\beta}{\partial r_{\mathbf{n}\upsilon\alpha}\,\partial r_{\mathbf{n}'\upsilon'\alpha'}}\right|_{\{\mathbf{T}_\mathbf{n}+\mathbf{d}_\upsilon\}} = \frac{1}{\sqrt{m_\upsilon\, m_{\upsilon'}}}\left.\frac{\partial^2 \mathcal{E}_\beta}{\partial r_{\mathbf{n}-\mathbf{n}'\upsilon\alpha}\,\partial r_{\mathbf{0}\upsilon'\alpha'}}\right|_{\{\mathbf{T}_\mathbf{n}+\mathbf{d}_\upsilon\}}.$$

■

Furthermore, Eq. (2.43) describes the components of the following vector equation:

$$\ddot{\mathbf{U}}(t) = -\,\mathbf{D}\cdot\mathbf{U}(t), \tag{2.45}$$

$$\Leftrightarrow \quad \mathbf{C}\cdot\ddot{\mathbf{U}}(t) = -\,\mathbf{C}\cdot\mathbf{D}\cdot\underbrace{\mathbf{C}^{\mathrm{t}}\cdot\mathbf{C}}_{=\mathbb{1}}\cdot\mathbf{U}(t),$$

$$\Leftrightarrow \quad \ddot{\mathcal{U}}(t) = -\,\mathbf{W}\cdot\mathcal{U}(t). \tag{2.46}$$

In this way, we found

$$\ddot{\mathcal{U}}_i(t) = -\omega_i^2\,\mathcal{U}_i(t), \qquad \forall\, i = 1,\ldots,3\,N_{\mathrm{at}}. \tag{2.47}$$

We just have $3\,N_{\mathrm{at}}$ differential equations of independent one dimensional harmonic oscillators, if the directions of the eigenmodes are used. These equations can be easily solved by

$$\mathcal{U}_i(t) = \mathcal{U}_i^{(0)}\cos\left(\omega_i\, t + \omega_i^{(0)}\right), \tag{2.48}$$

where $\mathcal{U}_i^{(0)}$ and $\omega_i^{(0)}$ are real constants defined from the initial conditions. Using Eq. (2.40) we obtain

$$\mathbf{U}(t) = \mathbf{C}^{\mathrm{t}}\cdot\mathcal{U}(t), \tag{2.49}$$

$$\overset{(2.39)}{\Leftrightarrow} \quad \mathbf{U}(t) = \sum_{i=1}^{3\,N_{\mathrm{at}}} \mathbf{e}^{(i)}\,\mathcal{U}_i(t), \tag{2.50}$$

$$\Leftrightarrow \quad U_{\upsilon\alpha}(\mathbf{T}_\mathbf{n}, t) = \sum_{i=1}^{3\,N_{\mathrm{at}}} e_{\upsilon\alpha}^{(i)}(\mathbf{T}_\mathbf{n})\,\mathcal{U}_i(t), \tag{2.51}$$

and finally

$$\mathbf{U}(t) = \sum_{i=1}^{3\,N_{\text{at}}} \mathbf{e}^{(i)}\, \mathscr{U}_i^{(0)} \cos\left(\omega_i\, t + \omega_i^{(0)}\right) \tag{2.52}$$

for the mass-normalized nuclei displacement vector. The main problem in this procedure is the diagonalization of the large dynamical matrix **D** to get the nuclei motion direction vectors $\mathbf{e}^{(i)}$ and angular frequencies ω_i of the eigenmodes. But one can circumvent the diagonalization of the full dynamical matrix, if one considers the translation invariance with respect to the Bravais-grid. Due to this symmetry, the eigenmodes correspond to plane waves with certain wave vectors $\mathbf{q_n}$ and the motions related to these wave vectors are independent to each other. Now we describe the construction of these wave vectors:

Definition 2.2 (*Reciprocal lattice vectors and grid*) Using the three-dimensional lattice vectors $\mathbf{a}_1, \mathbf{a}_2, \mathbf{a}_3$, one defines the reciprocal lattice vectors

$$\mathbf{b}_1 = 2\pi\, \frac{\mathbf{a}_2 \times \mathbf{a}_3}{\mathbf{a}_1^{\text{t}} \cdot (\mathbf{a}_2 \times \mathbf{a}_3)}, \quad \mathbf{b}_2 = 2\pi\, \frac{\mathbf{a}_3 \times \mathbf{a}_1}{\mathbf{a}_2^{\text{t}} \cdot (\mathbf{a}_3 \times \mathbf{a}_1)}, \quad \mathbf{b}_3 = 2\pi\, \frac{\mathbf{a}_1 \times \mathbf{a}_2}{\mathbf{a}_3^{\text{t}} \cdot (\mathbf{a}_1 \times \mathbf{a}_2)}, \tag{2.53}$$

where $\mathbf{a}_1 \times \mathbf{a}_2$ denotes the cross product between the vectors $\mathbf{a}_1$ and $\mathbf{a}_2$. The reciprocal lattice vectors form the so called reciprocal grid

$$\mathbf{G_n} = n_1\, \mathbf{b}_1 + n_2\, \mathbf{b}_2 + n_3\, \mathbf{b}_3, \tag{2.54}$$

with $\mathbf{n} = [n_1, n_2, n_3]^{\text{t}}$ and $n_1, n_2, n_3 \in \mathbb{Z}$.

By definition, the reciprocal lattice vectors obey

$$\mathbf{a}_i^{\text{t}} \cdot \mathbf{b}_j = 2\pi\, \delta_{ij}, \qquad \forall\, i, j = 1, 2, 3. \tag{2.55}$$

Due to the periodic boundary conditions, the wave vector **q** of a plane wave that is compatible to the simulation cell must fulfill

$$\begin{aligned} & e^{\mathrm{i}\,\mathbf{q}^{\text{t}}\cdot\mathbf{r}} = e^{\mathrm{i}\,\mathbf{q}^{\text{t}}\cdot(\mathbf{r}+N_i\,\mathbf{a}_i)}, && \forall\, i = 1, 2, 3 \ \text{ and } \ \forall\, \mathbf{r} \in \mathbb{R}^3, \\ \Leftrightarrow \quad & e^{\mathrm{i}\,\mathbf{q}^{\text{t}}\cdot N_i\,\mathbf{a}_i} = 1, && \forall\, i = 1, 2, 3. \end{aligned} \tag{2.56}$$

Therefore, the compatible wave vectors lie on a grid and are given by

$$\mathbf{q_n} = \frac{n_1}{N_1}\, \mathbf{b}_1 + \frac{n_2}{N_2}\, \mathbf{b}_2 + \frac{n_3}{N_3}\, \mathbf{b}_3 \tag{2.57}$$

with $\mathbf{n} = [n_1, n_2, n_3]^{\text{t}}$ and $n_1, n_2, n_3 \in \mathbb{Z}$. By construction of the reciprocal lattice vectors, if one adds a vector $\mathbf{G_{n'}}$ of the reciprocal grid to the wave vector $\mathbf{q_n}$, the

Bravais-grid points $\mathbf{T_n}$ remain invariant, since $e^{i\,\mathbf{G}^t_{\mathbf{n}'}\cdot\mathbf{T_n}} = 1$. Therefore in the following, we only need to consider the $N_1\,N_2\,N_3$ non-equivalent wave vectors $\mathbf{q_n}$ with

$$n_i = 1, \ldots, N_i, \qquad \forall i = 1, 2, 3. \tag{2.58}$$

Now we can transform the generalized nuclei displacements and momenta to the reciprocal space with the help of the Fourier transform:

Definition 2.3 (*Discrete Fourier transform*) Let $\mathbf{F} = [F_1, \ldots, F_N]^t \in \mathbb{C}^N$ be a vector of N complex numbers. The Fourier transform of $\mathbf{F}$ is the N-dimensional Vector $\mathfrak{F} = [\mathfrak{F}_1, \ldots, \mathfrak{F}_N]^t \in \mathbb{C}^N$ that components are defined by

$$\mathfrak{F}_k = \frac{1}{\sqrt{N}} \sum_{j=1}^{N} F_j\, e^{-i\frac{2\pi\, j\, k}{N}}. \tag{2.59}$$

To derive the inverse Fourier transform, we need the following theorem about the geometric sum, which proof was taken from Ref [6]:

Mathematical Theorem 2.1 (Geometric sum) *One obtains for any* $1 \neq x \in \mathbb{C}$ *and* $n \in \mathbb{N}_0$:

$$\sum_{i=0}^{n} x^i = \frac{1 - x^{n+1}}{1 - x}. \tag{2.60}$$

Proof We utilize mathematical induction to prove Eq. (2.60). For $n = 0$, we have

$$\sum_{i=0}^{0} x^i = x^0 = 1 = \frac{1-x}{1-x} = \frac{1 - x^{0+1}}{1 - x}.$$

Now we assume that Eq. (2.60) is valid up to a given $n \in \mathbb{N}_0$ and prove Eq. (2.60) for $n + 1$:

$$\begin{aligned}
\sum_{i=0}^{n+1} x^i &= x^{n+1} + \sum_{i=0}^{n} x^i \\
&= x^{n+1} + \frac{1 - x^{n+1}}{1 - x} \\
&= \frac{x^{n+1} - x^{(n+1)+1}}{1 - x} + \frac{1 - x^{n+1}}{1 - x}
\end{aligned}$$

$$= \frac{1 - x^{(n+1)+1}}{1 - x}.$$

■

Mathematical Theorem 2.2 (Inverse discrete Fourier transform) *Let* $\mathfrak{F} = [\mathfrak{F}_1, \ldots, \mathfrak{F}_N]^t \in \mathbb{C}^N$ *be the Fourier transform of* $\mathbf{F} = [F_1, \ldots, F_N]^t \in \mathbb{C}^N$. *Then one obtains*

$$F_j = \frac{1}{\sqrt{N}} \sum_{k=1}^{N} \mathfrak{F}_k \, e^{i \frac{2\pi j k}{N}}. \tag{2.61}$$

Proof This proof is taken from Ref. [7]. For $\ell = j$, we have

$$\sum_{k=1}^{N} e^{i \frac{2\pi k (\ell - j)}{N}} = \sum_{k=1}^{N} e^0 = \sum_{k=1}^{N} 1 = N.$$

For $\ell \neq j$, we have $e^{i \frac{2\pi (\ell - j)}{N}} \neq 1$. For $\ell, j \in \mathbb{Z}$, we have $e^{i \frac{2\pi\, 0 (\ell - j)}{N}} = 1 = e^{i 2\pi (\ell - j)} = e^{i \frac{2\pi N (\ell - j)}{N}}$. Therefore, we obtain for $\ell, j \in \mathbb{Z}$ and $\ell \neq j$ with the help of the geometric sum:

$$\begin{aligned} \sum_{k=1}^{N} e^{i \frac{2\pi k (\ell - j)}{N}} &= \sum_{k=0}^{N-1} \left(e^{i \frac{2\pi (\ell - j)}{N}} \right)^k \\ &\overset{(2.60)}{=} \frac{1 - \left(e^{i \frac{2\pi (\ell - j)}{N}} \right)^N}{1 - e^{i \frac{2\pi (\ell - j)}{N}}} \\ &= \frac{1 - e^{i 2\pi (\ell - j)}}{1 - e^{i \frac{2\pi (\ell - j)}{N}}} \\ &= \frac{1 - 1}{1 - e^{i \frac{2\pi (\ell - j)}{N}}} \\ &= 0. \end{aligned}$$

In summary, we proved for all $\ell, k \in \mathbb{Z}$:

$$\sum_{k=1}^{N} e^{i \frac{2\pi k (\ell - j)}{N}} = N \, \delta_{\ell j}. \tag{2.62}$$

Using this, we get

$$\frac{1}{\sqrt{N}}\sum_{k=1}^{N}\mathfrak{F}_k\,\mathrm{e}^{-\mathrm{i}\,\frac{2\pi\,j\,k}{N}} \overset{(2.59)}{=} \frac{1}{\sqrt{N}}\sum_{k=1}^{N}\left(\frac{1}{\sqrt{N}}\sum_{\ell=1}^{N}F_\ell\,\mathrm{e}^{\mathrm{i}\,\frac{2\pi\,\ell\,k}{N}}\right)\mathrm{e}^{-\mathrm{i}\,\frac{2\pi\,j\,k}{N}}$$

$$= \frac{1}{N}\sum_{\ell=1}^{N}F_\ell\underbrace{\sum_{k=1}^{N}\mathrm{e}^{\mathrm{i}\,\frac{2\pi\,k\,(\ell-j)}{N}}}_{\overset{(2.62)}{=}N\,\delta_{\ell j}}$$

$$= F_j.$$

■

Definition 2.4 (*Nuclei displacements and momenta in reciprocal space*) For any wave vector $\mathbf{q}_{\mathbf{n}'}$ that is compatible with the simulation cell, we define the nuclei displacement in reciprocal space by

$$\mathfrak{U}_{\upsilon\alpha}(\mathbf{q}_{\mathbf{n}'}) = \frac{1}{\sqrt{N_1\,N_2\,N_3}}\sum_{\mathbf{T}_\mathbf{n}}U_{\upsilon\alpha}(\mathbf{T}_\mathbf{n})\,\mathrm{e}^{-\mathrm{i}\,\mathbf{q}^{\mathrm{t}}_{\mathbf{n}'}\cdot\mathbf{T}_\mathbf{n}} \tag{2.63}$$

and the nuclei momenta in reciprocal space by

$$\mathfrak{P}_{\upsilon\alpha}(\mathbf{q}_{\mathbf{n}'}) = \frac{1}{\sqrt{N_1\,N_2\,N_3}}\sum_{\mathbf{T}_\mathbf{n}}P_{\upsilon\alpha}(\mathbf{T}_\mathbf{n})\,\mathrm{e}^{-\mathrm{i}\,\mathbf{q}^{\mathrm{t}}_{\mathbf{n}'}\cdot\mathbf{T}_\mathbf{n}}, \tag{2.64}$$

where the sum runs over all lattice points $\mathbf{T}_\mathbf{n}$ of the Bravais grid inside the simulation cell.

Theorem 2.3 (Nuclei displacements and momenta in real space from reciprocal space) *For any point of the Bravais lattice* $\mathbf{T}_\mathbf{n}$, *we obtain the nuclei displacement in real space by*

$$U_{\upsilon\alpha}(\mathbf{T}_\mathbf{n}) = \frac{1}{\sqrt{N_1\,N_2\,N_3}}\sum_{\mathbf{q}_{\mathbf{n}'}}\mathfrak{U}_{\upsilon\alpha}(\mathbf{q}_{\mathbf{n}'})\,\mathrm{e}^{\mathrm{i}\,\mathbf{q}^{t}_{\mathbf{n}'}\cdot\mathbf{T}_\mathbf{n}} \tag{2.65}$$

and the nuclei momenta in real space by

$$P_{\upsilon\alpha}(\mathbf{T}_\mathbf{n}) = \frac{1}{\sqrt{N_1\,N_2\,N_3}}\sum_{\mathbf{q}_{\mathbf{n}'}}\mathfrak{P}_{\upsilon\alpha}(\mathbf{q}_{\mathbf{n}'})\,\mathrm{e}^{\mathrm{i}\,\mathbf{q}^{t}_{\mathbf{n}'}\cdot\mathbf{T}_\mathbf{n}}, \tag{2.66}$$

where the sum runs over all non-equivalent wave vectors $\mathbf{q_{n'}}$ that are compatible to the simulation cell.

Proof We have

$$
\begin{aligned}
&\frac{1}{\sqrt{N_1\,N_2\,N_3}}\sum_{\mathbf{q_{n'}}}\mathfrak{U}_{\upsilon\alpha}(\mathbf{q_{n'}})\,\mathrm{e}^{\mathrm{i}\,\mathbf{q}^{\mathrm{t}}_{\mathbf{n'}}\cdot\mathbf{T_n}}\\
&\overset{(2.63)}{=}\frac{1}{\sqrt{N_1\,N_2\,N_3}}\sum_{\mathbf{q'_n}}\left(\frac{1}{\sqrt{N_1\,N_2\,N_3}}\sum_{\mathbf{T_{n''}}}U_{\upsilon\alpha}(\mathbf{T_{n''}})\,\mathrm{e}^{-\mathrm{i}\,\mathbf{q}^{\mathrm{t}}_{\mathbf{n'}}\cdot\mathbf{T_{n''}}}\right)\mathrm{e}^{\mathrm{i}\,\mathbf{q}^{\mathrm{t}}_{\mathbf{n'}}\cdot\mathbf{T_n}}\\
&=\frac{1}{N_1\,N_2\,N_3}\sum_{\mathbf{T_{n''}}}U_{\upsilon\alpha}(\mathbf{T_{n''}})\sum_{\mathbf{q_{n'}}}\mathrm{e}^{\mathrm{i}\,\mathbf{q}^{\mathrm{t}}_{\mathbf{n'}}\cdot(\mathbf{T_n}-\mathbf{T_{n''}})}.
\end{aligned}
$$

We get further using the condition Eq. (2.57) for non-equivalent simulation cell compatible wave vectors $\mathbf{q_{n'}}$ and the definition of the Bravais grid vectors

$$
\begin{aligned}
&\sum_{\mathbf{q_{n'}}}\mathrm{e}^{\mathrm{i}\,\mathbf{q}^{\mathrm{t}}_{\mathbf{n'}}\cdot(\mathbf{T_n}-\mathbf{T_{n''}})}\\
&=\sum_{n'_1}^{N_1}\sum_{n'_2}^{N_2}\sum_{n'_3}^{N_3}\mathrm{e}^{\mathrm{i}\left(\frac{n'_1}{N_1}\,\mathbf{b}_1+\frac{n'_2}{N_2}\,\mathbf{b}_2+\frac{n'_3}{N_3}\,\mathbf{b}_3\right)^{\mathrm{t}}\cdot\left((n_1-n''_1)\,\mathbf{a}_1+(n_2-n''_2)\,\mathbf{a}_2+(n_3-n''_3)\,\mathbf{a}_3\right)}\\
&\overset{(2.55)}{=}\sum_{n'_1}^{N_1}\sum_{n'_2}^{N_2}\sum_{n'_3}^{N_3}\mathrm{e}^{\mathrm{i}\,2\pi\left(\frac{n'_1\,(n_1-n''_1)}{N_1}+\frac{n'_2\,(n_2-n''_2)}{N_2}+\frac{n'_3\,(n_3-n''_3)}{N_3}\right)}\\
&=\sum_{n'_1}^{N_1}\mathrm{e}^{\mathrm{i}\,\frac{2\pi\,n'_1\,(n_1-n''_1)}{N_1}}\sum_{n'_2}^{N_2}\mathrm{e}^{\mathrm{i}\,\frac{2\pi\,n'_2\,(n_2-n''_2)}{N_2}}\sum_{n'_3}^{N_3}\mathrm{e}^{\mathrm{i}\,\frac{2\pi\,n'_3\,(n_3-n''_3)}{N_3}}\\
&\overset{(2.62)}{=}N_1\,\delta_{n_1n''_1}\,N_2\,\delta_{n_2n''_2}\,N_3\,\delta_{n_3n''_3}\\
&=N_1\,N_2\,N_3\,\delta_{\mathbf{T_{n''}}\mathbf{T_n}},
\end{aligned}
\tag{2.67}
$$

which proves Eq. (2.65). The proof of Eq. (2.66) for the momenta is performed analogously. ■

As in the proof, we can show for the sum over the Bravais grid vectors $\mathbf{T_n}$ localized in the simulation cell:

$$\begin{aligned}
\sum_{\mathbf{T_n}} e^{i(\mathbf{q_{n''}}-\mathbf{q_{n'}})^t\cdot\mathbf{T_n}} &= \sum_{n_1}^{N_1}\sum_{n_2}^{N_2}\sum_{n_3}^{N_3} e^{i\left(\frac{n''_1-n'_1}{N_1}\mathbf{b}_1+\frac{n''_2-n'_2}{N_2}\mathbf{b}_2+\frac{n''_3-n'_3}{N_3}\mathbf{b}_3\right)^t\cdot(n_1\mathbf{a}_1+n_2\mathbf{a}_2+n_3\mathbf{a}_3)} \\
&\overset{(2.55)}{=} \sum_{n_1}^{N_1}\sum_{n_2}^{N_2}\sum_{n_3}^{N_3} e^{i\,2\pi\left(\frac{n_1(n''_1-n'_1)}{N_1}+\frac{n_2(n''_2-n'_2)}{N_2}+\frac{n_3(n''_3-n'_3)}{N_3}\right)} \\
&= \sum_{n_1}^{N_1} e^{i\frac{2\pi n_1(n''_1-n'_1)}{N_1}} \sum_{n_2}^{N_2} e^{i\frac{2\pi n_2(n''_2-n'_2)}{N_2}} \sum_{n_3}^{N_3} e^{i\frac{2\pi n_3(n''_3-n'_3)}{N_3}} \\
&\overset{(2.62)}{=} N_1\,\delta_{n''_1 n'_1}\,N_2\,\delta_{n''_2 n'_2}\,N_3\,\delta_{n''_3 n'_3}, \\
&= N_1 N_2 N_3\,\delta_{\mathbf{q_{n''}}\mathbf{q_{n'}}}.
\end{aligned} \tag{2.68}$$

The nuclei displacement $U_{\upsilon\alpha}(\mathbf{T_n})$ are real in contrast to the complex nuclei displacements $\mathfrak{U}_{\upsilon\alpha}(\mathbf{q_{n''}})$ in reciprocal space. For every wave-vector $\mathbf{q_{n''}}$, we can write the corresponding $3\,N_b$ numbers $\mathfrak{U}_{\upsilon\alpha}(\mathbf{q_{n''}})$ into a vector $\boldsymbol{\mathfrak{U}}(\mathbf{q_{n''}}) \in \mathbb{C}^{3\,N_b}$ and the corresponding $3\,N_b$ numbers $\mathfrak{P}_{\upsilon\alpha}(\mathbf{q_{n''}})$ into a vector $\boldsymbol{\mathfrak{P}}(\mathbf{q_{n''}}) \in \mathbb{C}^{3\,N_b}$. Using these vectors, we can transform the harmonic Hamilton function Eq. (2.37) to the reciprocal space. To do this, we use

$$\begin{aligned}
\mathbf{P}^t\cdot\mathbf{P} &= \sum_{\upsilon\alpha}\sum_{\mathbf{T_n}} P^*_{\upsilon\alpha}(\mathbf{T_n})\,P_{\upsilon\alpha}(\mathbf{T_n}) \\
&\overset{(2.64)}{=} \frac{1}{N_1 N_2 N_3}\sum_{\upsilon\alpha}\sum_{\mathbf{T_n}}\left(\sum_{\mathbf{q_{n'}}}\mathfrak{P}^*_{\upsilon\alpha}(\mathbf{q_{n'}})\,e^{-i\,\mathbf{q}^t_{\mathbf{n'}}\cdot\mathbf{T_n}}\right)\left(\sum_{\mathbf{q_{n''}}}\mathfrak{P}_{\upsilon\alpha}(\mathbf{q_{n''}})\,e^{i\,\mathbf{q}^t_{\mathbf{n''}}\cdot\mathbf{T_n}}\right) \\
&= \frac{1}{N_1 N_2 N_3}\sum_{\upsilon\alpha}\sum_{\mathbf{q_{n'}}}\mathfrak{P}^*_{\upsilon\alpha}(\mathbf{q_{n'}})\sum_{\mathbf{q_{n''}}}\mathfrak{P}_{\upsilon\alpha}(\mathbf{q_{n''}})\underbrace{\sum_{\mathbf{T_n}} e^{i(\mathbf{q_{n''}}-\mathbf{q_{n'}})^t\cdot\mathbf{T_n}}}_{\overset{(2.68)}{=}N_1 N_2 N_3\,\delta_{\mathbf{q_{n''}}\mathbf{q_{n'}}}} \\
&= \sum_{\mathbf{q_{n'}}}\sum_{\upsilon\alpha}\mathfrak{P}^*_{\upsilon\alpha}(\mathbf{q_{n'}})\mathfrak{P}_{\upsilon\alpha}(\mathbf{q_{n'}}) \\
&= \sum_{\mathbf{q_{n'}}}\boldsymbol{\mathfrak{P}}^\dagger(\mathbf{q_{n'}})\cdot\boldsymbol{\mathfrak{P}}(\mathbf{q_{n'}}).
\end{aligned} \tag{2.69}$$

Here, $\boldsymbol{\mathfrak{P}}^\dagger(\mathbf{q_{n'}})$ denotes the complex conjugated of the vector $\boldsymbol{\mathfrak{P}}(\mathbf{q_{n'}})$. Furthermore, we have

$$\begin{aligned}
\mathbf{U}^t\cdot\mathbf{D}\cdot\mathbf{U} &= \sum_{\upsilon\alpha}\sum_{\mathbf{T_n}} U^*_{\upsilon\alpha}(\mathbf{T_n})\sum_{\upsilon'\alpha'}\sum_{\mathbf{T_{n'}}} D_{\mathbf{T_n}\upsilon\alpha,\mathbf{T_{n'}}\upsilon'\alpha'}\,U_{\upsilon'\alpha'}(\mathbf{T_{n'}}) \\
&\overset{(2.63)}{=} \frac{1}{N_1 N_2 N_3}\sum_{\upsilon\alpha}\sum_{\mathbf{T_n}}\left(\sum_{\mathbf{q_{n''}}}\mathfrak{U}^*_{\upsilon\alpha}(\mathbf{q_{n''}})\,e^{-i\,\mathbf{q}^t_{\mathbf{n''}}\cdot\mathbf{T_n}}\right)\times
\end{aligned}$$

$$
\begin{aligned}
&\times \sum_{\upsilon'\alpha'} \sum_{\mathbf{T}_{\mathbf{n}'}} D_{\mathbf{T}_{\mathbf{n}}\upsilon\alpha,\mathbf{T}_{\mathbf{n}'}\upsilon'\alpha'} \left(\sum_{\mathbf{q}_{\mathbf{n}'''}} \mathfrak{U}_{\upsilon'\alpha'}(\mathbf{q}_{\mathbf{n}'''})\, \mathrm{e}^{\mathrm{i}\, \mathbf{q}_{\mathbf{n}'''}^{\mathrm{t}} \cdot \mathbf{T}_{\mathbf{n}'}} \right) \\
&= \frac{1}{N_1 N_2 N_3} \sum_{\upsilon\alpha} \sum_{\mathbf{q}_{\mathbf{n}''}} \mathfrak{U}^*_{\upsilon\alpha}(\mathbf{q}_{\mathbf{n}''}) \sum_{\upsilon'\alpha'} \sum_{\mathbf{q}_{\mathbf{n}'''}} \mathfrak{U}_{\upsilon'\alpha'}(\mathbf{q}_{\mathbf{n}'''}) \times \\
&\quad \times \sum_{\mathbf{T}_{\mathbf{n}'}} \mathrm{e}^{\mathrm{i}\, \mathbf{q}_{\mathbf{n}'''}^{\mathrm{t}} \cdot \mathbf{T}_{\mathbf{n}'}} \sum_{\mathbf{T}_{\mathbf{n}}} D_{\mathbf{T}_{\mathbf{n}}\upsilon\alpha,\mathbf{T}_{\mathbf{n}'}\upsilon'\alpha'}\, \mathrm{e}^{-\mathrm{i}\, \mathbf{q}_{\mathbf{n}''}^{\mathrm{t}} \cdot \mathbf{T}_{\mathbf{n}}} \\
&= \frac{1}{N_1 N_2 N_3} \sum_{\upsilon\alpha} \sum_{\mathbf{q}_{\mathbf{n}''}} \mathfrak{U}^*_{\upsilon\alpha}(\mathbf{q}_{\mathbf{n}''}) \sum_{\upsilon'\alpha'} \sum_{\mathbf{q}_{\mathbf{n}'''}} \mathfrak{U}_{\upsilon'\alpha'}(\mathbf{q}_{\mathbf{n}'''}) \times \\
&\quad \times \sum_{\mathbf{T}_{\mathbf{n}'}} \mathrm{e}^{\mathrm{i}\, (\mathbf{q}_{\mathbf{n}'''} - \mathbf{q}_{\mathbf{n}''})^{\mathrm{t}} \cdot \mathbf{T}_{\mathbf{n}'}} \sum_{\mathbf{T}_{\mathbf{n}}} D_{\mathbf{T}_{\mathbf{n}}\upsilon\alpha,\mathbf{T}_{\mathbf{n}'}\upsilon'\alpha'}\, \mathrm{e}^{-\mathrm{i}\, \mathbf{q}_{\mathbf{n}''}^{\mathrm{t}} \cdot (\mathbf{T}_{\mathbf{n}} - \mathbf{T}_{\mathbf{n}'})} \\
&\overset{(2.44)}{=} \frac{1}{N_1 N_2 N_3} \sum_{\upsilon\alpha} \sum_{\mathbf{q}_{\mathbf{n}''}} \mathfrak{U}^*_{\upsilon\alpha}(\mathbf{q}_{\mathbf{n}''}) \sum_{\upsilon'\alpha'} \sum_{\mathbf{q}_{\mathbf{n}'''}} \mathfrak{U}_{\upsilon'\alpha'}(\mathbf{q}_{\mathbf{n}'''}) \times \\
&\quad \times \sum_{\mathbf{T}_{\mathbf{n}'}} \mathrm{e}^{\mathrm{i}\, (\mathbf{q}_{\mathbf{n}'''} - \mathbf{q}_{\mathbf{n}''})^{\mathrm{t}} \cdot \mathbf{T}_{\mathbf{n}'}} \sum_{\mathbf{T}_{\mathbf{n}}} D_{\mathbf{T}_{\mathbf{n}} - \mathbf{T}_{\mathbf{n}'}\upsilon\alpha,\mathbf{0}\upsilon'\alpha'}\, \mathrm{e}^{-\mathrm{i}\, \mathbf{q}_{\mathbf{n}''}^{\mathrm{t}} \cdot (\mathbf{T}_{\mathbf{n}} - \mathbf{T}_{\mathbf{n}'})} \\
&= \frac{1}{N_1 N_2 N_3} \sum_{\upsilon\alpha} \sum_{\mathbf{q}_{\mathbf{n}''}} \mathfrak{U}^*_{\upsilon\alpha}(\mathbf{q}_{\mathbf{n}''}) \sum_{\upsilon'\alpha'} \sum_{\mathbf{q}_{\mathbf{n}'''}} \mathfrak{U}_{\upsilon'\alpha'}(\mathbf{q}_{\mathbf{n}'''}) \times \\
&\quad \times \sum_{\mathbf{T}_{\mathbf{n}'}} \mathrm{e}^{\mathrm{i}\, (\mathbf{q}_{\mathbf{n}'''} - \mathbf{q}_{\mathbf{n}''})^{\mathrm{t}} \cdot \mathbf{T}_{\mathbf{n}'}} \sum_{\mathbf{T}_{\mathbf{n}}} D_{\mathbf{T}_{\mathbf{n}}\upsilon\alpha,\mathbf{0}\upsilon'\alpha'}\, \mathrm{e}^{-\mathrm{i}\, \mathbf{q}_{\mathbf{n}''}^{\mathrm{t}} \cdot \mathbf{T}_{\mathbf{n}}}. \qquad (2.70)
\end{aligned}
$$

In the last step, we use the translation invariance with respect to the Bravais-grid, so that summing over $\mathbf{T_n}$ or $\mathbf{T_n} - \mathbf{T_{n'}}$ is equivalent. This allows the following definition:

Definition 2.5 (*Fourier transform of the dynamical matrix*) One defines the matrix $\mathbf{D}(\mathbf{q}_{\mathbf{n}''}) \in \mathbb{C}^{3N_b \times 3N_b}$ by

$$
D_{\upsilon\alpha,\upsilon'\alpha'}(\mathbf{q}_{\mathbf{n}''}) = \sum_{\mathbf{T}_{\mathbf{n}}} D_{\mathbf{T}_{\mathbf{n}}\upsilon\alpha,\mathbf{0}\upsilon'\alpha'}\, \mathrm{e}^{-\mathrm{i}\, \mathbf{q}_{\mathbf{n}''}^{\mathrm{t}} \cdot \mathbf{T}_{\mathbf{n}}}, \qquad (2.71)
$$

which depends on the wave-vector $\mathbf{q}_{\mathbf{n}''}$ and is the Fourier transform of the dynamical matrix.

Theorem 2.4 (Properties of the Fourier transform of the dynamical matrix) *For any wave-vector $\mathbf{q}_{\mathbf{n}''}$ that is compatible with the simulation cell, the Fourier* *transform* $\mathbf{D}(\mathbf{q}_{\mathbf{n}''})$ *of the dynamical matrix is hermitian:*

$$\mathbf{D}^{\dagger}(\mathbf{q}_{\mathbf{n}''}) = \mathbf{D}(\mathbf{q}_{\mathbf{n}''}). \tag{2.72}$$

Furthermore, one has

$$\mathbf{D}(-\mathbf{q}_{\mathbf{n}''}) = \mathbf{D}^{*}(\mathbf{q}_{\mathbf{n}''}) \tag{2.73}$$

and the real eigenvalues $\omega_j^2(\mathbf{q}_{\mathbf{n}''})$ *obey*

$$\omega_j^2(-\mathbf{q}_{\mathbf{n}''}) = \omega_j^2(\mathbf{q}_{\mathbf{n}''}), \tag{2.74}$$

so that the eigenvector $\mathbf{e}^{(j)}(-\mathbf{q}_{\mathbf{n}''})$ *is related to the same eigenvalue* $\omega_j^2(-\mathbf{q}_{\mathbf{n}''}) = \omega_j^2(\mathbf{q}_{\mathbf{n}''})$ *as the eigenvector* $\mathbf{e}^{(j)}(\mathbf{q}_{\mathbf{n}''})$*.*

Proof We obtain

$$\begin{aligned}
D^{\dagger}_{\upsilon\alpha,\upsilon'\alpha'}(\mathbf{q}_{\mathbf{n}''}) &= D^{*}_{\upsilon'\alpha',\upsilon\alpha}(\mathbf{q}_{\mathbf{n}''}) \\
&\overset{(2.71)}{=} \sum_{\mathbf{T}_{\mathbf{n}'}} D_{\mathbf{T}_{\mathbf{n}'}\upsilon'\alpha',\mathbf{0}\upsilon\alpha}\, \mathrm{e}^{\mathrm{i}\,\mathbf{q}^{\mathrm{t}}_{\mathbf{n}''}\cdot\mathbf{T}_{\mathbf{n}'}} \\
&\overset{(2.44)}{=} \sum_{\mathbf{T}_{\mathbf{n}'}} D_{\mathbf{0}\upsilon'\alpha',-\mathbf{T}_{\mathbf{n}'}\upsilon\alpha}\, \mathrm{e}^{\mathrm{i}\,\mathbf{q}^{\mathrm{t}}_{\mathbf{n}''}\cdot\mathbf{T}_{\mathbf{n}'}} \\
&= \sum_{\mathbf{T}_{\mathbf{n}'}} D_{-\mathbf{T}_{\mathbf{n}'}\upsilon\alpha,\mathbf{0}\upsilon'\alpha'}\, \mathrm{e}^{-\mathrm{i}\,\mathbf{q}^{\mathrm{t}}_{\mathbf{n}''}\cdot(-\mathbf{T}_{\mathbf{n}'})} \\
&= \sum_{\mathbf{T}_{\mathbf{n}'}} D_{\mathbf{T}_{\mathbf{n}'}\upsilon\alpha,\mathbf{0}\upsilon'\alpha'}\, \mathrm{e}^{-\mathrm{i}\,\mathbf{q}^{\mathrm{t}}_{\mathbf{n}''}\cdot\mathbf{T}_{\mathbf{n}'}} \\
&\overset{(2.71)}{=} D_{\upsilon\alpha,\upsilon'\alpha'}(\mathbf{q}_{\mathbf{n}''}),
\end{aligned}$$

which proves that $\mathbf{D}(\mathbf{q}_{\mathbf{n}''})$ is hermitian. Since $\mathbf{q}_{\mathbf{n}''}$, $\mathbf{T}_{\mathbf{n}}$ and $D_{\mathbf{T}_{\mathbf{n}}\upsilon\alpha,\mathbf{0}\upsilon'\alpha'}$ are real, Eq. (2.73) follows directly from the $\mathbf{D}(\mathbf{q}_{\mathbf{n}''})$ definition (2.71). Since $\mathbf{D}(\mathbf{q}_{\mathbf{n}''})$ is hermitian, its eigenvalues $\omega_j^2(\mathbf{q}_{\mathbf{n}''})$ are real. Let $\omega_j^2(\mathbf{q}_{\mathbf{n}''})$ be an eigenvalue of $\mathbf{D}(\mathbf{q}_{\mathbf{n}''})$ with corresponding eigenvector $\mathbf{e}^{(j)}(\mathbf{q}_{\mathbf{n}''})$:

$$\omega_j^2(\mathbf{q}_{\mathbf{n}''})\,\mathbf{e}^{(j)}(\mathbf{q}_{\mathbf{n}''}) = \mathbf{D}(\mathbf{q}_{\mathbf{n}''})\cdot\mathbf{e}^{(j)}(\mathbf{q}_{\mathbf{n}''}). \tag{2.75}$$

Considering the complex conjugate of the above equation, we obtain

$$\begin{aligned}
&\omega_j^2(\mathbf{q}_{\mathbf{n}''})\,\mathbf{e}^{(j)*}(\mathbf{q}_{\mathbf{n}''}) = \mathbf{D}^*(\mathbf{q}_{\mathbf{n}''})\cdot\mathbf{e}^{(j)*}(\mathbf{q}_{\mathbf{n}''}),\\
\Leftrightarrow\quad &\omega_j^2(\mathbf{q}_{\mathbf{n}''})\,\mathbf{e}^{(j)*}(\mathbf{q}_{\mathbf{n}''}) = \mathbf{D}(-\mathbf{q}_{\mathbf{n}''})\cdot\mathbf{e}^{(j)*}(\mathbf{q}_{\mathbf{n}''}),\\
\Leftrightarrow\quad &\omega_j^2(-\mathbf{q}_{\mathbf{n}''})\,\mathbf{e}^{(j)}(-\mathbf{q}_{\mathbf{n}''}) = \mathbf{D}(-\mathbf{q}_{\mathbf{n}''})\cdot\mathbf{e}^{(j)}(-\mathbf{q}_{\mathbf{n}''}).
\end{aligned} \tag{2.76}$$

Hence, $\mathbf{e}^{(j)}(-\mathbf{q}_{\mathbf{n}''}) = \mathbf{e}^{(j)*}(\mathbf{q}_{\mathbf{n}''})$ is related to the same eigenvalue $\omega_j^2(-\mathbf{q}_{\mathbf{n}''}) = \omega_j^2(\mathbf{q}_{\mathbf{n}''})$ as $\mathbf{e}^{(j)}(\mathbf{q}_{\mathbf{n}''})$. ■

With the help of the Fourier transform of the dynamical matrix, we obtain further from Eq. (2.70):

$$\begin{aligned}
\mathbf{U}^{\mathrm{t}}\cdot\mathbf{D}\cdot\mathbf{U} =& \frac{1}{N_1\,N_2\,N_3}\sum_{\upsilon\alpha}\sum_{\mathbf{q}_{\mathbf{n}''}}\mathfrak{U}^*_{\upsilon\alpha}(\mathbf{q}_{\mathbf{n}''})\sum_{\upsilon'\alpha'}\sum_{\mathbf{q}_{\mathbf{n}'''}}\mathfrak{U}_{\upsilon'\alpha'}(\mathbf{q}_{\mathbf{n}'''})\times\\
&\times\sum_{\mathbf{T}_{\mathbf{n}'}}\mathrm{e}^{\mathrm{i}\,(\mathbf{q}_{\mathbf{n}'''}-\mathbf{q}_{\mathbf{n}''})^{\mathrm{t}}\cdot\mathbf{T}_{\mathbf{n}'}}\underbrace{\sum_{\mathbf{T}_{\mathbf{n}}}D_{\mathbf{T}_{\mathbf{n}}\upsilon\alpha,\mathbf{0}\upsilon'\alpha'}\,\mathrm{e}^{-\mathrm{i}\,\mathbf{q}^{\mathrm{t}}_{\mathbf{n}''}\cdot\mathbf{T}_{\mathbf{n}}}}_{\overset{(2.71)}{=}D_{\upsilon\alpha,\upsilon'\alpha'}(\mathbf{q}_{\mathbf{n}'''})}\\
=& \frac{1}{N_1\,N_2\,N_3}\sum_{\upsilon\alpha}\sum_{\mathbf{q}_{\mathbf{n}''}}\mathfrak{U}^*_{\upsilon\alpha}(\mathbf{q}_{\mathbf{n}''})\sum_{\upsilon'\alpha'}\sum_{\mathbf{q}_{\mathbf{n}'''}}D_{\upsilon\alpha,\upsilon'\alpha'}(\mathbf{q}_{\mathbf{n}'''})\,\mathfrak{U}_{\upsilon'\alpha'}(\mathbf{q}_{\mathbf{n}'''})\times\\
&\times\underbrace{\sum_{\mathbf{T}_{\mathbf{n}'}}\mathrm{e}^{\mathrm{i}\,(\mathbf{q}_{\mathbf{n}'''}-\mathbf{q}_{\mathbf{n}''})^{\mathrm{t}}\cdot\mathbf{T}_{\mathbf{n}'}}}_{\overset{(2.68)}{=}N_1\,N_2\,N_3\,\delta_{\mathbf{q}_{\mathbf{n}'''}\mathbf{q}_{\mathbf{n}''}}}\\
=& \sum_{\mathbf{q}_{\mathbf{n}''}}\sum_{\upsilon\alpha}\mathfrak{U}^*_{\upsilon\alpha}(\mathbf{q}_{\mathbf{n}''})\sum_{\upsilon'\alpha'}D_{\upsilon\alpha,\upsilon'\alpha'}(\mathbf{q}_{\mathbf{n}''})\,\mathfrak{U}_{\upsilon'\alpha'}(\mathbf{q}_{\mathbf{n}''})\\
=& \sum_{\mathbf{q}_{\mathbf{n}''}}\mathfrak{U}^\dagger(\mathbf{q}_{\mathbf{n}''})\cdot\mathbf{D}(\mathbf{q}_{\mathbf{n}''})\cdot\mathfrak{U}(\mathbf{q}_{\mathbf{n}''}).
\end{aligned}$$

In summary, we have for the harmonic Hamilton function of the nuclei

$$\begin{aligned}
H^{(\mathrm{harm})} =& \frac{1}{2}\Big(\mathbf{P}^{\mathrm{t}}\cdot\mathbf{P} + \mathbf{U}^{\mathrm{t}}\cdot\mathbf{D}\cdot\mathbf{U}\Big)\\
=& \frac{1}{2}\sum_{\mathbf{q}_{\mathbf{n}}}\Big(\mathfrak{P}^\dagger(\mathbf{q}_{\mathbf{n}})\cdot\mathfrak{P}(\mathbf{q}_{\mathbf{n}}) + \mathfrak{U}^\dagger(\mathbf{q}_{\mathbf{n}})\cdot\mathbf{D}(\mathbf{q}_{\mathbf{n}})\cdot\mathfrak{U}(\mathbf{q}_{\mathbf{n}})\Big).
\end{aligned} \tag{2.77}$$

There are $N_1\,N_2\,N_3$ non-equivalent wave vectors $\mathbf{q}_{\mathbf{n}}$. The motion corresponding to any wave vector $\mathbf{q}_{\mathbf{n}}$ is independent to the other wave vectors. For each wave vector $\mathbf{q}_{\mathbf{n}}$, we have $3\,N_{\mathrm{b}}$ coupled harmonic oscillators. To decouple these oscillators, we have to diagonalize the Fourier transform of the dynamical matrix $\mathbf{D}(\mathbf{q}_{\mathbf{n}})$ for each wave vector $\mathbf{q}_{\mathbf{n}}$. We are searching for the $3\,N_{\mathrm{b}}$ eigenvectors $\mathbf{e}^{(j)}(\mathbf{q}_{\mathbf{n}})$ and corresponding eigenvalues $\omega_j^2(\mathbf{q}_{\mathbf{n}})$ of $\mathbf{D}(\mathbf{q}_{\mathbf{n}})$. $\mathbf{e}^{(j)}(\mathbf{q}_{\mathbf{n}})$ describes the nuclei motion directions in the basis cell for the jth eigenmode in dependence of the wave vector $\mathbf{q}_{\mathbf{n}}$. An unitary matrix $\mathbf{C}(\mathbf{q}_{\mathbf{n}}) \in \mathbb{C}^{3\,N_{\mathrm{b}}\times 3\,N_{\mathrm{b}}}$ with $\mathbf{C}^\dagger(\mathbf{q}_{\mathbf{n}})\cdot\mathbf{C}(\mathbf{q}_{\mathbf{n}}) = \mathbb{1}$ exist that diagonalizes $\mathbf{D}(\mathbf{q}_{\mathbf{n}})$:

$$\mathbf{C}(\mathbf{q_n})\cdot\mathbf{D}(\mathbf{q_n})\cdot\mathbf{C}^\dagger(\mathbf{q_n})=\mathbf{W}(\mathbf{q_n})=\begin{bmatrix}\omega_1^2(\mathbf{q_n}) & 0 & \dots & 0\\ 0 & \omega_2^2(\mathbf{q_n}) & 0 & \vdots\\ \vdots & 0 & \ddots & 0\\ 0 & \dots & 0 & \omega_{3N_\mathrm{b}}^2(\mathbf{q_n})\end{bmatrix}. \tag{2.78}$$

Here, $\mathbf{C}^\dagger(\mathbf{q_n})$ denotes the complex conjugated and transposed matrix of $\mathbf{C}(\mathbf{q_n})$. The columns of $\mathbf{C}^\dagger(\mathbf{q_{n''}})$ are the searched eigenvectors:

$$\mathbf{C}^\dagger(\mathbf{q_n})=\left[\mathbf{e}^{(1)}(\mathbf{q_n}),\mathbf{e}^{(2)}(\mathbf{q_n}),\dots,\mathbf{e}^{(3N_\mathrm{b})}(\mathbf{q_n})\right]. \tag{2.79}$$

With the help of $\mathbf{C}(\mathbf{q_n})$, we can project the nuclei displacements and momenta in reciprocal space onto the eigenmodes for each wave vector $\mathbf{q_n}$:

$$\mathcal{U}(\mathbf{q_n})=\mathbf{C}(\mathbf{q_n})\cdot\mathfrak{U}(\mathbf{q_n}), \tag{2.80}$$

$$\mathcal{P}(\mathbf{q_n})=\mathbf{C}(\mathbf{q_n})\cdot\mathfrak{P}(\mathbf{q_n}). \tag{2.81}$$

Then we obtain for the harmonic Hamilton function

$$\begin{aligned}H^{(\mathrm{harm})}=&\frac{1}{2}\sum_{\mathbf{q_n}}\mathfrak{P}^\dagger(\mathbf{q_n})\cdot\underbrace{\mathbf{C}^\dagger(\mathbf{q_n})\cdot\mathbf{C}(\mathbf{q_n})}_{=\mathbb{1}}\cdot\mathfrak{P}(\mathbf{q_n})\\ &+\frac{1}{2}\sum_{\mathbf{q_n}}\mathfrak{U}^\dagger(\mathbf{q_n})\cdot\underbrace{\mathbf{C}^\dagger(\mathbf{q_n})\cdot\mathbf{C}(\mathbf{q_n})}_{=\mathbb{1}}\cdot\mathbf{D}(\mathbf{q_n})\cdot\underbrace{\mathbf{C}^\dagger(\mathbf{q_n})\cdot\mathbf{C}(\mathbf{q_n})}_{=\mathbb{1}}\cdot\mathfrak{U}(\mathbf{q_n})\\ =&\frac{1}{2}\sum_{\mathbf{q_n}}\Big(\mathbf{C}(\mathbf{q_n})\cdot\mathfrak{P}(\mathbf{q_n})\Big)^\dagger\cdot\mathbf{C}(\mathbf{q_n})\cdot\mathfrak{P}(\mathbf{q_n})\\ &+\frac{1}{2}\Big(\mathbf{C}(\mathbf{q_n})\cdot\mathfrak{U}(\mathbf{q_n})\Big)^\dagger\cdot\underbrace{\mathbf{C}(\mathbf{q_n})\cdot\mathbf{D}(\mathbf{q_n})\cdot\mathbf{C}^\dagger(\mathbf{q_n})}_{=\mathbf{W}(\mathbf{q_n})}\cdot\mathbf{C}(\mathbf{q_n})\cdot\mathfrak{U}(\mathbf{q_n})\\ =&\frac{1}{2}\sum_{\mathbf{q_n}}\Big(\mathcal{P}^\dagger(\mathbf{q_n})\cdot\mathcal{P}(\mathbf{q_n})+\mathcal{U}^\dagger(\mathbf{q_n})\cdot\mathbf{W}(\mathbf{q_n})\cdot\mathcal{U}(\mathbf{q_n})\Big)\end{aligned} \tag{2.82}$$

and finally

$$\boxed{H^{(\mathrm{harm})}=\frac{1}{2}\sum_{\mathbf{q_n}}\sum_{i=1}^{3N_\mathrm{b}}\Big(\mathcal{P}_i^*(\mathbf{q_n})\,\mathcal{P}_i(\mathbf{q_n})+\omega_i^2(\mathbf{q_n})\,\mathcal{U}_i^*(\mathbf{q_n})\,\mathcal{U}_i(\mathbf{q_n})\Big).} \tag{2.83}$$

Similar to Eq. (2.42), this is the Hamilton function of $3\,N_\mathrm{at}$ independent harmonic oscillators, but we have complex amplitudes and momenta. To get this Hamilton function, we only need to diagonalize the $3\,N_\mathrm{b}\times 3\,N_\mathrm{b}$ matrix $\mathbf{D}(\mathbf{q_n})$ for $N_1\,N_2\,N_3$

wave vectors $\mathbf{q_n}$, which circumvents the diagonalization of the large $3\,N_1\,N_2\,N_3\,N_b \times 3\,N_1\,N_2\,N_3\,N_b$ dynamical matrix $\mathbf{D}$.

Now we solve the equations of motion in reciprocal space. For this, we first transform the Newton equation (2.43) to reciprocal space. We have for every $\mathbf{T_n}$, υ and α:

$$\ddot{U}_{\upsilon\alpha}(\mathbf{T_n}) = -\sum_{\mathbf{T_{n'}}}\sum_{\upsilon'\alpha'} D_{\mathbf{T_n}\upsilon\alpha,\mathbf{T_{n'}}\upsilon'\alpha'}\, U_{\upsilon'\alpha'}(\mathbf{T_{n'}})$$

$$\overset{(2.65)}{\Leftrightarrow} \quad \frac{1}{\sqrt{N_1\,N_2\,N_3}}\sum_{\mathbf{q_{n''}}} \ddot{\mathfrak{U}}_{\upsilon\alpha}(\mathbf{q_{n''}})\, \mathrm{e}^{\mathrm{i}\,\mathbf{q}^{\mathrm{t}}_{\mathbf{n''}}\cdot\mathbf{T_n}} = -\sum_{\mathbf{T_{n'}}}\sum_{\upsilon',\alpha'} D_{\mathbf{T_n}\upsilon\alpha,\mathbf{T_{n'}}\upsilon'\alpha'} \times$$

$$\times\left(\frac{1}{\sqrt{N_1\,N_2\,N_3}}\sum_{\mathbf{q_{n''}}} \mathfrak{U}_{\upsilon'\alpha'}(\mathbf{q_{n''}})\, \mathrm{e}^{\mathrm{i}\,\mathbf{q}^{\mathrm{t}}_{\mathbf{n''}}\cdot\mathbf{T_{n'}}}\right),$$

$$\Leftrightarrow \quad \frac{1}{N_1\,N_2\,N_3}\sum_{\mathbf{q_{n''}}} \ddot{\mathfrak{U}}_{\upsilon\alpha}(\mathbf{q_{n''}})\, \mathrm{e}^{\mathrm{i}\,\mathbf{q}^{\mathrm{t}}_{\mathbf{n''}}\cdot\mathbf{T_n}} = -\frac{1}{N_1\,N_2\,N_3}\sum_{\mathbf{q_{n''}}}\sum_{\upsilon',\alpha'} \mathfrak{U}_{\upsilon'\alpha'}(\mathbf{q_{n''}}) \times$$

$$\times\sum_{\mathbf{T_{n'}}} D_{\mathbf{T_n}\upsilon\alpha,\mathbf{T_{n'}}\upsilon'\alpha'}\, \mathrm{e}^{\mathrm{i}\,\mathbf{q}^{\mathrm{t}}_{\mathbf{n''}}\cdot\mathbf{T_{n'}}}.$$

We take an arbitrary wave vector $\mathbf{q_{n'''}}$, multiply the above equation by $\mathrm{e}^{-\mathrm{i}\,\mathbf{q}^{\mathrm{t}}_{\mathbf{n'''}}\cdot\mathbf{T_n}}$ and sum over all vectors $\mathbf{T_n}$ of the Bravais grid in the simulation cell. Doing this, we obtain for the left side

$$\frac{1}{N_1\,N_2\,N_3}\sum_{\mathbf{q_{n''}}} \ddot{\mathfrak{U}}_{\upsilon\alpha}(\mathbf{q_{n''}}) \underbrace{\sum_{\mathbf{T_n}} \mathrm{e}^{\mathrm{i}\,(\mathbf{q_{n''}}-\mathbf{q_{n'''}})^{\mathrm{t}}\cdot\mathbf{T_n}}}_{\overset{(2.68)}{=} N_1\,N_2\,N_3\,\delta_{\mathbf{q_{n''}}\mathbf{q_{n'''}}}} = \ddot{\mathfrak{U}}_{\upsilon\alpha}(\mathbf{q_{n'''}})$$

and for the right side

$$-\frac{1}{N_1\,N_2\,N_3}\sum_{\mathbf{q_{n''}}}\sum_{\upsilon',\alpha'} \mathfrak{U}_{\upsilon'\alpha'}(\mathbf{q_{n''}}) \sum_{\mathbf{T_n}} \mathrm{e}^{-\mathrm{i}\,\mathbf{q}^{\mathrm{t}}_{\mathbf{n'''}}\cdot\mathbf{T_n}} \sum_{\mathbf{T_{n'}}} D_{\mathbf{T_n}\upsilon\alpha,\mathbf{T_{n'}}\upsilon'\alpha'}\, \mathrm{e}^{\mathrm{i}\,\mathbf{q}^{\mathrm{t}}_{\mathbf{n''}}\cdot\mathbf{T_{n'}}}$$

$$= -\frac{1}{N_1\,N_2\,N_3}\sum_{\mathbf{q_{n''}}}\sum_{\upsilon',\alpha'} \mathfrak{U}_{\upsilon'\alpha'}(\mathbf{q_{n''}}) \times$$

$$\times\sum_{\mathbf{T_n}} \mathrm{e}^{\mathrm{i}\,(\mathbf{q_{n''}}-\mathbf{q_{n'''}})^{\mathrm{t}}\cdot\mathbf{T_n}} \sum_{\mathbf{T_{n'}}} D_{\mathbf{T_n}\upsilon\alpha,\mathbf{T_{n'}}\upsilon'\alpha'}\, \mathrm{e}^{-\mathrm{i}\,\mathbf{q}^{\mathrm{t}}_{\mathbf{n''}}\cdot(\mathbf{T_n}-\mathbf{T_{n'}})}$$

$$\overset{(2.44)}{=} -\frac{1}{N_1\,N_2\,N_3}\sum_{\mathbf{q_{n''}}}\sum_{\upsilon',\alpha'} \mathfrak{U}_{\upsilon'\alpha'}(\mathbf{q_{n''}}) \times$$

$$\times\sum_{\mathbf{T_n}} \mathrm{e}^{\mathrm{i}\,(\mathbf{q_{n''}}-\mathbf{q_{n'''}})^{\mathrm{t}}\cdot\mathbf{T_n}} \sum_{\mathbf{T_{n'}}} D_{\mathbf{T_n}-\mathbf{T_{n'}}\upsilon\alpha,\mathbf{0}\upsilon'\alpha'}\, \mathrm{e}^{-\mathrm{i}\,\mathbf{q}^{\mathrm{t}}_{\mathbf{n''}}\cdot(\mathbf{T_n}-\mathbf{T_{n'}})}$$

$$
\begin{aligned}
&= -\frac{1}{N_1\,N_2\,N_3}\sum_{\mathbf{q}_{\mathbf{n}''}}\sum_{\upsilon',\alpha'}\mathfrak{U}_{\upsilon'\alpha'}(\mathbf{q}_{\mathbf{n}''})\times\\
&\quad\times\sum_{\mathbf{T}_\mathbf{n}}\mathrm{e}^{\mathrm{i}\,(\mathbf{q}_{\mathbf{n}''}-\mathbf{q}_{\mathbf{n}'''})^{\mathrm{t}}\cdot\mathbf{T}_\mathbf{n}}\underbrace{\sum_{\mathbf{T}_{\mathbf{n}'}}D_{\mathbf{T}_{\mathbf{n}'}\upsilon\alpha,\mathbf{0}\upsilon'\alpha'}\,\mathrm{e}^{-\mathrm{i}\,\mathbf{q}^{\mathrm{t}}_{\mathbf{n}''}\cdot\mathbf{T}_{\mathbf{n}'}}}_{\stackrel{(2.71)}{=}D_{\upsilon\alpha,\upsilon'\alpha'}(\mathbf{q}_{\mathbf{n}''})}\\
&= -\frac{1}{N_1\,N_2\,N_3}\sum_{\mathbf{q}_{\mathbf{n}''}}\sum_{\upsilon',\alpha'}D_{\upsilon\alpha,\upsilon'\alpha'}(\mathbf{q}_{\mathbf{n}''})\,\mathfrak{U}_{\upsilon'\alpha'}(\mathbf{q}_{\mathbf{n}''})\underbrace{\sum_{\mathbf{T}_\mathbf{n}}\mathrm{e}^{\mathrm{i}\,(\mathbf{q}_{\mathbf{n}''}-\mathbf{q}_{\mathbf{n}'''})^{\mathrm{t}}\cdot\mathbf{T}_\mathbf{n}}}_{\stackrel{(2.68)}{=}N_1\,N_2\,N_3\,\delta_{\mathbf{q}_{\mathbf{n}''}\mathbf{q}_{\mathbf{n}'''}}}\\
&= -\sum_{\upsilon',\alpha'}D_{\upsilon\alpha,\upsilon'\alpha'}(\mathbf{q}_{\mathbf{n}'''})\,\mathfrak{U}_{\upsilon'\alpha'}(\mathbf{q}_{\mathbf{n}'''})
\end{aligned}
$$

and, therefore, in total

$$
\ddot{\mathfrak{U}}_{\upsilon\alpha}(\mathbf{q}_{\mathbf{n}'''}) = -\sum_{\upsilon',\alpha'}D_{\upsilon\alpha,\upsilon'\alpha'}(\mathbf{q}_{\mathbf{n}'''})\,\mathfrak{U}_{\upsilon'\alpha'}(\mathbf{q}_{\mathbf{n}'''}).
$$

Since the above equation is valid for every υ and α, we obtain finally for every non-equivalent wave vector $\mathbf{q}_{\mathbf{n}'''}$ that is compatible with the simulation cell:

$$
\begin{aligned}
&& \ddot{\mathfrak{U}}(\mathbf{q}_{\mathbf{n}'''}) &= -\,\mathbf{D}(\mathbf{q}_{\mathbf{n}'''})\cdot\mathfrak{U}(\mathbf{q}_{\mathbf{n}'''})\\
\Leftrightarrow\quad & \mathbf{C}(\mathbf{q}_{\mathbf{n}'''})\cdot\ddot{\mathfrak{U}}(\mathbf{q}_{\mathbf{n}'''}) &&= -\,\mathbf{C}(\mathbf{q}_{\mathbf{n}'''})\cdot\mathbf{D}(\mathbf{q}_{\mathbf{n}'''})\cdot\underbrace{\mathbf{C}^\dagger(\mathbf{q}_{\mathbf{n}'''})\cdot\mathbf{C}(\mathbf{q}_{\mathbf{n}'''})}_{=\mathbb{1}}\cdot\mathfrak{U}(\mathbf{q}_{\mathbf{n}'''})\\
\Leftrightarrow\quad && \ddot{\mathcal{U}}(\mathbf{q}_{\mathbf{n}'''}) &= -\,\mathbf{W}(\mathbf{q}_{\mathbf{n}'''})\cdot\mathcal{U}(\mathbf{q}_{\mathbf{n}'''}).
\end{aligned}
\tag{2.84}
$$

This means

$$
\ddot{\mathcal{U}}_j(\mathbf{q}_{\mathbf{n}'''}) = -\omega_j^2(\mathbf{q}_{\mathbf{n}'''})\,\mathcal{U}_j(\mathbf{q}_{\mathbf{n}'''})\qquad\forall\,j=1,\ldots,3\,N_\mathrm{b}\quad\text{and}\quad\forall\,\mathbf{q}_{\mathbf{n}'''}.\tag{2.85}
$$

This uncoupled equations of independent harmonic oscillators can be solved by

$$
\mathcal{U}_j(\mathbf{q}_{\mathbf{n}'''}) = \mathcal{U}_j^{(0)}(\mathbf{q}_{\mathbf{n}'''})\,\cos\Big(\omega_j(\mathbf{q}_{\mathbf{n}'''})\,t+\omega_j^{(0)}(\mathbf{q}_{\mathbf{n}'''})\Big)\tag{2.86}
$$

with $\mathcal{U}_j^{(0)}(\mathbf{q}_{\mathbf{n}'''})\in\mathbb{C}$ and $\omega_j^{(0)}(\mathbf{q}_{\mathbf{n}'''})\in\mathbb{R}$ being constants determined by the initial conditions. Due to Eq. (2.80), we have

$$
\mathfrak{U}(\mathbf{q}_{\mathbf{n}'''}) = \mathbf{C}^\dagger(\mathbf{q}_{\mathbf{n}'''})\cdot\mathcal{U}(\mathbf{q}_{\mathbf{n}'''}),\tag{2.87}
$$

$$
\stackrel{(2.79)}{\Leftrightarrow}\qquad\mathfrak{U}(\mathbf{q}_{\mathbf{n}'''}) = \sum_{j=1}^{3\,N_\mathrm{b}}\mathbf{e}^{\,(j)}(\mathbf{q}_{\mathbf{n}'''})\,\mathcal{U}_j(\mathbf{q}_{\mathbf{n}'''}),\tag{2.88}
$$

$$\Leftrightarrow \qquad \mathfrak{U}_{\upsilon\alpha}(\mathbf{q}_{\mathbf{n}'''}) = \sum_{j=1}^{3\,N_b} e_{\upsilon\alpha}^{(j)}(\mathbf{q}_{\mathbf{n}'''})\,\mathcal{U}_j(\mathbf{q}_{\mathbf{n}'''}), \tag{2.89}$$

and get

$$\mathfrak{U}(\mathbf{q}_{\mathbf{n}'''}) = \sum_{j=1}^{3\,N_b} \mathbf{e}^{(j)}(\mathbf{q}_{\mathbf{n}'''})\,\mathcal{U}_j^{(0)}(\mathbf{q}_{\mathbf{n}'''})\,\cos\Big(\omega_j(\mathbf{q}_{\mathbf{n}'''})\,t + \omega_j^{(0)}(\mathbf{q}_{\mathbf{n}'''})\Big). \tag{2.90}$$

Using the inverse Fourier transform (2.65), we obtain further

$$U_{\upsilon\alpha}(\mathbf{T}_\mathbf{n}) = \frac{1}{\sqrt{N_1\,N_2\,N_3}} \sum_{\mathbf{q}_{\mathbf{n}'''}} \sum_{j=1}^{3\,N_b} e_{\upsilon\alpha}^{(j)}(\mathbf{q}_{\mathbf{n}'''})\,\mathrm{e}^{\mathrm{i}\,\mathbf{q}_{\mathbf{n}'''}^{t}\cdot\mathbf{T}_\mathbf{n}}\,\mathcal{U}_j^{(0)}(\mathbf{q}_{\mathbf{n}'''}) \times$$
$$\times \cos\Big(\omega_j(\mathbf{q}_{\mathbf{n}'''})\,t + \omega_j^{(0)}(\mathbf{q}_{\mathbf{n}'''})\Big). \tag{2.91}$$

Adding a vector $\mathbf{G}_\mathbf{n}$ of the reciprocal grid to any wave vector $\mathbf{q}_{\mathbf{n}'''}$ produces an equivalent wave vector. Therefore, it is equivalent for the above sum, if we use $-\mathbf{q}_{\mathbf{n}'''}$ instead of $\mathbf{q}_{\mathbf{n}'''}$. Due to $\omega_j^2(-\mathbf{q}_{\mathbf{n}'''}) = \omega_j^2(\mathbf{q}_{\mathbf{n}'''})$ and $e_{\upsilon\alpha}^{(j)}(-\mathbf{q}_{\mathbf{n}'''}) = e_{\upsilon\alpha}^{(j)*}(\mathbf{q}_{\mathbf{n}'''})$, we set $\omega_j(-\mathbf{q}_{\mathbf{n}'''}) = \omega_j(\mathbf{q}_{\mathbf{n}'''})$, $\omega_j^{(0)}(-\mathbf{q}_{\mathbf{n}'''}) = \omega_j^{(0)}(\mathbf{q}_{\mathbf{n}'''})$ and $\mathcal{U}_j^{(0)}(-\mathbf{q}_{\mathbf{n}'''}) = \mathcal{U}_j^{(0)*}(\mathbf{q}_{\mathbf{n}'''})$. Therefore, using $-\mathbf{q}_{\mathbf{n}'''}$ instead of $\mathbf{q}_{\mathbf{n}'''}$ in the sum contained in Eq. (2.91) produces the complex conjugated result. To get finally real displacements, we use in the sum the negative and positive non-equivalent wave vectors and perform a linear combination of both results that is real:

$$U_{\upsilon\alpha}(\mathbf{T}_\mathbf{n}) = \frac{1}{\sqrt{N_1\,N_2\,N_3}} \sum_{\mathbf{q}_{\mathbf{n}'''}} \sum_{j=1}^{3\,N_b} \frac{1}{2}\,\cos\Big(\omega_j(\mathbf{q}_{\mathbf{n}'''})\,t + \omega_j^{(0)}(\mathbf{q}_{\mathbf{n}'''})\Big) \times$$
$$\times \Big((1-\mathrm{i})\,e_{\upsilon\alpha}^{(j)}(\mathbf{q}_{\mathbf{n}'''})\,\mathrm{e}^{\mathrm{i}\,\mathbf{q}_{\mathbf{n}'''}^{t}\cdot\mathbf{T}_\mathbf{n}}\,\mathcal{U}_j^{(0)}(\mathbf{q}_{\mathbf{n}'''})$$
$$+(1+\mathrm{i})\,e_{\upsilon\alpha}^{(j)}(-\mathbf{q}_{\mathbf{n}'''})\,\mathrm{e}^{-\mathrm{i}\,\mathbf{q}_{\mathbf{n}'''}^{t}\cdot\mathbf{T}_\mathbf{n}}\,\mathcal{U}_j^{(0)}(-\mathbf{q}_{\mathbf{n}'''})\Big). \tag{2.92}$$

The term in the bracket is real due to

$$(1-\mathrm{i})\,(a+\mathrm{i}\,b) + (1+\mathrm{i})\,(a-\mathrm{i}\,b) = a+\mathrm{i}\,b-\mathrm{i}\,a-\mathrm{i}^2\,b+a-\mathrm{i}\,b+\mathrm{i}\,a-\mathrm{i}^2\,b$$
$$=2\,a+2\,b.$$

Other linear combinations of $(a+\mathrm{i}\,b)$ and $(a-\mathrm{i}\,b)$ would also work, but one has to take care that a and b both occur in the final result, so that there is no information loss.

In the harmonic approximation, the movement of the nuclei can be analytically described by independent oscillations in directions of phonon modes, which belong to a certain phonon branch j and have a certain wave vector $\mathbf{q}_\mathbf{n}$. The presented description of the nuclei utilizes the harmonic approximation of the PES $\mathcal{E}_\beta(\{\mathbf{r}_{\mathbf{n}\upsilon}\})$ generated by the electrons in the state β. To get this electronic PES, we have to solve the problem of interacting electrons in an external potential of fixed nuclei at coordinates $\{\mathbf{r}_{\mathbf{n}\upsilon}\}$. But, in contrast to the nuclei, the electrons cannot be treated as classical particles, so that a quantum mechanical description is needed.

2.3 Density Functional Theory

The great challenge in the quantum mechanical description of the electrons is finding the high dimensional electronic wave function $\phi_{\alpha,\mathbf{r}_1,\ldots,\mathbf{r}_{N_\mathrm{at}}}(\mathbf{R}_1,\ldots,\mathbf{R}_{N_\mathrm{e}})$ in order to describe and to calculate the properties of the interacting electrons in the external potential of fixed nuclei. This many-body wave function is found by solving the time-independent Schrödinger equation (2.19) with the electronic Hamiltonian

$$\hat{H}_\mathrm{e} = \underbrace{-\sum_{i=1}^{N_\mathrm{e}} \frac{\hbar^2}{2m_\mathrm{e}^2}\nabla^2_{\mathbf{R}_i}}_{=\hat{T}_\mathrm{e}} + \underbrace{\sum_{i=1}^{N_\mathrm{e}}\sum_{j=i+1}^{N_\mathrm{e}} \frac{e^2}{4\pi\varepsilon_0}\frac{1}{|\mathbf{R}_i-\mathbf{R}_j|}}_{=\hat{V}_\mathrm{int}} + \underbrace{\sum_{i=1}^{N_\mathrm{e}} V_\mathrm{ext}(\mathbf{R}_i)}_{=\hat{V}_\mathrm{ext}}$$
$$+ \underbrace{\sum_{k=1}^{N_\mathrm{at}}\sum_{\ell=k+1}^{N_\mathrm{at}} \frac{e^2}{4\pi\varepsilon_0}\frac{Z_k Z_\ell}{|\mathbf{r}_k-\mathbf{r}_\ell|}}_{=E_\mathrm{nn}}, \tag{2.93}$$

where

$$V_\mathrm{ext}(\mathbf{R}) = -\sum_{k=1}^{N_\mathrm{at}} \frac{e^2}{4\pi\varepsilon_0}\frac{Z_k}{|\mathbf{r}_k-\mathbf{R}|}. \tag{2.94}$$

E_nn in Eq. (2.93) is a constant and describes the interaction energy of the nuclei. In most of the cases, only the ground state of the electrons, which is described by the ground state wave function $\phi_{0,\mathbf{r}_1,\ldots,\mathbf{r}_{N_\mathrm{at}}}(\mathbf{R}_1,\ldots,\mathbf{R}_{N_\mathrm{e}})$, is needed. The main idea behind density functional theory (DFT) is using the electron density of the ground state $n_{\mathrm{e}_0}(\mathbf{r})$ as the main variable to characterize the ground state instead of using the ground state wave function $\phi_{0,\mathbf{r}_1,\ldots,\mathbf{r}_{N_\mathrm{at}}}(\mathbf{R}_1,\ldots,\mathbf{R}_{N_\mathrm{e}})$. The electron density $n_\mathrm{e}(\mathbf{r})$ can be calculated for any state $|\phi\rangle$ from the electron density operator

$$\hat{n}_\mathrm{e}(\mathbf{r}) = \sum_{i=1}^{N_\mathrm{e}} \delta(\mathbf{r}-\mathbf{R}_i) \tag{2.95}$$

as

$$
\begin{aligned}
n_e(\mathbf{r}) =& \langle \phi | \hat{n}_e(\mathbf{r}) | \phi \rangle \\
=& \int_{\mathbb{R}^3} d^3R_1 \cdots \int_{\mathbb{R}^3} d^3R_{N_e}\, \phi^*(\mathbf{R}_1, \dots, \mathbf{R}_{N_e}) \left(\sum_{i=1}^{N_e} \delta(\mathbf{r} - \mathbf{R}_i) \right) \phi(\mathbf{R}_1, \dots, \mathbf{R}_{N_e}) \\
=& \sum_{i=1}^{N_e} \int_{\mathbb{R}^3} d^3R_1 \cdots \int_{\mathbb{R}^3} d^3R_{i-1} \int_{\mathbb{R}^3} d^3R_{i+1} \cdots \int_{\mathbb{R}^3} d^3R_{N_e} \times \\
& \times \left| \phi(\mathbf{R}_1, \dots, \mathbf{R}_{i-1}, \mathbf{r}, \mathbf{R}_{i+1}, \dots, \mathbf{R}_{N_e}) \right|^2 \\
=& \sum_{i=1}^{N_e} \int_{\mathbb{R}^3} d^3R_1 \cdots \int_{\mathbb{R}^3} d^3R_{i-1} \int_{\mathbb{R}^3} d^3R_{i+1} \cdots \int_{\mathbb{R}^3} d^3R_{N_e} \times \\
& \times \left| \phi(\mathbf{r}, \dots, \mathbf{R}_{i-1}, \mathbf{R}_1, \mathbf{R}_{i+1}, \dots, \mathbf{R}_{N_e}) \right|^2 \\
=& N_e \int_{\mathbb{R}^3} d^3R_2 \cdots \int_{\mathbb{R}^3} d^3R_{N_e} \left| \phi(\mathbf{r}, \mathbf{R}_2, \dots, \mathbf{R}_{N_e}) \right|^2 .
\end{aligned}
\tag{2.96}
$$

Here we used that interchanging two electrons does not change $|\phi|^2$, since electrons are indistinguishable. By construction, the electron density $n_e(\mathbf{r})$ obeys

$$
N_e = \int_{\mathbb{R}^3} d^3r\, n_e(\mathbf{r}). \tag{2.97}
$$

In the electronic Hamiltonian (2.93), the terms $\hat{T}_e$ and $\hat{V}_{int}$ look always the same for any electronic system, whereas only $\hat{V}_{ext}$ and E_{nn} are specific for a given system. Fortunately, with E_{nn} being a constant $\langle \phi | \hat{V}_{ext} | \phi \rangle$ can be directly calculated from the electron density as:

$$
\begin{aligned}
\langle \phi | \hat{V}_{ext} | \phi \rangle =& \int_{\mathbb{R}^3} d^3R_1 \cdots \int_{\mathbb{R}^3} d^3R_{N_e} \phi^*(\mathbf{R}_1, \dots, \mathbf{R}_{N_e}) \left(\sum_{i=1}^{N_e} V_{ext}(\mathbf{R}_i) \right) \phi(\mathbf{R}_1, \dots, \mathbf{R}_{N_e}) \\
=& \int_{\mathbb{R}^3} d^3R_1 \cdots \int_{\mathbb{R}^3} d^3R_{N_e} \left(\sum_{i=1}^{N_e} V_{ext}(\mathbf{R}_1) \right) \times \\
& \times \left| \phi(\mathbf{R}_i, \mathbf{R}_2, \dots, \mathbf{R}_{i-1}, \mathbf{R}_1, \mathbf{R}_{i+1}, \dots, \mathbf{R}_{N_e}) \right|^2
\end{aligned}
$$

$$
\begin{aligned}
&= \int_{\mathbb{R}^3} d^3R_1 \cdots \int_{\mathbb{R}^3} d^3R_{N_e} \left(\sum_{i=1}^{N_e} V_{\text{ext}}(\mathbf{R}_1) \right) \left| \phi(\mathbf{R}_1, \ldots, \mathbf{R}_{N_e}) \right|^2 \\
&= \int_{\mathbb{R}^3} d^3R_1 \left(\sum_{i=1}^{N_e} V_{\text{ext}}(\mathbf{R}_1) \right) \int_{\mathbb{R}^3} d^3R_2 \cdots \int_{\mathbb{R}^3} d^3R_{N_e} \left| \phi(\mathbf{R}_1, \ldots, \mathbf{R}_{N_e}) \right|^2 \\
&= \int_{\mathbb{R}^3} d^3r \, V_{\text{ext}}(\mathbf{r}) \, N_e \int_{\mathbb{R}^3} d^3R_2 \cdots \int_{\mathbb{R}^3} d^3R_{N_e} \left| \phi(\mathbf{r}, \mathbf{R}_2, \ldots, \mathbf{R}_{N_e}) \right|^2 \\
&= \int_{\mathbb{R}^3} d^3r \, V_{\text{ext}}(\mathbf{r}) \, n_e(\mathbf{r}).
\end{aligned} \tag{2.98}
$$

Since the only system specific term $\hat{V}_{\text{ext}}$ in the electronic Hamiltonian $\hat{H}_e$ is directly related to the electron density, it seems possible to utilize the electron density as the main variable instead of the many-body wave function. The following derivation of DFT is based on Ref [8], but we provide further details where necessary for a better understanding.

Since in this book we are concerned only with non-magnetic materials and systems for which relativistic effects can be neglected, the electronic Hamiltonian (2.93) does not explicitly depend on spin, hence half of the electrons have spin up and the other half spin down. Therefore, the spin does not need to be considered explicitly and a total electron density is sufficient for describing both spin up and spin down electrons. In the other cases, one has to consider explicitly the spin by introducing an additional spin density that is the difference between the density of spin up and down electrons [8].

2.3.1 Hohenberg-Kohn Theorems

A high reduction of complexity is reached by using the electron density, which is just a function in three dimensions in contrast to the high dimensional electronic wave function. One may ask whether there is a great information loss along with using the electron density as the basis variable instead of the many-body electronic wave function for the description of a system of interacting electrons in an external potential. Fortunately, the answer is *no* for ground states, which follows from two theorems formulated by Hohenberg and Kohn in 1964 [9]:

Theorem 2.5 (Hohenberg-Kohn unique density theorem) *The ground state electron density $n_{e_0}(\mathbf{r})$ determines uniquely, except for a constant, the external potential $V_{ext}(\mathbf{r})$ in the Hamiltonian for any system of interacting electrons.*

Due to the fully determination of the Hamiltonian except for a constant shift of the energy, the many-body wave functions are determined for all states (ground and excited). Therefore all properties of the system are completely determined by the ground state electron density $n_{e_0}(\mathbf{r})$.

Proof We assume, that two external potentials $V_{\mathrm{ext}}^{(1)}(\mathbf{r})$ and $V_{\mathrm{ext}}^{(2)}(\mathbf{r})$ exist, which differ by more than a constant and which lead to the same ground state electron density $n_{\mathrm{e}_0}(\mathbf{r})$. These two external potentials lead to different Hamiltonians $\hat{H}_{\mathrm{e}}^{(1)}$ and $\hat{H}_{\mathrm{e}}^{(2)}$ with nuclei interaction energies $E_{\mathrm{nn}}^{(1)}$ and $E_{\mathrm{nn}}^{(2)}$, respectively, and same kinetic energy $\hat{T}_{\mathrm{e}}$ of the electrons. We assume that both Hamiltonians have got a non-degenerate ground state with associated normalized ground state wave function $\Psi^{(1)}$ and $\Psi^{(2)}$, respectively. The proof for degenerated ground states can be found in Ref. [10]. Since the ground state is not-degenerate and $\Psi^{(2)} \neq \Psi^{(1)}$ is not the ground state wave function of $\hat{H}_{\mathrm{e}}^{(1)}$, we obtain for the ground state energy $E^{(1)}$:

$$\begin{aligned}
E^{(1)} &= \left\langle \Psi^{(1)} \middle| \hat{H}_{\mathrm{e}}^{(1)} \middle| \Psi^{(1)} \right\rangle \\
&< \left\langle \Psi^{(2)} \middle| \hat{H}_{\mathrm{e}}^{(1)} \middle| \Psi^{(2)} \right\rangle \\
&= \left\langle \Psi^{(2)} \middle| \hat{H}_{\mathrm{e}}^{(2)} \middle| \Psi^{(2)} \right\rangle + \left\langle \Psi^{(2)} \middle| \hat{H}_{\mathrm{e}}^{(1)} - \hat{H}_{\mathrm{e}}^{(2)} \middle| \Psi^{(2)} \right\rangle \\
&= E^{(2)} + \left\langle \Psi^{(2)} \middle| E_{\mathrm{nn}}^{(1)} - E_{\mathrm{nn}}^{(2)} \middle| \Psi^{(2)} \right\rangle + \left\langle \Psi^{(2)} \middle| \hat{V}_{\mathrm{ext}}^{(1)} - \hat{V}_{\mathrm{ext}}^{(2)} \middle| \Psi^{(2)} \right\rangle \\
&= E^{(2)} + \left(E_{\mathrm{nn}}^{(1)} - E_{\mathrm{nn}}^{(2)}\right) + \left\langle \Psi^{(2)} \middle| \hat{V}_{\mathrm{ext}}^{(1)} - \hat{V}_{\mathrm{ext}}^{(2)} \middle| \Psi^{(2)} \right\rangle \\
&\overset{(2.98)}{=} E^{(2)} + \left(E_{\mathrm{nn}}^{(1)} - E_{\mathrm{nn}}^{(2)}\right) + \int_{\mathbb{R}^3} d^3r \left(V_{\mathrm{ext}}^{(1)}(\mathbf{r}) - V_{\mathrm{ext}}^{(2)}(\mathbf{r})\right) n_{\mathrm{e}_0}(\mathbf{r}).
\end{aligned} \tag{2.99}$$

If we consider the ground state energy $E^{(2)}$ of $\hat{H}^{(2)}$ in exactly the same way, we get

$$E^{(2)} < E^{(1)} + \left(E_{\mathrm{nn}}^{(2)} - E_{\mathrm{nn}}^{(1)}\right) + \int_{\mathbb{R}^3} d^3r \left(V_{\mathrm{ext}}^{(2)}(\mathbf{r}) - V_{\mathrm{ext}}^{(1)}(\mathbf{r})\right) n_{\mathrm{e}_0}(\mathbf{r}). \tag{2.100}$$

Adding the the two latter equations, we obtain the contradictory inequality

$$E^{(1)} + E^{(2)} < E^{(1)} + E^{(2)}. \tag{2.101}$$

Hence, our initial assumption is wrong and a ground state electron density $n_{\mathrm{e}_0}(\mathbf{r})$ corresponds to exactly one external potential $V_{\mathrm{ext}}(\mathbf{r})$ in the Hamiltonian $\hat{H}_{\mathrm{e}}$. Since the Hamiltonian is uniquely defined except for a constant by setting the external potential, the many-body wave functions of all states are fully determined by solving the Schrödinger equation with this Hamiltonian. ∎

Theorem 2.6 (Hohenberg-Kohn energy functional) *An electron density $n_e(\mathbf{r})$ that is the ground state density of a Hamiltonian with some external potential $V_{ext}(\mathbf{r})$ is called V-representable. On the space of V-representable electron densities, a functional for the energy $E_{HK}\big(n_e(\mathbf{r}), V_{ext}(\mathbf{r}), E_{nn}\big)$ in terms of the electron density $n_e(\mathbf{r})$, the external potential $V_{ext}(\mathbf{r})$, and the nuclei interaction energy E_{nn} can be defined. For any particular $V_{ext}(\mathbf{r})$ and E_{nn}, the exact ground state electron density $n_{e_0}(\mathbf{r})$ corresponds to the V-representable electron density $n_e(\mathbf{r})$ that minimizes this functional and the exact ground state energy of the system corresponds to the global minimum value of this functional. Hence, the functional $E_{HK}\big(n_e(\mathbf{r}), V_{ext}(\mathbf{r}), E_{nn}\big)$ is sufficient to determine the exact ground state energy and density.*

Proof We consider a particular Hamiltonian $\hat{H}_{\mathrm{e}}^{(1)}$ with external potential $V_{\mathrm{ext}}^{(1)}(\mathbf{r})$ and nuclei interaction energy $E_{\mathrm{nn}}^{(1)}$. Let $E^{(1)}$ be the energy, $\Psi^{(1)}$ the normalized wave function and $n_{\mathrm{e}}^{(1)}(\mathbf{r})$ the electronic density of the ground state of $\hat{H}_{\mathrm{e}}^{(1)}$. Since the Hohenberg-Kohn Theorem 1 states that the ground state electron density determines uniquely all properties of the system, especially the energy, there exists a one to one map between the ground state density and the ground state energy. Hence,

$$E_{\mathrm{HK}}\left(n_{\mathrm{e}}^{(1)}(\mathbf{r}),\, V_{\mathrm{ext}}^{(1)}(\mathbf{r}),\, E_{\mathrm{nn}}^{(1)}\right) = E^{(1)} \tag{2.102}$$

is well defined. Since the ground state energy $E^{(1)}$ obeys

$$\begin{aligned} E^{(1)} &= \langle\Psi^{(1)}|\hat{H}_{\mathrm{e}}^{(1)}|\Psi^{(1)}\rangle \\ &= \langle\Psi^{(1)}|\hat{T}_{\mathrm{e}} + \hat{V}_{\mathrm{int}}|\Psi^{(1)}\rangle + \langle\Psi^{(1)}|\hat{V}_{\mathrm{ext}}^{(1)}|\Psi^{(1)}\rangle + \langle\Psi^{(1)}|E_{\mathrm{nn}}^{(1)}|\Psi^{(1)}\rangle \\ &\overset{(2.98)}{=} \langle\Psi^{(1)}|\hat{T}_{\mathrm{e}} + \hat{V}_{\mathrm{int}}|\Psi^{(1)}\rangle + \int_{\mathbb{R}^3} d^3r\, V_{\mathrm{ext}}^{(1)}(\mathbf{r})\, n_{\mathrm{e}}^{(1)}(\mathbf{r}) + E_{\mathrm{nn}}^{(1)}, \end{aligned} \tag{2.103}$$

we get by equating Eq. (2.103) and Eq. (2.102)

$$\langle\Psi^{(1)}|\hat{T}_{\mathrm{e}} + \hat{V}_{\mathrm{int}}|\Psi^{(1)}\rangle = E_{\mathrm{HK}}\left(n_{\mathrm{e}}^{(1)}(\mathbf{r}),\, V_{\mathrm{ext}}^{(1)}(\mathbf{r}),\, E_{\mathrm{nn}}^{(1)}\right) - \int_{\mathbb{R}^3} d^3r\, V_{\mathrm{ext}}^{(1)}(\mathbf{r})\, n_{\mathrm{e}}^{(1)}(\mathbf{r}) - E_{\mathrm{nn}}^{(1)}. \tag{2.104}$$

Since the right hand side is a functional of the electron density $n_{\mathrm{e}}^{(1)}(\mathbf{r})$, the term

$$F_{\mathrm{HK}}\left(n_{\mathrm{e}}^{(1)}(\mathbf{r})\right) := \langle\Psi^{(1)}|\hat{T}_{\mathrm{e}} + \hat{V}_{\mathrm{int}}|\Psi^{(1)}\rangle \tag{2.105}$$

must be also a functional of the electron density $n_{\mathrm{e}}^{(1)}(\mathbf{r})$. This term F_{HK} describes the kinetic and interaction energy of the electrons as a function of the electron density and is universal, because it depends neither on the external potential $\hat{V}_{\text{ext}}^{(1)}(\mathbf{r})$ nor on the nuclei interaction energy E_{nn}. Hence, E_{HK} is also defined for any other V-representable electron density $n_{\text{e}}^{(2)}(\mathbf{r})$ that differs from $n_{\text{e}}^{(1)}(\mathbf{r})$ and that is, by definition, the ground state density of a Hamiltonian $\hat{H}_{\text{e}}^{(2)}$ with external potential $V_{\text{ext}}^{(2)}(\mathbf{r})$, nuclei interaction energy $E_{\text{nn}}^{(2)}$, and related normalized ground state wave function $\Psi^{(2)}$ by

$$\begin{aligned} E_{\text{HK}}\left(n_{\text{e}}^{(2)}(\mathbf{r}), V_{\text{ext}}^{(1)}(\mathbf{r}), E_{\text{nn}}^{(1)}\right) =& F_{\text{HK}}\left(n_{\text{e}}^{(2)}(\mathbf{r})\right) + \int_{\mathbb{R}^3} d^3r\, V_{\text{ext}}^{(1)}(\mathbf{r})\, n_{\text{e}}^{(2)}(\mathbf{r}) + E_{\text{nn}}^{(1)} \\ =& \left\langle \Psi^{(2)} \middle| \hat{T}_{\text{e}} + \hat{V}_{\text{int}} \middle| \Psi^{(2)} \right\rangle + \left\langle \Psi^{(2)} \middle| \hat{V}_{\text{ext}}^{(1)} \middle| \Psi^{(2)} \right\rangle \\ &+ \left\langle \Psi^{(2)} \middle| E_{\text{nn}}^{(1)} \middle| \Psi^{(2)} \right\rangle \\ =& \left\langle \Psi^{(2)} \middle| \hat{H}_{\text{e}}^{(1)} \middle| \Psi^{(2)} \right\rangle. \end{aligned} \tag{2.106}$$

The later denotes the energy of the interacting electrons described by the Hamiltonian $\hat{H}_{\text{e}}^{(1)}$ with external potential $V_{\text{ext}}^{(1)}(\mathbf{r})$ and nuclei interaction energy $E_{\text{nn}}^{(1)}$ in the state $|\Psi^{(2)}\rangle$ that belongs to the density $n_{\text{e}}^{(2)}(\mathbf{r})$. Since $n_{\text{e}}^{(2)}(\mathbf{r})$ differs from $n_{\text{e}}^{(1)}(\mathbf{r})$, the related wave functions $\Psi^{(2)}$ and $\Psi^{(1)}$ are also different. Since $\Psi^{(1)}$ is the unique ground state wave function of Hamiltonian $\hat{H}_{\text{e}}^{(1)}$, we obtain

$$\begin{aligned} E_{\text{HK}}\left(n_{\text{e}}^{(1)}(\mathbf{r}), V_{\text{ext}}^{(1)}(\mathbf{r}), E_{\text{nn}}^{(1)}\right) &\overset{(2.102)}{=} E^{(1)} \\ &= \left\langle \Psi^{(1)} \middle| \hat{H}_{\text{e}}^{(1)} \middle| \Psi^{(1)} \right\rangle \\ &< \left\langle \Psi^{(2)} \middle| \hat{H}_{\text{e}}^{(1)} \middle| \Psi^{(2)} \right\rangle \\ &\overset{(2.106)}{=} E_{\text{HK}}\left(n_{\text{e}}^{(2)}(\mathbf{r}), V_{\text{ext}}^{(1)}(\mathbf{r}), E_{\text{nn}}^{(1)}\right). \end{aligned} \tag{2.107}$$

So, the functional $E_{\text{HK}}\left(n_{\text{e}}(\mathbf{r}), V_{\text{ext}}^{(1)}(\mathbf{r}), E_{\text{nn}}^{(1)}\right)$ returns its global minimum, if $n_{\text{e}}(\mathbf{r})$ equals the ground state density $n_{\text{e}}^{(2)}(\mathbf{r})$, and the related minimal value of the functional corresponds to the ground state energy $E^{(1)}$. Hence, from the knowledge of the functional E_{HK}, one can exactly determine the ground state density and energy. ■

The Hohenberg-Kohn Theorem 1 justifies that the ground state of a system of interacting electrons in an external potential $V_{\text{ext}}(\mathbf{r})$ can be fully described by the ground state electron density $n_{\text{e}_0}(\mathbf{r})$. The Hohenberg-Kohn Theorem 2 provides the existence of a functional

$$E_{\text{HK}}\left(n_{\text{e}}(\mathbf{r}), V_{\text{ext}}(\mathbf{r}), E_{\text{nn}}\right) = F_{\text{HK}}\left(n_{\text{e}}(\mathbf{r})\right) + \int_{\mathbb{R}^3} d^3r\, V_{\text{ext}}(\mathbf{r})\, n_{\text{e}}(\mathbf{r}) + E_{\text{nn}},$$

from which the energy of the system can be directly calculated from the electron density $n_{\text{e}}(\mathbf{r})$ using the external potential $V_{\text{e}}(\mathbf{r})$ and the nuclei interaction energy E_{nn}. Moreover, the ground state density $n_{\text{e}_0}(\mathbf{r})$ can be found by minimizing E_{HK}

with respect to the electron density $n_e(\mathbf{r})$. However, the Hohenberg-Kohn Theorem 2 provides no practical guide to calculate E_{HK} in terms of the electron density, or more detailed, to calculate the universal functional

$$F_{HK}\Big(n_e(\mathbf{r})\Big) = \big\langle \Psi \big| \hat{T}_e + \hat{V}_{int} \big| \Psi \big\rangle$$

of the kinetic T_e and interaction energy V_{int} of the electrons in terms of the electron density $n_e(\mathbf{r})$ without using the related many-body electron wave function $\Psi(\mathbf{R}_1, \ldots, \mathbf{R}_{N_e})$. But this is essential to use the Hohenberg-Kohn theorems for practical calculations. Unfortunately, the construction of a proper approximation of the functional $F_{HK}(n_e(\mathbf{r}))$ is quite difficult. For example: If one wants to construct only the electron interaction energy in terms of the electron density and remembers the classical electrostatic interaction energy

$$V_{int}(\mathbf{R}_1, \ldots, \mathbf{R}_{N_e}) = \frac{1}{2} \sum_{i=1}^{N_e} \sum_{\substack{j=1 \\ j \neq i}}^{N_e} \frac{e^2}{4\pi\varepsilon_0} \frac{1}{\big|\mathbf{R}_i - \mathbf{R}_j\big|}$$

of N_e electrons considered as point charges at positions $\mathbf{R}_1, \ldots, \mathbf{R}_{N_e}$, one may think to use the following construction:

$$E_{Ha}\Big(n_e(\mathbf{r})\Big) = \frac{1}{2} \int_{\mathbb{R}^3} d^3r \int_{\mathbb{R}^3} d^3r' \, \frac{e^2}{4\pi\varepsilon_0} \frac{n_e(\mathbf{r})\, n_e(\mathbf{r}')}{|\mathbf{r} - \mathbf{r}'|}. \tag{2.108}$$

This is the so-called Hartree energy and is only an approximation of the true interaction energy of the electrons. It includes, for instance, the interaction energy of a single electron with its own density, which is unphysical. In addition, any state can only be filled by a maximum of one electron due to the Pauli exclusion principle. This leads to a drop of the electron density around the position of any electron in solids, which is also not included in the Hartree energy.

2.3.2 *Kohn–Sham Equations*

In 1965, just one year after the Hohenberg-Kohn theorems were published, Kohn and Sham introduced the following ansatz to be able to calculate the unknown kinetic and interaction energy of the electrons in terms of the electron density [11]:

Theorem 2.7 (Kohn–Sham Ansatz)

1. *The ground state density $n_{e_0}(\mathbf{r})$ of a system of interacting electrons in an external potential $V_{ext}(\mathbf{r})$ is identical to the ground state density of an auxiliary system of non-interacting electrons in an adapted effective potential $V_{eff}(\mathbf{r})$. If this is possible, $n_{e_0}(\mathbf{r})$ is called non-interacting-V-representable.*
2. *The total energy functional $E_{KS}(n_e(\mathbf{r}), V_{ext}(\mathbf{r}), E_{nn})$ of the system of interacting electrons in an external potential $V_{ext}(\mathbf{r})$ with nuclei interaction energy E_{nn} is constructed as followed:*

$$E_{KS}\Big(n_e(\mathbf{r}), V_{ext}(\mathbf{r}), E_{nn}\Big)=T_{aux}\Big(n_e(\mathbf{r})\Big) + E_{Ha}\Big(n_e(\mathbf{r})\Big) + \int_{\mathbb{R}^3} d^3r\, V_{ext}(\mathbf{r})\, n_e(\mathbf{r}) + E_{nn} + E_{xc}\Big(n_e(\mathbf{r})\Big), \tag{2.109}$$

where $T_{aux}(n_e(\mathbf{r}))$ is the exact kinetic energy functional of the auxiliary system of non-interacting electrons and $E_{xc}(n_e(\mathbf{r}))$ is the so-called exchange correlation functional that takes all errors into account that are introduced by using $T_{aux}(n_e(\mathbf{r}))$ and $E_{Ha}(n_e(\mathbf{r}))$, the Hartree energy.

Calculations in the auxiliary system of non-interacting electrons are much easier: With the help of single-electron wave functions $\varphi_i(\mathbf{r})$, which are orthonormal

$$\delta_{ij} = \langle \varphi_i | \varphi_j \rangle = \int_{\mathbb{R}^3} d^3r\, \varphi_i^*(\mathbf{r})\, \varphi_j(\mathbf{r}), \tag{2.110}$$

the many-body wave function $\Psi_{\mathrm{aux}}(\mathbf{R}_1, \ldots, \mathbf{R}_{N_e})$ of this auxiliary system is constructed as a slater determinant

$$\begin{aligned}\Psi_{\mathrm{aux}}(\mathbf{R}_1, \ldots, \mathbf{R}_{N_e}) &= \frac{1}{\sqrt{N_e!}} \det\left(\begin{bmatrix} \varphi_1(\mathbf{R}_1) & \cdots & \varphi_1(\mathbf{R}_{N_e}) \\ \vdots & & \vdots \\ \varphi_{N_e}(\mathbf{R}_1) & \cdots & \varphi_{N_e}(\mathbf{R}_{N_e}) \end{bmatrix}\right) \\ &= \frac{1}{\sqrt{N_e!}} \sum_{\mathfrak{s}\in\mathfrak{S}_{N_e}} \mathrm{sgn}(\mathfrak{s}) \prod_{i=1}^{N_e} \varphi_{\mathfrak{s}(i)}(\mathbf{R}_i).\end{aligned} \tag{2.111}$$

Here $\mathfrak{s}$ denotes a permutation and $\mathfrak{S}_{N_e}$ denotes the set of all permutations of the integer set $\{1, \ldots, N_e\}$. A permutation $\mathfrak{s} \in \mathfrak{S}_{N_e}$ is a function that reorders the integer set $\{1, \ldots, N_e\}$ and $\mathfrak{s}(i)$ denotes the number at the ith position in the integer set after reordering. For any permutation $\mathfrak{s}$, sig($\mathfrak{s}$) is defined, which is equal to $+1$, whenever $\mathfrak{s}$ can be achieved from $\{1, \ldots, N_e\}$ by successively interchanging two entries an even number of times, and which is equal to -1, whenever it can be achieved by an odd number of such interchanges.

Using Ψ_{aux}, the kinetic energy T_{aux} of the auxiliary system can be exactly calculated as:

$$\begin{aligned}
T_{\text{aux}} =& \langle \Psi_{\text{aux}} | \hat{T}_{\text{e}} | \Psi_{\text{aux}} \rangle \\
=& \int_{\mathbb{R}^3} d^3 R_1 \cdots \int_{\mathbb{R}^3} d^3 R_{N_{\text{e}}} \, \Psi^*_{\text{aux}}(\mathbf{R}_1, \ldots, \mathbf{R}_{N_{\text{e}}}) \left(\sum_{j=1}^{N_{\text{e}}} \frac{\hat{P}_j^2}{2 m_{\text{e}}} \right) \Psi_{\text{aux}}(\mathbf{R}_1, \ldots, \mathbf{R}_{N_{\text{e}}}) \\
=& \int_{\mathbb{R}^3} d^3 R_1 \cdots \int_{\mathbb{R}^3} d^3 R_{N_{\text{e}}} \left(\frac{1}{\sqrt{N_{\text{e}}!}} \sum_{\mathfrak{s} \in \mathfrak{S}_{N_{\text{e}}}} \operatorname{sgn}(\mathfrak{s}) \prod_{i=1}^{N_{\text{e}}} \varphi^*_{\mathfrak{s}(i)}(\mathbf{R}_i) \right) \left(\sum_{j=1}^{N_{\text{e}}} \frac{\hat{P}_j^2}{2 m_{\text{e}}} \right) \times \\
& \times \left(\frac{1}{\sqrt{N_{\text{e}}!}} \sum_{\mathfrak{s}' \in \mathfrak{S}_{N_{\text{e}}}} \operatorname{sgn}(\mathfrak{s}') \prod_{i'=1}^{N_{\text{e}}} \varphi_{\mathfrak{s}'(i')}(\mathbf{R}_{i'}) \right) \\
=& -\frac{\hbar^2}{2 m_{\text{e}}} \sum_{j=1}^{N_{\text{e}}} \sum_{\mathfrak{s} \in \mathfrak{S}_{N_{\text{e}}}} \sum_{\mathfrak{s}' \in \mathfrak{S}_{N_{\text{e}}}} \frac{\operatorname{sgn}(\mathfrak{s}) \operatorname{sgn}(\mathfrak{s}')}{N_{\text{e}}!} \int_{\mathbb{R}^3} d^3 R_1 \cdots \int_{\mathbb{R}^3} d^3 R_{N_{\text{e}}} \times \\
& \times \prod_{i=1}^{N_{\text{e}}} \varphi^*_{\mathfrak{s}(i)}(\mathbf{R}_i) \, \nabla^2_{\mathbf{R}_j} \varphi_{\mathfrak{s}'(i)}(\mathbf{R}_i) \\
=& -\frac{\hbar^2}{2 m_{\text{e}}} \sum_{j=1}^{N_{\text{e}}} \sum_{\mathfrak{s} \in \mathfrak{S}_{N_{\text{e}}}} \sum_{\mathfrak{s}' \in \mathfrak{S}_{N_{\text{e}}}} \frac{\operatorname{sgn}(\mathfrak{s}) \operatorname{sgn}(\mathfrak{s}')}{N_{\text{e}}!} \times \\
& \times \left(\int_{\mathbb{R}^3} d^3 R_j \, \varphi^*_{\mathfrak{s}(j)}(\mathbf{R}_j) \, \nabla^2_{\mathbf{R}_j} \varphi_{\mathfrak{s}'(j)}(\mathbf{R}_j) \right) \prod_{\substack{i=1 \\ i \neq j}}^{N_{\text{e}}} \int_{\mathbb{R}^3} d^3 R_i \, \varphi^*_{\mathfrak{s}(i)}(\mathbf{R}_i) \, \varphi_{\mathfrak{s}'(i)}(\mathbf{R}_i) \\
=& -\frac{\hbar^2}{2 m_{\text{e}}} \sum_{j=1}^{N_{\text{e}}} \sum_{\mathfrak{s} \in \mathfrak{S}_{N_{\text{e}}}} \frac{\operatorname{sgn}(\mathfrak{s})^2}{N_{\text{e}}!} \int_{\mathbb{R}^3} d^3 r \, \varphi^*_{\mathfrak{s}(j)}(\mathbf{r}) \, \nabla^2_{\mathbf{r}} \varphi_{\mathfrak{s}(j)}(\mathbf{r}) \\
=& -\frac{\hbar^2}{2 m_{\text{e}}} \sum_{j=1}^{N_{\text{e}}} \int_{\mathbb{R}^3} d^3 r \, \varphi^*_j(\mathbf{r}) \, \nabla^2_{\mathbf{r}} \varphi_j(\mathbf{r}).
\end{aligned} \tag{2.112}$$

Analogously, the electron density $n_{\text{e}}(\mathbf{r})$ can be obtained from $\Psi_{\text{aux}}(\mathbf{R}_1, \ldots, \mathbf{R}_{N_{\text{e}}})$ with the help of $\hat{n}_{\text{e}}(\mathbf{r})$ by

$$\boxed{n_{\text{e}}(\mathbf{r}) = \sum_{i=1}^{N_{\text{e}}} \left| \varphi_i(\mathbf{r}) \right|^2 .} \tag{2.113}$$

For the ground state, the above electron density of the auxiliary system of non-interacting electrons is equal to the electron density of our system of interacting electrons because of the Kohn–Sham Ansatz. As mentioned above, to calculate the unknown kinetic and interaction energy $F_{\mathrm{HK}}\big(n_{\mathrm{e}}(\mathbf{r})\big)$ of the system of interacting electrons in terms of the electron density $n_{\mathrm{e}}(\mathbf{r})$, Kohn and Sham gather all the differences between the complete functional and the sum of the kinetic energy of the auxiliary system of non-interacting electrons, T_{aux}, and the Hartree energy, $E_{\mathrm{Ha}}\big(n_{\mathrm{e}}(\mathbf{r})\big)$, into the exchange and correlation functional $E_{\mathrm{xc}}\big(n_{\mathrm{e}}(\mathbf{r})\big)$:

$$F_{\mathrm{HK}}\Big(n_{\mathrm{e}}(\mathbf{r})\Big)= T_{\mathrm{aux}}\Big(n_{\mathrm{e}}(\mathbf{r})\Big)+E_{\mathrm{Ha}}\Big(n_{\mathrm{e}}(\mathbf{r})\Big) + E_{\mathrm{xc}}\Big(n_{\mathrm{e}}(\mathbf{r})\Big). \tag{2.114}$$

T_{aux} can be considered as a functional of $n_{\mathrm{e}}(\mathbf{r})$, because $n_{\mathrm{e}}(\mathbf{r})$ can be always expressed by Eq. (2.113) from single-electron wave functions $\varphi_i(\mathbf{r})$ following the Kohn–Sham Ansatz and T_{aux} can be calculated from Eq. (2.112) using the single-electron wave functions $\varphi_i(\mathbf{r})$. In addition, $F_{\mathrm{HK}}\big(n_{\mathrm{e}}(\mathbf{r})\big)$ is well defined, since $n_{\mathrm{e}}(\mathbf{r})$ is as a non-interacting-V-representable density also a V-representable density. This verifies that $E_{\mathrm{xc}}\big(n_{\mathrm{e}}(\mathbf{r})\big)$ is indeed a functional, since all other terms appearing in Eq. (2.114) are functionals. The advantage of the above construction is that by explicitly separating out the independent kinetic and the long range Hartee interaction energy, the remaining exchange correlation functional $E_{\mathrm{xc}}\big(n_{\mathrm{e}}(\mathbf{r})\big)$ can be expected to be a nearly local functional of the electron density $n_{\mathrm{e}}(\mathbf{r})$:

$$E_{\mathrm{xc}}\Big(n_{\mathrm{e}}(\mathbf{r})\Big)= \int_{\mathbb{R}^3} d^3r'\, n_{\mathrm{e}}(\mathbf{r}')\, \varepsilon_{\mathrm{xc}}\big(n_{\mathrm{e}}(\mathbf{r}), \mathbf{r}'\big), \tag{2.115}$$

where $\varepsilon_{\mathrm{xc}}\big(n_{\mathrm{e}}(\mathbf{r}), \mathbf{r}'\big)$ is the exchange correlation energy density at point $\mathbf{r}'$ that depends only upon the electron density $n_{\mathrm{e}}(\mathbf{r})$ in some neighborhood of point $\mathbf{r}'$.

According to the Hohenberg-Kohn Theorem 2, the ground state electron density $n_{\mathrm{e}_0}(\mathbf{r})$ can be found by minimizing the total energy $E_{\mathrm{KS}}\big(n_{\mathrm{e}}(\mathbf{r}), V_{\mathrm{ext}}(\mathbf{r}), E_{\mathrm{nn}}\big)$ with respect to the electron density $n_{\mathrm{e}}(\mathbf{r})$. Following the Kohn–Sham Ansatz, every ground state electron density can be represented by the ground state electron density of an auxiliary system of non-interacting electrons. In this way, any ground state electron density can be constructed from single-electron wave functions $\varphi_i(\mathbf{r})$ using Eq. (2.113). Hence, the minimization of the total energy E_{KS} can be done with respect to the single-electron wave functions $\varphi_i(\mathbf{r})$. In the ground state, the variation of the total energy with respect to any of the single-electron wave functions $\varphi_k^*(\mathbf{R})$ must be zero with the constraints that every single-electron wave function is normalized:

$$\frac{\delta}{\delta\varphi_k^*(\mathbf{R})}\left(E_{\mathrm{KS}}\Big(n_{\mathrm{e}_0}(\mathbf{r}), V_{\mathrm{ext}}(\mathbf{r}), E_{\mathrm{nn}}\Big)-\sum_{i=1}^{N_{\mathrm{e}}}\varepsilon_i\Big(\langle\varphi_i|\varphi_i\rangle - 1\Big)\right) = 0. \tag{2.116}$$

ε_i are Lagrange multipliers that take the normalization constraints into account. For $E_{\mathrm{KS}}\big(n_{\mathrm{e}_0}(\mathbf{r}), V_{\mathrm{ext}}(\mathbf{r}), E_{\mathrm{nn}}\big)$, we put in (2.109) and obtain

$$0 = \frac{\delta}{\delta \varphi_k^*(\mathbf{R})} \Bigg(T_{\text{aux}}\Big(n_{e_0}(\mathbf{r})\Big) + E_{\text{Ha}}\Big(n_{e_0}(\mathbf{r})\Big) + \int\limits_{\mathbb{R}^3} d^3r\, V_{\text{ext}}(\mathbf{r})\, n_{e_0}(\mathbf{r})$$

$$+ E_{\text{nn}} + E_{\text{xc}}\Big(n_{e_0}(\mathbf{r})\Big) - \sum_{i=1}^{N_e} \varepsilon_i \Big(\langle \varphi_i | \varphi_i \rangle - 1 \Big) \Bigg).$$

Now we insert (2.112) for $T_{\text{aux}}\big(n_{e_0}(\mathbf{r})\big)$ and obtain further

$$0 = \frac{\delta}{\delta \varphi_k^*(\mathbf{R})} \left(-\frac{\hbar^2}{2m_e} \sum_{j=1}^{N_e} \int\limits_{\mathbb{R}^3} d^3r\, \varphi_j^*(\mathbf{r})\, \nabla_{\mathbf{r}}^2\, \varphi_j(\mathbf{r}) + E_{\text{Ha}}\Big(n_{e_0}(\mathbf{r})\Big) + \int\limits_{\mathbb{R}^3} d^3r\, V_{\text{ext}}(\mathbf{r})\, n_{e_0}(\mathbf{r}) \right.$$

$$\left. + E_{\text{nn}} + E_{\text{xc}}\Big(n_{e_0}(\mathbf{r})\Big) - \sum_{i=1}^{N_e} \varepsilon_i \left(\int\limits_{\mathbb{R}^3} d^3r\, \varphi_i^*(\mathbf{r})\, \varphi_i(\mathbf{r}) - 1 \right) \right)$$

$$= -\frac{\hbar^2}{2m_e} \nabla_{\mathbf{R}}^2\, \varphi_k(\mathbf{R})$$

$$+ \frac{\delta}{\delta n_{e_0}(\mathbf{R})} \left(E_{\text{Ha}}\Big(n_{e_0}(\mathbf{r})\Big) + \int\limits_{\mathbb{R}^3} d^3r\, V_{\text{ext}}(\mathbf{r})\, n_{e_0}(\mathbf{r}) + E_{\text{xc}}\Big(n_{e_0}(\mathbf{r})\Big) \right) \underbrace{\frac{\delta n_{e_0}(\mathbf{R})}{\delta \varphi_k^*(\mathbf{R})}}_{=\varphi_k(\mathbf{R})}$$

$$- \varepsilon_k\, \varphi_k(\mathbf{R}). \tag{2.117}$$

By using

$$\frac{\delta}{\delta n_{e_0}(\mathbf{R})} \int\limits_{\mathbb{R}^3} d^3r\, V_{\text{ext}}(\mathbf{r})\, n_{e_0}(\mathbf{r}) = V_{\text{ext}}(\mathbf{R}) \tag{2.118}$$

and

$$\frac{\delta E_{\text{Ha}}\Big(n_{e_0}(\mathbf{r})\Big)}{\delta n_{e_0}(\mathbf{R})} \overset{(2.108)}{=} \frac{1}{2} \frac{\delta}{\delta n_{e_0}(\mathbf{R})} \int\limits_{\mathbb{R}^3} d^3r \int\limits_{\mathbb{R}^3} d^3r'\, \frac{e^2}{4\pi\varepsilon_0} \frac{n_{e_0}(\mathbf{r})\, n_{e_0}(\mathbf{r}')}{|\mathbf{r} - \mathbf{r}'|}$$

$$= \frac{1}{2} \int\limits_{\mathbb{R}^3} d^3r'\, \frac{e^2}{4\pi\varepsilon_0} \frac{n_{e_0}(\mathbf{r}')}{|\mathbf{R} - \mathbf{r}'|} + \frac{1}{2} \int\limits_{\mathbb{R}^3} d^3r\, \frac{e^2}{4\pi\varepsilon_0} \frac{n_{e_0}(\mathbf{r})}{|\mathbf{r} - \mathbf{R}|}$$

$$= \int\limits_{\mathbb{R}^3} d^3r\, \frac{e^2}{4\pi\varepsilon_0} \frac{n_{e_0}(\mathbf{r})}{|\mathbf{r} - \mathbf{R}|}, \tag{2.119}$$

which is called Hartree potential, and the definition of the exchange and correlation potential

$$V_{\mathrm{xc}}(\mathbf{R}) := \frac{\delta E_{\mathrm{xc}}\Big(n_{\mathrm{e}_0}(\mathbf{r})\Big)}{\delta n_{\mathrm{e}_0}(\mathbf{R})}, \tag{2.120}$$

one gets

$$\boxed{\left(-\frac{\hbar^2}{2m_{\mathrm{e}}}\nabla^2_{\mathbf{R}} + \underbrace{V_{\mathrm{ext}}(\mathbf{R}) + \int\limits_{\mathbb{R}^3} d^3r\, \frac{e^2}{4\pi\varepsilon_0}\frac{n_{\mathrm{e}_0}(\mathbf{r})}{|\mathbf{r}-\mathbf{R}|} + V_{\mathrm{xc}}(\mathbf{R})}_{=V_{\mathrm{eff}}(\mathbf{R})}\right)\varphi_k(\mathbf{R}) = \varepsilon_k\,\varphi_k(\mathbf{R}).} \tag{2.121}$$

This equation correspond to a Schrödinger equation of a single electron with the Hamiltonian

$$\hat{H}_{\mathrm{KS}} = -\frac{\hbar^2}{2m_{\mathrm{e}}}\nabla^2_{\mathbf{r}} + V_{\mathrm{eff}}(\mathbf{r}) \tag{2.122}$$

and constitutes along with Eq. (2.113) the Kohn–Sham equations. Since the effective potential $V_{\mathrm{eff}}(\mathbf{r})$ depends also on the unknown ground state electron density $n_{\mathrm{e}_0}(\mathbf{r})$, the determination of the ground state single-electron wave functions $\varphi_i(\mathbf{r})$ must be done self-consistently, as it is discussed in more detail later in Sect. 2.3.6.

Once the single-electron wave functions $\varphi_i(\mathbf{r})$ have been found by solving Eq. (2.121), they are sorted in increasing order according to their eigenenergies ε_i. The single-electron wave functions do not take the spin into account, so that each orbital can be occupied by two electrons – one with spin up and one with spin down. Therefore, to take the spin into account, one constructs the list of single-electron wave functions just as $\varphi_1(\mathbf{r}), \varphi_1(\mathbf{r}), \varphi_2(\mathbf{r}), \varphi_2(\mathbf{r}), \ldots$ with corresponding energies $\varepsilon_1, \varepsilon_1, \varepsilon_2, \varepsilon_2, \ldots$. Finally, the N_{e} single-electron wave functions $\varphi_i(\mathbf{r})$'s with the lowest eigenenergies ε_i are chosen from this list to construct the ground state electron density $n_{\mathrm{e}_0}(\mathbf{r})$ via Eq. (2.113).

The Hohenberg-Kohn theorems ensure that the ground state electron density $n_{\mathrm{e}_0}(\mathbf{r})$ is unique, minimizes the total energy functional $E_{\mathrm{KS}} = E_{\mathrm{HK}}$ and is uniquely connected to the external potential $V_{\mathrm{ext}}(\mathbf{r})$. Hence, the corresponding effective potential $V_{\mathrm{eff}}(\mathbf{r})$ that is constructed from this unique ground state density $n_{\mathrm{e}_0}(\mathbf{r})$ is also unique. This potential is called Kohn–Sham potential $V_{\mathrm{KS}}(\mathbf{r})$ and, by solving the single-electron Schödinger equation (2.121) with $V_{\mathrm{KS}}(\mathbf{r})$, one gets the unique single-electron wave functions $\varphi_i(\mathbf{r})$ that represent the ground state density $n_{\mathrm{e}_0}(\mathbf{r})$ using Eq. (2.113). These single-electron wave functions $\varphi_i(\mathbf{r})$ are called Kohn–Sham orbitals and the related energy eigenvalues ε_i are called Kohn–Sham energies. Using the Kohn–Sham orbitals and Kohn–Sham energies, one can calculate the total energy of the ground state as follows:

$$
\begin{aligned}
E_{\mathrm{KS}} &\overset{(2.109)}{=} T_{\mathrm{aux}}\Big(n_{\mathrm{e}_0}(\mathbf{r})\Big) + E_{\mathrm{Ha}}\Big(n_{\mathrm{e}_0}(\mathbf{r})\Big) + \int_{\mathbb{R}^3} d^3r\, V_{\mathrm{ext}}(\mathbf{r})\, n_{\mathrm{e}_0}(\mathbf{r}) + E_{\mathrm{nn}} + E_{\mathrm{xc}}\Big(n_{\mathrm{e}_0}(\mathbf{r})\Big) \\
&= -\frac{\hbar^2}{2m_{\mathrm{e}}} \sum_{j=1}^{N_{\mathrm{e}}} \int_{\mathbb{R}^3} d^3r\, \varphi_j^*(\mathbf{r})\, \nabla_{\mathbf{r}}^2\, \varphi_j(\mathbf{r}) \\
&\quad + \frac{1}{2} \int_{\mathbb{R}^3} d^3r \int_{\mathbb{R}^3} d^3r'\, \frac{e^2}{4\pi\varepsilon_0}\, \frac{n_{\mathrm{e}_0}(\mathbf{r})\, n_{\mathrm{e}_0}(\mathbf{r}')}{|\mathbf{r}-\mathbf{r}'|} \\
&\quad + \int_{\mathbb{R}^3} d^3r\, V_{\mathrm{ext}}(\mathbf{r})\, n_{\mathrm{e}_0}(\mathbf{r}) + E_{\mathrm{nn}} + E_{\mathrm{xc}}\Big(n_{\mathrm{e}_0}(\mathbf{r})\Big) \\
&= -\frac{\hbar^2}{2m_{\mathrm{e}}} \sum_{j=1}^{N_{\mathrm{e}}} \int_{\mathbb{R}^3} d^3r\, \varphi_j^*(\mathbf{r})\, \nabla_{\mathbf{r}}^2\, \varphi_j(\mathbf{r}) + \int_{\mathbb{R}^3} d^3r\, V_{\mathrm{ext}}(\mathbf{r})\, n_{\mathrm{e}_0}(\mathbf{r}) \\
&\quad + \int_{\mathbb{R}^3} d^3r \int_{\mathbb{R}^3} d^3r'\, \frac{e^2}{4\pi\varepsilon_0}\, \frac{n_{\mathrm{e}_0}(\mathbf{r})\, n_{\mathrm{e}_0}(\mathbf{r}')}{|\mathbf{r}-\mathbf{r}'|} \\
&\quad + \int_{\mathbb{R}^3} d^3r\, V_{\mathrm{xc}}(\mathbf{r})\, n_{\mathrm{e}_0}(\mathbf{r}) + E_{\mathrm{nn}} + E_{\mathrm{xc}}\Big(n_{\mathrm{e}_0}(\mathbf{r})\Big) \\
&\quad - \frac{1}{2} \int_{\mathbb{R}^3} d^3r \int_{\mathbb{R}^3} d^3r'\, \frac{e^2}{4\pi\varepsilon_0}\, \frac{n_{\mathrm{e}_0}(\mathbf{r})\, n_{\mathrm{e}_0}(\mathbf{r}')}{|\mathbf{r}-\mathbf{r}'|} - \int_{\mathbb{R}^3} d^3r\, V_{\mathrm{xc}}(\mathbf{r})\, n_{\mathrm{e}_0}(\mathbf{r}).
\end{aligned}
$$

Now we insert Eq. (2.113) for the equilibrium electron density $n_{\mathrm{e}_0}(\mathbf{r})$ and get further

$$
\begin{aligned}
E_{\mathrm{KS}} &= -\frac{\hbar^2}{2m_{\mathrm{e}}} \sum_{j=1}^{N_{\mathrm{e}}} \int_{\mathbb{R}^3} d^3r\, \varphi_j^*(\mathbf{r})\, \nabla_{\mathbf{r}}^2\, \varphi_j(\mathbf{r}) + \int_{\mathbb{R}^3} d^3r\, V_{\mathrm{ext}}(\mathbf{r}) \left(\sum_{j=1}^{N_{\mathrm{e}}} |\varphi_j(\mathbf{r})|^2 \right) \\
&\quad + \int_{\mathbb{R}^3} d^3r \int_{\mathbb{R}^3} d^3r'\, \frac{e^2}{4\pi\varepsilon_0}\, \frac{\left(\sum_{j=1}^{N_{\mathrm{e}}} |\varphi_j(\mathbf{r})|^2 \right) n_{\mathrm{e}_0}(\mathbf{r}')}{|\mathbf{r}-\mathbf{r}'|} + \int_{\mathbb{R}^3} d^3r\, V_{\mathrm{xc}}(\mathbf{r}) \left(\sum_{j=1}^{N_{\mathrm{e}}} |\varphi_j(\mathbf{r})|^2 \right) \\
&\quad + E_{\mathrm{nn}} - \frac{1}{2} \int_{\mathbb{R}^3} d^3r \int_{\mathbb{R}^3} d^3r'\, \frac{e^2}{4\pi\varepsilon_0}\, \frac{n_{\mathrm{e}_0}(\mathbf{r})\, n_{\mathrm{e}_0}(\mathbf{r}')}{|\mathbf{r}-\mathbf{r}'|} - \int_{\mathbb{R}^3} d^3r\, V_{\mathrm{xc}}(\mathbf{r})\, n_{\mathrm{e}_0}(\mathbf{r}) \\
&\quad + E_{\mathrm{xc}}\Big(n_{\mathrm{e}_0}(\mathbf{r})\Big) \\
&= \sum_{j=1}^{N_{\mathrm{e}}} \int_{\mathbb{R}^3} d^3r\, \varphi_j^*(\mathbf{r})\, \underbrace{\left(-\frac{\hbar^2}{2m_{\mathrm{e}}}\, \nabla_{\mathbf{r}}^2 + V_{\mathrm{ext}}(\mathbf{r}) + \int_{\mathbb{R}^3} d^3r'\, \frac{e^2}{4\pi\varepsilon_0}\, \frac{n_{\mathrm{e}_0}(\mathbf{r}')}{|\mathbf{r}'-\mathbf{r}|} + V_{\mathrm{xc}}(\mathbf{r}) \right) \varphi_j(\mathbf{r})}_{\overset{(2.121)}{=} \varepsilon_j\, \varphi_j(\mathbf{r})} \\
&\quad + E_{\mathrm{nn}} - \frac{1}{2} \int_{\mathbb{R}^3} d^3r \int_{\mathbb{R}^3} d^3r'\, \frac{e^2}{4\pi\varepsilon_0}\, \frac{n_{\mathrm{e}_0}(\mathbf{r})\, n_{\mathrm{e}_0}(\mathbf{r}')}{|\mathbf{r}-\mathbf{r}'|} - \int_{\mathbb{R}^3} d^3r\, V_{\mathrm{xc}}(\mathbf{r})\, n_{\mathrm{e}_0}(\mathbf{r})
\end{aligned}
$$

$$+ E_{\mathrm{xc}}\Big(n_{\mathrm{e}_0}(\mathbf{r})\Big).$$

Finally we obtain

$$\boxed{\begin{aligned} E_{\mathrm{KS}} = &\sum_{j=1}^{N_{\mathrm{e}}} \varepsilon_j + E_{\mathrm{nn}} \\ &-\frac{1}{2}\int\limits_{\mathbb{R}^3} d^3r \int\limits_{\mathbb{R}^3} d^3r' \,\frac{e^2}{4\pi\varepsilon_0}\,\frac{n_{\mathrm{e}_0}(\mathbf{r})\,n_{\mathrm{e}_0}(\mathbf{r}')}{|\mathbf{r}-\mathbf{r}'|} - \int\limits_{\mathbb{R}^3} d^3r\; V_{\mathrm{xc}}(\mathbf{r})\,n_{\mathrm{e}_0}(\mathbf{r}) \\ &+E_{\mathrm{xc}}\Big(n_{\mathrm{e}_0}(\mathbf{r})\Big). \end{aligned}} \tag{2.123}$$

The fact that the total energy E_{KS} is not simply the sum of the Kohn–Sham energies, ε_i, makes more obvious that the Kohn–Sham orbitals $\varphi_i(\mathbf{r})$ are not physical single-electron wave functions. But the Kohn–Sham orbitals $\varphi_i(\mathbf{r})$ produce the correct ground state density $n_{\mathrm{e}_0}(\mathbf{r})$ from Eq. (2.113) and ground state energy from Eq. (2.123). To determine the Kohn-Sham orbitals, one just has to solve a Schrödinger equation (2.121) of a single electron in an effective Potential. In this way, the complicated many-body wave function of the interacting electrons is not needed within the Kohn–Sham ansatz, which allows for practical numerical calculations. Moreover, the found ground state electron density and the related ground state energy would be exact, if the exact form of the exchange and correlation functional $E_{\mathrm{xc}}\big(n_{\mathrm{e}}(\mathbf{r})\big)$ were known, which is not the case. Note, the interpretation of the Kohn–Sham eigenvalues ε_j with $j > N_{\mathrm{e}}$ as excitation energies is a contentious issue [12].

2.3.3 *Approximations to the Exchange Correlation Functional*

Several very successful approaches exist for the approximation of the exchange and correlation functional $E_{\mathrm{xc}}\big(n_{\mathrm{e}}(\mathbf{r})\big)$ as a function of the electron density $n_{\mathrm{e}}(\mathbf{r})$. We will now discuss those more relevant to our work.

2.3.3.1 Local Density Approximation (LDA)

One of the most popular ones is the local density approximation (LDA), which is based on the model system of the homogeneous electron gas in an uniform positively charged background that describes the nuclei. Using quantum Monte Carlo methods, it was possible to calculate numerically the exact ground state energy of this model system for various electronic densities in 1980 [13]. In these calculations, the homogeneous electronic density $n_{\mathrm{e}}(\mathbf{r})$ is characterized by the Wigner-Seitz sphere radius

$$r_s\big(n_e(\mathbf{r})\big) = \sqrt[3]{\frac{3}{4\pi\, n_e(\mathbf{r})}}, \tag{2.124}$$

which corresponds to the radius of a sphere that contains one electron on average. Thus, r_s can be interpreted as the average distance between electrons. In this way, the ground state energies for several values of the Wigner-Seitz sphere radius r_s were calculated and tabulated in Ref. [13]. Since the system is homogeneous, the corresponding ground state energy densities and also the related exact exchange and correlation energy densities can be derived for the studied r_s values. In LDA, the exchange and correlation functional is constructed as an integral over all space with the exchange-correlation energy density at each point assumed to be the same as in a homogeneous electron gas with that density:

$$E_{xc}^{(\mathrm{LDA})}\Big(n_e(\mathbf{r})\Big) = \int_{\mathbb{R}^3} d^3r'\, n_e(\mathbf{r}')\, \varepsilon_{xc}^{(\mathrm{hom})}\big(n_e(\mathbf{r}')\big), \tag{2.125}$$

whereas $\varepsilon_{xc}^{(\mathrm{hom})}\big(n_e(\mathbf{r}')\big)$ is the exchange-correlation energy density of the homogeneous electron gas with electron density $n_e(\mathbf{r}')$. This approximation is called local, since $\varepsilon_{xc}^{(\mathrm{hom})}\big(n_e(\mathbf{r}')\big)$ only depends on the electron density $n_e(\mathbf{r}')$ at the current integration point $\mathbf{r}'$. The exchange-correlation energy density of the homogeneous electron gas is expressed as a function of r_s

$$\varepsilon_{xc}^{(\mathrm{hom})}\big(n_e(\mathbf{r})\big) \equiv \varepsilon_{xc}^{(\mathrm{hom})}\Big(r_s\big(n_e(\mathbf{r})\big)\Big). \tag{2.126}$$

and was parametrized several times [14–17] using the exact exchange and correlation energy densities at the studied r_s values [13]. The most accurate parametrization was done by Perdew and Wang in 1992 [17]:

$$\begin{aligned}\varepsilon_{xc}^{(\mathrm{hom})}(r_s) = &-2\,A\,(1+\alpha_1\, r_s)\,\ln\left(1+\frac{1}{2\,A\left(\beta_1\, r_s^{\frac{1}{2}} + \beta_2\, r_s + \beta_3\, r_s^{\frac{3}{2}} + \beta_4\, r_s^2\right)}\right)\\ &-\frac{3}{4\pi}\left(\frac{9\pi}{4}\right)^{\frac{1}{3}}\frac{1}{r_s}.\end{aligned} \tag{2.127}$$

Here, $A, \alpha_1, \beta_1, \beta_2, \beta_3$, and β_4 are parameters, which are tabulated in Ref. [17]. The construction of the related exchange and correlation potential $V_{xc}^{(\mathrm{hom})}(\mathbf{r})$ is straight forward:

$$\begin{aligned} V_{\mathrm{xc}}^{(\mathrm{hom})}(\mathbf{r}) &= \frac{\delta}{\delta n_{\mathrm{e}}(\mathbf{r})}\left(n_{\mathrm{e}}(\mathbf{r})\,\varepsilon_{\mathrm{xc}}^{(\mathrm{hom})}\left(r_{\mathrm{s}}\left(n_{\mathrm{e}}(\mathbf{r})\right)\right)\right) \\ &= \varepsilon_{\mathrm{xc}}^{(\mathrm{hom})}(r_{\mathrm{s}}) + \frac{\partial \varepsilon_{\mathrm{xc}}^{(\mathrm{hom})}(r_{\mathrm{s}})}{\partial r_{\mathrm{s}}} \underbrace{\frac{-r_{\mathrm{s}}}{3}}_{=n_{\mathrm{e}}\frac{\delta r_{\mathrm{s}}}{\delta n_{\mathrm{e}}}} . \end{aligned} \tag{2.128}$$

Since the LDA functional was developed from purely theoretical insights, it can be considered as an ab-initio method. Based on the model system of the homogeneous electron gas, one expects that LDA will work best for solids with electrons close to a homogeneous gas like a metal with nearly free valence electrons. Since, for the electron densities typical found in solids, the range of the effects of exchange and correlation is rather short, LDA produces reasonable results for a wide range of solids including also semiconductors and isolators.

2.3.3.2 Generalized Gradient Approximation (GGA)

The success of LDA lead to the development of further functionals. A natural approach to extend LDA is to incorporate the gradient of the electron density in the integral over the exchange correlation energy density. Indeed, this leads to the very well-known generalized gradient approximation (GGA):

$$E_{\mathrm{xc}}^{(\mathrm{GGA})}\left(n_{\mathrm{e}}(\mathbf{r})\right) = \int_{\mathbb{R}^3} d^3r'\, n_{\mathrm{e}}(\mathbf{r}')\,\varepsilon_{\mathrm{xc}}^{(\mathrm{hom})}\left(n_{\mathrm{e}}(\mathbf{r}')\right)\, f_{\mathrm{xc}}\left(n_{\mathrm{e}}(\mathbf{r}'), \nabla n_{\mathrm{e}}(\mathbf{r}')\right). \tag{2.129}$$

Several approaches exist to determine the additional function $f_{\mathrm{xc}}\left(n_{\mathrm{e}}(\mathbf{r}'), \nabla n_{\mathrm{e}}(\mathbf{r}')\right)$. Since some of them use experimental data for the determination, these approaches cannot be called ab-initio any more. The most common approaches for GGA are BLYP [18, 19], PW91 [20] and PBE [21].

2.3.4 *Bloch Waves in Crystalline Systems*

To calculate the ground state density $n_{\mathrm{e}_0}(\mathbf{r})$ with the help of the Kohn–Sham Ansatz, one has to determine the Kohn–Sham orbitals $\varphi_i(\mathbf{r})$ from the Schrödinger equation (2.121) of a single electron described by the Hamilton operator $\hat{H}_{\mathrm{KS}}$. As discussed in Sect. 2.2.1, for crystalline solids one uses a basis cell, which is spanned by the three vectors $\mathbf{a}_1, \mathbf{a}_2, \mathbf{a}_3$, with the volume $\mathcal{V} = \left|\mathbf{a}_1 \cdot (\mathbf{a}_2 \times \mathbf{a}_3)\right|$ and contains N_{b} atoms. The calculations are done with a simulation cell, which is created by repeating the basis cell N_1 times in $\mathbf{a}_1$, N_2 times in $\mathbf{a}_2$ and N_3 times in $\mathbf{a}_3$ direction with $N_1, N_2, N_3 \in \mathbb{N}$. The simulation cell is spanned by the three vectors $N_1\,\mathbf{a}_1$, $N_2\,\mathbf{a}_2$, $N_3\,\mathbf{a}_3$ and contains $N_{\mathrm{at}} = N_1\,N_2\,N_3\,N_{\mathrm{b}}$ atoms. Commonly, one utilizes periodic boundary conditions, so that this simulation cell is periodically repeated to fill the whole space. Due to this, if one performs a translation with any translation vector $T_{\mathbf{n}}$ of the Bravais grid (2.28), the effective potential $V_{\text{eff}}(\mathbf{r})$ in the Hamiltonian $\hat{H}_{\text{KS}}$ is invariant, *i.e.*, $V_{\text{eff}}(\mathbf{r} + \mathbf{T_n}) = V_{\text{eff}}(\mathbf{r})$. If we denote by $\hat{\mathcal{T}}_{\mathbf{n}}$ the translation operator related to the translation vector $\mathbf{T_n}$, then the Hamiltonian $\hat{H}_{\text{KS}}$ commutes with the translation operator $\hat{\mathcal{T}}_{\mathbf{n}}$:

$$\hat{\mathcal{T}}_{\mathbf{n}}\, \hat{H}_{\text{KS}} = \hat{H}_{\text{KS}}\, \hat{\mathcal{T}}_{\mathbf{n}}, \tag{2.130}$$

since the derivative operator $\nabla_{\mathbf{r}}$ is always translation invariant. Due to this, the eigenfunctions of the Hamiltonian $\hat{H}_{\text{KS}}$, which are the Kohn–Sham orbitals $\varphi_i(\mathbf{r})$, can be chosen to be eigenfunctions of all translation operator $\hat{\mathcal{T}}_{\mathbf{n}}$ simultaneously:

$$\hat{\mathcal{T}}_{\mathbf{n}}\, \varphi_i(\mathbf{r}) = \mathsf{t}_{\mathbf{n}}\, \varphi_i(\mathbf{r}), \tag{2.131}$$

where $\mathsf{t}_{\mathbf{n}} \in \mathbb{C}$. Since the product of two translation operators $\hat{\mathcal{T}}_{\mathbf{n}_1}$ and $\hat{\mathcal{T}}_{\mathbf{n}_2}$ corresponds again to a translation operator

$$\hat{\mathcal{T}}_{\mathbf{n}_1}\, \hat{\mathcal{T}}_{\mathbf{n}_2} = \hat{\mathcal{T}}_{\mathbf{n}_1+\mathbf{n}_2}, \tag{2.132}$$

one obtains

$$\mathsf{t}_{\mathbf{n}_1+\mathbf{n}_2}\, \varphi_i(\mathbf{r}) = \hat{\mathcal{T}}_{\mathbf{n}_1+\mathbf{n}_2}\, \varphi_i(\mathbf{r}) = \hat{\mathcal{T}}_{\mathbf{n}_1}\, \hat{\mathcal{T}}_{\mathbf{n}_2}\, \varphi_i(\mathbf{r}) = \mathsf{t}_{\mathbf{n}_1}\, \mathsf{t}_{\mathbf{n}_2}\, \varphi_i(\mathbf{r}). \tag{2.133}$$

Due to this and since any translation can be broken into a product of primitive translations, one can always write any $\mathsf{t}_{\mathbf{n}}$ as

$$\mathsf{t}_{\mathbf{n}} = \Big(\mathsf{t}(\mathbf{a}_1)\Big)^{n_1} \Big(\mathsf{t}(\mathbf{a}_2)\Big)^{n_2} \Big(\mathsf{t}(\mathbf{a}_3)\Big)^{n_3}. \tag{2.134}$$

Here $\mathsf{t}(\mathbf{a}_i)$ denotes the eigenvalue of the translation operator that translate by the lattice vector $\mathbf{a}_i$. Due due the periodic boundary conditions,

$$\Big(\mathsf{t}(\mathbf{a}_1)\Big)^{N_1} = 1, \qquad \Big(\mathsf{t}(\mathbf{a}_2)\Big)^{N_2} = 1, \qquad \Big(\mathsf{t}(\mathbf{a}_3)\Big)^{N_3} = 1 \tag{2.135}$$

must be true, from which one obtains

$$\mathsf{t}(\mathbf{a}_1) = \mathrm{e}^{\frac{2\pi \mathrm{i}\, m_1}{N_1}}, \qquad \mathsf{t}(\mathbf{a}_2) = \mathrm{e}^{\frac{2\pi \mathrm{i}\, m_1}{N_2}}, \qquad \mathsf{t}(\mathbf{a}_3) = \mathrm{e}^{\frac{2\pi \mathrm{i}\, m_1}{N_3}} \tag{2.136}$$

with $m_1, m_2, m_3 \in \mathbb{Z}$. Using the definition (2.53) of the reciprocal lattice vectors, one can write

$$\mathsf{t}_{\mathbf{n}} = \mathrm{e}^{\mathrm{i}\,\mathbf{k}^{\mathrm{t}}\cdot\mathbf{T_n}} \tag{2.137}$$

with

$$\mathbf{k} = \frac{m_1}{N_1}\,\mathbf{b}_1 + \frac{m_2}{N_2}\,\mathbf{b}_2 + \frac{m_3}{N_3}\,\mathbf{b}_3 \tag{2.138}$$

being a vector in reciprocal space. Adding a vector $\mathbf{G}_{\mathbf{n}'}$ of the reciprocal grid (2.54) to $\mathbf{k}$ does not influence Eq. (2.137), since $\mathbf{G}_{\mathbf{n}'}^{t} \cdot \mathbf{T}_{\mathbf{n}} = 2\pi\, z, z \in \mathbb{Z}$ and one has $\mathrm{e}^{\mathrm{i}\,\mathbf{G}_{\mathbf{n}'}^{t} \cdot \mathbf{T}_{\mathbf{n}}} = 1$. Consequently, the range of $\mathbf{k}$ can be restricted to one primitive cell of the reciprocal lattice, *i.e.*,

$$m_1 \in \{0, 1, \ldots, N_1 - 1\}, \quad m_2 \in \{0, 1, \ldots, N_2 - 1\}, \quad m_3 \in \{0, 1, \ldots, N_3 - 1\}, \tag{2.139}$$

so that one has only to deal with $N_1\, N_2\, N_3$ different $\mathbf{k}$ vectors. In Summary, we found

$$\varphi_i(\mathbf{r} + \mathbf{T}_{\mathbf{n}}) = \hat{\mathcal{T}}_{\mathbf{n}}\, \varphi_i(\mathbf{r}) = \mathrm{e}^{\mathrm{i}\,\mathbf{k}^{t} \cdot \mathbf{T}_{\mathbf{n}}}\, \varphi_i(\mathbf{r}),$$

which is the famous Bloch theorem initially derived by Felix Bloch in 1929 [22].

Theorem 2.8 (Bloch functions) *In a periodic system, which is invariant under translations with vectors* $\mathbf{T}_{\mathbf{n}}$, *the eigenfunctions* $\varphi_i(\mathbf{r})$ *of any single-particle Hamiltonian can be chosen to be eigenfunctions of the translations along lattice vectors, with the eigenvalues determined by a given vector* $\mathbf{k}$*:*

$$\varphi_i(\mathbf{r} + \mathbf{T}_{\mathbf{n}}) = \mathrm{e}^{\mathrm{i}\,\mathbf{k}^{t} \cdot \mathbf{T}_{\mathbf{n}}}\, \varphi_i(\mathbf{r}). \tag{2.140}$$

Due to this, one can write the eigenfunctions as so called Bloch functions

$$\varphi_{\ell\mathbf{k}}(\mathbf{r}) = \mathrm{e}^{\mathrm{i}\,\mathbf{k}^{t} \cdot \mathbf{r}}\, \mathrm{u}_{\ell\mathbf{k}}(\mathbf{r}), \tag{2.141}$$

where the function $\mathrm{u}_{\ell\mathbf{k}}(\mathbf{r})$ *is periodic, i.e.,* $\mathrm{u}_{\ell\mathbf{k}}(\mathbf{r} + \mathbf{T}_{\mathbf{n}}) = \mathrm{u}_{\ell\mathbf{k}}(\mathbf{r})$.

Thanks to the Bloch theorem, one needs only to consider $N_1\, N_2\, N_3$ $\mathbf{k}$-points and derive, for each of these $\mathbf{k}$-points, the related Kohn–Sham orbitals

$$\varphi_{\ell\mathbf{k}}(\mathbf{r}) = \mathrm{e}^{\mathrm{i}\,\mathbf{k}^{t} \cdot \mathbf{r}}\, \mathrm{u}_{\ell\mathbf{k}}(\mathbf{r}).$$

In the Γ-point approximation, one considers only the $\mathbf{k} = \mathbf{0}$ point and the basis cell is equal to the simulation cell.

2.3.5 Using a Set of Basis Functions

Furthermore, the periodical functions $\mathrm{u}_{\ell\mathbf{k}}(\mathbf{r})$ are represented by a linear combination of N_{bs} basis functions $\chi_i(\mathbf{r}), i = 1, \ldots, N_{\mathrm{bs}}$:

$$\mathrm{u}_{\ell\mathbf{k}}(\mathbf{r}) = \sum_{i=1}^{N_{\mathrm{bs}}} c_{\ell i}(\mathbf{k})\, \chi_i(\mathbf{r}), \tag{2.142}$$

where $c_{\ell i}(\mathbf{k}) \in \mathbb{C}$ are the coefficients.. For the set of basis functions, availabe DFT codes use commonly plane waves or atom centered Gaußian functions. Now the task is to find the coefficients $c_{ki}(\mathbf{k}) \in \mathbb{C}$ for each considered **k**-point. One defines the N_{bs}-dimensional vector

$$\mathbf{c}^{(\ell)}(\mathbf{k}) = \begin{bmatrix} c_{\ell 1}(\mathbf{k}) \\ \vdots \\ c_{\ell N_{\mathrm{bs}}}(\mathbf{k}) \end{bmatrix}, \tag{2.143}$$

which contains the coefficients necessary to construct the periodical function $\mathrm{u}_{\ell\mathbf{k}}(\mathbf{r})$ of the ℓth Kohn–Sham orbital $\varphi_{\ell\mathbf{k}}(\mathbf{r})$. Taking the Kohn–Sham equation (2.121)

$$\begin{aligned} \hat{H}_{\mathrm{KS}}\, \varphi_{\ell\mathbf{k}}(\mathbf{r}) &= \varepsilon_{\ell\mathbf{k}}\, \varphi_{\ell\mathbf{k}}(\mathbf{r}), \\ \Leftrightarrow \qquad \underbrace{\mathrm{e}^{-\mathrm{i}\,\mathbf{k}\cdot r}\, \hat{H}_{\mathrm{KS}}\, \mathrm{e}^{\mathrm{i}\,\mathbf{k}\cdot r}}_{=:\hat{H}_{\mathrm{KS}}(\mathbf{k})}\, \mathrm{u}_{\ell\mathbf{k}}(\mathbf{r}) &= \varepsilon_{\ell\mathbf{k}}\, \mathrm{u}_{\ell\mathbf{k}}(\mathbf{r}), \end{aligned} \tag{2.144}$$

multiplying it with $u^*_{m,\mathbf{k}}(\mathbf{r})$ and integrating it over the whole space yields to

$$\begin{aligned} &\int_{\mathbb{R}^3} d^3r\, u^*_{m,\mathbf{k}}(\mathbf{r})\, \hat{H}_{\mathrm{KS}}(\mathbf{k})\, \mathrm{u}_{\ell\mathbf{k}}(\mathbf{r}) = \int_{\mathbb{R}^3} d^3r\, \mathrm{u}^*_{m\mathbf{k}}(\mathbf{r})\, \varepsilon_{\ell\mathbf{k}}\, \mathrm{u}_{\ell\mathbf{k}}(\mathbf{r}), \\ \Leftrightarrow \quad &\int_{\mathbb{R}^3} d^3r \left(\sum_{i=1}^{N_{\mathrm{bs}}} c^*_{mi}(\mathbf{k})\, \chi^*_i(\mathbf{r}) \right) \hat{H}_{\mathrm{KS}}(\mathbf{k}) \left(\sum_{j=1}^{N_{\mathrm{bs}}} c_{\ell j}(\mathbf{k})\, \chi_j(\mathbf{r}) \right) \\ &\qquad = \int_{\mathbb{R}^3} d^3r \left(\sum_{i=1}^{N_{\mathrm{bs}}} c^*_{mi}(\mathbf{k})\, \chi^*_i(\mathbf{r}) \right) \varepsilon_{\ell\mathbf{k}} \left(\sum_{j=1}^{N_{\mathrm{bs}}} c_{\ell j}(\mathbf{k})\, \chi_j(\mathbf{r}) \right), \\ \Leftrightarrow \quad &\sum_{i=1}^{N_{\mathrm{bs}}} c^*_{mi}(\mathbf{k}) \sum_{j=1}^{N_{\mathrm{bs}}} c_{\ell j}(\mathbf{k}) \int_{\mathbb{R}^3} d^3r\, \chi^*_i(\mathbf{r})\, \hat{H}_{\mathrm{KS}}(\mathbf{k})\, \chi_j(\mathbf{r}) \\ &\qquad = \sum_{i=1}^{N_{\mathrm{bs}}} c^*_{mi}(\mathbf{k}) \sum_{j=1}^{N_{\mathrm{bs}}} \varepsilon_{\ell\mathbf{k}}\, c_{\ell j}(\mathbf{k}) \int_{\mathbb{R}^3} d^3r\, \chi^*_i(\mathbf{r})\, \chi_j(\mathbf{r}), \\ \Leftrightarrow \quad &\sum_{i=1}^{N_{\mathrm{bs}}} c^*_{mi}(\mathbf{k}) \sum_{j=1}^{N_{\mathrm{bs}}} \big\langle \chi_i \big| \hat{H}_{\mathrm{KS}}(\mathbf{k}) \big| \chi_j \big\rangle\, c_{\ell j}(\mathbf{k}) = \sum_{i=1}^{N_{\mathrm{bs}}} c^*_{mi}(\mathbf{k}) \sum_{j=1}^{N_{\mathrm{bs}}} \varepsilon_{\ell\mathbf{k}}\, \langle \chi_i | \chi_j \rangle\, c_{\ell j}(\mathbf{k}). \end{aligned}$$

This equation is satisfied when

$$\sum_{j=1}^{N_{bs}} \langle \chi_i | \hat{H}_{KS}(\mathbf{k}) | \chi_j \rangle c_{\ell j}(\mathbf{k}) = \sum_{j=1}^{N_{bs}} \varepsilon_{\ell \mathbf{k}} \langle \chi_i | \chi_j \rangle c_{\ell j}(\mathbf{k}) \tag{2.145}$$

is fulfilled. If one defines the $N_{bs} \times N_{bs}$ Hamilton Matrix

$$\mathbf{H}_{KS}(\mathbf{k}) = \left[\langle \chi_i | \hat{H}_{KS}(\mathbf{k}) | \chi_j \rangle \right]_{i,j=1,\ldots,N_{bs}} \tag{2.146}$$

and the $N_{bs} \times N_{bs}$ overlap matrix

$$\mathbf{S} = \left[\langle \chi_i | \chi_j \rangle \right]_{i,j=1,\ldots,N_{bs}}, \tag{2.147}$$

Equation (2.145) can be written as

$$\boxed{\mathbf{H}_{KS}(\mathbf{k}) \cdot \mathbf{c}^{(\ell)}(\mathbf{k}) = \varepsilon_{\ell \mathbf{k}} \, \mathbf{S} \cdot \mathbf{c}^{(\ell)}(\mathbf{k}).} \tag{2.148}$$

In this way, solving the Kohn–Sham equations with a chosen set of basis functions leads to solving the generalized eigenvalue problem formulated above for each **k**-point considered. In general, this is the most time consuming part in DFT calculations, since its computational cost scales with $O(N_{bs}^3)$. Hence, the choice of the basis set functions controls the accuracy and the computational cost. The computation time can be reduced by solving Eq. (2.148) in parallel for each **k**-point. If one considers an ideal crystal structure, only a reduced number of so called irreducible **k**-points must be considered with particular weights, which reduces further the computational cost.

2.3.6 Solving the Kohn–Sham Equations Self Consistently

As previously mentioned, the Kohn–Sham orbitals $\varphi_i(\mathbf{r})$ cannot be directly obtained by solving the one-electron Schrödinger equation (2.121), because the effective potential $V_{\text{eff}}(\mathbf{r})$ depends on the unknown ground state electron density $n_{e_0}(\mathbf{r})$. This knot can be delt with performing an iteration of the following loop (see Fig. 2.1).

In any step k, one puts a density $n_e^{(\text{in},k)}(\mathbf{r})$ into the effective potential $V_{\text{eff}}(\mathbf{r})$, solves the one-electron Schrödinger equation (2.121) with this effective potential and, by using the obtained single-electron wave functions $\varphi_i(\mathbf{r})$ and Eq. (2.113), gets a new density $n_e^{(\text{out},k)}(\mathbf{r})$. If the output density differs from the input density, one calculates a new density from the previous calculated densities and uses it as input in a new iteration of the loop. The loop is initialized with an initial guess for the electronic density, maybe as a superposition of atomic electronic densities. The procedure ends when the input and output densities are equal up to the desired accuracy. At that point, the self-consistent solution of the electron density has been found. By construction of the Kohn–Sham equations and assuming the validity of the Kohn–Sham Ansatz, this

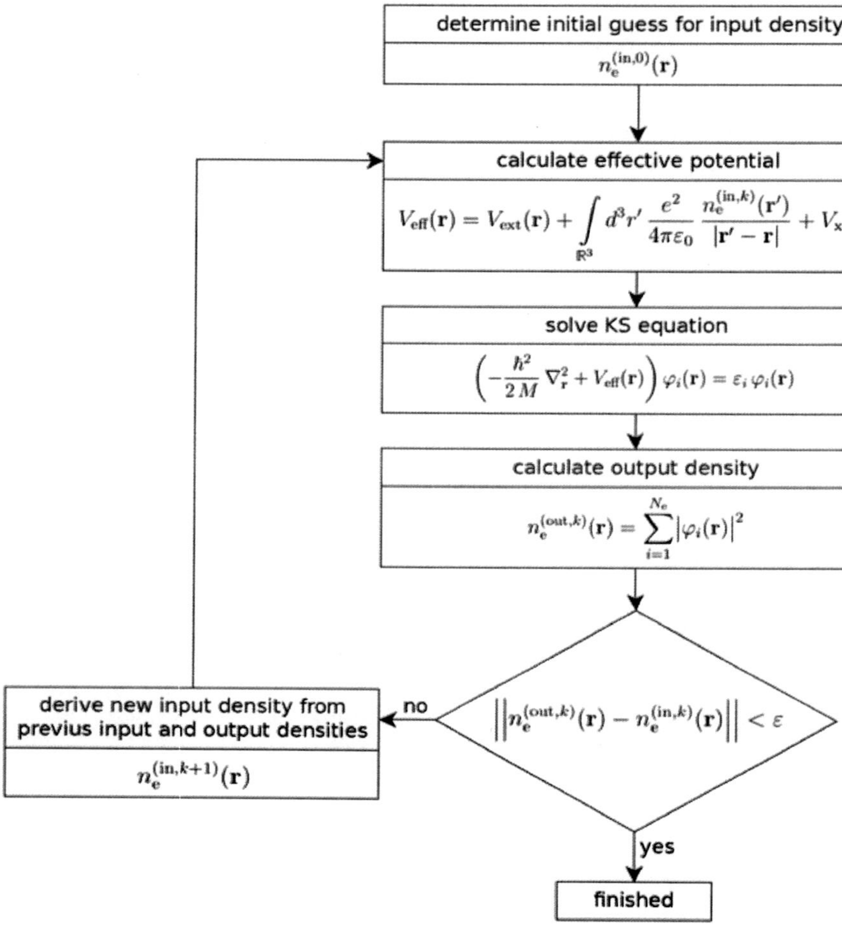

Fig. 2.1 The self-consistent loop to solve the Kohn–Sham equations is illustrated

self-consistent electron density is equal to the ground state density of the interacting electron system, since the ground state density is unique and, hence, no other density can be a self-consistent solution of the Kohn–Sham equations.

2.3.7 Density Mixing to Speed up the Solution of the Kohn–Sham Equations

The convergence of the iterative procedure described previously depends strongly on how the new input density is constructed from the previous output densities. If one just takes the previous output density $n_e^{(\mathrm{out},k)}(\mathbf{r})$ as the input density $n_e^{(\mathrm{in},k+1)}(\mathbf{r})$

of the next loop, the procedure will not converge in most of the cases due to spatial electron density sloshing. It is, therefore, convenient to mix the results of the last and the previous iterations in the construction of the next input density. The different techniques proposed for this step are known as mixing methods. In the development and analysis of mixing methods, one considers the electron density residual

$$\mathrm{Res}\left(n_{\mathrm{e}}^{(\mathrm{in},k)}(\mathbf{r})\right) = n_{\mathrm{e}}^{(\mathrm{out},k)}(\mathbf{r}) - n_{\mathrm{e}}^{(\mathrm{in},k)}(\mathbf{r}), \tag{2.149}$$

which depends explicitly only on the input density $n_{\mathrm{e}}^{(\mathrm{in},k)}(\mathbf{r})$, since the output density $n_{\mathrm{e}}^{(\mathrm{out},k)}(\mathbf{r})$ is calculated from the Kohn–Sham equations by using the input density $n_{\mathrm{e}}^{(\mathrm{in},k)}(\mathbf{r})$ in the effective potential $V_{\mathrm{eff}}(\mathbf{r})$. The electron density residual becomes zero for the self-consistent solution of the density.

2.3.7.1 Linear Mixing

The simplest method is adding a small amount of the electron density residual to the old input density $n_{\mathrm{e}}^{(\mathrm{in},k)}(\mathbf{r})$ for getting the new input density $n_{\mathrm{e}}^{(\mathrm{in},k+1)}(\mathbf{r})$, which corresponds just to linear mixing of the old input $n_{\mathrm{e}}^{(\mathrm{in},k)}(\mathbf{r})$ and output density $n_{\mathrm{e}}^{(\mathrm{out},k)}(\mathbf{r})$:

$$\begin{aligned} n_{\mathrm{e}}^{(\mathrm{in},k+1)}(\mathbf{r}) &= n_{\mathrm{e}}^{(\mathrm{in},k)}(\mathbf{r}) + \alpha\, \mathrm{Res}\left(n_{\mathrm{e}}^{(\mathrm{in},k)}(\mathbf{r})\right) \\ &= n_{\mathrm{e}}^{(\mathrm{in},k)}(\mathbf{r}) + \alpha\left(n_{\mathrm{e}}^{(\mathrm{out},k)}(\mathbf{r}) - n_{\mathrm{e}}^{(\mathrm{in},k)}(\mathbf{r})\right) \\ &= (1-\alpha)\, n_{\mathrm{e}}^{(\mathrm{in},k)}(\mathbf{r}) + \alpha\, n_{\mathrm{e}}^{(\mathrm{out},k)}(\mathbf{r}) \end{aligned} \tag{2.150}$$

with $\alpha \in \mathbb{R}$ and $0 < \alpha < 1$. This mixing schema works in almost all cases but converges slowly.

2.3.7.2 Pulay Mixing

To improve the linear mixing, Pulay [23] constructs the new input density as the linear combination of the last m input densities of the self-consistent loop:

$$n_{\mathrm{e}}^{(\mathrm{in},k+1)}(\mathbf{r}) = \sum_{i=1}^{m} \alpha_i\, n_{\mathrm{e}}^{(\mathrm{in},k+1-i)}(\mathbf{r}), \tag{2.151}$$

where the coefficients α_i fulfill the constraint

$$\sum_{i=1}^{m} \alpha_i = 1 \tag{2.152}$$

to ensure that the new input density also produces the right number of electrons. The coefficients are derived from the condition that the scalar product of the residual of the new input electron density should be minimal with the constraint (2.152). This leads to the following equations for $\ell = 1, \ldots, m$:

$$\frac{\partial}{\partial \alpha_\ell}\left(\left\langle \mathrm{Res}\left(n_{\mathrm{e}}^{(\mathrm{in},k+1)}(\mathbf{r})\right)\middle|\mathrm{Res}\left(n_{\mathrm{e}}^{(\mathrm{in},k+1)}(\mathbf{r})\right)\right\rangle - \lambda\left(\sum_{i=1}^{m}\alpha_i - 1\right)\right) = 0, \quad (2.153)$$

where λ is a Lagrangian multiplier which takes the constraint (2.152) into account. Since the electron density residual of the new electron density $n_{\mathrm{e}}^{(\mathrm{in},k+1)}(\mathbf{r})$ is unknown a priori, Pullay assumes that the electron density residual is linear with the density

$$\begin{aligned}\mathrm{Res}\left(\sum_{i=1}^{m}\alpha_i\, n_{\mathrm{e}}^{(\mathrm{in},k+1-i)}(\mathbf{r})\right) &\approx \sum_{i=1}^{m}\alpha_i\,\mathrm{Res}\left(n_{\mathrm{e}}^{(\mathrm{in},k+1-i)}(\mathbf{r})\right),\\ &= \sum_{i=1}^{m}\alpha_i\left(n_{\mathrm{e}}^{(\mathrm{out},k+1-i)}(\mathbf{r}) - n_{\mathrm{e}}^{(\mathrm{in},k+1-i)}(\mathbf{r})\right). \quad (2.154)\end{aligned}$$

Now the right hand side is already known. Then he utilizes the linearity of the scalar product and obtains

$$\frac{\partial}{\partial \alpha_\ell}\left(\sum_{i=1}^{m}\sum_{j=1}^{m}\alpha_i\,\alpha_j\left\langle \mathrm{Res}\left(n_{\mathrm{e}}^{(\mathrm{in},k+1-i)}(\mathbf{r})\right)\middle|\mathrm{Res}\left(n_{\mathrm{e}}^{(\mathrm{in},k+1-j)}(\mathbf{r})\right)\right\rangle - \lambda\sum_{i=1}^{m}\alpha_i\right) = 0. \quad (2.155)$$

If one defines

$$\begin{aligned}b_{ij} :=& \left\langle \mathrm{Res}\left(n_{\mathrm{e}}^{(\mathrm{in},k+1-i)}(\mathbf{r})\right)\middle|\mathrm{Res}\left(n_{\mathrm{e}}^{(\mathrm{in},k+1-j)}(\mathbf{r})\right)\right\rangle\\ =& \left\langle n_{\mathrm{e}}^{(\mathrm{out},k+1-i)}(\mathbf{r}) - n_{\mathrm{e}}^{(\mathrm{in},k+1-i)}(\mathbf{r})\middle| n_{\mathrm{e}}^{(\mathrm{out},k+1-j)}(\mathbf{r}) - n_{\mathrm{e}}^{(\mathrm{in},k+1-j)}(\mathbf{r})\right\rangle, \quad (2.156)\end{aligned}$$

then one gets from Eq. (2.155) the following system of $m+1$ linear equations

$$\begin{bmatrix} b_{11} & b_{12} & \ldots & b_{1m} & -1 \\ b_{21} & b_{22} & \ldots & b_{2m} & -1 \\ \vdots & \vdots & \ddots & \vdots & \vdots \\ b_{m1} & b_{m2} & \ldots & b_{mm} & -1 \\ -1 & -1 & \ldots & -1 & 0 \end{bmatrix} \cdot \begin{bmatrix} \alpha_1 \\ \alpha_2 \\ \vdots \\ \alpha_m \\ \lambda \end{bmatrix} = \begin{bmatrix} 0 \\ 0 \\ \vdots \\ 0 \\ -1 \end{bmatrix} \quad (2.157)$$

from which the parameters α_i are obtained. Here, a factor of 2 was divided out in the first m equations when performing the derivative with respect to α_ℓ in Eq. (2.155). The last equation is given by differentiating with respect to λ instead of α_ℓ.

This mixing method needs less steps to converge in contrast to the linear mixing, but there is no guarantee for convergence in general.

2.3.7.3 Kerker Mixing

The self-consistent loop converges slowly or even not at all in metals, at localized defects like vacancies or at surfaces, since the electron density sloshes around between several points. The reason behind this is an overestimation of the long range contributions to the electron density during mixing. To overcome this, Kerker [24] performs firstly a three-dimensional Fourier transformation of the $\mathbf{r}$-dependent electron density $n_e(\mathbf{r})$ to get the $\mathbf{k}$-dependent electron density in reciprocal space

$$\tilde{n}_e(\mathbf{k}) = \frac{1}{\mathcal{V}} \int_{\mathcal{V}} d^3r \, n_e(\mathbf{r}) \, e^{-i\mathbf{k}\cdot\mathbf{r}}, \tag{2.158}$$

where $\mathcal{V}$ denotes the volume of the simulation cell. Then, Kerker performs a linear mixing in reciprocal space with a $\mathbf{k}$-dependent mixing parameter:

$$\tilde{n}_e^{(\text{in},k+1)}(\mathbf{k}) = \tilde{n}_e^{(\text{in},k)}(\mathbf{k}) + \alpha \frac{|\mathbf{k}|^2}{|\mathbf{k}|^2 + k_0} \text{Res}\Big(\tilde{n}_e^{(\text{in},k)}(\mathbf{k})\Big). \tag{2.159}$$

In this way, the mixing parameter for short range contributions, $|\mathbf{k}| >> k_0$, equals α, while for long range contributions, $|\mathbf{k}| << k_0$, a damped $\frac{\alpha\,|\mathbf{k}|^2}{k_0^2}$ parameter is used. This mixing method prevents the electron density sloshing mentioned above.

2.3.7.4 Pulay–Kerker Mixing

To get the fast convergence of the Pulay mixer and to prevent the electron density sloshing, one combines the Pulay and Kerker mixing. In this way, first Pulay mixing is applied on the previous input and output densities. Then, Kerker mixing is applied on the new input density obtained by the Pulay mixer in the reciprocal space:

$$\begin{aligned}
\tilde{n}_e^{(\text{in},k+1)}(\mathbf{k}) =& \tilde{n}_e^{(\text{Pulay},\text{in},k+1)}(\mathbf{k}) + \alpha \frac{|\mathbf{k}|^2}{|\mathbf{k}|^2 + k_0} \text{Res}\Big(\tilde{n}_e^{(\text{Pulay},\text{in},k+1)}(\mathbf{k})\Big) \\
=& \tilde{n}_e^{(\text{Pulay},\text{in},k+1)}(\mathbf{k}) + \alpha \frac{|\mathbf{k}|^2}{|\mathbf{k}|^2 + k_0} \sum_{i=1}^{m} \alpha_i \, \text{Res}\Big(\tilde{n}_e^{(\text{in},k+1-i)}(\mathbf{k})\Big) \\
=& \tilde{n}_e^{(\text{Pulay},\text{in},k+1)}(\mathbf{k}) + \alpha \frac{|\mathbf{k}|^2}{|\mathbf{k}|^2 + k_0} \sum_{i=1}^{m} \alpha_i \Big(\tilde{n}_e^{(\text{out},k+1-i)}(\mathbf{k}) - \tilde{n}_e^{(\text{in},k+1-i)}(\mathbf{k})\Big).
\end{aligned} \tag{2.160}$$

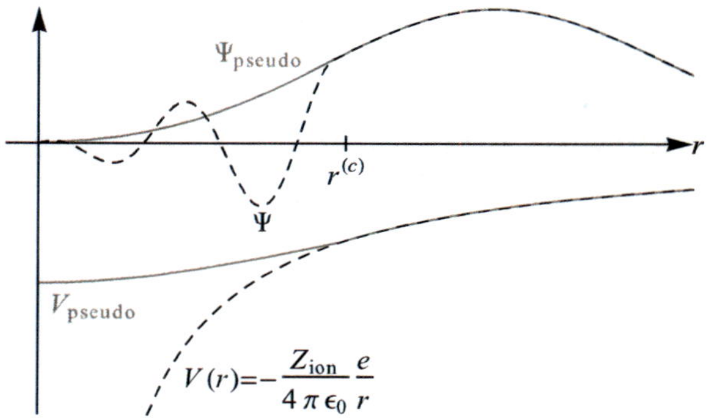

Fig. 2.2 The concept of a pseudopotential is illustrated

2.3.8 Pseudopotentials

The computational cost may be further reduced based on the following considerations: In most of the cases, the electrons can be separated in core and valence electrons. The valence electrons occupy higher energy levels and are delocalized between the nuclei. They are, thus, actively involved in the atomic bonding. The core electrons occupy lower energy levels and are tightly bound to the nucleus. Therefore, they are localized close to the nucleus and do not contribute considerably to atomic bonding. Their wave function varies rapidly on short distances, so that many basis functions are necessary for an accurate description. In a good approximation, the wave function of core electrons has no overlap with electronic orbitals from neighboring atoms, so that they are mainly identical for isolated or bonded atoms. Due to this, the core electrons can be treated as frozen, which is known as the frozen core approximation. Within this approximation, only the valence electrons are treated explicitly with wave functions and the core electrons together with the nuclei are replaced by an effective ion potential that acts on the valence electrons. This reduces significantly the computational cost, since less electrons need to be treated with wave functions.

But the wave function of the valence electrons still has got several nodes on short distances close to the core, which would need many basis functions for an accurate representation (See for this Fig. 2.2). Origin of this knots is the diverging effective ion potential $V(r)$ generated from the core electrons and the nucleus at short distances. Therefore, this potential is replaced by a pseudopotential $V_{\text{pseudo}}(r)$ for distances smaller than a given cutoff radius $r^{(c)}$ to the ion. The pseudopotential is constructed in such a way that the obtained pseudo wave function has no knots for distances smaller than $r^{(c)}$ and that it is identical to the original wave function for distances greater than $r^{(c)}$. The pseudo wave function can be accurately treated with less numerical effort than the original wave function. All relevant electronic properties can be obtained from the pseudo wave function, since atomic bonding mostly takes places at distances greater than $r^{(c)}$.

In 1979, Hamann et al. [25] introduced the following requirements for "good" pseudopotentials that yield approximately the same results compared to all-electron calculations for a wide range of atomic structures:

1. The valence eigenvalues of all-electron and pseudopotential calculations agree for the chosen atomic reference configuration.
2. All-electron and pseudo valence wave functions agree for distances greater than a chosen core radius $r^{(c)}$.
3. The integrated charge from 0 to $r^{(c)}$ agrees for each all-electron and pseudo valence wave function, which is called norm-conservation.
4. The logarithmic derivatives of the all-electron and pseudo valence wave functions agree at $r^{(c)}$. For a given wave function $\psi(\mathbf{r})$, the logarithmic derivative $D_{\ln}$ is defined as
$$D_{\ln}\Big(\psi(\mathbf{r})\Big) \equiv r\,\frac{d}{dr}\ln\Big(\psi(\mathbf{r})\Big), \tag{2.161}$$
where $r = |\mathbf{r}|$ and ln denotes the logarithmic function.

Pseudopotentials fullfilling the above requirements are called norm-conserving and are the basis for most modern pseudopotential calculations. Smooth pseudopotentials fulfill automatically point 4. Due to points 1 and 2, the pseudopotential agrees outside the core radius with the original potential, since the eigenfunctions and eigenvalues define uniquely the potential except for a constant shift of the energy, which is fixed, because the potential is defined to be zero at infinity. Point 3 ensures that the total charge inside the core region is correct. Therefore, points 1 to 4 ensure that the pseudo orbitals are equal to the true orbitals outside the core.

In the original publication of Hamann et al. [25], an additional requirement was added to the 4th point. We skip this, since later it was shown that this additional requirement is implied by the four points listed above [8].

2.3.8.1 Norm-Conserving Pseudopotentials of Goedecker, Teter, and Hutter

In 1996, Goedecker, Teter, and Hutter developed norm-conserving pseudopotentials for the elements of the first two rows of the periodic table [26]. Two years later in 1998, they included relativistic effects in their pseudopotentials and provided pseudopotentials for all elements from H to Rn [27]. The obtained pseudopotentials are highly efficient and reliable: In contrast to most other pseudopotentials, which are tabulated on a radial grid, these pseudopotentials utilize an analytical functional form with a few parameters. Gaußians are used for the radial functions, so that the analytical functional form is well localized and has got a smooth shape in both real and Fourier space. This allows to perform an efficient real space integration of the pseudopotential with only a quadratic scaling with respect to the system size in contrast to the cubic scaling of the commonly used Fourier space integration scheme [27]. In addition, these pseudopotentials are purely ab-initio, since all contained parameters were obtained by reproducing all-electron DFT calculations in the LDA-approximation of isolated atoms. The validity of the obtained pseudopotentials has been thoroughly tested, among other ways by comparing bond lengths of many small molecules with all- electron DFT calculations and they have been found to be very accurate [26, 27].

The pseudopotential consists of a so called local part that only depends on the distance r to the ion:

$$V_{\text{loc}}(r) = -\frac{e^2}{4\pi\varepsilon_0}\frac{Z_{\text{ion}}}{r}\,\text{erf}\left(\frac{r}{\sqrt{2}\,r_{\text{loc}}}\right) + \text{e}^{-\frac{1}{2}\left(\frac{r}{r_{\text{loc}}}\right)^2}\left(C_1 + C_2\left(\frac{r}{r_{\text{loc}}}\right)^2 + C_3\left(\frac{r}{r_{\text{loc}}}\right)^3 + C_4\left(\frac{r}{r_{\text{loc}}}\right)^6\right), \tag{2.162}$$

where $Z_{\text{ion}}\,e$ is the effective ion charge, *i.e.*, the charge of the nucleus minus the charge of the core electrons, erf denotes the error function and r_{loc}, C_1, C_2, C_3, C_4 are parameters. The error function is introduced to avoid the singularity at $r = 0$, thereby allowing for smoother pseudo wave function. The pseudopotential also contains a so called non-local part. In contrast to the local component, this part differs for wave functions with different angular momentum. The non-local part is a sum over separable terms:

$$V_{\text{nl}}(\mathbf{r}, \mathbf{r}') = \sum_{\ell=0}^{3}\sum_{i=1}^{3}\sum_{j=1}^{3}\sum_{m=-\ell}^{\ell} Y_{\ell m}(\hat{\mathbf{r}})\, \mathfrak{p}_i^{(\ell)}(r)\, h_{ij}^{(\ell)}\, \mathfrak{p}_i^{(\ell)}(r')\, Y_{\ell m}^*(\hat{\mathbf{r}}'). \tag{2.163}$$

Here, $Y_{\ell m}(\hat{\mathbf{r}})$ denotes the spherical harmonic as a function of the normalized vector $\hat{\mathbf{r}} = \frac{\mathbf{r}}{r}$, $h_{ij}^{(\ell)}$ are parameters and the radial projectors $\mathfrak{p}_i^{(\ell)}(r)$ are defined by

$$\mathfrak{p}_i^{(\ell)}(r) = \frac{\sqrt{2}\, r^{\ell+2(i-1)}\, \text{e}^{-\frac{r^2}{2r_\ell^2}}}{r_\ell^{\ell+\frac{4i-1}{2}}\sqrt{\Gamma\left(\ell + \frac{4i-1}{2}\right)}}. \tag{2.164}$$

Γ denotes the gamma function and r_ℓ is a parameter. The radial projectors are normalized:

$$\int_0^\infty dr\, \mathfrak{p}_i^{(\ell)}\, \mathfrak{p}_i^{(\ell)}\, r^2 = 1. \tag{2.165}$$

All pseudopotential parameters are tabulated in Ref. [26] for the non-relativistic and in Ref. [27] for the relativistic case.

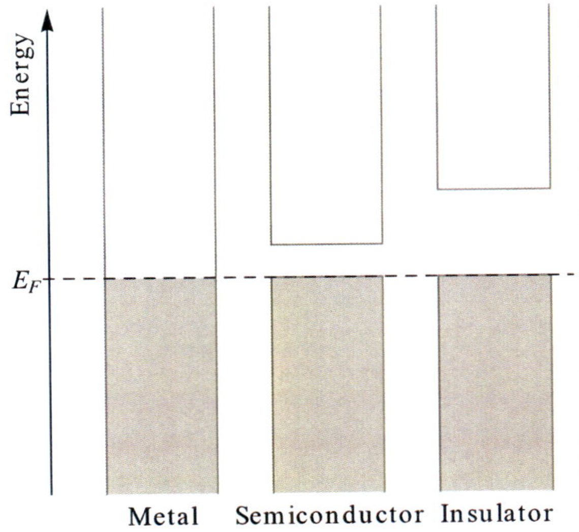

Fig. 2.3 Electronic band structures of metals, semiconductors and insulators are illustrated. Gray filled ares indicate the occupied energy levels at $T = 0$

2.3.9 Electronic Band Structure of Solids

Density functional theory allows to calculate the allowed energy levels of the electrons in a solid. Due to the huge number of electrons in a solid, these electronic energy levels lie closely together and form so called bands. Between the bands, there are forbidden energy ranges for the electrons. In the ground state, the N_e electrons fill up the N_e lowest energy levels. The highest occupied energy level E_F is called **Fermi energy**. The highest occupied band is called **valence band** and the lowest band that is not or not completely filled by the electrons is called **conduction band**. See for this Fig. 2.3.

The unoccupied states in the ground state are non-bonding, so that electrons occupying these states can freely move in the solid. Therefore, the electric conductivity depends strongly on the number of electrons in these non-bonding states. In the ground state at zero temperature $T = 0$, there are no electrons in non-bonding states. But under ambient conditions $T = 300$ K, the electrons have got thermal energy and, therefore, some electrons are excited in the non-bonding states contained in the conduction band. In an insulator, there is a huge gap between the valence and conduction band, so that the thermal energy is not sufficient to excite electrons into the conduction band and there is no electric conductivity. In a semiconductor, there is a small gap between the valence and conduction band, so that a few electrons are excited into the conduction band at $T = 300$ K and one observes a small conductivity. In a metal, there is no gap between the valence and conduction band, so that a significant amount of electrons is excited into the conduction band at $T = 300$ K, which explains the good conductivity of metals.

2.4 T_e-dependent Density Functional Theory

Ground state DFT cannot be used to describe excited electrons. However, as early as in 1965, only one year after the publication of the Hohenberg-Kohn theorems [9], Mermin developed an extension of DFT to electrons at a finite electronic temperature T_e [28]. The description of electrons at a finite T_e needs several tools of thermodynamics and statistical mechanics, which are introduced in the next two sections based on Ref. [29] and Ref. [30] before T_e-dependent DFT is derived.

2.4.1 *Basic Considerations of Thermodynamics*

A macroscopic system contains a huge number of particles and, therefore, a huge number of degrees of freedom. Due to this, the energy separation between the eigenstates of the Hamilton operator is very small. The smallest perturbation can induce a transfer of the system from one to another state which has practically the same energy, so that there are always fluctuations of the state. On the other hand, any macroscopic measurement takes place on a long timescale compared to the particle motions in the system. In this way, most of the degrees of freedom can be integrated out. Hence, the system can be described by a macro state that only depends on a few so called thermodynamical variables like energy E, temperature T, pressure p, volume $\mathcal{V}$, number of particles N. Also all properties of the macro state are functions of these few thermodynamical variables, which leads to far-reaching consequences. Thermodynamical potentials are introduced, which depend only on the macro state and not on how this state was reached. One thermodynamical potential is the internal energy E. The energy conservation leads to the formulation of the first law of thermodynamics, which states that the change of the internal energy of a system is given by

$$dE = \delta Q - p\,d\mathcal{V} + \mu\,dN, \tag{2.166}$$

where δQ denotes the heat flow into the system, $-p\,d\mathcal{V}$ the work performed on the system and $\mu\,dN$ the energy change due to particle exchange. μ is the chemical potential and describes the change in energy, when one particle is added or removed. Here, we only treat one type of particle, since we want to describe later the system of electrons at a finite T_e. Since the total exchanged heat Q depends not only on the macro state but also on the process leading to the macro state, the entropy S is defined, which is again a thermodynamical potential. The second law of thermodynamics defines the change in entropy as

$$dS \geq \frac{\delta Q}{T}, \tag{2.167}$$

where T is the absolute temperature and the equality holds for quasi-static processes, *i.e.*, processes during which the system is always in equilibrium. The entropy is properly defined, since the third law of thermodynamics states that $T = 0$ cannot be reached in any system during a finite time, so that $T > 0$ always. Solving Eq. (2.166) for δQ and inserting the result for δQ in Eq. (2.167) leads to

$$dS \geq \frac{1}{T} dE + \frac{p}{T} d\mathcal{V} - \frac{\mu}{T} dN. \tag{2.168}$$

In this way, $E, \mathcal{V}, N$ are the natural variables describing the entropy S and, if the macro state is characterized by these variables, the entropy can be easily used to describe transformations. Solving the above equation for dE yields

$$dE \leq T\, dS - p\, d\mathcal{V} + \mu\, dN \tag{2.169}$$

and indicates that $S, \mathcal{V}, N$ are the natural variables of the internal energy E.

Thermodynamical potentials can be used to determine the equilibrium macro state of a system at given conditions. If a system is isolated, *i.e.*, no heat exchange $\delta Q = 0$, no particle exchange $dN = 0$ and no work is performed $-p\, d\mathcal{V} = 0$, then $dE = 0$ is valid and the internal energy E keeps constant due to (2.166). Furthermore, one obtains $dS \geq 0$ from Eq. (2.168), so that the macro state corresponding to equilibrium is characterized by the maximum of the entropy S.

Now we consider a system which is kept at a given constant temperature T. This condition is achieved through heat exchange with a system so large that it does not change its temperature due to this exchange and, therefore, is called heat bath. Since the heat bath and the considered system are isolated from the environment, one obtains for the internal energy E of the considered system from the first law of thermodynamics:

$$\begin{aligned}
& & dE \overset{(2.166)}{=} \underbrace{\delta Q}_{\overset{(2.167)}{\leq} T\, dS} \underbrace{- p\, d\mathcal{V}}_{=0} + \mu \underbrace{dN}_{=0}, & \\
\Leftrightarrow & & dE - T\, dS \leq 0, & \\
\Leftrightarrow & & dE - T\, dS - S \underbrace{dT}_{=0} \leq 0, & \\
\Leftrightarrow & & dE - d(T\, S) \leq 0, & \\
\Leftrightarrow & & d(E - T\, S) \leq 0. & \qquad (2.170)
\end{aligned}$$

It follows that the so called Helmholtz free energy

$$F = E - T\, S \tag{2.171}$$

is minimized, if the system is in the equilibrium macro state. The Helmholtz free energy F is a thermodynamical potential with the natural variables $T, \mathcal{V}, N$, since one gets from the definition of F and Eq. (2.169):

$$dF \leq -S\,dT - p\,d\mathcal{V} + \mu\,dN. \tag{2.172}$$

Furthermore, the thermodynamical variables form two disjoint groups: The extensive variables like volume $\mathcal{V}$, internal energy E, Helmholtz free energy F, and entropy S depend on the number of particles N in the system, whereas intensive variables like temperature T, pressure p, chemical potential μ do not depend on the number of particles N. If one considers two identical systems and combines them, then the value of the extensive variables is doubled whereas the intensive variables are unchanged.

2.4.2 Basic Considerations of Statistical Mechanics

Now we give an overview about statistical mechanics based on Ref. [29]. But in contrast to Ref. [29] and most commonly used textbooks, we develop all statements from the definition of the entropy instead of performing an integration in the phase space of all degrees of freedom.

Statistical mechanics allows to get deeper insights compared to employing the rules of thermodynamics. For this, a micro state is introduced, which corresponds to a particular state of the system, which is characterized by the many-body wave function Ψ or the corresponding ket $|\Psi\rangle$. Usually, several micro states are compatible to or represent a macro state. Therefore, a macro state corresponds to the collection of all representing micro states together with their probability of occurrence. Such a collection is also called statistical ensemble. A macro state cannot be described by a wave function or a ket, since it is represented by several micro states. If the macro state is represented by Ω micro states, where the ith one is characterized by the ket $|\Psi_i\rangle$ and occurs with the probability $0 \leq \mathbb{p}_i \leq 1$, then the macro state is described by the density matrix

$$\hat{\rho} = \sum_{i=1}^{\Omega} \mathbb{p}_i\, |\Psi_i\rangle\langle\Psi_i|, \qquad \sum_{i=1}^{\Omega} \mathbb{p}_i = 1. \tag{2.173}$$

Here, the kets $|\Psi_i\rangle$ are orthonormal. We allow $\Omega = \infty$. To be as general as possible, we also allow $\mathbb{p}_i = 0$, but, obviously, $\mathbb{p}_i > 0$ for micro states compatible with the macro state. By definition, any density matrix is Hermitian. If the system is in the macro state described by $\hat{\rho}$, the corresponding average value of any operator $\hat{A}$ is given by

$$\langle \hat{A} \rangle_{\hat{\rho}} = \mathrm{Tr}\big(\hat{\rho}\,\hat{A}\big), \tag{2.174}$$

where Tr denotes the trace, which is defined for any operator $\hat{A}$ with the help of a complete set of orthonormal kets $\big\{|\phi_i\rangle, i \in \mathcal{N}\big\}$ by

$$\mathrm{Tr}(\hat{A}) = \sum_{i\in\mathcal{N}} \langle \phi_i | \hat{A} | \phi_i \rangle. \tag{2.175}$$

The probability that a micro state $|\psi\rangle$ occurs is given by $\langle\psi|\hat{\rho}|\psi\rangle$. The particular choice of the complete set does not influence the value of the trace, since one obtains by using an other complete set of orthonormal kets $\{|\varphi_j\rangle,\ j \in \mathcal{M}\}$:

$$\begin{aligned}
\sum_{i\in\mathcal{N}} \langle \phi_i | \hat{A} | \phi_i \rangle &= \sum_{i\in\mathcal{N}} \langle \phi_i | \underbrace{\left(\sum_{j\in\mathcal{M}} |\varphi_j\rangle\langle\varphi_j| \right)}_{=\hat{\mathbb{1}}} \hat{A} | \phi_i \rangle \\
&= \sum_{i\in\mathcal{N}} \sum_{j\in\mathcal{M}} \langle \phi_i | \varphi_j \rangle \langle \varphi_j | \hat{A} | \phi_i \rangle \\
&= \sum_{i\in\mathcal{N}} \sum_{j\in\mathcal{M}} \langle \varphi_j | \hat{A} | \phi_i \rangle \langle \phi_i | \varphi_j \rangle \\
&= \sum_{j\in\mathcal{M}} \langle \phi_j | \underbrace{\left(\sum_{i\in\mathcal{N}} |\varphi_i\rangle\langle\varphi_i| \right)}_{=\hat{\mathbb{1}}} \hat{A} | \phi_j \rangle \\
&= \sum_{j\in\mathcal{M}} \langle \phi_j | \hat{A} | \phi_j \rangle.
\end{aligned} \tag{2.176}$$

Hence, the trace is well defined.

Theorem 2.9 (Properties of the density matrix) *Let $\hat{\rho}$ be an arbitrary density matrix and let f be a real function that can be expressed in a Taylor series with vanishing constant term. Then one gets*

$$f(\hat{\rho}) = \sum_{i=1}^{\Omega} f(\mathbb{p}_i)\, |\Psi_i\rangle\langle\Psi_i| \tag{2.177}$$

and

$$Tr\Big(f(\hat{\rho})\Big) = \sum_{i=1}^{\Omega} f(\mathbb{p}_i). \tag{2.178}$$

Consequently, one has

$$Tr(\hat{\rho}) = 1. \tag{2.179}$$

If more than one $\mathbb{p}_i$ is greater than zero in the definition of $\hat{\rho}$, then one obtains $\hat{\rho}^2 \neq \hat{\rho}$ and $Tr(\hat{\rho}^2) < 1$.

Proof Let $\hat{\rho} = \sum_{i=1}^{\Omega} \mathbb{p}_i \, |\Psi_i\rangle\langle\Psi_i|$ be an arbitrary density matrix with $\sum_{i=1}^{\Omega} \mathbb{p}_i = 1$. To simplify things, we add further kets $|\Psi_i\rangle$ with $\mathbb{p}_i = 0$ to the sum in $\hat{\rho}$, so that $\{|\Psi_i\rangle, i = 1, 2, \ldots, \Omega'\}$ forms a complete set of orthonormal kets. First, we prove for all $n \in \mathbb{N}$

$$\hat{\rho}^n = \sum_{i=1}^{\Omega} \mathbb{p}_i^n \, |\Psi_i\rangle\langle\Psi_i| \tag{2.180}$$

by using mathematical induction. This statement is true for $n = 1$ by definition of $\hat{\rho}$. Now we assume that Eq. (2.180) is valid up to a given $n \in \mathbb{N}$ and prove it for $n + 1$:

$$\begin{aligned}
\hat{\rho}^{n+1} &= \hat{\rho}\,\hat{\rho}^n \\
&= \left(\sum_{i=1}^{\Omega} \mathbb{p}_i \, |\Psi_i\rangle\langle\Psi_i|\right)\left(\sum_{j=1}^{\Omega} \mathbb{p}_j^n \, |\Psi_j\rangle\langle\Psi_j|\right) \\
&= \sum_{i=1}^{\Omega}\sum_{j=1}^{\Omega} \mathbb{p}_i \, \mathbb{p}_j^n \, |\Psi_i\rangle \underbrace{\langle\Psi_i|\Psi_j\rangle}_{\delta_{ij}} \langle\Psi_j| \\
&= \sum_{i=1}^{\Omega} \mathbb{p}_i^{n+1} \, |\Psi_i\rangle\langle\Psi_i|.
\end{aligned}$$

Now we obtain

$$\begin{aligned}
f(\hat{\rho}) &= \sum_{n=1}^{\infty} c_n \, \hat{\rho}^n \\
&\overset{(2.180)}{=} \sum_{n=1}^{\infty} c_n \left(\sum_{i=1}^{\Omega} \mathbb{p}_i^n \, |\Psi_i\rangle\langle\Psi_i|\right) \\
&= \sum_{i=1}^{\Omega} \left(\sum_{n=1}^{\infty} c_n \, \mathbb{p}_i^n\right) |\Psi_i\rangle\langle\Psi_i| \\
&= \sum_{i=1}^{\Omega} f(\mathbb{p}_i) \, |\Psi_i\rangle\langle\Psi_i|.
\end{aligned}$$

Since $\{|\Psi_i\rangle, i \in \Omega'\}$ forms a complete set of orthonormal kets, we have further

$$\mathrm{Tr}\Big(f(\hat{\rho})\Big) = \sum_{j=1}^{\Omega'} \langle \Psi_j | \underbrace{\left(\sum_{i=1}^{\Omega} f(\mathbb{p}_i) \, |\Psi_i\rangle\langle\Psi_i| \right)}_{\overset{(2.177)}{=} f(\hat{\rho})} |\Psi_j\rangle$$

$$= \sum_{j=1}^{\Omega'} \sum_{i=1}^{\Omega} f(\mathbb{p}_i) \, \langle \Psi_j | \Psi_i \rangle \, \underbrace{\langle \Psi_i | \Psi_j \rangle}_{=\delta_{ij}}$$

$$= \sum_{i=1}^{\Omega} f(\mathbb{p}_i).$$

Equation (2.179) follows directly from this by using the identity map for f. Without loss of generality we assume $\mathbb{p}_i > 0$ for all $i = 1, \ldots, \Omega$. If more than one $\mathbb{p}_i$ is bigger than zero, then we have $\Omega \geq 2$ and due to $\sum_{i=1}^{\Omega} \mathbb{p}_i = 1$ and $\mathbb{p}_i > 0$ for all $i = 1, \ldots, \Omega$ we obtain $\mathbb{p}_i < 1$ for all $i = 1, \ldots, \Omega$. Consequently, we have $\mathbb{p}_i^2 \neq \mathbb{p}_i$ for all $i = 1, \ldots, \Omega$ and we obtain

$$\hat{\rho}^2 \overset{(2.177)}{=} \sum_{i=1}^{\Omega} \mathbb{p}_i^2 \, |\Psi_i\rangle\langle\Psi_i| \neq \sum_{i=1}^{\Omega} \mathbb{p}_i \, |\Psi_i\rangle\langle\Psi_i| = \hat{\rho}.$$

We get from $\mathbb{p}_i > 0$ for all $i = 1, \ldots, \Omega$:

$$\begin{aligned} \mathrm{Tr}\big(\hat{\rho}^2\big) &\overset{(2.178)}{=} \sum_{i=1}^{\Omega} \mathbb{p}_i^2 \\ &< \sum_{i=1}^{\Omega} \mathbb{p}_i^2 + \underbrace{\sum_{i=1}^{\Omega} \sum_{\substack{j=1 \\ j \neq i}}^{\Omega} \mathbb{p}_i \, \mathbb{p}_j}_{>0} \\ &= \sum_{i=1}^{\Omega} \sum_{j=1}^{\Omega} \mathbb{p}_i \, \mathbb{p}_j \\ &= \underbrace{\left(\sum_{i=1}^{\Omega} \mathbb{p}_i \right)}_{=1} \underbrace{\left(\sum_{j=1}^{\Omega} \mathbb{p}_j \right)}_{=1} \\ &= 1. \end{aligned} \tag{2.181}$$

■

The time evolution of the density matrix is given by

$$
\begin{aligned}
\frac{\partial \hat{\rho}}{\partial t} &= \sum_{i=1}^{\Omega} \mathbb{p}_i \left|\frac{\partial \Psi_i}{\partial t}\right\rangle\!\left\langle \Psi_i\right| + \sum_{i=1}^{\Omega} \mathbb{p}_i \left|\Psi_i\right\rangle\!\left\langle \frac{\partial \Psi_i}{\partial t}\right| \\
&\overset{(2.11)}{=} \sum_{i=1}^{\Omega} \mathbb{p}_i \left|-\frac{\mathrm{i}}{\hbar}\,\hat{H}\,\Psi_i\right\rangle\!\left\langle \Psi_i\right| + \sum_{i=1}^{\Omega} \mathbb{p}_i \left|\Psi_i\right\rangle\!\left\langle -\frac{\mathrm{i}}{\hbar}\,\hat{H}\,\Psi_i\right|. \\
&= -\frac{\mathrm{i}}{\hbar}\sum_{i=1}^{\Omega} \mathbb{p}_i\,\hat{H}\,\left|\Psi_i\right\rangle\!\left\langle \Psi_i\right| + \frac{\mathrm{i}}{\hbar}\sum_{i=1}^{\Omega} \mathbb{p}_i\,\left|\Psi_i\right\rangle\!\left\langle \Psi_i\right| \underbrace{\hat{H}^{\dagger}}_{=\hat{H}} \\
&= -\frac{\mathrm{i}}{\hbar}\left(\hat{H}\,\hat{\rho} - \hat{\rho}\,\hat{H}\right).
\end{aligned}
$$

$\hat{H}$ denotes the Hamilton operator and we assume, that the probabilities $\mathbb{p}_i$ does not depend on time. Finally, we obtain the so called **von Neumann equation**

$$
\boxed{\frac{\partial \hat{\rho}}{\partial t} = -\frac{\mathrm{i}}{\hbar}\left[\hat{H}, \hat{\rho}\right]_{-},} \tag{2.182}
$$

where $[.\,,\ .]_{-}$ denotes the commutator.

Now we want to define the averaged entropy $\langle S\rangle$ of a macro state. Firstly, the entropy of a micro state is a function of its probability of occurrence p. If one combines two systems A and B to a single system AB, then the probability $\mathbb{p}_{AB}$ of the microstate of the combined system is the product of the probabilities $\mathbb{p}_A$ and $\mathbb{p}_B$ of the microstates of the systems A and B, respectively: $\mathbb{p}_{AB} = \mathbb{p}_A\,\mathbb{p}_B$. Since the entropy is an extensive variable, the entropy of the combined system AB must be the sum of the entropies of the systems A and B. Therefore, the entropy of a microstate with probability p must be $S \propto \ln(\mathbb{p})$. One defines finally $S = -k_{\mathrm{B}}\,\ln(\mathbb{p})$, where k_{B} is the Boltzmann constant to set the correct unit. The minus sign is set to get always a non-negative entropy, since $\ln(\mathbb{p}) \leq 0$ is valid for $0 \leq \mathbb{p} \leq 1$. Therefore, the averaged entropy $\langle S\rangle$ of a macro state is defined from the corresponding density matrix $\hat{\rho}$ by

$$
\langle S\rangle_{\hat{\rho}} = -k_{\mathrm{B}} \sum_{i=1}^{\Omega} \mathbb{p}_i\,\ln(\mathbb{p}_i) \tag{2.183}
$$

Using Eq. (2.178), we obtain finally

$$
\boxed{\langle S\rangle_{\hat{\rho}} = -k_{\mathrm{B}}\,\mathrm{Tr}\Big(\hat{\rho}\,\ln\big(\hat{\rho}\big)\Big).} \tag{2.184}
$$

The entropy is well defined, since one yields for $\mathbb{p}_i = 0$:

$$
\lim_{\mathbb{p}_i \to 0} \mathbb{p}_i\,\ln(\mathbb{p}_i) = \lim_{\mathbb{p}_i \to 0} \frac{\ln(\mathbb{p}_i)}{\frac{1}{\mathbb{p}_i}} = \lim_{\mathbb{p}_i \to 0} \frac{\frac{1}{\mathbb{p}_i}}{-\frac{1}{\mathbb{p}_i^2}} = \lim_{\mathbb{p}_i \to 0} -\mathbb{p}_i = 0. \tag{2.185}
$$

Here we applied the rule of L' Hospital, since the nominator $\ln(\mathbb{p}_i)$ diverges to $-\infty$ and the denominator $\frac{1}{\mathbb{p}_i}$ diverges to ∞. One obtains the minimal value $\langle S\rangle_{\hat{\rho}} = 0$ of the entropy, if one $\mathbb{p}_i = 1$ and all other $\mathbb{p}_i$'s are zero, *i.e.*, only one micro state represents the macro state.

Now we want to continue with equilibrium ensembles at given conditions. For this, we firstly need several statements about convex functions, which are taken together with the proofs from appendix B of Ref. [30] except for the proof of (i) of the following theorem.

Mathematical Theorem 2.3 (Properties of convex functions) *A real function f is convex on an interval $I \subseteq \mathbb{R}$, if the following is valid:*

$$\forall x_1, x_2 \in I,\ \lambda \in [0, 1]: \ f\big(\lambda\, x_1 + (1-\lambda)\, x_2\big) \leq \lambda\, f(x_1) + (1-\lambda)\, f(x_2). \tag{2.186}$$

(i) Let be $\Omega \in \mathbb{N}$ and the numbers $0 \leq \mathbb{p}_i \leq 1$ for all $i = 1, 2, \ldots, \Omega$ given with $\sum_{i=1}^{\Omega} \mathbb{p}_i = 1$. Then a real function f that is convex on an interval I obeys

$$\forall x_1, \ldots, x_\Omega \in I: \quad f\left(\sum_{i=1}^{\Omega} \mathbb{p}_i\, x_i\right) \leq \sum_{i=1}^{\Omega} \mathbb{p}_i\, f(x_i), \tag{2.187}$$

which is called Jensens's inequality.

(ii) Let f be a differentiable function on I. Then one has

$$\forall x_1, x_2 \in I: \quad f(x_1) - f(x_2) - (x_1 - x_2)\, f'(x_2) \geq 0 \ \Leftrightarrow\ f \text{ is convex on } I. \tag{2.188}$$

(iii) Let f be a twice-differentiable function on I. Then one has

$$\forall x \in I: \quad f''(x) \geq 0 \ \Rightarrow\ f \text{ is convex on } I. \tag{2.189}$$

Proof (i) We use mathematical induction. Equation (2.187) is trivially true for $\Omega = 1$ and from the definition of convex functions (2.186) one obtains the validity for $\Omega = 2$. Now we assume that Eq. (2.187) is valid up to a given $\Omega \in \mathbb{N}$ and show the validity for $\Omega + 1$: We have $\sum_{i=1}^{\Omega+1} \mathbb{p}_i = 1$ and $x_1, \ldots, x_{\Omega+1} \in I$. If $\mathbb{p}_{\Omega+1} = 1$ is valid, then all other $\mathbb{p}_i$'s must be zero and (2.187) is trivially true. Let be $\mathbb{p}_{\Omega+1} < 1$. It follows that $1 = \sum_{i=1}^{\Omega} \frac{\mathbb{p}_i}{1-\mathbb{p}_{\Omega+1}}$ is true. Let be $x_{\min} = \min\{x_1, \ldots, x_\Omega\}$ and $x_{\max} = \max\{x_1, \ldots, x_\Omega\}$. Then we obtain

$$x_{\min} = \sum_{i=1}^{\Omega} \frac{\mathbb{p}_i}{1-\mathbb{p}_{\Omega+1}} x_{\min} \leq \sum_{i=1}^{\Omega} \frac{\mathbb{p}_i}{1-\mathbb{p}_{\Omega+1}} x_i \leq \sum_{i=1}^{\Omega} \frac{\mathbb{p}_i}{1-\mathbb{p}_{\Omega+1}} x_{\max} = x_{\max}$$

and, consequently, $\sum_{i=1}^{\Omega} \frac{\mathbb{p}_i}{1-\mathbb{p}_{\Omega+1}} x_i \in I$. Then we have

$$\begin{aligned}
f\left(\sum_{i=1}^{\Omega+1} \mathbb{p}_i\, x_i\right) &= f\left(\mathbb{p}_{\Omega+1}\, x_{\Omega+1} + (1-\mathbb{p}_{\Omega+1}) \sum_{i=1}^{\Omega} \frac{\mathbb{p}_i}{1-\mathbb{p}_{\Omega+1}}\, x_i\right) \\
&\overset{(2.186)}{\leq} \mathbb{p}_{\Omega+1}\, f(x_{\Omega+1}) + (1-\mathbb{p}_{\Omega+1})\, f\left(\sum_{i=1}^{\Omega} \frac{\mathbb{p}_i}{1-\mathbb{p}_{\Omega+1}}\, x_i\right) \\
&\leq \mathbb{p}_{\Omega+1}\, f(x_{\Omega+1}) + (1-\mathbb{p}_{\Omega+1}) \sum_{i=1}^{\Omega} \frac{\mathbb{p}_i}{1-\mathbb{p}_{\Omega+1}}\, f(x_i) \\
&= \sum_{i=1}^{\Omega+1} \mathbb{p}_i\, f(x_i).
\end{aligned}$$

(ii) First we prove $\forall x_1, x_2 \in I: \quad f$ is convex on $I \Rightarrow f(x_1) - f(x_2) - (x_1 - x_2)\, f'(x_2) \geq 0$. By the definition of a convex function, we have for all $\lambda \in (0, 1]$ and all $x_1, x_2 \in I$:

$$\begin{aligned}
& f\Big(\lambda\, x_1 + (1-\lambda)\, x_2\Big) \leq \lambda\, f(x_1) + (1-\lambda)\, f(x_2), \\
\Leftrightarrow \quad & 0 \leq \lambda\, f(x_1) + f(x_2) - \lambda\, f(x_2) - f\Big(\lambda\, x_1 + (1-\lambda)\, x_2\Big), \\
\Leftrightarrow \quad & 0 \leq f(x_1) - f(x_2) - \frac{1}{\lambda}\Big(f\Big(\lambda\, x_1 + (1-\lambda)\, x_2\Big) - f(x_2)\Big), \\
\Leftrightarrow \quad & 0 \leq f(x_1) - f(x_2) - (x_1 - x_2) \times \\
& \times \frac{f\Big(x_2 + \lambda\,(x_1 - x_2)\Big) - f(x_2)}{\lambda\,(x_1 - x_2)}.
\end{aligned}$$

Since f is differentiable, one can perform the limit $\lambda \to 0$ and obtains Eq. (2.188).

Now we prove $\forall x_1, x_2 \in I: \quad f(x_1) - f(x_2) - (x_1 - x_2)\, f'(x_2) \geq 0 \Rightarrow f$ is convex on I. We have for all $x_1, x_2 \in I$

$$f(x_1) - f(x_2) \geq (x_1 - x_2)\, f'(x_2). \tag{2.190}$$

It follows for $x_1, x_2 \in I$ and $\lambda \in [0, 1]$:

$$
\begin{aligned}
&\lambda\, f(x_1) + (1-\lambda)\, f(x_2) - f\Big(\lambda\, x_1 + (1-\lambda)\, x_2\Big) \\
=&\lambda\, f(x_1) + (1-\lambda)\, f(x_2) - f\Big(\lambda\, x_1 + (1-\lambda)\, x_2\Big) \\
&\underbrace{-\lambda\, f\Big(\lambda\, x_1 + (1-\lambda)\, x_2\Big) + \lambda\, f\Big(\lambda\, x_1 + (1-\lambda)\, x_2\Big)}_{=0} \\
=&\lambda \underbrace{\Big(f(x_1) - f\Big(\lambda\, x_1 + (1-\lambda)\, x_2\Big)\Big)}_{\overset{(2.190)}{\geq}\Big(x_1-\lambda\, x_1-(1-\lambda)\, x_2\Big)\, f'\Big(\lambda\, x_1+(1-\lambda)\, x_2\Big)} \\
&+ (1-\lambda) \underbrace{\Big(f(x_2) - f\Big(\lambda\, x_1 + (1-\lambda)\, x_2\Big)\Big)}_{\overset{(2.190)}{\geq}\Big(x_2-\lambda\, x_1-(1-\lambda)\, x_2\Big)\, f'\Big(\lambda\, x_1+(1-\lambda)\, x_2\Big)} \\
\geq&\lambda\, (1-\lambda)\, (x_1 - x_2)\, f'\Big(\lambda\, x_1 + (1-\lambda)\, x_2\Big) \\
&+ (1-\lambda)\, \lambda\, (x_2 - x_1)\, f'\Big(\lambda\, x_1 + (1-\lambda)\, x_2\Big) \\
=&\lambda\, (1-\lambda)\, (x_1 - x_2)\, f'\Big(\lambda\, x_1 \\
&+ (1-\lambda)\, x_2\Big) - \lambda\, (1-\lambda)\, (x_1 - x_2)\, f'\Big(\lambda\, x_1 + (1-\lambda)\, x_2\Big) \\
=&0.
\end{aligned}
$$

From this we obtain that f is convex in I by definition.

(iii) Since $f''(x) \geq 0$ is valid for all $x \in I$, we obtain for all $x_1, x_2 \in I$:

$$
\begin{aligned}
0 \leq& \int_{x_2}^{x_1} dt \int_{x_2}^{t} dx\, f''(x) \\
=& \int_{x_2}^{x_1} dt\, \Big(f'(t) - f'(x_2)\Big) \\
=& f(x_1) - f(x_2) - (x_1 - x_2)\, f'(x_2).
\end{aligned}
$$

It follows from (ii) that f is convex on I. ■

Mathematical Theorem 2.4 (Gibb's inequality) *Let be $\Omega \in \mathbb{N}$ and the numbers $0 \leq \mathbb{p}_i \leq 1$ and $0 \leq \mathbb{p}'_i \leq 1$ for $i = 1, \ldots, \Omega$ given with $\sum_{i=1}^{\Omega} \mathbb{p}_i = 1$ and $\sum_{i=1}^{\Omega} \mathbb{p}'_i = 1$. Then one obtains*

$$\sum_{i=1}^{\Omega} \mathbb{p}_i \left(\ln(\mathbb{p}_i) - \ln(\mathbb{p}'_i) \right) \geq 0. \tag{2.191}$$

Proof The function $f(x) = x\, \ln(x)$ is convex in $[0, \infty)$ due to Eq. (2.189), since one has $f'(x) = \ln(x) + 1$ and $f''(x) = \frac{1}{x} \geq 0$ for all $x \in [0, \infty)$. We further define $x_i := \frac{\mathbb{p}_i}{\mathbb{p}'_i} \in [0, \infty)$ for $i = 1, \ldots, \Omega$. Then we obtain from Jensen's inequality (2.187) applied to $f(x) = x\, \ln(x)$:

$$\begin{aligned}
0 &= \left(\sum_{i=1}^{\Omega} \mathbb{p}_i \right) \underbrace{\ln \left(\sum_{i=1}^{\Omega} \mathbb{p}_i \right)}_{=0} \\
&= \left(\sum_{i=1}^{\Omega} \mathbb{p}'_i \frac{\mathbb{p}_i}{\mathbb{p}'_i} \right) \ln \left(\sum_{i=1}^{\Omega} \mathbb{p}'_i \frac{\mathbb{p}_i}{\mathbb{p}'_i} \right) \\
&= \left(\sum_{i=1}^{\Omega} \mathbb{p}'_i x_i \right) \ln \left(\sum_{i=1}^{\Omega} \mathbb{p}'_i x_i \right) \\
&\overset{(2.187)}{\leq} \sum_{i=1}^{\Omega} \mathbb{p}'_i x_i \, \ln(x_i) \\
&= \sum_{i=1}^{\Omega} \mathbb{p}'_i \frac{\mathbb{p}_i}{\mathbb{p}'_i} \ln \left(\frac{\mathbb{p}_i}{\mathbb{p}'_i} \right) \\
&= \sum_{i=1}^{\Omega} \mathbb{p}_i \left(\ln(\mathbb{p}_i) - \ln(\mathbb{p}'_i) \right).
\end{aligned} \tag{2.192}$$

■

Now we can consider equilibrium ensembles under given conditions. The following discussion is based on Ref. [30], but more details are presented.

Theorem 2.10 (Microcanonical ensemble) *The equilibrium macro state of an isolated system is given by the microcanonical ensemble. The energy of any isolated system keeps constant at a given value E. Let* $\{|\Psi_{E_i}\rangle, i = 1, \ldots \Omega\}$ *be the set of all micro states of the isolated system with energy E, so that* $\hat{H}\,|\Psi_{E_i}\rangle = E\,|\Psi_{E_i}\rangle$ *is valid for all* $i = 1, \ldots, \Omega$*, where* $\hat{H}$ *is the Hamiltonian of the system. Then the microcanonical density matrix is given by*

$$\hat{\rho}^{(mc)} = \sum_{i=1}^{\Omega} \frac{1}{\Omega} |\Psi_{E_i}\rangle\langle\Psi_{E_i}|. \tag{2.193}$$

All representing micro sates occur with the same probability in the isolated system at equilibrium. The microcanonical density matrix maximizes the entropy $\langle S\rangle$ *among all of the system compatible density matrices* $\hat{\rho}$*:*

$$\langle S\rangle_{\hat{\rho}^{(mc)}} \geq \langle S\rangle_{\hat{\rho}}. \tag{2.194}$$

Proof Since any macro state that describes the isolated system must be represented by micro states of energy E, one obtains for the microcanonical density matrix

$$\hat{\rho}^{(\mathrm{mc})} = \sum_{i=1}^{\Omega} \mathbb{p}_i^{(\mathrm{mc})} |\Psi_{E_i}\rangle\langle\Psi_{E_i}|.$$

Furthermore, the microcanonical ensemble maximizes the entropy $\langle S\rangle$, since the entropy is maximized in an isolated system at equilibrium, as mentioned previously. Hence, the probabilities $\mathbb{p}_i^{(\mathrm{mc})}$ of the microcanonical density matrix can be obtained from the following variation principle:

$$\frac{\delta}{\delta \mathbb{p}_j}\left(\langle S\rangle_{\hat{\rho}^{(\mathrm{mc})}} + \lambda\left(\mathrm{Tr}\left(\hat{\rho}^{(\mathrm{mc})}\right) - 1\right)\right) = 0,$$

$$\Leftrightarrow \quad \frac{\delta}{\delta \mathbb{p}_j^{(\mathrm{mc})}}\left(-k_\mathrm{B}\sum_{i=1}^{\Omega} \mathbb{p}_i^{(\mathrm{mc})} \ln\left(\mathbb{p}_i^{(\mathrm{mc})}\right) + \lambda\left(\sum_{i=1}^{\Omega} \mathbb{p}_i^{(\mathrm{mc})} - 1\right)\right) = 0,$$

where λ is a Lagrangian multiplier, which takes the normalization of the $\mathbb{p}_i^{(\mathrm{mc})}$'s into account. For any $j \in \{1, \ldots, \Omega\}$, it follows:

$$-k_\mathrm{B} \ln\left(\mathbb{p}_j^{(\mathrm{mc})}\right) - k_\mathrm{B}\, \mathbb{p}_j^{(\mathrm{mc})} \frac{1}{\mathbb{p}_j^{(\mathrm{mc})}} + \lambda = 0,$$

$$\Leftrightarrow \quad \mathbb{p}_j^{(\mathrm{mc})} = e^{-\frac{k_\mathrm{B}-\lambda}{k_\mathrm{B}}}.$$

Consequently, all $\mathbb{p}_j^{(\mathrm{mc})}$ are equal and one obtains finally $\mathbb{p}_j^{(\mathrm{mc})} = \frac{1}{\Omega}$ from the normalization condition $\sum_{i=1}^{\Omega} \mathbb{p}_i^{(\mathrm{mc})} = 1$. In order to check that the corresponding entropy, S, is a maximum, let us consider an arbitrary density matrix

$$\hat{\rho} = \sum_{i=1}^{\Omega} \mathbb{p}_i \, |\Psi_{E_i}\rangle\langle\Psi_{E_i}|,$$

that is compatible with the isolated system at energy E. Then one obtains from the inequality of Gibbs (2.191):

$$\begin{aligned}
\langle S\rangle_{\hat{\rho}^{(\mathrm{mc})}} - \langle S\rangle_{\hat{\rho}} &= -k_\mathrm{B} \sum_{i=1}^{\Omega} \frac{1}{\Omega} \ln\left(\frac{1}{\Omega}\right) + k_\mathrm{B} \sum_{i=1}^{\Omega} \mathbb{p}_i \ln\left(\mathbb{p}_i\right) \\
&= -k_\mathrm{B} \ln\left(\frac{1}{\Omega}\right) + k_\mathrm{B} \sum_{i=1}^{\Omega} \mathbb{p}_i \ln\left(\mathbb{p}_i\right) \\
&= -k_\mathrm{B} \sum_{i=1}^{\Omega} \mathbb{p}_i \ln\left(\frac{1}{\Omega}\right) + k_\mathrm{B} \sum_{i=1}^{\Omega} \mathbb{p}_i \ln\left(\mathbb{p}_i\right) \\
&= k_\mathrm{B} \sum_{i=1}^{\Omega} \mathbb{p}_i \left(\ln\left(\mathbb{p}_i\right) - \ln\left(\frac{1}{\Omega}\right)\right) \\
&\overset{(2.191)}{\geq} 0.
\end{aligned}$$

■

In the microcanconical ensemble, all representing micro states occur with the same probability $\mathbb{p}_i^{(\mathrm{mc})} = \frac{1}{\Omega}$, where Ω is the number of the representing micro states. One obtains for the average of the entropy in the microcanonical ensemble

$$\langle S\rangle_{\hat{\rho}^{(\mathrm{mc})}} \overset{(2.184)}{=} -k_\mathrm{B} \sum_{i=1}^{\Omega} \frac{1}{\Omega} \ln\left(\frac{1}{\Omega}\right) = k_\mathrm{B} \ln(\Omega). \tag{2.195}$$

The entropy can be calculated from the logarithm of the number representing micro states. To count the representing micro states, the following asymptotic correct approximation of the factorial is very useful, since one deals commonly with large numbers in statistical physics. The proof is based on Ref. [31], but the proof of asymptotic correctness is added.

Mathematical Theorem 2.5 (Stirling approximation of the factorial) *One obtains for* $N \in \mathbb{N}$

$$\ln(N!) \approx N \ln(N) - N. \tag{2.196}$$

This approximation is asymptotically correct, i.e., the relative error of the above approximation converges to 0 for $N \to \infty$.

Proof The logarithm increases strict monotonously for $x \in (0, \infty)$, since one has $\frac{d}{dx} \ln(x) = \frac{1}{x} > 0$. From this, we obtain for $n \in \mathbb{N}$:

$$\int_{n-1}^{n} dx \underbrace{\ln(x)}_{<\ln(n)} < \ln(n) < \int_{n}^{n+1} dx \underbrace{\ln(x)}_{>\ln(n)}.$$

If we consider a fixed $N \in \mathbb{N}$ and add the above inequality for $n = 1, 2, \ldots, N$, we get

$$\begin{aligned} \int_0^N dx \ln(x) &< \ln(1) + \ln(2) + \ldots + \ln(N) < \int_1^{N+1} dx \ln(x), \\ x \ln(x) - x\Big|_{x=0}^{x=N} &< \ln(N!) < x \ln(x) - x\Big|_{x=1}^{x=N+1}, \\ N \ln(N) - N &< \ln(N!) < (N+1) \ln(N+1) - N. \end{aligned}$$

The integral on the left hand side is well defined due to Eq. (2.185). The approximation is given by the first inequality of the above equation. For $N \geq 3 > e \approx 2.718$, we have $\ln(N) > \ln(e) = 1$ and therefore $N \ln(N) - N > 0$. Consequently, we obtain as an upper limit for the relative error of the approximation for $N \geq 3$:

$$\begin{aligned} \frac{\ln(N!) - \left(N \ln(N) - N\right)}{N \ln(N) - N} &< \frac{\left((N+1) \ln(N+1) - N\right) - \left(N \ln(N) - N\right)}{N \ln(N) - N} \\ &= \frac{(N+1) \ln(N+1) - N \ln(N)}{N \ln(N) - N} \\ &= \frac{N \ln\left(1 + \frac{1}{N}\right) + \ln(N+1)}{N\left(\ln(N) - 1\right)}. \end{aligned}$$

Now we consider $N \to \infty$ and apply the rule of L' Hospital, since the nominator and the denominator converge to ∞:

$$\begin{aligned}\lim_{N\to\infty} \frac{N \ln\left(1+\frac{1}{N}\right)+\ln(N+1)}{N\left(\ln(N)-1\right)} &= \lim_{N\to\infty} \frac{\ln\left(1+\frac{1}{N}\right)+N\frac{1}{1+\frac{1}{N}}\frac{-1}{N^2}+\frac{1}{N+1}}{\left(\ln(N)-1\right)+N\frac{1}{N}} \\ &= \lim_{N\to\infty} \frac{\ln\left(1+\frac{1}{N}\right)}{\ln(N)} \\ &=0.\end{aligned}$$

Hence, the approximation is asymptotically correct. ■

Theorem 2.11 (Canonical ensemble) *The equilibrium macro state of a system kept at temperature T is given by the canonical ensemble. In order to fix the temperature, the system exchanges heat with a heath bath, where the heat bath and the system are isolated from the environment. Let $\left\{|\Psi_{E_i}\rangle, i = 1, \ldots \Omega\right\}$ be the complete set of orthonormal eigenkets of the Hamiltonian $\hat{H}$ of the system, so that $\hat{H}\,|\Psi_{E_i}\rangle = E_i\,|\Psi_{E_i}\rangle$ is valid for all $i = 1, \ldots, \Omega$. Then the canonical density matrix is given by*

$$\hat{\rho}^{(c)} = \sum_{i=1}^{\Omega} \frac{e^{-\frac{E_i}{k_B T}}}{\mathscr{Z}^{(c)}} |\Psi_{E_i}\rangle\langle\Psi_{E_i}|, \tag{2.197}$$

where

$$\mathscr{Z}^{(c)} = \sum_{i=1}^{\Omega} e^{-\frac{E_i}{k_B T}} \tag{2.198}$$

denotes the canonical partition function. The canonical density matrix minimizes the Helmholtz free energy $\langle F\rangle = \langle E\rangle - T\,\langle S\rangle$ among all density matrices $\hat{\rho}$:

$$\langle F\rangle_{\hat{\rho}^{(c)}} \leq \langle F\rangle_{\hat{\rho}}. \tag{2.199}$$

Furthermore, one obtains

$$\langle F\rangle_{\hat{\rho}^{(c)}} = -k_B T\,\ln\left(\mathscr{Z}^{(c)}\right). \tag{2.200}$$

Proof The equilibrium macro state, which is given by the canonical ensemble, minimizes the Helmholtz free energy $\langle F\rangle$, as previously mentioned. For a given density matrix $\hat{\rho}$, the Helmholtz free energy is defined as

$$
\begin{aligned}
\langle F\rangle_{\sum_{i=1}^{\Omega} \mathbb{p}_i |\Psi_i\rangle\langle\Psi_i|} &= \langle E\rangle_{\sum_{i=1}^{\Omega} \mathbb{p}_i |\Psi_i\rangle\langle\Psi_i|} - T\,\langle S\rangle_{\sum_{i=1}^{\Omega} \mathbb{p}_i |\Psi_i\rangle\langle\Psi_i|} \\
&= \mathrm{Tr}\big(\hat{\rho}\,\hat{H}\big) + T\,k_B\,\mathrm{Tr}\Big(\hat{\rho}\,\ln\big(\hat{\rho}\big)\Big) \\
&= \sum_{j=1}^{\Omega}\sum_{i=1}^{\Omega} \mathbb{p}_i \underbrace{\big\langle\Psi_j\big|\Psi_i\big\rangle}_{=\delta_{ij}} \big\langle\Psi_i\big|\hat{H}\big|\Psi_j\big\rangle + k_B\,T \sum_{i=1}^{\Omega} \mathbb{p}_i\,\ln\big(\mathbb{p}_i\big) \\
&= \sum_{i=1}^{\Omega} \mathbb{p}_i \big\langle\Psi_i\big|\hat{H}\big|\Psi_i\big\rangle + k_B\,T \sum_{i=1}^{\Omega} \mathbb{p}_i\,\ln\big(\mathbb{p}_i\big).
\end{aligned} \tag{2.201}
$$

Here, we assume that $\big\{|\Psi_i\rangle, i = 1, \ldots, \Omega\big\}$ forms a complete set of orthonormal kets. This is always possible by extending an initial set with additional normalized kets $|\Psi_i\rangle$ that have $\mathbb{p}_i = 0$ and are orthogonal to the other kets.

At first, we consider a fixed complete set $\big\{\big|\Psi_i\big\rangle, i = 1, \ldots, \Omega\big\}$ of arbitrary normalized kets $|\Psi_i\rangle$ and try to find the related optimal $\mathbb{p}_i^{(o)}$'s that minimize the free energy $\langle F\rangle$ under the condition $\sum_{i=1}^{\Omega} \mathbb{p}_i^{(o)} = 1$. These $\mathbb{p}_i^{(o)}$'s can be found from the following variation principle:

$$
\begin{aligned}
0 &= \frac{\delta}{\delta\mathbb{p}_j^{(o)}}\left(\langle F\rangle_{\sum_{i=1}^{\Omega} \mathbb{p}_i^{(o)} |\Psi_i\rangle\langle\Psi_i|} + \lambda\left(\sum_{i=1}^{\Omega} \mathbb{p}_i^{(o)} - 1\right)\right) \\
\Leftrightarrow\quad 0 &\overset{(2.201)}{=} \frac{\delta}{\delta\mathbb{p}_j^{(o)}}\left(\sum_{i=1}^{\Omega} \mathbb{p}_i^{(o)} \big\langle\Psi_i\big|\hat{H}\big|\Psi_i\big\rangle + k_B\,T \sum_{i=1}^{\Omega} \mathbb{p}_i^{(o)}\,\ln\big(\mathbb{p}_i^{(o)}\big)\right. \\
&\qquad \left. + \lambda\left(\sum_{i=1}^{\Omega} \mathbb{p}_i^{(o)} - 1\right)\right), \\
\Leftrightarrow\quad 0 &= \big\langle\Psi_j\big|\hat{H}\big|\Psi_j\big\rangle + k_B\,T\,\ln\big(\mathbb{p}_j^{(o)}\big) + k_B\,T\,\mathbb{p}_j^{(o)}\,\frac{1}{\mathbb{p}_j^{(o)}} + \lambda, \\
\Leftrightarrow\quad \mathbb{p}_j^{(o)} &= \mathrm{e}^{-\frac{\langle\Psi_j|\hat{H}|\Psi_j\rangle}{k_B T}}\,\underbrace{\mathrm{e}^{-\frac{k_B T + \lambda}{k_B T}}}_{=:\frac{1}{\mathscr{Z}}}.
\end{aligned} \tag{2.202}
$$

The constant $\mathscr{Z}$ can be derived from the normalization constraint of the $\mathbb{p}_i^{(o)}$'s:

$$
\begin{aligned}
1 &= \sum_{j=1}^{\Omega} \mathbb{p}_j^{(o)}, \\
\Leftrightarrow\quad 1 &= \sum_{j=1}^{\Omega} \frac{1}{\mathscr{Z}}\,\mathrm{e}^{-\frac{\langle\Psi_j|\hat{H}|\Psi_j\rangle}{k_B T}},
\end{aligned}
$$

$$\Leftrightarrow \qquad \mathscr{Z} = \sum_{j=1}^{\Omega} \mathrm{e}^{-\frac{\langle \Psi_j | \hat{H} | \Psi_j \rangle}{k_\mathrm{B}\, T}} . \tag{2.203}$$

Inserting Eq. (2.202) into the Helmholtz free energy (2.201) yields

$$\begin{aligned}
\langle F \rangle_{\sum\limits_{i=1}^{\Omega} \mathbb{p}_i^{(\mathrm{o})} |\Psi_i\rangle\langle\Psi_i|} &= \sum_{i=1}^{\Omega} \mathbb{p}_i^{(\mathrm{o})} \big\langle \Psi_i \big| \hat{H} \big| \Psi_i \big\rangle + k_\mathrm{B}\, T \sum_{i=1}^{\Omega} \mathbb{p}_i^{(\mathrm{o})} \ln\!\big(\mathbb{p}_i^{(\mathrm{o})}\big) \\
&= \sum_{i=1}^{\Omega} \mathrm{e}^{-\frac{\langle \Psi_i | \hat{H} | \Psi_i \rangle}{k_\mathrm{B}\, T}} \frac{1}{\mathscr{Z}} \big\langle \Psi_i \big| \hat{H} \big| \Psi_i \big\rangle \\
&\quad + k_\mathrm{B}\, T \sum_{i=1}^{\Omega} \mathrm{e}^{-\frac{\langle \Psi_i | \hat{H} | \Psi_i \rangle}{k_\mathrm{B}\, T}} \frac{1}{\mathscr{Z}} \ln\left(\mathrm{e}^{-\frac{\langle \Psi_i | \hat{H} | \Psi_i \rangle}{k_\mathrm{B}\, T}} \frac{1}{\mathscr{Z}} \right) \\
&= \sum_{i=1}^{\Omega} \mathrm{e}^{-\frac{\langle \Psi_i | \hat{H} | \Psi_i \rangle}{k_\mathrm{B}\, T}} \frac{1}{\mathscr{Z}} \big\langle \Psi_i \big| \hat{H} \big| \Psi_i \big\rangle \\
&\quad + k_\mathrm{B}\, T \sum_{i=1}^{\Omega} \mathrm{e}^{-\frac{\langle \Psi_i | \hat{H} | \Psi_i \rangle}{k_\mathrm{B}\, T}} \frac{1}{\mathscr{Z}} \frac{-\langle \Psi_i | \hat{H} | \Psi_i \rangle}{k_\mathrm{B}\, T} \\
&\quad - k_\mathrm{B}\, T \underbrace{\sum_{i=1}^{\Omega} \mathrm{e}^{-\frac{\langle \Psi_i | \hat{H} | \Psi_i \rangle}{k_\mathrm{B}\, T}}}_{\overset{(2.203)}{=} \mathscr{Z}} \frac{1}{\mathscr{Z}} \ln(\mathscr{Z}) \\
&= -\, k_\mathrm{B}\, T \ln(\mathscr{Z}).
\end{aligned} \tag{2.204}$$

The $\mathbb{p}_i^{(\mathrm{o})}$'s found from Eq. (2.202) indeed minimize the Helmholtz free energy $\langle F \rangle$ for given kets $|\Psi_i\rangle$, since one obtains for any arbitrary $\mathbb{p}_i$'s obeying $0 \le \mathbb{p}_i \le 1$ and $\sum\limits_{i=1}^{\Omega} \mathbb{p}_i = 1$:

$$\begin{aligned}
&\langle F \rangle_{\sum\limits_{i=1}^{\Omega} \mathbb{p}_i |\Psi_i\rangle\langle\Psi_i|} - \langle F \rangle_{\sum\limits_{i=1}^{\Omega} \mathbb{p}_i^{(\mathrm{o})} |\Psi_i\rangle\langle\Psi_i|} \\
&= \sum_{i=1}^{\Omega} \mathbb{p}_i \big\langle \Psi_i \big| \hat{H} \big| \Psi_i \big\rangle + k_\mathrm{B}\, T \sum_{i=1}^{\Omega} \mathbb{p}_i \ln(\mathbb{p}_i) + k_\mathrm{B}\, T \ln(\mathscr{Z}) \\
&= -\, k_\mathrm{B}\, T \sum_{i=1}^{\Omega} \mathbb{p}_i \frac{-\big\langle \Psi_i \big| \hat{H} \big| \Psi_i \big\rangle}{k_\mathrm{B}\, T} + k_\mathrm{B}\, T \sum_{i=1}^{\Omega} \mathbb{p}_i \ln(\mathbb{p}_i) + k_\mathrm{B}\, T \ln(\mathscr{Z}) \underbrace{\sum_{i=1}^{\Omega} \mathbb{p}_i}_{=1} \\
&= -\, k_\mathrm{B}\, T \sum_{i=1}^{\Omega} \mathbb{p}_i \ln\left(\mathrm{e}^{\frac{-\langle \Psi_i | \hat{H} | \Psi_i \rangle}{k_\mathrm{B}\, T}} \right) + k_\mathrm{B}\, T \sum_{i=1}^{\Omega} \mathbb{p}_i \ln(\mathbb{p}_i) - k_\mathrm{B}\, T \sum_{i=1}^{\Omega} \mathbb{p}_i \ln\left(\frac{1}{\mathscr{Z}} \right)
\end{aligned}$$

$$\begin{aligned}
&= -k_B T \sum_{i=1}^{\Omega} \mathbb{p}_i \ln \underbrace{\left(e^{\frac{-\langle\Psi_i|\hat{H}|\Psi_i\rangle}{k_B T}} \frac{1}{\mathscr{Z}} \right)}_{\overset{(2.202)}{=} \mathbb{p}_i^{(o)}} + k_B T \sum_{i=1}^{\Omega} \mathbb{p}_i \ln(\mathbb{p}_i) \\
&= k_B T \sum_{i=1}^{\Omega} \mathbb{p}_i \left(\ln(\mathbb{p}_i) - \ln\left(\mathbb{p}_i^{(o)}\right) \right) \\
&\overset{(2.191)}{\geq} 0.
\end{aligned} \tag{2.205}$$

The last step follows from the Gibb's inequality (2.191). Finally, we want to find the optimal kets that minimize the free energy. Let $\{|\psi_{E_i}\rangle, i = 1, \ldots, \Omega\}$ be the complete set of normalized eigenkets of the Hamilton operator $\hat{H}$. The exponential function e^x is convex on $\mathbb{R}$ due to Eq. (2.189), since one has $\frac{d^2 e^x}{dx^2} = e^x \geq 0$. We obtain further by applying Jensen's inequality (2.187) to the exponential function:

$$\begin{aligned}
\langle F \rangle_{\sum_{i=1}^{\Omega} \mathbb{p}_i^{(o)} |\Psi_i\rangle\langle\Psi_i|} &= -k_B T \ln(\mathscr{Z}) \\
&\overset{(2.203)}{=} -k_B T \ln \left(\sum_{j=1}^{\Omega} \exp\left(-\frac{\langle\Psi_j|\hat{H}|\Psi_j\rangle}{k_B T} \right) \right) \\
&= -k_B T \ln \left(\sum_{j=1}^{\Omega} \exp\left(\langle\Psi_j|\hat{H}| \underbrace{\left(\sum_{i=1}^{\Omega} |\Psi_{E_i}\rangle\langle\Psi_{E_i}| \right)}_{=\hat{\mathbb{1}}} |\Psi_j\rangle \frac{-1}{k_B T} \right) \right) \\
&= -k_B T \ln \left(\sum_{j=1}^{\Omega} \exp\left(\sum_{i=1}^{\Omega} \langle\Psi_j| \underbrace{\hat{H}|\Psi_{E_i}\rangle}_{=E_i |\Psi_{E_i}\rangle} \langle\Psi_{E_i}|\Psi_j\rangle \frac{-1}{k_B T} \right) \right) \\
&= -k_B T \ln \left(\sum_{j=1}^{\Omega} \underbrace{\exp\left(\sum_{i=1}^{\Omega} \left|\langle\Psi_j|\Psi_{E_i}\rangle\right|^2 \frac{-E_i}{k_B T} \right)}_{\overset{(2.187)}{\leq} \sum_{i=1}^{\Omega} |\langle\Psi_j|\Psi_{E_i}\rangle|^2 \exp\left(\frac{-E_i}{k_B T}\right)} \right) \\
&\geq -k_B T \ln \left(\sum_{j=1}^{\Omega} \sum_{i=1}^{\Omega} \langle\Psi_{E_i}|\Psi_j\rangle \langle\Psi_j|\Psi_{E_i}\rangle \exp\left(\frac{-E_i}{k_B T} \right) \right)
\end{aligned}$$

$$
\begin{aligned}
&= -k_B T \ln\left(\sum_{i=1}^{\Omega} \exp\left(\frac{-E_i}{k_B T}\right) \langle\Psi_{E_i}| \underbrace{\left(\sum_{j=1}^{\Omega} |\Psi_j\rangle\langle\Psi_j|\right)}_{=\hat{1}} |\Psi_{E_i}\rangle\right) \\
&= -k_B T \ln\left(\sum_{i=1}^{\Omega} \exp\left(\frac{-\langle\Psi_{E_i}|\hat{H}|\Psi_{E_i}\rangle}{k_B T}\right)\right) \\
&= -k_B T \ln\left(\mathscr{Z}^{(c)}\right) \\
&= \langle F\rangle_{\hat{\rho}^{(c)}} .
\end{aligned}
\tag{2.206}
$$

Indeed, Jensen's inequality (2.187) can be applied, since $0 \leq \left|\langle\Psi_j|\Psi_{E_i}\rangle\right|^2 \leq 1$ and

$$
\begin{aligned}
\sum_{i=1}^{\Omega} \left|\langle\Psi_j|\Psi_{E_i}\rangle\right|^2 &= \sum_{i=1}^{\Omega} \langle\Psi_j|\Psi_{E_i}\rangle \langle\Psi_{E_i}|\Psi_j\rangle \\
&= \langle\Psi_j| \underbrace{\left(\sum_{i=1}^{\Omega} |\Psi_{E_i}\rangle \langle\Psi_{E_i}|\right)}_{=\hat{1}} |\Psi_j\rangle \\
&= 1
\end{aligned}
\tag{2.207}
$$

are valid for any $j = 1, \ldots, \Omega$. In this way, we showed that for a given set of kets $\{|\Psi_i\rangle, i = 1, \ldots, \Omega\}$ the Helmholtz free energy is minimized, if the corresponding $\mathbb{p}_i^{(o)}$'s from Eq. (2.202) are used in the density matrix $\hat{\rho}$. Furthermore, if for a given set of kets $\{|\Psi_i\rangle, i = 1, \ldots, \Omega\}$ always the related $\mathbb{p}_i^{(o)}$'s are used, the Helmholtz free energy is minimized, if the set of eigenkets $\{|\Psi_{E_i}\rangle, i = 1, \ldots, \Omega\}$ of the Hamilton operator $\hat{H}$ is used. Hence, the canonical density matrix $\hat{\rho}^{(c)}$ leads to the global minimum of the Helmholtz free energy, which is given by (2.200) due to Eq. (2.204). ■

2.4.3 *Mermin's Theorems*

Now we have all tools required to describe the system of interacting electrons at a given electronic temperature T_e in the external potential $V_{ext}(\mathbf{r})$ of the fixed nuclei. The electronic Hamiltonian $\hat{H}_e$ is still given by Eq. (2.93). But now the electrons are described by a density matrix

$$\hat{\rho}_e = \sum_{\alpha=1}^{\Omega} \mathbb{p}_\alpha \, |\Psi_\alpha\rangle\langle\Psi_\alpha| \tag{2.208}$$

rather than a many-body wave function Ψ, since the electrons are considered at a given $T_e > 0$. From any given density matrix $\hat{\rho}_e$, the electron density can be calculated by

$$\begin{aligned} n_e(\mathbf{r}) &= \langle \hat{n}_e(\mathbf{r}) \rangle_{\hat{\rho}_e} \\ &= \sum_{\alpha=1}^{\Omega} \mathbb{p}_\alpha \left\langle \Psi_\alpha \middle| \hat{n}_e(\mathbf{r}) \middle| \Psi_\alpha \right\rangle \\ &= \sum_{\alpha=1}^{\Omega} \mathbb{p}_\alpha \, n_e^{(\alpha)}(\mathbf{r}). \end{aligned} \tag{2.209}$$

Here, we denote by $n_e^{(\alpha)}(\mathbf{r})$ the electron density obtained from the many-body wave function Ψ_α and we assume that $\{|\Psi_\alpha\rangle, \alpha = 1, \ldots, \Omega\}$ forms a complete set of orthonormal kets to simplify things. Using this, we obtain for the external potential

$$\begin{aligned} \left\langle \hat{V}_{ext} \right\rangle_{\hat{\rho}_e} &= \sum_{\alpha=1}^{\Omega} \mathbb{p}_\alpha \left\langle \Psi_\alpha \middle| \hat{V}_{ext} \middle| \Psi_\alpha \right\rangle \\ &= \sum_{\alpha=1}^{\Omega} \mathbb{p}_\alpha \int_{\mathbb{R}^3} d^3r \, V_{ext}(\mathbf{r}) \, n_e^{(\alpha)}(\mathbf{r}) \\ &= \int_{\mathbb{R}^3} d^3r \, V_{ext}(\mathbf{r}) \, n_e(\mathbf{r}). \end{aligned} \tag{2.210}$$

We have further for the Hamiltonian $\hat{H}_e$ of the electronic system

$$\begin{aligned} \left\langle \hat{H}_e \right\rangle_{\hat{\rho}_e} &= \left\langle \hat{T}_e + \hat{V}_{int} \right\rangle_{\hat{\rho}_e} + \left\langle \hat{V}_{ext} \right\rangle_{\hat{\rho}_e} + \left\langle E_{nn} \right\rangle_{\hat{\rho}_e} \\ &= \sum_{\alpha=1}^{\Omega} \mathbb{p}_\alpha \left\langle \Psi_\alpha \middle| \hat{T}_e + \hat{V}_{int} \middle| \Psi_\alpha \right\rangle + \int_{\mathbb{R}^3} d^3r \, V_{ext}(\mathbf{r}) \, n_e(\mathbf{r}) + E_{nn}. \end{aligned} \tag{2.211}$$

The operators for the kinetic energy $\hat{T}_e$ and for the interaction between the electrons $\hat{V}_{int}$ are universal for any system of N_e interacting electrons. Only $V_{ext}(\mathbf{r})$ and E_{nn} differ for particular electronic systems. E_{nn} is just a constant and the energy contribution of $V_{ext}(\mathbf{r})$ in the Hamiltonian can be directly calculated from the electronic density $n_e(\mathbf{r})$. This indicates that the electron density $n_e(\mathbf{r})$ could be used as the main variable instead of the complicated density operator $\hat{\rho}_e$.

Indeed, similar to density functional theory of the ground state, the electron density can be used as the main variable, if the electronic system is considered at equilibrium. This was found by Mermin in 1965 [28] just one year later after the Hohenberg-Kohn theorems were published:

Theorem 2.12 (Mermin's unique density theorem) *The equilibrium electron density $n_{e_0}(\mathbf{r})$ determines uniquely, except for a constant, the external potential $V_{ext}(\mathbf{r})$ in the Hamiltonian $\hat{H}_e$ for any system of interacting electrons at a given electronic temperature T_e. Due to the fully determination of the Hamiltonian except for a constant shift of the energy, the many-body wave functions are determined for all states (ground and excited) and the equilibrium density matrix is known. Therefore, all properties of the system can be derived from the equilibrium electron density.*

Proof The equilibrium state of an electronic system at a given T_e in an external potential $V_{ext}(\mathbf{r})$ is related to the canonical density matrix $\hat{\rho}_e^{(c)}$, which minimizes the Helmholtz free energy, as we proved in the previous section. For any density matrix $\hat{\rho}_e$, the Helmholtz free energy is given by

$$\begin{aligned}\langle F\rangle_{\hat{\rho}_e} =&\big\langle \hat{H}_e\big\rangle_{\hat{\rho}_e} - T_e\,\langle S_e\rangle_{\hat{\rho}_e}\\ =&\sum_{\alpha=1}^{\Omega} \mathbb{p}_\alpha \big\langle \Psi_\alpha\big|\hat{T}_e + \hat{V}_{int}\big|\Psi_\alpha\big\rangle + \int_{\mathbb{R}^3} d^3r\, V_{ext}(\mathbf{r})\, n_e(\mathbf{r}) + E_{nn}\\ &+ k_B\,T_e \sum_{\alpha=1}^{\Omega} \mathbb{p}_\alpha\, \ln\big(\mathbb{p}_\alpha\big). \end{aligned} \tag{2.212}$$

Now we assume that two external potentials $V_{ext}^{(1)}(\mathbf{r})$ and $V_{ext}^{(2)}(\mathbf{r})$ exist, which differ by more than a constant and lead to the same equilibrium electron density $n_{e_0}(\mathbf{r})$. Due to this, the Hamiltonians $\hat{H}_e^{(1)}$ and $\hat{H}_e^{(2)}$ differ by more than a constant and the related eigenfunctions of the Hamiltonians differ by more than a phase factor. Consequently, the related canonical density matrices $\hat{\rho}_e^{(c1)}$ and $\hat{\rho}_e^{(c2)}$ are different. Since $\hat{\rho}_e^{(c1)}$ leads to the global minimum of the Helmholtz free energy $F^{(1)}$ obtained from $\hat{H}_e^{(1)}$ and $\hat{\rho}_e^{(c2)}$ differs from $\hat{\rho}_e^{(c1)}$, we obtain

$$\begin{aligned} F^{(1)} &= \big\langle \hat{H}_e^{(1)}\big\rangle_{\hat{\rho}_e^{(c1)}} - T_e\,\langle S_e\rangle_{\hat{\rho}_e^{(c1)}}\\ &< \big\langle \hat{H}_e^{(1)}\big\rangle_{\hat{\rho}_e^{(c2)}} - T_e\,\langle S_e\rangle_{\hat{\rho}_e^{(c2)}}\\ &\overset{(2.212)}{=} \sum_{\alpha=1}^{\Omega} \mathbb{p}_\alpha^{(c2)} \big\langle \Psi_\alpha^{(2)}\big|\hat{T}_e + \hat{V}_{int}\big|\Psi_\alpha^{(2)}\big\rangle + \int_{\mathbb{R}^3} d^3r\, V_{ext}^{(1)}(\mathbf{r})\, n_{e_0}(\mathbf{r}) + E_{nn} \end{aligned}$$

$$+ k_B T_e \sum_{\alpha=1}^{\Omega} \mathbb{p}_\alpha^{(c2)} \ln\left(\mathbb{p}_\alpha^{(c2)}\right)$$

$$= \sum_{\alpha=1}^{\Omega} \mathbb{p}_\alpha^{(c2)} \left\langle \Psi_\alpha^{(2)} \middle| \hat{T}_e + \hat{V}_{int} \middle| \Psi_\alpha^{(2)} \right\rangle + \int_{\mathbb{R}^3} d^3r \, V_{ext}^{(2)}(\mathbf{r})\, n_{e_0}(\mathbf{r}) + E_{nn}$$

$$+ k_B T_e \sum_{\alpha=1}^{\Omega} \mathbb{p}_\alpha^{(c2)} \ln\left(\mathbb{p}_\alpha^{(c2)}\right)$$

$$+ \int_{\mathbb{R}^3} d^3r \left(V_{ext}^{(1)}(\mathbf{r}) - V_{ext}^{(2}(\mathbf{r}) \right) n_{e_0}(\mathbf{r})$$

$$\overset{(2.212)}{=} \left\langle \hat{H}_e^{(2)} \right\rangle_{\hat{\rho}_e^{(c2)}} - T_e \left\langle S_e \right\rangle_{\hat{\rho}_e^{(c2)}} + \int_{\mathbb{R}^3} d^3r \left(V_{ext}^{(1)}(\mathbf{r}) - V_{ext}^{(2)}(\mathbf{r}) \right) n_{e_0}(\mathbf{r})$$

$$= F^{(2)} + \int_{\mathbb{R}^3} d^3r \left(V_{ext}^{(1)}(\mathbf{r}) - V_{ext}^{(2)}(\mathbf{r}) \right) n_{e_0}(\mathbf{r}) \tag{2.213}$$

Here, T_e denotes the electronic temperature and $\hat{T}_e$ the kinetic energy operator of the electrons. If we treat the global minimum of the Helmholtz free energy $F^{(2)}$ obtained from $\hat{H}_e^{(2)}$ in the same way, we have

$$F^{(2)} < F^{(1)} + \int_{\mathbb{R}^3} d^3r \left(V_{ext}^{(2)}(\mathbf{r}) - V_{ext}^{(1)}(\mathbf{r}) \right) n_e(\mathbf{r}) \tag{2.214}$$

Adding the two latter inequalities yields

$$F^{(1)} + F^{(2)} < F^{(1)} + F^{(2)}, \tag{2.215}$$

which is a contradiction. Hence, our initial assumption is wrong and the equilibrium electron density $n_{e_0}(\mathbf{r})$ determines uniquely the external potential $V_{ext}(\mathbf{r})$ except for a constant. Due to the determination of the external potential, the Hamiltonian is completely defined except for a constant shift of the energy and, therefore, the many-body wave functions of all states are fully determined by solving the Schrödinger equation with this Hamiltonian. From the eigenfunctions and eigenvalues of the Hamiltonian, the related canonical density matrix $\hat{\rho}_e^{(c)}$ can be constructed from Eq. (2.197). In this way, all properties of the system can be calculated from the equilibrium electron density $n_{e_0}(\mathbf{r})$. ■

Theorem 2.13 (Mermin's Helmholtz free energy functional) *Any electron density $n_e(\mathbf{r})$ that can be constructed from a density operator $\hat{\rho}_e$ is called* *ρ-representable. On the space of ρ-representable electron densities, an universal functional for the Helmholtz free energy*

$$F_M\Big(n_e(\mathbf{r}), V_{ext}(\mathbf{r}), E_{nn}, T_e\Big) = \inf_{\hat{\rho}_e \to n_e(\mathbf{r})} Tr\Big(\hat{\rho}_e\,(\hat{T}_e + \hat{V}_{int} + \hat{V}_{ext}) + k_B\,T_e\,\hat{\rho}_e\,\ln(\hat{\rho}_e)\Big) + E_{nn} \tag{2.216}$$

in terms of the electron density $n_e(\mathbf{r})$, the external potential $V_{ext}(\mathbf{r})$, the nuclei interaction energy E_{nn} and the electronic temperature T_e can be defined, since the operators $\hat{T}_e$ and $\hat{V}_{int}$ are universal. For any particular $V_{ext}(\mathbf{r})$, E_{nn} and T_e, the exact equilibrium electron density $n_{e_0}(\mathbf{r})$ corresponds to the ρ-representable electron density $n_e(\mathbf{r})$ that minimizes this functional and the exact equilibrium state Helmholtz free energy of the system corresponds to the global minimum value of this functional. Hence, the functional $F_M\big(n_e(\mathbf{r}), V_{ext}(\mathbf{r}), E_{nn}, T_e\big)$ is sufficient to determine the exact equilibrium state energy and electron density.

Proof Let $n_\text{e}(\mathbf{r})$ be an arbitrary ρ-representable electron density. By definition, there exists at least one-electron density operator $\hat{\rho}_\text{e}$, that generates this electron density. Therefore, the Helmholtz free energy energy functional of Eq. (2.216) is well defined. For any given $V_\text{ext}(\mathbf{r})$, E_nn and T_e, the related canonical density matrix $\hat{\rho}_\text{e}^{(c)}$ corresponds to the global minimum of the Helmholtz free energy. Hence, the electron density $n_{\text{e}_0}(\mathbf{r})$, which is generated by the canonical density matrix, corresponds to the equilibrium electron density, is ρ-representable and the functional Eq. (2.216) takes its global minimum at this electron density, by construction. ∎

Mermin's theorems prove that the equilibrium electron density can be used as the main variable and that the Helmholtz free energy can be considered as a functional of the electron density. But to do so, one has to perform a minimization over a set of electron density operators, which must be constructed from the complicated many-body wave functions of the electronic system. Due to this, the Mermin's Helmholtz free energy functional is useless for practical calculations.

2.4.4 T_e-dependent Kohn–Sham Equations

Fortunately, the Kohn–Sham Ansatz from ground state DFT can be extended to finite T_e's:

Theorem 2.14 (T_e-dependent Kohn–Sham Ansatz)

1. *The equilibrium electron density $n_{e_0}(\mathbf{r})$ of a system of interacting electrons at a given temperature T_e in an external potential $V_{ext}(\mathbf{r})$ is identical to*

the equilibrium electron density of an auxiliary system of non-interacting electrons at the temperature T_e in an adapted effective potential $V_{eff}(\mathbf{r})$. If this is possible, $n_{e_0}(\mathbf{r})$ is called non-interacting-ρ-representable.

2. *The Helmholtz free energy functional $F_{KS}\big(n_e(\mathbf{r}), V_{ext}(\mathbf{r}), E_{nn}, T_e\big)$ of the system of interacting electrons at a temperature T_e in an external potential $V_{ext}(\mathbf{r})$ with nuclei interaction energy E_{nn} is constructed as follows:*

$$\begin{aligned} F_{KS}\big(n_e(\mathbf{r}), V_{ext}(\mathbf{r}), E_{nn}, T_e\big) = {} & T_{aux}\big(n_e(\mathbf{r})\big) + E_{Ha}\big(n_e(\mathbf{r})\big) \\ & + \int_{\mathbb{R}^3} d^3r\, V_{ext}(\mathbf{r})\, n_e(\mathbf{r}) + E_{nn} - T_e\, S_{aux}\big(n_e(\mathbf{r})\big) \\ & + E_{xc}\big(n_e(\mathbf{r}), T_e\big), \end{aligned} \tag{2.217}$$

where $T_{aux}\big(n_e(\mathbf{r})\big)$ is the exact kinetic energy functional and $S_{aux}\big(n_e(\mathbf{r})\big)$ is the exact entropy functional of the auxiliary system of non-interacting electrons and $E_{xc}\big(n_e(\mathbf{r}), T_e\big)$ is the so-called exchange correlation functional that takes all errors into account that are introduced by using $T_{aux}\big(n_e(\mathbf{r})\big)$, $S_{aux}\big(n_e(\mathbf{r})\big)$ and $E_{Ha}\big(n_e(\mathbf{r})\big)$.

First we consider the auxiliary system of N_e non-interacting electrons. We denote the eigenfunctions of the Hamiltonian $\hat{H}_e^{(\mathrm{aux})}$ by $\Psi_{E_\alpha^{(\mathrm{aux})}}(\mathbf{R}_1, \ldots, \mathbf{R}_{N_e})$, $\alpha = 1, 2, \ldots$ and the related energy eigenvalues by $E_\alpha^{(\mathrm{aux})}$. The canonical density matrix, which describes the equilibrium at temperature T_e, is given by

$$\hat{\rho}_e^{(\mathrm{aux})} = \sum_{\alpha=1}^{\infty} \mathbb{p}_\alpha^{(\mathrm{aux})} \left|\Psi_{E_\alpha}^{(\mathrm{aux})}\right\rangle\!\left\langle\Psi_{E_\alpha}^{(\mathrm{aux})}\right|. \tag{2.218}$$

Using the equilibrium Helmholtz free energy

$$\begin{aligned} F^{(\mathrm{aux})} & \overset{(2.204)}{=} - k_B\, T_e \ln\big(\mathcal{Z}^{(\mathrm{aux})}\big) \\ \Leftrightarrow \qquad \frac{1}{\mathcal{Z}^{(\mathrm{aux})}} & = e^{\frac{F^{(\mathrm{aux})}}{k_B T_e}}, \end{aligned} \tag{2.219}$$

we obtain for the probabilities within the canonical density matrix:

$$\mathbb{p}_\alpha^{(\mathrm{aux})} = \frac{e^{-\frac{E_\alpha^{(\mathrm{aux})}}{k_B T_e}}}{\mathcal{Z}^{(\mathrm{aux})}} = e^{-\frac{E_\alpha^{(\mathrm{aux})} - F^{(\mathrm{aux})}}{k_B T_e}}. \tag{2.220}$$

Due to this, these equilibrium probabilities can be assumed as a function of the energy and the number of electrons in the system:

$$\mathbb{p}_\alpha^{(\mathrm{aux})} \equiv \mathbb{p}\big(E_\alpha^{(\mathrm{aux})}, N_e\big) = e^{-\frac{E_\alpha^{(\mathrm{aux})} - F^{(\mathrm{aux})}(N_e)}{k_B T_e}}. \tag{2.221}$$

Here, we write $F^{(\mathrm{aux})}(N_\mathrm{e})$ to point out that this is the Helmholtz free energy of N_e non-interacting electrons. Any eigenfunction $\Psi_{E_\alpha^{(\mathrm{aux})}}(\mathbf{R}_1,\ldots,\mathbf{R}_{N_\mathrm{e}})$ of the Hamiltonian $\hat{H}_\mathrm{e}^{(\mathrm{aux})}$ is a slater determinant of N_e orthonormal single-electron wave functions / orbitals $\varphi_{\alpha(1)}(\mathbf{r}),\ldots,\varphi_{\alpha(N_\mathrm{e})}(\mathbf{r})$. Here, we denote by $\alpha(1),\ldots,\alpha(N_\mathrm{e})$ the indices of the single-electron wave functions that are used to construct $\Psi_{E_\alpha^{(\mathrm{aux})}}(\mathbf{R}_1,\ldots,\mathbf{R}_{N_\mathrm{e}})$. The indices $\alpha(1),\ldots,\alpha(N_\mathrm{e})$ are distinct due to the Pauli exclusion principle. If we denote the energy of the single-electron orbital $\varphi_\beta(\mathbf{r})$ by ε_β, then we get simply for the energy of $\Psi_{E_\alpha^{(\mathrm{aux})}}(\mathbf{R}_1,\ldots,\mathbf{R}_{N_\mathrm{e}})$:

$$E_\alpha^{(\mathrm{aux})}=\sum_{i=1}^{N_\mathrm{e}}\varepsilon_{\alpha(i)}. \tag{2.222}$$

Furthermore, for every single-electron orbital $\varphi_\beta(\mathbf{r})$, we introduce the equilibrium occupation number f_β, which is the probability of there being an electron in this orbital in equilibrium. By definition, f_β is calculated by summing up the probabilities $\mathbb{p}_\alpha^{(\mathrm{aux})}$ of all slater-determinant states $\Psi_{E_\alpha^{(\mathrm{aux})}}(\mathbf{R}_1,\ldots,\mathbf{R}_{N_\mathrm{e}})$ of N_e distinct single-electron orbitals containing the orbital $\varphi_\beta(\mathbf{r})$:

$$f_\beta=\sum_{\substack{\alpha=1\\ \beta\in\{\alpha(1),\ldots,\alpha(N_\mathrm{e})\}}}^{\infty}\mathbb{p}_\alpha^{(\mathrm{aux})}=\sum_{\substack{\alpha=1\\ \beta\in\{\alpha(1),\ldots,\alpha(N_\mathrm{e})\}}}^{\infty}p\big(E_\alpha^{(\mathrm{aux})},N_\mathrm{e}\big). \tag{2.223}$$

Now we derive f_β as a function of ε_β taken from on Ref. [32]. Since we have

$$1=\sum_{\alpha=1}^{\infty}\mathbb{p}_\alpha^{(\mathrm{aux})}=\sum_{\substack{\alpha=1\\ \beta\in\{\alpha(1),\ldots,\alpha(N_\mathrm{e})\}}}^{\infty}\mathbb{p}_\alpha^{(\mathrm{aux})}+\sum_{\substack{\alpha=1\\ \beta\notin\{\alpha(1),\ldots,\alpha(N_\mathrm{e})\}}}^{\infty}\mathbb{p}_\alpha^{(\mathrm{aux})} \tag{2.224}$$

for all $\beta=1,2,\ldots$, it follows

$$f_\beta=1-\sum_{\substack{\alpha=1\\ \beta\notin\{\alpha(1),\ldots,\alpha(N_\mathrm{e})\}}}^{\infty}\mathbb{p}_\alpha^{(\mathrm{aux})}=1-\sum_{\substack{\alpha=1\\ \beta\notin\{\alpha(1),\ldots,\alpha(N_\mathrm{e})\}}}^{\infty}p\Big(E_\alpha^{(\mathrm{aux})},N_\mathrm{e}\Big). \tag{2.225}$$

In this way, the equilibrium occupation number f_β can be obtained by summing over all slater-determinant states of N_e distinct single-electron orbitals not containing the orbital $\varphi_\beta(\mathbf{r})$ and using their probabilities. The set of all slater-determinant states of N_e distinct single-electron orbitals not containing $\varphi_\beta(\mathbf{r})$ is identical to the set of all slater-determinant states of $N_\mathrm{e}+1$ distinct single-electron orbitals containing $\varphi_\beta(\mathbf{r})$, from which the orbital $\varphi_\beta(\mathbf{r})$ was removed, so that the energy $E_\alpha^{(\mathrm{aux})}$ of the state containing $N_\mathrm{e}+1$ orbitals was reduced by ε_β. Hence, f_β can be derived by summation over of all slater-determinant states of $N_\mathrm{e}+1$ distinct single-electron orbitals containing $\varphi_\beta(\mathbf{r})$ and using the probabilities, which one obtains after removing from this states containing $N_e + 1$ orbitals the orbital $\varphi_\beta(\mathbf{r})$, which reduces their energy $E_\alpha^{(\mathrm{aux})}$ by ε_β:

$$f_\beta(N_e) = 1 - \sum_{\substack{\alpha=1 \\ \beta \in \{\alpha(1),\ldots,\alpha(N_e),\alpha(N_e+1)\}}}^{\infty} p\Big(E_\alpha^{(\mathrm{aux})} - \varepsilon_\beta, N_e\Big). \tag{2.226}$$

Here we write $f_\beta(N_e)$ to point out that is an occupation number of a system containing N_e electrons. The chemical potential μ_e of the electrons corresponds to the change in Helmholtz free energy, if one electron is added or removed from the system:

$$\mu_e = F^{(\mathrm{aux})}(N_e + 1) - F^{(\mathrm{aux})}(N_e). \tag{2.227}$$

Therefore, we obtain

$$\begin{aligned} p\Big(E_\alpha^{(\mathrm{aux})} - \varepsilon_\beta, N_e\Big) &= \mathrm{e}^{-\frac{E_\alpha^{(\mathrm{aux})} - \varepsilon_\beta - F^{(\mathrm{aux})}(N_e)}{k_B T_e}} \\ &= \mathrm{e}^{\frac{\varepsilon_\beta - \mu_e}{k_B T_e}}\, \mathrm{e}^{-\frac{E_\alpha^{(\mathrm{aux})} - F^{(\mathrm{aux})}(N_e+1)}{k_B T_e}} \\ &= \mathrm{e}^{\frac{\varepsilon_\beta - \mu_e}{k_B T_e}}\, p\Big(E_\alpha^{(\mathrm{aux})}, N_e + 1\Big). \end{aligned} \tag{2.228}$$

Inserting this into Eq. (2.226), we have

$$\begin{aligned} f_\beta(N_e) &= 1 - \mathrm{e}^{\frac{\varepsilon_\beta - \mu_e}{k_B T_e}} \sum_{\substack{\alpha=1 \\ \beta \in \{\alpha(1),\ldots,\alpha(N_e),\alpha(N_e+1)\}}}^{\infty} p\Big(E_\alpha^{(\mathrm{aux})}, N_e\Big) \\ &\overset{(2.223)}{=} 1 - \mathrm{e}^{\frac{\varepsilon_\beta - \mu_e}{k_B T_e}}\, f_\beta(N_e + 1). \end{aligned} \tag{2.229}$$

The above equation gives an exact relation between the occupation of the single-electron orbital $\varphi_\beta(\mathbf{r})$ being occupied at temperature T_e in an N_e-electron system and in an $(N_e + 1)$-electron system. For large N_e, the occupation of a single-electron orbital changes insignificantly, if one electron is added to the system. Therefore, we can replace $f_\beta(N_e + 1)$ by $f_\beta(N_e)$ in Eq. (2.229) and obtain

$$f_\beta(N_e) = 1 - \mathrm{e}^{\frac{\varepsilon_\beta - \mu_e}{k_B T_e}}\, f_\beta(N_e). \tag{2.230}$$

From this, we get the equilibrium occupation of the single-electron orbital $\varphi_\beta(\mathbf{r})$ at temperature T_e by

$$\boxed{f_\beta = \frac{1}{\mathrm{e}^{\frac{\varepsilon_\beta - \mu_e}{k_B T_e}} + 1}.} \tag{2.231}$$

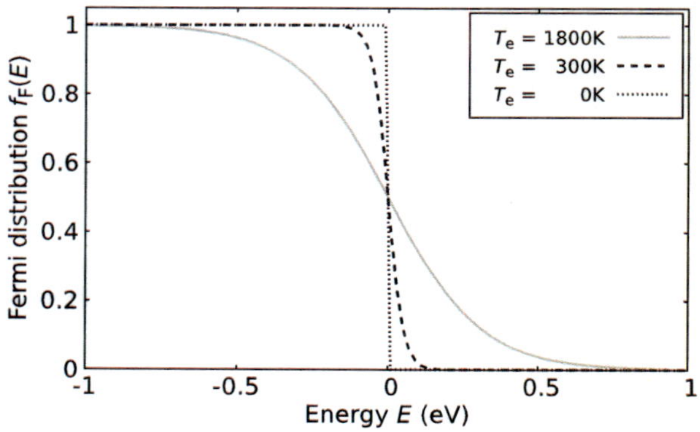

Fig. 2.4 The Fermi distribution with $\mu_e = 0$ is shown for selected temperatures T_e

This is the famous Fermi distribution, which is plotted for various temperatures T_e in Fig. 2.4. Here we removed the label N_e, since, from now on, we will be dealing with a fixed number of electrons.

Now we derive the entropy S_{aux} of the auxiliary system as a function of the single-electron orbital occupation numbers f_β. For our derivation we were inspired by Ref. [33]. The statistical ensemble used determines the equilibrium values of the occupation numbers, but it does not influence the functional dependency of the entropy with those occupation numbers. Therefore, we can use the microcanonical ensemble to calculate S_{aux}. We consider $M >> 1$ copies of the auxiliary system with given fixed occupation numbers f_β and calculate the total entropy $S_{aux}^{(M)}$ of this set of systems in the microcanonical ensemble using Eq. (2.195) by counting the number Ω of possible arrangements of the M systems compatible to the given occupation numbers. Since the entropy is an extensive variable, we have

$$S_{aux} = \frac{1}{M} S_{aux}^{(M)}. \tag{2.232}$$

A given single-electron orbital $\varphi_\beta(\mathbf{r})$ can be empty or can be filled by one electron. Let $0 \leq n_\beta \leq M$ be the number of systems in our set of M systems, in which the single-electron orbital $\varphi_\beta(\mathbf{r})$ is filled with one electron. The number Ω_β of arrangements that n_β systems show a filled and $M - n_\beta$ systems show an empty orbital $\varphi_\beta(\mathbf{r})$ is given by the binomial coefficient

$$\Omega_\beta = \binom{M}{n_\beta} = \frac{M!}{M!\,(M - n_\beta)!}, \tag{2.233}$$

since it yields in how many ways n_β things can be chosen from M things. Since all systems are compatible with the occupation numbers, the probability of the orbital $\varphi_\beta(\mathbf{r})$ being filled is given by f_β, so that we have

$$n_\beta = \lfloor f_\beta\, M \rfloor, \tag{2.234}$$

where $\lfloor x \rfloor$ denotes the biggest integer number that is smaller or equal to x. From this, we get

$$f_\beta \approx \frac{n_\beta}{M}. \tag{2.235}$$

Since the electrons do not interact with each other, the total number Ω of possible arrangements of the M systems is given by

$$\Omega = \prod_{\beta=1}^{\infty} \Omega_\beta. \tag{2.236}$$

Hence, we obtain

$$\begin{aligned}
S_{\text{aux}} &\overset{(2.232)}{=} \frac{1}{M} S_{\text{aux}}^{(M)} \\
&\overset{(2.195)}{=} \frac{k_B}{M} \ln(\Omega) \\
&\overset{(2.236)}{=} \frac{k_B}{M} \ln\left(\prod_{\beta=1}^{\infty} \Omega_\beta\right) \\
&= \frac{k_B}{M} \sum_{\beta=1}^{\infty} \ln(\Omega_\beta) \\
&\overset{(2.233)}{=} \frac{k_B}{M} \sum_{\beta=1}^{\infty} \Big(\ln(M!) - \ln(n_\beta!) - \ln\big((M - n_\beta)!\big)\Big).
\end{aligned} \tag{2.237}$$

Since we deal with $M >> 1$, we have also $n_\beta >> 1$ from Eq. (2.234) for $f_\beta > 0$. If $f_\beta = 0$ occurs, then we formally drop this β from the sum. In this way, we can use Stirling's approximation (2.196) of the logarithm of the factorial and obtain further

$$\begin{aligned}
&\frac{1}{M}\Big(\ln(M!) - \ln(n_\beta!) - \ln\big((M - n_\beta)!\big)\Big) \\
\overset{(2.196)}{\approx}\ & \frac{1}{M}\Big(M \ln(M) - M - n_\beta \ln(n_\beta) + n_\beta - (M - n_\beta) \ln(M - n_\beta) + (M - n_\beta)\Big) \\
=\ & \frac{1}{M}\Big(M \ln(M) - n_\beta \ln(n_\beta) - (M - n_\beta) \ln(M - n_\beta)\Big) \\
=\ & \ln(M) - \frac{n_\beta}{M} \ln(n_\beta) - \frac{M - n_\beta}{M} \ln(M - n_\beta) \\
=\ & \underbrace{\left(-\frac{n_\beta}{M} - \left(1 - \frac{n_\beta}{M}\right)\right)}_{=-1} \ln\left(\frac{1}{M}\right) - \frac{n_\beta}{M} \ln(n_\beta) - \left(1 - \frac{n_\beta}{M}\right) \ln(M - n_\beta) \\
=\ & -\frac{n_\beta}{M} \ln\left(\frac{n_\beta}{M}\right) - \left(1 - \frac{n_\beta}{M}\right) \ln\left(1 - \frac{n_\beta}{M}\right)
\end{aligned}$$

$$\overset{(2.235)}{\approx} - f_\beta \ln(f_\beta) - (1 - f_\beta) \ln(1 - f_\beta). \tag{2.238}$$

Putting this in (2.237) yields

$$S_{\text{aux}} \approx -k_{\text{B}} \sum_{\beta=1}^{\infty} \Big(f_\beta \ln(f_\beta) - (1 - f_\beta) \ln(1 - f_\beta) \Big). \tag{2.239}$$

Now let us consider the limit $M \to \infty$, so that Stirling's formula (2.196) and Eq. (2.235) become exact, and we obtain finally

$$\boxed{S_{\text{aux}} = -k_{\text{B}} \sum_{\beta=1}^{\infty} \Big(f_\beta \ln(f_\beta) - (1 - f_\beta) \ln(1 - f_\beta) \Big).} \tag{2.240}$$

Now we can add the eventually dropped β's with $f_\beta = 0$ to the sum, since $f_\beta = 0$ yields no contribution to the sum due to $f_\beta \ln(1 - f_\beta) = \ln(1) = 0$ and $f_\beta \ln(f_\beta) \overset{(2.185)}{=} 0$. Equation (2.240) is valid for any set of occupation numbers f_β, since we did not use the values from Eq. (2.231) corresponding to the equilibrium in the derivation. Furthermore, we obtain for the equilibrium electron density of the auxiliary system.

$$\begin{aligned} n_{\text{e}}(\mathbf{r}) &= \big\langle \hat{n}_{\text{e}}(\mathbf{r}) \big\rangle_{\hat{\rho}_{\text{e}}^{(\text{aux})}} \\ &= \sum_{\alpha=1}^{\infty} \mathbb{p}_{\alpha}^{(\text{aux})} \big\langle \Psi_{E_\alpha}^{(\text{aux})} \big| \hat{n}_{\text{e}}(\mathbf{r}) \big| \Psi_{E_\alpha}^{(\text{aux})} \big\rangle \\ &= \sum_{\alpha=1}^{\infty} \mathbb{p}_{\alpha}^{(\text{aux})} \sum_{i=1}^{N_{\text{e}}} \big| \varphi_{\alpha(i)}(\mathbf{r}) \big|^2 \\ &= \sum_{\beta=1}^{\infty} \big| \varphi_\beta(\mathbf{r}) \big|^2 \sum_{\substack{\alpha = 1 \\ \beta \in \{\alpha(1), \ldots, \alpha(N_{\text{e}})\}}}^{\infty} \mathbb{p}_{\alpha}^{(\text{aux})} \end{aligned} \tag{2.241}$$

and finally

$$\boxed{n_{\text{e}}(\mathbf{r}) = \sum_{\beta=1}^{\infty} f_\beta \big| \varphi_\beta(\mathbf{r}) \big|^2.} \tag{2.242}$$

From this, we get

$$N_{\text{e}} = \int_{\mathbb{R}^3} d^3r \, n_{\text{e}}(\mathbf{r}) = \sum_{\beta=1}^{\infty} f_\beta \int_{\mathbb{R}^3} d^3r \, \big| \varphi_\beta(\mathbf{r}) \big|^2 = \sum_{\beta=1}^{\infty} f_\beta. \tag{2.243}$$

Furthermore, we have for the kinetic energy in the auxiliary system

$$
\begin{aligned}
T_{\text{aux}} &= \left\langle \hat{T}_e \right\rangle_{\hat{\rho}_e^{(\text{aux})}} \\
&= \sum_{\alpha=1}^{\infty} \mathbb{p}_\alpha^{(\text{aux})} \left\langle \Psi_{E_\alpha}^{(\text{aux})} \middle| \hat{T}_e \middle| \Psi_{E_\alpha}^{(\text{aux})} \right\rangle \\
&= \sum_{\alpha=1}^{\infty} \mathbb{p}_\alpha^{(\text{aux})} \sum_{i=1}^{N_e} \frac{-\hbar^2}{2\,m_e} \int_{\mathbb{R}^3} d^3r\, \varphi^*_{\alpha(i)}(\mathbf{r})\, \nabla^2_{\mathbf{r}}\, \varphi_{\alpha(i)}(\mathbf{r}) \\
&= -\frac{\hbar^2}{2\,m_e} \sum_{\beta=1}^{\infty} \int_{\mathbb{R}^3} d^3r\, \varphi^*_\beta(\mathbf{r})\, \nabla^2_{\mathbf{r}}\, \varphi_\beta(\mathbf{r}) \sum_{\substack{\alpha=1 \\ \beta \in \{\alpha(1),\ldots,\alpha(N_e)\}}}^{\infty} \mathbb{p}_\alpha^{(\text{aux})} \\
&= -\frac{\hbar^2}{2\,m_e} \sum_{\beta=1}^{\infty} f_\beta \int_{\mathbb{R}^3} d^3r\, \varphi^*_\beta(\mathbf{r})\, \nabla^2_{\mathbf{r}}\, \varphi_\beta(\mathbf{r}).
\end{aligned}
\tag{2.244}
$$

Now we consider the system of interacting electrons. Let $\hat{\rho}_e^{(c)}$ be the equilibrium density matrix of this system and $\{|\Psi_{E_\alpha}\rangle, \alpha = 1, \ldots, \infty\}$ the set of eigenkets of the corresponding Hamiltonian. Then we obtain for the Helmholtz free energy of the system of interaction electrons

$$
\begin{aligned}
\langle F \rangle_{\hat{\rho}_e^{(c)}} =& \sum_{\alpha=1}^{\infty} \mathbb{p}_\alpha^{(c)} \left\langle \Psi_{E_\alpha} \middle| \hat{T}_e + \hat{V}_{\text{int}} \middle| \Psi_{E_\alpha} \right\rangle + \int_{\mathbb{R}^3} d^3r\, V_{\text{ext}}(\mathbf{r})\, n_e(\mathbf{r}) + E_{\text{nn}} \\
&+ k_B\, T_e \sum_{\alpha=1}^{\infty} \mathbb{p}_\alpha^{(c)} \ln\left(\mathbb{p}_\alpha^{(c)}\right) \\
=& T_{\text{aux}} + E_{\text{Ha}}\Big(n_e(\mathbf{r})\Big) + \int_{\mathbb{R}^3} d^3r\, V_{\text{ext}}(\mathbf{r})\, n_e(\mathbf{r}) + E_{\text{nn}} - T_e\, S_{\text{aux}} + E_{\text{xc}}.
\end{aligned}
\tag{2.245}
$$

Due to Mermins theorems, the Helmholtz free energy of the system of interaction electrons is a functional of the electron density, *i.e.*,

$$
\langle F \rangle_{\hat{\rho}_e^{(c)}} = F_M\Big(n_e(\mathbf{r}), V_{\text{ext}}(\mathbf{r}), E_{\text{nn}}, T_e\Big).
$$

In equilibrium, the electron density of our system of interacting electrons is equal to the electron density of the auxiliary system of non-interacting electrons due to the Kohn–Sham Ansatz. Hence, the electron density can be expressed by Eq. (2.242) from the single-electron wave functions and the occupation numbers, so that T_{aux} and S_{aux} can be considered as functionals of the electron density, since they can be calculated from the single-electron wave functions and their occupation numbers.

We get by rearranging Eq. (2.245)

$$
\begin{aligned}
E_{\mathrm{xc}} =& F_{\mathrm{M}}\Big(n_{\mathrm{e}}(\mathbf{r}),\, V_{\mathrm{ext}}(\mathbf{r}),\, E_{\mathrm{nn}},\, T_{\mathrm{e}}\Big) - T_{\mathrm{aux}}\Big(n_{\mathrm{e}}(\mathbf{r})\Big) - E_{\mathrm{Ha}}\Big(n_{\mathrm{e}}(\mathbf{r})\Big) \\
&- \int_{\mathbb{R}^3} d^3r\, V_{\mathrm{ext}}(\mathbf{r})\, n_{\mathrm{e}}(\mathbf{r}) - E_{\mathrm{nn}} + T_{\mathrm{e}}\, S_{\mathrm{aux}}\Big(n_{\mathrm{e}}(\mathbf{r})\Big).
\end{aligned} \tag{2.246}
$$

Since all terms on the right hand side are functionals of the electron density, E_{xc} is also a functional of the electron density. We obtain further

$$
\begin{aligned}
E_{\mathrm{xc}}\Big(n_{\mathrm{e}}(\mathbf{r}),\, T_{\mathrm{e}}\Big) =& \sum_{\alpha=1}^{\infty} \mathbb{p}_{\alpha}^{(\mathrm{c})} \left\langle \Psi_{E_\alpha} \middle| \hat{T}_{\mathrm{e}} + \hat{V}_{\mathrm{int}} \middle| \Psi_{E_\alpha} \right\rangle - T_{\mathrm{aux}}\Big(n_{\mathrm{e}}(\mathbf{r})\Big) - E_{\mathrm{Ha}}\Big(n_{\mathrm{e}}(\mathbf{r})\Big) \\
&+ k_{\mathrm{B}}\, T_{\mathrm{e}} \sum_{\alpha=1}^{\infty} \mathbb{p}_{\alpha}^{(\mathrm{c})} \ln\big(\mathbb{p}_{\alpha}^{(\mathrm{c})}\big) + T_{\mathrm{e}}\, S_{\mathrm{aux}}\Big(n_{\mathrm{e}}(\mathbf{r})\Big).
\end{aligned} \tag{2.247}
$$

If we denote by $n_{\mathrm{e}}^{(\alpha)}(\mathbf{r})$ the electron density generated from $|\Psi_{E_\alpha}\rangle$, then we have

$$
n_{\mathrm{e}}(\mathbf{r}) = \sum_{\alpha=1}^{\infty} \mathbb{p}_{\alpha}^{(\mathrm{c})}\, n_{\mathrm{e}}^{(\alpha)}(\mathbf{r}). \tag{2.248}
$$

Due to the definition of the exchange correlation functional $E_{\mathrm{xc}}\big(n_{\mathrm{e}}(\mathbf{r})\big)$ of the kinetic and interaction energy of the electrons in ground state DFT, we get further

$$
\begin{aligned}
&\sum_{\alpha=1}^{\infty} \mathbb{p}_{\alpha}^{(\mathrm{c})} \left\langle \Psi_{E_\alpha} \middle| \hat{T}_{\mathrm{e}} + \hat{V}_{\mathrm{int}} \middle| \Psi_{E_\alpha} \right\rangle \\
&= \sum_{\alpha=1}^{\infty} \mathbb{p}_{\alpha}^{(\mathrm{c})} \left(T_{\mathrm{aux}}\Big(n_{\mathrm{e}}^{(\alpha)}(\mathbf{r})\Big) + E_{\mathrm{Ha}}\Big(n_{\mathrm{e}}^{(\alpha)}(\mathbf{r})\Big) + E_{\mathrm{xc}}\Big(n_{\mathrm{e}}^{(\alpha)}(\mathbf{r})\Big) \right) \\
&= T_{\mathrm{aux}}\left(\sum_{\alpha=1}^{\infty} \mathbb{p}_{\alpha}^{(\mathrm{c})}\, n_{\mathrm{e}}^{(\alpha)}(\mathbf{r}) \right) + E_{\mathrm{Ha}}\left(\sum_{\alpha=1}^{\infty} \mathbb{p}_{\alpha}^{(\mathrm{c})}\, n_{\mathrm{e}}^{(\alpha)}(\mathbf{r}) \right) + E_{\mathrm{xc}}\left(\sum_{\alpha=1}^{\infty} \mathbb{p}_{\alpha}^{(\mathrm{c})}\, n_{\mathrm{e}}^{(\alpha)}(\mathbf{r}) \right) \\
&= T_{\mathrm{aux}}\Big(n_{\mathrm{e}}(\mathbf{r})\Big) + E_{\mathrm{Ha}}\Big(n_{\mathrm{e}}(\mathbf{r})\Big) + E_{\mathrm{xc}}\Big(n_{\mathrm{e}}(\mathbf{r})\Big).
\end{aligned} \tag{2.249}
$$

and finally

$$
E_{\mathrm{xc}}\Big(n_{\mathrm{e}}(\mathbf{r}),\, T_{\mathrm{e}}\Big) = E_{\mathrm{xc}}\Big(n_{\mathrm{e}}(\mathbf{r})\Big) + k_{\mathrm{B}}\, T_{\mathrm{e}} \sum_{\alpha=1}^{\infty} \mathbb{p}_{\alpha}^{(\mathrm{c})} \ln\big(\mathbb{p}_{\alpha}^{(\mathrm{c})}\big) + T_{\mathrm{e}}\, S_{\mathrm{aux}}\Big(n_{\mathrm{e}}(\mathbf{r})\Big). \tag{2.250}
$$

Consequently, the finite-temperature exchange-correlation functional corresponds to the exchange correlation functional of ground state DFT plus a correction term related to the entropy. Now we derive the single-electron orbitals $\varphi_\beta(\mathbf{r})$ at the equilibrium. These equilibrium orbitals minimize the Helmholtz free energy (2.217), if the equilibrium electron density $n_{\mathrm{e}_0}(\mathbf{r})$ is inserted and the equilibrium occupation numbers f_β from Eq. (2.231) are used, under the constraint that the single-electron orbitals are normalized. This corresponds to the following variation principle:

$$\frac{\delta}{\delta\varphi_\alpha^*(\mathbf{R})}\left(F_{\mathrm{KS}}\Big(n_{\mathrm{e}_0}(\mathbf{r}),\,V_{\mathrm{ext}}(\mathbf{r}),\,E_{\mathrm{nn}},\,T_\mathrm{e}\Big)-\sum_{\beta=1}^{\infty}\varepsilon_\beta\,f_\beta\,\Big(\langle\varphi_\beta|\varphi_\beta\rangle-1\Big)\right)=0, \tag{2.251}$$

where the Lagrangian multipliers ε_β are used to take the normalization condition into account. From this, we obtain from Eq. (2.217):

$$\begin{aligned}0=\frac{\delta}{\delta\varphi_\alpha^*(\mathbf{R})}\Bigg(&T_{\mathrm{aux}}\Big(n_{\mathrm{e}_0}(\mathbf{r})\Big)+E_{\mathrm{Ha}}\Big(n_{\mathrm{e}_0}(\mathbf{r})\Big)+\int_{\mathbb{R}^3}d^3r\,V_{\mathrm{ext}}(\mathbf{r})\,n_{\mathrm{e}_0}(\mathbf{r})+E_{\mathrm{nn}}\\&-T_\mathrm{e}\,S_{\mathrm{aux}}\Big(n_{\mathrm{e}_0}(\mathbf{r})\Big)+E_{\mathrm{xc}}\Big(n_{\mathrm{e}_0}(\mathbf{r}),\,T_\mathrm{e}\Big)\\&-\sum_{\beta=1}^{\infty}\varepsilon_\beta\,f_\beta\left(\int_{\mathbb{R}^3}d^3r\,\varphi_\beta^*(\mathbf{r})\,\varphi_\beta(\mathbf{r})-1\right)\Bigg).\end{aligned}$$

We have further by inserting T_{aux} from Eq. (2.244) and S_{aux} from (2.240):

$$\begin{aligned}0=\frac{\delta}{\delta\varphi_\alpha^*(\mathbf{R})}\Bigg(&-\frac{\hbar^2}{2\,m_\mathrm{e}}\sum_{\beta=1}^{\infty}f_\beta\int_{\mathbb{R}^3}d^3r\,\varphi_\beta^*(\mathbf{r})\,\nabla_\mathbf{r}^2\,\varphi_\beta(\mathbf{r})+E_{\mathrm{Ha}}\Big(n_{\mathrm{e}_0}(\mathbf{r})\Big)\\&+\int_{\mathbb{R}^3}d^3r\,V_{\mathrm{ext}}(\mathbf{r})\,n_{\mathrm{e}_0}(\mathbf{r})\\&+E_{\mathrm{nn}}+T_\mathrm{e}\,k_\mathrm{B}\sum_{\beta=1}^{\infty}\Big(f_\beta\,\ln(f_\beta)-(1-f_\beta)\,\ln(1-f_\beta)\Big)\\&+E_{\mathrm{xc}}\Big(n_{\mathrm{e}_0}(\mathbf{r}),\,T_\mathrm{e}\Big)-\sum_{\beta=1}^{\infty}\varepsilon_\beta\,f_\beta\left(\int_{\mathbb{R}^3}d^3r\,\varphi_\beta^*(\mathbf{r})\,\varphi_\beta(\mathbf{r})-1\right)\Bigg)\\=-&\frac{\hbar^2}{2\,m_\mathrm{e}}\,f_\alpha\,\nabla_\mathbf{R}^2\,\varphi_\alpha(\mathbf{R})\\&+\frac{\delta}{\delta n_{\mathrm{e}_0}(\mathbf{R})}\left(E_{\mathrm{Ha}}\Big(n_{\mathrm{e}_0}(\mathbf{r})\Big)+\int_{\mathbb{R}^3}d^3r\,V_{\mathrm{ext}}(\mathbf{r})\,n_{\mathrm{e}_0}(\mathbf{r})+E_{\mathrm{xc}}\Big(n_{\mathrm{e}_0}(\mathbf{r}),\,T_\mathrm{e}\Big)\right)\underbrace{\frac{\delta n_{\mathrm{e}_0}(\mathbf{R})}{\delta\varphi_\alpha^*(\mathbf{R})}}_{=f_\alpha\,\varphi_\alpha(\mathbf{R})}\\&-\varepsilon_\alpha\,f_\alpha\,\varphi_\alpha(\mathbf{R}).\end{aligned}$$

We can divide by f_α, since we have $f_\alpha > 0$ due to Eq. (2.231). If we define

$$V_{xc}(\mathbf{R}, T_e) := \frac{\delta E_{xc}\Big(n_{e_0}(\mathbf{r}), T_e\Big)}{\delta n_{e_0}(\mathbf{R})} \tag{2.252}$$

analogously to Eq. (2.120) of ground state DFT and use Eq. (2.119) for $\frac{\delta E_{Ha}\left(n_{e_0}(\mathbf{r})\right)}{\delta n_{e_0}(\mathbf{R})}$, we obtain finally

$$\boxed{\left(-\frac{\hbar^2}{2m_e}\nabla^2_{\mathbf{R}} + \underbrace{V_{ext}(\mathbf{R}) + \int\limits_{\mathbb{R}^3} d^3r\, \frac{e^2}{4\pi\varepsilon_0}\, \frac{n_{e0}(\mathbf{r})}{|\mathbf{r}-\mathbf{R}|} + V_{xc}(\mathbf{R}, T_e)}_{=V_{eff}(\mathbf{R}, T_e)}\right)\varphi_\alpha(\mathbf{R}) = \varepsilon_\alpha\, \varphi_\alpha(\mathbf{R}).} \tag{2.253}$$

This is exactly the same equation as Eq. (2.121) for ground state DFT, if one realizes that here the equilibrium electron density at T_e plays the same role as the ground state electron density in ground state DFT. The equilibrium single-electron orbitals $\varphi_\alpha(\mathbf{r})$ are obtained by solving the above equation, which corresponds to a Schrödinger equation of a single electron in an effective potential $V_{eff}(\mathbf{R}, T_e)$, which depends on the equilibrium electron density $n_{e_0}(\mathbf{r})$. There is only a small difference with the ground state DFT: Here, the exchange-correlation potential included in $V_{eff}(\mathbf{R}, T_e)$ depends additionally on T_e, since the exchange-correlation energy (2.250) contains an additional correction term that accounts for the entropy difference between the system of interacting and non-interacting electrons. But in most commonly used T_e-dependent DFT codes, the entropy of the system of interacting electrons is assumed to be equal to that of the non-interacting system, so that the exchange-correlation energy E_{xc} and potential V_{xc} are identical to those of the ground state DFT.

Due to this similarities, the equilibrium electron density $n_{e_0}(\mathbf{r})$, the equilibrium single-electron orbitals $\varphi_\alpha(\mathbf{r})$, which we already know as Kohn–Sham orbitals, and the related energy eigenvalues ε_β, which we already know as Kohn–Sham energies, are found in a self consistent way similar to ground state DFT described in Sect. 2.3.6. Again, since we do not take explicitly the spin into account, we consider the single-electron orbitals $\varphi_\beta(\mathbf{r})$'s with eigenenergies ε_β's twice – one for spin up and one for spin down. There is only one difference: After solving Eq. (2.253) for the single-electron orbitals $\varphi_\beta(\mathbf{r})$ and the related energy eigenvalues β, one calculates the output electron density from Eq. (2.242) with the help of occupation numbers f_β that form a Fermi distribution (2.231). In this Fermi distribution, the chemical potential μ_e of the electrons is not known *a priori*, but can be obtained from the known energy eigenvalues ε_β and T_e and the fact that there are N_e electrons in the system, using

$$N_e \overset{(2.243)}{=} \sum_{\beta=1}^{\infty} f_\beta \overset{(2.231)}{=} \sum_{\beta=1}^{\infty} \frac{1}{e^{\frac{\varepsilon_\beta - \mu_e}{k_B T_e}} + 1}. \tag{2.254}$$

In real life numerical calculations, the sum over β appearing in (2.254) is finite due to the finite size of the basis set, so that μ_e can be numerically obtained from the above equation.

After determining the equilibrium electron density $n_{e_0}(\mathbf{r})$, the Kohn–Sham orbitals $\varphi_\alpha(\mathbf{r})$ and energies ε_β, the equilibrium Helmholtz free energy is given by

$$\begin{aligned} F_{KS} \overset{(2.217)}{=} & T_{aux}\Big(n_{e_0}(\mathbf{r})\Big) + E_{Ha}\Big(n_{e_0}(\mathbf{r})\Big) + \int_{\mathbb{R}^3} d^3r \, V_{ext}(\mathbf{r})\, n_{e_0}(\mathbf{r}) + E_{nn} \\ & - T_e\, S_{aux}\Big(n_{e_0}(\mathbf{r})\Big) + E_{xc}\Big(n_{e_0}(\mathbf{r}), T_e\Big). \end{aligned}$$

Now we insert the kinetic energy T_{aux} from Eq. (2.244) and the Hartree energy E_{Ha} from Eq. (2.108) and obtain further

$$\begin{aligned} F_{KS} = & -\frac{\hbar^2}{2m_e} \sum_{\beta=1}^{\infty} f_\beta \int_{\mathbb{R}^3} d^3r \, \varphi_\beta^*(\mathbf{r})\, \nabla_{\mathbf{r}}^2\, \varphi_\beta(\mathbf{r}) \\ & + \frac{1}{2} \int_{\mathbb{R}^3} d^3r \int_{\mathbb{R}^3} d^3r' \, \frac{e^2}{4\pi\varepsilon_0} \frac{n_{e_0}(\mathbf{r})\, n_{e_0}(\mathbf{r}')}{|\mathbf{r}-\mathbf{r}'|} \\ & + \int_{\mathbb{R}^3} d^3r \, V_{ext}(\mathbf{r})\, n_{e_0}(\mathbf{r}) + E_{nn} - T_e\, S_{aux}\Big(n_{e_0}(\mathbf{r})\Big) + E_{xc}\Big(n_{e_0}(\mathbf{r}), T_e\Big) \\ = & -\frac{\hbar^2}{2m_e} f_\beta \sum_{\beta=1}^{\infty} \int_{\mathbb{R}^3} d^3r \, \varphi_\beta^*(\mathbf{r})\, \nabla_{\mathbf{r}}^2\, \varphi_\beta(\mathbf{r}) + \int_{\mathbb{R}^3} d^3r \, V_{ext}(\mathbf{r})\, n_{e_0}(\mathbf{r}) \\ & + \int_{\mathbb{R}^3} d^3r \int_{\mathbb{R}^3} d^3r' \, \frac{e^2}{4\pi\varepsilon_0} \frac{n_{e_0}(\mathbf{r})\, n_{e_0}(\mathbf{r}')}{|\mathbf{r}-\mathbf{r}'|} + \int_{\mathbb{R}^3} d^3r \, V_{xc}(\mathbf{r}, T_e)\, n_{e_0}(\mathbf{r}) \\ & + E_{nn} - T_e\, S_{aux}\Big(n_{e_0}(\mathbf{r})\Big) \\ & - \frac{1}{2} \int_{\mathbb{R}^3} d^3r \int_{\mathbb{R}^3} d^3r' \, \frac{e^2}{4\pi\varepsilon_0} \frac{n_{e_0}(\mathbf{r})\, n_{e_0}(\mathbf{r}')}{|\mathbf{r}-\mathbf{r}'|} - \int_{\mathbb{R}^3} d^3r \, V_{xc}(\mathbf{r}, T_e)\, n_{e_0}(\mathbf{r}) \\ & + E_{xc}\Big(n_{e_0}(\mathbf{r}), T_e\Big). \end{aligned}$$

By inserting Eq. (2.242) for the equilibrium electron density $n_{e_0}(\mathbf{r})$ we get further

$$
\begin{aligned}
F_{\mathrm{KS}} = & -\frac{\hbar^2}{2m_{\mathrm{e}}} f_\beta \sum_{\beta=1}^{\infty} \int\limits_{\mathbb{R}^3} d^3r\, \varphi_\beta^*(\mathbf{r})\, \nabla_{\mathbf{r}}^2\, \varphi_\beta(\mathbf{r}) + \int\limits_{\mathbb{R}^3} d^3r\, V_{\mathrm{ext}}(\mathbf{r}) \left(\sum_{\beta=1}^{\infty} f_\beta \left|\varphi_\beta(\mathbf{r})\right|^2 \right) \\
& + \int\limits_{\mathbb{R}^3} d^3r \int\limits_{\mathbb{R}^3} d^3r'\, \frac{e^2}{4\pi\varepsilon_0} \frac{\left(\sum\limits_{\beta=1}^{\infty} f_\beta \left|\varphi_\beta(\mathbf{r})\right|^2 \right) n_{\mathrm{e}_0}(\mathbf{r}')}{|\mathbf{r}-\mathbf{r}'|} \\
& + \int\limits_{\mathbb{R}^3} d^3r\, V_{\mathrm{xc}}(\mathbf{r}, T_{\mathrm{e}}) \left(\sum_{j=1}^{N_{\mathrm{e}}} f_\beta \left|\varphi_\beta(\mathbf{r})\right|^2 \right) \\
& + E_{\mathrm{nn}} - T_{\mathrm{e}}\, S_{\mathrm{aux}}\Big(n_{\mathrm{e}_0}(\mathbf{r})\Big) - \frac{1}{2} \int\limits_{\mathbb{R}^3} d^3r \int\limits_{\mathbb{R}^3} d^3r'\, \frac{e^2}{4\pi\varepsilon_0} \frac{n_{\mathrm{e}_0}(\mathbf{r})\, n_{\mathrm{e}_0}(\mathbf{r}')}{|\mathbf{r}-\mathbf{r}'|} \\
& - \int\limits_{\mathbb{R}^3} d^3r\, V_{\mathrm{xc}}(\mathbf{r}, T_{\mathrm{e}})\, n_{\mathrm{e}_0}(\mathbf{r}) + E_{\mathrm{xc}}\Big(n_{\mathrm{e}_0}(\mathbf{r}), T_{\mathrm{e}}\Big) \\
= & \sum_{\beta=1}^{\infty} f_\beta \int\limits_{\mathbb{R}^3} d^3r\, \varphi_\beta^*(\mathbf{r}) \times \\
& \times \underbrace{\left(-\frac{\hbar^2}{2m_{\mathrm{e}}} \nabla_{\mathbf{r}}^2 + V_{\mathrm{ext}}(\mathbf{r}) + \int\limits_{\mathbb{R}^3} d^3r'\, \frac{e^2}{4\pi\varepsilon_0} \frac{n_{\mathrm{e}_0}(\mathbf{r}')}{|\mathbf{r}'-\mathbf{r}|} + V_{\mathrm{xc}}(\mathbf{r}, T_{\mathrm{e}}) \right) \varphi_\beta(\mathbf{r})}_{\overset{(2.253)}{=} \varepsilon_\beta\, \varphi_\beta(\mathbf{r})} \\
& + E_{\mathrm{nn}} - T_{\mathrm{e}}\, S_{\mathrm{aux}}\Big(n_{\mathrm{e}_0}(\mathbf{r})\Big) - \frac{1}{2} \int\limits_{\mathbb{R}^3} d^3r \int\limits_{\mathbb{R}^3} d^3r'\, \frac{e^2}{4\pi\varepsilon_0} \frac{n_{\mathrm{e}_0}(\mathbf{r})\, n_{\mathrm{e}_0}(\mathbf{r}')}{|\mathbf{r}-\mathbf{r}'|} \\
& - \int\limits_{\mathbb{R}^3} d^3r\, V_{\mathrm{xc}}(\mathbf{r}, T_{\mathrm{e}})\, n_{\mathrm{e}_0}(\mathbf{r}) + E_{\mathrm{xc}}\Big(n_{\mathrm{e}_0}(\mathbf{r}), T_{\mathrm{e}}\Big).
\end{aligned}
$$

Finally, we obtain

$$
\boxed{
\begin{aligned}
F_{\mathrm{KS}} = & \sum_{\beta=1}^{\infty} f_\beta\, \varepsilon_\beta + E_{\mathrm{nn}} + k_{\mathrm{B}}\, T_{\mathrm{e}} \sum_{\beta=1}^{\infty} \Big(f_\beta\, \ln(f_\beta) - (1-f_\beta)\, \ln(1-f_\beta) \Big) \\
& - \frac{1}{2} \int\limits_{\mathbb{R}^3} d^3r \int\limits_{\mathbb{R}^3} d^3r'\, \frac{e^2}{4\pi\varepsilon_0} \frac{n_{\mathrm{e}_0}(\mathbf{r})\, n_{\mathrm{e}_0}(\mathbf{r}')}{|\mathbf{r}-\mathbf{r}'|} - \int\limits_{\mathbb{R}^3} d^3r\, V_{\mathrm{xc}}(\mathbf{r}, T_{\mathrm{e}})\, n_{\mathrm{e}_0}(\mathbf{r}) \\
& + E_{\mathrm{xc}}\Big(n_{\mathrm{e}_0}(\mathbf{r}), T_{\mathrm{e}}\Big).
\end{aligned}
}
\tag{2.255}
$$

If we consider $T_e \to 0$, the entropy contribution vanishes and the Fermi distribution (2.231) transforms to

$$f_\beta(\epsilon_\beta) = \begin{cases} 1 & \varepsilon_\beta < \mu_e \\ 0 & \varepsilon_\beta \geq \mu_e \end{cases}, \tag{2.256}$$

so that just the lowest N_e single-electron orbitals $\varphi_\beta(\mathbf{r})$ are used with an occupation of 1. Then the calculations are the same as for ground state DFT. Consequently, ground state DFT corresponds to the limit $T_e \to 0$ of T_e-dependent DFT, as expected.

Pseudopotentials derived from ground state DFT can also be used for T_e-dependent DFT calculations, since the Fermi distribution at increased T_e only changes significantly the occupation of the valence electrons (except for very, very high T_e's, which are seldom considered).

2.5 Summary

A proper theoretical description of a solid needs an accurate treatment of the electrons, since they are responsible for interatomic bonding. The nuclei can be treated classically. In contrast, a precise ab-initio description of the electrons is only possible with quantum mechanical methods.

Since electrons are much lighter than the nuclei, the movement of the electrons and the nuclei can be separated using the Born-Oppenheimer approximation. The nuclei move in an effective potential generated by the electrons, which are always assumed to be in equilibrium with the nuclei, since the electrons adapt immediately their state to the actual nuclei positions. If one approximates the effective electron potential up to the second order (harmonic approximation) with respect to the nuclei coordinates, the classical equations of motions of the nuclei can be analytically solved. This is one possibility to obtain the eigen oscillations of nuclei, which are called phonons.

For electrons in equilibrium at a given electronic temperature T_e, T_e-dependent density functional theory (DFT) allows to derive the equilibrium Helmholtz free energy of the electrons as a function of the nuclei coordinates and the temperature T_e. Numerical calculations can be performed within T_e-dependent DFT for systems of up to 1000 atoms, since the electron density $n_e(\mathbf{r})$ is used as the main variable instead of the complicated many-body electron wave function Ψ. The Mermin's theorems show that the Helmholtz free energy of the electrons is a functional of the electron density and the minimal value of this functional corresponds to the equilibrium and is reached, if the equilibrium electron density is inserted. Furthermore, the Kohn–Sham Ansatz is applied, which states that the equilibrium electron density is equal to the equilibrium electron density of an auxiliary system of non-interacting electrons in an adapted effective potential. This allows to perform the calculations in a much easier way in the system of non-interacting electrons. All errors, which are introduced by considering the system of non-interacting electrons, are put into the so called exchange and correlation term E_{xc}, which is also a functional of the electron density.

Since several very successful and numerically efficient approximations of the E_{xc} functional exists, density functional theory is the workhorse for accurately describing electrons in a solid.

References

1. F. Schwabl, *Quantenmechanik Eine Einführung*, 7th edn. (Springer, 2007). https://www.springer.com/de/book/9783540736745
2. P.A.M. Dirac, Math. Proc. Camb. Philos. Soc. **35**(3), 416 (1939). https://doi.org/10.1017/S0305004100021162
3. A.S. Eddington, *Philosophy of Physical Science* (Cambridge University Press, 1939)
4. M. Born, R. Oppenheimer, Annalen der Physik **389**(20), 457 (1927). https://doi.org/10.1002/andp.19273892002
5. G. Czycholl, *Theoretische Festkörper Physik Band 1*, 4th edn. (Springer, 2016). https://www.springer.com/de/book/9783662471401
6. O. Forster, *Analysis 1*, 9th edn. (Vieweg, 2008)
7. H.R. Schwarz, N. Koeckler, *Numerische Mathematik*, 6th edn. (Teubner, 2006)
8. R.M. Martin, *Electronic Structure: Basic Theory and Practical Methods* (Cambridge University Press, 2004). https://doi.org/10.1017/CBO9780511805769, https://www.cambridge.org/core/books/electronic-structure/DDFE838DED61D7A402FDF20D735BC63A
9. P. Hohenberg, W. Kohn, Phys. Rev. **136**, B864 (1964). https://doi.org/10.1103/PhysRev.136.B864
10. F. Bassani, F. Fumi, M.P. Tosi, *Highlights of condensed-matter theory* (North-Holland physics publ., 1985)
11. W. Kohn, L.J. Sham, Phys. Rev. **140**, A1133 (1965). https://doi.org/10.1103/PhysRev.140.A1133
12. R. van Meer, O.V. Gritsenko, E.J. Baerends, Journal of Chemical Theory and Computation **10**, 4432 (2014). https://doi.org/10.1021/ct500727c
13. D.M. Ceperley, B.J. Alder, Phys. Rev. Lett. **45**, 566 (1980). https://doi.org/10.1103/PhysRevLett.45.566
14. S.H. Vosko, L. Wilk, M. Nusair, Canadian Journal of Physics **58**(8), 1200 (1980). https://doi.org/10.1139/p80-159
15. J.P. Perdew, A. Zunger, Phys. Rev. B **23**, 5048 (1981). https://doi.org/10.1103/PhysRevB.23.5048
16. L.A. Cole, J.P. Perdew, Phys. Rev. A **25**, 1265 (1982). https://doi.org/10.1103/PhysRevA.25.1265
17. J.P. Perdew, Y. Wang, Phys. Rev. B **45**, 13244 (1992). https://doi.org/10.1103/PhysRevB.45.13244
18. A.D. Becke, Phys. Rev. A **38**, 3098 (1988). https://doi.org/10.1103/PhysRevA.38.3098
19. C. Lee, W. Yang, R.G. Parr, Phys. Rev. B **37**, 785 (1988). https://doi.org/10.1103/PhysRevB.37.785
20. J.P. Perdew, J.A. Chevary, S.H. Vosko, K.A. Jackson, M.R. Pederson, D.J. Singh, C. Fiolhais, Phys. Rev. B **46**, 6671 (1992). https://doi.org/10.1103/PhysRevB.46.6671
21. J.P. Perdew, K. Burke, M. Ernzerhof, Phys. Rev. Lett. **77**, 3865 (1996). https://doi.org/10.1103/PhysRevLett.77.3865
22. F. Bloch, Zeitschrift für Physik **52**(7), 555 (1929). https://doi.org/10.1007/BF01339455
23. P. Pulay, Chemical Physics Letters **73**(2), 393 (1980). https://doi.org/10.1016/0009-2614(80)80396-4
24. G.P. Kerker, Phys. Rev. B **23**, 3082 (1981). https://doi.org/10.1103/PhysRevB.23.3082
25. D.R. Hamann, M. Schlüter, C. Chiang, Phys. Rev. Lett. **43**, 1494 (1979). https://doi.org/10.1103/PhysRevLett.43.1494

26. S. Goedecker, M. Teter, J. Hutter, Phys. Rev. B **54**, 1703 (1996). https://doi.org/10.1103/PhysRevB.54.1703
27. C. Hartwigsen, S. Goedecker, J. Hutter, Phys. Rev. B **58**, 3641 (1998). https://doi.org/10.1103/PhysRevB.58.3641
28. N.D. Mermin, Phys. Rev. **137**, A1441 (1965). https://doi.org/10.1103/PhysRev.137.A1441
29. F. Schwabl, *Statistische Mechanik*, 3rd edn. (Springer, 2006). https://www.springer.com/de/book/9783540310952
30. R.G. Parr, W. Yang, *Density-Functinal Theory of Atoms and Molecules* (Oxford University Press, 1989)
31. R. Baierlein, *Thermal Physics* (Cambridge University Press, 1999)
32. N.W. Ashcroft, N.D. Mermin, *Solid State Physics* (Saunders College Publishing, 1976)
33. K. Huang, *Statistical Mechanics*, 2nd edn. (Wiley, 1987)

Chapter 3
Ab-initio Description of a Fs-laser Excitation

Abstract The ability to generate few femtosecond $\left(1\,\text{fs} = 10^{-15}\,\text{s}\right)$ laser-pules opened access to a wide range of applications in material research and processing far beyond the limitations of conventional thermal treatments. In this chapter, the ab-initio description of the processes in matter following a femtosecond laser-pulse excitation are presented. For this, first the classical description and generation of electromagnetic waves and their interaction with matter are reviewed. Since the electrons in a solid require ab-initio quantum mechanical methods, the second quantization is introduced, which provides an equivalent description of quantum mechanics for many body systems. Within second quantization, the effects of a femtosecond laser-interaction on matter are described ab-initio. Again, all important statements are derived and proved on strict physical grounds to provide the reader a deep understanding facilitating the connections between the different related fields of theoretical physics. Finally, the DFT code CHIVES [1–8] is presented, which was developed in the group of Prof. Dr. Garcia to calculate in a fast but accurately way the structural response of a solid after a femtosecond laser-pulse excitation.

3.1 Basic Considerations of Electrodynamics

First a review of Electrodynamics is provided, which is taken mostly from [9, 10]. For completeness, also the main statements of vector calculus are summarized in Sect. A.1 of the appendix. Certain particles like electrons or protons, for example, have a mass and also carry a charge. To describe the distribution and motion of charges, the charge density $\rho(\mathbf{r}, t)$ and the charge current density $\mathbf{J}(\mathbf{r}, t)$ are introduced. The charge density $\rho(\mathbf{r}, t)$ describes how much charge per volume is located at point $\mathbf{r} = [x, y, z]^t$ and time point t. The charge current density $\mathbf{J}(\mathbf{r}, t)$ describes how the charge density $\rho(\mathbf{r}, t)$ moves with which speed $\dot{\mathbf{r}}(\mathbf{r}, t)$ at point $\mathbf{r}$ and time t:

$$\mathbf{J}(\mathbf{r}, t) = \rho(\mathbf{r}, t)\,\dot{\mathbf{r}}(\mathbf{r}, t). \tag{3.1}$$

© The Author(s), under exclusive license to Springer Nature Switzerland AG 2021

B. Bauerhenne, *Materials Interaction with Femtosecond Lasers*,
https://doi.org/10.1007/978-3-030-85135-4_3

The total charge $\mathbb{q}_{\text{tot}}(t)$ inside any volume $\mathcal{V}$ is calculated by

$$\mathbb{q}_{\text{tot}}(t) = \int_{\mathcal{V}} d^3r\, \rho(\mathbf{r}, t). \tag{3.2}$$

The total current $J(t)$ through the surface $\mathbb{S}(\mathcal{V})$ of any volume $\mathcal{V}$ is calculated by

$$J(t) = \int_{\mathbb{S}(\mathcal{V})} d\mathbf{a}^{\mathrm{t}} \cdot \mathbf{J}(\mathbf{r}, t). \tag{3.3}$$

Here, $d\mathbf{a}$ denotes unit vectors that are orthogonal to the surface $\mathbb{S}(\mathcal{V})$ pointing outward of the volume $\mathcal{V}$. Due to charge conservation, the change of the total charge $\mathbb{q}_{\text{tot}}(t)$ in any volume $\mathcal{V}$ is always equal to the total charge current $J(t)$ through the surface $\mathbb{S}(\mathcal{V})$ into the volume $\mathcal{V}$:

$$\begin{aligned} -\frac{d}{dt}\,\mathbb{q}_{\text{tot}}(t) &= J(t), \\ \Leftrightarrow \qquad -\frac{d}{dt}\int_{\mathcal{V}} d^3r\, \rho(\mathbf{r}, t) &= \int_{\mathbb{S}(\mathcal{V})} d\mathbf{a}^{\mathrm{t}} \cdot \mathbf{J}(\mathbf{r}, t). \end{aligned}$$

The minus sign follows the convention that a positive charge current $J(t)$ out of the volume $\mathcal{V}$ produces a negative change of the total charge $\mathbb{q}_{\text{tot}}(t)$. Invoking the **divergence theorem** on the right hand side, we get further

$$-\int_{\mathcal{V}} d^3r\, \dot{\rho}(\mathbf{r}, t) = \int_{\mathcal{V}} d^3r \left(\nabla \cdot \mathbf{J}(\mathbf{r}, t)\right).$$

The dot on $\dot{\rho}(\mathbf{r}, t)$ denotes the time derivative similar to $\dot{\mathbf{r}}$. In the limit of zero volume we obtain at any point $\mathbf{r}$ and time t

$$\boxed{-\dot{\rho}(\mathbf{r}, t) = \nabla \cdot \mathbf{J}(\mathbf{r}, t).} \tag{3.4}$$

This is the charge **continuity equation**, which has its origin in the charge conservation.

3.1.1 *Maxwell Equations in Vacuum*

Charges are the sources of the electromagnetic interaction. To describe these long-range electromagnetic interactions between charges, the electric field $\mathbf{E}$ and the magnetic field $\mathbf{B}$ are introduced. The physics of electric and magnetic fields are summarized in the **Maxwell equations**:

Theorem 3.1 (Maxwell equations in vacuum) *Let ε_0 be the vacuum permittivity and μ_0 be the vacuum permeability. The electric field* $\mathbf{E}$ *and the magnetic field* $\mathbf{B}$ *are calculated from the charge density ρ and the charge current density* $\mathbf{J}$ *in vacuum by*

$$\nabla \cdot \mathbf{E} = \frac{\rho}{\varepsilon_0}, \tag{3.5}$$

$$\nabla \cdot \mathbf{B} = 0, \tag{3.6}$$

$$\nabla \times \mathbf{E} = -\frac{\partial \mathbf{B}}{\partial t}, \tag{3.7}$$

$$\nabla \times \mathbf{B} = \mu_0 \mathbf{J} + \mu_0 \varepsilon_0 \frac{\partial \mathbf{E}}{\partial t}. \tag{3.8}$$

A time-oscillating electric and magnetic field induce each other, so that an electromagnetic wave can propagate in space without presence of charges or currents of charges. To derive the corresponding wave equation for the electric field $\mathbf{E}$ in the case $\rho \equiv 0$ and $\mathbf{J} \equiv \mathbf{0}$, we apply the rotation $\nabla\times$ on Eq. (3.7):

$$\begin{aligned}
& \nabla \times \left(\nabla \times \mathbf{E}\right) = -\nabla \times \frac{\partial \mathbf{B}}{\partial t}, \\
\Leftrightarrow \quad & \nabla \underbrace{\left(\nabla \cdot \mathbf{E}\right)}_{\overset{(3.5)}{=} 0} - \nabla^2 \mathbf{E} = -\frac{\partial}{\partial t} \underbrace{\left(\nabla \times \mathbf{B}\right)}_{\overset{(3.8)}{=} \mu_0 \varepsilon_0 \frac{\partial \mathbf{E}}{\partial t}}, \\
\Leftrightarrow \quad & -\nabla^2 \mathbf{E} = -\mu_0 \varepsilon_0 \frac{\partial^2 \mathbf{E}}{\partial t^2}, \\
\Leftrightarrow \quad & \underbrace{\frac{1}{\mu_0 \varepsilon_0}}_{= \mathbb{c}^2} \nabla^2 \mathbf{E} = \frac{\partial^2 \mathbf{E}}{\partial t^2}.
\end{aligned} \tag{3.9}$$

This is the classical wave equation, from which we can directly read the wave propagation speed

$$\boxed{\mathbb{c} = \sqrt{\frac{1}{\mu_0 \varepsilon_0}}} \tag{3.10}$$

that equals the speed of light in vacuum. Analogously, we obtain the wave equation for the magnetic field $\mathbf{B}$ by applying the rotation $\nabla\times$ on Eq. (3.8):

$$\begin{aligned}
&\nabla \times (\nabla \times \mathbf{B}) = \mu_0\, \varepsilon_0\, \nabla \times \left(\frac{\partial \mathbf{E}}{\partial t}\right),\\
\Leftrightarrow\quad &\nabla \underbrace{(\nabla \cdot \mathbf{B})}_{\stackrel{(3.6)}{=}0} - \nabla^2 \mathbf{B} = \mu_0\, \varepsilon_0\, \frac{\partial}{\partial t} \underbrace{(\nabla \times \mathbf{E})}_{\stackrel{(3.7)}{=}-\frac{\partial \mathbf{B}}{\partial t}},\\
\Leftrightarrow\quad &-\nabla^2 \mathbf{B} = -\mu_0\, \varepsilon_0\, \frac{\partial^2 \mathbf{B}}{\partial t^2},\\
\Leftrightarrow\quad &\frac{1}{\mu_0\, \varepsilon_0}\, \nabla^2 \mathbf{B} = \frac{\partial^2 \mathbf{B}}{\partial t^2}.
\end{aligned} \tag{3.11}$$

It is convenient to introduce the vector and scalar potentials:

Definition 3.1 (*Vector and scalar potential*) A vector potential $\mathbf{A}(\mathbf{r}, t) \in \mathbb{R}^3$ and a scalar potential $\phi_s(\mathbf{r}, t) \in \mathbb{R}$ are defined, from which the magnetic field $\mathbf{B}$ and electric field $\mathbf{E}$ are calculated by

$$\mathbf{B} = \nabla \times \mathbf{A}, \tag{3.12}$$

$$\mathbf{E} = -\nabla \phi_s - \frac{\partial \mathbf{A}}{\partial t}. \tag{3.13}$$

These definitions not only simplify the solution of the Maxwell equations, but also makes possible a seamless transition to the language of quantum mechanics. Indeed, the definition of the vector and scalar potential are meaningful: Due to Eq. (3.6), a vector potential **A** must exist for any magnetic field **B**, so that **B** can be just calculated from **A** by Eq. (3.12). If we use this for **B** in Eq. (3.7), we get

$$\begin{aligned}
&\nabla \times \mathbf{E} = -\frac{\partial}{\partial t} \underbrace{(\nabla \times \mathbf{A})}_{\stackrel{(3.12)}{=}\mathbf{B}},\\
\Leftrightarrow\quad &\nabla \times \left(\mathbf{E} + \frac{\partial A}{\partial t}\right) = 0.
\end{aligned}$$

Because the rotation of the term in the bracket is zero, this term can be calculated from the gradient of a scalar potential $\phi_s(\mathbf{r}, t)$:

$$\mathbf{E} + \frac{\partial A}{\partial t} = -\nabla \phi_s.$$

From this, we obtain directly Eq. (3.13).

Theorem 3.2 (Calculation of the vector and scalar potentials) $\phi_s(\mathbf{r}, t)$ *and* $\mathbf{A}(\mathbf{r}, t)$ *are obtained from the Maxwell equations by*

$$\nabla^2\phi_s + \frac{\partial}{\partial t}(\nabla \cdot \mathbf{A}) = -\frac{\rho}{\varepsilon_0}, \tag{3.14}$$

$$\nabla^2\mathbf{A} - \mu_0\,\varepsilon_0\,\frac{\partial^2\mathbf{A}}{\partial t^2} - \nabla\left(\nabla\cdot\mathbf{A} + \mu_0\,\varepsilon_0\,\frac{\partial\phi_s}{\partial t}\right) = -\mu_0\,\mathbf{J}. \tag{3.15}$$

Proof If we use Eq. (3.13) for **E** in Eq. (3.5), we get

$$\begin{aligned} &\nabla\cdot\underbrace{\left(-\nabla\phi_s - \frac{\partial\mathbf{A}}{\partial t}\right)}_{\stackrel{(3.13)}{=}\mathbf{E}} = \frac{\rho}{\varepsilon_0},\\ \Leftrightarrow\quad & -\nabla^2\phi_s - \nabla\cdot\frac{\partial A}{\partial t} = \frac{\rho}{\varepsilon_0},\\ \Leftrightarrow\quad & -\nabla^2\phi_s - \frac{\partial}{\partial t}(\nabla\cdot\mathbf{A}) = \frac{\rho}{\varepsilon_0}, \end{aligned}$$

from which Eq. (3.14) follows. If we insert Eq. (3.12) for **B** and Eq. (3.13) for **E** in Eq. (3.8), we have

$$\begin{aligned} &\nabla\times\underbrace{(\nabla\times\mathbf{A})}_{\stackrel{(3.12)}{=}\mathbf{B}} = \mu_0\,\mathbf{J} + \mu_0\,\varepsilon_0\,\frac{\partial}{\partial t}\underbrace{\left(-\nabla\phi_s - \frac{\partial\mathbf{A}}{\partial t}\right)}_{\stackrel{(3.13)}{=}\mathbf{E}},\\ &\Leftrightarrow\nabla(\nabla\cdot\mathbf{A}) - \nabla^2\mathbf{A} = \mu_0\,\mathbf{J} - \mu_0\,\varepsilon_0\,\nabla\left(\frac{\partial\phi_s}{\partial t}\right) - \mu_0\,\varepsilon_0\,\frac{\partial^2\mathbf{A}}{\partial t^2},\\ &\Leftrightarrow\nabla\left(\nabla\cdot\mathbf{A} + \mu_0\,\varepsilon_0\frac{\partial\phi_s}{\partial t}\right) - \nabla^2\mathbf{A} + \mu_0\,\varepsilon_0\,\frac{\partial^2\mathbf{A}}{\partial t^2} = \mu_0\,\mathbf{J}, \end{aligned}$$

from which Eq. (3.15) follows. ■

The electric and magnetic field do not change, if one transforms **A** and ϕ_s with the help of a scalar $\xi(\mathbf{r}, t) \in \mathbb{R}$ by

$$\begin{aligned} \mathbf{A}' &= \mathbf{A} + \nabla\xi,\\ \phi_s' &= \phi_s - \frac{\partial\xi}{\partial t}, \end{aligned}$$

since

$$\mathbf{B}' \overset{(3.12)}{=} \nabla \times (\mathbf{A} + \nabla \xi) \qquad = \nabla \times \mathbf{A} + \underbrace{\nabla \times (\nabla \xi)}_{=\mathbf{0}} \qquad = \mathbf{B},$$

$$\mathbf{E}' \overset{(3.13)}{=} -\nabla \left(\phi_s - \frac{\partial \xi}{\partial t} \right) - \frac{\partial}{\partial t}(\mathbf{A} + \nabla \xi) \quad = -\nabla \phi_s - \frac{\partial \mathbf{A}}{\partial t} + \underbrace{\frac{\partial}{\partial t}\nabla \xi - \frac{\partial}{\partial t}\nabla \xi}_{=0} \quad = \mathbf{E}.$$

This allows to transform $\mathbf{A}$ and ϕ_s in such a way that both fulfill an additional condition to simplify Eq. (3.14) and Eq. (3.15). Such a procedure is called gauge. There are two famous gauges. In the Coulomb gauge, Eq. (3.14) for the calculation of ϕ_s is simplified:

Corollary 3.1 (Coulomb gauge of the vector and scalar potentials) *The vector potential $\mathbf{A}(\mathbf{r}, t)$ is chosen in the way that*

$$\nabla \cdot \mathbf{A} = 0. \tag{3.16}$$

Due to this, one obtains $\mathbf{A}(\mathbf{r}, t)$ and $\phi_s(\mathbf{r}, t)$ from

$$\nabla^2 \phi_s = -\frac{\rho}{\varepsilon_0}, \tag{3.17}$$

$$\nabla^2 \mathbf{A} - \mu_0 \varepsilon_0 \frac{\partial^2 \mathbf{A}}{\partial t^2} = -\mu_0 \mathbf{J} + \mu_0 \varepsilon_0 \nabla \left(\frac{\partial \phi_s}{\partial t} \right). \tag{3.18}$$

Proof Equation (3.17) follows directly from Eq. (3.14) due to

$$\nabla^2 \phi_s + \frac{\partial}{\partial t} \underbrace{(\nabla \cdot \mathbf{A})}_{\overset{(3.16)}{=} 0} = -\frac{\rho}{\varepsilon_0},$$

and (3.18) follows directly from Eq. (3.15) due to

$$\nabla^2 \mathbf{A} - \mu_0 \varepsilon_0 \frac{\partial^2 \mathbf{A}}{\partial t^2} - \nabla \Big(\underbrace{\nabla \cdot \mathbf{A}}_{\overset{(3.16)}{=} 0} + \mu_0 \varepsilon_0 \frac{\partial \phi_s}{\partial t} \Big) = -\mu_0 \mathbf{J}.$$

■

In the Coulomb gauge, Eq. (3.18) still looks complicated. Therefore, in Lorenz gauge, Eq. (3.14) and Eq. (3.15) are simplified in such a way, that both equations are symmetric:

Corollary 3.2 (Lorenz gauge of the vector and scalar potentials) *The vector potential $\mathbf{A}(\mathbf{r}, t)$ and the scalar potential $\phi_s(\mathbf{r}, t)$ are chosen in the way that*

$$\nabla \cdot \mathbf{A} = -\mu_0 \varepsilon_0 \frac{\partial \phi_s}{\partial t}. \tag{3.19}$$

Due to this, one obtains $\mathbf{A}(\mathbf{r}, t)$ *and* $\phi_s(\mathbf{r}, t)$ *from*

$$\nabla^2 \phi_s - \mu_0 \, \varepsilon_0 \, \frac{\partial^2 \phi_s}{\partial t^2} = -\frac{1}{\varepsilon_0} \, \rho, \tag{3.20}$$

$$\nabla^2 \mathbf{A} - \mu_0 \, \varepsilon_0 \, \frac{\partial^2 \mathbf{A}}{\partial t^2} = -\mu_0 \, \mathbf{J}. \tag{3.21}$$

Proof Equation (3.20) follows directly from Eq. (3.14) due to

$$\nabla^2 \phi_s + \underbrace{\frac{\partial}{\partial t} (\nabla \cdot \mathbf{A})}_{\overset{(3.19)}{=} -\mu_0 \, \varepsilon_0 \, \frac{\partial^2 \phi_s}{\partial t^2}} = -\frac{\rho}{\varepsilon_0},$$

and Eq. (3.21) follows directly from Eq. (3.15) due to

$$\nabla^2 \mathbf{A} - \mu_0 \, \varepsilon_0 \, \frac{\partial^2 \mathbf{A}}{\partial t^2} - \nabla \underbrace{\left(\nabla \cdot \mathbf{A} + \mu_0 \, \varepsilon_0 \, \frac{\partial \phi_s}{\partial t} \right)}_{\overset{(3.19)}{=} 0} = -\mu_0 \, \mathbf{J}.$$

■

The Lorenz gauge allows to derive general analytical integral equations for ϕ_s and $\mathbf{A}$:

Theorem 3.3 (Integral equations of the vector and scalar potentials) *In the Lorenz gauge, the vector potential* $\mathbf{A}(\mathbf{r}, t)$ *and the scalar potential* $\phi_s(\mathbf{r}, t)$ *can be calculated from the charge density* $\rho(\mathbf{r}, t)$ *and the charge current density* $\mathbf{J}(\mathbf{r}, t)$ *using the retarded time*

$$t_r = t - \frac{|\mathbf{r} - \mathbf{r}'|}{\mathbb{c}} \tag{3.22}$$

by

$$\phi_s(\mathbf{r}, t) = \frac{1}{4\pi \, \varepsilon_0} \int\limits_{\mathbb{R}^3} d^3 r' \, \frac{\rho(\mathbf{r}', t_r)}{|\mathbf{r} - \mathbf{r}'|}, \tag{3.23}$$

$$\mathbf{A}(\mathbf{r}, t) = \frac{\mu_0}{4\pi} \int\limits_{\mathbb{R}^3} d^3 r' \, \frac{\mathbf{J}(\mathbf{r}', t_r)}{|\mathbf{r} - \mathbf{r}'|}. \tag{3.24}$$

Proof *First, we show that the scalar potential from Eq. (3.23) and vector potential from Eq. (3.24) fullfill the Lorenz gauge Eq. (3.19):*

We have $\mathbf{r} = [x, y, z]^{\mathrm{t}}$ and $\mathbf{r}' = [x', y', z']^{\mathrm{t}}$ and get for the retarded time

$$\frac{\partial t_{\mathrm{r}}}{\partial x} = -\frac{1}{\mathbb{c}} \frac{\partial |\mathbf{r}-\mathbf{r}'|}{\partial x}, \quad \frac{\partial t_{\mathrm{r}}}{\partial y} = -\frac{1}{\mathbb{c}} \frac{\partial |\mathbf{r}-\mathbf{r}'|}{\partial y}, \quad \frac{\partial t_{\mathrm{r}}}{\partial z} = -\frac{1}{\mathbb{c}} \frac{\partial |\mathbf{r}-\mathbf{r}'|}{\partial z}. \tag{3.25}$$

The product rule says

$$\nabla_{\mathbf{r}} \cdot \left(\frac{\mathbf{J}(\mathbf{r}', t_{\mathrm{r}})}{|\mathbf{r}-\mathbf{r}'|} \right) = \frac{1}{|\mathbf{r}-\mathbf{r}'|} \left(\nabla_{\mathbf{r}} \cdot \mathbf{J}(\mathbf{r}', t_{\mathrm{r}}) \right) + \mathbf{J}(\mathbf{r}', t_{\mathrm{r}}) \cdot \left(\nabla_{\mathbf{r}} \frac{1}{|\mathbf{r}-\mathbf{r}'|} \right) \tag{3.26}$$

and

$$\begin{aligned} \nabla_{\mathbf{r}'} \cdot \left(\frac{\mathbf{J}(\mathbf{r}', t_{\mathrm{r}})}{|\mathbf{r}-\mathbf{r}'|} \right) &= \frac{1}{|\mathbf{r}-\mathbf{r}'|} \left(\nabla_{\mathbf{r}'} \cdot \mathbf{J}(\mathbf{r}', t_{\mathrm{r}}) \right) \\ &\quad + \mathbf{J}(\mathbf{r}', t_{\mathrm{r}}) \cdot \left(\nabla_{\mathbf{r}'} \frac{1}{|\mathbf{r}-\mathbf{r}'|} \right), \end{aligned} \tag{3.27}$$

$$\Leftrightarrow \quad -\mathbf{J}(\mathbf{r}', t_{\mathrm{r}}) \cdot \left(\nabla_{\mathbf{r}'} \frac{1}{|\mathbf{r}-\mathbf{r}'|} \right) = \frac{1}{|\mathbf{r}-\mathbf{r}'|} \left(\nabla_{\mathbf{r}'} \cdot \mathbf{J}(\mathbf{r}', t_{\mathrm{r}}) \right) - \nabla_{\mathbf{r}'} \cdot \left(\frac{\mathbf{J}(\mathbf{r}', t_{\mathrm{r}})}{|\mathbf{r}-\mathbf{r}'|} \right). \tag{3.28}$$

Since

$$\nabla_{\mathbf{r}} \frac{1}{|\mathbf{r}-\mathbf{r}'|} = -\nabla_{\mathbf{r}'} \frac{1}{|\mathbf{r}-\mathbf{r}'|} \tag{3.29}$$

is valid, we obtain from Eq. (3.26)

$$\begin{aligned} \nabla_{\mathbf{r}} \cdot \left(\frac{\mathbf{J}(\mathbf{r}', t_{\mathrm{r}})}{|\mathbf{r}-\mathbf{r}'|} \right) &= \frac{1}{|\mathbf{r}-\mathbf{r}'|} \left(\nabla_{\mathbf{r}} \cdot \mathbf{J}(\mathbf{r}', t_{\mathrm{r}}) \right) - \mathbf{J}(\mathbf{r}', t_{\mathrm{r}}) \cdot \left(\nabla_{\mathbf{r}'} \frac{1}{|\mathbf{r}-\mathbf{r}'|} \right) \\ &\overset{(3.28)}{=} \frac{1}{|\mathbf{r}-\mathbf{r}'|} \left(\nabla_{\mathbf{r}} \cdot \mathbf{J}(\mathbf{r}', t_{\mathrm{r}}) \right) + \frac{1}{|\mathbf{r}-\mathbf{r}'|} \left(\nabla_{\mathbf{r}'} \cdot \mathbf{J}(\mathbf{r}', t_{\mathrm{r}}) \right) \\ &\quad - \nabla_{\mathbf{r}'} \cdot \left(\frac{\mathbf{J}(\mathbf{r}', t_{\mathrm{r}})}{|\mathbf{r}-\mathbf{r}'|} \right). \end{aligned} \tag{3.30}$$

Furthermore, we obtain

$$\begin{aligned} \nabla_{\mathbf{r}} \cdot \mathbf{J}(\mathbf{r}', t_{\mathrm{r}}) &= \frac{dJ_x(\mathbf{r}', t_{\mathrm{r}})}{dx} + \frac{dJ_y(\mathbf{r}', t_{\mathrm{r}})}{dy} + \frac{dJ_z(\mathbf{r}', t_{\mathrm{r}})}{dz} \\ &= \frac{\partial J_x(\mathbf{r}', t_{\mathrm{r}})}{\partial t_{\mathrm{r}}} \frac{\partial t_{\mathrm{r}}}{\partial x} + \frac{\partial J_y(\mathbf{r}', t_{\mathrm{r}})}{\partial t_{\mathrm{r}}} \frac{\partial t_{\mathrm{r}}}{\partial y} + \frac{\partial J_z(\mathbf{r}', t_{\mathrm{r}})}{\partial t_{\mathrm{r}}} \frac{\partial t_{\mathrm{r}}}{\partial z} \\ &\overset{(3.25)}{=} -\frac{1}{\mathbb{c}} \left(\frac{\partial J_x(\mathbf{r}', t_{\mathrm{r}})}{\partial t_{\mathrm{r}}} \frac{\partial |\mathbf{r}-\mathbf{r}'|}{\partial x} + \frac{\partial J_y(\mathbf{r}', t_{\mathrm{r}})}{\partial t_{\mathrm{r}}} \frac{\partial |\mathbf{r}-\mathbf{r}'|}{\partial y} \right. \\ &\qquad \left. + \frac{\partial J_z(\mathbf{r}', t_{\mathrm{r}})}{\partial t_{\mathrm{r}}} \frac{\partial |\mathbf{r}-\mathbf{r}'|}{\partial z} \right) \end{aligned}$$

$$= -\frac{1}{\mathbb{c}}\frac{\partial \mathbf{J}(\mathbf{r}', t_r)}{\partial t_r} \cdot \left(\nabla_{\mathbf{r}} |\mathbf{r} - \mathbf{r}'|\right). \tag{3.31}$$

Analogously, we have

$$\begin{aligned}\nabla_{\mathbf{r}'} \cdot \mathbf{J}(\mathbf{r}', t_r) &= \frac{\partial J_x(\mathbf{r}', t_r)}{\partial x'} + \frac{\partial J_y(\mathbf{r}', t_r)}{\partial y'} + \frac{\partial J_z(\mathbf{r}', t_r)}{\partial z'} \\ &\quad + \frac{\partial J_x(\mathbf{r}', t_r)}{\partial t_r}\frac{\partial t_r}{\partial x'} + \frac{\partial J_y(\mathbf{r}', t_r)}{\partial t_r}\frac{\partial t_r}{\partial y'} + \frac{\partial J_z(\mathbf{r}', t_r)}{\partial t_r}\frac{\partial t_r}{\partial z'} \\ &\overset{(3.4)}{=} -\dot{\rho}(\mathbf{r}', t_r) - \frac{1}{\mathbb{c}}\frac{\partial \mathbf{J}(\mathbf{r}', t_r)}{\partial t_r} \cdot \left(\nabla_{\mathbf{r}'} |\mathbf{r} - \mathbf{r}'|\right),\end{aligned} \tag{3.32}$$

where we use the continuity equation (3.4) for the three first terms. If we utilize

$$\nabla_{\mathbf{r}} |\mathbf{r} - \mathbf{r}'| = -\nabla_{\mathbf{r}'} |\mathbf{r} - \mathbf{r}'| \tag{3.33}$$

we get further

$$\begin{aligned}\nabla_{\mathbf{r}} \cdot \left(\frac{\mathbf{J}(\mathbf{r}', t_r)}{|\mathbf{r} - \mathbf{r}'|}\right) &\overset{(3.30)}{=} \frac{1}{|\mathbf{r} - \mathbf{r}'|}\left(\nabla_{\mathbf{r}} \cdot \mathbf{J}(\mathbf{r}', t_r)\right) + \frac{1}{|\mathbf{r} - \mathbf{r}'|}\left(\nabla_{\mathbf{r}'} \cdot \mathbf{J}(\mathbf{r}', t_r)\right) \\ &\quad - \nabla_{\mathbf{r}'} \cdot \left(\frac{\mathbf{J}(\mathbf{r}', t_r)}{|\mathbf{r} - \mathbf{r}'|}\right) \\ &= \frac{1}{|\mathbf{r} - \mathbf{r}'|}\left(-\frac{1}{c}\frac{\partial \mathbf{J}(\mathbf{r}', t_r)}{\partial t_r} \cdot \left(\nabla_{\mathbf{r}} |\mathbf{r} - \mathbf{r}'|\right)\right) \\ &\quad + \frac{1}{|\mathbf{r} - \mathbf{r}'|}\left(-\dot{\rho}(\mathbf{r}', t_r) - \frac{1}{c}\frac{\partial \mathbf{J}(\mathbf{r}', t_r)}{\partial t_r} \cdot \left(\nabla_{\mathbf{r}'} |\mathbf{r} - \mathbf{r}'|\right)\right) \\ &\quad - \nabla_{\mathbf{r}'} \cdot \left(\frac{\mathbf{J}(\mathbf{r}', t_r)}{|\mathbf{r} - \mathbf{r}'|}\right) \\ &= -\frac{1}{|\mathbf{r} - \mathbf{r}'|}\dot{\rho}(\mathbf{r}', t_r) - \nabla_{\mathbf{r}'} \cdot \left(\frac{\mathbf{J}(\mathbf{r}', t_r)}{|\mathbf{r} - \mathbf{r}'|}\right).\end{aligned} \tag{3.34}$$

Therefore, we obtain

$$\begin{aligned}\nabla_{\mathbf{r}} \cdot \mathbf{A}(\mathbf{r}, t) &\overset{(3.24)}{=} \frac{\mu_0}{4\pi}\int_{\mathbb{R}^3} d^3r'\, \nabla_{\mathbf{r}} \cdot \left(\frac{\mathbf{J}(\mathbf{r}', t_r)}{|\mathbf{r} - \mathbf{r}'|}\right) \\ &= \frac{\mu_0}{4\pi}\left(-\int_{\mathbb{R}^3} d^3r'\, \frac{\dot{\rho}(\mathbf{r}', t_r)}{|\mathbf{r} - \mathbf{r}'|} - \int_{\mathbb{R}^3} d^3r'\, \nabla_{\mathbf{r}'} \cdot \left(\frac{\mathbf{J}(\mathbf{r}', t_r)}{|\mathbf{r} - \mathbf{r}'|}\right)\right)\end{aligned}$$

$$= -\mu_0\,\varepsilon_0\,\frac{\partial}{\partial t}\left(\frac{1}{4\pi\,\varepsilon_0}\int\limits_{\mathbb{R}^3} d^3r'\,\frac{\rho(\mathbf{r}',t_\mathrm{r})}{|\mathbf{r}-\mathbf{r}'|}\right) - \underbrace{\frac{\mu_0}{4\pi}\int\limits_{\mathbb{S}(\mathbb{R}^3)} d\mathbf{a}'^{\mathrm{t}}\cdot\frac{\mathbf{J}(\mathbf{r}',t_\mathrm{r})}{|\mathbf{r}-\mathbf{r}'|}}_{=0}$$

$$\overset{(3.23)}{=} -\mu_0\,\varepsilon_0\,\frac{\partial\phi_\mathrm{s}(\mathbf{r},t)}{\partial t} \tag{3.35}$$

and the condition (3.19) of the Lorenz gauge is proven. Here, we applied the divergence theorem and utilized that $\mathbf{J}=0$ is valid on the surface at infinity $\mathbb{S}(\mathbb{R}^3)$.

Now we prove that $\phi_\mathrm{s}(\mathbf{r},t)$ from Eq. (3.23) *fulfills Eq.* (3.20).

In the following, we just write $\nabla\cdot$ instead of $\nabla_\mathbf{r}\cdot$ and ∇ instead of $\nabla_\mathbf{r}$, since we are only dealing with derivatives with respect to $\mathbf{r}$. We have

$$\nabla\,t_\mathrm{r} = \nabla\left(t-\frac{|\mathbf{r}-\mathbf{r}'|}{\mathbb{c}}\right) = -\frac{1}{\mathbb{c}}\,\nabla\,|\mathbf{r}-\mathbf{r}'| = -\frac{1}{\mathbb{c}}\,\frac{\mathbf{r}-\mathbf{r}'}{|\mathbf{r}-\mathbf{r}'|} \tag{3.36}$$

and, therefore,

$$\nabla\rho(\mathbf{r}',t_\mathrm{r}) = \dot{\rho}(\mathbf{r}',t_\mathrm{r})\,\nabla t_\mathrm{r} \overset{(3.36)}{=} -\frac{\dot{\rho}(\mathbf{r}',t_\mathrm{r})}{\mathbb{c}}\,\frac{\mathbf{r}-\mathbf{r}'}{|\mathbf{r}-\mathbf{r}'|}. \tag{3.37}$$

Performing the time derivative, we obtain

$$\nabla\dot{\rho}(\mathbf{r}',t_\mathrm{r}) = -\frac{\ddot{\rho}(\mathbf{r}',t_\mathrm{r})}{\mathbb{c}}\,\frac{\mathbf{r}-\mathbf{r}'}{|\mathbf{r}-\mathbf{r}'|}. \tag{3.38}$$

Using

$$\nabla\left(\frac{1}{|\mathbf{r}-\mathbf{r}'|}\right) = -\frac{\mathbf{r}-\mathbf{r}'}{|\mathbf{r}-\mathbf{r}'|^3}, \tag{3.39}$$

we get

$$\begin{aligned}
\nabla\phi_\mathrm{s}(\mathbf{r},t) &\overset{(3.23)}{=} \frac{1}{4\pi\,\varepsilon_0}\int\limits_{\mathbb{R}^3} d^3r'\,\nabla\left(\frac{\rho(\mathbf{r}',t_\mathrm{r})}{|\mathbf{r}-\mathbf{r}'|}\right)\\
&= \frac{1}{4\pi\,\varepsilon_0}\int\limits_{\mathbb{R}^3} d^3r'\left(\left(\nabla\rho(\mathbf{r}',t_\mathrm{r})\right)\frac{1}{|\mathbf{r}-\mathbf{r}'|} + \rho(\mathbf{r}',t_\mathrm{r})\,\nabla\left(\frac{1}{|\mathbf{r}-\mathbf{r}'|}\right)\right)\\
&\overset{(3.37)}{=} \frac{1}{4\pi\,\varepsilon_0}\int\limits_{\mathbb{R}^3} d^3r'\left(-\frac{\dot{\rho}(\mathbf{r}',t_\mathrm{r})}{\mathbb{c}}\,\frac{\mathbf{r}-\mathbf{r}'}{|\mathbf{r}-\mathbf{r}'|^2} - \rho(\mathbf{r}',t_\mathrm{r})\,\frac{\mathbf{r}-\mathbf{r}'}{|\mathbf{r}-\mathbf{r}'|^3}\right).
\end{aligned} \tag{3.40}$$

Furthermore, we have

$$\nabla \cdot \left(\frac{\mathbf{r} - \mathbf{r}'}{|\mathbf{r} - \mathbf{r}'|^2} \right) = \frac{1}{|\mathbf{r} - \mathbf{r}'|^2} \tag{3.41}$$

and

$$\nabla \cdot \left(\frac{\mathbf{r} - \mathbf{r}'}{|\mathbf{r} - \mathbf{r}'|^3} \right) = 4\pi\, \delta^{(3)}(\mathbf{r} - \mathbf{r}'). \tag{3.42}$$

Here, $\delta^{(3)}(\mathbf{r})$ denotes the three-dimensional delta function of Dirac. Using this and $\frac{1}{c^2} = \mu_0\, \varepsilon_0$, we obtain

$$\begin{aligned}
\nabla^2 \phi_s(\mathbf{r}, t) &= \nabla \cdot \Big(\nabla \phi_s(\mathbf{r}, t) \Big) \\
&\overset{(3.40)}{=} -\frac{1}{4\pi\varepsilon_0} \int_{\mathbb{R}^3} d^3r' \left(\frac{1}{\mathbb{c}} \frac{\mathbf{r} - \mathbf{r}'}{|\mathbf{r} - \mathbf{r}'|^2} \cdot \Big(\nabla \dot{\rho}(\mathbf{r}', t_r) \Big) + \frac{\dot{\rho}(\mathbf{r}', t_r)}{\mathbb{c}} \nabla \cdot \left(\frac{\mathbf{r} - \mathbf{r}'}{|\mathbf{r} - \mathbf{r}'|^2} \right) \right) \\
&\quad - \frac{1}{4\pi\varepsilon_0} \int_{\mathbb{R}^3} d^3r' \left(\frac{\mathbf{r} - \mathbf{r}'}{|\mathbf{r} - \mathbf{r}'|^3} \cdot \Big(\nabla \rho(\mathbf{r}', t_r) \Big) + \rho(\mathbf{r}', t_r) \nabla \cdot \left(\frac{\mathbf{r} - \mathbf{r}'}{|\mathbf{r} - \mathbf{r}'|^3} \right) \right) \\
&= -\frac{1}{4\pi\varepsilon_0} \int_{\mathbb{R}^3} d^3r' \left(-\frac{\ddot{\rho}(\mathbf{r}', t_r)}{\mathbb{c}^2} \frac{1}{|\mathbf{r} - \mathbf{r}'|} + \frac{\dot{\rho}(\mathbf{r}', t_r)}{\mathbb{c}} \frac{1}{|\mathbf{r} - \mathbf{r}'|^2} \right) \\
&\quad - \frac{1}{4\pi\varepsilon_0} \int_{\mathbb{R}^3} d^3r' \left(-\frac{\dot{\rho}(\mathbf{r}', t_r)}{\mathbb{c}} \frac{1}{|\mathbf{r} - \mathbf{r}'|^2} + \rho(\mathbf{r}', t_r)\, 4\pi\, \delta^{(3)}(\mathbf{r} - \mathbf{r}') \right) \\
&= \frac{1}{\mathbb{c}^2} \frac{\partial^2}{\partial t^2} \left(\frac{1}{4\pi\varepsilon_0} \int_{\mathbb{R}^3} d^3r' \frac{\rho(\mathbf{r}', t_r)}{|\mathbf{r} - \mathbf{r}'|} \right) - \frac{1}{\varepsilon_0} \rho(\mathbf{r}, t_r) \\
&\overset{(3.23)}{=} \mu_0\, \varepsilon_0 \frac{\partial^2 \phi_s(\mathbf{r}, t)}{\partial t^2} - \frac{1}{\varepsilon_0} \rho(\mathbf{r}, t_r),
\end{aligned} \tag{3.43}$$

which proves that $\phi_s(\mathbf{r}, t)$ from (3.23) fulfills Eq. (3.20).

Analogously, the prove can be done for the components of $\mathbf{A}(\mathbf{r}, t)$, *since the equations* (3.20) *and* (3.21) *have the same form / structure.* ■

Equations (3.23) and (3.24) reflect that the fields propagate at the speed of light $\mathbb{c}$. The charge density ρ at point $\mathbf{r}'$ and retarded time $t_r = t - \frac{|\mathbf{r}-\mathbf{r}'|}{\mathbb{c}}$ is responsible for the scalar field ϕ_s at point $\mathbf{r}$ and time t, since the field needs the time $\frac{|\mathbf{r}-\mathbf{r}'|}{\mathbb{c}}$ to propagate from the point $\mathbf{r}'$ to the point $\mathbf{r}$, where the scalar field is considered. The charge density from the past affects the scalar field at the current time. The same holds for the charge current density $\mathbf{J}$ and the vector potential $\mathbf{A}$.

Furthermore, one can even derive a direct solution of the Maxwell equations for the electric and magnetic fields from the integral equations of $\mathbf{A}$ and ϕ_s:

Theorem 3.4 (Integral equations of the electric and magnetic fields) *Using the retarded time $t_r = t - \frac{|\mathbf{r}-\mathbf{r}'|}{\mathbb{c}}$, one can directly calculate the electric and magnetic field from the charge density $\rho(\mathbf{r}, t)$ and the charge current density $\mathbf{J}(\mathbf{r}, t)$ by*

$$\mathbf{E}(\mathbf{r}, t) = \frac{1}{4\pi\varepsilon_0} \int_{\mathbb{R}^3} d^3r' \left(\frac{\rho(\mathbf{r}', t_r)}{|\mathbf{r}-\mathbf{r}'|^2} \frac{\mathbf{r}-\mathbf{r}'}{|\mathbf{r}-\mathbf{r}'|} + \frac{\dot{\rho}(\mathbf{r}', t_r)}{\mathbb{c}\,|\mathbf{r}-\mathbf{r}'|} \frac{\mathbf{r}-\mathbf{r}'}{|\mathbf{r}-\mathbf{r}'|} - \frac{\dot{\mathbf{J}}(\mathbf{r}', t_r)}{\mathbb{c}^2\,|\mathbf{r}-\mathbf{r}'|} \right), \tag{3.44}$$

$$\mathbf{B}(\mathbf{r}, t) = \frac{\mu_0}{4\pi} \int_{\mathbb{R}^3} d^3r' \left(\frac{\mathbf{J}(\mathbf{r}', t_r)}{|\mathbf{r}-\mathbf{r}'|^2} + \frac{\dot{\mathbf{J}}(\mathbf{r}', t_r)}{\mathbb{c}\,|\mathbf{r}-\mathbf{r}'|} \right) \times \frac{\mathbf{r}-\mathbf{r}'}{|\mathbf{r}-\mathbf{r}'|}. \tag{3.45}$$

Proof We utilize the Lorenz gauge, since the fields are independent of the used gauge. In the Lorenz gauge, we have the integral equations Eqs. (3.23) and (3.24) for ϕ_s and $\mathbf{A}$, respectively. We obtain the electric field just by

$$\mathbf{E}(\mathbf{r}, t) \overset{(3.13)}{=} -\nabla\phi_s(\mathbf{r}, t) - \frac{\partial \mathbf{A}(\mathbf{r}, t)}{\partial t}.$$

We already calculated $\nabla\phi_s$ in Eq. (3.40). For the second term, we have

$$-\frac{\partial \mathbf{A}(\mathbf{r}, t)}{\partial t} \overset{(3.24)}{=} -\frac{\mu_0}{4\pi} \int_{\mathbb{R}^3} d^3r' \frac{\dot{\mathbf{J}}(\mathbf{r}', t_r)}{|\mathbf{r}-\mathbf{r}'|} = -\frac{1}{4\pi\varepsilon_0} \int_{\mathbb{R}^3} d^3r' \frac{\dot{\mathbf{J}}(\mathbf{r}', t_r)}{\mathbb{c}^2\,|\mathbf{r}-\mathbf{r}'|}, \tag{3.46}$$

if we use $\mathbb{c}^2 = \frac{1}{\mu_0\,\varepsilon_0}$. This proves Eq. (3.44). We obtain further

$$\begin{aligned} \mathbf{B}(\mathbf{r}, t) &\overset{(3.12)}{=} \nabla \times \mathbf{A}(\mathbf{r}, t) \\ &\overset{3.24}{=} \frac{\mu_0}{4\pi} \int_{\mathbb{R}^3} d^3r' \,\nabla \times \left(\frac{\mathbf{J}(\mathbf{r}', t_r)}{|\mathbf{r}-\mathbf{r}'|} \right) \\ &= \frac{\mu_0}{4\pi} \int_{\mathbb{R}^3} d^3r' \left(\frac{1}{|\mathbf{r}-\mathbf{r}'|} \left(\nabla \times \mathbf{J}(\mathbf{r}', t_r) \right) - \mathbf{J}(\mathbf{r}', t_r) \times \nabla \left(\frac{1}{|\mathbf{r}-\mathbf{r}'|} \right) \right). \end{aligned} \tag{3.47}$$

We have

$$\begin{aligned}
\nabla \times \mathbf{J}(\mathbf{r}', t_\mathrm{r}) &= \begin{bmatrix} \frac{\partial J_z(\mathbf{r}', t_\mathrm{r})}{\partial y} - \frac{\partial J_y(\mathbf{r}', t_\mathrm{r})}{\partial z} \\ \frac{\partial J_x(\mathbf{r}', t_\mathrm{r})}{\partial z} - \frac{\partial J_z(\mathbf{r}', t_\mathrm{r})}{\partial x} \\ \frac{\partial J_y(\mathbf{r}', t_\mathrm{r})}{\partial x} - \frac{\partial J_x(\mathbf{r}', t_\mathrm{r})}{\partial y} \end{bmatrix} \\
&= \begin{bmatrix} \dot{j}_z \frac{\partial t_\mathrm{r}}{\partial y} - \dot{j}_y \frac{\partial t_\mathrm{r}}{\partial z} \\ \dot{j}_x \frac{\partial t_\mathrm{r}}{\partial z} - \dot{j}_z \frac{\partial t_\mathrm{r}}{\partial x} \\ \dot{j}_y \frac{\partial t_\mathrm{r}}{\partial x} - \dot{j}_x \frac{\partial t_\mathrm{r}}{\partial y} \end{bmatrix} \\
&= -\dot{\mathbf{J}}(\mathbf{r}', t_\mathrm{r}) \times \nabla t_\mathrm{r} \\
&\overset{3.36}{=} \frac{1}{\mathbb{C}} \dot{\mathbf{J}}(\mathbf{r}', t_\mathrm{r}) \times \frac{\mathbf{r} - \mathbf{r}'}{|\mathbf{r} - \mathbf{r}'|}.
\end{aligned} \tag{3.48}$$

Due to

$$\nabla \left(\frac{1}{|\mathbf{r} - \mathbf{r}'|} \right) = -\frac{\mathbf{r} - \mathbf{r}'}{|\mathbf{r} - \mathbf{r}'|^3},$$

we get

$$-\mathbf{J}(\mathbf{r}', t_\mathrm{r}) \times \nabla \left(\frac{1}{|\mathbf{r} - \mathbf{r}'|} \right) = \frac{\mathbf{J}(\mathbf{r}', t_\mathrm{r})}{|\mathbf{r} - \mathbf{r}'|^2} \times \frac{\mathbf{r} - \mathbf{r}'}{|\mathbf{r} - \mathbf{r}'|}. \tag{3.49}$$

Thus, Eq. (3.45) is proven. ∎

In practice, the direct calculation of the electric and magnetic fields from the last theorem is too complicated, so that one commonly derives the scalar and vector potentials and calculates the fields from them. In the static case, where

$$\rho(\mathbf{r}', t) \equiv \rho(\mathbf{r}'), \qquad \dot{\rho} = 0, \qquad \mathbf{J}(\mathbf{r}', t) \equiv \mathbf{J}(\mathbf{r}'), \qquad \dot{\mathbf{J}} = \mathbf{0}$$

are valid, Eq. (3.23) transforms to

$$\phi_\mathrm{s}(\mathbf{r}) = \frac{1}{4\pi\varepsilon_0} \int\limits_{\mathbb{R}^3} d^3r' \, \frac{\rho(\mathbf{r}')}{|\mathbf{r} - \mathbf{r}'|}, \tag{3.50}$$

which is the well-known **Coulomb potential**, Eq. (3.44) transforms to

$$\mathbf{E}(\mathbf{r}, t) = \frac{1}{4\pi\varepsilon_0} \int\limits_{\mathbb{R}^3} d^3r' \frac{\rho(\mathbf{r}')}{|\mathbf{r} - \mathbf{r}'|^2} \frac{\mathbf{r} - \mathbf{r}'}{|\mathbf{r} - \mathbf{r}'|}, \tag{3.51}$$

which is the well-known **Coulomb law**, and Eq. (3.45) transforms to

$$\mathbf{B}(\mathbf{r}, t) = \frac{\mu_0}{4\pi} \int\limits_{\mathbb{R}^3} d^3r' \, \frac{\mathbf{J}(\mathbf{r}')}{|\mathbf{r} - \mathbf{r}'|^2} \times \frac{\mathbf{r} - \mathbf{r}'}{|\mathbf{r} - \mathbf{r}'|}, \tag{3.52}$$

which is the well-known **Biot-Savart law**.

Now we consider a charge density and charge current density located in a finite Volume $\mathcal{V}$ around the origin. We are interested in the scalar field $\phi_s(\mathbf{r}, t)$ far away from $\mathcal{V}$, i.e., $|\mathbf{r}| = r >> 1$ or $\frac{1}{r} << 1$. If we perform a Taylor series approximation of the denominator in the integral of Eq. (3.23) in $\mathbf{r}'$, we obtain

$$\frac{1}{|\mathbf{r} - \mathbf{r}'|} = \frac{1}{r} + \frac{\hat{\mathbf{r}}^t}{r^2} \cdot \mathbf{r}' + O\left(\frac{1}{r^3}\right). \tag{3.53}$$

The missing terms go with $\frac{1}{r^3}$ or faster to zero, which we denote by $O\left(\frac{1}{r^3}\right)$. Using this approximation for Eq. (3.23) and

$$t_r = t - \frac{|\mathbf{r} - \mathbf{r}'|}{\mathbb{c}} \approx t - \frac{r}{\mathbb{c}} =: t_0, \tag{3.54}$$

we have

$$\begin{aligned} \phi_s(\mathbf{r}, t) &= \frac{1}{4\pi\varepsilon_0 r} \int_{\mathcal{V}} d^3r' \, \rho(\mathbf{r}', \underbrace{t_r}_{\approx t_0}) + \frac{\hat{\mathbf{r}}^t}{4\pi\varepsilon_0 r^2} \cdot \int_{\mathcal{V}} d^3r' \, \mathbf{r}' \rho(\mathbf{r}', \underbrace{t_r}_{\approx t_0}) + O\left(\frac{1}{r^3}\right) \\ &\approx \frac{1}{4\pi\varepsilon_0 r} \int_{\mathcal{V}} d^3r' \, \rho(\mathbf{r}', t_0) + \frac{\hat{\mathbf{r}}^t}{4\pi\varepsilon_0 r^2} \cdot \int_{\mathcal{V}} d^3r' \, \mathbf{r}' \rho(\mathbf{r}', t_0) + O\left(\frac{1}{r^3}\right). \end{aligned} \tag{3.55}$$

The first term is related to the total charge $\mathbb{q}_{tot}$ (*c.f.* Eq. (3.2)). The second term motivates to define so called **electric dipole moment**

$$\boxed{\mathbf{d}(t) := \int_{\mathcal{V}} d^3r' \, \mathbf{r}' \rho(\mathbf{r}', t).} \tag{3.56}$$

The electric dipole moment is a vector. If we have only two point charges with charge $\mathbb{q}$ and $-\mathbb{q}$ in the volume $\mathcal{V}$, the electric dipole moment is $\mathbb{q}$ times the vector from the negative to the positive charge. Using the definition of the total charge and the electric dipole moment, we obtain

$$\phi_s(\mathbf{r}, t) \approx \frac{\mathbb{q}_{tot}(t_0)}{4\pi\varepsilon_0 r} + \frac{\hat{\mathbf{r}}^t}{4\pi\varepsilon_0 r^2} \cdot \mathbf{d}(t_0) + O\left(\frac{1}{r^3}\right). \tag{3.57}$$

Therefore, at large distances to $\mathcal{V}$, the first term dominates, which is just characterized by the total charge $\mathbb{q}_{tot}$. This term corresponds to a point charge $\mathbb{q}_{tot}$ located at the origin in $\mathcal{V}$. If the total charge is zero, the dominating term at large distances is the second one, which is characterized by the electric dipole moment $\mathbf{d}$.

Next we consider the time derivative of the electric dipole moment:

$$\dot{\mathbf{d}}(t) \overset{(3.56)}{=} \frac{d}{dt}\int_{\mathcal{V}} d^3r\,\mathbf{r}\,\rho(\mathbf{r},t) = \int_{\mathcal{V}} d^3r\,\mathbf{r}\,\dot{\rho}(\mathbf{r},t) \overset{(3.4)}{=} -\int_{\mathcal{V}} d^3r\,\mathbf{r}\left(\nabla\cdot\mathbf{J}(\mathbf{r},t)\right). \tag{3.58}$$

Using the product rule we obtain

$$\begin{aligned} & \nabla\cdot\left(x\,\mathbf{J}\right) = x\left(\nabla\cdot\mathbf{J}\right) + \mathbf{J}\cdot\underbrace{(\nabla x)}_{=\mathbf{e}_x}, \\ \Leftrightarrow\quad & \nabla\cdot\left(x\,\mathbf{J}\right) = x\left(\nabla\cdot\mathbf{J}\right) + J_x, \\ \Leftrightarrow\quad & -x\left(\nabla\cdot\mathbf{J}\right) = -\nabla\cdot\left(x\,\mathbf{J}\right) + J_x, \end{aligned} \tag{3.59}$$

where $\mathbf{e}_x$ denotes the unit vector that points in x-direction. Integration over the volume $\mathcal{V}$ of the above equation yields

$$\begin{aligned} -\int_{\mathcal{V}} d^3r\,x\left(\nabla\cdot\mathbf{J}(\mathbf{r},t)\right) &= -\int_{\mathcal{V}} d^3r\,\nabla\cdot\left(x\,\mathbf{J}(\mathbf{r},t)\right) + \int_{\mathcal{V}} d^3r\,J_x(\mathbf{r},t) \\ &= -\int_{\mathbb{S}(\mathcal{V})} d\mathbf{a}^{\prime}\cdot\Big(x\,\underbrace{\mathbf{J}(\mathbf{r},t)}_{=\mathbf{0}}\Big) + \int_{\mathcal{V}} d^3r\,J_x(\mathbf{r},t) \\ &= \int_{\mathcal{V}} d^3r\,J_x(\mathbf{r},t). \end{aligned} \tag{3.60}$$

For the second identity, we applied the divergence theorem and, for the third identity, we utilized that the current $\mathbf{J}$ is entirely inside the volume $\mathcal{V}$, so that the current is zero on the surface $\mathbb{S}(\mathcal{V})$ and outside of $\mathcal{V}$. Since the analogous of the above transformations are valid for the y- and z-directions, we have

$$-\int_{\mathcal{V}} d^3r\,\mathbf{r}\left(\nabla\cdot\mathbf{J}(\mathbf{r},t)\right) = \int_{\mathcal{V}} d^3r\,\mathbf{J}(\mathbf{r},t). \tag{3.61}$$

Using Eq. (3.58), we obtain

$$\boxed{\dot{\mathbf{d}}(t) = \int_{\mathcal{V}} d^3r\,\mathbf{J}(\mathbf{r},t).} \tag{3.62}$$

In this way, the integral of the charge current density in the volume $\mathcal{V}$ correspond to the time derivative of the electric dipole moment.

3.1.2 Radiation of Electromagnetic Waves

Electromagnetic waves spread in space even in the absence of charges or currents of charges. In this subsection we describe how electromagnetic waves are initially generated.

Such a wave is only formed, if a time oscillating electric and magnetic field induce each other. A stationary charge cannot form a wave, since such a charge only generates an electric field that is constant in time. A charge that moves with a constant velocity is seen by other inertial frame as static. Since all inertial frames are equivalent, such a charge also cannot form a wave. Therefore, only accelerated charges can generate electromagnetic waves, or, with other words, radiate.

To show this, we consider accelerated charges, which do not leave a small volume $\mathcal{V}$ around the origin. We calculate approximately the electric field $\mathbf{E}(\mathbf{r}, t)$ and the magnetic field $\mathbf{B}(\mathbf{r}, t)$ far away from $\mathcal{V}$ in the so called radiation zone, *i.e.*, $|\mathbf{r}| = r >> 1$. For this, we utilize the general integral equations of the fields in contrast to Ref. [9], where the vector and scalar potentials are calculated and from them the fields. Such a solution needs further assumptions and approximations. The accelerated charges are described by a time-dependent charge density $\rho(\mathbf{r}, t)$ and time-dependent charge current density $\mathbf{J}(\mathbf{r}, t)$ located in $\mathcal{V}$. We obtain for the magnetic field from the general integral Eq. (3.45)

$$\mathbf{B}(\mathbf{r}, t) = \frac{\mu_0}{4\pi} \int\limits_{\mathcal{V}} d^3r' \left(\frac{\mathbf{J}(\mathbf{r}', t_\mathrm{r})}{|\mathbf{r} - \mathbf{r}'|^2} + \frac{\dot{\mathbf{J}}(\mathbf{r}', t_\mathrm{r})}{\mathbb{c}\,|\mathbf{r} - \mathbf{r}'|} \right) \times \frac{\mathbf{r} - \mathbf{r}'}{|\mathbf{r} - \mathbf{r}'|}, \tag{3.63}$$

because the currents are completely located in $\mathcal{V}$. Since $\mathbf{r}' \in \mathcal{V}$, $\mathcal{V}$ is a small volume and we consider $|\mathbf{r}| >> 1$, we have

$$\mathbf{r} - \mathbf{r}' \approx \mathbf{r} \tag{3.64}$$

and obtain

$$t_\mathrm{r} \overset{(3.22)}{=} t - \frac{|\mathbf{r} - \mathbf{r}'|}{\mathbb{c}} \approx t - \frac{r}{\mathbb{c}} =: t_0 \tag{3.65}$$

for the retarded time. Using $|\mathbf{r} - \mathbf{r}'| \approx r$ and neglecting the term that decays with $\frac{1}{r^2}$, we can simplify Eq. (3.63) for the magnetic field

$$\begin{aligned} \mathbf{B}(\mathbf{r}, t) &\approx \frac{\mu_0}{4\pi} \int\limits_{\mathcal{V}} d^3r' \left(\frac{\mathbf{J}(\mathbf{r}', t_\mathrm{r})}{r^2} + \frac{\dot{\mathbf{J}}(\mathbf{r}', t_\mathrm{r})}{\mathbb{c}\,r} \right) \times \hat{\mathbf{r}} \\ &\approx \frac{\mu_0}{4\pi\,\mathbb{c}\,r} \int\limits_{\mathcal{V}} d^3r'\, \dot{\mathbf{J}}(\mathbf{r}', \underbrace{t_\mathrm{r}}_{\approx t_0}) \times \hat{\mathbf{r}} \end{aligned}$$

$$\approx -\frac{\mu_0}{4\pi\,\mathbb{c}\,r}\,\hat{\mathbf{r}}\times\underbrace{\int_{\mathcal{V}} d^3r'\,\dot{\mathbf{J}}(\mathbf{r}',t_0)}_{\overset{(3.62)}{=}\ddot{\mathbf{d}}(t_0)}$$

and obtain finally

$$\boxed{\mathbf{B}(\mathbf{r},t)\approx -\frac{\mu_0}{4\pi\,\mathbb{c}\,r}\,\hat{\mathbf{r}}\times\ddot{\mathbf{d}}(t_0)} \tag{3.66}$$

Performing similar approximations, we obtain for the electric field from the general integral Eq. (3.44):

$$\begin{aligned}\mathbf{E}(\mathbf{r},t) &\overset{(3.44)}{=} \frac{1}{4\pi\varepsilon_0}\int_{\mathcal{V}} d^3r'\left(\frac{\rho(\mathbf{r}',t_r)}{|\mathbf{r}-\mathbf{r}'|^2}\,\frac{\mathbf{r}-\mathbf{r}'}{|\mathbf{r}-\mathbf{r}'|}+\frac{\dot{\rho}(\mathbf{r}',t_r)}{\mathbb{c}\,|\mathbf{r}-\mathbf{r}'|}\,\frac{\mathbf{r}-\mathbf{r}'}{|\mathbf{r}-\mathbf{r}'|}-\frac{\dot{\mathbf{J}}(\mathbf{r}',t_r)}{\mathbb{c}^2\,|\mathbf{r}-\mathbf{r}'|}\right)\\ &\approx \frac{1}{4\pi\varepsilon_0}\int_{\mathcal{V}} d^3r'\left(\frac{\rho(\mathbf{r}',t_r)}{r^2}\,\hat{\mathbf{r}}+\frac{\dot{\rho}(\mathbf{r}',t_r)}{\mathbb{c}\,r}\,\hat{\mathbf{r}}-\frac{\dot{\mathbf{J}}(\mathbf{r}',t_r)}{\mathbb{c}^2\,r}\right)\\ &\approx \frac{\hat{\mathbf{r}}}{4\pi\varepsilon_0\,\mathbb{c}\,r}\int_{\mathcal{V}} d^3r'\,\dot{\rho}(\mathbf{r}',t_r)-\frac{1}{4\pi\varepsilon_0\,\mathbb{c}^2\,r}\int_{\mathcal{V}} d^3r'\,\dot{\mathbf{J}}(\mathbf{r}',t_r).\end{aligned} \tag{3.67}$$

To simplify further, we want to write the first integral with the help of the charge current density. For this, we utilize the continuity equation for the charge density, take into account that the retarded time t_r also depend on $\mathbf{r}'$ and obtain from Eq. (3.32):

$$\dot{\rho}(\mathbf{r}',t_r)=-\nabla_{\mathbf{r}'}\cdot\mathbf{J}(\mathbf{r}',t_r)-\frac{1}{\mathbb{c}}\,\dot{\mathbf{J}}^{t}(\mathbf{r}',t_r)\cdot\Big(\nabla_{\mathbf{r}'}\,|\mathbf{r}-\mathbf{r}'|\Big).$$

If we also utilize the approximation

$$\nabla_{\mathbf{r}'}\,|\mathbf{r}-\mathbf{r}'|=-\frac{\mathbf{r}-\mathbf{r}'}{|\mathbf{r}-\mathbf{r}'|}\approx-\hat{\mathbf{r}},$$

we have

$$\dot{\rho}(\mathbf{r}',t_r)=-\nabla_{\mathbf{r}'}\cdot\mathbf{J}(\mathbf{r}',t_r)+\frac{1}{\mathbb{c}}\,\dot{\mathbf{J}}^{t}(\mathbf{r}',t_r)\cdot\hat{\mathbf{r}}$$

and get from Eq. (3.67):

$$\begin{aligned}\mathbf{E}(\mathbf{r},t)\approx&-\frac{\hat{\mathbf{r}}}{4\pi\varepsilon_0\,\mathbb{c}\,r}\int_{\mathcal{V}} d^3r'\,\nabla_{\mathbf{r}'}\cdot\mathbf{J}(\mathbf{r}',t_r)+\frac{\hat{\mathbf{r}}}{4\pi\varepsilon_0\,\mathbb{c}^2\,r}\int_{\mathcal{V}} d^3r'\,\dot{\mathbf{J}}^{t}(\mathbf{r}',\underbrace{t_r}_{\approx t_0})\cdot\hat{\mathbf{r}}\\ &-\frac{1}{4\pi\varepsilon_0\,\mathbb{c}^2\,r}\int_{\mathcal{V}} d^3r'\,\dot{\mathbf{J}}(\mathbf{r}',\underbrace{t_r}_{\approx t_0})\end{aligned}$$

$$\approx -\frac{\hat{\mathbf{r}}}{4\pi\,\varepsilon_0\,\mathbb{c}\,r}\int\limits_{\mathbb{S}(\mathcal{V})} d\mathbf{a}^{\mathrm{t}}\cdot\underbrace{\mathbf{J}(\mathbf{r}',t_{\mathrm{r}})}_{=\mathbf{0}}+\frac{\hat{\mathbf{r}}}{4\pi\,\varepsilon_0\,\mathbb{c}^2\,r}\underbrace{\int\limits_{\mathcal{V}} d^3r'\,\dot{\mathbf{J}}^{\mathrm{t}}(\mathbf{r}',t_0)}_{\overset{(3.62)}{=}\ddot{\mathbf{d}}^{\mathrm{t}}(t_0)}\cdot\hat{\mathbf{r}}$$

$$-\frac{1}{4\pi\,\varepsilon_0\,\mathbb{c}^2\,r}\underbrace{\int\limits_{\mathcal{V}} d^3r'\,\dot{\mathbf{J}}(\mathbf{r}',t_0)}_{\overset{(3.62)}{=}\ddot{\mathbf{d}}(t_0)}$$

$$\approx \frac{1}{4\pi\,\varepsilon_0\,\mathbb{c}^2\,r}\,\hat{\mathbf{r}}\left(\ddot{\mathbf{d}}^{\mathrm{t}}(t_0)\cdot\hat{\mathbf{r}}\right)-\frac{1}{4\pi\,\varepsilon_0\,\mathbb{c}^2\,r}\,\ddot{\mathbf{d}}(t_0)$$

$$=\frac{\mu_0}{4\pi\,r}\left(\hat{\mathbf{r}}\left(\hat{\mathbf{r}}^{\mathrm{t}}\cdot\ddot{\mathbf{d}}(t_0)\right)-\ddot{\mathbf{d}}(t_0)\right). \tag{3.68}$$

Here, we used the divergence theorem, the identity $\mathbb{c}^2 \overset{(3.10)}{=} \frac{1}{\mu_0\,\varepsilon_0}$ and that the currents are completely contained in the volume $\mathcal{V}$. Taking the vector identity

$$\hat{\mathbf{r}}\times\left(\hat{\mathbf{r}}\times\ddot{\mathbf{d}}\right)=\hat{\mathbf{r}}\left(\hat{\mathbf{r}}^{\mathrm{t}}\cdot\ddot{\mathbf{d}}\right)-\ddot{\mathbf{d}}\underbrace{\left(\hat{\mathbf{r}}^{\mathrm{t}}\cdot\hat{\mathbf{r}}\right)}_{=1}$$

into account, we obtain finally

$$\boxed{\mathbf{E}(\mathbf{r},t)\approx\frac{\mu_0}{4\pi\,r}\left(\hat{\mathbf{r}}\times\left(\hat{\mathbf{r}}\times\ddot{\mathbf{d}}(\mathbf{r},t_0)\right)\right).} \tag{3.69}$$

The electric and magnetic fields of the electromagnetic wave are perpendicular to each other, obey $E/B = \mathbb{c}$ and decay with $\frac{1}{r}$ with respect to the source. There is no radiation of electromagnetic fields in or against the direction of the oscillating dipole moment $\ddot{\mathbf{d}}$, whereas the radiation is most efficient in the directions perpendicular to $\ddot{\mathbf{d}}$.

3.1.3 Energy in Electromagnetic Fields

The electromagnetic fields $\mathbf{E}$ and $\mathbf{B}$ act with the following force $\mathbf{f}$ on a charge $\mathbb{q}$ with velocity $\dot{\mathbf{r}}$:

$$\mathbf{f}=\mathbb{q}\left(\mathbf{E}-\dot{\mathbf{r}}\times\mathbf{B}\right). \tag{3.70}$$

Since this force moves the charge, the electromagnetic fields perform work on the charge. If the charge moves by the infinitesimal distance $d\mathbf{l} = \dot{\mathbf{r}}\,dt$, the infinitesimal work

$$
\begin{aligned}
dW &= \mathbf{f}^{\mathrm{t}} \cdot d\mathbf{l} \\
&\overset{(3.70)}{=} \mathfrak{q} \left(\mathbf{E} - \dot{\mathbf{r}} \times \mathbf{B}\right)^{\mathrm{t}} \cdot \dot{\mathbf{r}}\, dt \\
&= \mathfrak{q}\, \mathbf{E}^{\mathrm{t}} \cdot \dot{\mathbf{r}}\, dt
\end{aligned} \tag{3.71}
$$

is performed. The magnetic field $\mathbf{B}$ does not perform work on the charge, since the force $-\dot{\mathbf{r}} \times \mathbf{B}$ of the magnetic field is always perpendicular to the movement direction $\dot{\mathbf{r}}$ of the charge. Since the volume integral over the charge density ρ corresponds to the total charge $\mathfrak{q}$ (see Eq. (3.2)), the above equation can be generalized by using the charge density $\rho(\mathbf{r}, t)$ and the corresponding velocity $\dot{\mathbf{r}}(\mathbf{r}, t)$ of the charge density at point $\mathbf{r}$ and time t:

$$
\begin{aligned}
dW &= \int_{\mathcal{V}} d^3r \left(\rho(\mathbf{r}, t)\, \mathbf{E}^{\mathrm{t}}(\mathbf{r}, t) \cdot \dot{\mathbf{r}}(\mathbf{r}, t)\, dt\right) \\
&= \int_{\mathcal{V}} d^3r \left(\mathbf{E}^{\mathrm{t}}(\mathbf{r}, t) \cdot \underbrace{\rho(\mathbf{r}, t)\, \dot{\mathbf{r}}(\mathbf{r}, t)}_{\overset{(3.1)}{=} \mathbf{J}(\mathbf{r}, t)}\, dt\right) \\
&= \int_{\mathcal{V}} d^3r \left(\mathbf{E}^{\mathrm{t}}(\mathbf{r}, t) \cdot \mathbf{J}(\mathbf{r}, t)\, dt\right).
\end{aligned}
$$

Thus, the change of work on the charges in a volume $\mathcal{V}$ is given by

$$
\frac{dW}{dt} = \int_{\mathcal{V}} d^3r \left(\mathbf{E}^{\mathrm{t}}(\mathbf{r}, t) \cdot \mathbf{J}(\mathbf{r}, t)\right). \tag{3.72}
$$

The term $\mathbf{E}^{\mathrm{t}} \cdot \mathbf{J}$ describes the work that is performed per time and per volume. Using the Maxwell equations and the product rule, we can write this term just as a function of the electric and magnetic fields:

$$
\begin{aligned}
\mathbf{E}^{\mathrm{t}} \cdot \mathbf{J} &\overset{(3.8)}{=} \mathbf{E}^{\mathrm{t}} \cdot \left(\frac{1}{\mu_0} \left(\nabla \times \mathbf{B}\right) - \varepsilon_0 \frac{\partial \mathbf{E}}{\partial t}\right) \\
&= \frac{1}{\mu_0} \mathbf{E}^{\mathrm{t}} \cdot \left(\nabla \times \mathbf{B}\right) - \varepsilon_0\, \mathbf{E}^{\mathrm{t}} \cdot \frac{\partial \mathbf{E}}{\partial t} \\
&= \frac{1}{\mu_0} \left(\mathbf{B}^{\mathrm{t}} \cdot \underbrace{\left(\nabla \times \mathbf{E}\right)}_{\overset{(3.7)}{=} -\frac{\partial \mathbf{B}}{\partial t}} - \nabla \cdot \left(\mathbf{E} \times \mathbf{B}\right)\right) - \varepsilon_0\, \mathbf{E}^{\mathrm{t}} \cdot \frac{\partial \mathbf{E}}{\partial t} \\
&= -\varepsilon_0 \underbrace{\mathbf{E}^{\mathrm{t}} \cdot \frac{\partial \mathbf{E}}{\partial t}}_{= \frac{1}{2} \frac{\partial}{\partial t} \mathbf{E}^2} - \frac{1}{\mu_0} \underbrace{\mathbf{B}^{\mathrm{t}} \cdot \frac{\partial \mathbf{B}}{\partial t}}_{= \frac{1}{2} \frac{\partial}{\partial t} \mathbf{B}^2} - \frac{1}{\mu_0} \nabla \cdot \left(\mathbf{E} \times \mathbf{B}\right)
\end{aligned}
$$

$$= -\frac{1}{2}\frac{\partial}{\partial t}\left(\varepsilon_0\,\mathbf{E}^2 + \frac{1}{\mu_0}\mathbf{B}^2\right) - \frac{1}{\mu_0}\nabla\cdot\left(\mathbf{E}\times\mathbf{B}\right). \tag{3.73}$$

Inserting this in Eq. (3.72) yields

$$\begin{aligned}\frac{dW}{dt} &= -\frac{d}{dt}\int_{\mathcal{V}} d^3r\,\frac{1}{2}\left(\varepsilon_0\,\mathbf{E}^2(\mathbf{r},t) + \frac{1}{\mu_0}\mathbf{B}^2(\mathbf{r},t)\right) - \frac{1}{\mu_0}\int_{\mathcal{V}} d^3r\,\nabla\cdot\left(\mathbf{E}(\mathbf{r},t)\times\mathbf{B}(\mathbf{r},t)\right)\\ &= -\frac{d}{dt}\int_{\mathcal{V}} d^3r\,\frac{1}{2}\left(\varepsilon_0\,\mathbf{E}^2(\mathbf{r},t) + \frac{1}{\mu_0}\mathbf{B}^2(\mathbf{r},t)\right) - \frac{1}{\mu_0}\int_{\mathbb{S}(\mathcal{V})} d\mathbf{a}'\cdot\left(\mathbf{E}(\mathbf{r},t)\times\mathbf{B}(\mathbf{r},t)\right),\end{aligned} \tag{3.74}$$

where we used the divergence theorem in the last step. We can interpret the first integral as the total energy of the electromagnetic fields in the volume $\mathcal{V}$ and the second integral as the energy flow of the electromagnetic fields through the surface $\mathbb{S}(\mathcal{V})$ out of the volume $\mathcal{V}$. Therefore, we can make the following definitions:

Definition 3.2 (*Energy density*) The energy density of the electromagnetic fields is given by

$$u_{\mathrm{em}} = \frac{1}{2}\left(\varepsilon_0\,\mathbf{E}^2 + \frac{1}{\mu_0}\mathbf{B}^2\right). \tag{3.75}$$

Definition 3.3 (*Poynting vector*) The energy flow density of the electromagnetic fields is given by

$$\mathbf{S} = \frac{1}{\mu_0}\left(\mathbf{E}\times\mathbf{B}\right). \tag{3.76}$$

Theorem 3.5 (Poynting's theorem) *The work that the electromagnetic fields perform on the charges in a volume $\mathcal{V}$ is equal to the reduction of the energy stored in the fields minus the field energy that leaves the volume $\mathcal{V}$ through the surface $\mathbb{S}(\mathcal{V})$. Thus, the change of the work W performed on the charges in the volume $\mathcal{V}$ is given by*

$$\frac{dW}{dt} = -\frac{d}{dt}\int_{\mathcal{V}} d^3r\; u_{em}(\mathbf{r}) - \int_{\mathbb{S}(\mathcal{V})} d\mathbf{a}'\cdot\mathbf{S}(\mathbf{r}). \tag{3.77}$$

The work performed by the electromagnetic fields on the charges is equal to the increase of the mechanical energy of the charges. Therefore, if u_{mech} denotes the mechanical energy density of the charges, one obtains

$$\frac{\partial}{\partial t}\left(u_{mech} + u_{em}\right) = -\nabla \cdot \mathbf{S}. \tag{3.78}$$

Proof Equation (3.77) follows directly from Eq. (3.74), if one utilizes the definition of the energy density and the poynting vector. Equation (3.78) follows from

$$\frac{d}{dt}\int_{\mathcal{V}} d^3r \, u_{\rm mech}(\mathbf{r}) = -\frac{d}{dt}\int_{\mathcal{V}} d^3r \, u_{\rm em}(\mathbf{r}) - \int_{\mathcal{S}(\mathcal{V})} d\mathbf{a}^{\rm t} \cdot \mathbf{S}(\mathbf{r}),$$

$$\frac{d}{dt}\int_{\mathcal{V}} d^3r \left(u_{\rm mech}(\mathbf{r}) + u_{\rm em}(\mathbf{r})\right) = -\int_{\mathcal{V}} d^3r \left(\nabla \cdot \mathbf{S}(\mathbf{r})\right).$$

■

The last theorem verifies that energy is stored in the electromagnetic fields, so that an electromagnetic wave transports energy through the space. This energy flow is characterized by the poynting vector **S**.

3.1.4 Interaction of a Charged Particle with an Electromagnetic Wave

We remember that a free particle with charge $\mathfrak{q}$, mass m, and coordinate $\mathbf{r} = [x, y, z]^{\rm t}$ feels the following force in an electric **E** and magnetic field **B**:

$$\mathbf{f} = \mathfrak{q}\,(\mathbf{E} + \dot{\mathbf{r}} \times \mathbf{B}) = \mathfrak{q} \begin{bmatrix} E_x + \dot{y}\,B_z - \dot{z}\,B_y \\ E_y + \dot{z}\,B_x - \dot{x}\,B_z \\ E_z + \dot{x}\,B_y - \dot{y}\,B_x \end{bmatrix}. \tag{3.79}$$

We derive the Hamilton function of the charged particle based on appendix III of Ref. [11]. Firstly, the Lagrange function L of the charged particle is given by

$$L(\mathbf{r}, \dot{\mathbf{r}}, t) = \frac{m}{2}\,\dot{\mathbf{r}}^2 + \mathfrak{q}\,\dot{\mathbf{r}} \cdot \mathbf{A}(\mathbf{r}, t). \tag{3.80}$$

This Lagrange function yields indeed the correct electromagnetic force (3.79) on the charged particle: From the Lagrangian equation [12] in x-direction one gets

$$
\begin{aligned}
0 &= \frac{d}{dt}\left(\frac{\partial L}{\partial \dot{x}}\right) - \frac{\partial L}{\partial x} \\
&= \frac{d}{dt}\left(m\,\dot{x} + \mathrm{q}\,A_x\right) - \mathrm{q}\,\dot{\mathbf{r}}\cdot\frac{\partial \mathbf{A}}{\partial x} \\
&= m\,\ddot{x} + \mathrm{q}\left(\frac{\partial A_x}{\partial t} + \dot{x}\,\frac{\partial A_x}{\partial x} + \dot{y}\,\frac{\partial A_x}{\partial y} + \dot{z}\,\frac{\partial A_x}{\partial z}\right) - \mathrm{q}\left(\dot{x}\,\frac{\partial A_x}{\partial x} + \dot{y}\,\frac{\partial A_y}{\partial x} + \dot{z}\,\frac{\partial A_z}{\partial x}\right) \\
&= m\,\ddot{x} + \mathrm{q}\left(\frac{\partial A_x}{\partial t} + \dot{y}\left(\frac{\partial A_x}{\partial y} - \frac{\partial A_y}{\partial x}\right) + \dot{z}\left(\frac{\partial A_x}{\partial z} - \frac{\partial A_z}{\partial x}\right)\right).
\end{aligned}
\tag{3.81}
$$

From this, one obtains for the force f_x in x-direction due to Newton's law

$$
\begin{aligned}
f_x &= m\,\ddot{x} \\
&= \mathrm{q}\left(-\frac{\partial A_x}{\partial t} + \dot{y}\left(\frac{\partial A_y}{\partial x} - \frac{\partial A_x}{\partial y}\right) - \dot{z}\left(\frac{\partial A_x}{\partial z} - \frac{\partial A_z}{\partial x}\right)\right) \\
&= \mathrm{q}\,(E_x + \dot{y}\,B_z - \dot{z}\,B_y).
\end{aligned}
\tag{3.82}
$$

Analogously, one derives

$$
f_y = \mathrm{q}\,(E_y + \dot{y}\,B_z - \dot{z}\,B_y) \tag{3.83}
$$

$$
f_z = \mathrm{q}\,(E_z + \dot{x}\,B_y - \dot{y}\,B_x) \tag{3.84}
$$

and finally $\mathbf{f} = \mathrm{q}\,(\mathbf{E} + \dot{\mathbf{r}} \times \mathbf{B})$ in agreement with Eq. (3.79). Using the classical momentum $\mathbf{p} = m\,\dot{\mathbf{r}}$ of the particle, once can derive the Hamilton function H by [12]

$$
\begin{aligned}
H(\mathbf{r}, \mathbf{p}, t) &= \mathbf{p}\cdot\dot{\mathbf{r}} - L(\mathbf{r}, \dot{\mathbf{r}}, t) \\
&= \mathbf{p}\cdot\dot{\mathbf{r}} - \frac{m}{2}\,\dot{\mathbf{r}}^2 - q\,\dot{\mathbf{r}}\cdot\mathbf{A}(\mathbf{r}, t).
\end{aligned}
\tag{3.85}
$$

Using [12]

$$
\mathbf{p} = \frac{\partial L}{\partial \dot{\mathbf{r}}} = m\,\dot{\mathbf{r}} + \mathrm{q}\,\mathbf{A} \tag{3.86}
$$

one can write

$$
\dot{\mathbf{r}} = \frac{1}{m}\,(\mathbf{p} - \mathrm{q}\,\mathbf{A}). \tag{3.87}
$$

Inserting this into Eq. (3.85) one obtains

$$
\begin{aligned}
H(\mathbf{r}, \mathbf{p}, t) &= \frac{1}{m}\,\mathbf{p}\cdot(\mathbf{p} - \mathrm{q}\,\mathbf{A}) - \frac{1}{2m}\,(\mathbf{p} - \mathrm{q}\,\mathbf{A})^2 - \frac{\mathrm{q}}{m}\,(\mathbf{p} - \mathrm{q}\,\mathbf{A})\cdot\mathbf{A} \\
&= \frac{1}{m}\left(\mathbf{p}^2 - \mathrm{q}\,\mathbf{p}\cdot\mathbf{A} - \mathrm{q}\,\mathbf{p}\cdot\mathbf{A} + \mathrm{q}^2\,\mathbf{A}^2\right) - \frac{1}{2m}\,(\mathbf{p} - \mathrm{q}\,\mathbf{A})^2
\end{aligned}
$$

$$= \frac{1}{m}\,(\mathbf{p} - q\,\mathbf{A})^2 - \frac{1}{2\,m}\,(\mathbf{p} - q\,\mathbf{A})^2$$

and finally

$$\boxed{H(\mathbf{r}, \mathbf{p}, t) = \frac{1}{2\,m}\,\Big(\mathbf{p} - q\,\mathbf{A}(\mathbf{r}, t)\Big)^2.} \tag{3.88}$$

Next we switch to a Quantum mechanical description: In a first approximation, the vector potential can still be treated classically, but the momentum is described by the momentum operator $\hat{\mathbf{p}} = -\mathrm{i}\,\hbar\,\nabla_{\mathbf{r}}$ and the Hamilton operator is constructed from the classical Hamilton function H by

$$\begin{aligned}
\hat{H} &= \frac{1}{2\,m}\,(\hat{\mathbf{p}} - q\,\mathbf{A})^2 \\
&= \frac{\hat{\mathbf{p}}^2}{2\,m} - \frac{q}{2\,m}\,(\hat{\mathbf{p}}\cdot\mathbf{A} + \mathbf{A}\cdot\hat{\mathbf{p}}) + \frac{q^2}{2\,m}\,\mathbf{A}^2 \\
&= -\frac{\hbar^2}{2\,m}\,\nabla^2 + \frac{\mathrm{i}\,\hbar\,q}{2\,m}\,(\nabla\cdot\mathbf{A} + \mathbf{A}\cdot\nabla) + \frac{q^2}{2\,m}\,\mathbf{A}^2.
\end{aligned} \tag{3.89}$$

One obtains from the chain rule in the Coulomb gauge

$$\nabla\cdot\mathbf{A} = \underbrace{(\nabla\cdot\mathbf{A})}_{=0} + \mathbf{A}\cdot\nabla = \mathbf{A}\cdot\nabla, \tag{3.90}$$

so that the Hamiltonian transforms to

$$\hat{H} = -\frac{\hbar^2}{2\,m}\,\nabla^2 + \frac{\mathrm{i}\,\hbar\,q}{m}\,\mathbf{A}(\mathbf{r}, t)\cdot\nabla + \frac{q^2}{2\,m}\,\mathbf{A}^2(\mathbf{r}, t). \tag{3.91}$$

In summary, the electromagnetic field interaction yields to the additional term

$$\boxed{\hat{H}_{\mathrm{L}} = -\frac{q}{m}\,\mathbf{A}(\mathbf{r}, t)\cdot\hat{\mathbf{p}} + \frac{q^2}{2\,m}\,\mathbf{A}^2(\mathbf{r}, t).} \tag{3.92}$$

in the Hamiltonian.

To simplify things, one performs two approximations:

(i) For moderate field intensities, the second term can be neglected, since it is small compared to the first term.

(ii) The vector potential $\mathbf{A}$ is a function of time and place, i.e., $\mathbf{A} \equiv \mathbf{A}(\mathbf{r}, t)$. But electromagnetic waves with frequencies up to the optical or ultra violet range exhibit wavelengths much larger than the atomic dimensions. Therefore, one neglects the spatial dependency of $\mathbf{A}$.

Finally, one obtains for the Hamiltonian in the so called **electric dipole approximation**

$$\hat{H} = \frac{\hat{\mathbf{p}}^2}{2m} - \frac{\mathfrak{q}}{m}\mathbf{A}(t)\cdot\hat{\mathbf{p}}, \tag{3.93}$$

so that the interaction with the electromagnetic wave described by the vector potential **A** is approximately given by the following additional term in the Hamiltonian

$$\boxed{\hat{H}_{\mathrm{Ld}} = -\frac{\mathfrak{q}}{m}\mathbf{A}(t)\cdot\hat{\mathbf{p}}.} \tag{3.94}$$

3.2 Basic Considerations of Second Quantization

Electrons are described by a many-body electron wave function, which is obtained by solving the Schrödinger equation with the help of the Hamilton operator. To simplify things, one introduces the so called second quantization for describing many-body states, which is an equivalent description of quantum mechanics. Now a review of second quantization is given, which is taken from Ref. [13].

3.2.1 *Second Quantization for Electrons*

Any N_e-electron wave function can be constructed as a linear combination of slater determinants of single-electron wave functions. In a periodic solid, these single-electron wave functions can be chosen as Bloch functions $\varphi_{\ell\mathbf{k}}(\sigma, \mathbf{R})$ (see Sect. 2.3.4), which have a definite wave vector **k** and form a complete set $\{\varphi_{\ell\mathbf{k}}(\sigma, \mathbf{R})\}$. σ denotes the spin and **R** the coordinate. A N_e-electron wave function $\Psi_{N_\mathrm{e}}(\sigma_1, \mathbf{R}_1, \ldots, \sigma_{N_\mathrm{e}}, \mathbf{R}_{N_\mathrm{e}})$ that is a slater determinant of these single-electron Bloch functions can be characterized by which single-electron states are occupied.

Definition 3.4 *(Electron creation and annihilation operators)* The electron creation operator $\hat{c}^\dagger_{\ell\mathbf{k}}$ generates an electron in state $\ell\mathbf{k}$ and the electron annihilation operator $\hat{c}_{\ell\mathbf{k}}$ destroys the electron in state $\ell\mathbf{k}$.

If we denote the normalized vaccuum state without any electrons by $|0\rangle$, this normalized N_e-electron wave function $\Psi_{N_\mathrm{e}}(\sigma_1, \mathbf{R}_1, \ldots, \sigma_{N_\mathrm{e}}, \mathbf{R}_{N_\mathrm{e}})$ can be written as

$$\begin{aligned}
&\left\langle \sigma_1, \mathbf{R}_1, \ldots, \sigma_{N_e}, \mathbf{R}_{N_e} \middle| \frac{1}{\sqrt{N_e!}} \hat{c}^\dagger_{\ell_1\mathbf{k}_1} \cdots \hat{c}^\dagger_{\ell_{N_e}\mathbf{k}_{N_e}} \middle| 0 \right\rangle \\
&= \Psi_{N_e}(\sigma_1, \mathbf{R}_1, \ldots, \sigma_{N_e}, \mathbf{R}_{N_e}) \\
&= \frac{1}{\sqrt{N_e!}} \det\left(\begin{bmatrix} \varphi_{\ell_1\mathbf{k}_1}(\sigma_1, \mathbf{R}_1) & \cdots & \varphi_{\ell_1\mathbf{k}_1}(\sigma_{N_e}, \mathbf{R}_{N_e}) \\ \vdots & & \vdots \\ \varphi_{\ell_{N_e}\mathbf{k}_{N_e}}(\sigma_1, \mathbf{R}_1) & \cdots & \varphi_{\ell_{N_e}\mathbf{k}_{N_e}}(\sigma_{N_e}, \mathbf{R}_{N_e}) \end{bmatrix} \right).
\end{aligned} \tag{3.95}$$

$\hat{c}^\dagger_{\ell_1\mathbf{k}_1} \cdots \hat{c}^\dagger_{\ell_{N_e}\mathbf{k}_{N_e}} |0\rangle$ denotes a state in the so-called **Fock space**. The creation operator $\hat{c}^\dagger_{\ell_1\mathbf{k}_1}$ generates in the $(N_e - 1)$-electron state $\hat{c}^\dagger_{\ell_2\mathbf{k}_2} \cdots \hat{c}^\dagger_{\ell_{N_e}\mathbf{k}_{N_e}} |0\rangle$ an additional electron in the state $\ell_1\mathbf{k}_1$ to form the N_e-electron state $\hat{c}^\dagger_{\ell_1\mathbf{k}_1} \hat{c}^\dagger_{\ell_2\mathbf{k}_2} \cdots \hat{c}^\dagger_{\ell_{N_e}\mathbf{k}_{N_e}} |0\rangle$. The annihilation operator $\hat{c}_{\ell_1\mathbf{k}_1}$ destroys in the N_e-electron state $\hat{c}^\dagger_{\ell_1\mathbf{k}_1} \hat{c}^\dagger_{\ell_2\mathbf{k}_2} \cdots \hat{c}^\dagger_{\ell_{N_e}\mathbf{k}_{N_e}} |0\rangle$ the electron in the state $\ell_1\mathbf{k}_1$ to form the $(N_e - 1)$-electron state $\hat{c}^\dagger_{\ell_2\mathbf{k}_2} \cdots \hat{c}^\dagger_{\ell_{N_e}\mathbf{k}_{N_e}} |0\rangle$.

Now we show that the adjoint of the creation operator is, indeed, the annihilation operator:

$$\left(\hat{c}^\dagger_{\ell\mathbf{k}}\right)^\dagger = \hat{c}_{\ell\mathbf{k}}. \tag{3.96}$$

For this, let be $|\Psi_1\rangle = \hat{c}^\dagger_{\ell\mathbf{k}} |0\rangle$. Then we get

$$\begin{aligned}
1 &= \langle \Psi_1 | \Psi_1 \rangle \\
&= \left\langle \Psi_1 \middle| \hat{c}^\dagger_{\ell\mathbf{k}}\, 0 \right\rangle \\
&= \left\langle \left(\hat{c}^\dagger_{\ell\mathbf{k}}\right)^\dagger \Psi_1 \middle| 0 \right\rangle.
\end{aligned}$$

Only the bra $\langle \psi | = \langle 0 |$ fulfills $\langle \psi | 0 \rangle = 1$. Therefore, we obtain

$$\begin{aligned}
|0\rangle &= \left(\hat{c}^\dagger_{\ell\mathbf{k}}\right)^\dagger |\Psi_1\rangle \\
&= \left(\hat{c}^\dagger_{\ell\mathbf{k}}\right)^\dagger \hat{c}^\dagger_{\ell\mathbf{k}} |0\rangle.
\end{aligned}$$

This can only be fulfilled, if Eq. (3.96) is valid.

Electrons are fermions: If two electrons are interchanged, the many-body wave function changes the sign, which is realized by the determinant in Eq. (3.95). Therefore, one has

$$\hat{c}^\dagger_{\ell\mathbf{k}} \hat{c}^\dagger_{\ell'\mathbf{k}'} = -\hat{c}^\dagger_{\ell'\mathbf{k}'} \hat{c}^\dagger_{\ell\mathbf{k}}. \tag{3.97}$$

From this, it follows directly

$$\hat{c}^\dagger_{\ell\mathbf{k}} \hat{c}^\dagger_{\ell\mathbf{k}} = 0. \tag{3.98}$$

This indicates that a state can only be occupied by at most one electron, which is the well-known Pauli exclusion principle. The two above equations are also valid for the annihilation operator, which is just the adjoint of the creation operator. In addition, if there are no electrons, no electrons can be destroyed:

$$\hat{c}_{\ell\mathbf{k}}\,|0\rangle = 0. \tag{3.99}$$

For any slater-determinant state $|\Psi_{N_e}\rangle$, one has

$$\left(\hat{c}_{\ell\mathbf{k}}\,\hat{c}^{\dagger}_{\ell\mathbf{k}} + \hat{c}^{\dagger}_{\ell\mathbf{k}}\,\hat{c}_{\ell\mathbf{k}}\right)|\Psi_{N_e}\rangle = |\Psi_{N_e}\rangle,$$

since the state $\ell\mathbf{k}$ can be occupied or not in $|\Psi_{N_e}\rangle$. If this state is not occupied, the first term creates and then destroys the electron in this state, so that the state keeps unchanged in total, whereas the second term tries firstly to destroy the electron, which is not possible and yields zero. If this state is occupied, the second term destroys and then creates the electron in this state, so that the state keeps unchanged in total, whereas the first term tries firstly to create a second electron in this state, which is not possible and yields zero. In this way, we proved

$$\begin{aligned} \hat{c}_{\ell\mathbf{k}}\,\hat{c}^{\dagger}_{\ell\mathbf{k}} + \hat{c}^{\dagger}_{\ell\mathbf{k}}\,\hat{c}_{\ell\mathbf{k}} &= 1, \\ \Leftrightarrow \qquad \hat{c}_{\ell\mathbf{k}}\,\hat{c}^{\dagger}_{\ell\mathbf{k}} &= 1 - \hat{c}^{\dagger}_{\ell\mathbf{k}}\,\hat{c}_{\ell\mathbf{k}}, \end{aligned} \tag{3.100}$$

since any electronic many-body state can be constructed from a linear combination of slater-determinant states. In addition, we realized that the operator $\hat{c}^{\dagger}_{\ell\mathbf{k}}\,\hat{c}_{\ell\mathbf{k}}$ is one, if the state $\ell\mathbf{k}$ is occupied, and is zero, if not, which are the only two possibilities for an electronic state. This allows us to define the electron occupation number operator:

Definition 3.5 *(Electron occupation number operator)* The occupation number operator

$$\hat{n}_{\ell\mathbf{k}} = \hat{c}^{\dagger}_{\ell\mathbf{k}}\,\hat{c}_{\ell\mathbf{k}} \tag{3.101}$$

counts how many electrons occupy the state $\ell\mathbf{k}$.

Now we show

$$\hat{c}_{\ell\mathbf{k}}\,\hat{c}^{\dagger}_{\ell'\mathbf{k}'} = -\hat{c}^{\dagger}_{\ell'\mathbf{k}'}\,\hat{c}_{\ell\mathbf{k}} \tag{3.102}$$

for different states $\ell \neq \ell'$ or $\mathbf{k} \neq \mathbf{k}'$. For this, we write the slater determinant state as

$$|\Psi_{N_e}\rangle = \hat{c}^{\dagger}_{\ell_1\mathbf{k}_1}\,\hat{c}^{\dagger}_{\ell_2\mathbf{k}_2}\cdots\hat{c}^{\dagger}_{\ell_{n-1}\mathbf{k}_{n-1}}\,\hat{c}^{\dagger}_{\ell\mathbf{k}}\,\hat{c}^{\dagger}_{\ell_{n+1}\mathbf{k}_{n+1}}\cdots\hat{c}^{\dagger}_{\ell_{N_e}\mathbf{k}_{N_e}}\,|0\rangle,$$

where the state $\ell, \mathbf{k}$ occurs one time at position n. Then we obtain

$$
\begin{aligned}
&\hat{c}_{\ell\mathbf{k}}\,\hat{c}^\dagger_{\ell'\mathbf{k}'}\,|\Psi_{N_e}\rangle \\
&= \hat{c}_{\ell\mathbf{k}}\,\hat{c}^\dagger_{\ell'\mathbf{k}'}\,\hat{c}^\dagger_{\ell_1\mathbf{k}_1}\,\hat{c}^\dagger_{\ell_2\mathbf{k}_2}\cdots\hat{c}^\dagger_{\ell_{n-1}\mathbf{k}_{n-1}}\,\hat{c}^\dagger_{\ell\mathbf{k}}\,\hat{c}^\dagger_{\ell_{n+1}\mathbf{k}_{n+1}}\cdots\hat{c}^\dagger_{\ell_{N_e}\mathbf{k}_{N_e}}\,|0\rangle \\
&\overset{(3.97)}{=}(-1)^n\,\underbrace{\hat{c}_{\ell\mathbf{k}}\,\hat{c}^\dagger_{\ell\mathbf{k}}}_{\overset{(3.100)}{=}1-\hat{c}^\dagger_{\ell\mathbf{k}}\hat{c}_{\ell\mathbf{k}}}\,\hat{c}^\dagger_{\ell'\mathbf{k}'}\,\hat{c}^\dagger_{\ell_1\mathbf{k}_1}\,\hat{c}^\dagger_{\ell_2\mathbf{k}_2}\cdots\hat{c}^\dagger_{\ell_{n-1}\mathbf{k}_{n-1}}\,\hat{c}^\dagger_{\ell_{n+1}\mathbf{k}_{n+1}}\cdots\hat{c}^\dagger_{\ell_{N_e}\mathbf{k}_{N_e}}\,|0\rangle \\
&= (-1)^n\,\hat{c}^\dagger_{\ell'\mathbf{k}'}\,\hat{c}^\dagger_{\ell_1\mathbf{k}_1}\,\hat{c}^\dagger_{\ell_2\mathbf{k}_2}\cdots\hat{c}^\dagger_{\ell_{n-1}\mathbf{k}_{n-1}}\,\hat{c}^\dagger_{\ell_{n+1}\mathbf{k}_{n+1}}\cdots\hat{c}^\dagger_{\ell_{N_e}\mathbf{k}_{N_e}}\,|0\rangle
\end{aligned}
$$

and

$$
\begin{aligned}
&\hat{c}^\dagger_{\ell'\mathbf{k}'}\,\hat{c}_{\ell\mathbf{k}}\,|\Psi_{N_e}\rangle \\
&= \hat{c}^\dagger_{\ell'\mathbf{k}'}\,\hat{c}_{\ell\mathbf{k}}\,\hat{c}^\dagger_{\ell_1\mathbf{k}_1}\,\hat{c}^\dagger_{\ell_2\mathbf{k}_2}\cdots\hat{c}^\dagger_{\ell_{n-1}\mathbf{k}_{n-1}}\,\hat{c}^\dagger_{\ell\mathbf{k}}\,\hat{c}^\dagger_{\ell_{n+1}\mathbf{k}_{n+1}}\cdots\hat{c}^\dagger_{\ell_{N_e}\mathbf{k}_{N_e}}\,|0\rangle \\
&\overset{(3.97)}{=}(-1)^{n-1}\,\hat{c}^\dagger_{\ell'\mathbf{k}'}\,\underbrace{\hat{c}_{\ell\mathbf{k}}\,\hat{c}^\dagger_{\ell\mathbf{k}}}_{\overset{(3.100)}{=}1-\hat{c}^\dagger_{\ell\mathbf{k}}\hat{c}_{\ell\mathbf{k}}}\,\hat{c}^\dagger_{\ell_1\mathbf{k}_1}\,\hat{c}^\dagger_{\ell_2\mathbf{k}_2}\cdots\hat{c}^\dagger_{\ell_{n-1}\mathbf{k}_{n-1}}\,\hat{c}^\dagger_{\ell_{n+1}\mathbf{k}_{n+1}}\cdots\hat{c}^\dagger_{\ell_{N_e}\mathbf{k}_{N_e}}\,|0\rangle \\
&= (-1)^{n-1}\,\hat{c}^\dagger_{\ell'\mathbf{k}'}\,\hat{c}^\dagger_{\ell_1\mathbf{k}_1}\,\hat{c}^\dagger_{\ell_2\mathbf{k}_2}\cdots\hat{c}^\dagger_{\ell_{n-1}\mathbf{k}_{n-1}}\,\hat{c}^\dagger_{\ell_{n+1}\mathbf{k}_{n+1}}\cdots\hat{c}^\dagger_{\ell_{N_e}\mathbf{k}_{N_e}}\,|0\rangle.
\end{aligned}
$$

If the state $\ell\mathbf{k}$ is not occupied in $|\Psi_{N_e}\rangle$, we just have $\hat{c}_{\ell\mathbf{k}}\,\hat{c}^\dagger_{\ell'\mathbf{k}'}\,|\Psi_{N_e}\rangle = 0$ and $\hat{c}^\dagger_{\ell'\mathbf{k}'}\,\hat{c}_{\ell\mathbf{k}}\,|\Psi_{N_e}\rangle = 0$. Therefore, Eq. (3.102) is proven. Using the definition of the anti-commutator

$$
\left[\hat{c}^\dagger_{\ell\mathbf{k}}\,,\,\hat{c}^\dagger_{\ell'\mathbf{k}'}\right]_+ := \hat{c}^\dagger_{\ell\mathbf{k}}\,\hat{c}^\dagger_{\ell'\mathbf{k}'} + \hat{c}^\dagger_{\ell'\mathbf{k}'}\,\hat{c}^\dagger_{\ell\mathbf{k}}, \tag{3.103}
$$

we can compactly write the above presented properties of the creation and annihilation operators:

Theorem 3.6 (Properties of the electron creation and annihilation operators) *The electron creation and annihilation operators obey*

$$
\left[\hat{c}^\dagger_{\ell\mathbf{k}}\,,\,\hat{c}^\dagger_{\ell'\mathbf{k}'}\right]_+ = 0, \tag{3.104}
$$

$$
\left[\hat{c}_{\ell\mathbf{k}}\,,\,\hat{c}_{\ell'\mathbf{k}'}\right]_+ = 0, \tag{3.105}
$$

$$
\left[\hat{c}_{\ell\mathbf{k}}\,,\,\hat{c}^\dagger_{\ell'\mathbf{k}'}\right]_+ = \delta_{\ell\ell'}\,\delta_{\mathbf{k}\mathbf{k}'}. \tag{3.106}
$$

Using the properties of the creation- and annihilation operators, we can directly prove the following properties of the occupation number operator:

Theorem 3.7 (Properties of the electron occupation number operators) *The occupation number operator $\hat{n}_{\ell\mathbf{k}}$ is hermitian and its eigenvalues are just* 0 *or* 1. *Furthermore, one has*

$$\left[\hat{n}_{\ell\mathbf{k}}\,,\,\hat{n}_{\ell'\mathbf{k}'}\right]_{-} = 0 \tag{3.107}$$

for arbitrary $\ell, \ell', \mathbf{k}, \mathbf{k}'$ and for different states $\ell \neq \ell'$ or $\mathbf{k} \neq \mathbf{k}'$:

$$\left[\hat{n}_{\ell\mathbf{k}}\,,\,\hat{c}^{\dagger}_{\ell'\mathbf{k}'}\right]_{-} = 0, \tag{3.108}$$

$$\left[\hat{n}_{\ell\mathbf{k}}\,,\,\hat{c}_{\ell'\mathbf{k}'}\right]_{-} = 0. \tag{3.109}$$

Proof By definition, we get

$$\hat{n}^{\dagger}_{\ell\mathbf{k}} = \left(\hat{c}^{\dagger}_{\ell\mathbf{k}}\,\hat{c}_{\ell\mathbf{k}}\right)^{\dagger} = \hat{c}^{\dagger}_{\ell\mathbf{k}}\left(\hat{c}^{\dagger}_{\ell\mathbf{k}}\right)^{\dagger} = \hat{c}^{\dagger}_{\ell\mathbf{k}}\,\hat{c}_{\ell\mathbf{k}} = \hat{n}_{\ell\mathbf{k}}, \tag{3.110}$$

which proves that $\hat{n}_{\ell\mathbf{k}}$ is hermitian. Since we have

$$\left(\hat{n}_{\ell\mathbf{k}}\right)^{2} = \hat{n}_{\ell\mathbf{k}}\,\hat{n}_{\ell\mathbf{k}} = \hat{c}^{\dagger}_{\ell\mathbf{k}}\,\hat{c}_{\ell\mathbf{k}}\,\hat{c}^{\dagger}_{\ell\mathbf{k}}\,\hat{c}_{\ell\mathbf{k}} = \hat{c}^{\dagger}_{\ell\mathbf{k}}\left(1 - \hat{c}^{\dagger}_{\ell\mathbf{k}}\,\hat{c}_{\ell\mathbf{k}}\right)\hat{c}_{\ell\mathbf{k}} = \hat{c}^{\dagger}_{\ell\mathbf{k}}\,\hat{c}_{\ell\mathbf{k}} = \hat{n}_{\ell\mathbf{k}}, \tag{3.111}$$

the eigenvalue of $\hat{n}_{\ell\mathbf{k}}$ can only be 0 or 1. Furthermore, we obtain from the anticommutator rules of the creation and annihilation operators for different states $\ell \neq \ell'$ or $\mathbf{k} \neq \mathbf{k}'$:

$$\begin{aligned}
\left[\hat{n}_{\ell\mathbf{k}}\,,\,\hat{c}^{\dagger}_{\ell'\mathbf{k}'}\right]_{-} &= \hat{n}_{\ell\mathbf{k}}\,\hat{c}^{\dagger}_{\ell'\mathbf{k}'} - \hat{c}^{\dagger}_{\ell'\mathbf{k}'}\,\hat{n}_{\ell\mathbf{k}} \\
&= \hat{c}^{\dagger}_{\ell\mathbf{k}}\underbrace{\hat{c}_{\ell\mathbf{k}}\,\hat{c}^{\dagger}_{\ell'\mathbf{k}'}}_{=-\hat{c}^{\dagger}_{\ell'\mathbf{k}'}\,\hat{c}_{\ell\mathbf{k}}} - \hat{c}^{\dagger}_{\ell'\mathbf{k}'}\,\hat{c}^{\dagger}_{\ell\mathbf{k}}\,\hat{c}_{\ell\mathbf{k}} \\
&= -\underbrace{\hat{c}^{\dagger}_{\ell\mathbf{k}}\,\hat{c}^{\dagger}_{\ell'\mathbf{k}'}}_{=-\hat{c}^{\dagger}_{\ell'\mathbf{k}'}\,\hat{c}^{\dagger}_{\ell\mathbf{k}}}\,\hat{c}_{\ell\mathbf{k}} - \hat{c}^{\dagger}_{\ell'\mathbf{k}'}\,\hat{c}^{\dagger}_{\ell\mathbf{k}}\,\hat{c}_{\ell\mathbf{k}} \\
&= \hat{c}^{\dagger}_{\ell'\mathbf{k}'}\,\hat{c}^{\dagger}_{\ell\mathbf{k}}\,\hat{c}_{\ell\mathbf{k}} - \hat{c}^{\dagger}_{\ell'\mathbf{k}'}\,\hat{c}^{\dagger}_{\ell\mathbf{k}}\,\hat{c}_{\ell\mathbf{k}} \\
&= 0
\end{aligned}$$

and

$$
\begin{aligned}
\Big[\hat{n}_{\ell\mathbf{k}}\,,\,\hat{c}_{\ell'\mathbf{k}'}\Big]_- &= \hat{n}_{\ell\mathbf{k}}\,\hat{c}_{\ell'\mathbf{k}'} - \hat{c}_{\ell'\mathbf{k}'}\,\hat{n}_{\ell\mathbf{k}} \\
&= \hat{c}^\dagger_{\ell\mathbf{k}}\,\underbrace{\hat{c}_{\ell\mathbf{k}}\,\hat{c}_{\ell'\mathbf{k}'}}_{=-\hat{c}_{\ell'\mathbf{k}'}\,\hat{c}_{\ell\mathbf{k}}} - \hat{c}_{\ell'\mathbf{k}'}\,\hat{c}^\dagger_{\ell\mathbf{k}}\,\hat{c}_{\ell\mathbf{k}} \\
&= -\underbrace{\hat{c}^\dagger_{\ell\mathbf{k}}\,\hat{c}_{\ell'\mathbf{k}'}}_{=-\hat{c}_{\ell'\mathbf{k}'}\,\hat{c}^\dagger_{\ell\mathbf{k}}}\,\hat{c}_{\ell\mathbf{k}} - \hat{c}_{\ell'\mathbf{k}'}\,\hat{c}^\dagger_{\ell\mathbf{k}}\,\hat{c}_{\ell\mathbf{k}} \\
&= \hat{c}_{\ell'\mathbf{k}'}\,\hat{c}^\dagger_{\ell\mathbf{k}}\,\hat{c}_{\ell\mathbf{k}} - \hat{c}_{\ell'\mathbf{k}'}\,\hat{c}^\dagger_{\ell\mathbf{k}}\,\hat{c}_{\ell\mathbf{k}} \\
&= 0.
\end{aligned}
$$

■

The electronic Hamilton operator (2.93) contains one- and two-particle operators and can be written with the help of the creation and annihilation operators. For this, the contained one- and two-particle operators are transformed in the following way:

Theorem 3.8 (One-particle operator in second quantization) *A one particle operator*

$$\hat{V}^{(1)} = \sum_i v^{(1)}(\sigma_i, \mathbf{r}_i) \tag{3.112}$$

is written in second quantization as

$$\hat{V}^{(1)} = \sum_{\ell\mathbf{k},\ell'\mathbf{k}'} v^{(1)}_{\ell\mathbf{k},\ell'\mathbf{k}'}\,\hat{c}^\dagger_{\ell\mathbf{k}}\,\hat{c}_{\ell'\mathbf{k}'} \tag{3.113}$$

with the matrix element

$$V^{(1)}_{\ell\mathbf{k},\ell'\mathbf{k}'} = \sum_\sigma \int_{\mathbb{R}^3} d^3r\,\varphi^*_{\ell\mathbf{k}}(\sigma,\mathbf{r})\,v^{(1)}(\sigma,\mathbf{r})\,\varphi_{\ell'\mathbf{k}'}(\sigma,\mathbf{r}). \tag{3.114}$$

Theorem 3.9 (Two-particle operator in second quantization) *A two particle operator*

$$\hat{V}^{(2)} = \frac{1}{2}\sum_{i,j} v^{(2)}(\sigma_i, \mathbf{r}_i, \sigma_j, \mathbf{r}_j) \tag{3.115}$$

is written in second quantization as

$$\hat{V}^{(2)} = \frac{1}{2} \sum_{\ell\mathbf{k},\ell'\mathbf{k}',\ell''\mathbf{k}'',\ell'''\mathbf{k}'''} v^{(2)}_{\ell\mathbf{k},\ell'\mathbf{k}',\ell''\mathbf{k}'',\ell'''\mathbf{k}'''}\, \hat{c}^{\dagger}_{\ell\mathbf{k}}\, \hat{c}^{\dagger}_{\ell'\mathbf{k}'}\, \hat{c}_{\ell'''\mathbf{k}'''}\, \hat{c}_{\ell''\mathbf{k}''} \tag{3.116}$$

with the matrix element

$$V^{(2)}_{\ell\mathbf{k},\ell'\mathbf{k}',\ell''\mathbf{k}'',\ell'''\mathbf{k}'''} = \sum_{\sigma,\sigma'} \int_{\mathbb{R}^3} d^3r \int_{\mathbb{R}^3} d^3r'\, \varphi^*_{\ell\mathbf{k}}(\sigma,\mathbf{r})\, \varphi^*_{\ell'\mathbf{k}'}(\sigma',\mathbf{r}') \times \\ \times\, v^{(2)}(\sigma,\mathbf{r},\sigma',\mathbf{r}')\, \varphi_{\ell''\mathbf{k}''}(\sigma,\mathbf{r})\, \varphi_{\ell'''\mathbf{k}'''}(\sigma',\mathbf{r}'). \tag{3.117}$$

The proofs of the last two theorems are straight forward an can be found in Ref. [14] or Ref. [15]. A one-particle operator changes the state of one electron in the N_e-electron state and a two-particle operator of two electrons.

3.2.2 Second Quantization for Phonons

Also the motions of the nuclei can be treated quantum mechanically. The following description of second quantization for phonons is based on Ref. [16], but with more details at given points. In quantum mechanics, the nuclei displacements $u_{\mathbf{n}\upsilon\alpha}$ and momenta $p_{\mathbf{n}\upsilon\alpha}$ are described by the related operators $\hat{u}_{\mathbf{n}\upsilon\alpha}$ and $\hat{p}_{\mathbf{n}\upsilon\alpha}$, which are hermitian and obey the commutator relations

$$[\hat{u}_{\mathbf{n}\upsilon\alpha}\,,\ \hat{p}_{\mathbf{n}'\upsilon'\alpha'}]_- = \mathrm{i}\,\hbar\,\delta_{\mathbf{n}\mathbf{n}'}\,\delta_{\upsilon\upsilon'}\,\delta_{\alpha\alpha'}, \tag{3.118}$$

$$[\hat{u}_{\mathbf{n}\upsilon\alpha}\,,\ \hat{u}_{\mathbf{n}'\upsilon'\alpha'}]_- = 0, \tag{3.119}$$

$$[\hat{p}_{\mathbf{n}\upsilon\alpha}\,,\ \hat{p}_{\mathbf{n}'\upsilon'\alpha'}]_- = 0. \tag{3.120}$$

The generalized displacement and momenta operators $\hat{U}_{\upsilon\alpha}(\mathbf{T_n}) = \sqrt{m_\upsilon}\,\hat{u}_{\mathbf{n}\upsilon\alpha}$ and $\hat{P}_{\upsilon\alpha}(\mathbf{T_n}) = \frac{1}{\sqrt{m_\upsilon}}\,\hat{p}_{\mathbf{n}\upsilon\alpha}$ are just scaled by real constants. Therefore, this operators are also hermitian and fulfill the same commutator relations:

$$\left[\hat{U}_{\upsilon\alpha}(\mathbf{T_n})\,,\ \hat{P}_{\upsilon'\alpha'}(\mathbf{T_{n'}})\right]_- = \frac{\sqrt{m_\upsilon}}{\sqrt{m_{\upsilon'}}}\,[\hat{u}_{\mathbf{n}\upsilon\alpha}\,,\ \hat{p}_{\mathbf{n}'\upsilon'\alpha'}]_- = \mathrm{i}\,\hbar\,\delta_{\mathbf{T_n}\mathbf{T_{n'}}}\,\delta_{\upsilon\upsilon'}\,\delta_{\alpha\alpha'}, \tag{3.121}$$

$$\left[\hat{U}_{\upsilon\alpha}(\mathbf{T_n})\,,\ \hat{U}_{\upsilon'\alpha'}(\mathbf{T_{n'}})\right]_- = \sqrt{m_\upsilon}\,\sqrt{m_{\upsilon'}}\,[\hat{u}_{\mathbf{n}\upsilon\alpha}\,,\ \hat{u}_{\mathbf{n}'\upsilon'\alpha'}]_- = 0, \tag{3.122}$$

$$\left[\hat{P}_{\upsilon\alpha}(\mathbf{T_n})\,,\ \hat{P}_{\upsilon'\alpha'}(\mathbf{T_{n'}})\right]_- = \frac{1}{\sqrt{m_\upsilon}}\,\frac{1}{\sqrt{m_{\upsilon'}}}\,[\hat{u}_{\mathbf{n}\upsilon\alpha}\,,\ \hat{u}_{\mathbf{n}'\upsilon'\alpha'}]_- = 0. \tag{3.123}$$

The Fourier transforms $\hat{\mathfrak{U}}_{\upsilon\alpha}(\mathbf{q_n})$ and $\hat{\mathfrak{P}}_{\upsilon\alpha}(\mathbf{q_n})$ of the generalized displacement and momenta operators $\hat{U}_{\upsilon\alpha}(\mathbf{T_n})$ and $\hat{P}_{\upsilon\alpha}(\mathbf{T_n})$ are **not** hermitian, but we have

$$\hat{\mathfrak{U}}^{\dagger}_{\upsilon\alpha}(\mathbf{q_n}) \overset{(2.63)}{=} \left(\frac{1}{\sqrt{N_1\,N_2\,N_3}} \sum_{\mathbf{T_{n'}}} \hat{U}_{\upsilon\alpha}(\mathbf{T_{n'}})\,\mathrm{e}^{-\mathrm{i}\,\mathbf{q}^{\mathrm{t}}_{\mathbf{n}}\cdot\mathbf{T_{n'}}} \right)^{\dagger}$$
$$= \frac{1}{\sqrt{N_1\,N_2\,N_3}} \sum_{\mathbf{T_{n'}}} \hat{U}_{\upsilon\alpha}(\mathbf{T_{n'}})\,\mathrm{e}^{\mathrm{i}\,\mathbf{q}^{\mathrm{t}}_{\mathbf{n}}\cdot\mathbf{T_{n'}}}$$
$$\overset{(2.63)}{=} \hat{\mathfrak{U}}_{\upsilon\alpha}(-\mathbf{q_n}), \tag{3.124}$$

$$\hat{\mathfrak{P}}^{\dagger}_{\upsilon\alpha}(\mathbf{q_n}) \overset{(2.64)}{=} \left(\frac{1}{\sqrt{N_1\,N_2\,N_3}} \sum_{\mathbf{T_{n'}}} \hat{P}_{\upsilon\alpha}(\mathbf{T_{n'}})\,\mathrm{e}^{-\mathrm{i}\,\mathbf{q}^{\mathrm{t}}_{\mathbf{n}}\cdot\mathbf{T_{n'}}} \right)^{\dagger}$$
$$= \frac{1}{\sqrt{N_1\,N_2\,N_3}} \sum_{\mathbf{T_{n'}}} \hat{P}_{\upsilon\alpha}(\mathbf{T_{n'}})\,\mathrm{e}^{\mathrm{i}\,\mathbf{q}^{\mathrm{t}}_{\mathbf{n}}\cdot\mathbf{T_{n'}}}$$
$$\overset{(2.64)}{=} \hat{\mathfrak{P}}_{\upsilon\alpha}(-\mathbf{q_n}). \tag{3.125}$$

Furthermore, they fulfill the following commutator relation:

$$\left[\hat{\mathfrak{U}}_{\upsilon\alpha}(\mathbf{q_n})\,,\,\hat{\mathfrak{P}}^{\dagger}_{\upsilon'\alpha'}(\mathbf{q_{n'}})\right]_-$$
$$=\hat{\mathfrak{U}}_{\upsilon\alpha}(\mathbf{q_n})\,\hat{\mathfrak{P}}^{\dagger}_{\upsilon'\alpha'}(\mathbf{q_{n'}}) - \hat{\mathfrak{P}}^{\dagger}_{\upsilon'\alpha'}(\mathbf{q_{n'}})\,\hat{\mathfrak{U}}_{\upsilon\alpha}(\mathbf{q_n})$$
$$=\frac{1}{N_1\,N_2\,N_3} \left(\sum_{\mathbf{T_{n''}}} \hat{U}_{\upsilon\alpha}(\mathbf{T_{n''}})\,\mathrm{e}^{-\mathrm{i}\,\mathbf{q}^{\mathrm{t}}_{\mathbf{n}}\cdot\mathbf{T_{n''}}} \right) \left(\sum_{\mathbf{T_{n'''}}} \hat{P}_{\upsilon'\alpha'}(\mathbf{T_{n'''}})\,\mathrm{e}^{\mathrm{i}\,\mathbf{q}^{\mathrm{t}}_{\mathbf{n'}}\cdot\mathbf{T_{n'''}}} \right)$$
$$- \frac{1}{N_1\,N_2\,N_3} \left(\sum_{\mathbf{T_{n'''}}} \hat{P}_{\upsilon'\alpha'}(\mathbf{T_{n'''}})\,\mathrm{e}^{\mathrm{i}\,\mathbf{q}^{\mathrm{t}}_{\mathbf{n'}}\cdot\mathbf{T_{n'''}}} \right) \left(\sum_{\mathbf{T_{n''}}} \hat{U}_{\upsilon\alpha}(\mathbf{T_{n''}})\,\mathrm{e}^{-\mathrm{i}\,\mathbf{q}^{\mathrm{t}}_{\mathbf{n}}\cdot\mathbf{T_{n''}}} \right)$$
$$=\frac{1}{N_1\,N_2\,N_3} \sum_{\mathbf{T_{n''}}} \sum_{\mathbf{T_{n'''}}} \underbrace{\left(\hat{U}_{\upsilon\alpha}(\mathbf{T_{n''}})\,\hat{P}_{\upsilon'\alpha'}(\mathbf{T_{n'''}}) - \hat{P}_{\upsilon'\alpha'}(\mathbf{T_{n'''}})\,\hat{U}_{\upsilon\alpha}(\mathbf{T_{n''}}) \right)}_{=\left[\hat{U}_{\upsilon\alpha}(\mathbf{T_{n''}})\,,\,\hat{P}_{\upsilon'\alpha'}(\mathbf{T_{n'''}})\right]_- = \mathrm{i}\,\hbar\,\delta_{\mathbf{T_{n''}}\mathbf{T_{n'''}}}\,\delta_{\upsilon\upsilon'}\,\delta_{\alpha\alpha'}} \mathrm{e}^{-\mathrm{i}\,\mathbf{q}^{\mathrm{t}}_{\mathbf{n}}\cdot\mathbf{T_{n''}}}\,\mathrm{e}^{\mathrm{i}\,\mathbf{q}^{\mathrm{t}}_{\mathbf{n'}}\cdot\mathbf{T_{n'''}}}$$
$$=\mathrm{i}\,\hbar\,\delta_{\upsilon\upsilon'}\,\delta_{\alpha\alpha'}\,\frac{1}{N_1\,N_2\,N_3} \underbrace{\sum_{\mathbf{T_{n''}}} \mathrm{e}^{\mathrm{i}\,(\mathbf{q_{n'}}-\mathbf{q_n})^{\mathrm{t}}\cdot\mathbf{T_{n''}}}}_{\overset{(2.68)}{=} N_1\,N_2\,N_3\,\delta_{\mathbf{q_n}\mathbf{q_{n'}}}}$$
$$=\mathrm{i}\,\hbar\,\delta_{\mathbf{q_n}\mathbf{q_{n'}}}\,\delta_{\upsilon\upsilon'}\,\delta_{\alpha\alpha'}. \tag{3.126}$$

Analogously, one finds

$$\hat{\mathfrak{U}}^{\dagger}_{\upsilon\alpha}(\mathbf{q_n})\,\hat{\mathfrak{P}}_{\upsilon'\alpha'}(\mathbf{q_{n'}}) - \hat{\mathfrak{P}}^{\dagger}_{\upsilon'\alpha'}(\mathbf{q_{n'}})\,\hat{\mathfrak{U}}_{\upsilon\alpha}(\mathbf{q_n}) = \mathrm{i}\,\hbar\,\delta_{\mathbf{q_n}\mathbf{q_{n'}}}\,\delta_{\upsilon\upsilon'}\,\delta_{\alpha\alpha'}, \tag{3.127}$$
$$\hat{\mathfrak{U}}_{\upsilon\alpha}(\mathbf{q_n})\,\hat{\mathfrak{P}}^{\dagger}_{\upsilon'\alpha'}(\mathbf{q_{n'}}) - \hat{\mathfrak{P}}_{\upsilon'\alpha'}(\mathbf{q_{n'}})\,\hat{\mathfrak{U}}^{\dagger}_{\upsilon\alpha}(\mathbf{q_n}) = \mathrm{i}\,\hbar\,\delta_{\mathbf{q_n}\mathbf{q_{n'}}}\,\delta_{\upsilon\upsilon'}\,\delta_{\alpha\alpha'} \tag{3.128}$$

and

$$\left[\hat{\mathfrak{U}}_{\upsilon\alpha}(\mathbf{q_n})\,,\,\hat{\mathfrak{U}}_{\upsilon'\alpha'}(\mathbf{q_{n'}})\right]_- = \left[\hat{\mathfrak{U}}_{\upsilon\alpha}(\mathbf{q_n})\,,\,\hat{\mathfrak{U}}^\dagger_{\upsilon'\alpha'}(\mathbf{q_{n'}})\right]_- = 0, \tag{3.129}$$

$$\left[\hat{\mathfrak{P}}_{\upsilon\alpha}(\mathbf{q_n})\,,\,\hat{\mathfrak{P}}_{\upsilon'\alpha'}(\mathbf{q_{n'}})\right]_- = \left[\hat{\mathfrak{P}}_{\upsilon\alpha}(\mathbf{q_n})\,,\,\hat{\mathfrak{P}}^\dagger_{\upsilon'\alpha'}(\mathbf{q_{n'}})\right]_- = 0. \tag{3.130}$$

From the vectors $\hat{\mathfrak{U}}(\mathbf{q_n}) \in \mathbb{C}^{3\,N_b}$ and $\hat{\mathfrak{P}}(\mathbf{q_n}) \in \mathbb{C}^{3\,N_b}$, we obtain the displacements and momenta of the phonon eigenmodes in reciprocal space with the help of an unitary matrix $\mathbf{C}(\mathbf{q_n}) \in \mathbb{C}^{3\,N_b \times 3\,N_b}$ by $\hat{\mathcal{U}}(\mathbf{q_n}) = \mathbf{C}(\mathbf{q_n}) \cdot \hat{\mathfrak{U}}(\mathbf{q_n})$ and $\hat{\mathcal{P}}(\mathbf{q_n}) = \mathbf{C}(\mathbf{q_n}) \cdot \hat{\mathfrak{P}}(\mathbf{q_n})$. Since $\mathbf{C}(\mathbf{q_n})$ is unitary, one has $\mathbb{1} = \mathbf{C}(\mathbf{q_n}) \cdot \mathbf{C}^\dagger(\mathbf{q_n})$ or, for the matrix elements,

$$\delta_{jj'} = \sum_{k=1}^{3\,N_b} C_{jk}(\mathbf{q_n})\,C^\dagger_{kj'}(\mathbf{q_n}) = \sum_{k=1}^{3\,N_b} C_{jk}(\mathbf{q_n})\,C^*_{j'k}(\mathbf{q_n}), \qquad \forall\, j, j' = 1, \ldots, 3\,N_b. \tag{3.131}$$

Therefore, we obtain

$$\begin{aligned}
&\left[\hat{\mathcal{U}}_j(\mathbf{q_n})\,,\,\hat{\mathcal{P}}^\dagger_{j'}(\mathbf{q_{n'}})\right]_- \\
&= \left(\sum_{k=1}^{3\,N_b} C_{jk}(\mathbf{q_n})\,\hat{\mathfrak{U}}_k(\mathbf{q_n})\right)\left(\sum_{k'=1}^{3\,N_b} C_{j'k'}(\mathbf{q_{n'}})\,\hat{\mathfrak{P}}_{k'}(\mathbf{q_{n'}})\right)^\dagger \\
&\quad - \left(\sum_{k'=1}^{3\,N_b} C_{j'k'}(\mathbf{q_{n'}})\,\hat{\mathfrak{P}}_{k'}(\mathbf{q_{n'}})\right)^\dagger\left(\sum_{k=1}^{3\,N_b} C_{jk}(\mathbf{q_n})\,\hat{\mathfrak{U}}_k(\mathbf{q_n})\right) \\
&= \sum_{k=1}^{3\,N_b}\sum_{k'=1}^{3\,N_b} \underbrace{\left(\hat{\mathfrak{U}}_k(\mathbf{q_n})\,\hat{\mathfrak{P}}^\dagger_{k'}(\mathbf{q_{n'}}) - \hat{\mathfrak{P}}^\dagger_{k'}(\mathbf{q_{n'}})\,\hat{\mathfrak{U}}_k(\mathbf{q_n})\right)}_{=\left[\hat{\mathfrak{U}}_k(\mathbf{q_n})\,,\,\hat{\mathfrak{P}}^\dagger_{k'}(\mathbf{q_{n'}})\right]_- = \mathrm{i}\,\hbar\,\delta_{\mathbf{q_n}\mathbf{q_{n'}}}\,\delta_{kk'}} C_{jk}(\mathbf{q_n})\,C^*_{j'k'}(\mathbf{q_{n'}}) \\
&= \mathrm{i}\,\hbar\,\delta_{\mathbf{q_n}\mathbf{q_{n'}}} \underbrace{\sum_{k=1}^{3\,N_b} C_{jk}(\mathbf{q_n})\,C^*_{j'k}(\mathbf{q_n})}_{\overset{(3.131)}{=}\,\delta_{jj'}} \\
&= \mathrm{i}\,\hbar\,\delta_{jj'}\,\delta_{\mathbf{q_n}\mathbf{q_{n'}}}.
\end{aligned} \tag{3.132}$$

Analogously, one finds

$$\hat{\mathcal{U}}^\dagger_j(\mathbf{q_n})\,\hat{\mathcal{P}}_{j'}(\mathbf{q_{n'}}) - \hat{\mathcal{P}}^\dagger_{j'}(\mathbf{q_{n'}})\,\hat{\mathcal{U}}_j(\mathbf{q_n}) = \mathrm{i}\,\hbar\,\delta_{jj'}\,\delta_{\mathbf{q_n}\mathbf{q_{n'}}}, \tag{3.133}$$

$$\hat{\mathcal{U}}_j(\mathbf{q_n})\,\hat{\mathcal{P}}^\dagger_{j'}(\mathbf{q_{n'}}) - \hat{\mathcal{P}}_{j'}(\mathbf{q_{n'}})\,\hat{\mathcal{U}}^\dagger_j(\mathbf{q_n}) = \mathrm{i}\,\hbar\,\delta_{jj'}\,\delta_{\mathbf{q_n}\mathbf{q_{n'}}} \tag{3.134}$$

and

$$\left[\hat{\mathcal{U}}_j(\mathbf{q_n}),\, \hat{\mathcal{U}}_{j'}(\mathbf{q_{n'}})\right]_- = \left[\hat{\mathcal{U}}_j(\mathbf{q_n}),\, \hat{\mathcal{U}}^\dagger_{j'}(\mathbf{q_{n'}})\right]_- = 0, \tag{3.135}$$

$$\left[\hat{\mathcal{P}}_j(\mathbf{q_n}),\, \hat{\mathcal{P}}_{j'}(\mathbf{q_{n'}})\right]_- = \left[\hat{\mathcal{P}}_j(\mathbf{q_n}),\, \hat{\mathcal{P}}^\dagger_{j'}(\mathbf{q_{n'}})\right]_- = 0. \tag{3.136}$$

From now on, we write $\mathbf{q}$ instead of $\mathbf{q_n}$ to simplify the notation. We obtain for the nuclei Hamilton operator in the harmonic approximation (*c.f.* Eq. (2.83):

$$\hat{H}^{(\mathrm{harm})} = \frac{1}{2}\sum_{\mathbf{q}}\sum_{j=1}^{3N_b}\left(\hat{\mathcal{P}}^\dagger_j(\mathbf{q})\,\hat{\mathcal{P}}_j(\mathbf{q}) + \omega_j^2(\mathbf{q})\,\hat{\mathcal{U}}^\dagger_j(\mathbf{q})\,\hat{\mathcal{U}}_j(\mathbf{q})\right). \tag{3.137}$$

To get an hermitian Hamiltonian, we have to use $\hat{\mathcal{P}}^\dagger_j(\mathbf{q})\,\hat{\mathcal{P}}_j(\mathbf{q})$ instead of $\hat{\mathcal{P}}^2_j(\mathbf{q})$ and $\hat{\mathcal{U}}^\dagger_j(\mathbf{q})\,\hat{\mathcal{U}}_j(\mathbf{q})$ instead of $\hat{\mathcal{U}}^2_j(\mathbf{q})$, since both operators are not hermitian. The Hamiltonian can be transformed to second quantization with the help of the following phonon creation and annihilation operators using $\omega_j(\mathbf{q}) = \omega_j(-\mathbf{q})$ due to $\omega_j^2(\mathbf{q}) = \omega_j^2(-\mathbf{q})$:

Definition 3.6 (*Phonon creation and annihilation operators*) For the phonon mode characterized by $j\mathbf{q}$, one defines the creation operator

$$\hat{b}^\dagger_{j\mathbf{q}} := \sqrt{\frac{\omega_j(\mathbf{q})}{2\hbar}}\,\hat{\mathcal{U}}^\dagger_j(\mathbf{q}) - \mathrm{i}\sqrt{\frac{1}{2\hbar\,\omega_j(\mathbf{q})}}\,\hat{\mathcal{P}}^\dagger_j(\mathbf{q}) \tag{3.138}$$

$$= \sqrt{\frac{\omega_j(\mathbf{q})}{2\hbar}}\,\hat{\mathcal{U}}_j(-\mathbf{q}) - \mathrm{i}\sqrt{\frac{1}{2\hbar\,\omega_j(\mathbf{q})}}\,\hat{\mathcal{P}}_j(-\mathbf{q}) \tag{3.139}$$

and the annihilation operator

$$\hat{b}_{j\mathbf{q}} := \sqrt{\frac{\omega_j(\mathbf{q})}{2\hbar}}\,\hat{\mathcal{U}}_j(\mathbf{q}) + \mathrm{i}\sqrt{\frac{1}{2\hbar\,\omega_j(\mathbf{q})}}\,\hat{\mathcal{P}}_j(\mathbf{q}). \tag{3.140}$$

From the above definitions, we obtain directly

$$\hat{\mathcal{U}}_j(\mathbf{q}) = \sqrt{\frac{\hbar}{2\,\omega_j(\mathbf{q})}}\left(\hat{b}_{j\mathbf{q}} + \hat{b}^\dagger_{j-\mathbf{q}}\right), \tag{3.141}$$

$$\hat{\mathcal{P}}_j(\mathbf{q}) = -\,\mathrm{i}\sqrt{\frac{\hbar\,\omega_j(\mathbf{q})}{2}}\left(\hat{b}_{j\mathbf{q}} - \hat{b}^\dagger_{j-\mathbf{q}}\right). \tag{3.142}$$

Theorem 3.10 (Properties of the phonon creation and annihilation operators)
The phonon creation and annihilation operators obey

$$\left[\hat{b}_{j\mathbf{q}}, \hat{b}^{\dagger}_{j'\mathbf{q}'}\right]_{-} = \delta_{jj'}\,\delta_{\mathbf{q}\mathbf{q}'}, \tag{3.143}$$

$$\left[\hat{b}^{\dagger}_{j\mathbf{q}}, \hat{b}^{\dagger}_{j'\mathbf{q}'}\right]_{-} = 0, \tag{3.144}$$

$$\left[\hat{b}_{j\mathbf{q}}, \hat{b}_{j'\mathbf{q}'}\right]_{-} = 0 \tag{3.145}$$

and

$$\hat{b}^{\dagger}_{j\mathbf{q}}\,\hat{b}_{j\mathbf{q}} = \frac{\hat{H}^{(harm)}}{\hbar\,\omega_j(\mathbf{q})} - \frac{1}{2}. \tag{3.146}$$

Proof Considering the same phonon mode $j\mathbf{q}$, we obtain

$$\begin{aligned}
\left[\hat{b}_{j\mathbf{q}}, \hat{b}^{\dagger}_{j\mathbf{q}}\right]_{-} =& \Bigg[\sqrt{\frac{\omega_j(\mathbf{q})}{2\hbar}}\,\hat{\mathcal{U}}_j(\mathbf{q}) + \mathrm{i}\sqrt{\frac{1}{2\hbar\,\omega_j(\mathbf{q})}}\,\hat{\mathcal{P}}_j(\mathbf{q}), \sqrt{\frac{\omega_j(\mathbf{q})}{2\hbar}}\,\hat{\mathcal{U}}^{\dagger}_j(\mathbf{q}) \\
&- \mathrm{i}\sqrt{\frac{1}{2\hbar\,\omega_j(\mathbf{q})}}\,\hat{\mathcal{P}}^{\dagger}_j(\mathbf{q})\Bigg]_{-} \\
=& \frac{\omega_j(\mathbf{q})}{2\hbar}\underbrace{\left[\hat{\mathcal{U}}_j(\mathbf{q}), \hat{\mathcal{U}}^{\dagger}_j(\mathbf{q})\right]_{-}}_{=0} - \frac{\mathrm{i}}{2\hbar}\underbrace{\left[\hat{\mathcal{U}}_j(\mathbf{q}), \hat{\mathcal{P}}^{\dagger}_j(\mathbf{q})\right]_{-}}_{=\mathrm{i}\,\hbar} + \frac{\mathrm{i}}{2\hbar}\underbrace{\left[\hat{\mathcal{P}}_j(\mathbf{q}), \hat{\mathcal{U}}^{\dagger}_j(\mathbf{q})\right]_{-}}_{=-\mathrm{i}\,\hbar} \\
&+ \frac{1}{2\hbar\,\omega_j(\mathbf{q})}\underbrace{\left[\hat{\mathcal{P}}_j(\mathbf{q}), \hat{\mathcal{P}}^{\dagger}_j(\mathbf{q})\right]_{-}}_{=0} \\
=& 1,
\end{aligned}$$

$$\begin{aligned}
\left[\hat{b}_{j\mathbf{q}}, \hat{b}_{j\mathbf{q}}\right]_{-} =& \Bigg[\sqrt{\frac{\omega_j(\mathbf{q})}{2\hbar}}\,\hat{\mathcal{U}}_j(\mathbf{q}) + \mathrm{i}\sqrt{\frac{1}{2\hbar\,\omega_j(\mathbf{q})}}\,\hat{\mathcal{P}}_j(\mathbf{q}), \sqrt{\frac{\omega_j(\mathbf{q})}{2\hbar}}\,\underbrace{\hat{\mathcal{U}}_j(\mathbf{q})}_{=\hat{\mathcal{U}}^{\dagger}_j(-\mathbf{q})} \\
&+ \mathrm{i}\sqrt{\frac{1}{2\hbar\,\omega_j(\mathbf{q})}}\,\underbrace{\hat{\mathcal{P}}_j(\mathbf{q})}_{=\hat{\mathcal{P}}^{\dagger}_j(-\mathbf{q})}\Bigg]_{-} \\
=& \frac{\omega_j(\mathbf{q})}{2\hbar}\underbrace{\left[\hat{\mathcal{U}}_j(\mathbf{q}), \hat{\mathcal{U}}^{\dagger}_j(-\mathbf{q})\right]_{-}}_{=0} + \frac{\mathrm{i}}{2\hbar}\underbrace{\left[\hat{\mathcal{U}}_j(\mathbf{q}), \hat{\mathcal{P}}^{\dagger}_j(-\mathbf{q})\right]_{-}}_{=\mathrm{i}\,\hbar\,\delta_{\mathbf{q}-\mathbf{q}}=0}
\end{aligned}$$

$$+\frac{\mathrm{i}}{2\hbar}\underbrace{\left[\hat{\mathcal{P}}_j(\mathbf{q}),\hat{\mathcal{U}}_j^\dagger(-\mathbf{q})\right]_-}_{=-\mathrm{i}\hbar\,\delta_{\mathbf{q}-\mathbf{q}}=0}+\frac{1}{2\hbar\,\omega_j(\mathbf{q})}\underbrace{\left[\hat{\mathcal{P}}_j(\mathbf{q}),\hat{\mathcal{P}}_j^\dagger(-\mathbf{q})\right]_-}_{=0}$$

$$=0,$$

$$\left[\hat{b}_{j\mathbf{q}}^\dagger,\hat{b}_{j\mathbf{q}}^\dagger\right]_-=\Bigg[\sqrt{\frac{\omega_j(\mathbf{q})}{2\hbar}}\underbrace{\hat{\mathcal{U}}_j^\dagger(\mathbf{q})}_{=\hat{\mathcal{U}}_j(-\mathbf{q})}-\mathrm{i}\sqrt{\frac{1}{2\hbar\,\omega_j(\mathbf{q})}}\underbrace{\hat{\mathcal{P}}_j^\dagger(\mathbf{q})}_{=\hat{\mathcal{P}}_j(-\mathbf{q})},\sqrt{\frac{\omega_j(\mathbf{q})}{2\hbar}}\hat{\mathcal{U}}_j^\dagger(\mathbf{q})$$

$$-\mathrm{i}\sqrt{\frac{1}{2\hbar\,\omega_j(\mathbf{q})}}\hat{\mathcal{P}}_j^\dagger(\mathbf{q})\Bigg]$$

$$=\frac{\omega_j(\mathbf{q})}{2\hbar}\underbrace{\left[\hat{\mathcal{U}}_j(-\mathbf{q}),\hat{\mathcal{U}}_j^\dagger(\mathbf{q})\right]_-}_{=0}-\frac{\mathrm{i}}{2\hbar}\underbrace{\left[\hat{\mathcal{U}}_j(-\mathbf{q}),\hat{\mathcal{P}}_j^\dagger(\mathbf{q})\right]_-}_{=\mathrm{i}\hbar\,\delta_{-\mathbf{q}\mathbf{q}}=0}$$

$$-\frac{\mathrm{i}}{2\hbar}\underbrace{\left[\hat{\mathcal{P}}_j(-\mathbf{q}),\hat{\mathcal{U}}_j^\dagger(\mathbf{q})\right]_-}_{=-\mathrm{i}\hbar\,\delta_{-\mathbf{q}\mathbf{q}}=0}+\frac{1}{2\hbar\,\omega_j(\mathbf{q})}\underbrace{\left[\hat{\mathcal{P}}_j(-\mathbf{q}),\hat{\mathcal{P}}_j^\dagger(\mathbf{q})\right]_-}_{=0}$$

and $\quad =0.$

For different phonon modes, the above commutators are just zero. Equation (3.146) follows from

$$\hat{b}_{j\mathbf{q}}^\dagger\hat{b}_{j\mathbf{q}}=\left(\sqrt{\frac{\omega_j(\mathbf{q})}{2\hbar}}\hat{\mathcal{U}}_j^\dagger(\mathbf{q})-\mathrm{i}\sqrt{\frac{1}{2\hbar\,\omega_j(\mathbf{q})}}\hat{\mathcal{P}}_j^\dagger(\mathbf{q})\right)\times$$

$$\times\left(\sqrt{\frac{\omega_j(\mathbf{q})}{2\hbar}}\hat{\mathcal{U}}_j(\mathbf{q})+\mathrm{i}\sqrt{\frac{1}{2\hbar\,\omega_j(\mathbf{q})}}\hat{\mathcal{P}}_j(\mathbf{q})\right)$$

$$=\frac{\omega_j(\mathbf{q})}{2\hbar}\hat{\mathcal{U}}_j^\dagger(\mathbf{q})\hat{\mathcal{U}}_j(\mathbf{q})+\frac{\mathrm{i}}{2\hbar}\hat{\mathcal{U}}_j^\dagger(\mathbf{q})\hat{\mathcal{P}}_j(\mathbf{q})-\frac{\mathrm{i}}{2\hbar}\hat{\mathcal{P}}_j^\dagger(\mathbf{q})\hat{\mathcal{U}}_j(\mathbf{q})$$

$$+\frac{1}{2\hbar\,\omega_j(\mathbf{q})}\hat{\mathcal{P}}_j^\dagger(\mathbf{q})\hat{\mathcal{P}}_j(\mathbf{q})$$

$$=\frac{1}{2\hbar\,\omega_j(\mathbf{q})}\hat{\mathcal{P}}_j^\dagger(\mathbf{q})\hat{\mathcal{P}}_j(\mathbf{q})+\frac{\omega_j(\mathbf{q})}{2\hbar}\hat{\mathcal{U}}_j^\dagger(\mathbf{q})\hat{\mathcal{U}}_j(\mathbf{q})$$

$$+\frac{\mathrm{i}}{2\hbar}\underbrace{\left(\hat{\mathcal{U}}_j^\dagger(\mathbf{q}),\hat{\mathcal{P}}_j(\mathbf{q})-\hat{\mathcal{P}}_j^\dagger(\mathbf{q})\hat{\mathcal{U}}_j(\mathbf{q})\right)}_{=\mathrm{i}\hbar}$$

$$=\frac{1}{\hbar\,\omega_j(\mathbf{q})}\frac{\hat{\mathcal{P}}_j^\dagger(\mathbf{q})\hat{\mathcal{P}}_j(\mathbf{q})+\omega_j^2(\mathbf{q})\hat{\mathcal{U}}_j^\dagger(\mathbf{q})\hat{\mathcal{U}}_j(\mathbf{q})}{2}-\frac{1}{2}$$

$$=\frac{\hat{H}^{(\text{harm})}}{\hbar\omega_j(\mathbf{q})}-\frac{1}{2}.$$

■

The last theorem allows us to write the harmonic Hamiltonian (3.137) of the nuclei just as

$$\boxed{\hat{H}^{(\text{harm})}=\sum_j\sum_{\mathbf{q}}\hbar\omega_j(\mathbf{q})\left(\hat{b}^\dagger_{j\mathbf{q}}\hat{b}_{j\mathbf{q}}+\frac{1}{2}\right),} \tag{3.147}$$

which is the representation in second quantization. The factor $\frac{1}{2}$ accounts for the zero point motion of all phonon modes. A nuclei oscillation state is characterized by which phonon modes are discretely excited. $\hat{b}^\dagger_{j\mathbf{q}}$ generates one phonon in the state $j\mathbf{q}$ and $\hat{b}_{j\mathbf{q}}$ destroys one phonon in this state. We denote again by $|0\rangle$ the vacuum state and obtain

$$\hat{b}_{j\mathbf{q}}|0\rangle=0. \tag{3.148}$$

In contrast to electrons, a single phonon mode can be occupied by more than one phonon and there is no phonon number conservation. In this way, we can write the nuclei oscillation state in the Fock space as

$$\left(\hat{b}^\dagger_{j_1\mathbf{q}_1}\right)^{n_1}\left(\hat{b}^\dagger_{j_2\mathbf{q}_2}\right)^{n_2}\cdots\left(\hat{b}^\dagger_{j_{n_{N_\text{p}}}\mathbf{q}_{n_{N_\text{p}}}}\right)^{n_{N_\text{p}}}|0\rangle$$

with the help of non-negative integer numbers $n_1, n_2, \ldots, n_{N_\text{p}}$ that are the occupation numbers of the N_p different phonon modes. Similar to electrons, we can define a phonon occupation number operator:

Definition 3.7 (*Phonon occupation number operator*) The occupation number operator

$$\hat{n}_{j\mathbf{q}}=\hat{b}^\dagger_{j\mathbf{q}}\hat{b}_{j\mathbf{q}} \tag{3.149}$$

counts how many phonons occupy the state $j\mathbf{q}$.

Using the properties of the phonon creation and annihilation operators we can prove the following properties of the occupation number operator:

Theorem 3.11 (Properties of the phonon occupation number operator) *The occupation number operator is hermitian and obeys*

$$\left[\hat{n}_{j\mathbf{q}}, \hat{n}_{j'\mathbf{q}'}\right]_- = 0, \tag{3.150}$$

$$\left[\hat{n}_{j\mathbf{q}}, \hat{b}^\dagger_{j'\mathbf{q}'}\right]_- = \hat{b}^\dagger_{j\mathbf{q}}\, \delta_{jj'}\, \delta_{\mathbf{q}\mathbf{q}'}, \tag{3.151}$$

$$\left[\hat{n}_{j\mathbf{q}}, \hat{b}_{j'\mathbf{q}'}\right]_- = -\hat{b}_{j\mathbf{q}}\, \delta_{jj'}\, \delta_{\mathbf{q}\mathbf{q}'} \tag{3.152}$$

and has got the eigenvalues $0, 1, 2, \ldots$.

Proof The phonon occupation number operator is hermitian due to

$$\hat{n}^\dagger_{j\mathbf{q}} = \left(\hat{b}^\dagger_{j\mathbf{q}}\, \hat{b}_{j\mathbf{q}}\right)^\dagger = \hat{b}^\dagger_{j\mathbf{q}} \left(\hat{b}^\dagger_{j\mathbf{q}}\right)^\dagger = \hat{b}^\dagger_{j\mathbf{q}}\, \hat{b}_{j\mathbf{q}} = \hat{n}_{j\mathbf{q}}.$$

We obtain for $j \neq j'$ or $\mathbf{q} \neq \mathbf{q}'$:

$$\begin{aligned}
\left[\hat{n}_{j\mathbf{q}}, \hat{n}_{j'\mathbf{q}'}\right]_- &= \hat{n}_{j\mathbf{q}}\, \hat{n}_{j'\mathbf{q}'} - \hat{n}_{j'\mathbf{q}'}\, \hat{n}_{j\mathbf{q}} \\
&= \hat{b}^\dagger_{j\mathbf{q}} \underbrace{\hat{b}_{j\mathbf{q}}\, \hat{b}^\dagger_{j'\mathbf{q}'}}_{=\hat{b}^\dagger_{j'\mathbf{q}'}\, \hat{b}_{j\mathbf{q}}} \hat{b}_{j'\mathbf{q}'} - \hat{b}^\dagger_{j'\mathbf{q}'}\, \hat{b}_{j'\mathbf{q}'}\, \hat{b}^\dagger_{j\mathbf{q}}\, \hat{b}_{j\mathbf{q}} \\
&= \underbrace{\hat{b}^\dagger_{j\mathbf{q}}\, \hat{b}^\dagger_{j'\mathbf{q}'}}_{=\hat{b}^\dagger_{j'\mathbf{q}'}\, \hat{b}^\dagger_{j\mathbf{q}}} \underbrace{\hat{b}_{j\mathbf{q}}\, \hat{b}_{j'\mathbf{q}'}}_{=\hat{b}_{j'\mathbf{q}'}\, \hat{b}_{j\mathbf{q}}} - \hat{b}^\dagger_{j'\mathbf{q}'}\, \hat{b}_{j'\mathbf{q}'}\, \hat{b}^\dagger_{j\mathbf{q}}\, \hat{b}_{j\mathbf{q}} \\
&= \hat{b}^\dagger_{j'\mathbf{q}'} \underbrace{\hat{b}^\dagger_{j\mathbf{q}}\, \hat{b}_{j'\mathbf{q}'}}_{=\hat{b}_{j'\mathbf{q}'}\, \hat{b}^\dagger_{j\mathbf{q}}} \hat{b}_{j\mathbf{q}} - \hat{b}^\dagger_{j'\mathbf{q}'}\, \hat{b}_{j'\mathbf{q}'}\, \hat{b}^\dagger_{j\mathbf{q}}\, \hat{b}_{j\mathbf{q}} \\
&= \hat{b}^\dagger_{j'\mathbf{q}'}\, \hat{b}_{j'\mathbf{q}'}\, \hat{b}^\dagger_{j\mathbf{q}}\, \hat{b}_{j\mathbf{q}} - \hat{b}^\dagger_{j'\mathbf{q}'}\, \hat{b}_{j'\mathbf{q}'}\, \hat{b}^\dagger_{j\mathbf{q}}\, \hat{b}_{j\mathbf{q}} \\
&= 0
\end{aligned}$$

For $j = j'$ and $\mathbf{q} = \mathbf{q}'$, the above statement is trivial. We get further for arbitrary $j, j', \mathbf{q}, \mathbf{q}'$:

$$\begin{aligned}
\left[\hat{n}_{j\mathbf{q}}, \hat{b}^\dagger_{j'\mathbf{q}'}\right]_- &= \hat{n}_{j\mathbf{q}}\, \hat{b}^\dagger_{j'\mathbf{q}'} - \hat{b}^\dagger_{j'\mathbf{q}'}\, \hat{n}_{j\mathbf{q}} \\
&= \hat{b}^\dagger_{j\mathbf{q}} \underbrace{\hat{b}_{j\mathbf{q}}\, \hat{b}^\dagger_{j'\mathbf{q}'}}_{=\delta_{jj'}\, \delta_{\mathbf{q}\mathbf{q}'} + \hat{b}^\dagger_{j'\mathbf{q}'}\, \hat{b}_{j\mathbf{q}}} - \hat{b}^\dagger_{j'\mathbf{q}'}\, \hat{b}^\dagger_{j\mathbf{q}}\, \hat{b}_{j\mathbf{q}} \\
&= \hat{b}^\dagger_{j\mathbf{q}}\, \delta_{jj'}\, \delta_{\mathbf{q}\mathbf{q}'} + \underbrace{\hat{b}^\dagger_{j\mathbf{q}}\, \hat{b}^\dagger_{j'\mathbf{q}'}}_{=\hat{b}^\dagger_{j'\mathbf{q}'}\, \hat{b}^\dagger_{j\mathbf{q}}} \hat{b}_{j\mathbf{q}} - \hat{b}^\dagger_{j'\mathbf{q}'}\, \hat{b}^\dagger_{j\mathbf{q}}\, \hat{b}_{j\mathbf{q}} \\
&= \hat{b}^\dagger_{j\mathbf{q}}\, \delta_{jj'}\, \delta_{\mathbf{q}\mathbf{q}'} + \hat{b}^\dagger_{j'\mathbf{q}'}\, \hat{b}^\dagger_{j\mathbf{q}}\, \hat{b}_{j\mathbf{q}} - \hat{b}^\dagger_{j'\mathbf{q}'}\, \hat{b}^\dagger_{j\mathbf{q}}\, \hat{b}_{j\mathbf{q}} \\
&= \hat{b}^\dagger_{j\mathbf{q}}\, \delta_{jj'}\, \delta_{\mathbf{q}\mathbf{q}'}
\end{aligned}$$

and

$$\begin{aligned}\left[\hat{n}_{j\mathbf{q}},\hat{b}_{j'\mathbf{q}'}\right]_- &= \hat{n}_{j\mathbf{q}}\hat{b}_{j'\mathbf{q}'} - \hat{b}_{j'\mathbf{q}'}\hat{n}_{j\mathbf{q}} \\ &= \hat{b}^\dagger_{j\mathbf{q}}\underbrace{\hat{b}_{j\mathbf{q}}\hat{b}_{j'\mathbf{q}'}}_{=\hat{b}_{j'\mathbf{q}'}\hat{b}_{j\mathbf{q}}} - \hat{b}_{j'\mathbf{q}'}\hat{b}^\dagger_{j\mathbf{q}}\hat{b}_{j\mathbf{q}} \\ &= \underbrace{\hat{b}^\dagger_{j\mathbf{q}}\hat{b}_{j'\mathbf{q}'}}_{=-\delta_{jj'}\delta_{\mathbf{q}\mathbf{q}'}+\hat{b}_{j'\mathbf{q}'}\hat{b}^\dagger_{j\mathbf{q}}}\hat{b}_{j\mathbf{q}} - \hat{b}_{j'\mathbf{q}'}\hat{b}^\dagger_{j\mathbf{q}}\hat{b}_{j\mathbf{q}} \\ &= -\hat{b}_{j\mathbf{q}}\delta_{jj'}\delta_{\mathbf{q}\mathbf{q}'} + \hat{b}_{j'\mathbf{q}'}\hat{b}^\dagger_{j\mathbf{q}}\hat{b}_{j\mathbf{q}} - \hat{b}_{j'\mathbf{q}'}\hat{b}^\dagger_{j\mathbf{q}}\hat{b}_{j\mathbf{q}} \\ &= -\hat{b}_{j\mathbf{q}}\delta_{jj'}\delta_{\mathbf{q}\mathbf{q}'}.\end{aligned}$$

Now we show that the eigenvalues of $\hat{n}_{j\mathbf{q}}$ are all non-negative. For this, we consider a normalized eingenstate $|\psi_n\rangle$ of $\hat{n}_{j\mathbf{q}}$, i.e.,

$$\hat{n}_{j\mathbf{q}}|\psi_n\rangle = n\,|\psi_n\rangle,$$

and obtain

$$\begin{aligned}n &= n\,\langle\psi_n|\psi_n\rangle \\ &= \langle\psi_n|n\,\psi_n\rangle \\ &= \langle\psi_n|\hat{n}_{j\mathbf{q}}\,\psi_n\rangle \\ &= \left\langle\psi_n\middle|\hat{b}^\dagger_{j\mathbf{q}}\hat{b}_{j\mathbf{q}}\,\psi_n\right\rangle \\ &= \left\langle\hat{b}_{j\mathbf{q}}\psi_n\middle|\hat{b}_{j\mathbf{q}}\,\psi_n\right\rangle \\ &= \left|\hat{b}_{j\mathbf{q}}\psi_n\right|^2.\end{aligned}$$

Furthermore, we have

$$\begin{aligned}\hat{n}_{j\mathbf{q}}\hat{b}^\dagger_{j\mathbf{q}}|\psi_n\rangle &= \hat{b}^\dagger_{j\mathbf{q}}\underbrace{\hat{b}_{j\mathbf{q}}\hat{b}^\dagger_{j\mathbf{q}}}_{=1+\hat{b}^\dagger_{j\mathbf{q}}\hat{b}_{j\mathbf{q}}}|\psi_n\rangle \\ &= \hat{b}^\dagger_{j\mathbf{q}}\left(1+\hat{b}^\dagger_{j\mathbf{q}}\hat{b}_{j\mathbf{q}}\right)|\psi_n\rangle \\ &= \hat{b}^\dagger_{j\mathbf{q}}\left(1+\hat{n}_{j\mathbf{q}}\right)|\psi_n\rangle \\ &= (1+n)\,\hat{b}^\dagger_{j\mathbf{q}}|\psi_n\rangle\end{aligned}$$

and

$$\begin{aligned}
\hat{n}_{j\mathbf{q}}\hat{b}_{j\mathbf{q}}|\psi_n\rangle = & \underbrace{\hat{b}^\dagger_{j\mathbf{q}}\hat{b}_{j\mathbf{q}}}_{=-1+\hat{b}_{j\mathbf{q}}\hat{b}^\dagger_{j\mathbf{q}}}\hat{b}_{j\mathbf{q}}|\psi_n\rangle \\
=&\hat{b}_{j\mathbf{q}}\left(-1+\hat{b}^\dagger_{j\mathbf{q}}\hat{b}_{j\mathbf{q}}\right)|\psi_n\rangle \\
=&\hat{b}_{j\mathbf{q}}\left(-1+\hat{n}_{j\mathbf{q}}\right)|\psi_n\rangle \\
=&(-1+n)\,\hat{b}_{j\mathbf{q}}|\psi_n\rangle.
\end{aligned}$$

Therefore, we can increase the eigenvalue of the eigenstate by 1 with the help of the creation operator and we can decrease the eigenvalue of the eigenstate by 1 with the help of the annihilation operator. Since the eigenvalues of $\hat{n}_{j\mathbf{q}}$ are non-negative and $\hat{b}_{j\mathbf{q}}|0\rangle = 0$, we get that all eigenvalues of $\hat{n}_{j\mathbf{q}}$ are non-negative integers. ∎

With the help of the last theorem, we can verify that the occupation number operator $\hat{n}_{j\mathbf{q}}$ counts, indeed, the number of phonons in the state $j\mathbf{q}$. For this, we consider the state

$$|\Psi\rangle = \left(\hat{b}^\dagger_{j_1\mathbf{q}_1}\right)^{n_1}\left(\hat{b}^\dagger_{j_2\mathbf{q}_2}\right)^{n_2}\cdots\left(\hat{b}^\dagger_{j_{nN_\mathrm{p}}\mathbf{q}_{nN_\mathrm{p}}}\right)^{nN_\mathrm{p}}|0\rangle$$

and obtain

$$\begin{aligned}
\hat{n}_{j_i\mathbf{q}_i}|\Psi\rangle =& \underbrace{\hat{n}_{j_i\mathbf{q}_i}\hat{b}^\dagger_{j_1\mathbf{q}_1}}_{=\hat{b}^\dagger_{j_1\mathbf{q}_1}\hat{n}_{j_i\mathbf{q}_i}}\left(\hat{b}^\dagger_{j_1\mathbf{q}_1}\right)^{n_1-1}\cdots\left(\hat{b}^\dagger_{j_i\mathbf{q}_i}\right)^{n_i}\cdots\left(\hat{b}^\dagger_{j_{nN_\mathrm{p}}\mathbf{q}_{nN_\mathrm{p}}}\right)^{nN_\mathrm{p}}|0\rangle \\
=&\ldots \\
=&\left(\hat{b}^\dagger_{j_1\mathbf{q}_1}\right)^{n_1}\cdots\underbrace{\hat{n}_{j_i\mathbf{q}_i}\hat{b}^\dagger_{j_i\mathbf{q}_i}}_{=\hat{b}^\dagger_{j_i\mathbf{q}_i}+\hat{b}^\dagger_{j_i\mathbf{q}_i}\hat{n}_{j_i\mathbf{q}_i}}\left(\hat{b}^\dagger_{j_i\mathbf{q}_i}\right)^{n_i-1}\cdots\left(\hat{b}^\dagger_{j_{nN_\mathrm{p}}\mathbf{q}_{nN_\mathrm{p}}}\right)^{nN_\mathrm{p}}|0\rangle \\
=&\left(\hat{b}^\dagger_{j_1\mathbf{q}_1}\right)^{n_1}\cdots\hat{b}_{j_i\mathbf{q}_i}\left(\hat{b}^\dagger_{j_i\mathbf{q}_i}\right)^{n_i-1}\cdots\left(\hat{b}^\dagger_{j_{nN_\mathrm{p}}\mathbf{q}_{nN_\mathrm{p}}}\right)^{nN_\mathrm{p}}|0\rangle \\
&+\left(\hat{b}^\dagger_{j_1\mathbf{q}_1}\right)^{n_1}\cdots\hat{b}^\dagger_{j_i\mathbf{q}_i}\hat{n}_{j_i\mathbf{q}_i}\hat{b}^\dagger_{j_i\mathbf{q}_i}\left(\hat{b}^\dagger_{j_i\mathbf{q}_i}\right)^{n_i-2}\cdots\left(\hat{b}^\dagger_{j_{nN_\mathrm{p}}\mathbf{q}_{nN_\mathrm{p}}}\right)^{nN_\mathrm{p}}|0\rangle \\
=&\ldots \\
=&n_i\left(\hat{b}^\dagger_{j_1\mathbf{q}_1}\right)^{n_1}\cdots\left(\hat{b}^\dagger_{j_i\mathbf{q}_i}\right)^{n_i}\cdots\left(\hat{b}^\dagger_{j_{nN_\mathrm{p}}\mathbf{q}_{nN_\mathrm{p}}}\right)^{nN_\mathrm{p}}|0\rangle \\
&+\left(\hat{b}^\dagger_{j_1\mathbf{q}_1}\right)^{n_1}\cdots\left(\hat{b}^\dagger_{j_i\mathbf{q}_i}\right)^{n_i}\hat{n}_{j_i\mathbf{q}_i}\cdots\left(\hat{b}^\dagger_{j_{nN_\mathrm{p}}\mathbf{q}_{nN_\mathrm{p}}}\right)^{nN_\mathrm{p}}|0\rangle \\
=&n_i|\Psi\rangle \\
&+\left(\hat{b}^\dagger_{j_1\mathbf{q}_1}\right)^{n_1}\cdots\left(\hat{b}^\dagger_{j_i\mathbf{q}_i}\right)^{n_i}\cdots\left(\hat{b}^\dagger_{j_{nN_\mathrm{p}}\mathbf{q}_{nN_\mathrm{p}}}\right)^{nN_\mathrm{p}}\underbrace{\hat{n}_{j_i\mathbf{q}_i}}_{=\hat{b}^\dagger_{j_i\mathbf{q}_i}\hat{b}_{j_i\mathbf{q}_i}}|0\rangle \\
=&n_i|\Psi\rangle.
\end{aligned}$$

If only n phonons of the mode $j\mathbf{q}$ are occupied in the state $\left(\hat{b}^{\dagger}_{j\mathbf{q}}\right)^n \left|0\right\rangle$, we need the factor $\frac{1}{\sqrt{n!}}$ for the normalization due to

$$\left\langle 0\right|\left(\hat{b}_{j\mathbf{q}}\right)^n \left(\hat{b}^{\dagger}_{j\mathbf{q}}\right)^n \left|0\right\rangle = n!. \tag{3.153}$$

We prove this by mathematical induction over n. We have for $n = 0$: $\langle 0|0\rangle = 1$, by definition of the vacuum state, and for $n = 1$:

$$\left\langle 0\right| \underbrace{\hat{b}_{j\mathbf{q}}\hat{b}^{\dagger}_{j\mathbf{q}}}_{=1+\hat{b}^{\dagger}_{j\mathbf{q}}\hat{b}_{j\mathbf{q}}} \left|0\right\rangle = \langle 0|0\rangle = 1. \tag{3.154}$$

Now we assume, that Eq. (3.153) is valid up to a given $n \in \mathbb{N}_0$ and prove Eq. (3.153) for $n+1$:

$$\begin{aligned}
&\left\langle 0\right|\left(\hat{b}_{j\mathbf{q}}\right)^{n+1} \left(\hat{b}^{\dagger}_{j\mathbf{q}}\right)^{n+1} \left|0\right\rangle \\
=&\left\langle 0\right|\left(\hat{b}_{j\mathbf{q}}\right)^{n} \underbrace{\hat{b}_{j\mathbf{q}}\hat{b}^{\dagger}_{j\mathbf{q}}}_{=1+\hat{b}^{\dagger}_{j\mathbf{q}}\hat{b}_{j\mathbf{q}}} \left(\hat{b}^{\dagger}_{j\mathbf{q}}\right)^{n} \left|0\right\rangle \\
=&\left\langle 0\right|\left(\hat{b}_{j\mathbf{q}}\right)^{n} \left(\hat{b}^{\dagger}_{j\mathbf{q}}\right)^{n} \left|0\right\rangle + \left\langle 0\right|\left(\hat{b}_{j\mathbf{q}}\right)^{n} \hat{b}^{\dagger}_{j\mathbf{q}} \underbrace{\hat{b}_{j\mathbf{q}}\hat{b}^{\dagger}_{j\mathbf{q}}}_{=1+\hat{b}^{\dagger}_{j\mathbf{q}}\hat{b}_{j\mathbf{q}}} \left(\hat{b}^{\dagger}_{j\mathbf{q}}\right)^{n-1} \left|0\right\rangle \\
=&2\left\langle 0\right|\left(\hat{b}_{j\mathbf{q}}\right)^{n} \left(\hat{b}^{\dagger}_{j\mathbf{q}}\right)^{n} \left|0\right\rangle + \left\langle 0\right|\left(\hat{b}_{j\mathbf{q}}\right)^{n} \left(\hat{b}^{\dagger}_{j\mathbf{q}}\right)^{2} \underbrace{\hat{b}_{j\mathbf{q}}\hat{b}^{\dagger}_{j\mathbf{q}}}_{=1+\hat{b}^{\dagger}_{j\mathbf{q}}\hat{b}_{j\mathbf{q}}} \left(\hat{b}^{\dagger}_{j\mathbf{q}}\right)^{n-2} \left|0\right\rangle \\
=&\ldots \\
=&(n+1) \underbrace{\left\langle 0\right|\left(\hat{b}_{j\mathbf{q}}\right)^{n} \left(\hat{b}^{\dagger}_{j\mathbf{q}}\right)^{n} \left|0\right\rangle}_{=n!} \\
=&(n+1)!.
\end{aligned}$$

Now we derive the average values of the occupation numbers of the phonon modes in equilibrium at temperature T taken from Ref. [16]. For this, we need the following theorem about the geometric series, which proof is taken from Ref [17]:

Theorem 3.12 (Geometric series) *One obtains for* $|x| < 1$*:*

$$\sum_{i=0}^{\infty} x^i = \frac{1}{1-x}. \tag{3.155}$$

Proof Since one has for $|x| < 1$: $\lim_{n\to\infty} x^{n+1} = 0$, it follows

$$\lim_{n\to\infty} x^i \overset{(2.60)}{=} \lim_{n\to\infty} \frac{1-x^{n+1}}{1-x} = \frac{1}{1-x}. \tag{3.156}$$

■

Theorem 3.13 (Equilibrium phonon occupation numbers at temperature T_i) *In the harmonic approximation, the average value of the phonon occupation numbers in equilibrium at temperature* T_i *is given by the* **Bose-Einstein distribution**

$$\langle \hat{n}_{j\mathbf{q}} \rangle = \frac{1}{\mathrm{e}^{\frac{\hbar\omega_j(\mathbf{q})}{k_B T_i}} - 1} \tag{3.157}$$

where k_B *denotes the Boltzmann constant.*

Proof In the harmonic approximation, the different phonon modes are independent to each other. The total Hamilton operator $\hat{H}^{(\mathrm{harm})}$ in Eq. (3.147) is just the sum of the Hamilton operators of the different phonon modes. Therefore, we only need to consider a single arbitrary phonon mode $j\mathbf{q}$. The internal energy contained in this phonon mode is given by

$$E_{j\mathbf{q}} = \hbar\omega_j(\mathbf{q}) \left(n_{j\mathbf{q}} + \frac{1}{2} \right),$$

where $n_{j\mathbf{q}}$ is the occupation number of this phonon mode. We denote by $|\Psi_{n_{j\mathbf{q}}}\rangle$ the state, in which only this phonon mode is $n_{j\mathbf{q}}$ times occupied. In this way, we get

$$\left\langle \Psi_{n_{j\mathbf{q}}} \middle| \hat{H}^{(\mathrm{harm})} \middle| \Psi_{n_{j\mathbf{q}}} \right\rangle = E_{j\mathbf{q}}. \tag{3.158}$$

In thermodynamical equilibrium at temperature T_i, the phonon modes form a canonical ensemble. Therefore, the probability of an occupation $n_{j\mathbf{q}}$ is given by

$$\mathbb{p}(n_{j\mathbf{q}}) = \frac{\mathrm{e}^{-\frac{E_{j\mathbf{q}}}{k_\mathrm{B} T_\mathrm{i}}}}{\mathscr{Z}^{(\mathrm{c})}} = \frac{\mathrm{e}^{-\frac{\hbar\omega_j(\mathbf{q})\left(n_{j\mathbf{q}}+\frac{1}{2}\right)}{k_\mathrm{B} T_\mathrm{i}}}}{\mathscr{Z}^{(\mathrm{c})}} \tag{3.159}$$

with the partition function

$$\mathscr{Z}^{(c)} = \sum_{n_{j\mathbf{q}}=0}^{\infty} \mathrm{e}^{-\frac{E_{j\mathbf{q}}}{k_B T_i}} = \sum_{n_{j\mathbf{q}}=0}^{\infty} \mathrm{e}^{-\frac{\hbar\omega_j(\mathbf{q})\left(n_{j\mathbf{q}}+\frac{1}{2}\right)}{k_B T_i}} = \mathrm{e}^{-\frac{\hbar\omega_j(\mathbf{q})}{2k_B T_i}} \sum_{n_{j\mathbf{q}}=0}^{\infty} \left(\mathrm{e}^{-\frac{\hbar\omega_j(\mathbf{q})}{k_B T_i}}\right)^{n_{j\mathbf{q}}}. \tag{3.160}$$

Using this, we obtain for the average occupation number

$$\begin{aligned}
\langle n_{j\mathbf{q}}\rangle &= \sum_{n_{j\mathbf{q}}=0}^{\infty} n_{j\mathbf{q}}\, \mathbb{p}(n_{j\mathbf{q}}) \\
&= \sum_{n_{j\mathbf{q}}=0}^{\infty} \left(n_{j\mathbf{q}} \frac{\mathrm{e}^{-\frac{\hbar\omega_j(\mathbf{q})\left(n_{j\mathbf{q}}+\frac{1}{2}\right)}{k_B T_i}}}{\mathscr{Z}^{(c)}} \right) \\
&= \underbrace{\frac{1}{\mathrm{e}^{-\frac{\hbar\omega_j(\mathbf{q})}{2k_B T_i}} \sum_{n_{j\mathbf{q}}=0}^{\infty} \left(\mathrm{e}^{-\frac{\hbar\omega_j(\mathbf{q})}{k_B T_i}}\right)^{n_{j\mathbf{q}}}}}_{=\frac{1}{\mathscr{Z}^{(c)}}} \mathrm{e}^{-\frac{\hbar\omega_j(\mathbf{q})}{2k_B T_i}} \sum_{n_{j\mathbf{q}}=0}^{\infty} \left(n_{j\mathbf{q}} \mathrm{e}^{-\frac{\hbar\omega_j(\mathbf{q}) n_{j\mathbf{q}}}{k_B T_i}} \right) \\
&= \frac{1}{\sum_{n_{j\mathbf{q}}=0}^{\infty} \left(\mathrm{e}^{-\frac{\hbar\omega_j(\mathbf{q})}{k_B T_i}}\right)^{n_{j\mathbf{q}}}} \sum_{n_{j\mathbf{q}}=0}^{\infty} \left(n_{j\mathbf{q}} \mathrm{e}^{-\frac{\hbar\omega_j(\mathbf{q}) n_{j\mathbf{q}}}{k_B T_i}} \right) \\
&= -\frac{k_B T_i}{\hbar} \frac{\partial}{\partial\omega_j(\mathbf{q})} \ln\left(\sum_{n_{j\mathbf{q}}=0}^{\infty} \mathrm{e}^{-\frac{\hbar\omega_j(\mathbf{q}) n_{j\mathbf{q}}}{k_B T_i}} \right) \\
&\overset{(3.155)}{=} -\frac{k_B T_i}{\hbar} \frac{\partial}{\partial\omega_j(\mathbf{q})} \ln\left(\frac{1}{1-\mathrm{e}^{-\frac{\hbar\omega_j(\mathbf{q})}{k_B T_i}}} \right) \\
&= \frac{k_B T_i}{\hbar} \frac{\partial}{\partial\omega_j(\mathbf{q})} \ln\left(1-\mathrm{e}^{-\frac{\hbar\omega_j(\mathbf{q})}{k_B T_i}} \right) \\
&= \frac{\mathrm{e}^{-\frac{\hbar\omega_j(\mathbf{q})}{k_B T_i}}}{1-\mathrm{e}^{-\frac{\hbar\omega_j(\mathbf{q})}{k_B T_i}}} \\
&= \frac{1}{\mathrm{e}^{\frac{\hbar\omega_j(\mathbf{q})}{k_B T_i}}-1}.
\end{aligned} \tag{3.161}$$

Here, we utilized the theorem about the geometric series (3.155), which can be applied, since $\mathrm{e}^{-\frac{\hbar\omega_j(\mathbf{q})}{k_B T_i}} < 1$ is valid due to $\frac{\hbar\omega_j(\mathbf{q})}{k_B T_i} > 0$. ■

We present in Fig. 3.1 the Bose-Einstein distribution of the phonons for various temperatures. In contrast to the Fermi distribution of the electrons (see Fig. 2.4, the Bose-Einstein distribution diverges to infinity at zero energy.

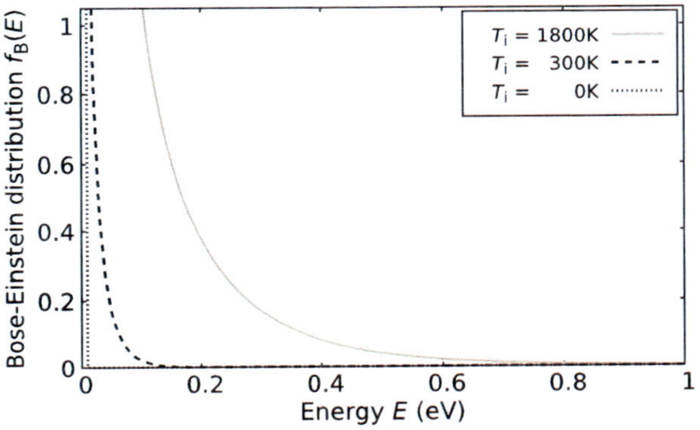

Fig. 3.1 The Bose-Einstein distribution is shown for selected temperatures T_i

3.3 Reduced Electron Density Matrices

Under ambient conditions, the electrons form a canonical ensemble at $T_e = 300$ K. Therefore, the electrons are entirely described by the N_e-electron density matrix

$$\hat{\rho}_e = \sum_{\ell} p_\ell \, |\Psi_\ell\rangle\langle\Psi_\ell|,$$

where $|\Psi_\ell\rangle$ are N_e-electron states. The time propagation of the N_e-electron density matrix is given by the von Neumann equation (2.182) with the help of the electronic Hamiltonian $\hat{H}_e$:

$$\frac{\partial \hat{\rho}_e}{\partial t} = -\frac{i}{\hbar}\left[\hat{H}_e \, , \, \hat{\rho}_e\right]_- .$$

Within the Born-Oppenheimer approximation, the electronic Hamiltonian $\hat{H}_e$ is just a sum of one- and two-particle operators:

$$\hat{H}_e = \sum_{i=1}^{N_e}\left(-\frac{\hbar^2}{2\,M^2}\nabla^2_{\mathbf{R}_i} - \sum_{k=1}^{N_{at}} \frac{e^2}{4\pi\,\varepsilon_0}\frac{Z_k}{|\mathbf{r}_k - \mathbf{R}_i|} + \frac{E_{nn}}{N_e}\right) + \frac{1}{2}\sum_{i=1}^{N_e}\sum_{\substack{j=1 \\ j\neq i}}^{N_e} \frac{e^2}{4\pi\,\varepsilon_0}\frac{1}{|\mathbf{R}_i - \mathbf{R}_j|}$$

$$= \sum_{i=1}^{N_e} \hat{H}_i^{(1)} + \frac{1}{2}\sum_{i=1}^{N_e}\sum_{\substack{j=1 \\ j\neq i}}^{N_e} \hat{V}_{ij}^{(int)} \tag{3.162}$$

with $\hat{V}_{ij}^{(int)} = \hat{V}_{ji}^{(int)}$. Hence, we do not need to work with the complicated full N_e-electron density matrix. Any N_e-electron state is a linear combination of slater determinants build from one-electron states. We consider a periodic crystal structure, so that these one-electron states are Bloch states $|\ell_i\mathbf{k}_i\rangle$ with definite wave vector $\mathbf{k}_i$. These Bloch states form a complete set of orthonormal one-electron states and the slater determinant states $|\ell_1\mathbf{k}_1, \ldots, \ell_{N_e}\mathbf{k}_{N_e}\rangle$ that are build from the one-electron

Bloch states $|\ell_1\mathbf{k}_1\rangle, \ldots, |\ell_{N_e}\mathbf{k}_{N_e}\rangle$ form a complete set of N_e-electron states:

$$\sum_{\ell_1\mathbf{k}_1,\ldots,\ell_{N_e}\mathbf{k}_{N_e}} \left|\ell_1\mathbf{k}_1, \ldots, \ell_{N_e}\mathbf{k}_{N_e}\right\rangle\!\left\langle\ell_1\mathbf{k}_1, \ldots, \ell_{N_e}\mathbf{k}_{N_e}\right| = \hat{\mathbb{1}}. \tag{3.163}$$

Now one defines reduced density matrices:

Definition 3.8 *(One-electron and two-electron reduced density matrices)* The one-electron reduced density matrix is defined as

$$\begin{aligned}&\left\langle\ell_1\mathbf{k}_1\middle|\hat{\rho}_e^{(1)}\middle|\ell_1'\mathbf{k}_1'\right\rangle\\ &= \sum_{\ell_2\mathbf{k}_2,\ldots,\ell_{N_e}\mathbf{k}_{N_e}} N_e\left\langle\ell_1\mathbf{k}_1, \ldots, \ell_{N_e}\mathbf{k}_{N_e}\middle|\hat{\rho}_e\middle|\ell_1'\mathbf{k}_1', \ell_2\mathbf{k}_2, \ldots, \ell_{N_e}\mathbf{k}_{N_e}\right\rangle\end{aligned} \tag{3.164}$$

and the two-electron reduced density matrix is defined as

$$\begin{aligned}&\left\langle\ell_1\mathbf{k}_1, \ell_2\mathbf{k}_2\middle|\hat{\rho}_e^{(2)}\middle|\ell_1'\mathbf{k}_1', \ell_2'\mathbf{k}_2'\right\rangle\\ &= \sum_{\ell_3\mathbf{k}_3,\ldots,\ell_{N_e}\mathbf{k}_{N_e}} N_e\,(N_e-1)\left\langle\ell_1\mathbf{k}_1, \ldots, \ell_{N_e}\mathbf{k}_{N_e}\middle|\hat{\rho}_e\middle|\ell_1'\mathbf{k}_1', \ell_2'\mathbf{k}_2', \ell_3\mathbf{k}_3 \ldots, \ell_{N_e}\mathbf{k}_{N_e}\right\rangle.\end{aligned} \tag{3.165}$$

By definition, one has

$$\begin{aligned}\operatorname*{Tr}_1\left(\hat{\rho}_e^{(1)}\right) &= \sum_{\ell_1\mathbf{k}_1,\ldots,\ell_{N_e}\mathbf{k}_{N_e}} N_e\left\langle\ell_1\mathbf{k}_1, \ldots, \ell_{N_e}\mathbf{k}_{N_e}\middle|\hat{\rho}_e\middle|\ell_1\mathbf{k}_1, \ldots, \ell_{N_e}\mathbf{k}_{N_e}\right\rangle\\ &= N_e\,\mathrm{Tr}\left(\hat{\rho}_e\right)\\ &= N_e,\end{aligned} \tag{3.166}$$

where $\operatorname*{Tr}_1$ denotes that the trace just runs over the first electron states. Furthermore, $\left\langle\ell_1\mathbf{k}_1\middle|\hat{\rho}_e^{(1)}\middle|\ell_1\mathbf{k}_1\right\rangle$ yields the occupation of the one-electron state $|\ell_1\mathbf{k}_1\rangle$.

Theorem 3.14 (Average value of the electronic Hamiltonian) *Let the Hamiltonian*

$$\hat{H} = \sum_{i=1}^{N_e} \hat{H}_i^{(1)} + \frac{1}{2}\sum_{i=1}^{N_e}\sum_{\substack{j=1\\ j\neq i}}^{N_e} \hat{V}_{ij}^{(int)}$$

be a sum of single-particle operators $\hat{H}_i^{(1)}$ *and of two-particle operators* $\hat{V}_{ij}^{(int)} = \hat{V}_{ji}^{(int)}$. *Then the average value of the Hamiltonian is given by*

$$\begin{aligned}\langle \hat{H} \rangle_{\rho_e} &= Tr(\hat{\rho}_e \hat{H}) \\ &= \underset{1}{Tr}\left(\hat{\rho}_e^{(1)} \hat{H}_1^{(1)}\right) + \frac{1}{2} \underset{1,2}{Tr}\left(\hat{\rho}_e^{(2)} \hat{V}_{12}^{(int)}\right). \end{aligned} \tag{3.167}$$

Here, $\underset{1}{Tr}$ *denotes the trace over the first electron states and* $\underset{1,2}{Tr}$ *over the first and second electron states.*

Proof The trace of the single-particle part of the Hamiltonian is given by

$$\begin{aligned}
&\mathrm{Tr}\left(\hat{\rho}_e \sum_{i=1}^{N_e} \hat{H}_i^{(1)}\right) \\
&= \sum_{\ell_1 \mathbf{k}_1, \ldots, \ell_{N_e} \mathbf{k}_{N_e}} \left\langle \ell_1 \mathbf{k}_1, \ldots, \ell_{N_e} \mathbf{k}_{N_e} \middle| \hat{\rho}_e \sum_{i=1}^{N_e} \hat{H}_i^{(1)} \middle| \ell_1 \mathbf{k}_1, \ldots, \ell_{N_e} \mathbf{k}_{N_e} \right\rangle \\
&= \sum_{i=1}^{N_e} \sum_{\ell_1 \mathbf{k}_1, \ldots, \ell_{N_e} \mathbf{k}_{N_e}} \left\langle \ell_1 \mathbf{k}_1, \ldots, \ell_{N_e} \mathbf{k}_{N_e} \middle| \hat{\rho}_e \hat{H}_i^{(1)} \middle| \ell_1 \mathbf{k}_1, \ldots, \ell_{N_e} \mathbf{k}_{N_e} \right\rangle \\
&= N_e \sum_{\ell_1 \mathbf{k}_1, \ldots, \ell_{N_e} \mathbf{k}_{N_e}} \left\langle \ell_1 \mathbf{k}_1, \ldots, \ell_{N_e} \mathbf{k}_{N_e} \middle| \hat{\rho}_e \hat{H}_1^{(1)} \middle| \ell_1 \mathbf{k}_1, \ldots, \ell_{N_e} \mathbf{k}_{N_e} \right\rangle \\
&= \sum_{\ell_1 \mathbf{k}_1, \ldots, \ell_{N_e} \mathbf{k}_{N_e}} \sum_{\ell'_1 \mathbf{k}'_1, \ldots, \ell'_{N_e} \mathbf{k}'_{N_e}} N_e \left\langle \ell_1 \mathbf{k}_1, \ldots, \ell_{N_e} \mathbf{k}_{N_e} \middle| \hat{\rho}_e \middle| \ell'_1 \mathbf{k}'_1, \ldots, \ell'_{N_e} \mathbf{k}'_{N_e} \right\rangle \times \\
&\qquad \times \left\langle \ell'_1 \mathbf{k}'_1, \ldots, \ell'_{N_e} \mathbf{k}'_{N_e} \middle| \hat{H}_1^{(1)} \middle| \ell_1 \mathbf{k}_1, \ldots, \ell_{N_e} \mathbf{k}_{N_e} \right\rangle \\
&= \sum_{\ell_1 \mathbf{k}_1} \sum_{\ell'_1 \mathbf{k}'_1} \left\langle \ell'_1 \mathbf{k}'_1 \middle| \hat{H}_1^{(1)} \middle| \ell_1 \mathbf{k}_1 \right\rangle \times \\
&\qquad \times \underbrace{\sum_{\ell_2 \mathbf{k}_2, \ldots, \ell_{N_e} \mathbf{k}_{N_e}} N_e \left\langle \ell_1 \mathbf{k}_1, \ldots, \ell_{N_e} \mathbf{k}_{N_e} \middle| \hat{\rho}_e \middle| \ell'_1 \mathbf{k}'_1, \ell_2 \mathbf{k}_2, \ldots, \ell_{N_e} \mathbf{k}_{N_e} \right\rangle}_{=\left\langle \ell_1 \mathbf{k}_1 \middle| \hat{\rho}_e^{(1)} \middle| \ell'_1 \mathbf{k}'_1 \right\rangle} \\
&= \sum_{\ell_1 \mathbf{k}_1} \sum_{\ell'_1 \mathbf{k}'_1} \left\langle \ell_1 \mathbf{k}_1 \middle| \hat{\rho}_e^{(1)} \middle| \ell'_1 \mathbf{k}'_1 \right\rangle \left\langle \ell'_1 \mathbf{k}'_1 \middle| \hat{H}_1^{(1)} \middle| \ell_1 \mathbf{k}_1 \right\rangle \\
&= \sum_{\ell_1 \mathbf{k}_1} \left\langle \ell_1 \mathbf{k}_1 \middle| \hat{\rho}_e^{(1)} \hat{H}_1^{(1)} \middle| \ell_1 \mathbf{k}_1 \right\rangle \\
&= \underset{1}{\mathrm{Tr}}\left(\hat{\rho}_e^{(1)} \hat{H}_1^{(1)}\right)
\end{aligned}$$

and the trace of the two-particle part of the Hamiltonian is given by

$$
\begin{aligned}
&\mathrm{Tr}\left(\hat{\rho}_{\mathrm{e}}\,\frac{1}{2}\sum_{i=1}^{N_{\mathrm{e}}}\sum_{\substack{j=1\\ j\neq i}}^{N_{\mathrm{e}}}\hat{V}_{ij}^{(\mathrm{int})}\right)\\
&=\sum_{\ell_1\mathbf{k}_1,\ldots,\ell_{N_{\mathrm{e}}}\mathbf{k}_{N_{\mathrm{e}}}}\left\langle \ell_1\mathbf{k}_1,\ldots,\ell_{N_{\mathrm{e}}}\mathbf{k}_{N_{\mathrm{e}}}\middle|\hat{\rho}_{\mathrm{e}}\,\frac{1}{2}\sum_{i=1}^{N_{\mathrm{e}}}\sum_{\substack{j=1\\ j\neq i}}^{N_{\mathrm{e}}}\hat{V}_{ij}^{(\mathrm{int})}\middle|\ell_1\mathbf{k}_1,\ldots,\ell_{N_{\mathrm{e}}}\mathbf{k}_{N_{\mathrm{e}}}\right\rangle\\
&=\frac{1}{2}\sum_{i=1}^{N_{\mathrm{e}}}\sum_{\substack{j=1\\ j\neq i}}^{N_{\mathrm{e}}}\sum_{\ell_1\mathbf{k}_1,\ldots,\ell_{N_{\mathrm{e}}}\mathbf{k}_{N_{\mathrm{e}}}}\left\langle \ell_1\mathbf{k}_1,\ldots,\ell_{N_{\mathrm{e}}}\mathbf{k}_{N_{\mathrm{e}}}\middle|\hat{\rho}_{\mathrm{e}}\,\hat{V}_{ij}^{(\mathrm{int})}\middle|\ell_1\mathbf{k}_1,\ldots,\ell_{N_{\mathrm{e}}}\mathbf{k}_{N_{\mathrm{e}}}\right\rangle\\
&=\frac{N_{\mathrm{e}}\left(N_{\mathrm{e}}-1\right)}{2}\sum_{\ell_1\mathbf{k}_1,\ldots,\ell_{N_{\mathrm{e}}}\mathbf{k}_{N_{\mathrm{e}}}}\left\langle \ell_1\mathbf{k}_1,\ldots,\ell_{N_{\mathrm{e}}}\mathbf{k}_{N_{\mathrm{e}}}\middle|\hat{\rho}_{\mathrm{e}}\,\hat{V}_{12}^{(\mathrm{int})}\middle|\ell_1\mathbf{k}_1,\ldots,\ell_{N_{\mathrm{e}}}\mathbf{k}_{N_{\mathrm{e}}}\right\rangle\\
&=\frac{1}{2}\sum_{\ell_1\mathbf{k}_1,\ldots,\ell_{N_{\mathrm{e}}}\mathbf{k}_{N_{\mathrm{e}}}}\;\sum_{\ell_1'\mathbf{k}_1',\ldots,\ell_{N_{\mathrm{e}}}'\mathbf{k}_{N_{\mathrm{e}}}'}N_{\mathrm{e}}\left(N_{\mathrm{e}}-1\right)\left\langle \ell_1\mathbf{k}_1,\ldots,\ell_{N_{\mathrm{e}}}\mathbf{k}_{N_{\mathrm{e}}}\middle|\hat{\rho}_{\mathrm{e}}\middle|\ell_1'\mathbf{k}_1',\ldots,\ell_{N_{\mathrm{e}}}'\mathbf{k}_{N_{\mathrm{e}}}'\right\rangle\times\\
&\qquad\qquad\times\left\langle \ell_1'\mathbf{k}_1',\ldots,\ell_{N_{\mathrm{e}}}'\mathbf{k}_{N_{\mathrm{e}}}'\middle|\hat{V}_{12}^{(\mathrm{int})}\middle|\ell_1\mathbf{k}_1,\ldots,\ell_{N_{\mathrm{e}}}\mathbf{k}_{N_{\mathrm{e}}}\right\rangle\\
&=\frac{1}{2}\sum_{\ell_1\mathbf{k}_1,\ell_2\mathbf{k}_2}\;\sum_{\ell_1'\mathbf{k}_1',\ell_2'\mathbf{k}_2'}\left\langle \ell_1'\mathbf{k}_1',\ell_2'\mathbf{k}_2'\middle|\hat{V}_{12}^{(\mathrm{int})}\middle|\ell_1\mathbf{k}_1,\ell_2\mathbf{k}_2\right\rangle\times\\
&\quad\times\underbrace{\sum_{\ell_3\mathbf{k}_3,\ldots,\ell_{N_{\mathrm{e}}}\mathbf{k}_{N_{\mathrm{e}}}}N_{\mathrm{e}}\left(N_{\mathrm{e}}-1\right)\left\langle \ell_1\mathbf{k}_1,\ldots,\ell_{N_{\mathrm{e}}}\mathbf{k}_{N_{\mathrm{e}}}\middle|\hat{\rho}_{\mathrm{e}}\middle|\ell_1'\mathbf{k}_1',\ell_2'\mathbf{k}_2',\ell_3\mathbf{k}_3\ldots,\ell_{N_{\mathrm{e}}}\mathbf{k}_{N_{\mathrm{e}}}\right\rangle}_{=\left\langle \ell_1\mathbf{k}_1,\ell_2\mathbf{k}_2\middle|\hat{\rho}_{\mathrm{e}}^{(2)}\middle|\ell_1'\mathbf{k}_1',\ell_2'\mathbf{k}_2'\right\rangle}\\
&=\frac{1}{2}\sum_{\ell_1\mathbf{k}_1,\ell_2\mathbf{k}_2}\;\sum_{\ell_1'\mathbf{k}_1',\ell_2'\mathbf{k}_2'}\left\langle \ell_1\mathbf{k}_1,\ell_2\mathbf{k}_2\middle|\hat{\rho}_{\mathrm{e}}^{(2)}\middle|\ell_1'\mathbf{k}_1',\ell_2'\mathbf{k}_2'\right\rangle\left\langle \ell_1'\mathbf{k}_1',\ell_2'\mathbf{k}_2'\middle|\hat{V}_{12}^{(\mathrm{int})}\middle|\ell_1\mathbf{k}_1,\ell_2\mathbf{k}_2\right\rangle\\
&=\frac{1}{2}\sum_{\ell_1\mathbf{k}_1,\ell_2\mathbf{k}_2}\left\langle \ell_1\mathbf{k}_1,\ell_2\mathbf{k}_2\middle|\hat{\rho}_{\mathrm{e}}^{(2)}\,\hat{V}_{12}^{(\mathrm{int})}\middle|\ell_1\mathbf{k}_1,\ell_2\mathbf{k}_2\right\rangle\\
&=\frac{1}{2}\,\underset{1,2}{\mathrm{Tr}}\left(\hat{\rho}_{\mathrm{e}}^{(2)}\,\hat{V}_{12}^{(\mathrm{int})}\right),
\end{aligned}
$$

which finishes the proof. ■

To calculate the average value of the electronic Hamiltonian, we only need the one-electron and two-electron reduced density matrices. To perform the time propagation of the one-electron reduced density matrix, we also do not need the full N_{e}-electron density matrix:

Theorem 3.15 (Time propagation of the one-electron reduced density matrix) *For an electronic Hamiltonian that is the sum of single-particle operators $\hat{H}_i^{(1)}$ and of two-particle operators $\hat{V}_{ij}^{(int)}$, the time propagation of the one-electron reduced density matrix is given by*

$$\left\langle \ell_1 \mathbf{k}_1 \left| \frac{\partial \hat{\rho}_e^{(1)}}{\partial t} \right| \ell_1' \mathbf{k}_1' \right\rangle = -\frac{\mathrm{i}}{\hbar} \left\langle \ell_1 \mathbf{k}_1 \left| \left[\hat{H}_1^{(1)}, \hat{\rho}_e^{(1)} \right]_- \right| \ell_1' \mathbf{k}_1' \right\rangle - \frac{\mathrm{i}}{2\hbar} \sum_{\ell_2 \mathbf{k}_2} \left\langle \ell_1 \mathbf{k}_1, \ell_2 \mathbf{k}_2 \left| \left[\hat{V}_{12}^{(int)}, \hat{\rho}_e^{(2)} \right]_- \right| \ell_1' \mathbf{k}_1', \ell_2 \mathbf{k}_2 \right\rangle . \tag{3.168}$$

Proof Using the definition of the one-electron reduced density matrix $\hat{\rho}_e^{(1)}$ and the time derivative of the full N_e-electron density matrix $\hat{\rho}_e$, we obtain

$$\begin{aligned}
&\left\langle \ell_1 \mathbf{k}_1 \left| \frac{\partial \hat{\rho}_e^{(1)}}{\partial t} \right| \ell_1' \mathbf{k}_1' \right\rangle \\
&= \sum_{\ell_2 \mathbf{k}_2, \ldots, \ell_{N_e} \mathbf{k}_{N_e}} \left\langle \ell_1 \mathbf{k}_1, \ldots, \ell_{N_e} \mathbf{k}_{N_e} \left| \frac{\partial \hat{\rho}_e}{\partial t} \right| \ell_1' \mathbf{k}_1', \ell_2 \mathbf{k}_2, \ldots, \ell_{N_e} \mathbf{k}_{N_e} \right\rangle \\
&= -\frac{\mathrm{i}}{\hbar} \sum_{\ell_2 \mathbf{k}_2, \ldots, \ell_{N_e} \mathbf{k}_{N_e}} \left\langle \ell_1 \mathbf{k}_1, \ldots, \ell_{N_e} \mathbf{k}_{N_e} \left| [\hat{H}, \hat{\rho}_e]_- \right| \ell_1' \mathbf{k}_1', \ell_2 \mathbf{k}_2, \ldots, \ell_{N_e} \mathbf{k}_{N_e} \right\rangle \\
&= -\frac{\mathrm{i}}{\hbar} \sum_{\ell_2 \mathbf{k}_2, \ldots, \ell_{N_e} \mathbf{k}_{N_e}} N_e \left\langle \ell_1 \mathbf{k}_1, \ldots, \ell_{N_e} \mathbf{k}_{N_e} \left| \hat{H}_1^{(1)} \hat{\rho}_e \right| \ell_1' \mathbf{k}_1', \ell_2 \mathbf{k}_2, \ldots, \ell_{N_e} \mathbf{k}_{N_e} \right\rangle \\
&\quad + \frac{\mathrm{i}}{\hbar} \sum_{\ell_2 \mathbf{k}_2, \ldots, \ell_{N_e} \mathbf{k}_{N_e}} N_e \left\langle \ell_1 \mathbf{k}_1, \ldots, \ell_{N_e} \mathbf{k}_{N_e} \left| \hat{\rho}_e \hat{H}_1^{(1)} \right| \ell_1' \mathbf{k}_1', \ell_2 \mathbf{k}_2, \ldots, \ell_{N_e} \mathbf{k}_{N_e} \right\rangle \\
&\quad - \frac{\mathrm{i}}{2\hbar} \sum_{\ell_2 \mathbf{k}_2, \ldots, \ell_{N_e} \mathbf{k}_{N_e}} N_e (N_e - 1) \left\langle \ell_1 \mathbf{k}_1, \ldots, \ell_{N_e} \mathbf{k}_{N_e} \left| \hat{V}_{12}^{(\mathrm{int})} \hat{\rho}_e \right| \ell_1' \mathbf{k}_1', \ell_2 \mathbf{k}_2, \ldots, \ell_{N_e} \mathbf{k}_{N_e} \right\rangle \\
&\quad + \frac{\mathrm{i}}{2\hbar} \sum_{\ell_2 \mathbf{k}_2, \ldots, \ell_{N_e} \mathbf{k}_{N_e}} N_e (N_e - 1) \left\langle \ell_1 \mathbf{k}_1, \ldots, \ell_{N_e} \mathbf{k}_{N_e} \left| \hat{\rho}_e \hat{V}_{12}^{(\mathrm{int})} \right| \ell_1' \mathbf{k}_1', \ell_2 \mathbf{k}_2, \ldots, \ell_{N_e} \mathbf{k}_{N_e} \right\rangle \\
&= -\frac{\mathrm{i}}{\hbar} \sum_{\ell_2 \mathbf{k}_2, \ldots, \ell_{N_e} \mathbf{k}_{N_e}} \sum_{\ell_1'' \mathbf{k}_1'', \ldots, \ell_{N_e}'' \mathbf{k}_{N_e}''} N_e \left\langle \ell_1 \mathbf{k}_1, \ldots, \ell_{N_e} \mathbf{k}_{N_e} \left| \hat{H}_1^{(1)} \right| \ell_1'' \mathbf{k}_1'', \ldots, \ell_{N_e}'' \mathbf{k}_{N_e}'' \right\rangle \times \\
&\qquad \times \left\langle \ell_1'' \mathbf{k}_1'', \ldots, \ell_{N_e}'' \mathbf{k}_{N_e}'' \left| \hat{\rho}_e \right| \ell_1' \mathbf{k}_1', \ell_2 \mathbf{k}_2, \ldots, \ell_{N_e} \mathbf{k}_{N_e} \right\rangle \\
&\quad + \frac{\mathrm{i}}{\hbar} \sum_{\ell_2 \mathbf{k}_2, \ldots, \ell_{N_e} \mathbf{k}_{N_e}} \sum_{\ell_1'' \mathbf{k}_1'', \ldots, \ell_{N_e}'' \mathbf{k}_{N_e}''} N_e \left\langle \ell_1 \mathbf{k}_1, \ldots, \ell_{N_e} \mathbf{k}_{N_e} \left| \hat{\rho}_e \right| \ell_1'' \mathbf{k}_1'', \ldots, \ell_{N_e}'' \mathbf{k}_{N_e}'' \right\rangle \times \\
&\qquad \times \left\langle \ell_1'' \mathbf{k}_1'', \ldots, \ell_{N_e}'' \mathbf{k}_{N_e}'' \left| \hat{H}_1^{(1)} \right| \ell_1' \mathbf{k}_1', \ell_2 \mathbf{k}_2, \ldots, \ell_{N_e} \mathbf{k}_{N_e} \right\rangle
\end{aligned}$$

$$
\begin{aligned}
&-\frac{\mathrm{i}}{2\hbar}\sum_{\ell_2\mathbf{k}_2,\ldots,\ell_{N_e}\mathbf{k}_{N_e}}\;\sum_{\ell''_1\mathbf{k}''_1,\ldots,\ell''_{N_e}\mathbf{k}''_{N_e}}\left\langle\ell_1\mathbf{k}_1,\ldots,\ell_{N_e}\mathbf{k}_{N_e}\middle|\hat{V}_{12}^{(\mathrm{int})}\middle|\ell''_1\mathbf{k}''_1,\ldots,\ell''_{N_e}\mathbf{k}''_{N_e}\right\rangle\times\\
&\qquad\times N_e\left(N_e-1\right)\left\langle\ell''_1\mathbf{k}''_1,\ldots,\ell''_{N_e}\mathbf{k}''_{N_e}\middle|\hat{\rho}_e\middle|\ell'_1\mathbf{k}'_1,\ell_2\mathbf{k}_2,\ldots,\ell_{N_e}\mathbf{k}_{N_e}\right\rangle\\
&+\frac{\mathrm{i}}{2\hbar}\sum_{\ell_2\mathbf{k}_2,\ldots,\ell_{N_e}\mathbf{k}_{N_e}}\;\sum_{\ell''_1\mathbf{k}''_1,\ldots,\ell''_{N_e}\mathbf{k}''_{N_e}}N_e\left(N_e-1\right)\left\langle\ell_1\mathbf{k}_1,\ldots,\ell_{N_e}\mathbf{k}_{N_e}\middle|\hat{\rho}_e\middle|\ell''_1\mathbf{k}''_1,\ldots,\ell''_{N_e}\mathbf{k}''_{N_e}\right\rangle\times\\
&\qquad\times\left\langle\ell''_1\mathbf{k}''_1,\ldots,\ell''_{N_e}\mathbf{k}''_{N_e}\middle|\hat{V}_{12}^{(\mathrm{int})}\middle|\ell'_1\mathbf{k}'_1,\ell_2\mathbf{k}_2,\ldots,\ell_{N_e}\mathbf{k}_{N_e}\right\rangle\\
=&-\frac{\mathrm{i}}{\hbar}\sum_{\ell''_1\mathbf{k}''_1}\left\langle\ell_1\mathbf{k}_1\middle|\hat{H}_1^{(1)}\middle|\ell''_1\mathbf{k}''_1\right\rangle\underbrace{\sum_{\ell_2\mathbf{k}_2,\ldots,\ell_{N_e}\mathbf{k}_{N_e}}N_e\left\langle\ell''_1\mathbf{k}''_1,\ell_2\mathbf{k}_2,\ldots,\ell_{N_e}\mathbf{k}_{N_e}\middle|\hat{\rho}_e\middle|\ell'_1\mathbf{k}'_1,\ell_2\mathbf{k}_2,\ldots,\ell_{N_e}\mathbf{k}_{N_e}\right\rangle}_{=\left\langle\ell''_1\mathbf{k}''_1\middle|\hat{\rho}_e^{(1)}\middle|\ell'_1\mathbf{k}'_1\right\rangle}\\
&+\frac{\mathrm{i}}{\hbar}\sum_{\ell''_1\mathbf{k}''_1}\left\langle\ell''_1\mathbf{k}''_1\middle|\hat{H}_1^{(1)}\middle|\ell'_1\mathbf{k}'_1\right\rangle\underbrace{\sum_{\ell_2\mathbf{k}_2,\ldots,\ell_{N_e}\mathbf{k}_{N_e}}N_e\left\langle\ell_1\mathbf{k}_1,\ldots,\ell_{N_e}\mathbf{k}_{N_e}\middle|\hat{\rho}_e\middle|\ell''_1\mathbf{k}''_1,\ell_2\mathbf{k}_2,\ldots,\ell_{N_e}\mathbf{k}_{N_e}\right\rangle}_{=\left\langle\ell_1\mathbf{k}_1\middle|\hat{\rho}_e^{(1)}\middle|\ell''_1\mathbf{k}''_1\right\rangle}\\
&-\frac{\mathrm{i}}{2\hbar}\sum_{\ell_2\mathbf{k}_2}\sum_{\ell''_1\mathbf{k}''_1,\ell''_2\mathbf{k}''_2}\left\langle\ell_1\mathbf{k}_1,\ell_2\mathbf{k}_2\middle|\hat{V}_{12}^{(\mathrm{int})}\middle|\ell''_1\mathbf{k}''_1,\ell''_2\mathbf{k}''_2\right\rangle\times\\
&\qquad\times\underbrace{\sum_{\ell_3\mathbf{k}_3,\ldots,\ell_{N_e}\mathbf{k}_{N_e}}N_e\left(N_e-1\right)\left\langle\ell''_1\mathbf{k}''_1,\ell''_2\mathbf{k}''_2,\ell_3\mathbf{k}_3\ldots,\ell_{N_e}\mathbf{k}_{N_e}\middle|\hat{\rho}_e\middle|\ell'_1\mathbf{k}'_1,\ell_2\mathbf{k}_2,\ldots,\ell_{N_e}\mathbf{k}_{N_e}\right\rangle}_{=\left\langle\ell''_1\mathbf{k}''_1,\ell''_2\mathbf{k}''_2\middle|\hat{\rho}_e^{(2)}\middle|\ell'_1\mathbf{k}'_1,\ell_2\mathbf{k}_2\right\rangle}\\
&+\frac{\mathrm{i}}{2\hbar}\sum_{\ell_2\mathbf{k}_2}\sum_{\ell''_1\mathbf{k}''_1,\ell''_2\mathbf{k}''_2}\left\langle\ell''_1\mathbf{k}''_1,\ell''_2\mathbf{k}''_2\middle|\hat{V}_{12}^{(\mathrm{int})}\middle|\ell'_1\mathbf{k}'_1,\ell_2\mathbf{k}_2\right\rangle\times\\
&\qquad\times\underbrace{\sum_{\ell_3\mathbf{k}_3,\ldots,\ell_{N_e}\mathbf{k}_{N_e}}N_e\left(N_e-1\right)\left\langle\ell_1\mathbf{k}_1,\ldots,\ell_{N_e}\mathbf{k}_{N_e}\middle|\hat{\rho}_e\middle|\ell''_1\mathbf{k}''_1,\ell''_2\mathbf{k}''_2,\ell_3\mathbf{k}_3,\ldots,\ell_{N_e}\mathbf{k}_{N_e}\right\rangle}_{=\left\langle\ell_1\mathbf{k}_1,\ell_2\mathbf{k}_2\middle|\hat{\rho}_e^{(2)}\middle|\ell''_1\mathbf{k}''_1,\ell''_2\mathbf{k}''_2\right\rangle}\\
=&-\frac{\mathrm{i}}{\hbar}\sum_{\ell''_1\mathbf{k}''_1}\left\langle\ell_1\mathbf{k}_1\middle|\hat{H}_1^{(1)}\middle|\ell''_1\mathbf{k}''_1\right\rangle\left\langle\ell''_1\mathbf{k}''_1\middle|\hat{\rho}_e^{(1)}\middle|\ell'_1\mathbf{k}'_1\right\rangle+\frac{\mathrm{i}}{\hbar}\sum_{\ell''_1\mathbf{k}''_1}\left\langle\ell_1\mathbf{k}_1\middle|\hat{\rho}_e^{(1)}\middle|\ell''_1\mathbf{k}''_1\right\rangle\left\langle\ell''_1\mathbf{k}''_1\middle|\hat{H}_1^{(1)}\middle|\ell'_1\mathbf{k}'_1\right\rangle\\
&-\frac{\mathrm{i}}{2\hbar}\sum_{\ell_2\mathbf{k}_2}\sum_{\ell''_1\mathbf{k}''_1,\ell''_2\mathbf{k}''_2}\left\langle\ell_1\mathbf{k}_1,\ell_2\mathbf{k}_2\middle|\hat{V}_{12}^{(\mathrm{int})}\middle|\ell''_1\mathbf{k}''_1,\ell''_2\mathbf{k}''_2\right\rangle\left\langle\ell''_1\mathbf{k}''_1,\ell''_2\mathbf{k}''_2\middle|\hat{\rho}_e^{(2)}\middle|\ell'_1\mathbf{k}'_1,\ell_2\mathbf{k}_2\right\rangle\\
&+\frac{\mathrm{i}}{2\hbar}\sum_{\ell_2\mathbf{k}_2}\sum_{\ell''_1\mathbf{k}''_1,\ell''_2\mathbf{k}''_2}\left\langle\ell_1\mathbf{k}_1,\ell_2\mathbf{k}_2\middle|\hat{\rho}_e^{(2)}\middle|\ell''_1\mathbf{k}''_1,\ell''_2\mathbf{k}''_2\right\rangle\left\langle\ell''_1\mathbf{k}''_1,\ell''_2\mathbf{k}''_2\middle|\hat{V}_{12}^{(\mathrm{int})}\middle|\ell'_1\mathbf{k}'_1,\ell_2\mathbf{k}_2\right\rangle\\
=&-\frac{\mathrm{i}}{\hbar}\left\langle\ell_1\mathbf{k}_1\middle|\hat{H}_1^{(1)}\,\hat{\rho}_e^{(1)}-\hat{\rho}_e^{(1)}\,\hat{H}_1^{(1)}\middle|\ell'_1\mathbf{k}'_1\right\rangle-\frac{\mathrm{i}}{2\hbar}\sum_{\ell_2\mathbf{k}_2}\left\langle\ell_1\mathbf{k}_1,\ell_2\mathbf{k}_2\middle|\hat{V}_{12}^{(\mathrm{int})}\,\hat{\rho}_e^{(2)}-\hat{\rho}_e^{(2)}\,\hat{V}_{12}^{(\mathrm{int})}\middle|\ell'_1\mathbf{k}'_1,\ell_2\mathbf{k}_2\right\rangle.\\
=&-\frac{\mathrm{i}}{\hbar}\left\langle\ell_1\mathbf{k}_1\middle|\left[\hat{H}_1^{(1)},\hat{\rho}_e^{(1)}\right]_-\middle|\ell'_1\mathbf{k}'_1\right\rangle-\frac{\mathrm{i}}{2\hbar}\sum_{\ell_2\mathbf{k}_2}\left\langle\ell_1\mathbf{k}_1,\ell_2\mathbf{k}_2\middle|\left[\hat{V}_{12}^{(\mathrm{int})},\hat{\rho}_e^{(2)}\right]_-\middle|\ell'_1\mathbf{k}'_1,\ell_2\mathbf{k}_2\right\rangle.
\end{aligned}
$$

■

The time derivative of the one-electron reduced density matrix is related to the two-electron reduced density matrix. In general, one can prove that the time derivative of the n-electron reduced density matrix is related to the $(n+1)$-electron reduced density matrix. The later is called **BBGKY-hierarchy** [18] and has the drawback that the equation for n-particles is coupled to that for $n+1$-particles. To overcome this problem, one can use, for example, the mean field approximation. In this approximation, to which DFT belongs, the two-body Coulomb interaction term is described by an effective one-particle operator, so that the electronic Hamiltonian is just a sum over one-particle operators and the term in Eq. (3.168) with the two-electron reduced density matrix is not present.

3.4 Effects of a Fs-laser Interaction on Matter

Lasers usually emit highly coherent and directed electromagnetic waves. Such a beam can be focused on a small spot size and easily directed on the material surface. Particularly, femtosecond laser allow to generate short bursts of waves with a time duration of a few femtoseconds, so that very high field intensities can be reached in the laser spot during these few femtoseconds. In this short interaction time, only the light electrons are influenced by the strong electromagnetic fields whereas the heavy nuclei cannot react. Using the Born-Oppenheimer approximation (see Sect. 2.2), where the motion of electrons and nuclei is decoupled, the Hamilton operator of the electrons without laser excitation is given by

$$\hat{H}_{\mathrm{e}} = -\sum_{i=1}^{N_{\mathrm{e}}} \frac{\hbar^2}{2m_{\mathrm{e}}^2} \nabla_{\mathbf{R}_i}^2 + \sum_{i=1}^{N_{\mathrm{e}}} V_{\mathrm{ext}}(\mathbf{R}_i) + E_{\mathrm{nn}} + \sum_{i=1}^{N_{\mathrm{e}}} \sum_{\substack{j=1 \\ j \neq i}}^{N_{\mathrm{e}}} \frac{e^2}{4\pi\varepsilon_0} \frac{1}{|\mathbf{R}_i - \mathbf{R}_j|}, \tag{3.169}$$

where

$$V_{\mathrm{ext}}(\mathbf{R}_i) = -\sum_{k=1}^{N_{\mathrm{at}}} \frac{e^2}{4\pi\varepsilon_0} \frac{Z_k}{|\mathbf{r}_k - \mathbf{R}_i|}$$

is the external potential given by the nuclei and

$$E_{\mathrm{nn}} = \sum_{k=1}^{N_{\mathrm{at}}} \sum_{\ell=k+1}^{N_{\mathrm{at}}} \frac{e^2}{4\pi\varepsilon_0} \frac{Z_k Z_\ell}{|\mathbf{r}_k - \mathbf{r}_\ell|}$$

is the nuclei interaction energy, which is just a constant in the electronic Hamiltonian.

3.4.1 Effects of the Fs-Laser Field

Since radiation only interacts with the electrons during the short interaction time, the laser interaction term is given by (see Eq. (3.92))

$$\hat{H}_{\mathrm{L}} = \sum_{i=1}^{N_{\mathrm{e}}} \frac{\mathrm{i}\,\hbar\, e}{m_{\mathrm{e}}}\,\mathbf{A}(\mathbf{R}_i, t)\cdot\nabla_{\mathbf{R}_i} + \sum_{i=1}^{N_{\mathrm{e}}} \frac{e^2}{2\,m_{\mathrm{e}}}\,\mathbf{A}^2(\mathbf{R}_i, t). \tag{3.170}$$

Therefore, the Hamiltonian of the electrons in the femtosecond laser-field denoted

$$\hat{H} = \sum_{i=1}^{N_{\mathrm{e}}} \hat{H}_i^{(1)} + \sum_{i=1}^{N_{\mathrm{e}}} \hat{H}_i^{(\mathrm{L})} + \frac{1}{2}\sum_{i=1}^{N_{\mathrm{e}}} \sum_{\substack{j=1\\ j\neq i}}^{N_{\mathrm{e}}} \hat{V}_{ij}^{(\mathrm{int})} \tag{3.171}$$

contains the sum of the one-particle operators

$$\hat{H}_i^{(1)} = \frac{\hbar^2}{2\,m_{\mathrm{e}}^2}\nabla_{\mathbf{R}_i}^2 + V_{\mathrm{ext}}(\mathbf{R}_i) + \frac{E_{\mathbf{nn}}}{N_{\mathrm{e}}}, \tag{3.172}$$

$$\hat{H}_i^{(\mathrm{L})} = \frac{\mathrm{i}\,\hbar\, e}{m_{\mathrm{e}}}\,\mathbf{A}(\mathbf{R}_i, t)\cdot\nabla_{\mathbf{R}_i} + \frac{e^2}{2\,m_{\mathrm{e}}}\,\mathbf{A}^2(\mathbf{R}_i, t), \tag{3.173}$$

and the sum of the two-particle operators

$$\hat{V}_{ij}^{(\mathrm{int})} = \frac{e^2}{4\pi\varepsilon_0}\,\frac{1}{\left|\mathbf{R}_i - \mathbf{R}_j\right|}, \tag{3.174}$$

which obey $\hat{V}_{ij}^{(\mathrm{int})} = \hat{V}_{ji}^{(\mathrm{int})}$. The electronic Hamiltonian

$$\hat{H}_0 = \sum_{i=1}^{N_{\mathrm{e}}} \hat{H}_i^{(1)} \tag{3.175}$$

without two-electron Coulomb interaction $\hat{V}_{ij}^{(\mathrm{int})}$ and laser-interaction $\hat{H}_i^{(\mathrm{L})}$ can be diagonalized. Since the solid forms a periodic crystal structure, the eigenstates of $\hat{H}_0$ with eigenenergies $\varepsilon_{\ell\mathbf{k}}$ are one-electron Bloch states $|\ell\mathbf{k}\rangle$. Using these complete set of one-electron eigenstates, we transform the full electronic Hamilton operator (3.171) with femtosecond laser-field interaction to second quantization:

$$\begin{aligned}\hat{H} =& \sum_{\ell\mathbf{k}} \varepsilon_{\ell\mathbf{k}}\,\hat{c}_{\ell\mathbf{k}}^{\dagger}\,\hat{c}_{\ell\mathbf{k}} + \sum_{\ell\mathbf{k},\ell'\mathbf{k}'} H_{\ell\mathbf{k},\ell'\mathbf{k}'}^{(\mathrm{L})}\,\hat{c}_{\ell\mathbf{k}}^{\dagger}\,\hat{c}_{\ell'\mathbf{k}'} \\ &+ \frac{1}{2}\sum_{\ell\mathbf{k},\ell'\mathbf{k}',\ell''\mathbf{k}'',\ell'''\mathbf{k}'''} V_{\ell\mathbf{k},\ell'\mathbf{k}',\ell''\mathbf{k}'',\ell'''\mathbf{k}'''}^{(\mathrm{int})}\,\hat{c}_{\ell\mathbf{k}}^{\dagger}\,\hat{c}_{\ell'\mathbf{k}'}^{\dagger}\,\hat{c}_{\ell'''\mathbf{k}'''}\,\hat{c}_{\ell''\mathbf{k}''},\end{aligned} \tag{3.176}$$

with

$$H^{(\mathrm{L})}_{\ell\mathbf{k},\ell'\mathbf{k}'} = \sum_{\sigma} \int_{\mathbb{R}^3} d^3R\, \varphi^*_{\ell\mathbf{k}}(\sigma,\mathbf{R}) \left(\frac{\mathrm{i}\,\hbar\, e}{m_\mathrm{e}} \mathbf{A}(\mathbf{R},t)\cdot\nabla_\mathbf{R} + \frac{e^2}{2\,m_\mathrm{e}} \mathbf{A}^2(\mathbf{R},t) \right) \varphi_{\ell'\mathbf{k}'}(\sigma,\mathbf{R}) \tag{3.177}$$

and

$$V^{(\mathrm{int})}_{\ell\mathbf{k},\ell'\mathbf{k}',\ell''\mathbf{k}'',\ell'''\mathbf{k}'''} = \sum_{\sigma,\sigma'} \int_{\mathbb{R}^3} d^3R \int_{\mathbb{R}^3} d^3R'\, \varphi^*_{\ell\mathbf{k}}(\sigma,\mathbf{R})\, \varphi^*_{\ell'\mathbf{k}'}(\sigma',\mathbf{R}') \times \\ \times \frac{e^2}{4\pi\varepsilon_0} \frac{1}{|\mathbf{R}-\mathbf{R}'|} \varphi_{\ell''\mathbf{k}''}(\mathbf{R})\, \varphi_{\ell'''\mathbf{k}'''}(\sigma',\mathbf{R}'). \tag{3.178}$$

The time propagation of the electronic system is described by the electron density matrix $\hat{\rho}_\mathrm{e}$. Before the femtosecond laser-excitation, the electrons form a canonical ensemble at ambient temperature, here $T_\mathrm{e} = 300$ K. First, we consider the influences of the femtosecond laser-excitation neglecting the electron-electron interaction, so that the electronic Hamiltonian is just a sum over one particle operators. Therefore, we only need to work with the one-electron density matrix $\hat{\rho}^{(1)}_\mathrm{e}$, which time propagation is, in this case, given by (*c.f.* Eq. (3.168))

$$\frac{\partial \hat{\rho}^{(1)}_\mathrm{e}}{\partial t} = -\frac{\mathrm{i}}{\hbar}\left[\hat{H}^{(1)}_1 + \hat{H}^{(\mathrm{L})}_1\,,\,\hat{\rho}^{(1)}_\mathrm{e}\right]_- . \tag{3.179}$$

To get an intuitive understanding of the femtosecond laser-driven changes in the electronic system, we subsequently introduce several approximations, which will allow to derive simple analytic equations for the change of the electronic occupations. For this, we follow Kuznetsov's model study in Ref. [19], but we take the freedom to provide more mathematical details. In addition, Kuznetsov works in the Heisenberg picture of quantum mechanics whereas we choose to work in the Schrödinger picture. Since the electronic Hamiltonian does not explicitly depend on the spin, the spin up and down electrons behave in a similar way and we do not need to take explicitly the spin into account. Kuznetsov neglects the Coulomb interaction $\hat{V}^{(\mathrm{int})}_{ij}$ between the electrons and considers for each **k**-point only two bands—the valence band ℓ_1 and the conduction band ℓ_2. Furthermore, he performs the dipole approximation (3.94) for the femtosecond laser interaction and assumes that the vector potential $\mathbf{A}(t) = A(t)\,\mathbf{e}_x$ is just oscillating in x-direction:

$$\hat{H}^{(\mathrm{L})}_1 = \frac{\mathrm{i}\,\hbar\,e}{m_\mathrm{e}} A(t) \frac{\partial}{\partial R_x}. \tag{3.180}$$

The electronic Hamiltonians reads in second quantization

$$\hat{H} = \sum_{\mathbf{k}} \left(\varepsilon_{\ell_1 \mathbf{k}} \, \hat{c}^\dagger_{\ell_1 \mathbf{k}} \, \hat{c}_{\ell_1 \mathbf{k}} + \varepsilon_{\ell_2 \mathbf{k}} \, \hat{c}^\dagger_{\ell_2 \mathbf{k}} \, \hat{c}_{\ell_2 \mathbf{k}} \right)$$
$$- \sum_{\mathbf{k}} \left(\mu_{\ell_1 \ell_2} \, A(t) \, \hat{c}^\dagger_{\ell_1 \mathbf{k}} \, \hat{c}_{\ell_2 \mathbf{k}} + \mu_{\ell_2 \ell_1} \, A(t) \, \hat{c}^\dagger_{\ell_2 \mathbf{k}} \, \hat{c}_{\ell_1 \mathbf{k}} \right), \tag{3.181}$$

where $\mu_{\ell_1 \ell_2}$ is the interband optical matrix element, which diagonal elements $\mu_{\ell_1 \ell_1}$ and $\mu_{\ell_2 \ell_2}$ are neglected as well as the **k** dependence. Therefore, any **k** state evolve independently with respect to other **k** states. Furthermore, we obtain

$$\begin{aligned}
\mu_{\ell_2 \ell_1} \, A(t) &= \int_{\mathbb{R}^3} d^3 R \, \varphi^*_{\ell_2 \mathbf{k}}(\mathbf{R}) \, \hat{H}_1^{(\mathrm{L})} \, \varphi_{\ell_1 \mathbf{k}}(\mathbf{R}) \\
&= \frac{\mathrm{i} \hbar e}{m_\mathrm{e}} \, A(t) \int_{\mathbb{R}^3} d^3 R \, \varphi^*_{\ell_2 \mathbf{k}}(\mathbf{R}) \, \frac{\partial}{\partial R_x} \varphi_{\ell_1 \mathbf{k}}(\mathbf{R}) \\
&= - \frac{\mathrm{i} \hbar e}{m_\mathrm{e}} \, A(t) \int_{\mathbb{R}^3} d^3 R \, \varphi_{\ell_1 \mathbf{k}}(\mathbf{R}) \, \frac{\partial}{\partial R_x} \varphi^*_{\ell_2 \mathbf{k}}(\mathbf{R}) \\
&= \int_{\mathbb{R}^3} d^3 R \, \varphi_{\ell_1 \mathbf{k}}(\mathbf{R}) \, \hat{H}_1^{(\mathrm{L})*} \, \varphi^*_{\ell_2 \mathbf{k}}(\mathbf{R}) \\
&= \mu^*_{\ell_1 \ell_2} \, A(t),
\end{aligned} \tag{3.182}$$

Here, we performed a partial integration and utilized that the Bloch wave functions $\varphi_{\ell \mathbf{k}}(\mathbf{R})$ decay fast enough to zero for $|\mathbf{R}| \to \infty$. In this way, we proved

$$\mu_{\ell_2 \ell_1} = \mu^*_{\ell_1 \ell_2}. \tag{3.183}$$

Hence, for any **k** state, the Hamiltonian can be written as a 2×2 matrix:

$$\hat{H}(\mathbf{k}) = \begin{bmatrix} \varepsilon_{\ell_1 \mathbf{k}} & -\mu_{\ell_1 \ell_2} \, A(t) \\ -\mu^*_{\ell_1 \ell_2} \, A(t) & \varepsilon_{\ell_2 \mathbf{k}} \end{bmatrix}. \tag{3.184}$$

We define

$$\hat{\mathbb{p}}_{\mathbf{k}} = \hat{c}^\dagger_{\ell_1 \mathbf{k}} \, \hat{c}_{\ell_2 \mathbf{k}} \tag{3.185}$$

and write the one-electron reduced electron density matrix also as a 2×2 matrix for each **k** state:

$$\hat{\rho}_\mathrm{e}^{(1)}(\mathbf{k}) = \begin{bmatrix} \hat{c}^\dagger_{\ell_1 \mathbf{k}} \, \hat{c}_{\ell_1 \mathbf{k}} & \hat{c}^\dagger_{\ell_1 \mathbf{k}} \, \hat{c}_{\ell_2 \mathbf{k}} \\ \hat{c}^\dagger_{\ell_2 \mathbf{k}} \, \hat{c}_{\ell_1 \mathbf{k}} & \hat{c}^\dagger_{\ell_2 \mathbf{k}} \, \hat{c}_{\ell_2 \mathbf{k}} \end{bmatrix} = \begin{bmatrix} \hat{n}_{\ell_1 \mathbf{k}} & \hat{\mathbb{p}}_{\mathbf{k}} \\ \hat{\mathbb{p}}^\dagger_{\mathbf{k}} & \hat{n}_{\ell_2 \mathbf{k}} \end{bmatrix}. \tag{3.186}$$

The diagonal elements correspond to the electron distribution functions within the bands and the non-diagonal elements describe the quantum-mechanical coherence between the bands. Due to Eq. (3.168), we obtain for the time derivative of $\hat{\rho}_\mathrm{e}^{(1)}(\mathbf{k})$:

$$
\begin{aligned}
&\frac{\partial \hat{\rho}_{\mathrm{e}}^{(1)}(\mathbf{k})}{\partial t}\\
&= -\frac{\mathrm{i}}{\hbar}\left[\hat{H}(\mathbf{k}),\, \hat{\rho}_{\mathrm{e}}^{(1)}(\mathbf{k})\right]_-\\
&= -\frac{\mathrm{i}}{\hbar}\,\hat{H}(\mathbf{k})\cdot\hat{\rho}_{\mathrm{e}}^{(1)}(\mathbf{k}) + \frac{\mathrm{i}}{\hbar}\,\hat{\rho}_{\mathrm{e}}^{(1)}(\mathbf{k})\cdot\hat{H}(\mathbf{k})\\
&= -\frac{\mathrm{i}}{\hbar}\begin{bmatrix} \varepsilon_{\ell_1\mathbf{k}} & -\mu_{\ell_1\ell_2}\,A(t)\\ -\mu^*_{\ell_1\ell_2}\,A(t) & \varepsilon_{\ell_2\mathbf{k}}\end{bmatrix}\cdot\begin{bmatrix}\hat{n}_{\ell_1\mathbf{k}} & \hat{\mathbb{p}}_{\mathbf{k}}\\ \hat{\mathbb{p}}^\dagger_{\mathbf{k}} & \hat{n}_{\ell_2\mathbf{k}}\end{bmatrix}\\
&\quad+\frac{\mathrm{i}}{\hbar}\begin{bmatrix}\hat{n}_{\ell_1\mathbf{k}} & \hat{\mathbb{p}}_{\mathbf{k}}\\ \hat{\mathbb{p}}^\dagger_{\mathbf{k}} & \hat{n}_{\ell_2\mathbf{k}}\end{bmatrix}\cdot\begin{bmatrix} \varepsilon_{\ell_1\mathbf{k}} & -\mu_{\ell_1\ell_2}\,A(t)\\ -\mu^*_{\ell_1\ell_2}\,A(t) & \varepsilon_{\ell_2\mathbf{k}}\end{bmatrix}\\
&= -\frac{\mathrm{i}}{\hbar}\begin{bmatrix} \varepsilon_{\ell_1\mathbf{k}}\,\hat{n}_{\ell_1\mathbf{k}} - \mu_{\ell_1\ell_2}\,A(t)\,\hat{\mathbb{p}}^\dagger_{\mathbf{k}} & \varepsilon_{\ell_1\mathbf{k}}\,\hat{\mathbb{p}}_{\mathbf{k}} - \mu_{\ell_1\ell_2}\,A(t)\,\hat{n}_{\ell_2\mathbf{k}}\\ -\mu^*_{\ell_1\ell_2}\,A(t)\,\hat{n}_{\ell_1\mathbf{k}} + \varepsilon_{\ell_2\mathbf{k}}\,\hat{\mathbb{p}}^\dagger_{\mathbf{k}} & -\mu^*_{\ell_1\ell_2}\,A(t)\,\hat{\mathbb{p}}^\dagger_{\mathbf{k}} + \varepsilon_{\ell_2\mathbf{k}}\,\hat{n}_{\ell_2\mathbf{k}}\end{bmatrix}\\
&\quad+\frac{\mathrm{i}}{\hbar}\begin{bmatrix} \varepsilon_{\ell_1\mathbf{k}}\,\hat{n}_{\ell_1\mathbf{k}} - \mu^*_{\ell_1\ell_2}\,A(t)\,\hat{\mathbb{p}}^\dagger_{\mathbf{k}} & -\mu_{\ell_1\ell_2}\,A(t)\,\hat{n}_{\ell_1\mathbf{k}} + \varepsilon_{\ell_2\mathbf{k}}\,\hat{\mathbb{p}}_{\mathbf{k}}\\ \varepsilon_{\ell_1\mathbf{k}}\,\hat{\mathbb{p}}^\dagger_{\mathbf{k}} - \mu^*_{\ell_1\ell_2}\,A(t)\,\hat{n}_{\ell_2\mathbf{k}} & -\mu_{\ell_1\ell_2}\,A(t)\,\hat{\mathbb{p}}^\dagger_{\mathbf{k}} + \varepsilon_{\ell_2\mathbf{k}}\,\hat{n}_{\ell_2\mathbf{k}}\end{bmatrix}\\
&=\frac{\mathrm{i}}{\hbar}\begin{bmatrix} \left(\mu_{\ell_1\ell_2} - \mu^*_{\ell_1\ell_2}\right)A(t)\,\hat{\mathbb{p}}^\dagger_{\mathbf{k}} & \mu_{\ell_1\ell_2}\,A(t)\left(\hat{n}_{\ell_2\mathbf{k}} - \hat{n}_{\ell_1\mathbf{k}}\right) + \left(\varepsilon_{\ell_2\mathbf{k}} - \varepsilon_{\ell_1\mathbf{k}}\right)\hat{\mathbb{p}}_{\mathbf{k}}\\ \mu^*_{\ell_1\ell_2}\,A(t)\left(\hat{n}_{\ell_1\mathbf{k}} - \hat{n}_{\ell_2\mathbf{k}}\right) + \left(\varepsilon_{\ell_1\mathbf{k}} - \varepsilon_{\ell_2\mathbf{k}}\right)\hat{\mathbb{p}}^\dagger_{\mathbf{k}} & \left(\mu^*_{\ell_1\ell_2} - \mu_{\ell_1\ell_2}\right)A(t)\,\hat{\mathbb{p}}^\dagger_{\mathbf{k}}\end{bmatrix}.
\end{aligned}
$$

From this, we get

$$
\frac{\partial \hat{n}_{\ell_2\mathbf{k}}}{\partial t} = \frac{2}{\hbar}\,\mathrm{Im}\left(\mu_{\ell_1\ell_2}\right)A(t)\,\hat{\mathbb{p}}^\dagger_{\mathbf{k}}, \tag{3.187}
$$

$$
\frac{\partial \hat{n}_{\ell_1\mathbf{k}}}{\partial t} = -\,\frac{\partial \hat{n}_{\ell_2\mathbf{k}}}{\partial t}, \tag{3.188}
$$

$$
\frac{\partial \hat{\mathbb{p}}_{\mathbf{k}}}{\partial t} = \frac{\mathrm{i}}{\hbar}\,\mu_{\ell_1\ell_2}\,A(t)\left(\hat{n}_{\ell_2\mathbf{k}} - \hat{n}_{\ell_1\mathbf{k}}\right) + \frac{\mathrm{i}}{\hbar}\left(\varepsilon_{\ell_2\mathbf{k}} - \varepsilon_{\ell_1\mathbf{k}}\right)\hat{\mathbb{p}}_{\mathbf{k}}. \tag{3.189}
$$

These are the well-known **Bloch equations** that describe the optical response of two-level systems. For times t before the laser excitation, the non-diagonal elements $\hat{\mathbb{p}}_{\mathbf{k}}$, $\hat{\mathbb{p}}^\dagger_{\mathbf{k}}$ of $\hat{\rho}_{\mathrm{e}}^{(1)}(\mathbf{k})$ are zero and the diagonal elements $\hat{n}_{\ell_1\mathbf{k}}$, $\hat{n}_{\ell_2\mathbf{k}}$ form a Fermi distribution with an electronic temperature of $T_{\mathrm{e}} = 300$ K. The femtosecond laser interaction causes non-zero values of the non-diagonal elements, increases the occupation $\hat{n}_{\ell_2\mathbf{k}}$ in the conduction band and decreases the occupation $\hat{n}_{\ell_1\mathbf{k}}$ in the valence band.

These effects of the femtosecond laser on the electronic system occur also in general: Electrons are excited to higher energy levels leaving unoccupied states at low energy levels, which are called holes. In addition, non-diagonal elements of the one-electron density matrix become non-zero, which correspond to polarization.

3.4.2 Electron Relaxation

To describe the relaxation processes of the electrons after the femtosecond laser-excitation, one has to take the electron-electron interactions $\hat{V}_{ij}^{(\mathrm{int})}$ into account. Therefore, the time propagation of the one-electron density matrix $\hat{\rho}_{\mathrm{e}}^{(1)}$ has to be modified to

$$\frac{\partial \hat{\rho}_{\mathrm{e}}^{(1)}}{\partial t} = -\frac{\mathrm{i}}{\hbar}\left[\hat{H}_1^{(1)} + \hat{H}_1^{(\mathrm{L})}, \hat{\rho}_{\mathrm{e}}^{(1)}\right]_- \underbrace{- \frac{\mathrm{i}}{2\hbar}\left[\hat{V}_{12}^{(\mathrm{int})}, \hat{\rho}_{\mathrm{e}}^{(2)}\right]_-}_{=: \left.\frac{\partial \hat{\rho}_{\mathrm{e}}^{(1)}}{\partial t}\right|_{\mathrm{collisions}}} . \tag{3.190}$$

After the femtosecond laser-excitation, the second term sets rapidly the non-diagonal elements of the one-electron density matrix $\hat{\rho}_{\mathrm{e}}^{(1)}$ to zero within a time $\tau_{\mathrm{dephasing}}$, which is extremely short for high laser fluencies. In good approximation one can assume that directly after the femtosecond laser-pulse excitation the non-diagonal elements become zero. Then, the diagonal elements of $\hat{\rho}_{\mathrm{e}}^{(1)}$, which correspond to the occupation numbers of the different bands, thermalize to a Fermi distribution within a time τ_{therm}, which is also very short. This time is known to be of the order of 100 fs [20] for low fluences. For high fluences, it was found to be shorter than 25 fs in GaAs [21]. In Summary, the electron-electron interaction causes a fast thermalization of the electrons to a Fermi distribution with a high electronic temperature T_{e}, which reaches values of several thousand Kelvins due to the high energy pumped in the electronic system by the femtosecond laser. Beside this, the femtosecond laser-pulse excites the electrons to very high energy levels in metals. Due to the high kinetic energy, these electrons can travel longer distances in the material compared to the other electrons and do not contribute to the thermalization to the Fermi distribution [22]. These so called ballistic electrons cause a ballistic energy transport in the order of 100 nm for noble metals. For transition metals with a strong, yet to be discussed, electron-phonon coupling, the ballistic energy transport is only of the order of 10 nm, since there are more scattering events between the ballistic electrons and the phonons.

3.4.3 *Electron-Phonon Relaxation*

On a timescale of picoseconds or tens of picoseconds, the hot electrons transfer their energy to the cold nuclei. To describe this relaxation process of the electrons and the nuclei, one has to go beyond the Born-Oppenheimer approximation (see Sect. 2.2) and has to include the coupling between the electron and nuclei motions, which is called electron-phonon coupling. The defects of the periodic crystal structure cause scattering processes of the hot Bloch electrons.

We will derive now the electron-phonon coupling (EPC) representation in second quantization based on Ref. [16]. But we perform a general derivation in contrast to Ref. [16], which only deals with one atom per basis cell. In addition, we also provide more mathematical details. The nuclei coordinates are given by (see Sect. 2.2.1)

$$\mathbf{r}_{\mathbf{n}\upsilon}(t) = \mathbf{T}_{\mathbf{n}} + \mathbf{d}_{\upsilon} + \mathbf{u}_{\mathbf{n}\upsilon}(t),$$

where the vectors $\mathbf{T_n}$ form a Bravais grid, $\mathbf{d}_\upsilon$ describes the relative position in the basis cell and $\mathbf{u}_{\mathbf{n}\upsilon}(t)$ is the displacement from the equilibrium position. In the electronic Hamiltonian (3.169), the external potential $\hat{V}_{\text{ext}}$ generated by the nuclei is given by

$$V_{\text{ext}}(\mathbf{R}_i) = \sum_{\mathbf{n}\upsilon} V^{(\upsilon)}(\mathbf{R}_i - \mathbf{r}_{\mathbf{n}\upsilon}), \tag{3.191}$$

$$V^{(\upsilon)}(\mathbf{R}_i - \mathbf{r}_{\mathbf{n}\upsilon}) = \frac{e^2}{4\pi\varepsilon_0} \frac{Z_\upsilon}{|\mathbf{R}_i - \mathbf{r}_{\mathbf{n}\upsilon}|}. \tag{3.192}$$

We perform a Taylor series expansion of the single nucleus potential $V^{(\upsilon)}(\mathbf{R}_i - \mathbf{r}_{\mathbf{n}\upsilon})$ with respect to the equilibrium position $\mathbf{T_n} + \mathbf{d}_\upsilon$ of the considered nucleus:

$$V^{(\upsilon)}(\mathbf{R}_i - \mathbf{r}_{\mathbf{n}\upsilon}) = V^{(\upsilon)}(\mathbf{R}_i - \mathbf{T_n} - \mathbf{d}_\upsilon) - \nabla V^{(\upsilon)}(\mathbf{R}_i - \mathbf{T_n} - \mathbf{d}_\upsilon)^{\text{t}} \cdot \mathbf{u}_{\mathbf{n}\upsilon} + \ldots \tag{3.193}$$

In the Born-Oppenheimer approximation, only the first term, which does not consider the nuclei movement described by the displacement $\mathbf{u}_{\mathbf{n}\upsilon}$, is taken into account. The second term and all following terms are responsible for the electron-phonon coupling. If one only takes the second term into account, the electron-phonon interacting term is given by

$$\hat{V}^{(\text{ep})} = \sum_{i=1}^{N_e} \hat{V}_i^{(\text{ep})}, \tag{3.194}$$

$$\hat{V}_i^{(\text{ep})} = -\sum_{\mathbf{n}\upsilon} \nabla V^{(\upsilon)}(\mathbf{R}_i - \mathbf{T_n} - \mathbf{d}_\upsilon)^{\text{t}} \cdot \mathbf{u}_{\mathbf{n}\upsilon}. \tag{3.195}$$

In second quantization, this term reads

$$\hat{V}^{(\text{ep})} = \sum_{\ell\mathbf{k}} \sum_{\ell'\mathbf{k}'} V^{(\text{ep})}_{\ell\mathbf{k},\ell'\mathbf{k}'} \hat{c}^\dagger_{\ell\mathbf{k}} \hat{c}_{\ell'\mathbf{k}'}, \tag{3.196}$$

$$V^{(\text{ep})}_{\ell\mathbf{k},\ell'\mathbf{k}'} = -\sum_\sigma \int_{\mathbb{R}^3} d^3r\, \varphi^*_{\ell\mathbf{k}}(\sigma, \mathbf{r}) \sum_{\mathbf{n}\upsilon} \nabla V^{(\upsilon)}(\mathbf{r} - \mathbf{T_n} - \mathbf{d}_\upsilon)^{\text{t}} \cdot \mathbf{u}_{\mathbf{n}\upsilon}\, \varphi_{\ell'\mathbf{k}'}(\sigma, \mathbf{r}). \tag{3.197}$$

Since the potential $\nabla V^{(\upsilon)}$ does not depend on the spin, $V^{(\text{ep})}_{\ell\mathbf{k},\ell'\mathbf{k}'}$ is spin diagonal, or with other words, the electron-phonon coupling does not change the spin. We therefore skip the spin to simplify any further analysis and perform a Fourier expansion of the single nuclei potential $V^{(\upsilon)}(\mathbf{r})$:

$$V^{(\upsilon)}(\mathbf{r}) = \sum_{\mathbf{k}''} V^{(\upsilon)}_{\mathbf{k}''} e^{i\,\mathbf{k}''^{\text{t}}\cdot\mathbf{r}} \tag{3.198}$$

with Fourier coefficients

$$V_{\mathbf{k}''}^{(\upsilon)} = \frac{1}{\mathcal{V}_\mathrm{s}} \int\limits_{\mathcal{V}_\mathrm{s}} d^3r\, V^{(\upsilon)}(\mathbf{r})\, \mathrm{e}^{-\mathrm{i}\,\mathbf{k}''^{\mathrm{t}}\cdot\mathbf{r}} \tag{3.199}$$

and $\mathcal{V}_\mathrm{s}$ denotes the volume of the simulation cell. Due to the nuclei displacements, the single nuclei potential $V^{(\upsilon)}(\mathbf{r})$ is not periodic with respect to the simulation cell in contrast to the total nuclei potential that is obtained from the sum over all nuclei contained in the simulation cell. Therefore, the sum over $\mathbf{k}''$ is not restricted to reciprocal lattice vectors or to the set of non-equivalent wave vectors. Due to Eq. (3.198), we obtain further

$$\nabla_\mathbf{r} V^{(\upsilon)}(\mathbf{r}) = \sum_{\mathbf{k}''} V_{\mathbf{k}''}^{(\upsilon)}\, \mathrm{e}^{\mathrm{i}\,\mathbf{k}''^{\mathrm{t}}\cdot\mathbf{r}}\, \mathrm{i}\,\mathbf{k}''. \tag{3.200}$$

Inserting this in Eq. (3.197) yields

$$V_{\ell\mathbf{k},\ell'\mathbf{k}'}^{(\mathrm{ep})} = \sum_{\mathbf{k}''} \int\limits_{\mathbb{R}^3} d^3r\, \varphi_{\ell\mathbf{k}}^*(\mathbf{r}) \sum_{\mathbf{n}\upsilon} V_{\mathbf{k}''}^{(\upsilon)}\, \mathrm{e}^{\mathrm{i}\,\mathbf{k}''^{\mathrm{t}}\cdot(\mathbf{r}-\mathbf{T_n}-\mathbf{d}_\upsilon)}\, \mathrm{i}\,\mathbf{k}''^{\mathrm{t}} \cdot \mathbf{u}_{\mathbf{n}\upsilon}\, \varphi_{\ell'\mathbf{k}'}(\mathbf{r}). \tag{3.201}$$

We transform the nuclei displacement $u_{\mathbf{n}\upsilon\alpha}$ in direction α to the reciprocal space and denote it in second quantization:

$$\begin{aligned} u_{\mathbf{n}\upsilon\alpha} &\overset{(2.36)}{=} \frac{1}{\sqrt{m_\upsilon}}\, U_{\upsilon\alpha}(\mathbf{T_n}) \\ &\overset{(2.65)}{=} \frac{1}{\sqrt{m_\upsilon}} \frac{1}{\sqrt{N_1\,N_2\,N_3}} \sum_{\mathbf{q_{n'}}} \mathfrak{U}_{\upsilon\alpha}(\mathbf{q_{n'}})\, \mathrm{e}^{\mathrm{i}\,\mathbf{q}_{\mathbf{n}'}^{\mathrm{t}}\cdot\mathbf{T_n}} \\ &\overset{(2.89)}{=} \frac{1}{\sqrt{m_\upsilon}} \frac{1}{\sqrt{N_1\,N_2\,N_3}} \sum_{\mathbf{q_{n'}}} \sum_{j=1}^{3\,N_\mathrm{b}} \mathcal{U}_j(\mathbf{q_{n'}})\, e_{\upsilon\alpha}^{(j)}(\mathbf{q_{n'}})\, \mathrm{e}^{\mathrm{i}\,\mathbf{q}_{\mathbf{n}'}^{\mathrm{t}}\cdot\mathbf{T_n}} \\ &\overset{(3.141)}{=} \frac{1}{\sqrt{m_\upsilon}} \frac{1}{\sqrt{N_1\,N_2\,N_3}} \sum_{\mathbf{q_{n'}}} \sum_{j=1}^{3\,N_\mathrm{b}} \sqrt{\frac{\hbar}{2\,\omega_j(\mathbf{q_{n'}})}} \left(\hat{b}_{j\mathbf{q_{n'}}} + \hat{b}_{j-\mathbf{q_{n'}}}^\dagger\right) e_{\upsilon\alpha}^{(j)}(\mathbf{q_{n'}})\, \mathrm{e}^{\mathrm{i}\,\mathbf{q}_{\mathbf{n}'}^{\mathrm{t}}\cdot\mathbf{T_n}}. \end{aligned} \tag{3.202}$$

Putting the final transformed nuclei displacement vector

$$\mathbf{u}_{\mathbf{n}\upsilon} = \frac{1}{\sqrt{N_1\,N_2\,N_3}} \sum_{\mathbf{q_{n'}}} \sum_{j=1}^{3\,N_\mathrm{b}} \sqrt{\frac{\hbar}{2\,m_\upsilon\,\omega_j(\mathbf{q_{n'}})}} \left(\hat{b}_{j\mathbf{q_{n'}}} + \hat{b}_{j-\mathbf{q_{n'}}}^\dagger\right) \mathbf{e}_\upsilon^{(j)}(\mathbf{q_{n'}})\, \mathrm{e}^{\mathrm{i}\,\mathbf{q}_{\mathbf{n}'}^{\mathrm{t}}\cdot\mathbf{T_n}} \tag{3.203}$$

into Eq. (3.201) and utilizing that $\mathbf{k}''$ is not restricted to the reciprocal lattice vectors or to the set of non-equivalent wave vectors, yields:

$$
\begin{aligned}
&V^{(\mathrm{ep})}_{\ell\mathbf{k},\ell'\mathbf{k}'} \\
&= -\frac{1}{\sqrt{N_1 N_2 N_3}} \sum_{\mathbf{q_{n'}}} \sum_{\mathbf{k}''} \sum_{j=1}^{3N_\mathrm{b}} \mathrm{i}\,\mathbf{k}''^{\mathrm{t}} \cdot \sum_{\upsilon} \mathbf{e}^{(j)}_{\upsilon}(\mathbf{q_{n'}})\, V^{(\upsilon)}_{\mathbf{k}''}\, \mathrm{e}^{-\mathrm{i}\,\mathbf{k}''^{\mathrm{t}}\cdot\mathbf{d}_\upsilon} \underbrace{\sum_{\mathbf{T_n}} \mathrm{e}^{\mathrm{i}\,(\mathbf{q_{n'}}-\mathbf{k}'')^{\mathrm{t}}\cdot\mathbf{T_n}}}_{\stackrel{(2.68)}{=} N_1 N_2 N_3\, \delta_{\mathbf{k}''\mathbf{q_{n'}}+\mathbf{G}}} \times \\
&\quad \times \int_{\mathbb{R}^3} d^3r\, \varphi^*_{\ell\mathbf{k}}(\mathbf{r})\, \mathrm{e}^{\mathrm{i}\,\mathbf{k}''^{\mathrm{t}}\cdot\mathbf{r}}\, \varphi_{\ell'\mathbf{k}'}(\mathbf{r}) \sqrt{\frac{\hbar}{2\,m_\upsilon\,\omega_j(\mathbf{q_{n'}})}} \left(\hat{b}_{j\mathbf{q_{n'}}} + \hat{b}^\dagger_{j-\mathbf{q_{n'}}}\right) \\
&= -\sqrt{N_1 N_2 N_3} \sum_{\mathbf{q_{n'}}} \sum_{\mathbf{G}} \sum_{j=1}^{3N_\mathrm{b}} \mathrm{i}\,(\mathbf{q_{n'}}+\mathbf{G})^{\mathrm{t}} \cdot \sum_{\upsilon} \mathbf{e}^{(j)}_{\upsilon}(\mathbf{q_{n'}})\, V^{(\upsilon)}_{\mathbf{q_{n'}}+\mathbf{G}}\, \mathrm{e}^{-\mathrm{i}\,(\mathbf{q_{n'}}+\mathbf{G})^{\mathrm{t}}\cdot\mathbf{d}_\upsilon} \times \\
&\quad \times \int_{\mathcal{V}_\mathrm{s}} d^3r\, \varphi^*_{\ell\mathbf{k}}(\mathbf{r})\, \mathrm{e}^{\mathrm{i}\,(\mathbf{q_{n'}}+\mathbf{G})^{\mathrm{t}}\cdot\mathbf{r}}\, \varphi_{\ell'\mathbf{k}'}(\mathbf{r}) \sqrt{\frac{\hbar}{2\,m_\upsilon\,\omega_j(\mathbf{q_{n'}})}} \left(\hat{b}_{j\mathbf{q_{n'}}} + \hat{b}^\dagger_{j-\mathbf{q_{n'}}}\right).
\end{aligned}
\tag{3.204}
$$

We only need the Fourier coefficients $V^{(\upsilon)}_{\mathbf{k}''}$ at $\mathbf{k}'' = \mathbf{q_{n'}} + \mathbf{G}$ with $\mathbf{G}$ being a reciprocal lattice vector and $\mathbf{q_{n'}}$ being a non-equivalent wave vector. The $\varphi_{\ell\mathbf{k}}(\mathbf{r})$ are Bloch functions (*c.f.* Eq. (2.141)),

$$
\varphi_{\ell\mathbf{k}}(\mathbf{r}) = \mathrm{u}_{\ell\mathbf{k}}(\mathbf{r})\, \mathrm{e}^{\mathrm{i}\,\mathbf{k}^{\mathrm{t}}\cdot\mathbf{r}}
$$

with $\mathrm{u}_{\ell\mathbf{k}}(\mathbf{r}+\mathbf{T_n}) = \mathrm{u}_{\ell\mathbf{k}}(\mathbf{r})$ for any vector $\mathbf{T_n}$ of the Bravais grid. Therefore, we obtain for the in Eq. (3.204) contained integral using that $\mathcal{V}_\mathrm{b}$ is the volume of the basis cell:

$$
\begin{aligned}
&\int_{\mathcal{V}_\mathrm{s}} d^3r\, \varphi^*_{\ell\mathbf{k}}(\mathbf{r})\, \mathrm{e}^{\mathrm{i}\,(\mathbf{q_{n'}}+\mathbf{G})^{\mathrm{t}}\cdot\mathbf{r}}\, \varphi_{\ell'\mathbf{k}'}(\mathbf{r}) \\
&= \sum_{\mathbf{T_n}} \int_{\mathcal{V}_\mathrm{b}} d^3r\, \varphi^*_{\ell\mathbf{k}}(\mathbf{r}+\mathbf{T_n})\, \mathrm{e}^{\mathrm{i}\,(\mathbf{q_{n'}}+\mathbf{G})^{\mathrm{t}}\cdot(\mathbf{r}+\mathbf{T_n})}\, \varphi_{\ell'\mathbf{k}'}(\mathbf{r}+\mathbf{T_n}) \\
&= \sum_{\mathbf{T_n}} \int_{\mathcal{V}_\mathrm{b}} d^3r\, \mathrm{u}^*_{\ell\mathbf{k}}(\mathbf{r}+\mathbf{T_n})\, \mathrm{e}^{-\mathrm{i}\,\mathbf{k}^{\mathrm{t}}\cdot(\mathbf{r}+\mathbf{T_n})}\, \mathrm{e}^{\mathrm{i}\,(\mathbf{q_{n'}}+\mathbf{G})^{\mathrm{t}}\cdot(\mathbf{r}+\mathbf{T_n})}\, \mathrm{u}_{\ell'\mathbf{k}'}(\mathbf{r}+\mathbf{T_n})\, \mathrm{e}^{\mathrm{i}\,\mathbf{k}'^{\mathrm{t}}\cdot(\mathbf{r}+\mathbf{T_n})} \\
&= \int_{\mathcal{V}_\mathrm{b}} d^3r\, \mathrm{e}^{\mathrm{i}\,(\mathbf{k}'-\mathbf{k}+\mathbf{q_{n'}}+\mathbf{G})^{\mathrm{t}}\cdot\mathbf{r}}\, \mathrm{u}^*_{\ell\mathbf{k}}(\mathbf{r})\, \mathrm{u}_{\ell'\mathbf{k}'}(\mathbf{r}) \underbrace{\sum_{\mathbf{T_n}} \mathrm{e}^{\mathrm{i}\,(\mathbf{k}'-\mathbf{k}+\mathbf{q_{n'}}+\mathbf{G})^{\mathrm{t}}\cdot\mathbf{T_n}}}_{\stackrel{(2.68)}{=} N_1 N_2 N_3\, \delta_{\mathbf{k}\mathbf{k}'+\mathbf{q_{n'}}+\mathbf{G}}} \\
&= N_1 N_2 N_3\, \delta_{\mathbf{k}\mathbf{k}'+\mathbf{q_{n'}}+\mathbf{G}} \int_{\mathcal{V}_\mathrm{b}} d^3r\, \mathrm{u}^*_{\ell\mathbf{k}'+\mathbf{q_{n'}}+\mathbf{G}}(\mathbf{r})\, \mathrm{u}_{\ell'\mathbf{k}'}(\mathbf{r}).
\end{aligned}
\tag{3.205}
$$

Inserting this into Eq. (3.204) produces

Fig. 3.2 Illustration of the two basis processes occurring during electron-phonon coupling

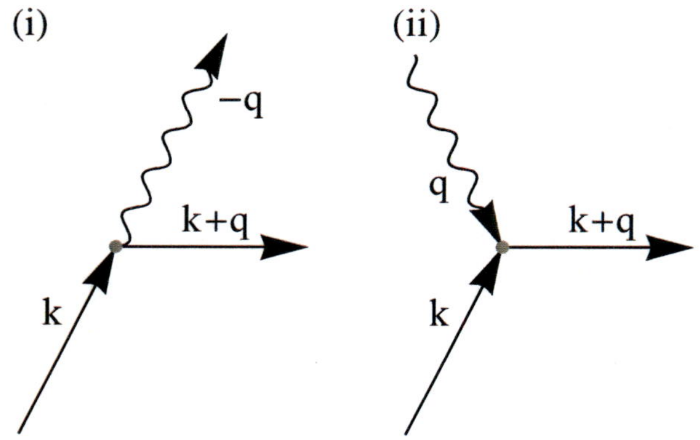

$$V^{(\mathrm{ep})}_{\ell\mathbf{k},\ell'\mathbf{k}'} = -\,\delta_{\mathbf{k}\mathbf{k}'+\mathbf{q}_{\mathbf{n}'}+\mathbf{G}}\sqrt{\frac{\hbar\,N_1\,N_2\,N_3}{2\,m_\upsilon\,\omega_j(\mathbf{q}_{\mathbf{n}'})}}\;N_1\,N_2\,N_3\times$$
$$\times\sum_{\mathbf{q}_{\mathbf{n}'}}\sum_{\mathbf{G}}\sum_{j=1}^{3\,N_\mathrm{b}}\mathrm{i}\,(\mathbf{q}_{\mathbf{n}'}+\mathbf{G})^\mathrm{t}\cdot\sum_\upsilon \mathbf{e}^{(j)}_\upsilon(\mathbf{q}_{\mathbf{n}'})\,V^{(\upsilon)}_{\mathbf{q}_{\mathbf{n}'}+\mathbf{G}}\,\mathrm{e}^{-\mathrm{i}\,(\mathbf{q}_{\mathbf{n}'}+\mathbf{G})^\mathrm{t}\cdot\mathbf{d}_\upsilon}\times$$
$$\times\int_{\mathcal{V}_\mathrm{b}} d^3r\,\mathrm{u}^*_{\ell\mathbf{k}'+\mathbf{q}_{\mathbf{n}'}+\mathbf{G}}(\mathbf{r})\,\mathrm{u}_{\ell'\mathbf{k}'}(\mathbf{r})\left(\hat{b}_{j\mathbf{q}_{\mathbf{n}'}}+\hat{b}^\dagger_{j-\mathbf{q}_{\mathbf{n}'}}\right). \tag{3.206}$$

Therefore, we obtain finally for the electron-phonon interacting term in second quantization (now we write $\mathbf{q}$ instead of $\mathbf{q}_{\mathbf{n}'}$ to simplify the notation)

$$\boxed{\hat{V}^{(\mathrm{ep})} = \sum_{\ell\ell'}\sum_{\mathbf{k}}\sum_{\mathbf{q}}\sum_{\mathbf{G}}\sum_{j=1}^{3\,N_\mathrm{b}} M^{(j\mathbf{q})}_{\ell\mathbf{k}+\mathbf{q}+\mathbf{G},\ell'\mathbf{k}}\left(\hat{b}_{j\mathbf{q}}+\hat{b}^\dagger_{j-\mathbf{q}}\right)\hat{c}^\dagger_{\ell\mathbf{k}+\mathbf{q}+\mathbf{G}}\,\hat{c}_{\ell'\mathbf{k}}} \tag{3.207}$$

with the electron-phonon coupling matrix element

$$\boxed{\begin{aligned} M^{(j\mathbf{q})}_{\ell\mathbf{k}+\mathbf{q}+\mathbf{G},\ell'\mathbf{k}} = &-\sqrt{\frac{\hbar\,N_1\,N_2\,N_3}{2\,m_\upsilon\,\omega_j(\mathbf{q})}}\;N_1\,N_2\,N_3\,\mathrm{i}\,(\mathbf{q}+\mathbf{G})^\mathrm{t}\cdot\sum_\upsilon \mathbf{e}^{(j)}_\upsilon(\mathbf{q})\,V^{(\upsilon)}_{\mathbf{q}+\mathbf{G}}\,\mathrm{e}^{-\mathrm{i}\,(\mathbf{q}+\mathbf{G})^\mathrm{t}\cdot\mathbf{d}_\upsilon}\times \\ &\times\sum_\sigma\int_{\mathcal{V}_\mathrm{b}} d^3r\,\mathrm{u}^*_{\ell\mathbf{k}+\mathbf{q}+\mathbf{G}}(\sigma,\mathbf{r})\,\mathrm{u}_{\ell'\mathbf{k}}(\sigma,\mathbf{r}).\end{aligned}} \tag{3.208}$$

Here, we again include the spin for completeness. The electron-phonon coupling is responsible for the following two basis processes (see Fig. 3.2):

(i) $\hat{b}^\dagger_{j-\mathbf{q}}\,\hat{c}^\dagger_{\ell\mathbf{k}+\mathbf{q}+\mathbf{G}}\,\hat{c}_{\ell'\mathbf{k}}$: A Bloch electron with wave-vector $\mathbf{k}$ scatters and gets the wave vector $\mathbf{k}+\mathbf{q}$ by emission of a phonon with wave vector $-\mathbf{q}$.
(ii) $\hat{b}_{j\mathbf{q}}\,\hat{c}^\dagger_{\ell\mathbf{k}+\mathbf{q}+\mathbf{G}}\,\hat{c}_{\ell'\mathbf{k}}$: A Bloch electron with wave-vector $\mathbf{k}$ scatters and gets the wave vector $\mathbf{k}+\mathbf{q}$ by absorption of a phonon with wave vector $\mathbf{q}$.

We can restrict $\mathbf{k}$ to the set of non-equivalent wave vectors. But $\mathbf{k}+\mathbf{q}$ can leave this set, so that a reciprocal lattice vector $\mathbf{G}$ must be added to keep $\mathbf{k}+\mathbf{q}+\mathbf{G}$ in the set of non-equivalent wave vectors. Such a process is called **Umklapp process**. In total, we have a quasi momentum conservation. During the scattering processes, the Bloch electron can change its state from ℓ to ℓ'. But the spin of the electron is not influenced, since $\hat{V}^{(\mathrm{ep})}$ does not act on the spin.

3.4.4 Electron-Phonon Coupling Strength

Energy is transfered from the hot electrons to the ions due to the electron-phonon coupling:

$$\frac{\partial E_{\mathrm{e}}}{\partial t} = \dot{E}_{\mathrm{ep}}, \tag{3.209}$$

$$\frac{\partial E_{\mathrm{i}}}{\partial t} = -\dot{E}_{\mathrm{ep}}. \tag{3.210}$$

Here, E_{e} denotes the energy of the electrons, E_{i} the energy of the ions and $\dot{E}_{\mathrm{ep}}$ is the energy transfer rate. The first ab-initio derivation of $\dot{E}_{\mathrm{ep}}$ was done by Allen et al. [23] in 1987 for metals. This method was improved by Wang et al.. [24] in 1994. However, this improved method was first detailed explained by Lin et al. [25] in 2008. Waldecker et al. extended this method to calculate different electron-phonon coupling constants for individual phonon branches in aluminium [26] and in antimony [27]. In 2017, Sadasivam et al. [28] extended Allens originally work to derive the electron-phonon coupling constants for twelve semiconductors also containing silicon. Here we derive the electron-phonon coupling rate $\dot{E}_{\mathrm{ep}}$ from ab-initio by summarizing the above studies, but provide more details where necessary. We start with need Fermis golden rule, which is proven in Ref [29]:

Theorem 3.16 (Fermis golden rule) *Let $|i\rangle$ and $|j\rangle$ be eigenstates of the unperturbed Hamiltonian $\hat{H}_0$. Due to the acting of the perturbation $\hat{V}$, the system changes from the state $|i\rangle$ with energy E_i to the state $|j\rangle$ with energy E_j. Then, the transition rate from $|i\rangle$ to $|j\rangle$ is given within first oder perturbation theory by*

$$\Gamma(i \to j) = \frac{2\pi}{\hbar} \left|\langle j|\hat{V}|i\rangle\right|^2 \delta(E_j - E_i), \tag{3.211}$$

where $\langle j|\hat{V}|i\rangle$ is the transition matrix element. The transition is only possible, if $E_i = E_j$ is valid, which is described by the delta function obeying $\delta(E_j - E_i) = \delta(E_i - E_j)$.

We consider the nuclei in the harmonic approximation together with the electrons described in the mean field approximation by one particle operators as the unperturbed Hamiltonian, which reads in second quantization

$$\hat{H}_0 = \underbrace{2\sum_{\ell\mathbf{k}} \varepsilon_{\ell\mathbf{k}}\, \hat{c}^\dagger_{\ell\mathbf{k}}\, \hat{c}_{\ell\mathbf{k}}}_{=E_\mathrm{e}} + \underbrace{\sum_{j\mathbf{q}} \hbar\,\omega_j(\mathbf{q}) \left(\hat{b}^\dagger_{j\mathbf{q}}\, \hat{b}_{j\mathbf{q}} + \frac{1}{2}\right)}_{=E_\mathrm{i}} . \tag{3.212}$$

We put a factor of 2 in the first term, since the sum over ℓ only runs over spin up states. $\hat{c}^\dagger_{\ell\mathbf{k}}$ creates one electron in the Bloch state $|\ell\mathbf{k}\rangle$ and $\hat{b}^\dagger_{j\mathbf{q}}$ creates one phonon in the phonon mode $j\mathbf{q}$. We denote by

$$E_0 = \sum_{j\mathbf{q}} \frac{\hbar\,\omega_j(\mathbf{q})}{2}$$

the contribution of the zero point motion. The electron-phonon coupling $\hat{V}^{(\mathrm{ep})}$ is considered as the perturbation, which does not change the spin state of the Bloch electrons. We consider the phonon mode $j\mathbf{q}$ with energy $\hbar\,\omega_j(\mathbf{q})$ and a single Bloch electron that is in the initial state $\ell'\mathbf{k}'$. Due to the electron phonon coupling, the Bloch electron scatters into the final state $\ell\mathbf{k}$, so that we have $n_{\ell'\mathbf{k}'} = 1$ and $n_{\ell\mathbf{k}} = 0$ for the electronic occupations at the beginning. There are two possibilities for the change of the phonon mode occupation $n_{j\mathbf{q}}$ due to the electron-phonon coupling:

(i) The Bloch electron scattering process generates an additional phonon in the state $j\mathbf{q}$, so that the phonon mode occupation changes from $n_{j\mathbf{q}}$ to $n_{j\mathbf{q}} + 1$. Then we have for the initial and final energies and normalized states of the system:

$$\begin{aligned}
E_\mathrm{initial} &= \varepsilon_{\ell'\mathbf{k}'} + n_{j\mathbf{q}}\,\hbar\,\omega_j(\mathbf{q}) + E_0, \\
E_\mathrm{final} &= \varepsilon_{\ell\mathbf{k}} + (n_{j\mathbf{q}} + 1)\,\hbar\,\omega_j(\mathbf{q}) + E_0, \\
|\mathrm{initial}\rangle &= \frac{1}{\sqrt{n_{j\mathbf{q}}!}} \left(\hat{b}^\dagger_{j\mathbf{q}}\right)^{n_{j\mathbf{q}}} \left(\hat{c}^\dagger_{\ell'\mathbf{k}'}\right)^{n_{\ell'\mathbf{k}'}} \Big|0\Big\rangle, \\
\langle\mathrm{final}| &= \Big\langle 0\Big| \frac{1}{\sqrt{(n_{j\mathbf{q}} + 1)!}} \left(\hat{c}_{\ell\mathbf{k}}\right)^{1-n_{\ell\mathbf{k}}} \left(\hat{b}_{j\mathbf{q}}\right)^{n_{j\mathbf{q}}+1} .
\end{aligned}$$

We obtain for the transition matrix element by considering the only non-vanishing term in $\hat{V}^{(\mathrm{ep})}$:

$$\begin{aligned}
&\Big\langle \mathrm{final}\Big| \hat{V}^{(\mathrm{ep})} \Big| \mathrm{initial}\Big\rangle \\
&= \frac{M^{(j\mathbf{q})}_{\ell\mathbf{k},\ell'\mathbf{k}'}}{\sqrt{n_{j\mathbf{q}}!}\,\sqrt{(n_{j\mathbf{q}} + 1)!}} \times
\end{aligned}$$

$$
\begin{aligned}
&\times \left\langle 0 \left| \left(\hat{c}_{\ell\mathbf{k}}\right)^{1-n_{\ell\mathbf{k}}} \left(\hat{b}_{j\mathbf{q}}\right)^{n_{j\mathbf{q}}+1} \hat{b}^{\dagger}_{j\mathbf{q}} \hat{c}^{\dagger}_{\ell\mathbf{k}} \hat{c}_{\ell'\mathbf{k}'} \left(\hat{b}^{\dagger}_{j\mathbf{q}}\right)^{n_{j\mathbf{q}}} \left(\hat{c}^{\dagger}_{\ell'\mathbf{k}'}\right)^{n_{\ell'\mathbf{k}'}} \right| 0 \right\rangle \\
=&\frac{M^{(j\mathbf{q})}_{\ell\mathbf{k},\ell'\mathbf{k}'}}{n_{j\mathbf{q}}!\,\sqrt{(n_{j\mathbf{q}}+1)}} \times \\
&\times \left\langle 0 \left| \left(\hat{b}_{j\mathbf{q}}\right)^{n_{j\mathbf{q}}+1} \hat{b}^{\dagger}_{j\mathbf{q}} \left(\hat{b}^{\dagger}_{j\mathbf{q}}\right)^{n_{j\mathbf{q}}} \left(\hat{c}_{\ell\mathbf{k}}\right)^{1-n_{\ell\mathbf{k}}} \hat{c}^{\dagger}_{\ell\mathbf{k}} \hat{c}_{\ell'\mathbf{k}'} \left(\hat{c}^{\dagger}_{\ell'\mathbf{k}'}\right)^{n_{\ell'\mathbf{k}'}} \right| 0 \right\rangle \\
=&\frac{M^{(j\mathbf{q})}_{\ell\mathbf{k},\ell'\mathbf{k}'}\sqrt{(n_{j\mathbf{q}}+1)}}{(n_{j\mathbf{q}}+1)!} n_{\ell'\mathbf{k}'}\left(1-n_{\ell\mathbf{k}}\right) \underbrace{\left\langle 0 \left| \left(\hat{b}_{j\mathbf{q}}\right)^{n_{j\mathbf{q}}+1} \left(\hat{b}^{\dagger}_{j\mathbf{q}}\right)^{n_{j\mathbf{q}}+1} \right| 0 \right\rangle}_{\overset{(3.153)}{=}(n_{j\mathbf{q}}+1)!} \\
=&M^{(j\mathbf{q})}_{\ell\mathbf{k},\ell'\mathbf{k}'}\sqrt{(n_{j\mathbf{q}}+1)}\, n_{\ell'\mathbf{k}'}\left(1-n_{\ell\mathbf{k}}\right)
\end{aligned}
$$

Since electrons are Ferminons and obey the Pauli exclusion principle, we have $n_{\ell'\mathbf{k}'}(1-n_{\ell\mathbf{k}}) \in \{0, 1\}$ and obtain, therefore, $n^2_{\ell'\mathbf{k}'}(1-n_{\ell\mathbf{k}})^2 = n_{\ell'\mathbf{k}'}(1-n_{\ell\mathbf{k}})$. Using this, we obtain finally for the transfer rate of this process from Fermis golden rule:

$$
\begin{aligned}
&\Gamma\left(\ell'\mathbf{k}' \to \ell\mathbf{k}, n_{j\mathbf{q}} \to n_{j\mathbf{q}}+1\right) \\
&= \frac{2\pi}{\hbar}\left|M^{(j\mathbf{q})}_{\ell\mathbf{k},\ell'\mathbf{k}'}\right|^2 (n_{j\mathbf{q}}+1)\, n_{\ell'\mathbf{k}'}\left(1-n_{\ell\mathbf{k}}\right)\delta\Big(\varepsilon_{\ell\mathbf{k}}-\varepsilon_{\ell'\mathbf{k}'}+\hbar\,\omega_j(\mathbf{q})\Big).
\end{aligned}
\tag{3.213}
$$

(ii) The Bloch electron scattering process destroys a phonon from the state $j\mathbf{q}$, so that the phonon mode occupation changes from $n_{j\mathbf{q}}$ to $n_{j\mathbf{q}}-1$. In this way, we get

$$
\begin{aligned}
E_{\text{initial}} =& \varepsilon_{\ell'\mathbf{k}'} + n_{j\mathbf{q}}\,\hbar\,\omega_j(\mathbf{q}) + E_0, \\
E_{\text{final}} =& \varepsilon_{\ell\mathbf{k}} + (n_{j\mathbf{q}}-1)\,\hbar\,\omega_j(\mathbf{q}) + E_0, \\
|\text{initial}\rangle =& \frac{1}{\sqrt{n_{j\mathbf{q}}!}}\left(\hat{b}^{\dagger}_{j\mathbf{q}}\right)^{n_{j\mathbf{q}}}\left(\hat{c}^{\dagger}_{\ell'\mathbf{k}'}\right)^{n_{\ell'\mathbf{k}'}}\left|0\right\rangle, \\
\langle\text{final}| =& \left\langle 0\right| \frac{1}{\sqrt{(n_{j\mathbf{q}}-1)!}}\left(\hat{c}_{\ell\mathbf{k}}\right)^{1-n_{\ell\mathbf{k}}}\left(\hat{b}_{j\mathbf{q}}\right)^{n_{j\mathbf{q}}-1}.
\end{aligned}
$$

Therefore, we obtain for the matrix element by considering the non-vanishing term:

$$
\begin{aligned}
&\left\langle\text{final}\left|\hat{V}^{(\text{ep})}\right|\text{initial}\right\rangle \\
=&\frac{M^{(j\mathbf{q})}_{\ell\mathbf{k},\ell'\mathbf{k}'}}{\sqrt{n_{j\mathbf{q}}!}\,\sqrt{(n_{j\mathbf{q}}-1)!}} \times
\end{aligned}
$$

$$
\begin{aligned}
&\times \left\langle 0 \left| \left(\hat{c}_{\ell\mathbf{k}}\right)^{1-n_{\ell\mathbf{k}}} \left(\hat{b}_{j\mathbf{q}}\right)^{n_{j\mathbf{q}}-1} \hat{b}_{j\mathbf{q}}\, \hat{c}^{\dagger}_{\ell\mathbf{k}}\, \hat{c}_{\ell'\mathbf{k}'} \left(\hat{b}^{\dagger}_{j\mathbf{q}}\right)^{n_{j\mathbf{q}}} \left(\hat{c}^{\dagger}_{\ell'\mathbf{k}'}\right)^{n_{\ell'\mathbf{k}'}} \right| 0 \right\rangle \\
=&\frac{M^{(j\mathbf{q})}_{\ell\mathbf{k},\ell'\mathbf{k}'}}{\sqrt{n_{j\mathbf{q}}}\,(n_{j\mathbf{q}}-1)!} \times \\
&\times \left\langle 0 \left| \left(\hat{b}_{j\mathbf{q}}\right)^{n_{j\mathbf{q}}-1} \hat{b}_{j\mathbf{q}} \left(\hat{b}^{\dagger}_{j\mathbf{q}}\right)^{n_{j\mathbf{q}}} \left(\hat{c}_{\ell\mathbf{k}}\right)^{1-n_{\ell\mathbf{k}}} \hat{c}^{\dagger}_{\ell\mathbf{k}}\, \hat{c}_{\ell'\mathbf{k}'} \left(\hat{c}^{\dagger}_{\ell'\mathbf{k}'}\right)^{n_{\ell'\mathbf{k}'}} \right| 0 \right\rangle \\
=&\frac{M^{(j\mathbf{q})}_{\ell\mathbf{k},\ell'\mathbf{k}'}\sqrt{n_{j\mathbf{q}}}}{n_{j\mathbf{q}}!}\, n_{\ell'\mathbf{k}'}\,(1-n_{\ell\mathbf{k}}) \underbrace{\left\langle 0 \left| \left(\hat{b}_{j\mathbf{q}}\right)^{n_{j\mathbf{q}}} \left(\hat{b}^{\dagger}_{j\mathbf{q}}\right)^{n_{j\mathbf{q}}} \right| 0 \right\rangle}_{\overset{(3.153)}{=} n_{j\mathbf{q}}!} \\
=&M^{(j\mathbf{q})}_{\ell\mathbf{k},\ell'\mathbf{k}'}\sqrt{n_{j\mathbf{q}}}\, n_{\ell'\mathbf{k}'}\,(1-n_{\ell\mathbf{k}})
\end{aligned}
$$

and find for the transfer rate of this process from Fermis golden rule:

$$
\begin{aligned}
&\Gamma\big(\ell'\mathbf{k}' \to \ell\mathbf{k}, n_{j\mathbf{q}} \to n_{j\mathbf{q}}-1\big) \\
=&\frac{2\pi}{\hbar}\left|M^{(j\mathbf{q})}_{\ell\mathbf{k},\ell'\mathbf{k}'}\right|^2 n_{j\mathbf{q}}\, n_{\ell'\mathbf{k}'}\,(1-n_{\ell\mathbf{k}})\,\delta\Big(\varepsilon_{\ell\mathbf{k}} - \varepsilon_{\ell'\mathbf{k}'} - \hbar\omega_j(\mathbf{q})\Big).
\end{aligned} \tag{3.214}
$$

From the definition (3.208) follows

$$
M^{(j\mathbf{q})}_{\ell'\mathbf{k}',\ell\mathbf{k}} = M^{(j\mathbf{q})*}_{\ell\mathbf{k},\ell'\mathbf{k}'} \quad \Rightarrow \quad \left|M^{(j\mathbf{q})}_{\ell'\mathbf{k}',\ell\mathbf{k}}\right|^2 = \left|M^{(j\mathbf{q})}_{\ell\mathbf{k},\ell'\mathbf{k}'}\right|^2. \tag{3.215}
$$

We now can consider the transition of the Bloch electron from the initial state $\ell\mathbf{k}$ to the final state $\ell'\mathbf{k}'$ and obtain analogously

$$
\begin{aligned}
&\Gamma\big(\ell\mathbf{k} \to \ell'\mathbf{k}', n_{j\mathbf{q}} \to n_{j\mathbf{q}}+1\big) \\
=&\frac{2\pi}{\hbar}\left|M^{(j\mathbf{q})}_{\ell'\mathbf{k}',\ell\mathbf{k}}\right|^2 (n_{j\mathbf{q}}+1)\, n_{\ell\mathbf{k}}\,(1-n_{\ell'\mathbf{k}'})\,\delta\Big(\varepsilon_{\ell'\mathbf{k}'} - \varepsilon_{\ell\mathbf{k}} + \hbar\omega_j(\mathbf{q})\Big) \\
=&\frac{2\pi}{\hbar}\left|M^{(j\mathbf{q})}_{\ell\mathbf{k},\ell'\mathbf{k}'}\right|^2 (n_{j\mathbf{q}}+1)\, n_{\ell\mathbf{k}}\,(1-n_{\ell'\mathbf{k}'})\,\delta\Big(\varepsilon_{\ell\mathbf{k}} - \varepsilon_{\ell'\mathbf{k}'} - \hbar\omega_j(\mathbf{q})\Big)
\end{aligned} \tag{3.216}
$$

and

$$
\begin{aligned}
&\Gamma\big(\ell\mathbf{k} \to \ell'\mathbf{k}', n_{j\mathbf{q}} \to n_{j\mathbf{q}}-1\big) \\
=&\frac{2\pi}{\hbar}\left|M^{(j\mathbf{q})}_{\ell'\mathbf{k}',\ell\mathbf{k}}\right|^2 n_{j\mathbf{q}}\, n_{\ell\mathbf{k}}\,(1-n_{\ell'\mathbf{k}'})\,\delta\Big(\varepsilon_{\ell'\mathbf{k}'} - \varepsilon_{\ell\mathbf{k}} - \hbar\omega_j(\mathbf{q})\Big) \\
=&\frac{2\pi}{\hbar}\left|M^{(j\mathbf{q})}_{\ell\mathbf{k},\ell'\mathbf{k}'}\right|^2 n_{j\mathbf{q}}\, n_{\ell\mathbf{k}}\,(1-n_{\ell'\mathbf{k}'})\,\delta\Big(\varepsilon_{\ell\mathbf{k}} - \varepsilon_{\ell'\mathbf{k}'} + \hbar\omega_j(\mathbf{q})\Big).
\end{aligned} \tag{3.217}
$$

With the help of the transfer rates, we can construct the rate equation for the electron occupation $n_{\ell\mathbf{k}}$ of the state $\ell\mathbf{k}$:

$$\frac{\partial n_{\ell\mathbf{k}}}{\partial t} = \sum_{j\mathbf{q}}\sum_{\ell'\mathbf{k}'}\Big(-\Gamma(\ell\mathbf{k}\to\ell'\mathbf{k}', n_{j\mathbf{q}}\to n_{j\mathbf{q}}+1) - \Gamma(\ell\mathbf{k}\to\ell'\mathbf{k}', n_{j\mathbf{q}}\to n_{j\mathbf{q}}-1)$$
$$+\Gamma(\ell'\mathbf{k}'\to\ell\mathbf{k}, n_{j\mathbf{q}}\to n_{j\mathbf{q}}+1) + \Gamma(\ell'\mathbf{k}'\to\ell\mathbf{k}, n_{j\mathbf{q}}\to n_{j\mathbf{q}}-1)\Big).$$

Putting in the transfer rates obtained from Fermis golden rule yields

$$\boxed{\begin{aligned}&\frac{\partial n_{\ell\mathbf{k}}}{\partial t}\\ &= \frac{2\pi}{\hbar}\sum_{j\mathbf{q}}\sum_{\ell'\mathbf{k}'}\left|M^{(j\mathbf{q})}_{\ell\mathbf{k},\ell'\mathbf{k}'}\right|^2\times\\ &\times\Big(-n_{\ell\mathbf{k}}\,(1-n_{\ell'\mathbf{k}'})\times\\ &\quad\times\Big((n_{j\mathbf{q}}+1)\,\delta\big(\varepsilon_{\ell\mathbf{k}}-\varepsilon_{\ell'\mathbf{k}'}-\hbar\,\omega_j(\mathbf{q})\big)+n_{j\mathbf{q}}\,\delta\big(\varepsilon_{\ell\mathbf{k}}-\varepsilon_{\ell'\mathbf{k}'}+\hbar\,\omega_j(\mathbf{q})\big)\Big)\\ &\quad+n_{\ell'\mathbf{k}'}\,(1-n_{\ell\mathbf{k}})\times\\ &\quad\times\Big((n_{j\mathbf{q}}+1)\,\delta\big(\varepsilon_{\ell\mathbf{k}}-\varepsilon_{\ell'\mathbf{k}'}+\hbar\,\omega_j(\mathbf{q})\big)+n_{j\mathbf{q}}\,\delta\big(\varepsilon_{\ell\mathbf{k}}-\varepsilon_{\ell'\mathbf{k}'}-\hbar\,\omega_j(\mathbf{q})\big)\Big)\Big).\end{aligned}} \tag{3.218}$$

In a similar way, we can construct the rate equation for the phonon occupation $n_{j\mathbf{q}}$ of the state $j\mathbf{q}$:

$$\begin{aligned}&\frac{\partial n_{j\mathbf{q}}}{\partial t}\\ &=\sum_{\ell\mathbf{k}}\sum_{\ell'\mathbf{k}'}\Big(\Gamma(\ell\mathbf{k}\to\ell'\mathbf{k}', n_{j\mathbf{q}}\to n_{j\mathbf{q}}+1) - \Gamma(\ell\mathbf{k}\to\ell'\mathbf{k}', n_{j\mathbf{q}}\to n_{j\mathbf{q}}-1)\\ &\qquad+\Gamma(\ell'\mathbf{k}'\to\ell\mathbf{k}, n_{j\mathbf{q}}\to n_{j\mathbf{q}}+1) - \Gamma(\ell'\mathbf{k}'\to\ell\mathbf{k}, n_{j\mathbf{q}}\to n_{j\mathbf{q}}-1)\Big)\\ &=\frac{2\pi}{\hbar}\sum_{j\mathbf{q}}\sum_{\ell'\mathbf{k}'}\left|M^{(j\mathbf{q})}_{\ell\mathbf{k},\ell'\mathbf{k}'}\right|^2\times\\ &\quad\times\Big(n_{\ell\mathbf{k}}\,(1-n_{\ell'\mathbf{k}'})\times\\ &\qquad\times\Big((n_{j\mathbf{q}}+1)\,\delta\big(\varepsilon_{\ell\mathbf{k}}-\varepsilon_{\ell'\mathbf{k}'}-\hbar\,\omega_j(\mathbf{q})\big)-n_{j\mathbf{q}}\,\delta\big(\varepsilon_{\ell\mathbf{k}}-\varepsilon_{\ell'\mathbf{k}'}+\hbar\,\omega_j(\mathbf{q})\big)\Big)\\ &\qquad+n_{\ell'\mathbf{k}'}\,(1-n_{\ell\mathbf{k}})\times\end{aligned}$$

$$\times\Bigg(\left(n_{j\mathbf{q}}+1\right)\underbrace{\delta\Big(\varepsilon_{\ell\mathbf{k}}-\varepsilon_{\ell'\mathbf{k}'}+\hbar\,\omega_j(\mathbf{q})\Big)}_{=\delta\left(\varepsilon_{\ell'\mathbf{k}'}-\varepsilon_{\ell\mathbf{k}}-\hbar\,\omega_j(\mathbf{q})\right)}-n_{j\mathbf{q}}\underbrace{\delta\Big(\varepsilon_{\ell\mathbf{k}}-\varepsilon_{\ell'\mathbf{k}'}-\hbar\,\omega_j(\mathbf{q})\Big)}_{\delta\left(\varepsilon_{\ell'\mathbf{k}'}-\varepsilon_{\ell\mathbf{k}}+\hbar\,\omega_j(\mathbf{q})\right)}\Bigg)\Bigg). \tag{3.219}$$

If we use $\left|M^{(j\mathbf{q})}_{\ell'\mathbf{k}',\ell\mathbf{k}}\right|^2=\left|M^{(j\mathbf{q})}_{\ell\mathbf{k},\ell'\mathbf{k}'}\right|^2$, we obtain finally

$$\boxed{\begin{aligned}\frac{\partial n_{j\mathbf{q}}}{\partial t}=\frac{4\pi}{\hbar}\sum_{j\mathbf{q}}\sum_{\ell'\mathbf{k}'}\left|M^{(j\mathbf{q})}_{\ell\mathbf{k},\ell'\mathbf{k}'}\right|^2 n_{\ell\mathbf{k}}\left(1-n_{\ell'\mathbf{k}'}\right)\times\\ \times\Big(\left(n_{j\mathbf{q}}+1\right)\delta\Big(\varepsilon_{\ell\mathbf{k}}-\varepsilon_{\ell'\mathbf{k}'}-\hbar\,\omega_j(\mathbf{q})\Big)-n_{j\mathbf{q}}\,\delta\Big(\varepsilon_{\ell\mathbf{k}}-\varepsilon_{\ell'\mathbf{k}'}+\hbar\,\omega_j(\mathbf{q})\Big)\Big).\end{aligned}} \tag{3.220}$$

Equation (3.218) and Eq. (3.220) form the **Bloch-Boltzmann-Peierls equations** [23]. These rate equations describe the changes in the electron and phonon occupations due to the electron-phonon coupling. They also fulfill the energy conservation

$$2\sum_{\ell\mathbf{k}}\varepsilon_{\ell\mathbf{k}}\,\frac{\partial n_{\ell\mathbf{k}}}{\partial t}+\sum_{j\mathbf{q}}\hbar\,\omega_j(\mathbf{q})\,\frac{\partial n_{j\mathbf{q}}}{\partial t}=0. \tag{3.221}$$

We can prove the energy conservation with the help of Eq. (3.218) and Eq. (3.219):

$$\begin{aligned}&2\sum_{\ell\mathbf{k}}\varepsilon_{\ell\mathbf{k}}\,\frac{\partial n_{\ell\mathbf{k}}}{\partial t}+\sum_{j\mathbf{q}}\hbar\,\omega_j(\mathbf{q})\,\frac{\partial n_{j\mathbf{q}}}{\partial t}\\
=&\frac{2\pi}{\hbar}\sum_{j\mathbf{q}}\sum_{\ell\mathbf{k}}\sum_{\ell'\mathbf{k}'}\left|M^{(j\mathbf{q})}_{\ell\mathbf{k},\ell'\mathbf{k}'}\right|^2\Big(-2\,\varepsilon_{\ell\mathbf{k}}\,n_{\ell\mathbf{k}}\left(1-n_{\ell'\mathbf{k}'}\right)\left(n_{j\mathbf{q}}+1\right)\delta\Big(\varepsilon_{\ell\mathbf{k}}-\varepsilon_{\ell'\mathbf{k}'}-\hbar\,\omega_j(\mathbf{q})\Big)\\
&-2\,\varepsilon_{\ell\mathbf{k}}\,n_{\ell\mathbf{k}}\left(1-n_{\ell'\mathbf{k}'}\right)n_{j\mathbf{q}}\,\delta\Big(\varepsilon_{\ell\mathbf{k}}-\varepsilon_{\ell'\mathbf{k}'}+\hbar\,\omega_j(\mathbf{q})\Big)\\
&+2\,\varepsilon_{\ell\mathbf{k}}\,n_{\ell'\mathbf{k}'}\left(1-n_{\ell\mathbf{k}}\right)\left(n_{j\mathbf{q}}+1\right)\delta\Big(\varepsilon_{\ell\mathbf{k}}-\varepsilon_{\ell'\mathbf{k}'}+\hbar\,\omega_j(\mathbf{q})\Big)\\
&+2\,\varepsilon_{\ell\mathbf{k}}\,n_{\ell'\mathbf{k}'}\left(1-n_{\ell\mathbf{k}}\right)n_{j\mathbf{q}}\,\delta\Big(\varepsilon_{\ell\mathbf{k}}-\varepsilon_{\ell'\mathbf{k}'}-\hbar\,\omega_j(\mathbf{q})\Big)\\
&+\hbar\,\omega_j(\mathbf{q})\,n_{\ell\mathbf{k}}\left(1-n_{\ell'\mathbf{k}'}\right)\left(n_{j\mathbf{q}}+1\right)\delta\Big(\varepsilon_{\ell\mathbf{k}}-\varepsilon_{\ell'\mathbf{k}'}-\hbar\,\omega_j(\mathbf{q})\Big)\\
&-\hbar\,\omega_j(\mathbf{q})\,n_{\ell\mathbf{k}}\left(1-n_{\ell'\mathbf{k}'}\right)n_{j\mathbf{q}}\,\delta\Big(\varepsilon_{\ell\mathbf{k}}-\varepsilon_{\ell'\mathbf{k}'}+\hbar\,\omega_j(\mathbf{q})\Big)\\
&+\hbar\,\omega_j(\mathbf{q})\,n_{\ell'\mathbf{k}'}\left(1-n_{\ell\mathbf{k}}\right)\left(n_{j\mathbf{q}}+1\right)\delta\Big(\varepsilon_{\ell\mathbf{k}}-\varepsilon_{\ell'\mathbf{k}'}+\hbar\,\omega_j(\mathbf{q})\Big)\\
&-\hbar\,\omega_j(\mathbf{q})\,n_{\ell'\mathbf{k}'}\left(1-n_{\ell\mathbf{k}}\right)n_{j\mathbf{q}}\,\delta\Big(\varepsilon_{\ell\mathbf{k}}-\varepsilon_{\ell'\mathbf{k}'}-\hbar\,\omega_j(\mathbf{q})\Big)\Big)\end{aligned}$$

$$
\begin{aligned}
&= \frac{2\pi}{\hbar} \sum_{j\mathbf{q}} \sum_{\ell\mathbf{k}} \sum_{\ell'\mathbf{k}'} \left| M^{(j\mathbf{q})}_{\ell\mathbf{k},\ell'\mathbf{k}'} \right|^2 \Bigg(\Big(-2\,\varepsilon_{\ell\mathbf{k}} + \hbar\,\omega_j(\mathbf{q})\Big)\, n_{\ell\mathbf{k}}\,(1 - n_{\ell'\mathbf{k}'})\,(n_{j\mathbf{q}} + 1)\, \underbrace{\delta\Big(\varepsilon_{\ell\mathbf{k}} - \varepsilon_{\ell'\mathbf{k}'} - \hbar\,\omega_j(\mathbf{q})\Big)}_{=\delta\left(\varepsilon_{\ell'\mathbf{k}'} - \varepsilon_{\ell\mathbf{k}} + \hbar\,\omega_j(\mathbf{q})\right)} \\
&\qquad + \Big(-2\,\varepsilon_{\ell\mathbf{k}} - \hbar\,\omega_j(\mathbf{q})\Big)\, n_{\ell\mathbf{k}}\,(1 - n_{\ell'\mathbf{k}'})\, n_{j\mathbf{q}}\, \underbrace{\delta\Big(\varepsilon_{\ell\mathbf{k}} - \varepsilon_{\ell'\mathbf{k}'} + \hbar\,\omega_j(\mathbf{q})\Big)}_{=\delta\left(\varepsilon_{\ell'\mathbf{k}'} - \varepsilon_{\ell\mathbf{k}} - \hbar\,\omega_j(\mathbf{q})\right)} \\
&\qquad + \Big(2\,\varepsilon_{\ell\mathbf{k}} + \hbar\,\omega_j(\mathbf{q})\Big)\, n_{\ell'\mathbf{k}'}\,(1 - n_{\ell\mathbf{k}})\,(n_{j\mathbf{q}} + 1)\, \delta\Big(\varepsilon_{\ell\mathbf{k}} - \varepsilon_{\ell'\mathbf{k}'} + \hbar\,\omega_j(\mathbf{q})\Big) \\
&\qquad + \Big(2\,\varepsilon_{\ell\mathbf{k}} - \hbar\,\omega_j(\mathbf{q})\Big)\, n_{\ell'\mathbf{k}'}\,(1 - n_{\ell\mathbf{k}})\, n_{j\mathbf{q}}\, \delta\Big(\varepsilon_{\ell\mathbf{k}} - \varepsilon_{\ell'\mathbf{k}'} - \hbar\,\omega_j(\mathbf{q})\Big) \Bigg) \\
&= \frac{2\pi}{\hbar} \sum_{j\mathbf{q}} \sum_{\ell\mathbf{k}} \sum_{\ell'\mathbf{k}'} \left| M^{(j\mathbf{q})}_{\ell\mathbf{k},\ell'\mathbf{k}'} \right|^2 \Bigg(\underbrace{\Big(-2\,\varepsilon_{\ell'\mathbf{k}'} + \hbar\,\omega_j(\mathbf{q})\Big)}_{=-\varepsilon_{\ell\mathbf{k}} - \varepsilon_{\ell'\mathbf{k}'}}\, n_{\ell'\mathbf{k}'}\,(1 - n_{\ell\mathbf{k}})\,(n_{j\mathbf{q}} + 1)\, \delta\Big(\varepsilon_{\ell\mathbf{k}} - \varepsilon_{\ell'\mathbf{k}'} + \hbar\,\omega_j(\mathbf{q})\Big) \\
&\qquad + \underbrace{\Big(-2\,\varepsilon_{\ell'\mathbf{k}'} - \hbar\,\omega_j(\mathbf{q})\Big)}_{=-\varepsilon_{\ell\mathbf{k}} - \varepsilon_{\ell'\mathbf{k}'}}\, n_{\ell'\mathbf{k}'}\,(1 - n_{\ell\mathbf{k}})\, n_{j\mathbf{q}}\, \delta\Big(\varepsilon_{\ell\mathbf{k}} - \varepsilon_{\ell'\mathbf{k}'} - \hbar\,\omega_j(\mathbf{q})\Big) \\
&\qquad + \Big(\varepsilon_{\ell\mathbf{k}} + \varepsilon_{\ell'\mathbf{k}'}\Big)\, n_{\ell'\mathbf{k}'}\,(1 - n_{\ell\mathbf{k}})\,(n_{j\mathbf{q}} + 1)\, \delta\Big(\varepsilon_{\ell\mathbf{k}} - \varepsilon_{\ell'\mathbf{k}'} + \hbar\,\omega_j(\mathbf{q})\Big) \\
&\qquad + \Big(\varepsilon_{\ell\mathbf{k}} + \varepsilon_{\ell'\mathbf{k}'}\Big)\, n_{\ell'\mathbf{k}'}\,(1 - n_{\ell\mathbf{k}})\, n_{j\mathbf{q}}\, \delta\Big(\varepsilon_{\ell\mathbf{k}} - \varepsilon_{\ell'\mathbf{k}'} - \hbar\,\omega_j(\mathbf{q})\Big) \Bigg) \\
&= 0.
\end{aligned}
$$

To derive the electron-phonon energy exchange rate

$$
\dot{E}_{\mathrm{ep}} = -\frac{\partial E_{\mathrm{i}}}{\partial t} = -\sum_{j\mathbf{q}} \hbar\,\omega_j(\mathbf{q})\, \frac{\partial n_{j\mathbf{q}}}{\partial t},
$$

we consider $\frac{\partial n_{j\mathbf{q}}}{\partial t}$ from Eq. (3.220):

$$
\begin{aligned}
-\frac{\partial n_{j\mathbf{q}}}{\partial t} &= \frac{4\pi}{\hbar} \sum_{\ell\mathbf{k}} \sum_{\ell'\mathbf{k}'} \left| M^{(j\mathbf{q})}_{\ell\mathbf{k},\ell'\mathbf{k}'} \right|^2 \Big(-(n_{j\mathbf{q}} + 1)\, n_{\ell\mathbf{k}}\,(1 - n_{\ell'\mathbf{k}'})\, \delta\Big(\varepsilon_{\ell\mathbf{k}} - \varepsilon_{\ell'\mathbf{k}'} - \hbar\,\omega_j(\mathbf{q})\Big) \\
&\qquad + n_{j\mathbf{q}}\, n_{\ell\mathbf{k}}\,(1 - n_{\ell'\mathbf{k}'})\, \delta\Big(\varepsilon_{\ell\mathbf{k}} - \varepsilon_{\ell'\mathbf{k}'} + \hbar\,\omega_j(\mathbf{q})\Big) \Big) \\
&= \frac{4\pi}{\hbar} \sum_{\ell\mathbf{k}} \sum_{\ell'\mathbf{k}'} \left| M^{(j\mathbf{q})}_{\ell\mathbf{k},\ell'\mathbf{k}'} \right|^2 \Big(-(n_{j\mathbf{q}} + 1)\, n_{\ell'\mathbf{k}'}\,(1 - n_{\ell\mathbf{k}})\, \underbrace{\delta\Big(\varepsilon_{\ell'\mathbf{k}'} - \varepsilon_{\ell\mathbf{k}} - \hbar\,\omega_j(\mathbf{q})\Big)}_{=\delta\left(\varepsilon_{\ell\mathbf{k}} - \varepsilon_{\ell'\mathbf{k}'} + \hbar\,\omega_j(\mathbf{q})\right)} \\
&\qquad + n_{j\mathbf{q}}\, n_{\ell\mathbf{k}}\,(1 - n_{\ell'\mathbf{k}'})\, \delta\Big(\varepsilon_{\ell\mathbf{k}} - \varepsilon_{\ell'\mathbf{k}'} + \hbar\,\omega_j(\mathbf{q})\Big) \Big) \\
&= \frac{4\pi}{\hbar} \sum_{\ell\mathbf{k}} \sum_{\ell'\mathbf{k}'} \left| M^{(j\mathbf{q})}_{\ell\mathbf{k},\ell'\mathbf{k}'} \right|^2 \Big(-(n_{j\mathbf{q}} + 1)\, n_{\ell'\mathbf{k}'}\,(1 - n_{\ell\mathbf{k}}) + n_{j\mathbf{q}}\, n_{\ell\mathbf{k}}\,(1 - n_{\ell'\mathbf{k}'}) \Big) \times \\
&\qquad \times \delta\Big(\varepsilon_{\ell\mathbf{k}} - \varepsilon_{\ell'\mathbf{k}'} + \hbar\,\omega_j(\mathbf{q})\Big).
\end{aligned}
$$

Due to

$$\begin{aligned}
& -\left(n_{j\mathbf{q}}+1\right) n_{\ell'\mathbf{k}'}\left(1-n_{\ell\mathbf{k}}\right)+n_{j\mathbf{q}} n_{\ell\mathbf{k}}\left(1-n_{\ell'\mathbf{k}'}\right) \\
= & -n_{j\mathbf{q}} n_{\ell'\mathbf{k}'}\left(1-n_{\ell\mathbf{k}}\right)-n_{\ell'\mathbf{k}'}\left(1-n_{\ell\mathbf{k}}\right)+n_{j\mathbf{q}} n_{\ell\mathbf{k}}-n_{j\mathbf{q}} n_{\ell\mathbf{k}} n_{\ell'\mathbf{k}'} \\
= & -n_{j\mathbf{q}} n_{\ell'\mathbf{k}'}+n_{j\mathbf{q}} n_{\ell\mathbf{k}} n_{\ell'\mathbf{k}'}-n_{\ell'\mathbf{k}'}+n_{\ell\mathbf{k}} n_{\ell'\mathbf{k}'}+n_{j\mathbf{q}} n_{\ell\mathbf{k}}-n_{j\mathbf{q}} n_{\ell\mathbf{k}} n_{\ell'\mathbf{k}'} \\
= & \left(n_{\ell\mathbf{k}}-n_{\ell'\mathbf{k}'}\right) n_{j\mathbf{q}}-n_{\ell'\mathbf{k}'}\left(1-n_{\ell\mathbf{k}}\right)
\end{aligned}$$

we get

$$\begin{aligned}
-\frac{\partial n_{j\mathbf{q}}}{\partial t}= & \frac{4\pi}{\hbar} \sum_{\ell\mathbf{k}} \sum_{\ell'\mathbf{k}'}\left|M_{\ell\mathbf{k},\ell'\mathbf{k}'}^{(j\mathbf{q})}\right|^2\Big(\left(n_{\ell\mathbf{k}}-n_{\ell'\mathbf{k}'}\right) n_{j\mathbf{q}}-n_{\ell'\mathbf{k}'}\left(1-n_{\ell\mathbf{k}}\right)\Big) \times \\
& \times \delta\Big(\varepsilon_{\ell\mathbf{k}}-\varepsilon_{\ell'\mathbf{k}'}+\hbar\,\omega_j(\mathbf{q})\Big).
\end{aligned}$$

Using this, we obtain for the electron-phonon coupling energy transfer rate

$$\begin{aligned}
\dot{E}_{\mathrm{ep}}= & -\sum_{j\mathbf{q}} \hbar\,\omega_j(\mathbf{q}) \frac{\partial n_{j\mathbf{q}}}{\partial t} \\
= & \frac{4\pi}{\hbar} \sum_{j\mathbf{q}} \sum_{\ell\mathbf{k}} \sum_{\ell'\mathbf{k}'} \hbar\,\omega_j(\mathbf{q})\left|M_{\ell\mathbf{k},\ell'\mathbf{k}'}^{(j\mathbf{q})}\right|^2\Big(\left(n_{\ell\mathbf{k}}-n_{\ell'\mathbf{k}'}\right) n_{j\mathbf{q}}-n_{\ell'\mathbf{k}'}\left(1-n_{\ell\mathbf{k}}\right)\Big) \times \\
& \times \delta\Big(\varepsilon_{\ell\mathbf{k}}-\varepsilon_{\ell'\mathbf{k}'}+\hbar\,\omega_j(\mathbf{q})\Big).
\end{aligned} \tag{3.222}$$

If we assume thermal equilibrium of the electrons and phonons, the electronic occupations $n_{\ell\mathbf{k}}$, $n_{\ell'\mathbf{k}'}$ follow a Fermi distribution $f_{\mathrm{F}}(\varepsilon_{\ell\mathbf{k}}, T_{\mathrm{e}})$, $f_{\mathrm{F}}(\varepsilon_{\ell'\mathbf{k}'}, T_{\mathrm{e}})$ with the electronic temperature T_{e} and the phonon occupation $n_{j\mathbf{q}}$ follows a Bose-Einstein distribution $f_{\mathrm{B}}\big(\hbar\,\omega_j(\mathbf{q}), T_{\mathrm{i}}\big)$ with nuclei temperature T_{i}. Then we obtain

$$\boxed{\begin{aligned}
\dot{E}_{\mathrm{ep}}= & \frac{2\pi}{\hbar} \sum_{j\mathbf{q}} \sum_{\ell\mathbf{k}} \sum_{\ell'\mathbf{k}'} \hbar\,\omega_j(\mathbf{q})\left|M_{\ell\mathbf{k},\ell'\mathbf{k}'}^{(j\mathbf{q})}\right|^2 \delta\Big(\varepsilon_{\ell\mathbf{k}}-\varepsilon_{\ell'\mathbf{k}'}+\hbar\,\omega_j(\mathbf{q})\Big) \times \\
& \times\Bigg(\Big(f_{\mathrm{F}}(\varepsilon_{\ell\mathbf{k}}, T_{\mathrm{e}})-f_{\mathrm{F}}(\varepsilon_{\ell'\mathbf{k}'}, T_{\mathrm{e}})\Big) f_{\mathrm{B}}\big(\hbar\,\omega_j(\mathbf{q}), T_{\mathrm{i}}\big) \\
& -f_{\mathrm{F}}(\varepsilon_{\ell'\mathbf{k}'}, T_{\mathrm{e}})\Big(1-f_{\mathrm{F}}(\varepsilon_{\ell\mathbf{k}}, T_{\mathrm{e}})\Big)\Bigg).
\end{aligned}} \tag{3.223}$$

Sadasivam et al. [28] use the above equation to derive the electron-phonon coupling constant for semiconductors. The drawback of this equation is given by the huge amount of different phonon modes and electronic states that must be taken into account in the sums. Therefore, further approximations were performed to reduce the calculation effort for metals [23–25]. We have

$$f_{\mathrm{F}}(\varepsilon', T)\left(1 - f_{\mathrm{F}}(\varepsilon, T)\right) = \left(f_{\mathrm{F}}(\varepsilon, T) - f_{\mathrm{F}}(\varepsilon', T)\right) f_{\mathrm{B}}(\varepsilon' - \varepsilon, T)$$

due to

$$\begin{aligned} f_{\mathrm{F}}(\varepsilon', T)\left(1 - f_{\mathrm{F}}(\varepsilon, T)\right) &= \frac{1}{\mathrm{e}^{\frac{\varepsilon'-\mu}{k_{\mathrm{B}}T}} + 1}\left(1 - \frac{1}{\mathrm{e}^{\frac{\varepsilon-\mu}{k_{\mathrm{B}}T}} + 1}\right) \\ &= \frac{1}{\mathrm{e}^{\frac{\varepsilon'-\mu}{k_{\mathrm{B}}T}} + 1}\left(\frac{\mathrm{e}^{\frac{\varepsilon-\mu}{k_{\mathrm{B}}T}} + 1}{\mathrm{e}^{\frac{\varepsilon-\mu}{k_{\mathrm{B}}T}} + 1} - \frac{1}{\mathrm{e}^{\frac{\varepsilon-\mu}{k_{\mathrm{B}}T}} + 1}\right) \\ &= \frac{1}{\mathrm{e}^{\frac{\varepsilon'-\mu}{k_{\mathrm{B}}T}} + 1}\,\frac{\mathrm{e}^{\frac{\varepsilon-\mu}{k_{\mathrm{B}}T}} + 1 - 1}{\mathrm{e}^{\frac{\varepsilon-\mu}{k_{\mathrm{B}}T}} + 1} \\ &= \frac{\mathrm{e}^{\frac{\varepsilon-\mu}{k_{\mathrm{B}}T}}}{\left(\mathrm{e}^{\frac{\varepsilon'-\mu}{k_{\mathrm{B}}T}} + 1\right)\left(\mathrm{e}^{\frac{\varepsilon-\mu}{k_{\mathrm{B}}T}} + 1\right)} \end{aligned}$$

and

$$\begin{aligned} &\left(f_{\mathrm{F}}(\varepsilon, T) - f_{\mathrm{F}}(\varepsilon', T)\right) f_{\mathrm{B}}(\varepsilon' - \varepsilon, T) \\ &= \left(\frac{1}{\mathrm{e}^{\frac{\varepsilon-\mu}{k_{\mathrm{B}}T}} + 1} - \frac{1}{\mathrm{e}^{\frac{\varepsilon'-\mu}{k_{\mathrm{B}}T}} + 1}\right)\frac{1}{\mathrm{e}^{\frac{\varepsilon'-\varepsilon}{k_{\mathrm{B}}T}} - 1} \\ &= \left(\frac{\mathrm{e}^{\frac{\varepsilon'-\mu}{k_{\mathrm{B}}T}} + 1}{\left(\mathrm{e}^{\frac{\varepsilon-\mu}{k_{\mathrm{B}}T}} + 1\right)\left(\mathrm{e}^{\frac{\varepsilon'-\mu}{k_{\mathrm{B}}T}} + 1\right)} - \frac{\mathrm{e}^{\frac{\varepsilon-\mu}{k_{\mathrm{B}}T}} + 1}{\left(\mathrm{e}^{\frac{\varepsilon'-\mu}{k_{\mathrm{B}}T}} + 1\right)\left(\mathrm{e}^{\frac{\varepsilon-\mu}{k_{\mathrm{B}}T}} + 1\right)}\right)\frac{1}{\mathrm{e}^{\frac{\varepsilon'-\varepsilon}{k_{\mathrm{B}}T}} - 1} \\ &= \frac{\mathrm{e}^{\frac{\varepsilon'-\mu}{k_{\mathrm{B}}T}} - \mathrm{e}^{\frac{\varepsilon-\mu}{k_{\mathrm{B}}T}}}{\left(\mathrm{e}^{\frac{\varepsilon'-\mu}{k_{\mathrm{B}}T}} + 1\right)\left(\mathrm{e}^{\frac{\varepsilon-\mu}{k_{\mathrm{B}}T}} + 1\right)}\,\frac{1}{\mathrm{e}^{\frac{\varepsilon'-\varepsilon}{k_{\mathrm{B}}T}} - 1} \\ &= \frac{\mathrm{e}^{\frac{\varepsilon-\mu}{k_{\mathrm{B}}T}}\left(\mathrm{e}^{\frac{\varepsilon'-\varepsilon}{k_{\mathrm{B}}T}} - 1\right)}{\left(\mathrm{e}^{\frac{\varepsilon'-\mu}{k_{\mathrm{B}}T}} + 1\right)\left(\mathrm{e}^{\frac{\varepsilon-\mu}{k_{\mathrm{B}}T}} + 1\right)\left(\mathrm{e}^{\frac{\varepsilon'-\varepsilon}{k_{\mathrm{B}}T}} - 1\right)} \\ &= \frac{\mathrm{e}^{\frac{\varepsilon-\mu}{k_{\mathrm{B}}T}}}{\left(\mathrm{e}^{\frac{\varepsilon'-\mu}{k_{\mathrm{B}}T}} + 1\right)\left(\mathrm{e}^{\frac{\varepsilon-\mu}{k_{\mathrm{B}}T}} + 1\right)}. \end{aligned}$$

Therefore, we get

$$\begin{aligned}\dot{E}_{\mathrm{ep}} =& \frac{2\pi}{\hbar} \sum_{j\mathbf{q}} \sum_{\ell\mathbf{k}} \sum_{\ell'\mathbf{k}'} \hbar\,\omega_j(\mathbf{q}) \left| M^{(j\mathbf{q})}_{\ell\mathbf{k},\ell'\mathbf{k}'} \right|^2 \delta\Big(\varepsilon_{\ell\mathbf{k}} - \varepsilon_{\ell'\mathbf{k}'} + \hbar\,\omega_j(\mathbf{q})\Big) \times \\ & \times \Big(f_{\mathrm{F}}(\varepsilon_{\ell\mathbf{k}}, T_{\mathrm{e}}) - f_{\mathrm{F}}(\varepsilon_{\ell'\mathbf{k}'}, T_{\mathrm{e}})\Big) \times \\ & \times \Big(f_{\mathrm{B}}\big(\hbar\,\omega_j(\mathbf{q}), T_{\mathrm{i}}\big) - f_{\mathrm{B}}\big(\varepsilon_{\ell'\mathbf{k}'} - \varepsilon_{\ell\mathbf{k}}, T_{\mathrm{e}}\big)\Big). \end{aligned} \tag{3.224}$$

In a metal, the density of states $\mathfrak{D}(E_{\mathrm{F}})$ at the Fermi energy E_{F} is bigger than zero. Therefore, Allen defines the **electron-phonon spectral function** [23]

$$f_{\mathrm{epsf}}(\varepsilon, \varepsilon', \omega) = \frac{2}{\hbar\,\mathfrak{D}(E_{\mathrm{F}})} \sum_{\ell\mathbf{k}} \sum_{\ell'\mathbf{k}'} \left| M^{(j\mathbf{q})}_{\ell\mathbf{k},\ell'\mathbf{k}'} \right|^2 \delta\big(\varepsilon_{\ell\mathbf{k}} - \varepsilon\big)\,\delta\big(\varepsilon_{\ell'\mathbf{k}'} - \varepsilon'\big)\,\delta\Big(\omega_j(\mathbf{q}) - \omega\Big) \tag{3.225}$$

and uses this definition for the electron-phonon coupling rate:

$$\begin{aligned}\dot{E}_{\mathrm{ep}} =& \pi\,\mathfrak{D}(E_{\mathrm{F}}) \int_0^\infty d\omega \int_{-\infty}^\infty d\varepsilon \int_{-\infty}^\infty d\varepsilon'\, f_{\mathrm{epsf}}(\varepsilon, \varepsilon', \omega)\,\hbar\,\omega\,\delta(\varepsilon - \varepsilon' + \hbar\,\omega) \times \\ & \times \Big(f_{\mathrm{F}}(\varepsilon, T_{\mathrm{e}}) - f_{\mathrm{F}}(\varepsilon', T_{\mathrm{e}})\Big)\Big(f_{\mathrm{B}}\big(\hbar\,\omega, T_{\mathrm{i}}\big) - f_{\mathrm{B}}\big(\varepsilon' - \varepsilon, T_{\mathrm{e}}\big)\Big) \\ =& \pi\,\mathfrak{D}(E_{\mathrm{F}}) \int_0^\infty d\omega \int_{-\infty}^\infty d\varepsilon\, f_{\mathrm{epsf}}(\varepsilon, \varepsilon + \hbar\,\omega, \omega)\,\hbar\,\omega \times \\ & \times \Big(f_{\mathrm{F}}(\varepsilon, T_{\mathrm{e}}) - f_{\mathrm{F}}(\varepsilon + \hbar\,\omega, T_{\mathrm{e}})\Big)\Big(f_{\mathrm{B}}\big(\hbar\,\omega, T_{\mathrm{i}}\big) - f_{\mathrm{B}}\big(\hbar\,\omega, T_{\mathrm{e}}\big)\Big). \end{aligned} \tag{3.226}$$

One defines the **Eliashberg function**

$$f_{\mathrm{El}}(\omega) = f_{\mathrm{epsf}}(E_{\mathrm{F}}, E_{\mathrm{F}}, \omega). \tag{3.227}$$

Since the electron energies vary in the eV range and the phonon energies in the meV range, one performs the approximations [24, 25]

$$f_{\mathrm{epsf}}(\varepsilon, \varepsilon + \hbar\,\omega, \omega) \approx \frac{\mathfrak{D}(\varepsilon)\,\mathfrak{D}(\varepsilon + \hbar\,\omega)}{\mathfrak{D}(E_{\mathrm{F}})^2}\, f_{\mathrm{El}}(\omega) \tag{3.228}$$

$$\begin{aligned} f_{\mathrm{F}}(\varepsilon, T_{\mathrm{e}}) - f_{\mathrm{F}}(\varepsilon + \hbar\,\omega, T_{\mathrm{e}}) \approx& f_{\mathrm{F}}(\varepsilon, T_{\mathrm{e}}) - \left(f_{\mathrm{F}}(\varepsilon, T_{\mathrm{e}}) + \frac{\partial f_{\mathrm{F}}(\varepsilon, T_{\mathrm{e}})}{\partial \varepsilon}\,\hbar\,\omega \right) \\ =& -\hbar\,\omega\,\frac{\partial f_{\mathrm{F}}(\varepsilon, T_{\mathrm{e}})}{\partial \varepsilon} \end{aligned} \tag{3.229}$$

and

$$\mathfrak{D}(\varepsilon + \hbar\,\omega) \approx \mathfrak{D}(\varepsilon). \tag{3.230}$$

Then the electron-phonon coupling rate is given by

$$\boxed{\begin{aligned}\dot{E}_{\text{ep}} = -\frac{\pi}{\mathfrak{D}(E_{\text{F}})} & \int_0^\infty d\omega\, f_{\text{El}}(\omega)\,(\hbar\,\omega)^2 \Big(f_{\text{B}}\big(\hbar\,\omega, T_{\text{i}}\big) - f_{\text{B}}\big(\hbar\,\omega, T_{\text{e}}\big)\Big) \times \\ & \times \int_{-\infty}^{\infty} d\varepsilon\, \mathfrak{D}(\varepsilon)^2\, \frac{\partial f_{\text{F}}(\varepsilon, T_{\text{e}})}{\partial \varepsilon}.\end{aligned}} \tag{3.231}$$

Waldecker et al. [26, 27] divides the phonon modes in different groups of phonon branches. For each branch, Waldecker defines an individual ionic temperature $T_{\text{i}}^{(b)}$. Using this temperature, Waldecker defines an individual electron-phonon coupling rate $E_{\text{ep}}^{(b)}$ for each phonon branch by integrating in the above equation only over the related phonon modes.

Furthermore, one defines the **electron-phonon coupling constant** G_{ep} from

$$\dot{E}_{\text{ep}} = -G_{\text{ep}}\, N_{\text{at}}\, (T_{\text{e}} - T_{\text{i}}), \tag{3.232}$$

where N_{at} is the number of atoms in the simulation cell.

3.5 Physical Picture of the Fs-laser Excitation

Utilizing the Born-Oppenheimer approximation, one can separate the motion of the ions and the electrons (see Sect. 2.2). The ions are exposed to an effective potential energy surface generated by the electrons. Before the onset of the laser excitation, the electrons fill the available states with a Fermi distribution at some ambient condition, e.g. at room temperature $T_{\text{e}} = 300$ K. The femtosecond laser-pulse excites mostly the electrons, which occupy higher energy levels and leave unoccupied hole states at low energies. It further creates a macroscopic polarization, that is represented in the non-zero values of the off-diagonal elements of the one-electron density matrix (see Sect. 3.4.1). Due to very fast dephasing processes, the off-diagonal elements vanish shortly for times after the laser pulse. On a timescale of typically 10 - 100 fs, the diagonal elements of the one-electron density matrix converge to a Fermi distribution with a higher electronic temperature T_{e}, so that the electrons can be described by a canonical ensemble (see Sect. 3.4.2). During these fast processes the ions remain cold. This state of hot electrons and cold ions far away from thermodynamic equilibrium continues to exist on a timescale of several picoseconds. Due to the electron-phonon coupling, energy is transferred from the hot electrons to the ions which decreases T_{e} and increases the ionic temperature T_{i} on a picosecond timescale (see Sect. 3.4.3).

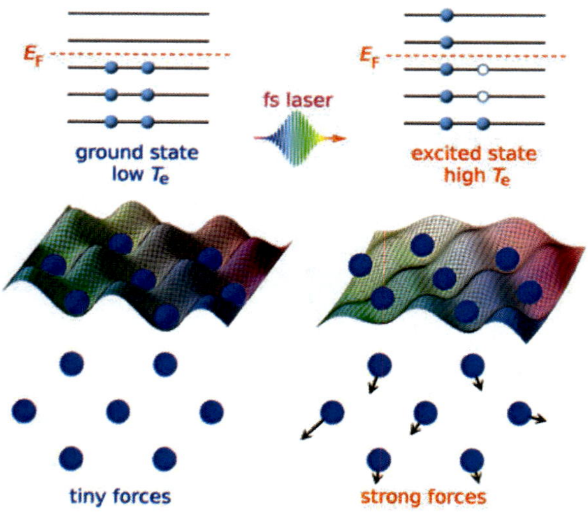

Fig. 3.3 Physical picture of a femtosecond laser-excitation

Since the electronic system thermalize fast to a high T_e after the femtosecond laser-pulse, one can neglect the electronic thermalization process and can assume, that the electronic temperature jumps instantaneous from $T_e = 300$ K to a high value. If one is just interested in the effects occurring in the first picosecond after the laser excitation, one can keep this high T_e constant neglecting the slow electron-phonon couping. This leads to the following physical picture of a femtosecond laser-excitation, which can be used, in principle, for any material and is illustrated in Fig. 3.3.

Before the femtosecond laser-excitation, the electrons form a Fermi-distribution with $T_e = 300$ K, so that there are no significant occupations above the Fermi energy E_F. The ions are located in the minima of the potential energy surface generated by the electrons and move slowly around their equilibrium positions due to the small thermal energy.

After the femtosecond laser-excitation, the electrons form a Fermi-distribution with a high constant T_e. The dramatic change of the electronic occupations induce significant changes in the potential energy surface of the ions. These changes manifest themselves in remarkable effects like bond hardening or softening, as well as lattice instabilities. The ions are not any more located in the minima of the potential energy surface but at the slopes, so that strong forces accelerate the ions far away from their initial positions. The changed potential energy surface leads to measurable ultrafast phenomena such as solid-to-solid and solid-to-liquid phase transitions, phonon squeezing and coherent phonons. These effects are discussed in the next chapter in detail.

3.6 Code for Highly Excited Valence Electron Systems (CHIVES)

A possible avenue to describe electrons at high T_e from ab-initio is given by temperature-dependent density functional theory (see Sect. 2.4). There are many implementations of DFT algorithms available, which all have their different advantages related to speed, accuracy and properties implemented. The usability is mainly related to the purpose for which the corresponding computer code was primarily developed. In the last years, the **C**ode for **H**ighly exc**I**ted **V**alence **E**lectron **S**ystems (**CHIVES**) [1–8] was developed in the group of Prof. Dr. Garcia at the university of Kassel to simulate the structural response of a solid or a nanostructure to a femtosecond laser-pulse excitation at affordable computational cost combined with reasonable accuracy. CHIVES is optimized to perform molecular dynamics (MD) simulations of several hundreds up to thousand of atoms before and after femtosecond laser-excitation using T_e-dependent DFT (see Sect. 2.4).

The nuclei are treated as classical objects, which obey Newton's equations of motion. These equations are integrated using the Velocity Verlet algorithm (see Sect. 4.1.1). In addition, the FIRE algorithm [30] is implemented for structural relaxation. The forces $\mathbf{f}_i$ on the nuclei are calculated from the potential energy surface $F(\mathbf{r}_1, \ldots, \mathbf{r}_{N_{at}}, T_e)$ generated from the electrons at temperature T_e by

$$\mathbf{f}_i = \nabla_{\mathbf{r}_i} F(\mathbf{r}_1, \ldots, \mathbf{r}_{N_{at}}, T_e). \tag{3.233}$$

The electrons are described within T_e-dependent DFT and are separated into core and valence electrons. The core electrons are treated by the norm-conserving pseudopotentials of Goedecker, Teter, and Hutter [31, 32] (see Sect. 2.3.8). CHIVES utilizes the non-relativistic pseudopotentials [31] for the elements H, B, C, N, O, F, Na, Mg, Al, Si, P, S, Cl and the relativistic pseudopotentials [32] for the elements He, Li, Be, Ne, and the elements from Ar up to Xe. The valence electrons are explicitly described by atom-centered Gaussian basis set functions of different angular momentum. A separate basis set is required for each individual element of the periodic table. To construct the basis set for an element, several exponents a for the Gaussian functions are used. For each exponent, the single s-type atom-centered Gaussian function

$$f_s(\mathbf{r}) = \mathrm{e}^{-a\,r^2}, \tag{3.234}$$

the three p-type atom-centered Gaussian functions

$$f_{p_x}(\mathbf{r}) = x\,\mathrm{e}^{-a\,r^2}, \qquad f_{p_y}(\mathbf{r}) = y\,\mathrm{e}^{-a\,r^2}, \qquad f_{p_z}(\mathbf{r}) = z\,\mathrm{e}^{-a\,r^2}, \tag{3.235}$$

and the five d-type atom-centered Gaussian functions

$$f_{d_{xy}}(\mathbf{r}) = x\,y\,\mathrm{e}^{-a\,r^2}, \qquad f_{d_{xz}}(\mathbf{r}) = x\,z\,\mathrm{e}^{-a\,r^2}, \qquad f_{d_{yz}}(\mathbf{r}) = y\,z\,\mathrm{e}^{-a\,r^2},$$
$$f_{d_{x^2-y^2}}(\mathbf{r}) = \left(x^2 - y^2\right)\mathrm{e}^{-a\,r^2}, \qquad f_{d_{z^2}}(\mathbf{r}) = z^2\,\mathrm{e}^{-a\,r^2}, \tag{3.236}$$

can be included into the basis set within CHIVES. Since the basis set controls the accuracy and the computational effort, one has to find a basis set that is as small as possible but large enough to describes all physical effects of interest with sufficient accuracy in order to perform fast, large scale MD simulations. Determining the basis set is a nontrivial and cumbersome task. As of today, optimized basis sets were developed and implemented in CHIVES for the following twenty three elements: H, He, Li, Be, B, N, C, O, F, Al, Si, P, S, Cl, Ti, V, MN, Cu, Ge, Sr, In, Sb and Te. Table A.50 in the appendix lists the basis set exponents together with the used angular momenta for the elements studied in this book.

If one utilizes plane waves instead of Gaussian functions, the construction of the basis set is straight forward for each element and the implementation is much simpler. But one needs much more basis set functions to reach the same accuracy compared to Gaussian basis functions.

CHIVES utilizes for the exchange and correlation potential the local density approximation (LDA) (see Sec 2.3.3). The exchange, correlation and Hartree potential and the electronic charge density are calculated on a fine mesh of grid points without further approximations. This allows to calculate the exchange, correlation and Hartree potential with order (N) methods [33], where computing time scales linearly with the number of atoms N used in the simulation cell. The density mixing in the self-consistent loop is done with the Pulay-Kerker mixer (see Sect. 2.3.7), which allows also the treatment of surfaces [8]. Due to this mixing scheme, CHIVES can only treat periodic boundary conditions. Furthermore, CHIVES can deal with arbitrary orthorhombic simulation cells, i.e., the simulation cell is spanned by the vectors $\mathbf{a}_1 = a_1\,\mathbf{e}_x, \mathbf{a}_2 = a_2\,\mathbf{e}_y, \mathbf{a}_3 = a_3\,\mathbf{e}_z$ with $a_1, a_2, a_3 \in \mathbb{R}^+$. To increase the accuracy, one can use a $N_1 \times N_2 \times N_3$ grid of the **k**-points with $N_1, N_2, N_3 \in \{1, 2\}$, so that the $N_1\,N_2\,N_3$ irreducible **k** points $\mathbf{k} = \frac{m_1}{N_1}\frac{2\pi}{a_1}\,\mathbf{e}_x + \frac{m_2}{N_2}\frac{2\pi}{a_2}\,\mathbf{e}_y + \frac{m_3}{N_3}\frac{2\pi}{a_3}\,\mathbf{e}_z$ with $m_1 \in \{0, N_1 - 1\}, m_2 \in \{0, N_2 - 1\}\, m_3 \in \{0, N_3 - 1\}$ are used to construct the basis set functions of the valence electrons (see Sect. 2.3.4). In summary, CHIVES treats 1, 2, 4 or 8 irreducible **k** points. Solving the Kohn-Sham equations for each **k** point is almost independent from the other **k** points. Thus, these calculations are parallelized over the different **k** points using the message passing interface (MPI). One open source implementation of MPI is OPEN MPI [1] [34]. This library allows to run a program on thousands of processors distributed on many computers. The Computation of the Hamilton matrix elements, charge density and density matrix is parallelized with open multi-processing (OpenMP), which allows to run a program on several processors of one computer with shared memory. To solve the Kohn-Sham equations, the efficient diagonalization of the optimized linear algebra package LAPACK [2] is used. In addition, the diagonalization can be performed on graphic processing units

---

1 https://www.open-mpi.org.

2 http://www.netlib.org/lapack.

(GPUs) using the MAGMA [3] library, which further speeds up the calculations. In the self-consistent loop, the charge density is efficiently initialized by an extrapolation from the previous time steps.

The above described implementations make CHIVES to one of the fastest available T_e-dependent DFT codes for performing MD simulations of laser excited solids and nanostructures. To evaluate the speed of CHIVES, its performance was compared with the freely available DFT code ABINIT [35]. This code can be executed using the same pseudopoentials as CHIVES. ABINIT uses plane waves for the basis functions with a selectable cut-off energy. Hence, one can compare directly the energies obtained from ABINIT and CHIVES. The benchmark [5] was performed with a simulation cell containing 288 Si atoms, which were randomly displaced from their equilibrium positions reproducing a thermal distribution at room temperature [2]. In the calculation of energies and forces of the atoms in this simulation cell, CHIVES is 200 times faster than ABINIT. The main reason for this is that CHIVES needs 4896 and ABINIT 47439 basis functions to calculate the energy with the same accuracy. We want to note, however, that ABNINT is faster compared to CHIVES for atomic configurations with high symmetry.

3.7 Summary

Femtosecond laser-pulses provide ultra high field intensities within a small spatial area. Such pulses can excite electrons from the valence band and / or conduction band to higher energy levels in the conduction band and create a macroscopic polarization with empty states in the valence band, the holes, leading to electron-hole pairs. The ions remain mainly unaffected because of the short interaction time. The macroscopic polarization vanishes immediately after the laser-interaction acting due to ultrafast dephasing processes. Then inelastic electron-electron scattering is responsible for a thermalization of the electrons to a Fermi distribution within tens of femtoseconds after the laser-pulse. This state of cold ions and hot electrons, which reach electronic temperatures T_e's of several 1000–10000 K, has a typical life time of several picoseconds, since the energy transfer from the electrons to the ions due to electron-phonon coupling proceeds slowly on a picosecond timescale.

The presence of hot electrons has dramatic influences on the potential energy surface of the ions, which are responsible for many ultrafast phenomena like solid-to-solid and solid-to-liquid phase transitions, generation of coherent phonons or phonon squeezing. A solid in a state with hot electrons can be accurately described by T_e-dependent DFT. The **C**ode for **H**ighly exc**I**ted **V**alence **E**lectron **S**ystems (**CHIVES**) allows to simulate these femtosecond laser-driven effects by utilizing a very efficient implementation of T_e-dependent DFT.

---

[3] https://icl.cs.utk.edu/magma.

References

1. E.S. Zijlstra, N. Huntemann, A. Kalitsov, M.E. Garcia, U. von Barth, Modell. Simul. Mater. Sci. Eng. **17**(1), 015009 (2009). http://stacks.iop.org/0965-0393/17/i=1/a=015009
2. E.S. Zijlstra, A. Kalitsov, T. Zier, M.E. Garcia, Phys. Rev. X **3**, 011005 (2013). https://doi.org/10.1103/PhysRevX.3.011005
3. E.S. Zijlstra, F. Cheenicode Kabeer, B. Bauerhenne, T. Zier, N. Grigoryan, M.E. Garcia, Appl. Phys. A **110**(3), 519 (2013). https://doi.org/10.1007/s00339-012-7183-0
4. E.S. Zijlstra, A. Kalitsov, T. Zier, M.E. Garcia, Adv. Mater. **25**(39), 5605 (2013). https://doi.org/10.1002/adma201302559
5. E.S. Zijlstra, T. Zier, B. Bauerhenne, S. Krylow, P.M. Geiger, M.E. Garcia, Appl. Phys. A **114**(1), 1 (2014). https://doi.org/10.1007/s00339-013-8080-x
6. B. Bauerhenne, E.S. Zijlstra, A. Kalitsov, M.E. Garcia, Nanotechnology **25**(14), 145701 (2014). https://doi.org/10.1088/0957-4484/25/14/145701
7. N.S. Grigoryan, T. Zier, M.E. Garcia, E.S. Zijlstra, J. Opt. Soc. Am. B **31**(11), C22 (2014). https://doi.org/10.1364/JOSAB.31.000C22
8. T. Zier, E.S. Zijlstra, S. Krylow, M.E. Garcia, Appl. Phys. A **123**(10), 625 (2017). https://doi.org/10.1007/s00339-017-1230-9
9. D.J. Griffiths, *Introduction to Electrodynamics*, 4th edn. (Pearson, 2013)
10. D.J. Griffiths, *Instructor's Solution Manual Introduction to Electrodynamics*, 3rd edn. (Pearson, 2009)
11. C.T. C., D. B., L. F., *Quantum mechanics Volume 2* (Wiley, 1978)
12. L.D. Landau, E.M. Lifschitz, *Mechanik*, 7th edn. (Vieweg + Sohn, Braunschweig, 1970)
13. P.R. Surjan, *Second Quantized Approach to Quantum Chemistry*, 1st edn. (Springer, 1989)
14. E.K.U. Gross, E. Runge, O. Heinonen, *Many-Particle Theory* (Adam Hilger, Bristol, 1991)
15. F. Schwabl, *Quantenmechanik für Fortgeschrittene*, 5th edn. (Springer, 2008). https://www.springer.com/de/book/9783540850755
16. G. Czycholl, *Theoretische Festkörper Physik Band 1*, 4th edn. (Springer, 2016). https://www.springer.com/de/book/9783662471401
17. O. Forster, *Analysis 1*, 9th edn. (Vieweg, 2008)
18. M. Bonitz, *Quantum kinetic theory*, 2nd edn. (Springer, 2016). https://link.springer.com/book/10.1007/978-3-319-24121-0
19. A.V. Kuznetsov, Phys. Rev. B **44**, 13381 (1991). https://doi.org/10.1103/PhysRevB.44.13381
20. T. Elsaesser, J. Shah, L. Rota, P. Lugli, Phys. Rev. Lett. **66**, 1757 (1991). https://doi.org/10.1103/PhysRevLett.66.1757
21. W.H. Knox, D.S. Chemla, G. Livescu, J.E. Cunningham, J.E. Henry, Phys. Rev. Lett. **61**, 1290 (1988). https://doi.org/10.1103/PhysRevLett.61.1290
22. J. Hohlfeld, S.S. Wellershoff, J. Güdde, U. Conrad, V. Jähnke, E. Matthias, Chemical Physics **251**(1), 237 (2000). https://doi.org/10.1016/S0301-0104(99)00330-4
23. P.B. Allen, Phys. Rev. Lett. **59**, 1460 (1987). https://doi.org/10.1103/PhysRevLett.59.1460
24. X.Y. Wang, D.M. Riffe, Y.S. Lee, M.C. Downer, Phys. Rev. B **50**, 8016 (1994). https://doi.org/10.1103/PhysRevB.50.8016
25. Z. Lin, L.V. Zhigilei, V. Celli, Phys. Rev. B **77**, 075133 (2008). https://doi.org/10.1103/PhysRevB.77.075133
26. L. Waldecker, R. Bertoni, R. Ernstorfer, J. Vorberger, Phys. Rev. X **6**, 021003 (2016). https://doi.org/10.1103/PhysRevX.6.021003
27. L. Waldecker, T. Vasileiadis, R. Bertoni, R. Ernstorfer, T. Zier, F.H. Valencia, M.E. Garcia, E.S. Zijlstra, Phys. Rev. B **95**, 054302 (2017). https://doi.org/10.1103/PhysRevB.95.054302
28. S. Sadasivam, M.K.Y. Chan, P. Darancet, Phys. Rev. Lett. **119**, 136602 (2017). https://doi.org/10.1103/PhysRevLett.119.136602
29. W. Greiner, *Quantum Mechanics An Introduction*, 4th edn. (Springer, 2000)
30. E. Bitzek, P. Koskinen, F. Gähler, M. Moseler, P. Gumbsch, Phys. Rev. Lett. **97**, 170201 (2006). https://doi.org/10.1103/PhysRevLett.97.170201

31. S. Goedecker, M. Teter, J. Hutter, Phys. Rev. B **54**, 1703 (1996). https://doi.org/10.1103/PhysRevB.54.1703
32. C. Hartwigsen, S. Goedecker, J. Hutter, Phys. Rev. B **58**, 3641 (1998). https://doi.org/10.1103/PhysRevB.58.3641
33. S. Goedecker, Rev. Mod. Phys. **71**, 1085 (1999). https://doi.org/10.1103/RevModPhys.71.1085
34. E. Gabriel, G.E. Fagg, G. Bosilca, T. Angskun, J.J. Dongarra, J.M. Squyres, V. Sahay, P. Kambadur, B. Barrett, A. Lumsdaine, R.H. Castain, D.J. Daniel, R.L. Graham, T.S. Woodall, in *Proceedings, 11th European PVM/MPI Users' Group Meeting* (Budapest, Hungary, 2004), pp. 97–104. https://www.open-mpi.org//papers/euro-pvmmpi-2004-overview
35. X. Gonze, G.M. Rignanese, M. Verstraete, J.M. Beuken, Y. Pouillon, R. Caracas, F. Jollet, M. Torrent, G. Zerah, M. Mikami, P. Ghosez, J.Y. Veithen, M.; Raty, V. Olevano, F. Bruneval, L. Reining, R.W. Godby, G. Onida, D.R. Hamann, D.C. Allan, Zeit. Kristallogr. **220**, 558 (2005). https://doi.org/10.1524/zkri.220.5.558.65066

Chapter 4
Ab-Initio MD Simulations of the Excited Potential Energy Surface

Abstract Within molecular dynamics (MD) simulations, we can account for the influences of a femtosecond laser-excitation on matter by increasing the electronic temperature T_e, if we neglect the fast thermalization processes of the electrons following the femtosecond laser-excitation. To perform ab-initio calculations with hot electrons, T_e-dependent DFT can be used. In such calculations [1–13], T_e is increased and commonly kept constant neglecting the electron-phonon coupling. The latter assumption may be acceptable, since these studies simulate only the first one or two picoseconds after the femtosecond laser-pulse, whereas the relaxation due to electron-phonon coupling takes place on a longer timescale. These studies with increased T_e and electron-phonon coupling neglected reveal the effects that are produced by the high T_e-induced changes in the potential energy surface of the ions. T_e-dependent DFT allows to calculate the forces on the atoms under the presence of hot electrons, so that the atomic motions following the femtosecond-laser excitation can be simulated. This provides a deep understanding of the induced processes like solid-to-solid and solid-to-liquid phase transitions, excitation of coherent phonons or phonon squeezing. In this chapter, it is first shown how MD simulations are performed and how to calculate the experimentally measurable diffraction peak intensities from the simulated atomic coordinates, which are not directly accessible in the experiments. Also the effect of laser-induced thermal phonon squeezing or antisqueezing is explained. Then high T_e-induced effects in selected interesting materials are studied by means of T_e-dependent DFT calculations. Silicon is the most used semiconductor for electronic devices like computers, since high purity silicon monocrystalline wafers can be produced and patterned at the nanoscale, for which femtosecond laser-pulse processing is a promising tool. Measurements in silicon showed that an increase of the temperature causes a reduction of the electronic band gap [14–17]. In experiments using a free electron laser it was observed that the excitation by an intense femtosecond laser induces a closure of band gap [18], which was explained by different phase transitions [18]. In 2014, again a femtosecond laser was utilized to apply high excitations and measurements with attosecond pulses showed that the direct band gap decreases due to the femtosecond-laser excitation [19]: The electronic excitation decreases the direct band gap immediately after the femtosecond-laser excitation. Then, a further decrease was measured and related to the atomic disordering. In 1994, Stampfli and Bennemann predicted a laser-induced melting and metalization of silicon using tight

© The Author(s), under exclusive license to Springer Nature Switzerland AG 2021

B. Bauerhenne, *Materials Interaction with Femtosecond Lasers*,
https://doi.org/10.1007/978-3-030-85135-4_4

binding theory [20]. Also the T_e-dependent DFT calculations of Silvestrelli et al. showed, that silicon melts and becomes metallic after an intense femtosecond-laser excitation [1]. Recently in 2019, Medvedev et al. calculated the band gap behavior following a femtosecond x-ray excitation from a x-ray free-electron laser [21] with similar conclusions. Thus, the high T_e-induced effects in silicon are intensively studied. It is shown that a femtosecond laser-pulse can induce thermal phonon antisqueezing and an ultrafast solid-to-liquid phase transition called non-thermal melting combined with metallization. There are, of course, more ultrafast phenomena than one can observe in silicon. An interesting additional one occurs when a femtosecond laser-pulse only excites a single phonon mode in a solid at moderate intensities. Then the atoms perform a coherent oscillation with a large amplitude in direction of this single phonon mode. This excitation of a coherent phonon can occur in the semimetal antimony and could already experimentally be observed [22–26]. An even more interesting ultrafast phenomenon is a solid-to-solid phase transition besides the solid-to-liquid phase transition, which occurs, as discussed in Si during non-thermal melting. Antimony crystallizes in the A7 structure under normal conditions similar to bismuth and arsenic. At a pressure of 26.3 GPa, arsenic transforms into the sc structure [27]. The necessary static pressure for the transition can be strongly reduced by additional excitation of the electrons with a femtosecond laser-pulse [28, 29]. Under this tailored conditions antimony performs a transition into the sc structure already at 8 GPa [30]. Therefore, it is of great scientific interest, if a femtosecond laser-pulse alone is capable of introducing an A7 to sc transition in antimony at least temporally. Thus, the laser excitation of a coherent phonon and the laser-induced solid-to-solid phase transition in antimony are studied at high T_e. Beside this, it is shown that a femtosecond laser-pulse can induce in antimony thermal phonon antisqueezing and non-thermal melting similar to silicon. Phonons oscillate in the THz frequency range and coherent phonons generated by a femtosecond laser can reach high amplitudes. Thus, in suitable materials, the excitation of coherent phonons can be used to create THz radiation [31]. Sources for THz radiation are of practical importance, since they allow advanced sensing applications, like concealed weapon detection, chemical and biological agent detection, and medical diagnostics [32]. Boron nitride nanotubes (BNNTs) are a promising material, especially for biological applications, since they exhibit exceptional mechanical properties [33] and are nontoxic [34]. They exhibit bonds with a permanent electric dipole moment due to nitrogen's larger electronegativity compared to boron. A femtosecond laser-pulse excites three coherent phonon modes in thin BNNTs [35]. Thus in the end of the chapter, it is analyzed by means of MD simulations if the high T_e-induced atomic oscillations in BNNTs emit THz radiation.

4.1 Molecular Dynamics Simulation Setup

The nuclei are treated classically as small particles with masses m_j, coordinates $\mathbf{r}_j$ and velocities $\mathbf{v}_j$. Following the Born-Oppenheimer approximation, they move on the potential energy surface $F(\mathbf{r}_1, \ldots, \mathbf{r}_{N_{at}}, T_e)$ generated by the electrons at temperature T_e. Therefore, the Hamilton function of the nuclei for fixed T_e is given by

$$H = \sum_{j=1}^{N_{\mathrm{at}}} \frac{(m_j \mathbf{v}_j)^2}{2 m_j} + F(\mathbf{r}_1, \ldots, \mathbf{r}_{N_{\mathrm{at}}}, T_{\mathrm{e}}). \tag{4.1}$$

The total energy described by H is conserved, if T_{e} is kept constant. T_{e}-dependent DFT allows to calculate ab-initio the potential energy surface $F(\mathbf{r}_1, \ldots, \mathbf{r}_{N_{\mathrm{at}}}, T_{\mathrm{e}})$ and the forces on the nuclei via

$$\mathbf{f}_j = -\nabla_{\mathbf{r}_j} F(\mathbf{r}_1, \ldots, \mathbf{r}_{N_{\mathrm{at}}}, T_{\mathrm{e}}). \tag{4.2}$$

The knowledge of the forces provides the possibility to perform molecular dynamics (MD) simulations of the nuclei by solving the Newton's equations of motions

$$\mathbf{f}_j(t) = m_j\, \dot{\mathbf{v}}_j(t). \tag{4.3}$$

For this, one utilizes initial conditions $\mathbf{r}_j(t_0)$ and $\mathbf{v}_j(t_0)$ at a given start time t_0 and calculates numerically the coordinates and velocities at the discrete times $t = t_0 + n\,\Delta t,\ n \in \mathbb{N}$ with the time step $\Delta t > 0$. The velocities are necessary to calculate the kinetic and total energy of the system to check the energy conservation during the MD simulation.

4.1.1 Velocity Verlet Algorithm

To perform a proper MD simulation, one should choose a numerical integrator of Newton's equations that fulfills the following properties [36]:

- The error of the integrator should be of high order in Δt.
- The total energy during a MD simulation with the integrator should keep constant and should not show a shift, especially on long times.
- Newton's equations are time reversible. Hence, the integrator should show time reversal symmetry, i.e., if all velocities are inverted at a given time, the system traces back to the initial conditions.
- The phase space contains all possible states of the system. If one considers a volume in phase space and propagates this volume in time with Hamilton dynamics, i.e., one takes all contained states as initial conditions and propagates this states, then the size of the volume is preserved. Hence, the integrator should preserve the volume in phase space.

Swope et al. [37] presented 1982 the following algorithm, which is easy to implement and fulfills the above criteria [36]:

Theorem 4.1 (Velocity Verlet algorithm) *Let* $\mathbf{r}_j(t)$ *be the position,* $\mathbf{v}_j(t)$ *the velocity,* m_j *the mass and* $\mathbf{f}_j(t)$ *the force on atom* j *at time* t*. The positions and velocities of the atoms at the next time step* $t + \Delta t$ *are calculated from*

$$\mathbf{r}_j(t+\Delta t) = \mathbf{r}_j(t) + \Delta t\, \mathbf{v}_j(t) + \frac{\Delta t^2}{2\,m_j}\,\mathbf{f}_j(t) \qquad + O\big(\Delta t^4\big), \tag{4.4}$$

$$\mathbf{v}_j(t+\Delta t) = \mathbf{v}_j(t) + \frac{\Delta t}{2\,m_j}\Big(\mathbf{f}_j(t) + \mathbf{f}_j(t+\Delta t)\Big) \quad + O\big(\Delta t^2\big). \tag{4.5}$$

Proof We will show that the error of this algorithm exhibits the presented orders in Δt for coordinates and velocities following Ref. [37], but with more mathematical details. At first, we perform a Taylor series approximation of the coordinates with respect to t:

$$\begin{aligned}
\mathbf{r}_j(t+\Delta t) &= \mathbf{r}_j(t) + \dot{\mathbf{r}}_j(t)\,\Delta t + \ddot{\mathbf{r}}_j(t)\,\frac{\Delta t^2}{2} + \dddot{\mathbf{r}}_j(t)\,\frac{\Delta t^3}{6} + O\big(\Delta t^4\big),\\
\mathbf{r}_j(t-\Delta t) &= \mathbf{r}_j(t) - \dot{\mathbf{r}}_j(t)\,\Delta t + \ddot{\mathbf{r}}_j(t)\,\frac{\Delta t^2}{2} - \dddot{\mathbf{r}}_j(t)\,\frac{\Delta t^3}{6} + O\big(\Delta t^4\big).
\end{aligned}$$

Using

$$\dot{\mathbf{r}}_j = \mathbf{v}_j, \qquad \ddot{\mathbf{r}}_j = \dot{\mathbf{v}}_j = \frac{1}{m_j}\,\mathbf{f}_j, \qquad \dddot{\mathbf{r}}_j = \ddot{\mathbf{v}}_j = \frac{1}{m_j}\,\dot{\mathbf{f}}_j,$$

we get

$$\mathbf{r}_j(t+\Delta t) = \mathbf{r}_j(t) + \Delta t\,\mathbf{v}_j(t) + \frac{\Delta t^2}{2\,m_j}\,\mathbf{f}_j(t) + \frac{\Delta t^3}{6\,m_j}\,\dot{\mathbf{f}}_j(t) + O\big(\Delta t^4\big), \tag{4.6}$$

$$\mathbf{r}_j(t-\Delta t) = \mathbf{r}_j(t) - \Delta t\,\mathbf{v}_j(t) + \frac{\Delta t^2}{2\,m_j}\,\mathbf{f}_j(t) - \frac{\Delta t^3}{6\,m_j}\,\dot{\mathbf{f}}_j(t) + O\big(\Delta t^4\big). \tag{4.7}$$

Adding Eq. (4.6) to Eq. (4.7) and solving for $\mathbf{r}_j(t+\Delta t)$ yields

$$\mathbf{r}_j(t+\Delta t) = 2\,\mathbf{r}_j(t) - \mathbf{r}_j(t-\Delta t) + \frac{\Delta t^2}{m_j}\,\mathbf{f}_j(t) + O\big(\Delta t^4\big). \tag{4.8}$$

Subtracting Eq. (4.7) from Eq. (4.6) yields

$$\begin{aligned}
\mathbf{r}_j(t+\Delta t) - \mathbf{r}_j(t-\Delta t) &= 2\,\Delta t\,\mathbf{v}_j(t) + O\big(\Delta t^3\big),\\
\Leftrightarrow \qquad \mathbf{v}_j(t) &= \frac{\mathbf{r}_j(t+\Delta t) - \mathbf{r}_j(t-\Delta t)}{2\,\Delta t} + O\big(\Delta t^2\big).
\end{aligned} \tag{4.9}$$

Equations (4.8) and (4.9) form the so called **Verlet** algorithm [38], which error is forth order in the coordinates and second order in the velocities. The Verlet algorithm is not self starting, since one needs the coordinates at two different time steps for the initialization. In addition, the calculation of the velocities from Eq. (4.9) is numerically instable, because the difference is formed between two almost identical numbers and the result is divided by a small number. The Velocity Verlet algorithm do not exhibit this drawbacks. Now we show that the Velocity Verlet algorithm is equivalent to the Verlet algorithm, so that both exhibit the same error orders. For this, we solve Eq. (4.9) for $-\mathbf{r}_j(t-\Delta t)$ and obtain

$$-\mathbf{r}_j(t-\Delta t) = -\mathbf{r}_j(t+\Delta t) + 2\,\Delta t\,\mathbf{v}_j(t).$$

Inserting this into Eq. (4.8) yields

$$\begin{aligned}
&\mathbf{r}_j(t+\Delta t) = 2\,\mathbf{r}_j(t)\ \overbrace{-\mathbf{r}_j(t+\Delta t) + 2\,\Delta t\,\mathbf{v}_j(t)}^{=-\mathbf{r}_j(t-\Delta t)}\ + \frac{\Delta t^2}{m_j}\,\mathbf{f}_j(t),\\
\Leftrightarrow\quad &\mathbf{r}_j(t+\Delta t) = \mathbf{r}_j(t) + \Delta t\,\mathbf{v}_j(t) + \frac{\Delta t^2}{2\,m_j}\,\mathbf{f}_j(t),
\end{aligned}$$

which, indeed, corresponds to the coordinate propagation of the Velocity Verlet algorithm. From the above equation, we obtain further

$$\frac{\mathbf{r}_j(t+\Delta t) - \mathbf{r}_j(t)}{\Delta t} = \mathbf{v}_j(t) + \frac{\Delta t}{2\,m_j}\,\mathbf{f}_j(t). \tag{4.10}$$

We get from Eq. (4.8)

$$\mathbf{r}_j(t+2\,\Delta t) = 2\,\mathbf{r}_j(t+\Delta t) - \mathbf{r}_j(t) + \frac{\Delta t^2}{m_j}\,\mathbf{f}_j(t+\Delta t) \tag{4.11}$$

and from Eq. (4.9)

$$\mathbf{v}_j(t+\Delta t) = \frac{\mathbf{r}_j(t+2\,\Delta t) - \mathbf{r}_j(t)}{2\,\Delta t}. \tag{4.12}$$

Inserting Eq. (4.11) into Eq. (4.12) yields

$$\begin{aligned}
\mathbf{v}_j(t+\Delta t) &= \frac{2\,\mathbf{r}_j(t+\Delta t) - \mathbf{r}_j(t) + \frac{\Delta t^2}{m_j}\,\mathbf{f}_j(t+\Delta t)\ -\ \mathbf{r}_j(t)}{2\,\Delta t}\\
&= \frac{\mathbf{r}_j(t+\Delta t) - \mathbf{r}_j(t)}{\Delta t}\ +\ \frac{\Delta t}{2\,m_j}\,\mathbf{f}_j(t+\Delta t)\\
&\overset{(4.10)}{=} \mathbf{v}_j(t) + \frac{\Delta t}{2\,m_j}\,\mathbf{f}_j(t)\ +\ \frac{\Delta t}{2\,m_j}\,\mathbf{f}_j(t+\Delta t)\\
&= \mathbf{v}_j(t) + \frac{\Delta t}{2\,m_j}\Big(\mathbf{f}_j(t) + \mathbf{f}_j(t+\Delta t)\Big),
\end{aligned}$$

which, indeed, corresponds to the velocity propagation of the Velocity Verlet algorithm. ■

In the Velocity Verlet algorithm, the coordinates $\mathbf{r}_j(t + \Delta t)$ of the next time step can be calculated with the help of Eq. (4.4) from the known quantities $\mathbf{r}_j(t)$, $\mathbf{v}_j(t)$, and $\mathbf{f}_j(t)$ of the actual time step t. Then, one has to calculate $\mathbf{f}_j(t + \Delta t)$ with the help of $\mathbf{r}_j(t + \Delta t)$, before one can calculate the velocities $\mathbf{v}_j(t + \Delta t)$ of the new time step with the help of Eq. (4.5).

Beside the Velocity Verlet algorithm, also so called predictor-corrector algorithms are commonly used to perform MD simulations [36, 39]. One famous example is the Nordsieck fifth-order predictor-corrector algorithm. All of these algorithms work as following:

Predictor: They use the coordinates $\mathbf{r}_j(t)$ and their time derivatives $\dot{\mathbf{r}}_j(t)$, $\ddot{\mathbf{r}}_j(t)$, ... up to a certain order at time t and predict the same quantities at time $t + \Delta t$.

Force calculation: They calculate the forces $\mathbf{f}_j(t + \Delta t)$ at time $t + \Delta t$ using the predicted coordinates $\mathbf{r}_j(t + \Delta t)$ and derive the discrepancy between $\frac{\mathbf{f}_j(t+\Delta t)}{m_j}$ and $\ddot{\mathbf{r}}_j(t + \Delta t)$.

Corrector: With the help of the previously mentioned discrepancy, they correct the coordinates $\mathbf{r}_j(t + \Delta t)$ and their time derivatives $\dot{\mathbf{r}}_j(t + \Delta t)$, $\ddot{\mathbf{r}}_j(t + \Delta t)$, ...

Compared to the Velocity Verlet algorithm, a predictor-corrector algorithm exhibits, on the one hand, a smaller error during one time step, which allows to simulate short processes with a high precision, and, on the other hand, several disadvantages:

- The computational costs are much higher for performing a time step.
- It does not conserve the volume in phase space and does not show time reversal symmetry, which both leads to a significant energy drift in MD simulations over long times.

Accordingly, we chose the Velocity Verlet algorithm, since it is more computational efficient and exhibits no energy drift in MD simulations over long times.

4.1.2 *Preparation of Initial Conditions*

Even at absolute zero but certainly at ambient conditions, the nuclei oscillate around their equilibrium positions, according to a related nuclei temperature $T_\mathrm{i} > 0$. In contrast to the electronic temperature T_e, which is just an input parameter of the potential energy surface F, T_i has to be determined from the velocities and coordinates of the nuclei. This is necessary to prepare initial coordinates and velocities compatible with normal conditions. Here, the following well-known theorem of classical statistical physics provides support (proof taken from Ref. [40]):

Theorem 4.2 (Equipartition theorem) $H(q_1, \ldots, q_N, p_1, \ldots, p_N)$ *denotes the Hamilton function of a classical system of N coordinates* q_i *and momenta* p_i. *This system is in contact with a heat bath of temperature* T. *The heat bath and the system are isolated. Let it be* $x_i = q_i$ *or* $x_i = p_i$. *In equilibrium, one obtains*

$$\left\langle x_i \frac{\partial H}{\partial x_j} \right\rangle = \delta_{ij}\, k_B\, T, \tag{4.13}$$

where k_B *denotes the Boltzmann constant and* $\langle \cdot \rangle$ *the thermodynamical average.*

Proof In equilibrium, the system can be described by the canonical ensemble. The classical partition function of the canonical ensemble is given by [41]

$$Z^{(c)} = \frac{1}{\zeta} \int_{-\infty}^{\infty} dq_1 \ldots \int_{-\infty}^{\infty} dq_N \int_{-\infty}^{\infty} dp_1 \ldots \int_{-\infty}^{\infty} dp_N \, \mathrm{e}^{-\frac{H}{k_B T}}. \tag{4.14}$$

Compared to the quantum mechanical partition function (2.198), which we already derived, the summation over all eigenstates of the system must be replaced by the integration over all variables and everything must be divided by the constant ζ, which value is not relevant for us here. Furthermore, the classical thermodynamical average in the canonical ensemble is given by the integration over all variables using the weight $\frac{\mathrm{e}^{-\frac{H}{k_B T}}}{Z^{(c)}}$ analogously to the quantum mechanical case using the canonical density matrix (2.197), which we also derived previously. We have to divide again the integral by the constant ζ similar to the calculation of $Z^{(c)}$. In this way, we obtain

$$\begin{aligned}
\left\langle x_i \frac{\partial H}{\partial x_j} \right\rangle &= \frac{1}{\zeta} \int_{-\infty}^{\infty} dq_1 \ldots \int_{-\infty}^{\infty} dq_N \int_{-\infty}^{\infty} dp_1 \ldots \int_{-\infty}^{\infty} dp_N \, \frac{\mathrm{e}^{-\frac{H}{k_B T}}}{Z^{(c)}} \, x_i \, \frac{\partial H}{\partial x_j} \\
&= -\frac{k_B T}{Z^{(c)}} \frac{1}{\zeta} \int_{-\infty}^{\infty} dq_1 \ldots \int_{-\infty}^{\infty} dq_N \int_{-\infty}^{\infty} dp_1 \ldots \int_{-\infty}^{\infty} dp_N \, x_i \, \frac{\partial}{\partial x_j} \left(\mathrm{e}^{-\frac{H}{k_B T}} \right) \\
&= 0 + \frac{k_B T}{Z^{(c)}} \frac{1}{\zeta} \int_{-\infty}^{\infty} dq_1 \ldots \int_{-\infty}^{\infty} dq_N \int_{-\infty}^{\infty} dp_1 \ldots \int_{-\infty}^{\infty} dp_N \, \mathrm{e}^{-\frac{H}{k_B T}} \underbrace{\frac{\partial x_i}{\partial x_j}}_{=\delta_{ij}} \\
&= \delta_{ij}\, k_B\, T.
\end{aligned}$$

Here we utilized partial integration and assumed that $\mathrm{e}^{-\frac{H}{k_B T}}$ decays fast enough to zero for $q_i \to \pm\infty$ or $p_i \to \pm\infty$. ■

Since the potential energy surface $F(\mathbf{r}_1, \dots, \mathbf{r}_{N_{at}}, T_e)$ does not depend on the velocities and the velocities occur quadratically in the Hamilton function (4.1), we obtain for the nuclei temperature T_i in equilibrium from the equipartition theorem:

$$\left\langle \sum_{j=1}^{N_{at}} \frac{m_j}{2} \mathbf{v}_j^2 \right\rangle = \frac{1}{2} \sum_{j=1}^{N_{at}} \sum_{\alpha=x,y,z} \langle m_j v_{j\alpha}^2 \rangle = \frac{1}{2} \sum_{j=1}^{N_{at}} \sum_{\alpha=x,y,z} \left\langle v_{j\alpha} \frac{\partial H}{\partial v_{j\alpha}} \right\rangle \stackrel{(4.13)}{=} \frac{3\, N_{at}}{2} k_B\, T_i. \tag{4.15}$$

This allows to use the following definition of the nuclei temperature T_i by assuming always thermodynamical equilibrium:

$$\boxed{T_i = \frac{2 \sum_{j=1}^{N_{at}} \frac{m_j}{2} \mathbf{v}_j^2}{3\, N_{at}\, k_B}.} \tag{4.16}$$

Within the harmonic approximation, we can even derive the exact equilibrium distributions of the momenta and displacements. For this, we first need the following theorem, which proof we took from Ref. [42]:

Mathematical Theorem 4.1 (Integral of the Gaußian function)

$$\int_{-\infty}^{\infty} dx\, e^{-x^2} = \sqrt{\pi}. \tag{4.17}$$

Proof We obtain from Fubini's theorem:

$$\left(\int_{-\infty}^{\infty} dx\, e^{-x^2} \right)^2 = \int_{-\infty}^{\infty} dx \int_{-\infty}^{\infty} dy\, e^{-x^2} e^{-y^2} = \int_{-\infty}^{\infty} dx \int_{-\infty}^{\infty} dy\, e^{-(x^2+y^2)}.$$

Now we utilize spherical coordinates $r = \sqrt{x^2 + y^2}$, $\phi = \arg(x, y)$ for the two dimensional integration:

$$\left(\int_{-\infty}^{\infty} dx\, e^{-x^2} \right)^2 = \int_{0}^{\infty} dr \int_{0}^{2\pi} d\phi\, r\, e^{-r^2} = 2\pi \int_{0}^{\infty} dr\, r\, e^{-r^2}.$$

We substitute $u = r^2$, $du = 2r$ and get

$$\left(\int_{-\infty}^{\infty} dx\, e^{-x^2}\right)^2 = \pi \int_0^{\infty} du\, e^{-u} = \left[-\pi\, e^{-u}\right]_{u=0}^{u=\infty} = \pi,$$

which proves Eq. (4.17). ■

Mathematical Theorem 4.2 (Gaußian probability distribution)

$$f_G(x) = \frac{1}{\sqrt{2\pi\,\sigma^2}}\, e^{-\frac{(x-\mu)^2}{2\sigma^2}} \tag{4.18}$$

is called Gaußian distribution and has got the mean value μ and the standard deviation σ. The Gaußian distribution with the mean value zero and the standard deviation of one is called normal Gaußian distribution.

Proof First we show that $f_G(x)$ is normalized and, therefore, is a probability distribution. For this, we substitute $u = \frac{x-\mu}{\sqrt{2\sigma^2}}$, $du = \frac{1}{\sqrt{2\sigma^2}}\, dx$ and obtain

$$\int_{-\infty}^{\infty} dx\, f_G(x) = \int_{-\infty}^{\infty} dx\, \frac{1}{\sqrt{2\pi\,\sigma^2}}\, e^{-\frac{(x-\mu)^2}{2\sigma^2}} = \int_{-\infty}^{\infty} du\, \frac{1}{\sqrt{\pi}}\, e^{-u^2} \overset{(4.17)}{=} 1. \tag{4.19}$$

We get for the mean value, if we use the substitution $u = x - \mu$, $du = dx$:

$$\begin{aligned}\int_{-\infty}^{\infty} dx\, f_G(x)\, x &= \int_{-\infty}^{\infty} dx\, \frac{1}{\sqrt{2\pi\,\sigma^2}}\, e^{-\frac{(x-\mu)^2}{2\sigma^2}}\, (x-\mu) + \mu \underbrace{\int_{-\infty}^{\infty} dx\, \frac{1}{\sqrt{2\pi\,\sigma^2}}\, e^{-\frac{(x-\mu)^2}{2\sigma^2}}}_{\overset{(4.19)}{=}1} \\ &= \underbrace{\int_{-\infty}^{\infty} du\, \frac{1}{\sqrt{2\pi\,\sigma^2}}\, e^{-\frac{u^2}{2\sigma^2}}\, u}_{=0} + \mu \\ &= \mu. \end{aligned} \tag{4.20}$$

The last step is valid, since the function in the integral is antisymmetric. We obtain further for any constant $\kappa \in \mathbb{R}$ with $\kappa > 0$ by substituting $u = \sqrt{\kappa}\, x$, $du = \sqrt{\kappa}\, dx$:

$$\int_{-\infty}^{\infty} dx \, \frac{1}{\sqrt{\pi}} \, \mathrm{e}^{-\kappa x^2} = \frac{1}{\sqrt{\kappa}} \underbrace{\int_{-\infty}^{\infty} du \, \frac{1}{\sqrt{\pi}} \, \mathrm{e}^{-u^2}}_{\overset{(4.17)}{=} 1} = \frac{1}{\sqrt{\kappa}}.$$

Now we differentiate both side with respect to κ:

$$-\int_{-\infty}^{\infty} dx \, \frac{1}{\sqrt{\pi}} \, \mathrm{e}^{-\kappa x^2} \, x^2 = -\frac{1}{2} \kappa^{-\frac{3}{2}},$$

$$\Leftrightarrow \quad \int_{-\infty}^{\infty} dx \, \frac{\sqrt{\kappa}}{\sqrt{\pi}} \, \mathrm{e}^{-\kappa x^2} \, x^2 = \frac{1}{2\kappa}. \tag{4.21}$$

Utilizing this, we obtain for the variance of the Gaußian distribution by substituting $u = x - \mu$, $du = dx$:

$$\begin{aligned} \int_{-\infty}^{\infty} dx \, f_{\mathrm{G}}(x) \, (x-\mu)^2 &= \int_{-\infty}^{\infty} dx \, \frac{1}{\sqrt{2\pi \, \sigma^2}} \, \mathrm{e}^{-\frac{(x-\mu)^2}{2\sigma^2}} \, (x-\mu)^2 \\ &= \int_{-\infty}^{\infty} du \, \frac{1}{\sqrt{2\pi \, \sigma^2}} \, \mathrm{e}^{-\frac{u^2}{2\sigma^2}} \, u^2 \\ &\overset{(4.21)}{=} \sigma^2. \end{aligned} \tag{4.22}$$

The last step follows by using $\kappa = \frac{1}{2\sigma^2}$ in Eq. (4.21). Since the variance yields σ^2, the standard deviation of the Gaußian distribution is σ. ■

Theorem 4.3 (Equilibrium distributions for a harmonic Hamilton function)
Let

$$H = \sum_{j=1}^{N} \frac{p_j^2}{2\,m_j} + \sum_{j=1}^{N} \frac{m_j \, \omega_j^2 \, q_j^2}{2}$$

be the Hamilton function of N independent harmonic oscillators. In equilibrium at temperature T, the momenta p_j show a Gaußian distribution with zero mean value and a standard deviation of $\sqrt{m_j \, k_B \, T}$

$$\mathbb{P}_{p_j}(p) = \frac{1}{\sqrt{2\pi \, m_j \, k_B \, T}} \, \mathrm{e}^{-\frac{p^2}{2 m_j k_B T}} \tag{4.23}$$

and the displacements q_j show a Gaußian distribution with zero mean value and a standard deviation of $\sqrt{\frac{k_B T}{m_j \omega_j^2}}$

$$\mathbb{p}_{q_j}(q) = \sqrt{\frac{m_j \omega_j^2}{2\pi k_B T}} \, \mathrm{e}^{-\frac{m_j \omega_j^2 q^2}{2 k_B T}} . \tag{4.24}$$

Proof The equilibrium is described by the classical canonical ensemble. Therfore, we obtain for the probability that the momentum p_k exhibits the value p:

$$\begin{aligned}
\mathbb{p}_{p_k}(p) &= \frac{1}{\zeta} \int\limits_{-\infty}^{\infty} dq_1 \ldots \int\limits_{-\infty}^{\infty} dq_N \int\limits_{-\infty}^{\infty} dp_1 \ldots \int\limits_{-\infty}^{\infty} dp_N \, \delta(p_k - p) \, \frac{\mathrm{e}^{-\frac{H}{k_B T}}}{Z^{(c)}} \\
&= \frac{1}{Z^{(c)}} \frac{1}{\zeta} \mathrm{e}^{-\frac{p^2}{2 m_k k_B T}} \times \\
&\quad \times \int\limits_{-\infty}^{\infty} dq_1 \ldots \int\limits_{-\infty}^{\infty} dq_N \int\limits_{-\infty}^{\infty} dp_1 \ldots \int\limits_{-\infty}^{\infty} dp_{k-1} \int\limits_{-\infty}^{\infty} dp_{k+1} \ldots \int\limits_{-\infty}^{\infty} dp_N \times \\
&\quad \times \mathrm{e}^{-\frac{\sum\limits_{j=1, j\neq k}^{N} \frac{p_j^2}{2 m_j} + \sum\limits_{j=1}^{N} \frac{m_j \omega_j^2 q_j^2}{2}}{k_B T}} .
\end{aligned}$$

Using

$$\begin{aligned}
Z^{(c)} &= \frac{1}{\zeta} \int\limits_{-\infty}^{\infty} dp_k \, \mathrm{e}^{-\frac{p_k^2}{2 m_k k_B T}} \times \\
&\quad \times \int\limits_{-\infty}^{\infty} dq_1 \ldots \int\limits_{-\infty}^{\infty} dq_N \int\limits_{-\infty}^{\infty} dp_1 \ldots \int\limits_{-\infty}^{\infty} dp_{k-1} \int\limits_{-\infty}^{\infty} dp_{k+1} \ldots \int\limits_{-\infty}^{\infty} dp_N \times \\
&\quad \times \mathrm{e}^{-\frac{\sum\limits_{j=1, j\neq k}^{N} \frac{p_j^2}{2 m_j} + \sum\limits_{j=1}^{N} \frac{m_j \omega_j^2 q_j^2}{2}}{k_B T}}
\end{aligned}$$

we get

$$\mathbb{p}_{p_k}(p) = \frac{\mathrm{e}^{-\frac{p^2}{2 m_k k_B T}}}{\int\limits_{-\infty}^{\infty} dp_k \, \mathrm{e}^{-\frac{p_k^2}{2 m_k k_B T}}} .$$

We substitute $u = \frac{p_k}{\sqrt{2 m_k k_B T}}$, $du = \frac{dp_k}{\sqrt{2 m_k k_B T}}$ and get

$$\mathbb{P}_{p_k}(p) = \frac{\mathrm{e}^{-\frac{p^2}{2m_k k_\mathrm{B} T}}}{\sqrt{2 m_k k_\mathrm{B} T} \int\limits_{-\infty}^{\infty} du\, \mathrm{e}^{-u^2}} \overset{(4.17)}{=} \frac{1}{\sqrt{2\pi\, m_k k_\mathrm{B} T}} \mathrm{e}^{-\frac{p^2}{2m_k k_\mathrm{B} T}},$$

which proves Eq. (4.23). The proof of Eq. (4.24) is done analogously, since the momenta p_j and displacements q_j are independent of each other in H and occur both quadratically. ■

At room temperature $T_\mathrm{i} = 300$ K, the nuclei only oscillate slightly around their equilibrium positions, so that the harmonic approximation of the potential energy surface F is acceptable. Utilizing the last theorem, we can initiate the nuclei velocities and coordinates at normal conditions within the harmonic approximation of the potential energy surface F. For this, we first have to derive the phonon eigenmodes of the system by numerically calculating the dynamical matrix $\mathbf{D} \in \mathbb{R}^{3N_\mathrm{at} \times 3N_\mathrm{at}}$:

We consider the equilibrium positions of the nuclei and denote by $u_{j\alpha}$ the displacement of atom j in direction α from its equilibrium position. We denote by

$$U_{j\alpha} = \sqrt{m_j}\, u_{j\alpha}$$

the mass-normalized displacements and by

$$P_{j\alpha} = \sqrt{m_j}\, v_{j\alpha}$$

the mass-normalized momenta (See Eqs. (2.36) and (2.35). To simplify things, here j already characterizes the desired atom wheres in Eq. (2.36) the atom is characterized by υ and $\mathbf{T_n}$). All mass-normalized displacements are written in the vector $\mathbf{U} \in \mathbb{R}^{3N_\mathrm{at}}$ and all mass-normalized momenta in the vector $\mathbf{P} \in \mathbb{R}^{3N_\mathrm{at}}$.

Theorem 4.4 (Numerical calculation of the dynamical matrix) *Let us define $f_{j\alpha}(0, \ldots, 0, u_{j'\alpha'}, 0 \ldots, 0)$ as the force on atom j in direction α, if all atoms are located at their equilibrium position except for atom j', which is displaced by $u_{j'\alpha'}$ in the α' direction. Then we can calculate the dynamical matrix numerically from*

$$\begin{aligned} D_{j\alpha, j'\alpha'} &= \frac{1}{\sqrt{m_j m_{j'}}} \frac{\partial^2 F}{\partial u_{j\alpha}\, \partial u_{j'\alpha'}} \\ &= -\frac{1}{\sqrt{m_j m_{j'}}} \frac{\partial f_{j\alpha}(0, \ldots, 0)}{\partial u_{j'\alpha'}} \\ &\approx -\frac{1}{\sqrt{m_j m_{j'}}} \frac{f_{j\alpha}(0, \ldots, 0, u_{j'\alpha'}, 0, \ldots, 0) - f_{j\alpha}(0, \ldots, 0, -u_{j'\alpha'}, 0, \ldots, 0)}{2u_{j'\alpha'}}. \end{aligned} \tag{4.25}$$

By displacing one atom by $u_{j'\alpha'}$ and calculating the corresponding forces on all atoms and by calculating the corresponding forces for the negative displacement we can set one row of the dynamical matrix from Eq. (4.25). To calculate the complete dynamical matrix, we have to displace formally all atoms in all directions. In a periodic crystal structure, the number of such calculations can be reduced due to symmetry arguments.

Furthermore, we have to determine the matrix $\mathbf{C}$ that diagonalizes the dynamical matrix $\mathbf{D}$ (See Eq. (2.38)). The columns of $\mathbf{C}^{\mathrm{t}}$ correspond to the nuclei motions of the different eigenmodes $\mathbf{e}^{(1)}, \ldots, \mathbf{e}^{(3N_{\mathrm{at}})} \in \mathbb{R}^{3N_{\mathrm{at}}}$ (See Eq. (2.39)). Utilizing $\mathbf{C}$, we also derive the mass-normalized displacements in direction of the phonon eigenmodes from

$$\mathcal{U} = \mathbf{C} \cdot \mathbf{U}$$

and the mass-normalized momenta in direction of the phonon eigenmodes from

$$\mathcal{P} = \mathbf{C} \cdot \mathbf{P}$$

(See Eqs. (2.40) and (2.41)). The Hamilton function in the harmonic approximation reads

$$H^{(\mathrm{harm})} = \sum_{k=1}^{3N_{\mathrm{at}}} \frac{1}{2} \mathcal{P}_k^2 + \sum_{k=1}^{3N_{\mathrm{at}}} \frac{1}{2} \omega_k^2 \, \mathcal{U}_k^2 \tag{4.26}$$

(See Eq. (2.42)). From Eqs. (4.23) and (4.24) we know that the values of the momenta form a Gaußian distribution with zero mean and a standard deviation of $\sqrt{k_{\mathrm{B}} T_{\mathrm{i}}}$ and that the values of the displacements form a Gaußian distribution with zero mean and a standard deviation of $\sqrt{\frac{k_{\mathrm{B}} T_{\mathrm{i}}}{\omega_k^2}}$. For this, we realized that we have to set $m_j = 1$ due to the construction of the related Hamilton function (4.26). Therefore, we can set vales of the momenta and displacements compatible to equilibrium conditions utilizing $6N_{\mathrm{at}}$ random numbers $g_1, \ldots, g_{6N_{\mathrm{at}}}$ that are realizations of a normal Gaußian distribution (the mean value is zero and the standard deviation is one):

$$\sum_{j=1}^{6N_{\mathrm{at}}} g_j = 0 \quad \text{and} \quad \sum_{j=1}^{6N_{\mathrm{at}}} g_j^2 = 1. \tag{4.27}$$

For the final setting, we have to multiply the random number with the correct factor to get the desired standard deviation of the considered quantity:

$$\mathcal{U}_k = \sqrt{\frac{k_{\mathrm{B}} T_{\mathrm{i}}}{\omega_k^2}} \, g_{k+3N_{\mathrm{at}}}, \tag{4.28}$$

$$\mathcal{P}_k = \sqrt{k_{\mathrm{B}} T_{\mathrm{i}}} \, g_k. \tag{4.29}$$

Due to

$$u_{j\alpha} = \frac{1}{\sqrt{m_j}} U_{j\alpha} = \sum_{k=1}^{3N_{\mathrm{at}}} e_{j\alpha}^{(k)} \frac{1}{\sqrt{m_j}} \mathscr{U}_k,$$

$$v_{j\alpha} = \frac{1}{\sqrt{m_j}} P_{j\alpha} = \sum_{k=1}^{3N_{\mathrm{at}}} e_{j\alpha}^{(k)} \frac{1}{\sqrt{m_j}} \mathscr{P}_k,$$

we obtain the corresponding displacements and velocities of the nuclei by

$$\boxed{u_{j\alpha} = \sum_{k=1}^{3N_{\mathrm{at}}} e_{j\alpha}^{(k)} \sqrt{\frac{k_{\mathrm{B}} T_{\mathrm{i}}}{m_j \omega_k^2}} g_{k+3N_{\mathrm{at}}},} \tag{4.30}$$

$$\boxed{v_{j\alpha} = \sum_{k=1}^{3N_{\mathrm{at}}} e_{j\alpha}^{(k)} \sqrt{\frac{k_{\mathrm{B}} T_{\mathrm{i}}}{m_j}} g_k.} \tag{4.31}$$

Using the above equations, we can initiate displacements and velocities of the nuclei that correspond to equilibrium conditions at ionic temperature T_{i} within the harmonic approximation of the potential energy surface F. The drawback of this method is that we have to calculate the dynamical matrix **D**, which is quite computational expensive for non symmetric atomic structures, where symmetry arguments cannot be used to reduce the number of necessary calculations. In addition, we have to diagonalize the dynamical matrix, which is also computational expensive for large numbers of atoms. Furthermore, the influence of higher than harmonic orders in the potential energy surface F is neglected.

To overcome these drawbacks, one can utilize a so called **thermostat** for the initialization. A MD simulation at constant electronic temperature T_{e} using the Hamilton function (4.1) corresponds to a simulation in a microcanonical ensemble, which is also called $N\mathcal{V}E$ ensemble, since the number of atoms N, the volume $\mathcal{V}$ and the total energy E are kept constant. To perform a MD simulation in a canonical ensemble, which corresponds to a $N\mathcal{V}T$ ensemble, we have to utilize a thermostat to control the temperature T. In 1980, Andersen presented the following thermostat [43]:

Theorem 4.5 (Andersen thermostat) *A MD simulation is performed, in which after every N_t time steps the velocities are reset by*

$$v_{j\alpha} = \sqrt{\frac{k_B T_i}{m_j}} g_{3(j-1)+\alpha}. \tag{4.32}$$

Here, the direction α exhibits the values 1, 2, 3 *instead of* x, y, z *to simplify things. For the resetting of the velocities, one uses always a different set* $\{g_1, \ldots, g_{3N_{at}}\}$ *of normal Gaußian distributed random numbers, which obey*

$$\sum_{j=1}^{3N_{at}} g_j = 0 \quad and \quad \sum_{j=1}^{3N_{at}} g_j^2 = 1. \tag{4.33}$$

This procedure generates coordinates and velocities corresponding to thermodynamical equilibrium at temperature T_i *for the Hamilton function* (4.1).

Proof In the Hamilton function (4.1), the potential energy surface F does not depend on the velocities. Therefore, the resetting of the velocities agrees to thermal equilibrium at T_i. After the resetting, both, velocities and coordinates equilibrate to thermal equilibrium in the N_t MD simulation steps, if N_t is sufficient large enough, since a MD simulation in the $N\mathcal{V}E$ ensemble converges always to thermal equilibrium at the given total energy E. The total energy E is distributed between the kinetic and the potential energy. If the potential energy during the velocity resetting is not compatible to T_i, kinetic energy is transformed to potential energy or the other way around in the N_t MD simulation steps to reach equilibrium. In this case, the reached temperature differs from the target T_i. Now the recurring resetting of the velocities and continuing of the MD simulation induces a convergence of the potential energy to the right value compatible to T_i. Hence, after resetting the velocities enough times, one obtains velocities and coordinates that are compatible the equilibrium at the temperature T_i. ■

To generate the initial conditions, we set the equilibrium coordinates of the nuclei and perform a MD simulation with the Anderson thermostat, which is also called **thermalization**. We check T_i obtained from Eq. (4.16) using the velocities during the thermalization and take the velocities and coordinates before the next velocity reset as initial conditions, if the target T_i is approximately reached. These initialization takes the complete potential energy surface F into account in contrast to Eqs. (4.30) and (4.31), where the harmonic approximation of F is used.

In this book, we utilized the thermalization with the Andersen thermostat or the initialization within the harmonic approximation to get the initial conditions. We took always the necessary random numbers from www.random.org, where real random numbers r_k generated from atmospheric noise are provided. Since these random numbers r_k form a uniform distribution in the interval [0, 1], we have to transform them to a normal Gaußian distribution. For this, we define for $x \in \mathbb{R}$ the integral of the normal Gaußian distribution

$$\mathcal{G}(x) := \int_{-\infty}^{x} du \, \frac{1}{\sqrt{2\pi}} \, e^{-\frac{u^2}{2}}, \tag{4.34}$$

which takes values in the interval [0, 1], and transform the random numbers r_k using the inverse of the above function by

$$g_k = \mathcal{G}^{-1}(r_k). \tag{4.35}$$

To calculate numerically the function $\mathcal{G}^{-1}(x)$, we utilized the function PPND16 from Ref. [44], which is implemented in the FORTRAN language.

4.2 Calculation of the Diffraction Peak Intensities

It is impossible presently and for the foreseeable future to measure the positions of a sufficiently large set of atoms in a solid on a few femtosecond time-scale. Techniques with sufficient spatial resolution lack temporal resolution by more than 10 orders of magnitude. On the other hand, techniques with sufficient temporal resolution lack spatial resolution at all (spectroscopic techniques) or measure the positions as an ensemble average.

However, the latter gives insights into the atomic structure by measuring the diffraction image generated from the diffraction of electrons or x-rays. Advanced experimental setups allow to measure the time-dependent diffraction image with a time resolution of 100 fs or better [26, 45–48]. Theoretically, the diffraction image can be directly calculated as a function of the atomic positions [49, 50], which allows to compare simulations with experiments. We consider the diffraction of a plane wave, since the incident wave on the solid can be always approximated by a plane wave, if the solid is located far enough away from the radiation source. Furthermore, we consider the diffraction by a solid with a periodic crystal structure. Let **k** be the wave vector of the plane wave with wavelength λ, so that we have

$$|\mathbf{k}| = \frac{2\pi}{\lambda}. \tag{4.36}$$

As we already described in Sect. 2.2.1, the atomic equilibrium positions in a periodic crystal structure are described by a Bravais grid, where at each grid point N_b atoms are identically placed. Each of these atoms is shifted by the vector $\mathbf{d}_\upsilon$ with respect to the Bravais grid point $\mathbf{T_n}$. Now the incident plane wave induces at the sites of the Bravais grid a coherent radiation of spherical waves in all directions. In special directions, the radiated beams show a constructive interference.

At first, we consider the scattering of the incoming wave by two sites of the Bravais lattice, one located at **0** and one located at $\mathbf{T_n}$. See for this Fig. 4.1. Since the diffraction pattern is commonly measured far away from the solid, the two rays that are emitted from the two sides and interfere in point B of the diffraction pattern can be assumed to be parallel to each other, so that they both have the same wave vector $\mathbf{k}'$. Furthermore, we consider only elastic scattering, so that we have

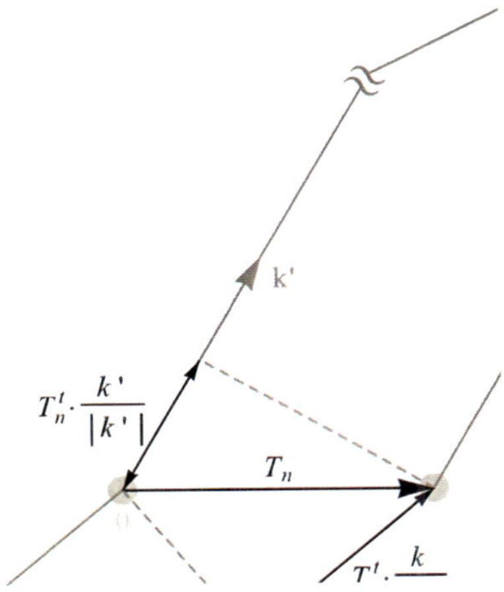

Fig. 4.1 The scattering by two sides of the Bravais lattice is illustrated

$$|\mathbf{k}'| = |\mathbf{k}|. \tag{4.37}$$

To get a constructive interference, the path difference between the two emitted rays must be a multiple of the wavelength λ:

$$\mathbf{T}^t_\mathbf{n} \cdot \frac{\mathbf{k}}{|\mathbf{k}|} - \mathbf{T}^t_\mathbf{n} \cdot \frac{\mathbf{k}'}{|\mathbf{k}'|} = n\,\lambda. \tag{4.38}$$

If we use Eqs. (4.36) and (4.37), we obtain the condition

$$\mathbf{T}^t_\mathbf{n} \cdot \left(\mathbf{k} - \mathbf{k}'\right) = 2\pi\, n,$$
$$\Leftrightarrow \quad e^{\mathrm{i}\,\mathbf{T}^t_\mathbf{n} \cdot (\mathbf{k}-\mathbf{k}')} = 1. \tag{4.39}$$

Next, we consider the scattering of the incoming wave by not just two, but by all sites of the Bravais lattice. Since all emitted rays must show a constructive interference, Eq. (4.39) must hold for all lattice points T_n of the Bravais grid. This is only possible, if the difference of the wave vectors corresponds to a vector of the reciprocal grid:

$$\mathbf{k} - \mathbf{k}' = \mathbf{G}_{\mathbf{n}'}. \tag{4.40}$$

This is the well-known **Laue condition**. Therefore, the diffraction image shows discrete peaks, which are called **Bragg peaks**. To represent the reciprocal lattice points related to the Bragg peaks, one commonly utilizes the so called Miller indices (h, k, l) with h, k, l being integers:

$$\mathbf{G}_{hkl} = h\,\mathbf{b}_1 + k\,\mathbf{b}_2 + l\,\mathbf{b}_3. \tag{4.41}$$

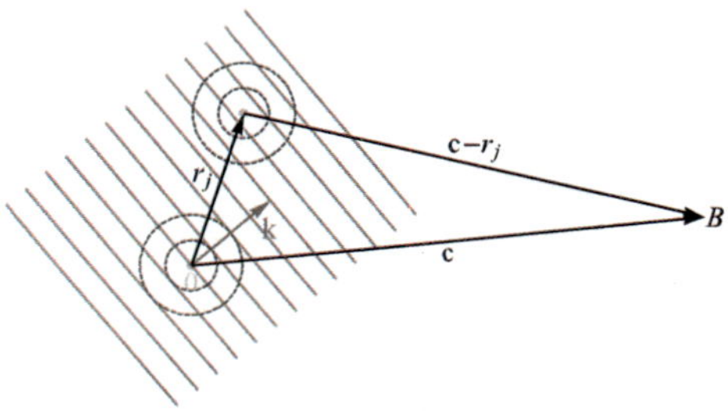

Fig. 4.2 The interference of two emitted rays is illustrated

Here $\mathbf{b}_1$, $\mathbf{b}_2$, $\mathbf{b}_3$ are the reciprocal grid vectors. To derive finally the intensity of the Bragg peaks, we have to consider the N_b atoms located at each site of the Bravais grid, since the incident wave is scattered by the atoms. The atomic coordinates are given by

$$\mathbf{r}_{\mathbf{n}\upsilon}(t) = \mathbf{T}_{\mathbf{n}} + \mathbf{d}_{\upsilon} + \mathbf{u}_{\mathbf{n}\upsilon}(t),$$

where $\mathbf{u}_{\mathbf{n}\upsilon}(t)$ is the time-dependent displacement of the atom from its equilibrium position. To simplify things, we introduce a counting order of the N_{at} atoms located in the simulation cell and write $\mathbf{r}_j(t)$ for the atomic coordinate of the j-th atom. The amplitude of the incident plane wave at the position of the j-th reads (See for this Fig. 4.2)

$$\Psi\big(\mathbf{r}_j(t), t\big) = \Psi_0\, \mathrm{e}^{\mathrm{i}\,(\mathbf{k}^{\mathrm{t}}\cdot\mathbf{r}_j(t) - \omega t)}, \tag{4.42}$$

where Ψ_0 is a constant and ω is the angular frequency. Here, we assume that Ψ_0 is constant inside the crystal, since the radiation source is located far away from the crystal and the dimensions of the crystal are small. The j-th atom emits a spherical wave coherently to the incident wave with an amplitude damped by the factor ξ with respect to the incident wave. For simplification, we assume, that this damping factor ξ is equal for all atoms. Therefore, we have for the emitted wave amplitude at point B

$$\Psi_B\big(\mathbf{r}_j(t), t\big) = \xi\; \Psi\big(\mathbf{r}_j(t), t\big)\, \frac{\mathrm{e}^{\mathrm{i}\,\mathbf{k}'^{\mathrm{t}}\cdot(\mathbf{c}-\mathbf{r}_j(t))}}{\big|\mathbf{c}-\mathbf{r}_j(t)\big|}. \tag{4.43}$$

Since the point B is far away from the crystal, i.e. $|\mathbf{c}| >> \big|\mathbf{r}_j(t)\big|$, we have $\big|\mathbf{c} - \mathbf{r}_j(t)\big| \approx |\mathbf{c}|$ and we obtain further by inserting Eq. (4.42):

$$\Psi_B\big(\mathbf{r}_j(t), t\big) \overset{(4.42)}{=} \xi\; \Psi_0\, \mathrm{e}^{\mathrm{i}\,(\mathbf{k}^{\mathrm{t}}\cdot\mathbf{r}_j(t)-\omega t)}\, \frac{\mathrm{e}^{\mathrm{i}\,\mathbf{k}'^{\mathrm{t}}\cdot(\mathbf{c}-\mathbf{r}_j(t))}}{\big|\mathbf{c}-\mathbf{r}_j(t)\big|} \approx \frac{\xi\,\Psi_0}{|\mathbf{c}|}\, \mathrm{e}^{\mathrm{i}\,\mathbf{k}'^{\mathrm{t}}\cdot\mathbf{c}}\, \mathrm{e}^{-\mathrm{i}\,\omega t}\, \mathrm{e}^{\mathrm{i}\,(\mathbf{k}-\mathbf{k}')^{\mathrm{t}}\cdot\mathbf{r}_j(t)}. \tag{4.44}$$

The above equation is valid for any atom j of the crystal. Therefore, at point B, we obtain for the total wave amplitude that is generated from the waves emitted from all atoms:

$$\Psi_B(t) = \sum_{j=1}^{N_{\text{at}}} \Psi_B\big(\mathbf{r}_j(t), t\big) \approx \frac{\xi\,\Psi_0}{|\mathbf{c}|}\,\mathrm{e}^{\mathrm{i}\,\mathbf{k}'^{\mathrm{t}}\cdot\mathbf{c}}\,\mathrm{e}^{-\mathrm{i}\omega t}\sum_{j=1}^{N_{\text{at}}}\mathrm{e}^{\mathrm{i}\,(\mathbf{k}-\mathbf{k}')^{\mathrm{t}}\cdot\mathbf{r}_j(t)}. \tag{4.45}$$

In the here relevant diffraction experiments, only the diffraction intensities are experimentally accessible:

$$I_B(t) = \left|\Psi_B(t)\right|^2 \approx \frac{\xi^2\,|\Psi_0|^2}{|\mathbf{c}|^2}\left|\sum_{j=1}^{N_{\text{at}}}\mathrm{e}^{\mathrm{i}\,(\mathbf{k}-\mathbf{k}')^{\mathrm{t}}\cdot\mathbf{r}_j(t)}\right|^2. \tag{4.46}$$

From the Laue condition, we already know, that a constructive interference is only possible, if $\mathbf{k}-\mathbf{k}'$ corresponds to a vector of the reciprocal grid. Thus, the elastic diffraction peak intensity related to a given reciprocal vector $\mathbf{q}$ has the following proportionality:

$$\boxed{I_{\mathbf{q}}(t) \propto \left|\sum_{j=1}^{N_{\text{at}}}\mathrm{e}^{\mathrm{i}\,\mathbf{q}^{\mathrm{t}}\cdot\mathbf{r}_j(t)}\right|^2.} \tag{4.47}$$

If one characterize the wave vector by the Miller indices (hkl), the corresponding intensity is denoted as $I_{hkl}(t)$. The appearing factor

$$SF_{\mathbf{q}} = \sum_{j=1}^{N_{\text{at}}}\mathrm{e}^{\mathrm{i}\,\mathbf{q}^{\mathrm{t}}\cdot\mathbf{r}_j(t)} \tag{4.48}$$

is called **structure factor**. In a real solid, there are always atomic displacements from the ideal symmetric crystal structure, which cause also a nonzero intensity for wave vectors $\mathbf{q}$ not belonging to the reciprocal grid. This is the so called **background** beside the Bragg peaks, to which also multi scattering processes contribute. In addition, the atomic displacements have a direct influence on the peak intensity within elastic diffraction. If the system is in thermal equilibrium, one can derive the so called Debye-Waller factor for the intensity dependence on the atomic displacements. Utilizing the following measure for the atomic displacements, the derivation of the Debye-Waller factor is presented based on Ref. [50].

Definition 4.1 (*Mean-square displacements of the atoms*) Let $\mathbf{u}_j(t)$ be the displacement of the j-atom in the simulation cell from its equilibrium position at time t. Then one defines the mean-square displacements of the atoms by

$$\mathrm{MSD}(t) = \frac{1}{N_{\text{at}}}\sum_{j=1}^{N_{\text{at}}}\left|\mathbf{u}_j(t)\right|^2, \tag{4.49}$$

where N_{at} is the number of atoms in the simulation cell.

Theorem 4.6 (Debye-Waller factor) *Let us consider a system of identical atoms that is in thermal equilibrium. Then the intensity of the diffraction peak related to the reciprocal vector* $\mathbf{q}$ *reads*

$$\langle I_{\mathbf{q}}(t)\rangle_t = I_0(\mathbf{q})\, \mathrm{e}^{-\frac{1}{3}|\mathbf{q}|^2 \langle MSD(t)\rangle_t}, \tag{4.50}$$

where

$$\langle f(t)\rangle_t := \lim_{\tau\to\infty} \frac{1}{\tau} \int_0^{\tau} dt\, f(t) \tag{4.51}$$

denotes the time average of a time dependent function $f(t)$ *and* $I_0(\mathbf{q})$ *denotes the intensity for an ideal crystal structure without atomic displacements. The occurring exponential factor is called Debye-Waller factor.*

Proof We denote the positions of the atoms by

$$\mathbf{r}_{\mathbf{n}\upsilon}(t) = \mathbf{T}_{\mathbf{n}} + \mathbf{d}_{\upsilon} + \mathbf{u}_{\mathbf{n}\upsilon}(t),$$

where $\mathbf{u}_{\mathbf{n}\upsilon}(t)$ denotes the displacement from the equilibrium position $\mathbf{T}_{\mathbf{n}} + \mathbf{d}_{\upsilon}$ at time t (c.f. Eq. (2.29)). Then the structure factor reads

$$SF_{\mathbf{q}}(t) = \sum_{\mathbf{n}\upsilon} \mathrm{e}^{\mathrm{i}\,\mathbf{q}^{\mathrm{t}}\cdot(\mathbf{T}_{\mathbf{n}\upsilon}+\mathbf{d}_{\upsilon})}\, \mathrm{e}^{\mathrm{i}\,\mathbf{q}^{\mathrm{t}}\cdot\mathbf{u}_{\mathbf{n}\upsilon}(t)},$$

where the sum runs over all atoms in the simulation cell. Now we perform the time average of the structure factor and obtain

$$\langle SF_{\mathbf{q}}(t)\rangle_t = \left\langle \sum_{\mathbf{n}\upsilon} \mathrm{e}^{\mathrm{i}\,\mathbf{q}^{\mathrm{t}}\cdot(\mathbf{T}_{\mathbf{n}\upsilon}+\mathbf{d}_{\upsilon})}\, \mathrm{e}^{\mathrm{i}\,\mathbf{q}^{\mathrm{t}}\cdot\mathbf{u}_{\mathbf{n}\upsilon}(t)} \right\rangle_t = \sum_{\mathbf{n}\upsilon} \mathrm{e}^{\mathrm{i}\,\mathbf{q}^{\mathrm{t}}\cdot(\mathbf{T}_{\mathbf{n}\upsilon}+\mathbf{d}_{\upsilon})} \left\langle \mathrm{e}^{\mathrm{i}\,\mathbf{q}^{\mathrm{t}}\cdot\mathbf{u}_{\mathbf{n}\upsilon}(t)} \right\rangle_t. \tag{4.52}$$

If the displacements of the atoms are not to large, we can perform a Taylor series expansion of the exponential function

$$\mathrm{e}^{\mathrm{i}\,\mathbf{q}^{\mathrm{t}}\cdot\mathbf{u}_{\mathbf{n}\upsilon}(t)} \approx 1 + \mathrm{i}\,\mathbf{q}^{\mathrm{t}}\cdot\mathbf{u}_{\mathbf{n}\upsilon}(t) - \frac{1}{2}\left(\mathbf{q}^{\mathrm{t}}\cdot\mathbf{u}_{\mathbf{n}\upsilon}(t)\right)^2$$

and get

$$\left\langle \mathrm{e}^{\mathrm{i}\,\mathbf{q}^{\mathrm{t}}\cdot\mathbf{u}_{\mathbf{n}\upsilon}(t)} \right\rangle_t \approx 1 + \mathrm{i}\langle \mathbf{q}^{\mathrm{t}}\cdot\mathbf{u}_{\mathbf{n}\upsilon}(t)\rangle_t - \frac{1}{2}\left\langle \left(\mathbf{q}^{\mathrm{t}}\cdot\mathbf{u}_{\mathbf{n}\upsilon}(t)\right)^2 \right\rangle_t. \tag{4.53}$$

The thermal energy induces a stochastic motion of the atoms. Consequently, the direction of $\mathbf{q}$ and $\mathbf{u}_{\mathbf{n}\upsilon}(t)$ is not correlated during the time propagation and we have

$$\left\langle \mathbf{q}^{\mathrm{t}} \cdot \mathbf{u}_{\mathbf{n}\upsilon}(t) \right\rangle_t = 0. \tag{4.54}$$

If we define $\vartheta(t)$ as the angle between $\mathbf{q}$ and $\mathbf{u}_{\mathbf{n}\upsilon}(t)$, we obtain for the quadratic term occurring in Eq. (4.53):

$$\begin{aligned} \left\langle \left(\mathbf{q}^{\mathrm{t}} \cdot \mathbf{u}_{\mathbf{n}\upsilon}(t)\right)^2 \right\rangle_t &= \left\langle |\mathbf{q}|^2 \left|\mathbf{u}_{\mathbf{n}\upsilon}(t)\right|^2 \cos\left(\vartheta(t)\right)^2 \right\rangle_t \\ &= |\mathbf{q}|^2 \left\langle \left|\mathbf{u}_{\mathbf{n}\upsilon}(t)\right|^2 \right\rangle_t \left\langle \cos\left(\vartheta(t)\right)^2 \right\rangle_t . \end{aligned} \tag{4.55}$$

All atoms are of the same type. Therefore, one can assume that the time average of the square displacement of a single atom is equal to the time average of the mean-square displacements of all atoms:

$$\left\langle \left|\mathbf{u}_{\mathbf{n}\upsilon}(t)\right|^2 \right\rangle_t = \left\langle \frac{1}{N_{\mathrm{at}}} \sum_{\mathbf{n}\upsilon} \left|\mathbf{u}_{\mathbf{n}\upsilon}(t)\right|^2 \right\rangle_t = \left\langle \mathrm{MSD}(t) \right\rangle_t. \tag{4.56}$$

Since the direction of $\mathbf{q}$ and $\mathbf{u}_{\mathbf{n}\upsilon}(t)$ is not correlated during the time propagation, we can average $\cos\left(\vartheta(t)\right)^2$ over all possible directions, i.e., we can average $\cos\left(\vartheta(t)\right)^2$ over a sphere using spherical coordinates:

$$\left\langle \cos\left(\vartheta(t)\right)^2 \right\rangle_t = \frac{\int_0^{2\pi} d\phi \int_0^{\pi} d\vartheta \, \sin(\vartheta) \cos(\vartheta)^2}{\int_0^{2\pi} d\phi \int_0^{\pi} d\vartheta \, \sin(\vartheta)} = \frac{2\pi \int_0^{\pi} d\vartheta \, \sin(\vartheta)\cos(\vartheta)^2}{4\pi}.$$

Now we substitute $u = -\cos(\vartheta)$, $du = \sin(\vartheta)\, d\vartheta$ and obtain

$$\left\langle \cos\left(\vartheta(t)\right)^2 \right\rangle_t = \frac{1}{2} \int_{-1}^{1} du\, u^2 = \left.\frac{u^3}{6}\right|_{u=-1}^{u=1} = \frac{1}{3}. \tag{4.57}$$

Putting all together, we find

$$\left\langle \mathrm{e}^{\mathrm{i}\,\mathbf{q}^{\mathrm{t}} \cdot \mathbf{u}_{\mathbf{n}\upsilon}(t)} \right\rangle_t \approx 1 - \frac{1}{6} |\mathbf{q}|^2 \left\langle \mathrm{MSD}(t) \right\rangle_t \approx \mathrm{e}^{-\frac{1}{6} |\mathbf{q}|^2 \left\langle \mathrm{MSD}(t) \right\rangle_t}, \tag{4.58}$$

where we utilize for the second approximation again the Taylor series approximation of the exponential function. Utilizing this, the time average of the structure factor reads

$$\left\langle SF_{\mathbf{q}}(t) \right\rangle_t = \sum_{\mathbf{n}\upsilon} \mathrm{e}^{\mathrm{i}\,\mathbf{q}^{\mathrm{t}} \cdot (\mathbf{T}_{\mathbf{n}\upsilon} + \mathbf{d}_\upsilon)}\, \mathrm{e}^{-\frac{1}{6} |\mathbf{q}|^2 \left\langle \mathrm{MSD}(t) \right\rangle_t} = \mathrm{e}^{-\frac{1}{6} |\mathbf{q}|^2 \left\langle \mathrm{MSD}(t) \right\rangle_t} \sum_{\mathbf{n}\upsilon} \mathrm{e}^{\mathrm{i}\,\mathbf{q}^{\mathrm{t}} \cdot (\mathbf{T}_{\mathbf{n}\upsilon} + \mathbf{d}_\upsilon)} \tag{4.59}$$

and we obtain finally for the time average of the diffraction peak intensity

$$\left\langle I_{\mathbf{q}}(t)\right\rangle_t \propto \left|\left\langle SF_{\mathbf{q}}(t)\right\rangle_t\right|^2 = \left|\sum_{\mathbf{n}\upsilon} \mathrm{e}^{\mathrm{i}\,\mathbf{q}^{\mathrm{t}}\cdot(\mathbf{T}_{\mathbf{n}\upsilon}+\mathbf{d}_\upsilon)}\right|^2 \mathrm{e}^{-\frac{1}{3}|\mathbf{q}|^2\left\langle \mathrm{MSD}(t)\right\rangle_t}, \tag{4.60}$$

which finishes the proof. ■

The last theorem indicates the following: The diffraction peak intensity decreases exponentially with the mean-square displacements of the atoms. Diffraction peaks with bigger wave vector **q** or higher integer numbers of the Miller indices decay stronger with respect to the mean-square displacements.

In the last theorem, the time average is formally carried out over an infinite period of time. But, to carry out a proper thermodynamical average, a finite period of time is enough. Since time-resolved diffraction experiments have a finite time resolution, which is about ~100 fs today, one practically reduces the period of time for the average to the time resolution of the experiments. By considering an infinitely short period of time, one obtains formally for the time-dependent relative peak intensity

$$\frac{I_{\mathbf{q}}(t)}{I_{\mathbf{q}}(t_0)} = \mathrm{e}^{-\frac{1}{3}|\mathbf{q}|^2\left(\mathrm{MSD}(t)-\mathrm{MSD}(t_0)\right)} \tag{4.61}$$

in the so-called **time-dependent Debye-Waller theory**. Here, t_0 denotes the initial time.

4.3 Fs-Laser Induced Thermal Phonon Squeezing and Antisqueezing

A femtosecond-laser excitation changes dramatically the PES. If the laser excitation is moderate, the interatomic bonding stays intact, so that the crystal structure survives. In order to check whether the crystal structure is preserved, one can test whether the atomic mean-square displacements remains below a certain limit:

Definition 4.2 (*Lindemann stability limit*) If the atomic mean-square displacements stays below 15% of the nearest-neighbor distance, the crystal structure remains intact.

Although the crystal structure survives, the femtosecond-laser excitation nevertheless affects the interatomic bonds. They can become softer (lower binding energy) or even harder (higher binding energy), so that the phonon eigenfrequencies decrease or increase, respectively. This femtosecond-laser induced change of the phonon eigenfrequencies occurs within a few femtoseconds and can lead to thermal **squeezing** or **antisqueezing**:

Definition 4.3 (*Thermal phonon squeezing and antisqueezing*) An ensemble of identical classical oscillators is thermally squeezed, if the variance of the distribution of displacements becomes temporally smaller compared to the value of that variance at room temperature (while the variance of the distribution of momenta increases). An ensemble of identical classical oscillators is thermally antisqueezed, if the variance of the distribution of displacements grows larger compared to the value of the variance at room temperature (while the variance of the distribution of momenta becomes smaller).

We want to note that no distinction is made between squeezing and antisqueezing in the common literature describing femtosecond laser-excitations of matter. The time-dependent oscillation of the variance of the displacement or momentum distribution is just called squeezing. We consider the atomic motions within the harmonic approximation (see Sect. 2.2.1) and the phonon eigenmodes at $T_\mathrm{e} = 300$ K before the laser excitation. We denote the mass-normalized displacements in direction of the $T_\mathrm{e} = 300$ K eigenmodes by $\tilde{\mathscr{U}}_i$ (c.f. Eq. (2.40)) and the mass-normalized momenta in direction of the $T_\mathrm{e} = 300$ K eigenmodes by $\tilde{\mathscr{P}}_i$ (c.f. Eq. (2.41)). Then, the Hamiltonian function reads

$$H^{(\mathrm{harm})} = \sum_{k=1}^{3\,N_\mathrm{at}} \frac{1}{2}\,\tilde{\mathscr{P}}_k^2 + \sum_{k=1}^{3\,N_\mathrm{at}} \frac{1}{2}\,\tilde{\omega}_k^2\,\tilde{\mathscr{U}}_k^2$$

In thermodynamical equilibrium at an atomic temperature T_i, the displacements and momenta follow a Gaußian distribution at any time:

$$\mathbb{p}_{\tilde{\mathscr{U}}_i}(u) \overset{(4.24)}{=} \frac{\tilde{\omega}_i}{\sqrt{2\pi\,k_\mathrm{B}\,T_\mathrm{i}}}\mathrm{e}^{-\frac{\tilde{\omega}_i^2\,u^2}{2\,k_\mathrm{B}\,T_\mathrm{i}}}, \tag{4.62}$$

$$\mathbb{p}_{\tilde{\mathscr{P}}_i}(p) \overset{(4.23)}{=} \frac{1}{\sqrt{2\pi\,k_\mathrm{B}\,T_\mathrm{i}}}\mathrm{e}^{-\frac{p^2}{2\,k_\mathrm{B}\,T_\mathrm{i}}}. \tag{4.63}$$

The laser excitation increases T_e to a new higher value at $t = 0$. Due to this, we obtain new phonon eigenmodes with related displacements $\mathscr{U}_i$, momenta $\mathscr{P}_i$ and eigenfrequencies ω_i. If we denote by $\mathscr{U}_i^{(0)}$ the displacement and by $\mathscr{P}_i^{(0)}$ the momentum at $t = 0$, the time dependent displacement of the i-th phonon eigenmode is given by

$$\mathscr{U}_i(t) = \mathscr{U}_i^{(0)}\,\cos(\omega_i\,t) + \mathscr{P}_i^{(0)}\,\frac{\sin(\omega_i\,t)}{\omega_i}, \tag{4.64}$$

since one has $\mathscr{P}_i(t) = \frac{d\mathscr{U}_i}{dt} = -\omega_i\, \mathscr{U}_i^{(0)} \sin(\omega_i\, t) + \mathscr{P}_i^{(0)} \cos(\omega_i\, t)$, so that $\mathscr{P}_i(0) = \mathscr{P}_i^{(0)}$. We obtain the displacements $\mathscr{U}$ in direction of the phonon eigenmodes from the vector $\mathbf{U}$ of all mass-normalized atomic displacements with the help of a orthonormal matrix $\mathbf{C}$ with $\mathbf{C}^{\mathrm{t}} \cdot \mathbf{C} = \mathbb{1}$:

$$\mathscr{U} = \mathbf{C} \cdot \mathbf{U}, \qquad \tilde{\mathscr{U}} = \tilde{\mathbf{C}} \cdot \mathbf{U} \quad \Rightarrow \quad \mathscr{U} = \underbrace{\mathbf{C} \cdot \tilde{\mathbf{C}}^{\mathrm{t}}}_{=:\mathbf{A}} \cdot \tilde{\mathscr{U}} \tag{4.65}$$

Due to this, the amplitudes of displacements and momenta at $t = 0$ can be obtained from the displacements and momenta of the phonon eigenmodes at the initial $T_{\mathrm{e}} = 300$ K by $\mathscr{U}^{(0)} = \mathbf{A} \cdot \tilde{\mathscr{U}}^{(0)}$ and $\mathscr{P}^{(0)} = \mathbf{A} \cdot \tilde{\mathscr{P}}^{(0)}$, or

$$\mathscr{U}_i^{(0)} = \sum_{j=1}^{3\,N_{\mathrm{at}}} A_{ij}\, \tilde{\mathscr{U}}_j^{(0)}, \qquad \mathscr{P}_i^{(0)} = \sum_{j=1}^{3\,N_{\mathrm{at}}} A_{ij}\, \tilde{\mathscr{P}}_j^{(0)}.$$

Inserting this into Eq. (4.64) yields

$$\mathscr{U}_i(t) = \cos(\omega_i\, t) \sum_{j=1}^{3\,N_{\mathrm{at}}} A_{ij}\, \tilde{\mathscr{U}}_j^{(0)} + \frac{\sin(\omega_i\, t)}{\omega_i} \sum_{j=1}^{3\,N_{\mathrm{at}}} A_{ij}\, \tilde{\mathscr{P}}_j^{(0)} \tag{4.66}$$

and

$$\begin{aligned}
\mathscr{U}_i^2(t) = {} & \cos^2(\omega_i\, t) \sum_{j=1}^{3\,N_{\mathrm{at}}} A_{ij}^2 \left(\tilde{\mathscr{U}}_j^{(0)}\right)^2 + \frac{\sin^2(\omega_i\, t)}{\omega_i^2} \sum_{j=1}^{3\,N_{\mathrm{at}}} A_{ij}^2 \left(\tilde{\mathscr{P}}_j^{(0)}\right)^2 \\
& + \cos^2(\omega_i\, t) \sum_{j\,=\,1}^{3\,N_{\mathrm{at}}} \sum_{\substack{k\,=\,1 \\ k \neq j}}^{3\,N_{\mathrm{at}}} A_{ij}\, A_{ik}\, \tilde{\mathscr{U}}_j^{(0)}\, \tilde{\mathscr{U}}_k^{(0)} \\
& + \frac{\sin^2(\omega_i\, t)}{\omega_i^2} \sum_{j\,=\,1}^{3\,N_{\mathrm{at}}} \sum_{\substack{k\,=\,1 \\ k \neq j}}^{3\,N_{\mathrm{at}}} A_{ij}\, A_{ik}\, \tilde{\mathscr{P}}_j^{(0)}\, \tilde{\mathscr{P}}_k^{(0)} \\
& + 2\, \frac{\cos(\omega_i\, t)\, \sin(\omega_i\, t)}{\omega_i} \sum_{j=1}^{3\,N_{\mathrm{at}}} \sum_{k=1}^{3\,N_{\mathrm{at}}} A_{ij}\, A_{ik}\, \tilde{\mathscr{U}}_j^{(0)}\, \tilde{\mathscr{P}}_k^{(0)}.
\end{aligned} \tag{4.67}$$

We perform a thermodynamic average by considering each individual oscillator i as an ensemble of independent and identical oscillators following a distribution like Eqs. (4.62) and (4.63) for displacement and momentum, respectively. In this way, we obtain, at time t, for the thermodynamic average $\langle \mathscr{U}_i^2(t) \rangle$ of the square of the displacement $\mathscr{U}_i^2(t)$ by integrating over all possible initial displacements and momenta and using that all oscillators i are independent to each other:

$$\left\langle \mathscr{U}_i^2(t) \right\rangle = \int_{-\infty}^{\infty} d\tilde{\mathscr{U}}_1^{(0)} \dots \int_{-\infty}^{\infty} d\tilde{\mathscr{U}}_{3N_{\text{at}}}^{(0)} \int_{-\infty}^{\infty} d\tilde{\mathscr{P}}_1^{(0)} \dots \int_{-\infty}^{\infty} d\tilde{\mathscr{P}}_{3N_{\text{at}}}^{(0)} \times$$
$$\times\, \mathbb{p}_{\tilde{\mathscr{U}}_1^{(0)}}\left(\tilde{\mathscr{U}}_1^{(0)}\right) \cdots \mathbb{p}_{\tilde{\mathscr{U}}_{3N_{\text{at}}}^{(0)}}\left(\tilde{\mathscr{U}}_{3N_{\text{at}}}^{(0)}\right) \mathbb{p}_{\tilde{\mathscr{P}}_1^{(0)}}\left(\tilde{\mathscr{P}}_1^{(0)}\right) \cdots \mathbb{p}_{\tilde{\mathscr{P}}_{3N_{\text{at}}}^{(0)}}\left(\tilde{\mathscr{P}}_{3N_{\text{at}}}^{(0)}\right) \times$$
$$\times\, \mathscr{U}_i^2(t). \tag{4.68}$$

If we insert Eq. (4.67) for $\mathscr{U}_i^2(t)$ and use

$$\int_{-\infty}^{\infty} d\tilde{\mathscr{U}}_j^{(0)}\, \mathbb{p}_{\tilde{\mathscr{U}}_j^{(0)}}\left(\tilde{\mathscr{U}}_j^{(0)}\right) = \int_{-\infty}^{\infty} d\tilde{\mathscr{U}}_j^{(0)} \frac{\tilde{\omega}_j}{\sqrt{2\pi\, k_B\, T_i}}\, e^{-\frac{\tilde{\omega}_j^2 \left(\tilde{\mathscr{U}}_j^{(0)}\right)^2}{2 k_B T_i}} \overset{(4.19)}{=} 1,$$

$$\int_{-\infty}^{\infty} d\tilde{\mathscr{U}}_j^{(0)}\, \mathbb{p}_{\tilde{\mathscr{U}}_j^{(0)}}\left(\tilde{\mathscr{U}}_j^{(0)}\right) \tilde{\mathscr{U}}_j^{(0)} = \int_{-\infty}^{\infty} d\tilde{\mathscr{U}}_j^{(0)} \frac{\tilde{\omega}_j}{\sqrt{2\pi\, k_B\, T_i}}\, e^{-\frac{\tilde{\omega}_j^2 \left(\tilde{\mathscr{U}}_j^{(0)}\right)^2}{2 k_B T_i}}\, \tilde{\mathscr{U}}_j^{(0)} \overset{(4.20)}{=} 0,$$

$$\int_{-\infty}^{\infty} d\tilde{\mathscr{U}}_j^{(0)}\, \mathbb{p}_{\tilde{\mathscr{U}}_j^{(0)}}\left(\tilde{\mathscr{U}}_j^{(0)}\right) \left(\tilde{\mathscr{U}}_j^{(0)}\right)^2 = \int_{-\infty}^{\infty} d\tilde{\mathscr{U}}_j^{(0)} \frac{\tilde{\omega}_j}{\sqrt{2\pi\, k_B\, T_i}}\, e^{-\frac{\tilde{\omega}_j^2 \left(\tilde{\mathscr{U}}_j^{(0)}\right)^2}{2 k_B T_i}} \left(\tilde{\mathscr{U}}_j^{(0)}\right)^2 \overset{(4.22)}{=} \frac{k_B\, T_i}{\tilde{\omega}_j^2},$$

$$\int_{-\infty}^{\infty} d\tilde{\mathscr{P}}_j^{(0)}\, \mathbb{p}_{\tilde{\mathscr{P}}_j^{(0)}}\left(\tilde{\mathscr{P}}_j^{(0)}\right) = \int_{-\infty}^{\infty} d\tilde{\mathscr{P}}_j^{(0)} \frac{1}{\sqrt{2\pi\, k_B\, T_i}}\, e^{-\frac{\left(\tilde{\mathscr{P}}_j^{(0)}\right)^2}{2 k_B T_i}} \overset{(4.19)}{=} 1,$$

$$\int_{-\infty}^{\infty} d\tilde{\mathscr{P}}_j^{(0)}\, \mathbb{p}_{\tilde{\mathscr{P}}_j^{(0)}}\left(\tilde{\mathscr{P}}_j^{(0)}\right) \tilde{\mathscr{P}}_j^{(0)} = \int_{-\infty}^{\infty} d\tilde{\mathscr{P}}_j^{(0)} \frac{1}{\sqrt{2\pi\, k_B\, T_i}}\, e^{-\frac{\left(\tilde{\mathscr{P}}_j^{(0)}\right)^2}{2 k_B T_i}}\, \tilde{\mathscr{P}}_j^{(0)} \overset{(4.20)}{=} 0,$$

$$\int_{-\infty}^{\infty} d\tilde{\mathscr{P}}_j^{(0)}\, \mathbb{p}_{\tilde{\mathscr{P}}_j^{(0)}}\left(\tilde{\mathscr{P}}_j^{(0)}\right) \left(\tilde{\mathscr{P}}_j^{(0)}\right)^2 = \int_{-\infty}^{\infty} d\tilde{\mathscr{P}}_j^{(0)} \frac{1}{\sqrt{2\pi\, k_B\, T_i}}\, e^{-\frac{\left(\tilde{\mathscr{P}}_j^{(0)}\right)^2}{2 k_B T_i}} \left(\tilde{\mathscr{P}}_j^{(0)}\right)^2 \overset{(4.22)}{=} k_B\, T_i,$$

we obtain

$$\left\langle \mathscr{U}_i^2(t) \right\rangle = \cos^2(\omega_i\, t) \sum_{j=1}^{3\, N_{\text{at}}} A_{ij}^2\, \frac{k_B\, T_i}{\tilde{\omega}_j^2} + \frac{\sin^2(\omega_i\, t)}{\omega_i^2} \sum_{j=1}^{3\, N_{\text{at}}} A_{ij}^2\, k_B\, T_i. \tag{4.69}$$

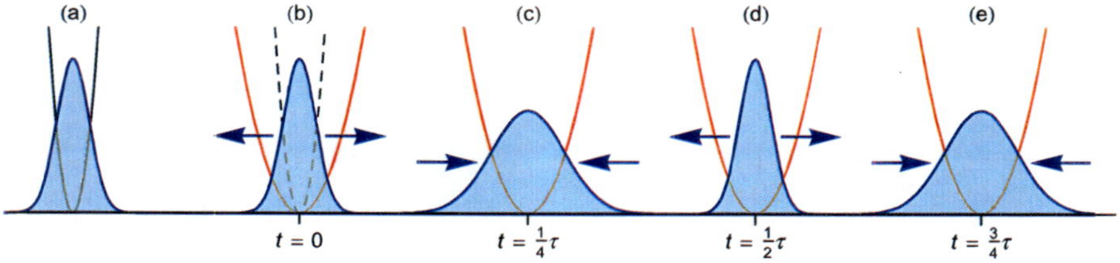

Fig. 4.3 Cartoon of antisqueezing is illustrated adapted from Fig. 2a–e of Ref. [5]. The blue curve indicates the distribution of the atomic displacements, the black curve indicates the PES before and the red curve after the laser interaction

Averaging over all phonon modes yields

$$\boxed{\langle\langle \mathcal{U}_i^2(t)\rangle\rangle_i = \frac{1}{3\,N_{\text{at}}}\sum_{i=1}^{3\,N_{\text{at}}}\left(\cos^2(\omega_i\,t)\sum_{j=1}^{3\,N_{\text{at}}}A_{ij}^2\,\frac{k_{\text{B}}\,T_{\text{i}}}{\tilde{\omega}_j^2}+\sin^2(\omega_i\,t)\sum_{j=1}^{3\,N_{\text{at}}}A_{ij}^2\,\frac{k_{\text{B}}\,T_{\text{i}}}{\omega_i^2}\right).} \tag{4.70}$$

This is the classical time-dependent variance[1] of the displacements of all phonon eigenmodes, which derivation we took from Ref. [5], but with more mathematical details.

A peak in the phonon density of states corresponds to an ensemble of nearly identical oscillators. We consider the distribution of the oscillator displacements of this ensemble. Now we assume, that the laser excitation makes the atomic bonds softer, so that the eigenfrequency of the oscillators decreases. This induces thermal antisqueezing, as illustrated in Fig. 4.3.

(a) Before the laser excitation, the displacement distribution corresponds to a Gaußian function with a standard deviation according to the atomic temperature $T_{\text{i}} = 300$ K and the eigenfrequency $\tilde{\omega}$.
(b) At time $t = 0$, the femtosecond-laser excitation nearly instantaneously changes the phonon mode frequency to a lower value ω with new oscillation period $\tau = \frac{2\pi}{\omega}$, which corresponds to a softening of the potential. Now, the displacement distribution is too narrow for the new potential, so that its width starts to increase over time.
(c) At time $t = \frac{1}{4}\tau$, due to the $\cos^2(\omega_i\,t)$ and $\sin^2(\omega_i\,t)$ terms in Eq. (4.69), the width of the displacement distribution reaches a maximum value and starts to decrease over time.
(d) At time $t = \frac{1}{2}\tau$, the width of the displacement distribution reaches a minimal value again, which would be equal to the initial width at $t = 0$ for an unperturbed ideal harmonic potential. But the reached minimal width is bigger due to anharmonicities and phonon-phonon couplings. After reaching this minimal value, the width starts to increase again.

[1] The mean value of the displacement of a harmonic oscillator is zero.

(e) Over time, the displacement distribution shows further oscillations of the width at values higher than the initial one.

The momentum distribution of the oscillators will show the opposite time behavior, i.e., the distribution will first decrease and then increase as a function of time. If the laser-excitation increases the eigenfrequency of the oscillators, a thermal squeezing will occur.

We can relate the average variance of the displacement of a phonon eigenmode to the average variance of the atomic displacements: If there is only one kind of atom with mass m in the simulation cell, we have for the vector of all atomic displacements $\mathbf{u} = \frac{1}{\sqrt{m}}\mathbf{U}$ and obtain

$$\mathbf{u}^{\mathrm{t}} \cdot \mathbf{u} = \frac{1}{m}\mathbf{U}^{\mathrm{t}} \cdot \mathbf{U} = \frac{1}{m}\left(\mathbf{C}^{\mathrm{t}} \cdot \mathscr{U}\right)^{\mathrm{t}} \cdot \mathbf{C}^{\mathrm{t}} \cdot \mathscr{U} = \frac{1}{m}\mathscr{U}^{\mathrm{t}} \cdot \mathbf{C} \cdot \mathbf{C}^{\mathrm{t}} \cdot \mathscr{U} = \frac{1}{m}\mathscr{U}^{\mathrm{t}} \cdot \mathscr{U}. \tag{4.71}$$

Due to this, we get for the variance of the atomic displacement on average

$$\begin{aligned}\left\langle \mathbf{u}_i^2(t)\right\rangle_i &= \frac{3}{m}\left\langle\left\langle \mathscr{U}_i^2(t)\right\rangle\right\rangle_i \\ &= \frac{1}{m\,N_{\mathrm{at}}}\sum_{i=1}^{3\,N_{\mathrm{at}}}\left(\cos^2(\omega_i\,t)\sum_{j=1}^{3\,N_{\mathrm{at}}} A_{ij}^2\,\frac{k_{\mathrm{B}}\,T_{\mathrm{i}}}{\tilde{\omega}_j^2} + \sin^2(\omega_i\,t)\sum_{j=1}^{3\,N_{\mathrm{at}}} A_{ij}^2\,\frac{k_{\mathrm{B}}\,T_{\mathrm{i}}}{\omega_i^2}\right),\end{aligned} \tag{4.72}$$

where $\mathbf{u}_i \in \mathbb{R}^3$ denotes the displacement of atom i from its equilibrium position. By definition, $\left\langle \mathbf{u}_i^2(t)\right\rangle_i$ is nothing else than the atomic mean-square displacements at time t:

$$\mathrm{MSD}(t) \overset{(4.49)}{=} \left\langle \mathbf{u}_i^2(t)\right\rangle_i . \tag{4.73}$$

Therefore, the squeezing or antisqueezing can be directly observed from the atomic mean-square displacements.

4.4 DFT Calculations and MD Simulations of Si at Various T_{e}'s

In the group of Prof. Dr. Garcia, the influence of heated electrons in Si was amply studied [5–13] by applying an increased T_{e} in electronic temperature-dependent density functional theory (DFT). Especially, many MD simulations of Si with increased T_{e} were performed. These calculations are all based on the CHIVES code (see Sect. 3.6).

4.4.1 Equilibrium Structure

Under normal conditions, Si crystallizes in the diamond-like structure, which can be constructed from a face centered cubic (fcc) structure that contains two atoms in the basis. One basis atom is located at fractional coordinates $[0, 0, 0]^t$ the other at $\left[\frac{a}{4}, \frac{a}{4}, \frac{a}{4}\right]^t$, where a denotes the lattice parameter. The fcc structure is generated by the three lattice vectors

$$\mathbf{a}_1 = \frac{a}{2}\begin{bmatrix}1\\1\\0\end{bmatrix}, \quad \mathbf{a}_2 = \frac{a}{2}\begin{bmatrix}1\\0\\1\end{bmatrix}, \quad \mathbf{a}_3 = \frac{a}{2}\begin{bmatrix}0\\1\\1\end{bmatrix}. \tag{4.74}$$

To generate the structure, we utilize the cubic conventional cell that is shown in Fig. 4.4a. This cell contains 8 atoms, that are drawn in green in Fig. 4.4a and which coordinates are listed in Table 4.1. The atoms that are attached to the conventional cell due to periodic boundary conditions are drawn in light green in Fig. 4.4a. By performing a lattice parameter optimization utilizing CHIVES at $T_e = 316$ K, we obtained the optimized value $a = 0.539872$ nm, which we used for all presented DFT calculations. This value corresponds to the equilibrium lattice parameter (see Fig. 4.5) within the DFT framework and differs slightly from the experimental value of $a = 0.54305$ nm [51]. In Table 4.2, we present the number of neighbors in the Si-diamond like structure as a function of the distance for the optimized lattice parameter.

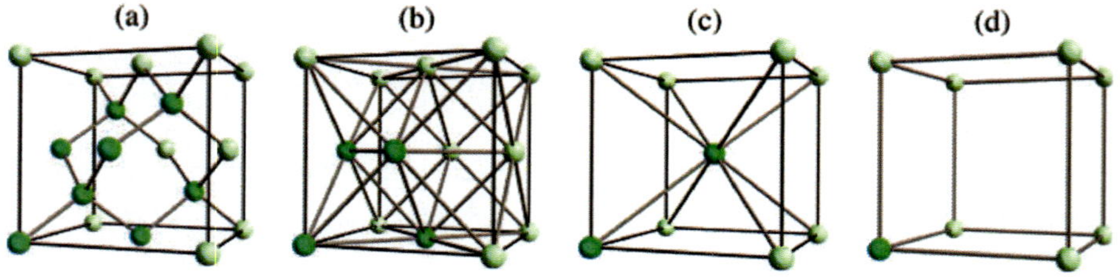

Fig. 4.4 Conventional cell of the diamond-like (**a**), fcc (**b**), bcc (**c**), and sc (**d**) structure is shown. The atoms belonging to the conventional cell are drawn in green, whereas atoms that are attached to the conventional cell due to periodic boundary conditions are drawn in light green. In addition, bonds between neighboring atoms are drawn

Table 4.1 Coordinates of the eight atoms contained in the Si conventional cell with length a

$\mathbf{r}_1$	$\mathbf{r}_2$	$\mathbf{r}_3$	$\mathbf{r}_4$	$\mathbf{r}_5$	$\mathbf{r}_6$	$\mathbf{r}_7$	$\mathbf{r}_8$
$\begin{bmatrix}0\\0\\0\end{bmatrix}$	$\begin{bmatrix}0\\\frac{a}{2}\\\frac{a}{2}\end{bmatrix}$	$\begin{bmatrix}\frac{a}{2}\\0\\\frac{a}{2}\end{bmatrix}$	$\begin{bmatrix}\frac{a}{2}\\\frac{a}{2}\\0\end{bmatrix}$	$\begin{bmatrix}\frac{a}{4}\\\frac{a}{4}\\\frac{a}{4}\end{bmatrix}$	$\begin{bmatrix}\frac{a}{4}\\\frac{3a}{4}\\\frac{3a}{4}\end{bmatrix}$	$\begin{bmatrix}\frac{3a}{4}\\\frac{a}{4}\\\frac{3a}{4}\end{bmatrix}$	$\begin{bmatrix}\frac{3a}{4}\\\frac{3a}{4}\\\frac{a}{4}\end{bmatrix}$

Table 4.2 Number of neighbors with respect to the distance in the ideal diamond-like structure of Si

Neighbor group	1	2	3	4	5	6	7	8	9	10
Dist. (nm)	0.2338	0.3817	0.4476	0.5399	0.5883	0.6612	0.7013	0.7635	0.7985	0.8536
Number at dist.	4	12	12	6	12	24	16	12	24	24
Number up to dist.	4	16	28	34	46	70	86	98	122	146
Neighbor group	11	12	13	14	15	16	17	18	19	20
Dist. (nm)	0.8850	0.9351	0.9639	1.0100	1.0367	1.0797	1.1048	1.1452	1.1689	1.2072
Number at dist.	12	8	24	48	36	6	12	36	28	24
Number up to dist.	158	166	190	238	274	280	292	328	356	380

4.4.2 *Cohesive Energies at Various T_e's*

Now we analyze the influence of an increased T_e on the bonding energy/cohesive energy of several Si bulk crystal structures.

Definition 4.4 (*Cohesive energy at electronic temperature T_e*) The cohesive energy is gained if initially isolated atoms bond to each other and form a structure. The cohesive energy $E_c(T_e)$ of a lattice structure at an electronic temperature T_e is defined as

$$E_c(T_e) = F(T_e) - N_{at}\, F_{at}(T_e), \tag{4.75}$$

where $F(T_e)$ denotes the Helmholtz Free energy of the structure containing N_{at} identical atoms and $F_{at}(T_e)$ is the Helmholtz Free energy of a single isolated atom. By definition, the cohesive energy $E_c(T_e)$ is negative.

We derived the Helmholtz Free energy $F_{at}(T_e)$ of an isolated Si atom for the 21 electronic temperatures 316 K (1 mHa[2]), 1578 K (5 mHa), 3158 K (10 mHa), ..., 31577 K (100 mHa) utilizing CHIVES. Then we calculated the Helmholtz Free energy $F(T_e)$ and the cohesive energy $E_c(T_e)$ of the diamond-like (dia), face centered cubic (fcc), body centered cubic (bcc), and simple cubic (sc) structure with the help of $F_{at}(T_e)$ for various lattice parameters a at the considered T_e's. In Fig. 4.4, we show the used conventional cells of these bulk crystal structures. In Fig. 4.5, we show the resulting cohesive energy curves as a function of the lattice parameter for these structures at $T_e = 316$ K and $T_e = 15789$ K.

Due to the increase of T_e, more anti-bonding orbitals at higher energy levels are occupied and bonding-orbitals at lower energy levels are emptied, so that the absolute value of the cohesive energy decreases with increasing T_e. At $T_e = 316$ K,

[2] 1 mHa = 0.001 Ha.

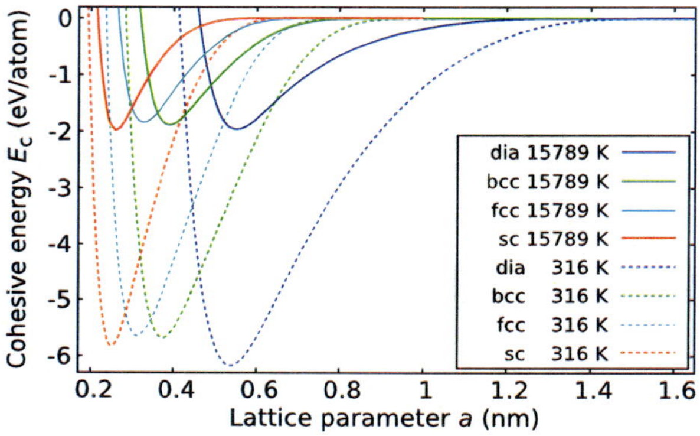

Fig. 4.5 DFT cohesive energies of the diamond-like, fcc, bcc and sc structures as a function of the lattice parameter are shown at $T_e = 316$ K and $T_e = 15789$ K

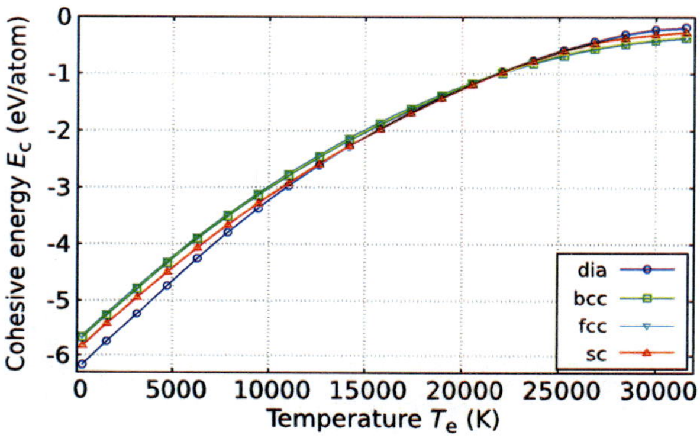

Fig. 4.6 Lattice parameter optimized DFT cohesive energies of the diamond-like, fcc, bcc, and sc structures as a function of T_e are shown

we obtained a cohesive energy $E_c = -6.17$ eV/atom of the diamond-like structure with the ideal lattice parameter. Our DFT value lies below the experimental value of (-4.62 ± 0.08) eV/atom [52] as expected, since LDA tends to overestimate the absolute value of the cohesive energy [53].

In Fig. 4.6, we present the DFT cohesive energies at the optimal lattice parameter for the four studied crystal structures as a function of T_e. One can clearly see that the diamond-like structure exhibits the biggest absolute value of the cohesive energy for $T_e < 15789$ K. But above $T_e = 15789$ K, the sc structure and than the bcc and fcc structures become more favorable. This behavior was also found by other T_e-dependent DFT calculations [54] and give us a first indication that the increase of T_e changes significantly the PES.

4.4.3 *Phonon Band Structure at Various T_e's*

In order to get a deeper insight into the increased T_e-induced changes in the PES, we studied the T_e-dependent phonon band structure of the diamond-like crystal structure.

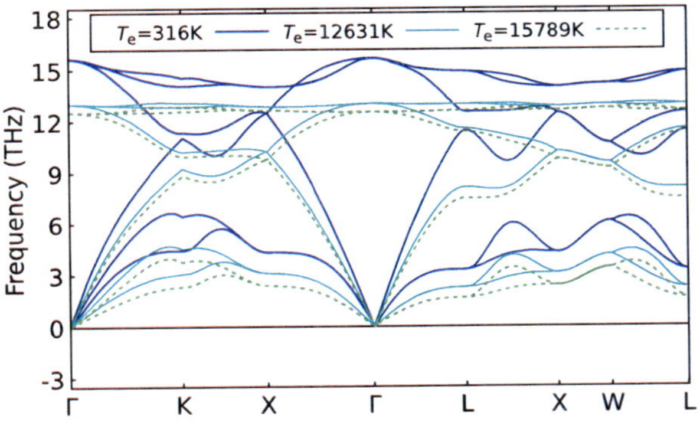

Fig. 4.7 DFT phonon bandstructure is shown at different T_e's

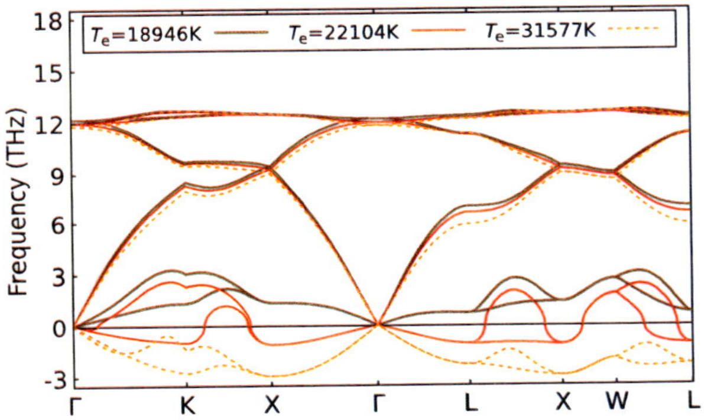

Fig. 4.8 DFT phonon bandstructure is shown at different T_e's

For this, we utilized a supercell that contains 512 Si atoms and consists of $4 \times 4 \times 4$ conventional cells. We applied periodic boundary conditions and studied various T_e's ranging from 316 K (1 mHa) up to 31577 K (100 mHa). We displaced one atom in the positive and in the negative x-direction and calculated the corresponding inter-atomic forces on all atoms. With the help of these forces and symmetry operations (see Theorem 4.4), we calculated the **q**-dependent dynamical matrix $\mathbf{D}(\mathbf{q})$ and obtained the **q**-dependent phonon frequencies by diagonalizing $\mathbf{D}(\mathbf{q})$ (see Sect. 2.2.1). In Figs. 4.7 and 4.8, we present the phonon bandstructure of the diamond-like Si structure at various T_e's. As usual in the literature, negative numbers are used to represent imaginary values of the phonon frequencies.

The frequencies of all phonon modes decrease with increasing electronic temperature T_e indicating a bond softening. The optical phonon frequencies decrease significantly, if T_e is increased from 316 K to 12631 K. Above $T_e = 12631$ K, the optical phonon frequency decrease very weakly with increasing T_e. In contrast, the acoustic phonon frequencies decrease uniformly with increasing T_e. Starting from $T_e = 19578$ K (63 mHa), the first branches of the acoustic phonon frequencies become negative and at $T_e = 31577$ K (100 mHa) all acoustic phonon branches are negative. This occurrence of negative phonon frequencies indicates possible crystal instabilities, which are induced by the increased T_e.

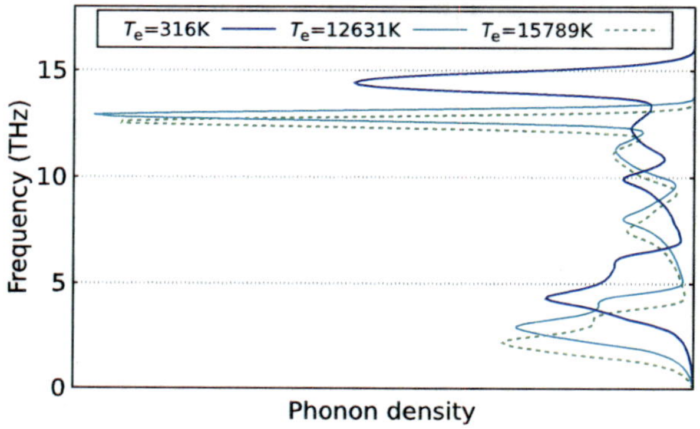

Fig. 4.9 DFT phonon density is shown at different T_e's

4.4.4 MD Simulations of Thermal Phonon Antisqueezing at Moderate T_e's

At a moderate increase of T_e, the phonon frequencies in Si decrease, but they still keep positive. The phonon density exhibits two main peaks: a localized one at optical phonon frequencies and a broader one at acoustic phonon frequencies. Due to the increase of T_e, the whole phonon density is shifted to smaller frequencies. This can be seen in Fig. 4.9, where the phonon density is shown for different T_e's. A peak in the phonon density corresponds to a large set of harmonic oscillators with approximately the same eigenfrequency ω. Due to the laser excitation, the eigenfrequency of these nearly degenerate oscillators is shifted to a smaller value. This indicates that thermal phonon antisqueezing may occur in femtosecond-laser excited Si.

Indeed, thermal phonon antisqueezing was observed in MD simulations with CHIVES performed by Zijlstra et al. [5]. Now a review of this study is presented.[3] For this, a supercell containing $N_{at} = 640$ Si atoms was used, which consists of $4 \times 4 \times 5$ conventional cells and is shown in Fig. 4.10. Periodic boundary conditions were applied and the atomic coordinates and velocities were initialized at an ionic temperature of $T_i = 316$ K (1 mHa). This initialization was performed within the harmonic approximation of the potential energy surface with the help of real random numbers (See Sect. 4.1.2). Using different sets of random numbers, 10 independent initializations were generated and used for the MD simulations to get always 10 independent runs. $T_e = 12631$ K (40 mHa) and $T_e = 15789$ K (50 mHa) were simulated for 1 ps using a time step of 2 fs.

To further analyze the antisqueezing, the atomic displacements from the equilibrium positions were calculated and denoted in the vector $\mathbf{u}(t) \in \mathbb{R}^{3N_{at}}$. To project onto the directions of selected phonon modes, the projection operators or matrices

[3] I took the atomic coordinates of these MD simulations and repeated the analysis of thermal phonon antisqueezing.

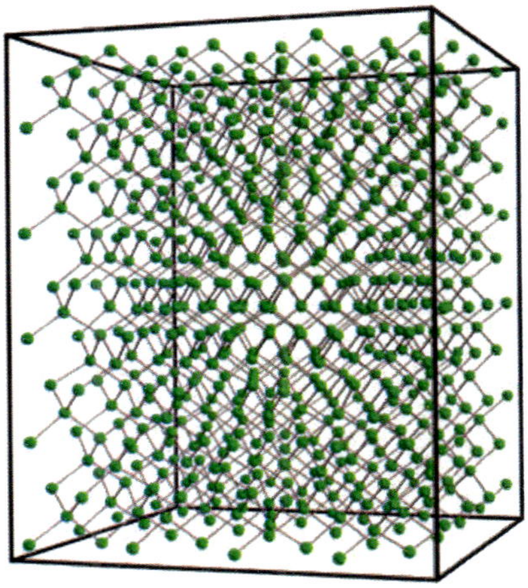

Fig. 4.10 Supercell containg 640 Si atoms

$$\mathbf{P}_{\mathcal{M}} = \sum_{i \in \mathcal{M}} \mathbf{e}^{(i)} \cdot \mathbf{e}^{(i)\mathrm{t}}, \qquad \mathbf{P}_{\overline{\mathcal{M}}} = \mathbb{1} - \mathbf{P}_{\mathcal{M}}. \tag{4.76}$$

are defined from the orthonormal eigenvectors $\{\mathbf{e}^{(i)}\}$ from Eq. (2.39) of the $3\,N_{\mathrm{at}} \times 3\,N_{\mathrm{at}}$ dynamical matrix $\mathbf{D}$. Here, $\mathbb{1}$ denotes the $3\,N_{\mathrm{at}} \times 3\,N_{\mathrm{at}}$ unit matrix, the indices of the selected phonon modes belong to the set $\mathcal{M}$ and the ones of the rest phonon modes to $\overline{\mathcal{M}}$, so that $\mathcal{M} \cup \overline{\mathcal{M}} = \{1, 2, \ldots, 3\,N_{\mathrm{at}}\}$. To observe the thermal phonon squeezing, the phonons were selected that correspond to the first peak in the phonon density (see Fig. 4.9), i.e.,

$$\mathcal{M}(T_e = 12631\,\mathrm{K}) = \{i : \nu_i \leq 5.1\,\mathrm{THz}\},$$
$$\mathcal{M}(T_e = 15789\,\mathrm{K}) = \{i : \nu_i \leq 4.7\,\mathrm{THz}\},$$

whereas ν_i denotes the frequency of the i-th phonon mode. Finally, $\mathbf{P}_{\mathcal{M}} \cdot \mathbf{u}$ defines the atomic displacements projected onto the directions of the selected phonon modes. Due to Eq. (4.49), the total atomic mean-square displacements is calculated from $\mathbf{u}$ by $\mathrm{MSD} = \frac{1}{N_{\mathrm{at}}} \mathbf{u}^{\mathrm{t}} \cdot \mathbf{u}$. The atomic mean-square displacements $\mathrm{MSD}_{\mathcal{M}}$ in direction of the selected phonon modes is calculated by inserting $\mathbf{P}_{\mathcal{M}} \cdot \mathbf{u}$ instead of $\mathbf{u}$:

$$\mathrm{MSD}_{\mathcal{M}}(t) = \frac{1}{N_{\mathrm{at}}} \sum_{i \in \mathcal{M}} \left(\mathbf{e}_i^{\mathrm{t}} \cdot \mathbf{u}(t)\right)^2. \tag{4.77}$$

Figure 4.11 shows the obtained atomic mean-square displacements in total and projected onto the selected and the rest phonon modes at $T_e = 12631$ K and $T_e = 15789$ K as a function of time t derived from the MD simulations with CHIVES. Indeed, the acoustic phonon modes corresponding to the first peak in the phonon density are mainly affected by the laser excitation and show the above mentioned thermal phonon antisqueezing. After the laser excitation or the increase of T_e, the

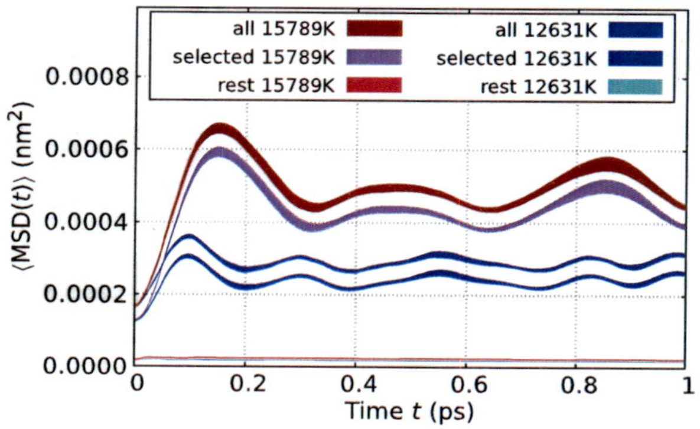

Fig. 4.11 DFT time-dependent atomic mean-square displacements in total and projected onto the selected and rest phonon modes averaged over 10 runs is shown at $T_e = 12631$ K and $T_e = 15789$ K adapted from Fig. 2g of Ref. [5]

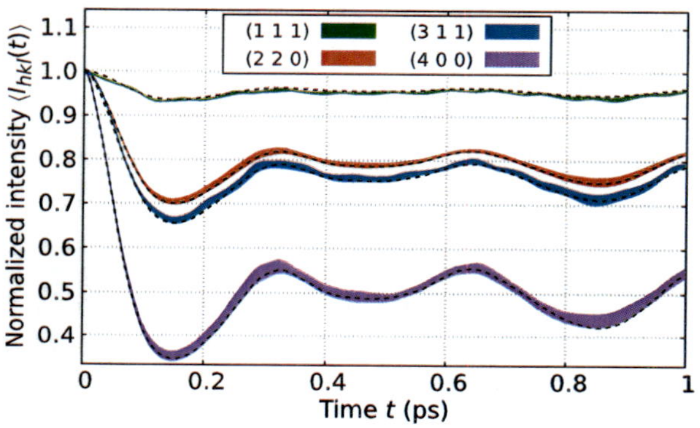

Fig. 4.12 DFT time-dependent normalized intensity $\langle I_{hkl}(t) \rangle$ of selected diffraction peaks averaged over 10 runs is shown at $T_e = 15789$ K. The dashed curves represent the normalized intensities derived from the total atomic mean-square displacements using time-dependent Debye-Waller theory

atomic mean-square displacements in direction of the selected phonon modes first increases and then oscillates at increased values as a function of time. In contrast, the atomic mean-square displacements in direction of the remain phonon modes nearly keeps unchanged after the increase of T_e.

At the studied T_e's of 12631 K and 15789 K, the crystal structure keeps intact in the MD simulations. This can be seen from the total atomic mean-square displacements which still stays below the Lindemann stability limit of 0.001225 nm^2 after the increase of T_e (see Fig. 4.11). For Si, the nearest-neighbor distance equals 0.234 nm (see Table 4.2), so that, following Lindemann, the critical atomic mean displacement yields 0.035 nm and, consequently, the critical atomic mean-square displacements yields 0.001225 nm^2.

Figure 4.12 shows the normalized intensity $\langle I_{hkl}(t) \rangle$ of selected diffraction peaks averaged over 10 runs as a function of time t derived from the MD simulations with CHIVES at $T_e = 15789$ K. Indeed, the diffraction peak intensities decrease moderately after the laser excitation and then oscillate at reduced values as a function of time, which can be fully explained using time-dependent Debye-Waller theory from the atomic mean-square displacements. In addition, from the time-dependent oscillation of the peak intensities, phonon squeezing was experimentally observed in the semimetal bismuth using femtosecond x-ray diffraction [55].

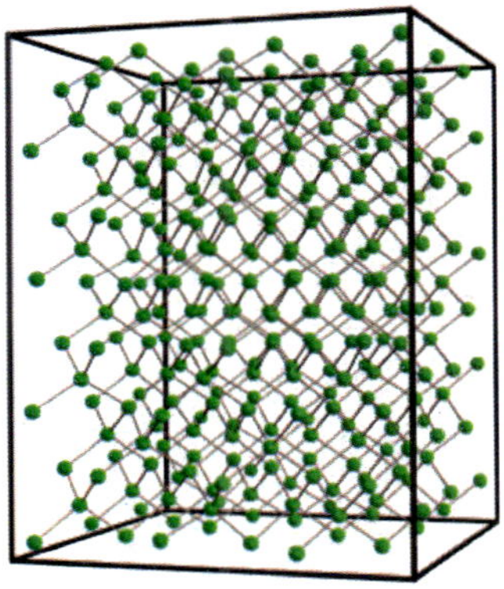

Fig. 4.13 Supercell containg 288 Si atoms

4.4.5 MD Simulations of Non-thermal Melting at High T_e's

The Si crystal structure stays intact in bulk MD simulations up to $T_e = 16736$ K (53 mHa) [8]. But, if T_e exceeds the critical value of 17052 K (54 mHa) [9], the T_e-induced deformation of the PES causes atomic forces stronger than the atomic bonding and leads to a destruction of the crystal structure within a few hundred femtoseconds.

This so called non-thermal melting of Si was observed in MD simulations with CHIVES performed by Zijlstra et al. [6, 8]. Now a review of this study is presented.[4] A bulk supercell containing 288 Si atoms was utilized, which consists of $3 \times 3 \times 5$ conventional cells and is shown in Fig. 4.13. Periodic boundary conditions were applied and the atoms were initialized at an ionic temperature of $T_i = 316$ K using the harmonic approximation of the PES similar to the bulk supercell containing 640 atoms. 40 independent initializations were generated to get always 40 independent MD simulation runs. The MD simulations were performed at the five T_e's 18946 K (60 mHa), 22104 K (70 mHa), 25262 K (80 mHa), 28420 K (90 mHa), 31577 K (100 mHa), and were stopped, when the atomic mean-square displacements reaches $0.2\,\text{nm}^2$. For this, again a time step of 2 fs was used.

With the help of the averaged atomic mean-square displacements $\langle \text{MSD}(t) \rangle$, the non-thermal melting of Si obtained from DFT can be divided into three stages as follows [6, 8].

1. **Super diffusive:**
Shortly after the laser excitation, the atoms move rapidly away from their equilibrium positions. The reason for this are strong atomic forces caused by the dramatical deformation of the PES due to the excited electrons. This movement is only hindered by a few atomic scattering events, since the atoms move in the existing voids of the diamond-like structure. (see Fig. 7.56). During this stage,

[4] I took the atomic coordinates of the MD simulations and repeated the analysis of non-thermal melting.

the average mean-square displacements and it's first derivative increase rapidly. Consequently, the time course of the averaged mean-square displacements is given by

$$\big\langle \mathrm{MSD}(t) \big\rangle = C\,(t - t_0)^{\alpha_s} \tag{4.78}$$

with constants C, t_0 and exponent $\alpha_s > 1$.

2. **Fractional diffusive:**
At this stage, the atoms start to scatter dominantly with each other after they have moved into the voids during the super diffusive stage. Due to the increased scattering rate, the increase of the mean-square displacements slows down, i.e., it's first derivative decreases with time. The time course of the average mean-square displacements is now given by

$$\big\langle \mathrm{MSD}(t) \big\rangle = C'\,(t - t_0')^{\alpha_f} \tag{4.79}$$

with constants C', t_0' and exponent $\alpha_f < 1$. Since the derivative of the average mean-square displacements increases during super diffusive and decreases during fractional diffusive motion, the transition time t_1 between both states is given by $\frac{d^2\langle \mathrm{MSD}(t_1)\rangle}{dt^2} = 0$.

3. **Normally diffusive:**
In the final stage, which is reached after the fractional diffusive atomic motion, the average mean-square displacements just increases linearly with time, i.e., the atomic motion is normally diffusive like in a liquid. Now the time course of the average mean-square displacements is given by

$$\big\langle \mathrm{MSD}(t) \big\rangle = 6\,D\,(t - t_0'') \tag{4.80}$$

with time constant t_0'' and diffusion coefficient D.

Figure 4.14 shows the atomic mean-square displacements averaged over 40 runs $\big\langle \mathrm{MSD}(t) \big\rangle$ at the five studied T_e's above the non-thermal melting threshold. In addition, the passing times between the three stages are marked. With increasing T_e, the passing times become smaller, since all processes occur faster. Figure 4.15 shows the super diffusion exponent α_s, the fractional diffusion exponent α_f, and the diffusion coefficient D at the five studied T_e's. With increasing T_e, the fractional diffusion exponent α_f decreases and the diffusion coefficient D increases. All values of D lie in the same range as the theoretical value $D = 0.02\,\mathrm{nm}^2/\mathrm{ps}$ of normal liquid Si at $T_e = T_i = 1800$ K, where the electrons and ions are in thermodynamic equilibrium [56].

The averaged atomic mean-square displacements exceeds the Lindemann stability limit in 100 to 200 fs after the increase of T_e, as it is shown in Fig. 4.14. The diffraction peak intensities decrease within a few hundred femtoseconds down to the background after the increase of T_e. This is shown for selected diffraction peak intensities in Fig. 4.16 at $T_e = 18946$ K. In addition, the diffraction peaks decay in slightly different times, which is a signature that this process is not thermal. For

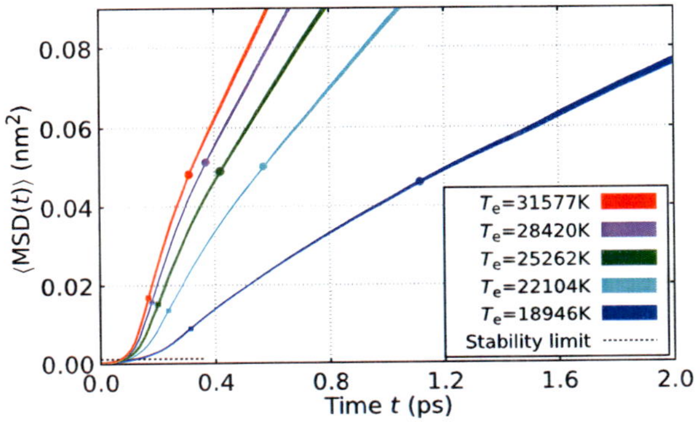

Fig. 4.14 DFT time-dependent atomic mean-square displacements $\langle \mathrm{MSD}(t) \rangle$ averaged over 40 runs are shown at the studied T_e's above the non-thermal melting threshold adapted from Fig. 3 of Ref. [8]. The dashed line indicates the Lindemann stability limit. The filled square indicates the transition time from super to fractional diffusive and the filled circle from fractional to normal diffusive

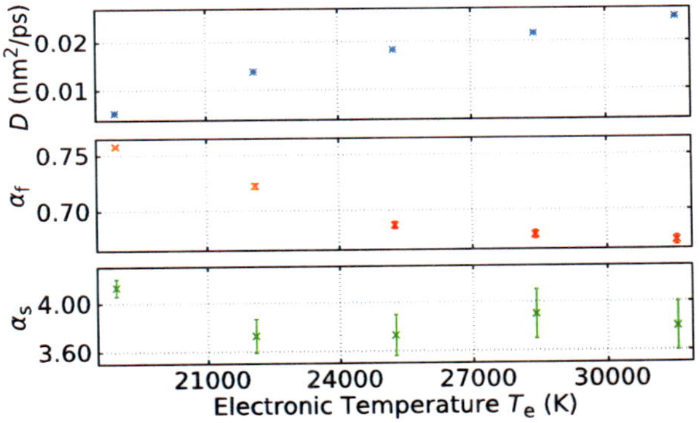

Fig. 4.15 Super diffusion exponent α_s, fractional diffusion exponent α_f, and diffusion coefficient D derived from the DFT MD simulations are presented at the studied T_e's. The error bars correspond to the fit error

melting under thermodynamic conditions, the melting starts at the surface and/or at crystal defects within the material due to the weaker bonding of the atoms at these locations. Then starting from these locations, a melting front proceeds through the material. (see Fig. 7.32). Consequently, the diffraction pattern would consist of a contribution from the solid part plus a contribution from the already molten part, so that the diffraction peaks decay concertedly [10]. Time-dependent Debye-Waller theory describes the Bragg peak decay only accurate for relative intensities above 0.4 (see Fig. 4.16). It is remarkable, that the phonon bandstructure does not show negative phonon frequencies at $T_e = 18946$ K (see Fig. 4.8) although Si melts non-thermally in MD simulations.

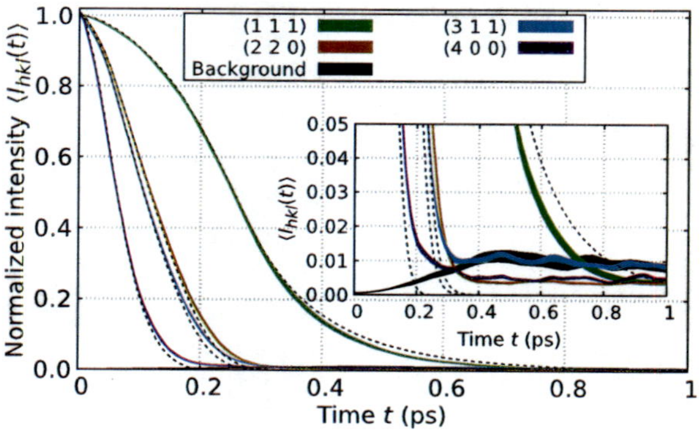

Fig. 4.16 DFT time-dependent normalized intensity $\langle I_{hkl}(t) \rangle$ of selected diffraction peaks and the background averaged over 40 runs is shown at $T_e = 18946$ K adapted from Fig. 2a of Ref. [10]. To study the background, we used the **q** point $\left(1\ 1\ \frac{5}{4}\right)$. The dashed curves represent the normalized intensities derived from the total atomic mean-square displacements using time-dependent Debye-Waller theory

Fig. 4.17 Supercell containg 432 Si atoms

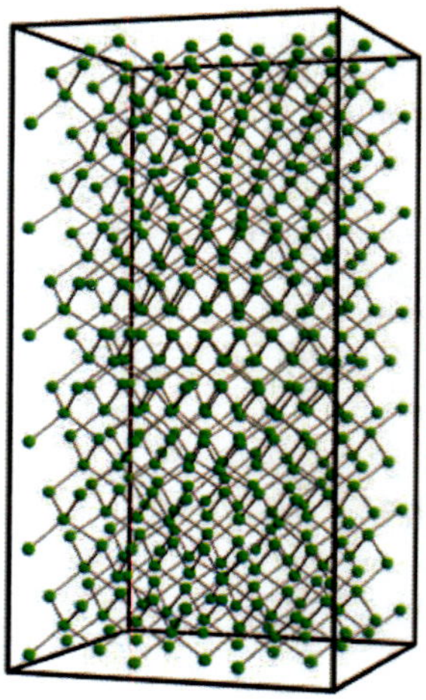

4.4.6 Behavior of the Electronic Indirect Band Gap

Si exhibits an indirect band gap between the Γ and the X points of the Brillouin zone, as one can see in Fig. 4.18, where the electronic bandstructure of Si derived from the DFT-code WIEN2k [57] is shown. In order to study the behavior of the electronic indirect band gap after an increase of T_e, we performed further MD simulations with CHIVES [13]. We utilized a bulk supercell containing 432 atoms and consisting of $3 \times 3 \times 6$ conventional cells. We show this supercell in Fig. 4.17. We applied periodic boundary conditions and initialized the atoms at an ionic temperature of $T_i = 316$ K (1 mHa) within the harmonic approximation similar to the previous cal-

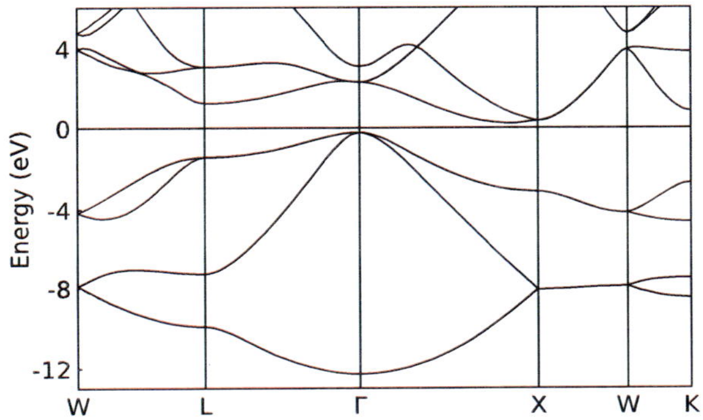

Fig. 4.18 Electronic band structure of Si at $T_e = 316$ K is shown. The Fermi energy is located at 0

culations. We generated 12 independent initializations to get always 12 independent MD simulation runs. We performed all calculations with a **k** point grid of $2 \times 2 \times 2$ that corresponds to 8 irreducible **k**-points in the first Brillouin zone in contrast to all other Si calculations with CHIVES presented in this section, which were done in the Γ-point approximation with a $1 \times 1 \times 1$ grid (see Sect. 3.6). We simulated the four T_e's, namely 12631 K (40 mHa), 18947 K (60 mHa), 25262 K (80 mHa), and 31578 K (100 mHa), for 0.25 ps using a time step of 2 fs.

At $T_e = 0$, the highest occupied Kohn-Sham energy corresponds to the highest energy of the valence band and the first unoccupied Kohn-Sham energy corresponds to the lowest energy of the conduction band. Thus, to determine the indirect electronic band gap E_{gap}, we sorted the Kohn-Sham energies ε_j ascending and calculate

$$E_{\text{gap}} = \varepsilon_{\frac{1}{2}N_e+1} - \varepsilon_{\frac{1}{2}N_e}. \tag{4.81}$$

The factor $\frac{1}{2}$ accounts that two electrons can fill a Kohn-Sham level due to the spin. N_e denotes the total number of electrons. $\frac{1}{2}N_e$ is indeed an integer number, since a Si atom has got an even number of electrons.

In Fig. 4.19, we show the behavior of the electronic indirect band gap $\langle E_{\text{gap}}(t)\rangle$ averaged over the 12 runs as a function of time t. At the beginning, the indirect band gap increases slightly with increasing T_e. Then, the band gap decreases monotonously to zero for T_e's above the non-thermal melting threshold. For $T_e = 12631$ K below the non-thermal melting threshold, it decreases slightly at first and then remains almost constant.

To analyze, whether the disorder of the crystal structure during melting is responsible for the disappearance of the band gap, we plot the averaged band gap $\langle E_{\text{gap}}\rangle$ as a function of the atomic mean-square displacements ⟨MSD⟩ in Fig. 4.20. One can clearly see that the band gap decreases as a universal function of the atomic mean-square displacements approximately independently of T_e. Furthermore, this function is approximately linear for a wide range of atomic mean-square displacements. We conclude that the reduction and final disappearance of the band gap is triggered by the laser-driven atomic displacements.

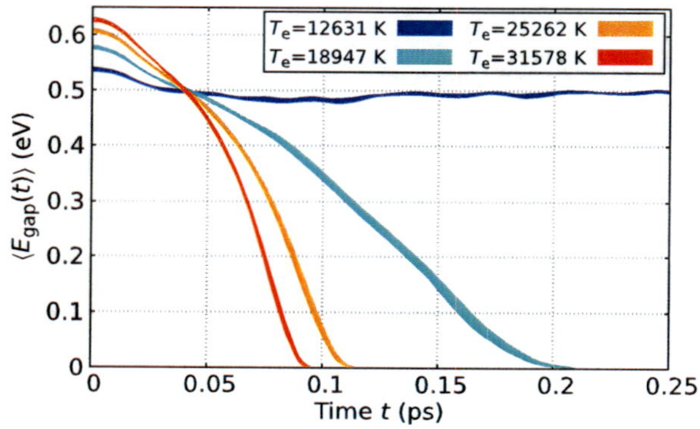

Fig. 4.19 Electronic band gap is shown as a function of time t averaged over 12 runs at various T_e's adapted from Fig. 4 of Ref. [13]

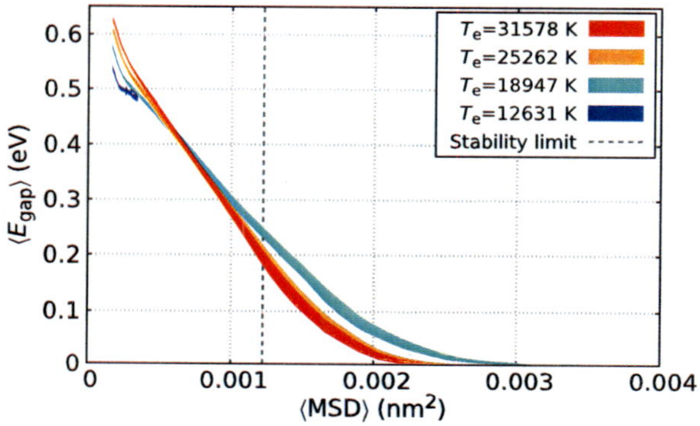

Fig. 4.20 Electronic band gap is shown as a function of atomic mean-square displacements averaged over 12 runs at various T_e's adapted from Fig. 5 of Ref. [13]

The band gap is very sensitive to atomic displacements from the equilibrium crystal structure, since it disappears already at an atomic mean-square displacements of $0.0030\,\text{nm}^2$, where the crystal structure is still recognizable. This can be clearly seen in Fig. 4.21, where the Si supercell is shown for an atomic mean-square displacements of $0.0030\,\text{nm}^2$ and $0.0002\,\text{nm}^2$. The latter corresponds to the atomic structure before the excitation. We also present the Si supercell for an atomic mean-square displacements of $0.0200\,\text{nm}^2$ where the crystal structure is clearly molten. Indeed, the increase of the atomic mean-square displacements from $0.0002\,\text{nm}^2$ to $0.0030\,\text{nm}^2$ has got a significant influence on the electronic density of states. We can clearly see this in Fig. 4.22, where the electronic density of states is shown for both mean-square displacements at $T_e = 18947$ K.

In experiments, direct access to the atomic mean-square displacements is not possible, but one can measure the time-resolved relative intensity of various Bragg peaks [45, 46]. Therefore, we plot the indirect band gap $\langle E_{gap}\rangle$ as a function of the relative Bragg peak intensity $\langle I_{hkl}\rangle$ averaged over 12 runs in Fig. 4.23 for the four Bragg peaks (111), (220), (311), and (400). We found again a strong correlation of the band gap with the relative Bragg peak intensity almost independent of T_e. This dependency is again approximately linear for a wide range of relative Bragg peak intensities. Due to time-dependent Debye-Waller theory (4.61), Bragg peaks related

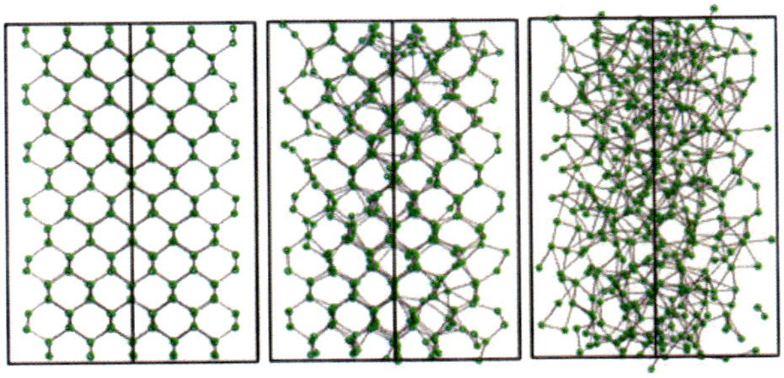

Fig. 4.21 432-atom Si supercell projected in the [110]-direction is shown for an atomic mean-square displacements MSD of 0.0002 nm^2 (left), 0.0030 nm^2 (middle), and 0.0200 nm^2 (right) adapted from Fig. 6 of Ref. [13]

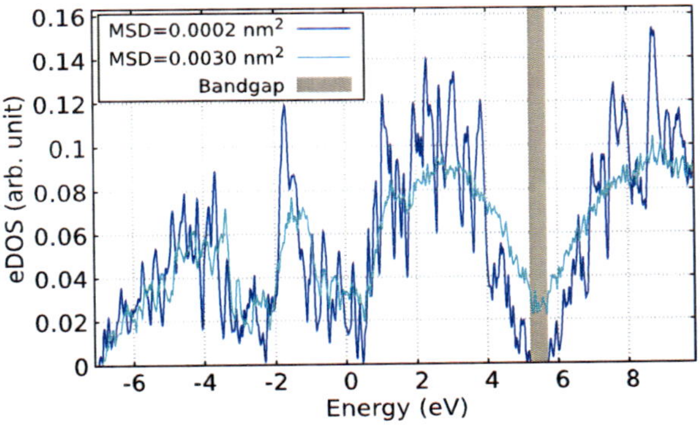

Fig. 4.22 Electronic density of states eDOS is shown at $T_e = 18947$ K for an atomic mean-square displacements MSD of 0.0002 nm^2 and 0.0030 nm^2 adapted from Fig. 7 of Ref. [13]. The band gap at MSD= 0.0002 nm^2 is highlighted in gray

to a smaller scattering vector $\mathbf{q}_{hkl}$ decay slower with increasing atomic mean-square displacements. Therefore, the indirect band gap decays faster to zero as a function of the relative peak intensity for Bragg peaks with smaller scattering vectors compared to ones with bigger scattering vectors.

We derived an indirect band gap of 0.48 eV at 316 K, in contrast to the experimental value, which yields 1.12 eV [17]. This is not surprising, since our simulations utilize DFT in the LDA approximation and LDA is known to generate too small band gaps in semiconductors compared to experiments [58]. As one can see in Fig. 4.19 at $t = 0$, the band gap of Si increases with increasing T_e for the diamond-like structure with small atomic displacements. This is also a known artifact of LDA [59], since the build-up of screening of the excited carries is not accurately described by LDA in semiconductors. In contrast, metals are well described by LDA, since LDA is based on the free electron gas, which is a good approximation of the electrons in metals. Since Si becomes metallic during melting, we strongly believe that the behavior of the band gap due to the increased T_e is correctly described by LDA. In addition, experiments also indicate a reduction of the band gap in Si with increasing temperature [14–19], as described in the beginning of the chapter.

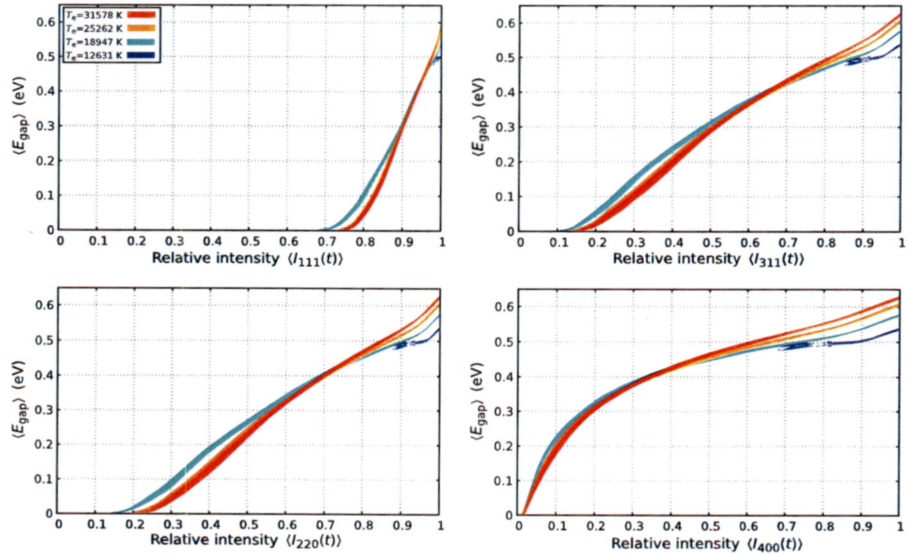

Fig. 4.23 Electronic band gap is shown as a function of the relative Bragg peak intensity averaged over 12 runs at various T_e's adapted from Fig. 8 of Ref. [13]

4.4.7 MD Simulations of a Thin-Film at Various T_e's

To analyze the influence of a surface, additional MD simulations with CHIVES were performed by Zier et al. [12]. Now a review of this study is presented.[5] A supercell containing 320 Si atoms in thin-film geometry was used. This supercell corresponds to $2 \times 2 \times 10$ conventional cells. Similar to the previous mentioned calculations, periodic boundary conditions were applied, but the simulation volume was doubled in z-direction ([$\bar{1}$11] direction of the crystal structure) to finally get a film of 5.3 nm thickness (see Fig. 4.24). To initialize the atomic positions and velocities, at first the equilibrium positions were calculated by applying the FIRE algorithm [60] for relaxation. In the equilibrium structure, there are some deformations from the ideal diamond-like structure at the surface, as it can be seen in Fig. 4.24 or Fig. 4.27 at $t = 0$ ps. Then, starting from the equilibrium positions, the atoms were initialized at an ionic temperature of $T_i = 316$ K (1 mHa) by applying the Andersen thermostat (see Sect. 4.1.2). But, in contrast to the previous mentioned bulk supercells, only one initialization was generated and used for the MD simulations. Furthermore, the twelve T_e's, namely 316 K (1 mHa), 3158 K (10 mHa), 6315 K (20 mHa), 9473 K (30 mHa), 12631 K (40 mHa), 15789 K (50 mHa), 17052 K (54 mHa), 18946 K (60 mHa), 22104 K (70 mHa), 25262 K (80 mHa), 28420 K (90 mHa), and 31577 K (100 mHa) were simulated using a time step of 2 fs for at least 1 ps, selected T_e's even for 10 ps. We want to note that the influence of the electron-phonon coupling may become

[5] I took the atomic coordinates of the MD simulations and repeated the analysis.

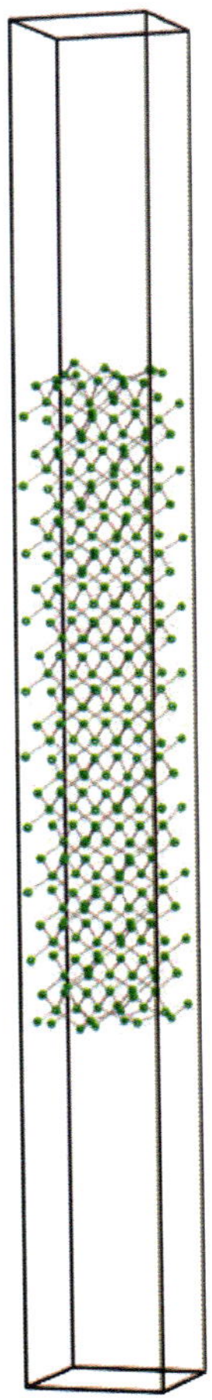

Fig. 4.24 Supercell with 320 Si atoms

relevant on the timescale of 10 ps with the consequence to significantly decrease T_e and heating up the lattice. However, here we are only interested in the effects that are caused by the change of the PES due to the increased T_e. To analyze the film expansion in z-direction, the atomic root mean-square displacements in z-direction

$$\mathrm{RMSD}_z(t) = \sqrt{\frac{1}{N_{\mathrm{at}}} \sum_{i=1}^{N_{\mathrm{at}}} u_{iz}^2} \tag{4.82}$$

was calculated at each time step t for all studied T_e's. To analyze, whether the crystal structure keeps intact or melts, the in plane atomic root mean-square displacements

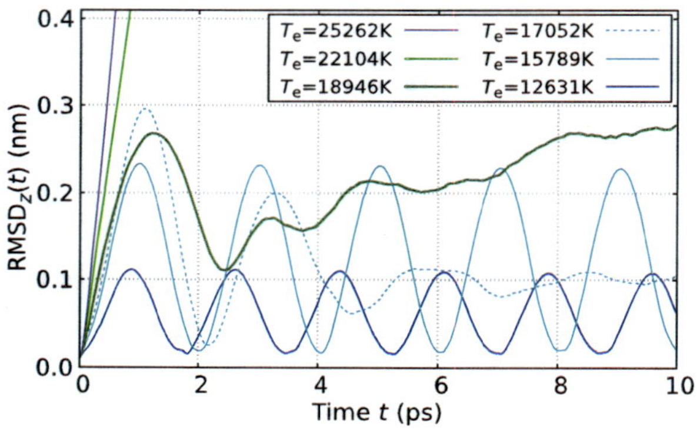

Fig. 4.25 DFT time-dependent atomic root mean-square displacements in z-direction $\mathrm{RMSD}_z(t)$ is shown for the thin film at various T_e's

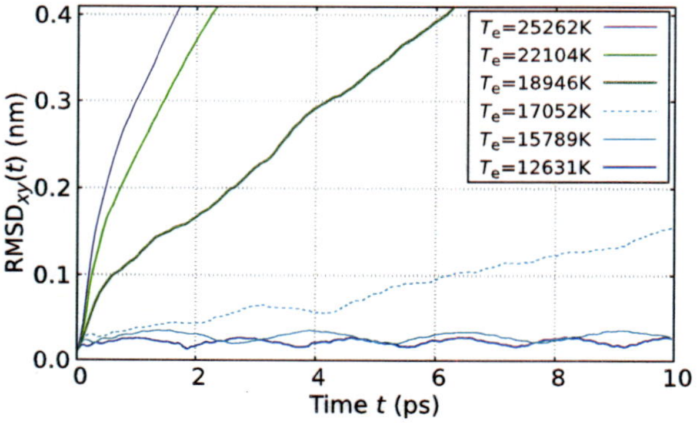

Fig. 4.26 DFT time-dependent in plane atomic root mean-square displacements $\mathrm{RMSD}_{xy}(t)$ is shown for the thin film at various T_e's

$$\mathrm{RMSD}_{xy}(t) = \sqrt{\frac{1}{N_{\mathrm{at}}} \sum_{i=1}^{N_{\mathrm{at}}} \left(u_{ix}^2 + u_{iy}^2 \right)} \tag{4.83}$$

was additional calculated. The resulting curves are presented in Figs. 4.25 and 4.26 at selected T_e's. At T_e's below the non-thermal melting threshold of $T_e = 17052$ K, the atomic root mean-square displacements in z-direction exhibits an harmonic oscillation and the one in xy-direction stays at small values during time. Here, the crystal structure keeps intact during the whole simulation time and the thin-film shows an harmonic expansion and shrinking in z-direction: A breathing mode perpendicular to the surface is induced [12]. This can be clearly seen in the averaged z-positions of the atomic planes with respect to the film center, which are shown in Fig. 4.28. The planes keep intact and the ones above the center move periodically up- and downwards and the ones below move periodically down- and upwards.

At the non-thermal melting threshold of $T_e = 17052$ K, the $\mathrm{RMSD}_z(t)$-value shows a damped oscillation as a function of time, which vanishes roughly at 5 ps, whereas the in plane atomic root mean-square displacements is increasing during time. Here, a part of the crystal starts to melt inside the film, as it can be seen in Fig. 4.27. This molten part surrounded by crystalline Si is formed 2.2 ps after the increase of T_e and than begins to grow during time. The reason for this is that

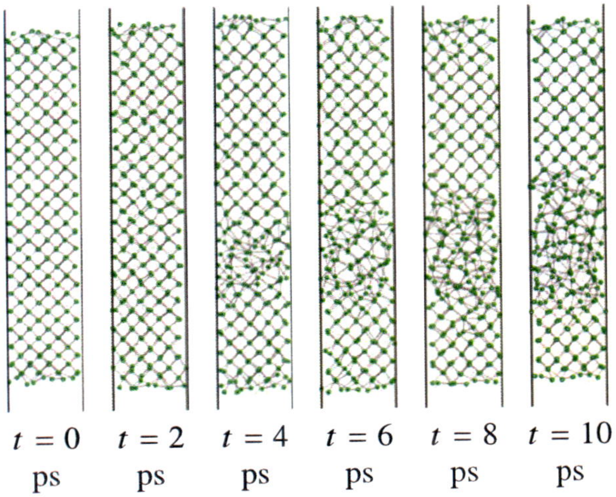

Fig. 4.27 Snapshots of the DFT MD simulation of the thin film containing 320 Si atoms at $T_e = 17052$ K projected in the [010]-direction. Bonds between neighboring atoms are drawn

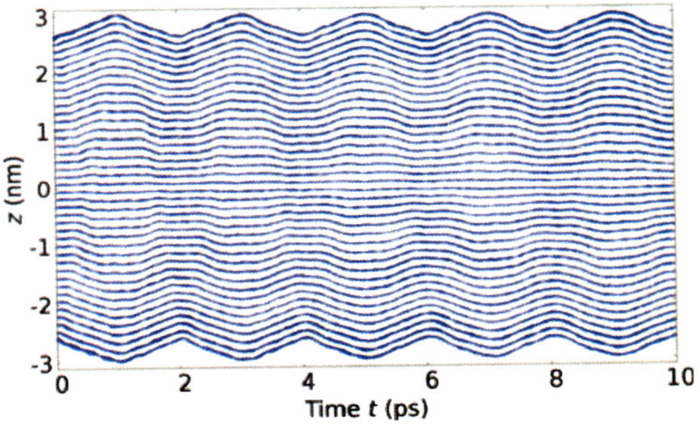

Fig. 4.28 Averaged z-position of the crystal planes with respect to the film center are shown as a function of time at $T_e = 15789$ K adapted from Fig. 3 of Ref. [12]. The line width corresponds to the standard deviations of the atomic z-coordinates in each plane

the expansion of the film firstly stabilized the crystal structure, since the pressure decreases in the film during expansion and the melting temperature T_m of Si increases with decreasing pressure. But, when the film shrinks again and reaches at 2.2 ps it's minimum thickness, the pressure increases inside the film and the crystal starts to melt in this high pressure region inside the film. The growing molten part inside the film slows down the further film expansions, since molten Si has got a smaller volume than crystalline Si.

At $T_e = 18946$ K, the entire film starts to melt from the very beginning and, after one expansion and shrinking, it starts to continuously expand with some fluctuations. This behavior can also be seen from the atomic root mean-square displacement in z-direction and from the in plane atomic root mean-square displacements. For higher T_e's, the entire film just melts and expands after the increase of T_e and ablation occurs. This can be seen in Fig. 4.29 at $T_e = 25262$ K as an example. Here, the atomic root mean-square displacements in z- and x, y-direction just increase as a function of time, as one can see in Figs. 4.25 and 4.26.

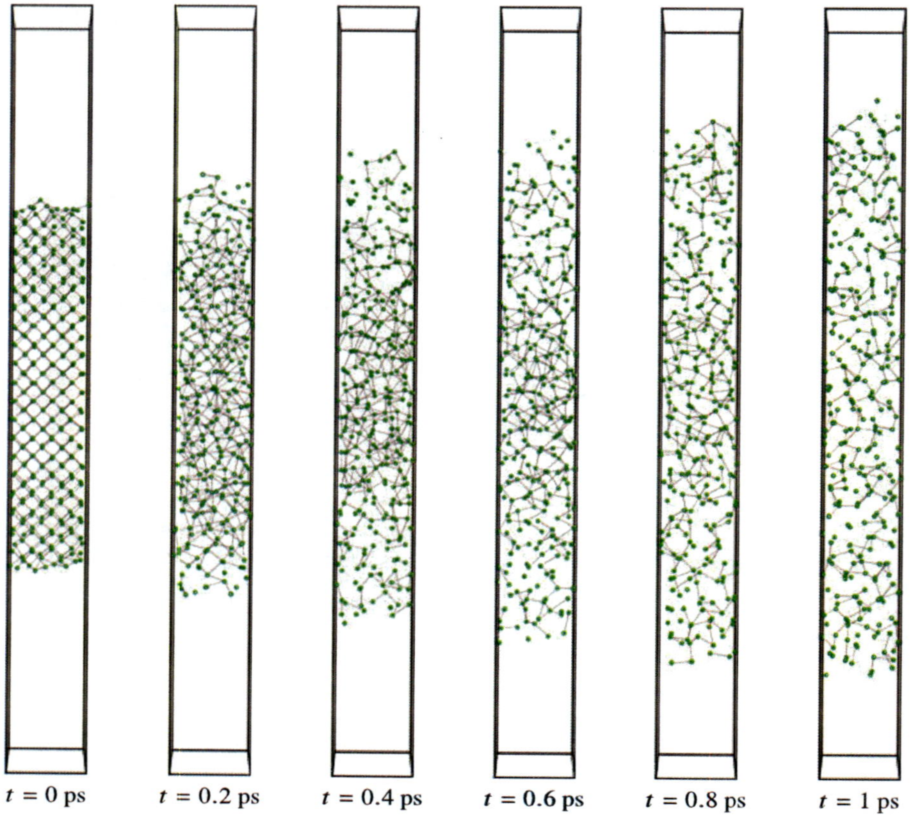

Fig. 4.29 Snapshots of the DFT MD simulation of the thin film containing 320 Si atoms at $T_e = 25262$ K. Bonds between neighboring atoms are drawn

4.4.8 *Summary of the Effects Induced by an Increased T_e*

In the preceding section, we studied the effects that are caused by the high-T_e-deformed PES in crystalline Si. Now we aim to summarize all revealed relevant effects in a single Fig. 4.30. For this, we consider representatively for T_e beside 316 K the two temperatures 12631 K and 15789 K below and the two temperatures 18946 K and 22104 K above the non-thermal melting threshold. We present four horizontal series of figures, each one describing one quantity of interest:

1. The fist row shows the atomic root mean-square displacement in z-direction $\mathrm{RMSD}_z(t)$ of the thin-film as a function of time t. Below the non-thermal melting threshold, $\mathrm{RMSD}_z(t)$ oscillates harmonically indicating the breathing mode of the thin-film. At $T_e = 18946$ K, the oscillation of $\mathrm{RMSD}_z(t)$ is damped, since the thin-film melts below the surface. At $T_e = 22104$ K, $\mathrm{RMSD}_z(t)$ simply increases rapidly as a function of time, while the thin-film expands and melts.

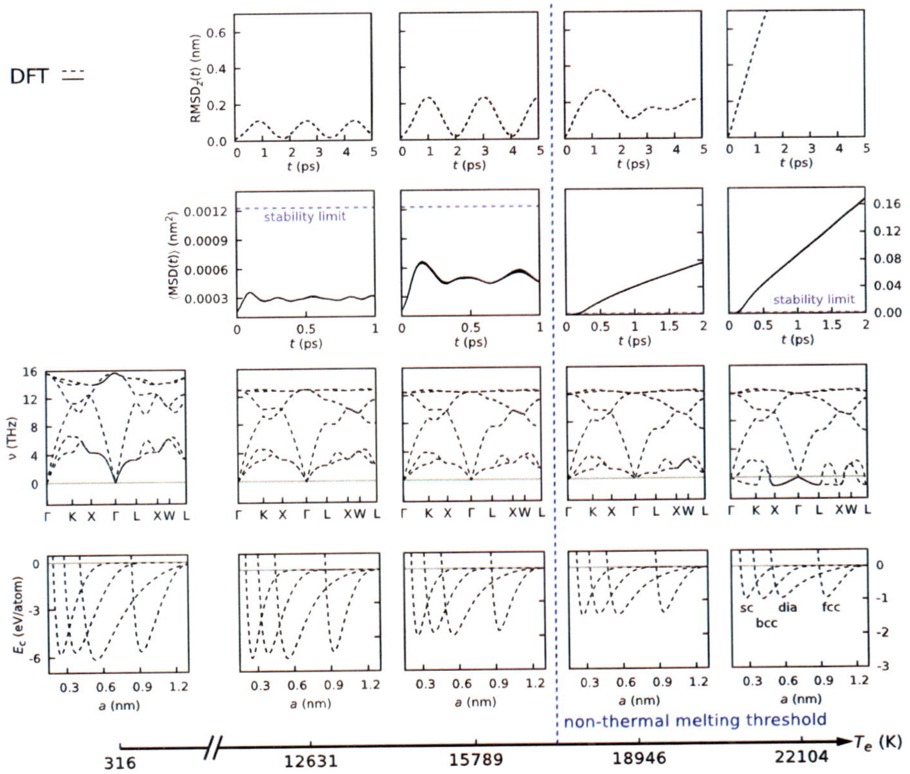

Fig. 4.30 Summary of the effects in Si induced by the increased T_e deformed PES. Description see text

2. The second row shows the average atomic mean-square displacement $\langle \mathrm{MSD}(t) \rangle$ as a function of time t for the bulk. Below the non-thermal melting threshold, $\langle \mathrm{MSD}(t) \rangle$ increases a bit after the increase of T_e and oscillates below the Lindemann stability limit indicating thermal phonon antisqueezing. Above the non-thermal melting threshold, $\langle \mathrm{MSD}(t) \rangle$ exceeds very fast the Lindemann stability limit and shows the three stages of non-thermal melting: First the time-derivative of $\langle \mathrm{MSD}(t) \rangle$ increases rapidly, decreases and then keeps constant, so that $\langle \mathrm{MSD}(t) \rangle$ increases linearly with time.
3. In the third row, the phonon bandstructure of the diamond-like structure is presented. The optical phonon branches decrease significantly from 316 K to 12631 K. Then, they decrease very weakly. The acoustic phonon branches decrease uniformly with increasing T_e. At 18946 K above the non-thermal melting threshold, all phonon branches are still positive, whereas at 22104 K, parts of the acoustic phonon branches become negative.
4. Lastly, in the forth or bottom series of figures the cohesive energies of the diamond-like, fcc, bcc and sc structures are presented at the considered T_e's.

To improve the visibility, we shifted the lattice parameter of the cohesive energy curve of the fcc structure by 0.6 nm. The increase of T_e reduces the absolute value of the cohesive energy curves and the diamond-like structure is not energetically preferred any more.

We skip the figure for $\text{RMSD}_z(t)$ and $\langle \text{MSD}(t) \rangle$ at $T_e = 316$ K, since both quantities show irrelevant fluctuations around the equilibrium value at this T_e.

4.5 DFT Calculations and MD Simulations of Sb at Various T_e's

In this section, we study extensively the influence of an increased T_e in Sb using T_e-dependent DFT with the help of our code CHIVES (see Sect. 3.6). Some of the presented results are already published [61–63].

4.5.1 Equilibrium Structure

Under normal conditions, Sb crystallizes in the A7 structure, which is stabilized by a so called Peierls distortion [64]. The atoms form hexagonal layers, which are placed on the top of each other (see Fig. 4.31). The distances between these planes alternate between short and long and the difference is characterized by the so-called Peierls parameter $\mathbf{z}$.

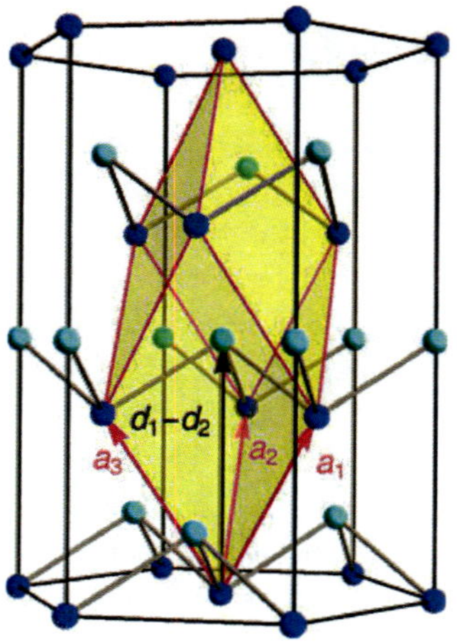

Fig. 4.31 The crystal structure of Sb is shown for $\mathbf{z} = 0.2336$. Atoms belonging to the basis $\mathbf{d}_1$ are drawn in blue, atoms belonging to the basis $\mathbf{d}_2$ are drawn in cyan, and bonds between neighboring atoms are drawn in gray. The cuboid spanned by the lattice vectors $\mathbf{a}_1$, $\mathbf{a}_2$, and $\mathbf{a}_3$ is also presented

The A7 structure is generated by the three lattice vectors

$$\mathbf{a}_1 = \begin{bmatrix} -\frac{1}{2}a \\ -\frac{\sqrt{3}}{6}a \\ \frac{1}{3}c \end{bmatrix}, \quad \mathbf{a}_2 = \begin{bmatrix} \frac{1}{2}a \\ -\frac{\sqrt{3}}{6}a \\ \frac{1}{3}c \end{bmatrix}, \quad \mathbf{a}_3 = \begin{bmatrix} 0 \\ \frac{\sqrt{3}}{3}a \\ \frac{1}{3}c \end{bmatrix}, \tag{4.84}$$

where a and c denote the two hexagonal lattice constants, which were experimentally found to be $a = 0.43007$ nm and $c = 1.12221$ nm at $T = 4.2$ K [64]. The corresponding reciprocal lattice vectors are

$$\mathbf{b}_1 = \begin{bmatrix} -\frac{2\pi}{a} \\ -\frac{2\pi\sqrt{3}}{3a} \\ \frac{2\pi}{c} \end{bmatrix}, \quad \mathbf{b}_2 = \begin{bmatrix} \frac{2\pi}{a} \\ -\frac{2\pi\sqrt{3}}{3a} \\ \frac{2\pi}{c} \end{bmatrix}, \quad \mathbf{b}_3 = \begin{bmatrix} 0 \\ \frac{4\pi\sqrt{3}}{3a} \\ \frac{2\pi}{c} \end{bmatrix}. \tag{4.85}$$

There are two atoms in the basis with the related basis vectors

$$\mathbf{d}_1 = \begin{bmatrix} 0 \\ 0 \\ \mathbb{z}c \end{bmatrix}, \quad \mathbf{d}_2 = \begin{bmatrix} 0 \\ 0 \\ -\mathbb{z}c \end{bmatrix}. \tag{4.86}$$

At $T = 4.2$ K $\mathbb{z} = 0.2336$ was found experimentally [64]. In the so-called A_{1g} phonon, $\mathbb{z}$ is oscillating, which means a motion of the hexagonal planes against each other. One can see this by comparing the Figs. 4.31, 4.32 and 4.33. If the distance between all hexagonal planes is identical, the Peierls distortion is removed and $\mathbb{z} = 0.25$. This is the first step to the sc structure, since the A7 structure can be obtained by distorting a sc structure. The later will be finally reached, if the ratio $\frac{a}{c}$ of the hexagonal lattice constants become $\sqrt{6} \approx 2.449$, which is close to the value 2.609

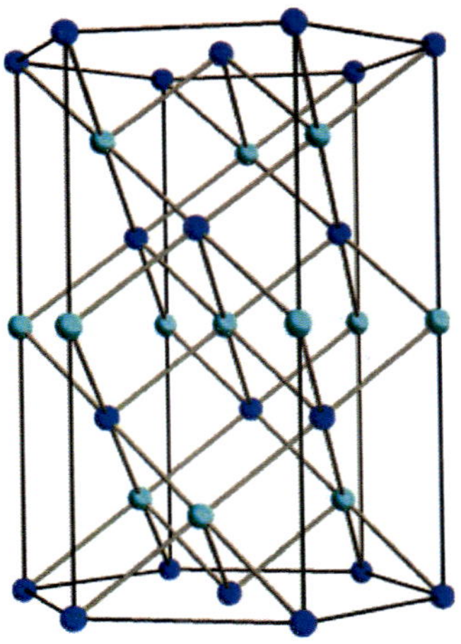

Fig. 4.32 The crystal structure of Sb is shown for $\mathbb{z} = 0.25$. Atoms belonging to the basis $\mathbf{d}_1$ are drawn in blue, atoms belonging to the basis $\mathbf{d}_2$ are drawn in cyan, and bonds between neighboring atoms are drawn in gray. Now the crystal structure corresponds to a sc structure that is distorted in direction of the body diagonal

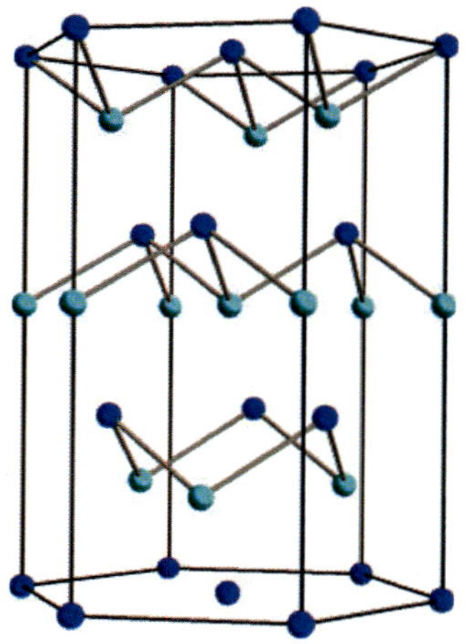

Fig. 4.33 The crystal structure of Sb is shown for $z = 0.2664$. Atoms belonging to the basis $\mathbf{d}_1$ are drawn in blue, atoms belonging to the basis $\mathbf{d}_2$ are drawn in cyan, and bonds between neighboring atoms are drawn in gray. The alternating distances are now exactly the other way around as with $z = 0.2336$

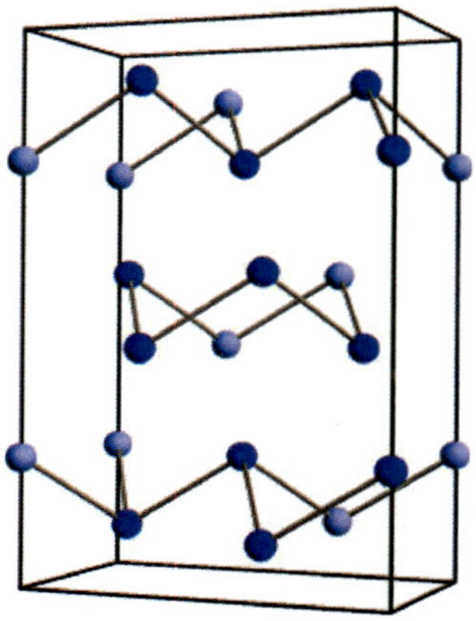

Fig. 4.34 Used minimal cell containing 12 Sb atoms

of Sb at $T = 4.2$ K [64]. For $z > 0.25$ the opposite alternating distances between the hexagonal planes occur compared with $z < 0.25$.

For our calculations, we utilized an orthorhombic minimal cell that is spanned by the three vectors

$$\mathbf{a}_2 - \mathbf{a}_1 = \begin{bmatrix} a \\ 0 \\ 0 \end{bmatrix}, \quad 2\,\mathbf{a}_3 - \mathbf{a}_1 - \mathbf{a}_2 = \begin{bmatrix} 0 \\ \sqrt{3}\,a \\ 0 \end{bmatrix}, \quad \mathbf{a}_1 + \mathbf{a}_1 + \mathbf{a}_2 = \begin{bmatrix} 0 \\ 0 \\ c \end{bmatrix} \tag{4.87}$$

and contains 12 atoms. This cell is shown in Fig. 4.34. The atoms belonging to the minimal cell, which are listed in Table 4.3, are drawn in blue, whereas atoms that are attached due to periodic boundary conditions are drawn in light blue.

To construct the simulation cell, the minimal cell is repeated N_x times in x-direction, N_y times in y-direction and N_z times in z-direction. The atoms contained

Table 4.3 Coordinates of the 12 atoms contained in the Sb minimal cell with lengths a, $\sqrt{3}\,a$, and c

$\mathbf{r}_1$	$\mathbf{r}_2$	$\mathbf{r}_3$	$\mathbf{r}_4$	$\mathbf{r}_5$	$\mathbf{r}_6$
$\begin{bmatrix} 0 \\ 0 \\ \mathbb{z}\,c \end{bmatrix}$	$\begin{bmatrix} 0 \\ 0 \\ (1-\mathbb{z})\,c \end{bmatrix}$	$\begin{bmatrix} \frac{1}{2}\,a \\ \frac{\sqrt{3}}{2}\,a \\ \mathbb{z}\,c \end{bmatrix}$	$\begin{bmatrix} \frac{1}{2}\,a \\ \frac{\sqrt{3}}{2}\,a \\ (1-\mathbb{z})\,c \end{bmatrix}$	$\begin{bmatrix} 0 \\ \frac{\sqrt{3}}{3}\,a \\ \left(\frac{1}{3}+\mathbb{z}\right)c \end{bmatrix}$	$\begin{bmatrix} 0 \\ \frac{\sqrt{3}}{3}\,a \\ \left(\frac{1}{3}-\mathbb{z}\right)c \end{bmatrix}$
$\mathbf{r}_7$	$\mathbf{r}_8$	$\mathbf{r}_9$	$\mathbf{r}_{10}$	$\mathbf{r}_{11}$	$\mathbf{r}_{12}$
$\begin{bmatrix} \frac{1}{2}\,a \\ \frac{5\sqrt{3}}{6}\,a \\ \left(\frac{1}{3}+\mathbb{z}\right)c \end{bmatrix}$	$\begin{bmatrix} \frac{1}{2}\,a \\ \frac{5\sqrt{3}}{6}\,a \\ \left(\frac{1}{3}-\mathbb{z}\right)c \end{bmatrix}$	$\begin{bmatrix} 0 \\ \frac{2\sqrt{3}}{3}\,a \\ \left(\frac{2}{3}+\mathbb{z}\right)c \end{bmatrix}$	$\begin{bmatrix} 0 \\ \frac{2\sqrt{3}}{3}\,a \\ \left(\frac{2}{3}-\mathbb{z}\right)c \end{bmatrix}$	$\begin{bmatrix} \frac{1}{2}\,a \\ \frac{\sqrt{3}}{6}\,a \\ \left(\frac{2}{3}+\mathbb{z}\right)c \end{bmatrix}$	$\begin{bmatrix} \frac{1}{2}\,a \\ \frac{\sqrt{3}}{6}\,a \\ \left(\frac{2}{3}-\mathbb{z}\right)c \end{bmatrix}$

in one minimal cell belong to six hexagonal planes, which have got in the simulation cell the following z-coordinates $z_1(i_z) < z_2(i_z) < z_3(i_z) < z_4(i_z) < z_5(i_z) < z_6(i_z)$:

$$z_1(i_z) = \left(\frac{1}{3} - \mathbb{z} + i_z\right)c, \quad z_2(i_z) = (\mathbb{z} + i_z)\,c, \quad z_3(i_z) = \left(\frac{2}{3} - \mathbb{z} + i_z\right)c, \tag{4.88}$$

$$z_4(i_z) = \left(\frac{1}{3} + \mathbb{z} + i_z\right)c, \quad z_5(i_z) = (1 - \mathbb{z} + i_z)\,c, \quad z_6(i_z) = \left(\frac{2}{3} + \mathbb{z} + i_z\right)c, \tag{4.89}$$

with $i_z \in \{0, 1, \ldots, N_z - 1\}$. In the minimal cell, two atoms belong to one hexagonal plane, as one can see in Table 4.3. Therefore, one has to average the z-coordinates of the two related atoms in the simulation cell to obtain the z-coordinates of the hexagonal planes. The Peierls parameter describes the difference between the two occurring alternating distances between the hexagonal planes. Due to this, one needs the relative z-position of three neighboring hexagonal planes to derive one Peierls parameter $\mathbb{z}$. Therefore, one gets two Peierls parameter from the above six hexagonal plane z-coordinates:

$$\mathbb{z}_1(i_z) = \frac{2\,z_1(i_z) - z_2(i_z) - z_3(i_z)}{6\left(z_1(i_z) - z_3(i_z)\right)}, \tag{4.90}$$

$$\mathbb{z}_2(i_z) = \frac{z_4(i_z) + z_5(i_z) - 2\,z_6(i_z)}{6\left(z_4(i_z) - z_6(i_z)\right)}. \tag{4.91}$$

Either one can derive a z-dependent Peierls parameter for the simulation cell or one just averages the obtained values to obtain one averaged Peierls parameter. Utilizing the experimental values of a, c and $\mathbb{z}$ at normal conditions, one obtains for Sb in the ideal A7 structure the number of neighbors as a function of the distance that is listed in Table 4.4.

Table 4.4 Number of neighbors with respect to the distance in the ideal A7 structure of Sb

Neighbor group	1	2	3	4	5	6	7	8	9	10
Dist. (nm)	0.2907	0.3337	0.4301	0.4490	0.5191	0.5252	0.5443	0.5970	0.6217	0.6741
Number at dist.	3	6	12	18	21	22	25	26	32	38
Number up to dist.	3	9	21	39	60	82	107	133	165	203
Neighbor group	11	12	13	14	15	16	17	18	19	20
Dist. (nm)	0.6788	0.6937	0.7358	0.7449	0.7560	0.7883	0.8601	0.8980	0.9079	0.9114
Number at dist.	44	50	56	62	74	80	86	92	98	104
Number up to dist.	247	297	353	415	498	569	655	747	845	949

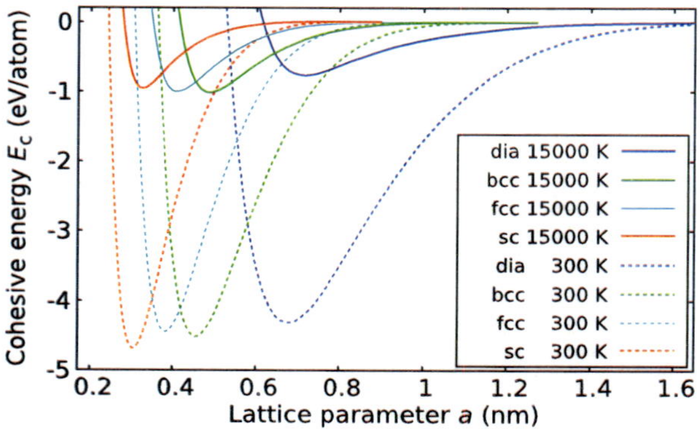

Fig. 4.35 DFT cohesive energies of the diamond-like, fcc, bcc and sc structures as a function of the lattice parameter are shown at $T_e = 300$ K and $T_e = 15000$ K

4.5.2 *Cohesive Energies at Various T_e's*

To get a first insight into the changes of the PES at increased T_e, we calculated the cohesive energies E_c of the fcc, bcc, sc and diamond-like crystal structures as a function of the lattice parameter a at the eleven T_e's 300 K, 3000 K, 6000 K, ..., 27000 K, and 30000 K. For the electrons, we used a $2 \times 2 \times 2$ grid of the **k** points containing 8 irreducible **k**-points in the first Brillouin zone (see Sect. 3.6) as for all Sb bulk calculations presented in this section. We present the resulting cohesive energy curves for $T_e = 300$ K and $T_e = 15000$ K, as an example, in Fig. 4.35.

Larger lattice parameters are preferred and the absolute values of the cohesive energies decrease at increasing T_e. In Fig. 4.36, we show the cohesive energy at the optimal lattice parameter as a function of T_e for the four considered structures. At $T_e < 15000$ K, the sc structure is the favorite one among the considered structures, whereas the bcc structure becomes more favorite at $T_e \geq 15000$ K. Under normal conditions, Sb crystallizes in the A7 structure which is a distortion of the sc structure. It is very reasonable that the sc structure is preferred among the considered structures at low T_e's.

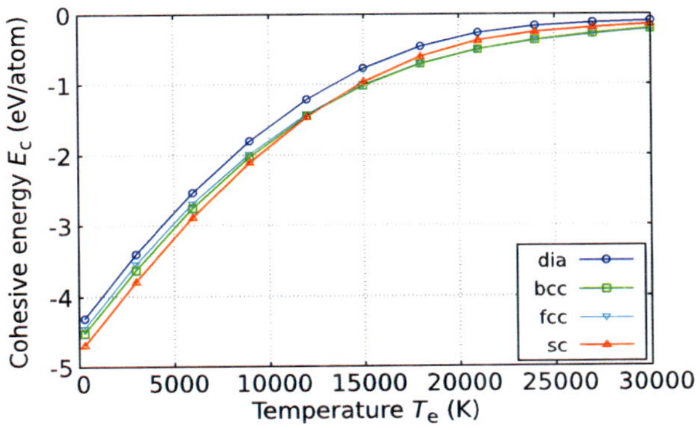

Fig. 4.36 DFT cohesive energies at the optimized lattice parameter of the diamond-like, fcc, bcc, and sc structures as a function of T_e are shown

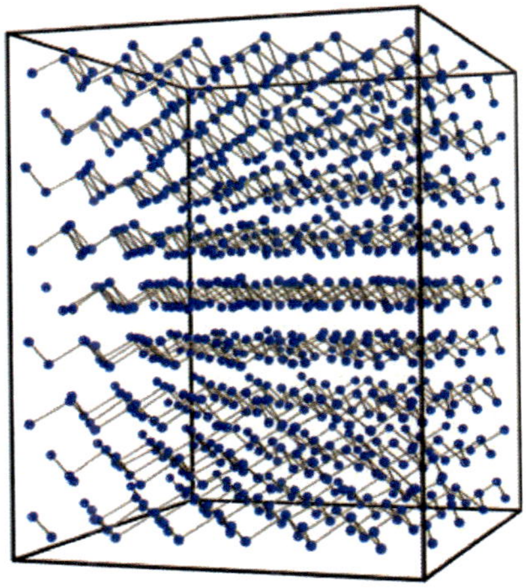

Fig. 4.37 Supercell containg 864 Sb atoms

4.5.3 *Potential Energy Surface and Displacive Excitation of the* A_{1g} *Phonon*

For a closer look at the A7 structure, we utilized a supercell that contains 864 Sb atoms and 18 hexagonal planes from $6 \times 4 \times 3$ minimal cells (Fig. 4.37). The value of the Peierls parameter $\mathbb{z}$ is not a consequence of crystal symmetries. Therefore, we varied $\mathbb{z}$ from 0.23 up to 0.25 and calculated the corresponding Helmholtz free energy F at various T_e's with CHIVES using the experimental lattice parameters $a = 0.43007$ nm and $c = 1.12221$ nm [64]. The resulting Helmholtz free energies $F_0(\mathbb{z}, T_e)$ of the ideal crystal structure as a function of the Peierls parameter $\mathbb{z}$ are presented in Fig. 4.38 at various T_e's.

To derive the energy ΔE, which is necessary to increase T_e from T_{e_1} to T_{e_2} at a given $\mathbb{z}$, we first calculated the internal energy

$$E(\mathbb{z}, T_e) = F_0(\mathbb{z}, T_e) + T_e\, S_e(\mathbb{z}, T_e) \tag{4.92}$$

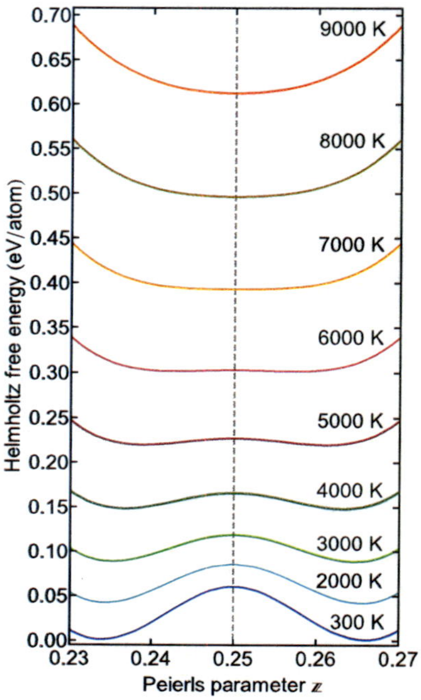

Fig. 4.38 DFT PES of the ideal A7 structure with the hexagonal lattice parameter ratio $\frac{c}{a} = 2.609$ is shown as a function of the Peierls parameter z at various T_e's adapted from Fig. 3 of Ref. [62]. The shift between the curves corresponds to the absorbed energy at $z = 0.25$ that is necessary to reach the next higher curve

with the help of the electronic entropy $S_e(z, T_e)$ obtained from DFT. Then we obtain ΔE just from

$$\Delta E = E(z, T_{e_2}) - E(z, T_{e_1}). \tag{4.93}$$

In Fig. 4.38, the shift between the curves correspond to the energy ΔE, which one needs to reach the next higher curve at $z = 0.25$.

On can clearly see that the increase of T_e dramatically changes the PES, which is described by the Helmholtz free energy $F_0(z, T_e)$. At $T_e = 300$ K, the PES exhibits the two minima at $z = 0.2340$ and at $z = 0.2660$, which are separated by a barrier with maximum at $z = 0.25$. The values $z = 0.2340$ and $z = 0.2660$ are equivalent, since $z = 0.2660$ describes just the opposite alternating distances compared to $z = 0.2340$. If T_e is increased, the height of the barrier becomes smaller and the two equivalent minima move closer to $z = 0.25$. At $T_e = 6600$ K, the barrier is vanished and the only minimum is located at $z = 0.25$. We present the T_e-dependent z-value of the minimum with $z \leq 0.25$ in the top panel of Fig. 4.40.

With the help of the PES, we can explain the laser-excitation of the A_{1g} phonon, which corresponds to an oscillation of the Peierls parameter z or the hexagonal planes against each other. Before the laser excitation at $T_e = 300$ K, the crystal structure stays at the minimum $z = 0.2340$ of the related PES (see Figs. 4.41 and 4.42). The femtosecond laser-pulse increases T_e to a higher value, so that the system is now localized on the PES of the higher T_e. The excitation to the higher T_e can be assumed so fast, that the crystal structure is still located at $z = 0.2340$ directly after

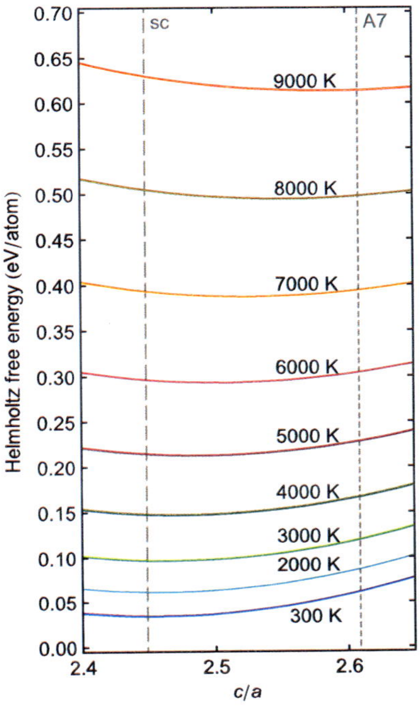

Fig. 4.39 DFT PES of the ideal crystal structure for $z = 0.25$ is shown as a function of the hexagonal lattice parameter ratio $\frac{c}{a}$ at various T_e's adapted from Fig. 4a of Ref. [62]. The shift between the curves corresponds to the absorbed energy at $\frac{c}{a} = 2.609$ that is necessary to reach the next higher curve

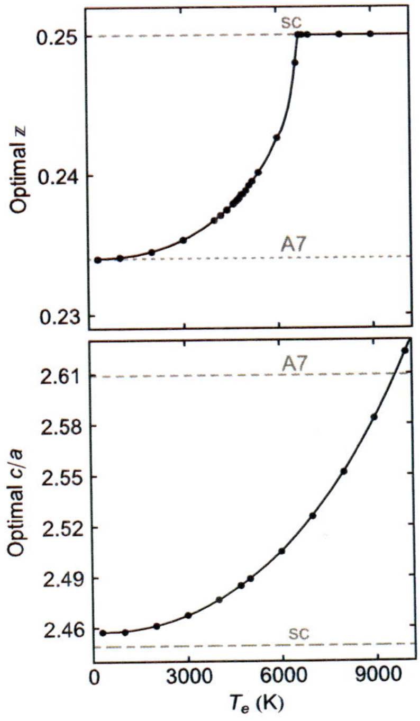

Fig. 4.40 Top: DFT optimal Peierls parameter z is shown as a function of T_e for the hexagonal lattice parameter ratio $\frac{c}{a} = 2.609$. Bottom: DFT optimal hexagonal lattice parameter ratio $\frac{c}{a}$ is shown as a function of T_e for the Peierls parameter $z = 0.25$ adapted from Fig. 4b of Ref. [62]. Solid lines are guides to the eye

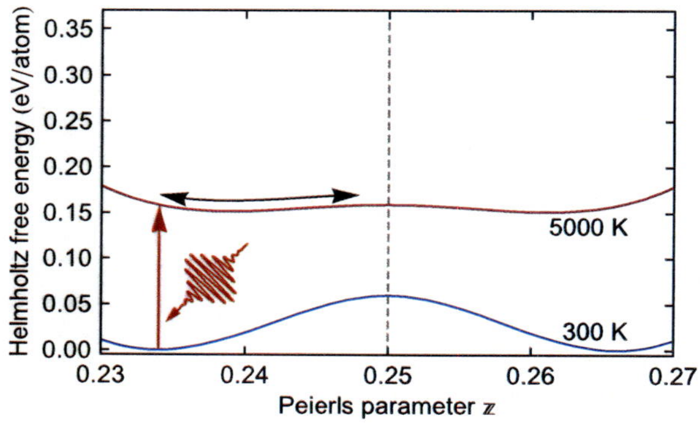

Fig. 4.41 DFT PES of the ideal A7 structure with $\frac{c}{a} = 2.609$ is shown as a function of z at $T_e = 300$ K and $T_e = 5000$ K. The shift between the two curves corresponds to the absorbed energy at $z = 0.2340$ to reach the curve at $T_e = 5000$ K. The excitation of the A_{1g} phonon is illustrated (see text)

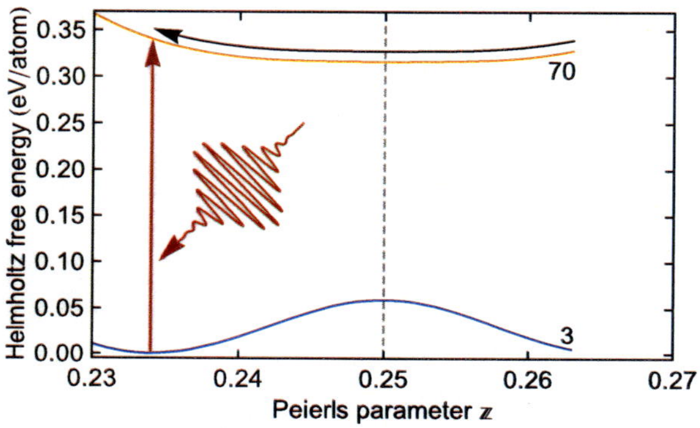

Fig. 4.42 DFT PES of the ideal A7 structure with $\frac{c}{a} = 2.609$ is shown as a function of z at $T_e = 300$ K and $T_e = 7000$ K. The shift between the two curves corresponds to the absorbed energy at $z = 0.2340$ to reach the curve at $T_e = 7000$ K. The excitation of the A_{1g} phonon is illustrated (see text)

the jump on the higher PES. In this state $z = 0.2340$ is not any more the minimum, so that this value corresponds to a displacement from the new equilibrium position. Therefore, the Peierls parameter z starts to oscillate around the new equilibrium position. For $T_e < 5100$ K, the amplitude of the oscillation does not exceed the barrier at $z = 0.25$, as it is illustrated in Fig. 4.41. For higher T_e, the oscillation overcomes the barrier, as it is illustrated in Fig. 4.42 as an example. The fundamentals of the excitation mechanism is the laser induced shift of the equilibrium position of the phonon coordinate in the PES. This mechanism is called **displacive excitation of a coherent phonon** and was first described in 1992 by Zeiger et al. [65].

Since we consider the ideal A7 structure without displacements or movements in directions of any other than the A_{1g} mode, this corresponds to an ionic temperature of $T_i = 0$ K and the influence of other phonon modes is neglected. If other phonon modes are present, the oscillation of the A_{1g} phonon mode will be damped, since energy is transferred from the A_{1g} phonon to the other phonon modes due to the phonon-phonon coupling. We will intensively study the influences of the other phonon modes in Sect. 4.5.5.

It is worth to analyze if an increase of T_e can induce an A7 to sc transition in Sb. For $T_e = 6600$ K, the only minimum of the PES is located at $z = 0.25$. Therefore, if the system is excited to this T_e, the Peierls parameter will oscillate around $z = 0.25$

and, due to the phonon-phonon coupling, the oscillation will be damped and, finally, the Peierls parameter will oscillate slightly around $z = 0.25$. At $z = 0.25$, Peierls distortion is removed and the first step towards the sc structure is performed. To reach finally the sc structure, the ratio $\frac{c}{a}$ of the hexagonal lattice constants a and c has to change from $\frac{c}{a} = 2.609$ to $\frac{c}{a} = \sqrt{6} \approx 2.449$. To check, if this may occur, we calculated the Helmholtz free energy of the ideal crystal structure as a function of $\frac{c}{a}$ for $z = 0.25$ at various T_e's with CHIVES. From the obtained curves, which are shown in Fig. 4.39, we derived the optimal $\frac{c}{a}$ ratio as a function of T_e, which we show in the bottom panel of Fig. 4.40. For $z = 0.25$, the optimal $\frac{c}{a}$ ratio yields $\frac{c}{a} = 2.458$ at $T_e = 300$ K closely to $\frac{c}{a} = \sqrt{6} \approx 2.449$ of the sc structure and increases with increasing T_e. At $T_e = 9700$ K, the optimal $\frac{c}{a}$ ratio reaches $\frac{c}{a} = 2.609$, which corresponds to the A7 structure. Therefore, for $T_e < 9700$ K, the $\frac{c}{a}$ ratio will decrease in direction of the value of the sc structure, but will not reach it. This indicates, that an increase of T_e may remove the Peierls distortion and, on a longer timescale, the hexagonal lattice parameters a and c will transform closer to the sc structure, so that a transition from A7 in direction to the sc structure may be possible if effects currently overlooked are included. A further analysis certainly needs to take the other phonon modes into account.

4.5.4 *Phonon Band Structure at Various T_e's*

To get insights into the individual phonon modes, we studied the T_e-dependent phonon band structure of the ideal A7 crystal structure with Peierls parameter $z = 0.2336$. For this, we utilized again the supercell containing 864 Sb atoms and studied various T_e's ranging from 300 K up to 30000 K. We displaced one atom in the positive and in the negative x, y, and z-direction and calculated the corresponding interatomic forces on all atoms. With the help of these forces and symmetry operations (see Theorem 4.4), we calculated the **q**-dependent dynamical matrix $\mathbf{D(q)}$ and obtained the **q**-dependent phonon frequencies and related atomic motions by diagonalizing $\mathbf{D(q)}$ (see Sect. 2.2.1). In Figs. 4.43 and 4.44, we present representatively the obtained phonon bandstructures at selected T_e's.

It becomes apparent from Figs. 4.43 and 4.44 that the optical phonon frequencies decrease, if T_e is increased from 300 K up to 3000 K. Then, the higher ones decrease and the lower ones increase when T_e is increased to 9000 K, so that the differences between the optical phonon frequencies becomes smaller. If T_e is further increased up to 30000 K, all optical phonon frequencies increase. The acoustic phonon frequencies just decrease, if T_e is increased. Starting from $T_e = 10000$ K, the first branch of the acoustic phonon modes becomes negative. At $T_e = 30000$ K, most of the acoustic branches are negative, but there are still positive acoustic branches.

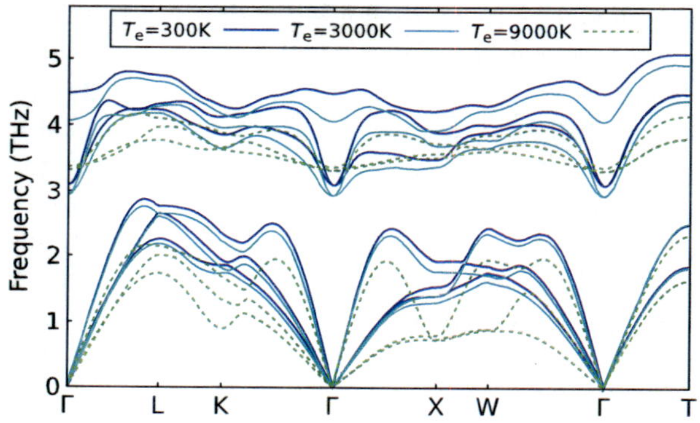

Fig. 4.43 DFT phonon bandstructure is shown at different T_e's

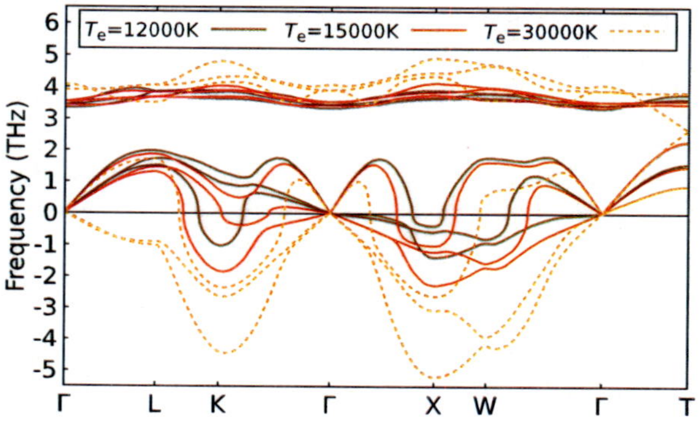

Fig. 4.44 DFT phonon bandstructure is shown at different T_e's

4.5.5 MD Simulations of the A_{1g}-Phonon Excitation at Various T_e's

To study the excitation of the A_{1g}-phonon mode under the influence of the other phonon modes we performed MD simulations at increased T_e of Sb initialized at an ionic temperature of $T_i = 300$ K.

For this, we utilized again the supercell containing 864 Sb atoms. To initialize the atoms at $T_i = 300$ K in six independent runs, we first set atomic displacements and velocities compatible to $T_i = 300$ K within the harmonic approximation as described in Sect. 4.1.2. In order to consider also higher than harmonic orders in the PES, we then performed an MD simulation at $T_e = 300$ K using a time step of 5 fs. After simulating 20 ps in this way, we took the atomic coordinates and velocities as the initial condition for the first run. We further continued this MD simulation at $T_e = 300$ K and took every 5 ps the atomic coordinates and velocities as initial conditions for the five other runs. Using these initial conditions, we performed MD simulations of a femtosecond-laser excitation by suddenly increasing T_e to a constant higher value. We simulated various T_e's for 3 ps – 5 ps.

The presence of oscillations from other phonons apart from the A_{1g} phonon causes significant changes compared to the excitation at $T_i = 0$ K studied in Sect. 4.5.3.

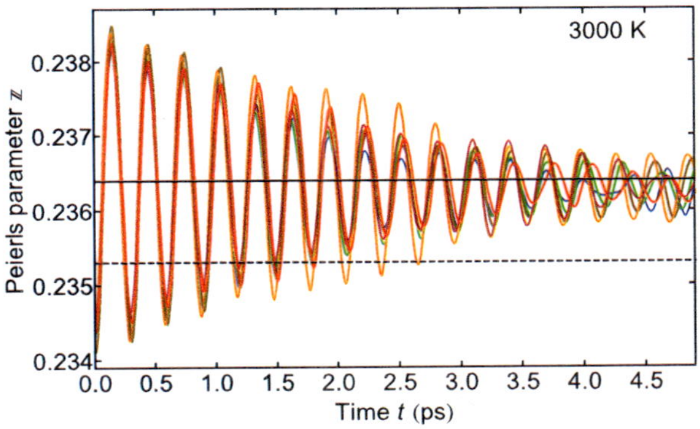

Fig. 4.45 Oscillation of the Peierls parameter z is shown for the six DFT MD simulations at $T_e = 3000$ K. The dashed line indicates the minimum z of the PES at $T_e = 3000$ K and $T_i = 0$ K and the solid one at $T_e = 3000$ K and $T_i = 300$ K

Already at $T_e = 300$ K, the Peierls parameter z oscillates around $z = 0.2345$ instead of $z = 0.2340$, which is the minimum of the $T_e = 300$ K and $T_i = 0$ K PES derived in Sect. 4.5.3. This shift of the equilibrium position becomes even bigger in the MD simulations with slightly increased T_e. In Fig. 4.45, we present the time-dependent Peierls parameter z of the six runs at $T_e = 3000$ K. The mean value of the oscillation differs significantly from the equilibrium Peierls parameter of the $T_e = 3000$ K and $T_i = 0$ K PES, which is indicated by a dashed line.

With the help of Theorem 4.7 we are able to include the influence of the other phonon modes on the PES within the harmonic approximation assuming thermal equilibrium:

Theorem 4.7 (Helmholtz free energy of N independent harmonic oscillators in equilibrium) *Let the Hamilton function be given as*

$$H = \sum_{j=1}^{N} \frac{1}{2} p_j^2 + \sum_{j=1}^{N} \frac{1}{2} \omega_j^2 q_j^2. \tag{4.94}$$

with coordinates q_j and momenta p_j and frequencies $\omega_j > 0$. In thermodynamical equilibrium, the Helmholtz free energy F at temperature T is given by

$$F = \frac{1}{2} k_B T \sum_{j=1}^{N} \ln\left(\omega_j^2\right) + f(T, N) \tag{4.95}$$

within the classical description, where k_B denotes the Boltzmann constant and $f(T, N)$ is only a function depending on T, N and natural constants.

Proof The thermodynamical equilibrium is described by the canonical ensemble. Analogously to the quantum mechanical case described in Eq. (2.200), the Helmholtz free energy is classical given by

$$F = -k_\mathrm{B}\,T\,\ln\left(Z^{(c)}\right), \tag{4.96}$$

where $Z^{(c)}$ denotes the classical partition function from Eq. (4.14). We obtain by inserting $Z^{(c)}$:

$$\begin{aligned}
F =& -k_\mathrm{B}\,T\,\ln\left(\frac{1}{\zeta}\int_{-\infty}^{\infty} dq_1 \ldots \int_{-\infty}^{\infty} dq_N \int_{-\infty}^{\infty} dp_1 \ldots \int_{-\infty}^{\infty} dp_N\, \mathrm{e}^{-\frac{\sum_{j=1}^{N}\frac{1}{2}p_j^2 + \sum_{j=1}^{N}\frac{1}{2}\omega_j^2 q_j^2}{k_\mathrm{B}\,T}}\right) \\
=& -k_\mathrm{B}\,T\,\ln\left(\frac{1}{\zeta}\prod_{j=1}^{N}\left(\int_{-\infty}^{\infty} dp_j\,\mathrm{e}^{-\frac{p_j^2}{2k_\mathrm{B}\,T}}\right)\left(\int_{-\infty}^{\infty} dq_j\,\mathrm{e}^{-\frac{\omega_j^2 q_j^2}{2k_\mathrm{B}\,T}}\right)\right) \\
=& k_\mathrm{B}\,T\,\ln(\zeta) - k_\mathrm{B}\,T\sum_{j=1}^{N}\ln\left(\int_{-\infty}^{\infty} dp_j\,\mathrm{e}^{-\frac{p_j^2}{2k_\mathrm{B}\,T}}\right) - k_\mathrm{B}\,T\sum_{j=1}^{N}\ln\left(\int_{-\infty}^{\infty} dq_j\, - \frac{\omega_j^2 q_j^2}{2\,k_\mathrm{B}\,T}\right).
\end{aligned} \tag{4.97}$$

We get by substituting $u = \frac{1}{\sqrt{2\,k_\mathrm{B}\,T}}\,p_j$, $du = \frac{1}{\sqrt{2\,k_\mathrm{B}\,T}}\,dp_j$:

$$\int_{-\infty}^{\infty} dp_j\,\mathrm{e}^{-\frac{p_j^2}{2k_\mathrm{B}\,T}} = \sqrt{2\,k_\mathrm{B}\,T}\underbrace{\int_{-\infty}^{\infty} du\,\mathrm{e}^{-u^2}}_{\overset{(4.17)}{=}\sqrt{\pi}} = \sqrt{2\,k_\mathrm{B}\,T\,\pi}$$

and by substituting $u = \frac{\omega_j}{\sqrt{2\,k_\mathrm{B}\,T}}\,q_j$, $du = \frac{\omega_j}{\sqrt{2\,k_\mathrm{B}\,T}}\,dq_j$:

$$\int_{-\infty}^{\infty} dq_j\,\mathrm{e}^{-\frac{\omega_j^2 q_j^2}{2k_\mathrm{B}\,T}} = \frac{\sqrt{2\,k_\mathrm{B}\,T}}{\omega_j}\underbrace{\int_{-\infty}^{\infty} du\,\mathrm{e}^{-u^2}}_{\overset{(4.17)}{=}\sqrt{\pi}} = \frac{\sqrt{2\,k_\mathrm{B}\,T\,\pi}}{\omega_j}.$$

Utilizing the above two equations, we obtain for Eq. (4.97):

$$\begin{aligned}
F =& k_\mathrm{B}\,T\,\ln(\zeta) - \frac{1}{2}k_\mathrm{B}\,T\,N\,\ln(2\,k_\mathrm{B}\,T\,\pi) - \frac{1}{2}k_\mathrm{B}\,T\,N\,\ln(2\,k_\mathrm{B}\,T\,\pi) - k_\mathrm{B}\,T\sum_{j=1}^{N}\ln\left(\frac{1}{\omega_j}\right) \\
=& \frac{1}{2}k_\mathrm{B}\,T\sum_{j=1}^{N}\ln\left(\omega_j^2\right)\underbrace{-k_\mathrm{B}\,T\,N\,\ln(2\,k_\mathrm{B}\,T\,\pi) - k_\mathrm{B}\,T\,\ln(\zeta)}_{=:f(T,N)}.
\end{aligned}$$

■

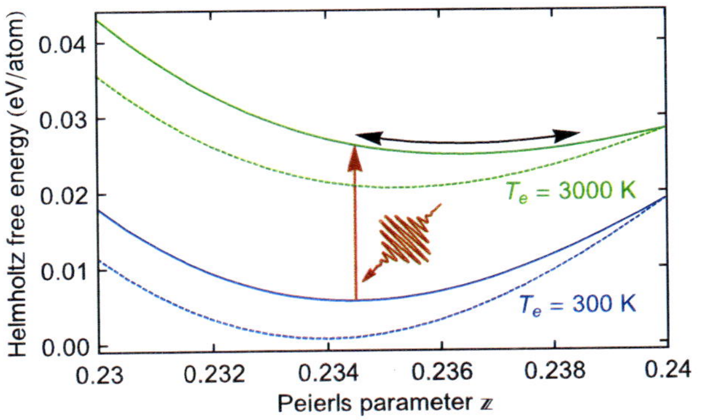

Fig. 4.46 DFT PES is shown as a function of the Peierls parameter $\mathbb{z}$ is shown at $T_e = 300$ K and $T_e = 3000$ K. The solid curves takes the other phonon modes into account whereas the dashed curves ignore them. The shift between the curves is selected to improve visibility. The excitation of the A_{1g} phonon is illustrated (see text)

In Sect. 4.5.3, we derived the T_e-dependent PES $F_0(\mathbb{z}, T_e)$ for the ideal crystal structure as a function of the Peierls parameter $\mathbb{z}$. Now we take the other phonon modes within the harmonic approximation into account. Using Theorem 4.7 and the assumption that the A_{1g} phonon is independent to the other phonon modes, we obtain

$$\boxed{F(\mathbb{z}, T_e, T_i) = F_0(\mathbb{z}, T_e) + \frac{1}{2} k_B T_i \sum_{j=1}^{N} \ln\left(\omega_j^2(\mathbb{z})\right),} \tag{4.98}$$

where $\omega_j(\mathbb{z})$ denotes the frequency of the j-th phonon mode for the ideal crystal structure with Peierls parameter $\mathbb{z}$. The sum over j runs over all phonon modes except the A_{1g} mode and the three translations. We skip the term $f(T, N)$ occurring Eq. (4.95), since we are only interested in the PES at $T_i = 300$ K, so that $f(T, N)$ is just a constant. At $T_e = 3000$ K, the Peierls parameter oscillates approximately around the equilibrium $\mathbb{z}$-value of the PES at $T_i = 300$ K (Fig. 4.46), which is indicated as a solid line in Fig. 4.45. The still occurring small mismatch has its origin in the higher than harmonic orders in the PES, which are not taken into account in Eq. (4.98).

In addition, after the increase of T_e or the laser excitation, the energy in the A_{1g} phonon mode is transferred to other phonon modes due to the phonon-phonon coupling. Consequently, the A_{1g} phonon oscillation is damped. This oscillation can overcome the barrier at $\mathbb{z} = 0.25$, if the initial oscillation amplitude is big enough. But, due to the damping, there is always a point in time at which the oscillation cannot overcome the barrier any more and will stay at one side. There must be ranges for T_e where the oscillation of the Peierls parameter $\mathbb{z}$ exactly overcomes the barrier 0 times, 1 times, 2 times, etc.

Indeed, our MD simulations show (see Fig. 4.47) that for $T_e \leq 4700$ K the oscillation do not reach the barrier, and for $T_e = 4800$ K the oscillation overcomes the

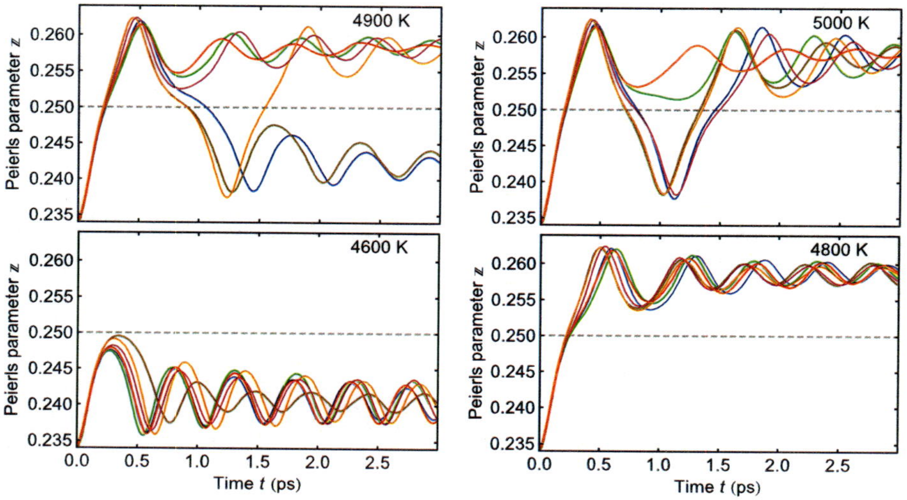

Fig. 4.47 Oscillation of the Peierls parameter z is shown for the six DFT MD simulations at various T_e's adapted from Fig. 6 of Ref. [62]

barrier one time. But, from our simulations, we cannot detect the interval for T_e, where the oscillation overcomes 2 times or 3 times the barrier: At $T_e = 4900$ K, three runs pass the barrier 1 time, two runs 2 times and one run 3 times, whereas at $T_e = 5000$ K, two runs pass the barrier still one time and the other four runs three times. This indicates that the difference in the initializations of the atoms at $T_i = 300$ K significantly influences the dynamical evolution of the system, which is a consequence of a finite size effect and, of course, should not occur. Thus, the simulation cell containing only 864 atoms is too small to reliably identify T_e's, where the Peierls parameter overcomes the barrier only a few times and a A7 to sc transition may be possible without melting of the crystal structure.

For $T_e \geq 6000$ K, all six runs show a similar behavior within the simulation time of 3 ps, since the laser-induced oscillation of Peierls parameter becomes so big that the influence of the different initializations of atoms at $T_i = 300$ K on the time evolution is minimal. At these T_e's, the Peierls parameter just oscillates around $z = 0.25$, as one can see in Fig. 4.48 for $T_e = 9000$ K as an example. To illustrate the corresponding vertical movement of the hexagonal planes against each other, we show snapshots of the supercell from run 4 at the three selected times t_0, t_1, and t_2 in Fig. 4.49. At t_1, the Peierls distortion is removed and the hexagonal planes show the same vertical distance between each other. At t_2, the alternating distances between the hexagonal planes are inverted compared to the initial time t_0.

At $T_e = 12000$ K, the laser excitation becomes so big that the crystals structure melts ultrafastly, which we will study intensively in Sect. 4.5.7.

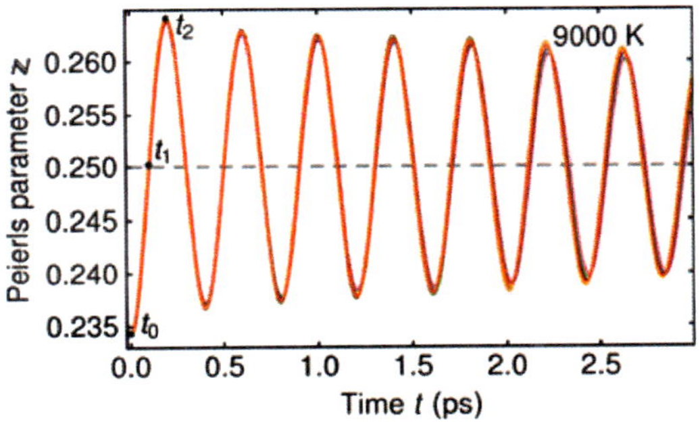

Fig. 4.48 Oscillation of the Peierls parameter z is shown for the six DFT MD simulations at $T_e = 9000$ K. The circles indicate the times $t_0 = 0$ ps, $t_1 = 0.1$ ps, and $t_2 = 0.2$ ps where snapshots were taken from run 4 for Fig. 4.49

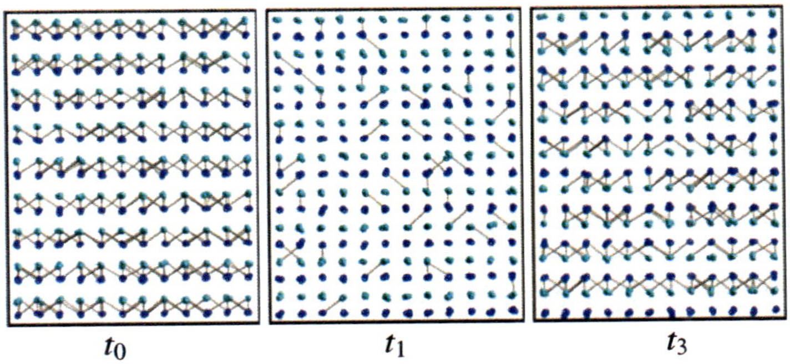

Fig. 4.49 Snapshots of the supercell containing 864 Sb atoms from run 4 are shown at the times t_0, t_1, and t_2 after the increase of T_e to 9000 K. A bond between two atoms is drawn if they are closer than 0.29 nm to each other. Atoms belonging to basis d_1 are drawn in blue and atoms belonging to basis d_2 are drawn in cyan

4.5.6 MD Simulations of Thermal Phonon Antisqueezing at Moderate T_e's

If T_e is moderately increased, the acoustic phonon frequencies decrease, but they keep still positive. In Fig. 4.50, we show the phonon density of Sb at $T_e = 300$ K, 6000 K, and 9000 K. There are two main peaks: one localized at the optical phonon frequencies and a more broader one at the acoustic phonon frequencies. The broader acoustic peak splits noticeably into two peaks when T_e is increased. We conclude: The frequency decrease and the presence of peaks in the phonon density indicate that thermal phonon antisqueezing may occur in Sb at moderate T_e's.

To verify this statement, we analyzed the atomic coordinates of the six MD simulation runs from the previous section at $T_e = 3000$ K, 6000 K and 9000 K. For this, we projected the atomic mean-square displacements from the equilibrium positions onto the directions of selected phonon modes as described in Sect. 4.4.4. We

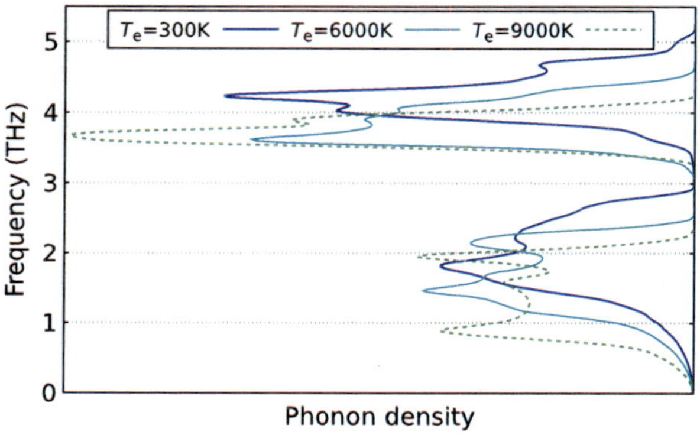

Fig. 4.50 DFT phonon density of Sb is shown at $T_e = 300$ K, 6000 K and 9000 K

selected the phonon modes that correspond to the peak located at low frequencies in the phonon density: At $T_e = 3000$ K, we selected the modes with frequencies $\nu \leq 2.0$ THz and, at $T_e = 6000$ K, we selected the modes with frequencies $\nu \leq 1.9$ THz and, at $T_e = 9000$ K, we selected the modes with frequencies $\nu \leq 1.3$ THz. Almost all acoustic phonon branches correspond to the selected phonon modes, as one can see in Fig. 4.43.

We show the time-dependent atomic mean-square displacements in total and projected onto the selected and rest phonon modes averaged over 6 runs in Fig. 4.51 at $T_e = 6000$ K as an example. The selected acoustic phonon modes show indeed a thermal antisqueezing: The corresponding atomic mean-square displacements first increases and then decreases back nearly to the initial value and shows then further oscillations. The atomic mean-square displacements of the rest phonon modes show a dominant oscillation, which has its origin in the excitation of the A_{1g} phonon mode, which belongs to the rest phonon modes. The frequency of the A_{1g} phonon mode is approximately twice times bigger than the central frequency of the acoustic peak (see Fig. 4.50), which shows antisqueezing in its related phonon modes. Since the oscillation corresponding to antisqueezing has the double frequency, as one can see in Eq. (4.70) due to the $\cos^2$ and the $\sin^2$, the atomic mean-square displacements of the selected and rest phonon modes show the same frequency although they have different origins: thermal phonon antisqueezing and coherent excitation of the A_{1g} phonon mode.

To detect the thermal phonon antisqueezing in a possible experimental setup, one can measure the time-resolved intensity of a Bragg peak that is not influenced by the A_{1g} phonon mode. The $(0\ 1\ -1)$ peak has the wave vector $\mathbf{q}_{01-1} = \left[\frac{2\pi}{a}, \frac{2\sqrt{3}\pi}{a}, 0\right]^t$ and is therefore unaffected by the A_{1g} phonon. In Fig. 4.52, we present the time-dependent normalized intensity $\langle I_{hkl}(t)\rangle$ of selected diffraction peaks averaged over 6 runs at $T_e = 6000$ K. The intensity of the $(0\ 1\ -1)$ peak decreases and then oscillates slightly at reduced values, which indicates that one may measure the thermal antisqueezing. The (1 1 0), (2 0 0), and, especially, the (1 2 0) peak intensities oscillate significantly due to the A_{1g} phonon excitation. Debye-Waller theory can only

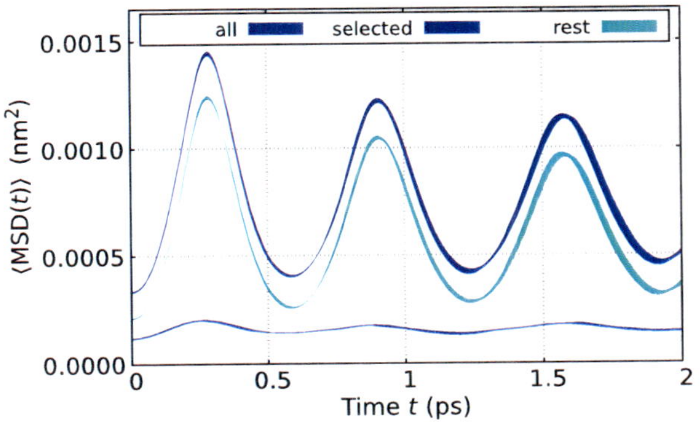

Fig. 4.51 DFT time-dependent atomic mean-square displacements in total and projected onto the selected and rest phonon modes averaged over 6 runs is shown at $T_e = 6000$ K

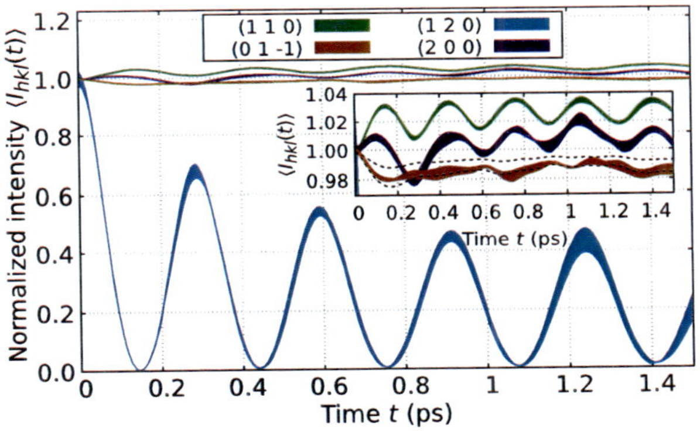

Fig. 4.52 DFT time-dependent normalized intensity $\langle I_{hkl}(t)\rangle$ of selected diffraction peaks averaged over 6 runs is shown at $T_e = 6000$ K. The dashed curves in the inset represent the normalized intensities derived from the total atomic mean-square displacements using time-dependent Debye-Waller theory for the (1 1 0) and (1 0 −1) peak

explain the time-dependent intensity of Bragg peaks, that are not influenced by the A_{1g} phonon. One can clearly see this in the inset of Fig. 4.52: The dashed curves represent the time-dependent intensity behavior of the (1 1 0) and (1 0 −1) peak obtained from Debye-Waller theory. Only the (1 0 −1) peak is proper described in contrast to the (1 1 1) peak, which is sensitive to the A_{1g} phonon oscillation.

At $T_e = 3000$ K and 9000 K, one observes a similar behavior.

4.5.7 MD Simulations of Non-thermal Melting at High T_e's

The crystal structure of Sb keeps intact in bulk MD simulations for T_e's below 12000 K. At $T_e \geq 12000$ K, the crystal structure melts within a few 100 fs. To study this ultrafast melting of Sb, we performed further MD simulations with CHIVES. For this, we utilized a supercell that contains $N_{at} = 432$ Sb atoms and 12 hexagonal planes and consists of $6 \times 3 \times 2$ minimal cells (see Fig. 4.53). Using the harmonic approximation and a long time MD simulation at $T_e = 300$ K, we initialized 30 initial conditions of atomic displacements and velocities compatible to $T_i = 300$ K with the same method as for the supercell with 864 atoms described in Sect. 4.5.5. Taking these initial conditions, we simulated the laser excitation by suddenly increasing T_e to a high constant value. We considered $T_e = 12000$ K, 15000 K, 18000 K, 24000 K, and 30000 K. We used a time step of 5 fs and simulated 2.5 ps in total, except for $T_e =$ 12000 K, where we simulated 7.5 ps. To analyze the non-thermal melting, we studied the atomic mean-square displacements. Since the z-direction is not equivalent to the x- and y-directions, we determined the individual atomic mean-square displacements in x-, y- and z-direction by

$$\mathrm{MSD}_x(t) = \frac{1}{N_{at}} \sum_{i=1}^{N_{at}} u_{ix}(t)^2, \ \mathrm{MSD}_y(t) = \frac{1}{N_{at}} \sum_{i=1}^{N_{at}} u_{iy}(t)^2, \ \mathrm{MSD}_z(t) = \frac{1}{N_{at}} \sum_{i=1}^{N_{at}} u_{iz}(t)^2, \tag{4.99}$$

where $\mathbf{u}_i(t) = \left[u_{ix}(t), u_{iy}(t), u_{iz}(t)\right]^t$ denotes the displacement vector of the i-th atom from its equilibrium position. We averaged the obtained values over the 30 runs and realized that

$$\langle \mathrm{MSD}_x(t) \rangle \approx \langle \mathrm{MSD}_y(t) \rangle \approx \langle \mathrm{MSD}_z(t) \rangle \tag{4.100}$$

is valid for the studied T_e's within the simulation time. Therefore, from now on, we only consider the averaged total atomic mean-square displacements $\langle \mathrm{MSD}(t) \rangle$. In Fig. 4.54, we present the latter for the studied T_e's.

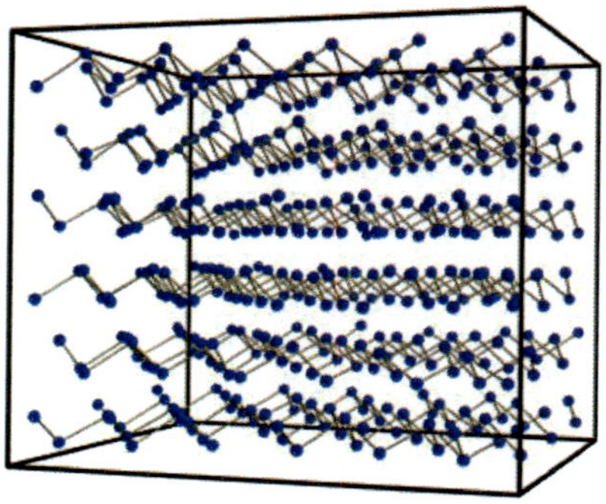

Fig. 4.53 Supercell containg 432 Sb atoms

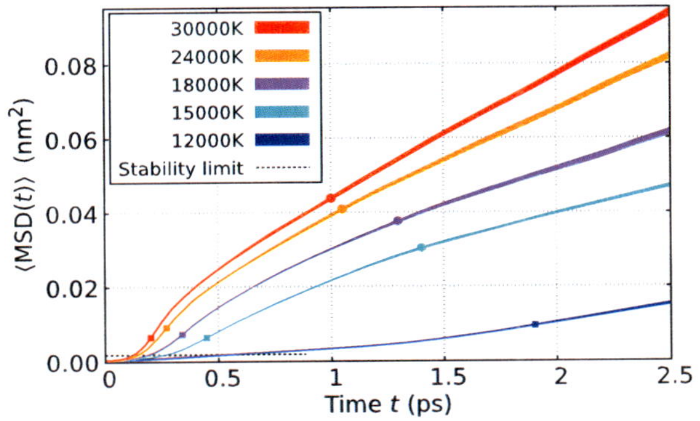

Fig. 4.54 DFT time-dependent atomic mean-square displacements $\langle \mathrm{MSD}(t) \rangle$ averaged over 30 runs are shown at the studied T_e's above the non-thermal melting threshold. The dashed line indicates the Lindemann stability limit. The filled square indicates the transition time from super to fractional diffusive and the filled circle from fractional to normal diffusive

A few hundred femtoseconds after the increase of T_e, the atomic mean square displacement exceed the Lindemann stability limit. Furthermore, Sb melts also in the three stages super diffusive, fractional diffusive, and normally diffusive similar to Si as described in Sect. 4.4.5. Reason for this is that the A7 structure of Sb is not closed packed and there is room for the atoms to move, as it is also the case in the diamond-like structure of Si. Therefore, after the increase of T_e, the Sb atoms accelerate into the voids during the super diffusive stage, so that the mean-square displacements rapidly increases. Then, in the fractional diffusive stage, the atoms start to scatter dominantly with each other, so that the time-dependent increase of the mean-square displacements decreases. In the third and last stage, the mean-square displacements increases linearly with the time as for a normal liquid. In Fig. 4.54, we present the transition time between super diffusive and the fractional diffusive state with a filled square and the transition time between fractional diffusive and normal diffusive with a filled circle. At $T_e = 12000$ K, which is slightly above the non-thermal melting threshold, the melting occurs slower compared to the other studied T_e's. Thus, one can only see the super diffusive stage and parts of the fractional diffusive stage in Fig. 4.54. We show the super diffusion exponent α_s, the fractional diffusive exponent α_f, and the diffusion coefficient D obtained at the studied T_e's in Fig. 4.55.

In available experimental setups, one can study the ultrafast melting of Sb by measuring the time-dependent intensity decay of the Bragg peaks. In Fig. 4.56, we show representatively the normalized intensity $\langle I_{hkl}(t) \rangle$ of selected Bragg peaks and the background as a function of time t at $T_e = 1800$ K averaged over 30 runs. The dashed lines indicate the normalized intensity obtained from Debye-Waller theory for the (1 1 0), (0 1 −1) and (2 0 0) peak. A few hundred femtoseconds after the increase of T_e, the intensities of the Bragg peaks decrease below the intensity of the background. The intensity of the (1 2 0) shows first a strong oscillation, which is

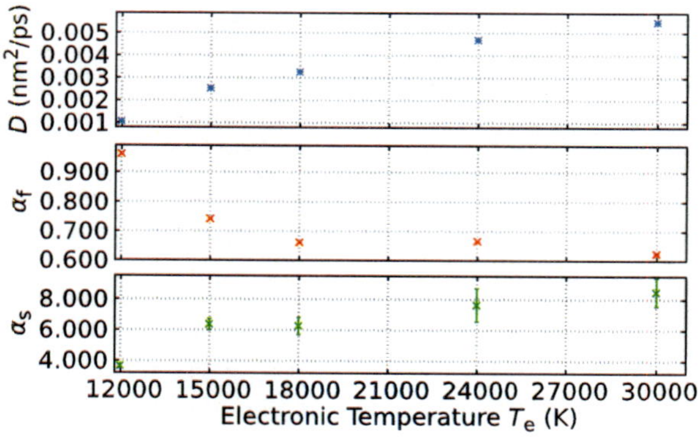

Fig. 4.55 Super diffusion exponent α_s, fractional diffusion exponent α_f, and diffusion coefficient D derived from the DFT MD simulations are presented at the studied T_e's. The error bars correspond to the fit error

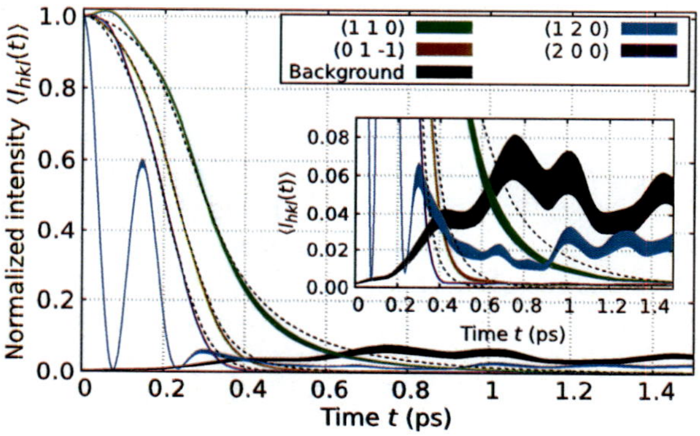

Fig. 4.56 DFT time-dependent normalized intensity $\langle I_{hkl}(t)\rangle$ of selected diffraction peaks and the background averaged over 30 runs is shown at $T_e = 1800$ K. For the background, the **q** point $(1\ 1\ \frac{7}{6})$ was considered. The dashed curves represent the normalized intensities derived from the total atomic mean-square displacements using time-dependent Debye-Waller theory (except for the (1 2 0) peak and the background)

related to the A_{1g} phonon excitation similar to the weak increase of the (1 1 0) and (2 0 0) peak at the beginning. These influences of the A_{1g} phonon on the bragg peak intensities cannot be described by Debye-Waller theory. Only the intensity behavior of the $(0\ 1\ -1)$ peak, which is not affected by the A_{1g} phonon, is accurately described by Debye-Waller theory for normalized intensities above 0.3.

4.5.8 MD Simulations of a Thin-Film at Various T_e's

4.5.8.1 Thin Film in z-Direction

To analyze the influence of the presence of a surface, we performed further MD simulations with CHIVES on a supercell containing 384 Sb atoms in thin-film geometry. This supercell consists of 24 hexagonal planes and corresponds to $4 \times 2 \times 4$ minimal cells. We applied periodic boundary conditions and doubled the simulation volume in z-direction ([111] direction of the crystal structure) to finally get a film of 4.5 nm thickness (see Fig. 4.57). For the electrons, we used a $2 \times 2 \times 1$ grid for the **k** points containing 4 irreducible **k**-points in the first Brillouin zone (see Sect. 3.6). To initialize the atoms at room temperature, we first performed the FIRE relaxation algorithm [60] to derive the equilibrium atomic positions. Secondly, we applied the Andersen thermostat (see Sect. 4.1.2) to derive one initialization of atomic coordinates and velocities at $T_i = 300$ K. Then we simulated the nine T_e's, namely 300 K, 3000 K, 6000 K, 9000 K, 12000 K, 15000 K, 18000 K, 21000 K, and 24000 K using a time step of 5 fs for at least 2.5 ps, and for T_e's below 15000 K even till 10 ps. We want to note that the influence of the electron-phonon coupling may become relevant on the timescale of 10 ps and will significantly decrease T_e and heating up the lattice. However, here we are only interested in the effects that are caused by the change of the PES due to the increased T_e.

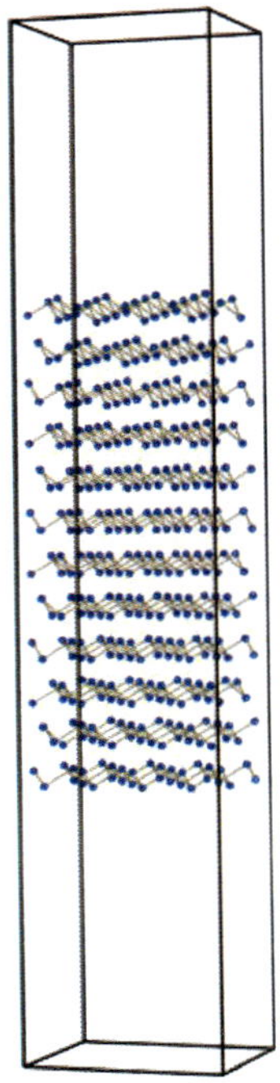

Fig. 4.57 Supercell with 384 Sb atoms

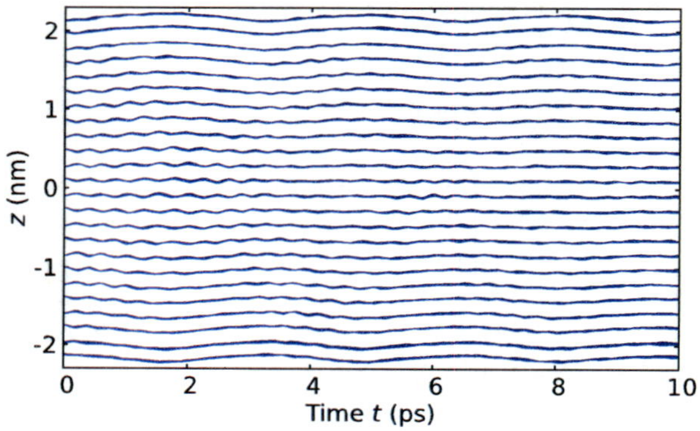

Fig. 4.58 DFT z-positions of the hexagonal planes with respect to the film center as a function of time at $T_e = 9000$ K. The line width corresponds to the standard deviations of the atomic z-coordinates in each plane

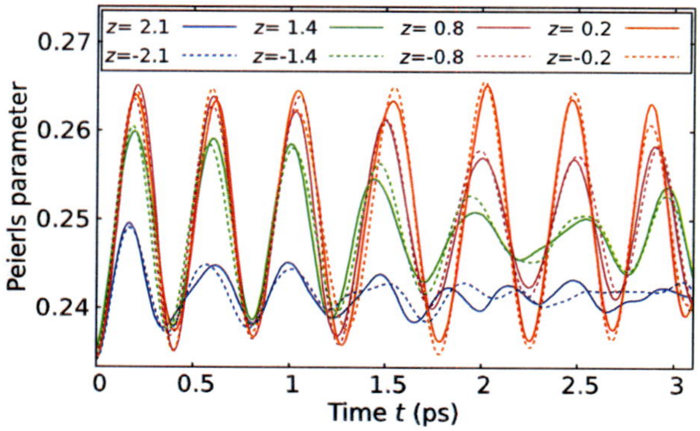

Fig. 4.59 DFT z-dependent Peierls parameters are shown as a function of time for the thin film at $T_e = 9000$ K. z denotes the distance from the film center

At T_e's below the non-thermal melting threshold 12000 K, the crystal structure stays intact and the thin film shows a breathing mode, i.e., the film thickness oscillates as a function of time. To get deeper insights, we derived the time-dependent z-positions of the hexagonal planes by averaging the z-coordinates of the corresponding atoms. In Fig. 4.58, we present the resulting curves at $T_e = 9000$ K. One can clearly see the breathing mode and the oscillation of the hexagonal planes against each other, which occurs due to the A_{1g}-phonon excitation. The A_{1g}-phonon oscillations appears mainly in the center of the film. This can be seen even more clearly in Fig. 4.59, where we show the z-dependent Peierls parameter as a function of time at $T_e = 12000$ K. We can define a z-dependent Peierls parameter, since we can derive from the z-coordinates of three hexagonal planes one Peierls parameter (see Sect. 4.5.1). Indeed, the Peierls parameter oscillation at the center of the film has a much bigger amplitude than near the surface.

At $T_e = 12000$ K, the thin film still shows the breathing mode, but it starts to melt approximately 4 ps after the increase of T_e, as one can see in Fig. 4.61, where the corresponding z-positions of the hexagonal planes are shown as a function of time. The initial expansion of the thin film prevents the melting, which only starts when the film contracts again. In addition, the oscillation of the A_{1g}-phonon mode is only visible during the first ps. At higher T_e's, the thin film just expands and melts

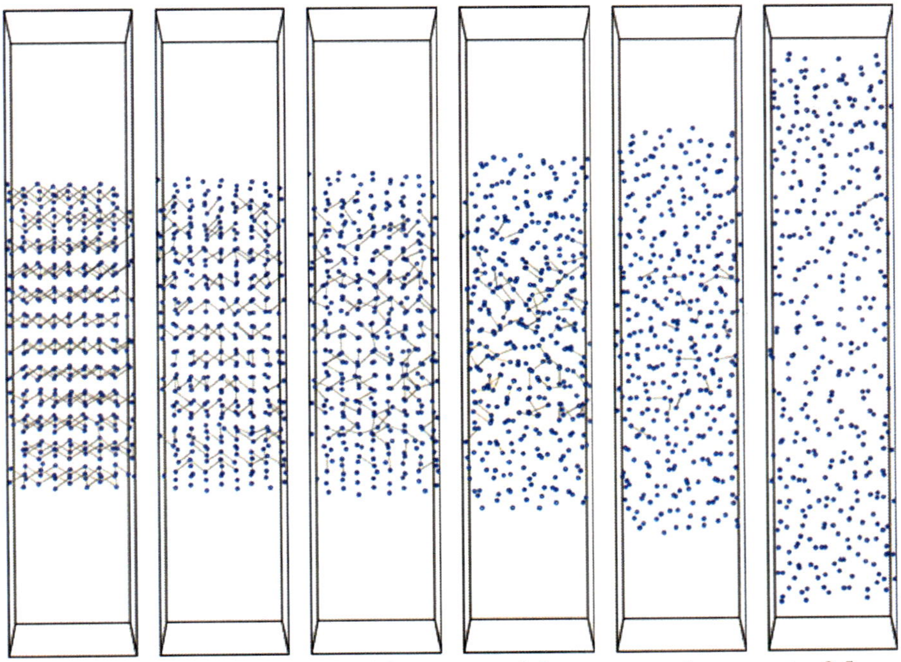

Fig. 4.60 Snapshots of the DFT-MD simulation of the thin film in z-direction containing 384 Sb atoms at $T_e = 18000$ K. Bonds between neighboring atoms are drawn

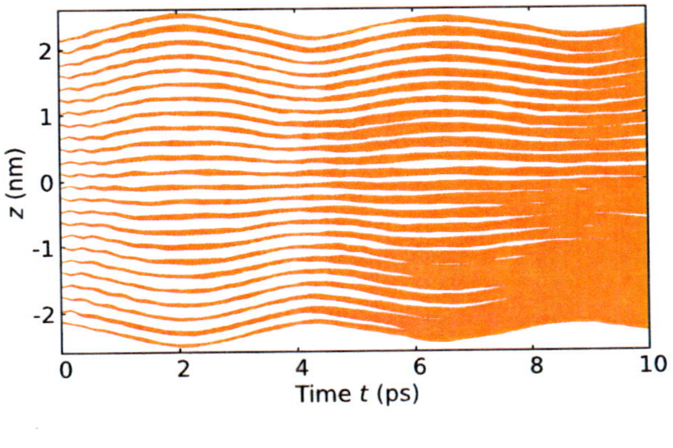

Fig. 4.61 Averaged z-position of the hexagonal planes with respect to the film center as a function of time at $T_e = 12000$ K. The line width corresponds to the standard deviations of the atomic z-coordinates in each plane

immediately after the increase of T_e. We present snapshots of the supercell of the DFT MD simulation at $T_e = 18000$ K in Fig. 4.60 as an example.

The above described effects of an increased T_e on the thin film can also be seen from the atomic root mean-square displacement in z-direction $\text{RMSD}_z(t)$ from Eq. (4.82) and from the in plane atomic root mean-square displacements $\text{RMSD}_{xy}(t)$ from Eq. (4.83). In Fig. 4.62, we present the $\text{RMSD}_z(t)$ curves for various T_e's. At

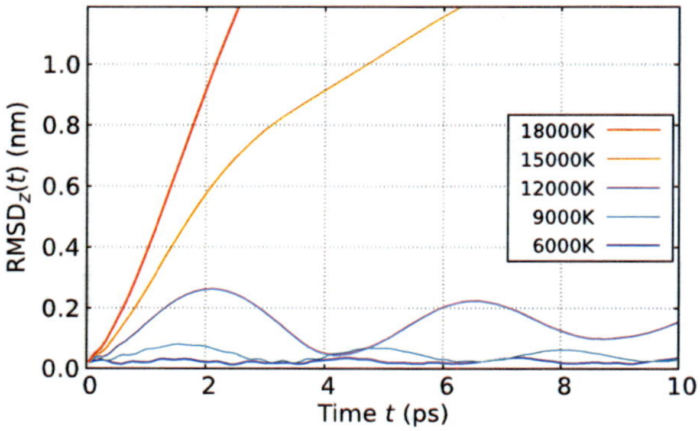

Fig. 4.62 DFT time-dependent atomic root mean-square displacements in z-direction RMSD$_z(t)$ is shown for the thin film in z-direction at various T_e's

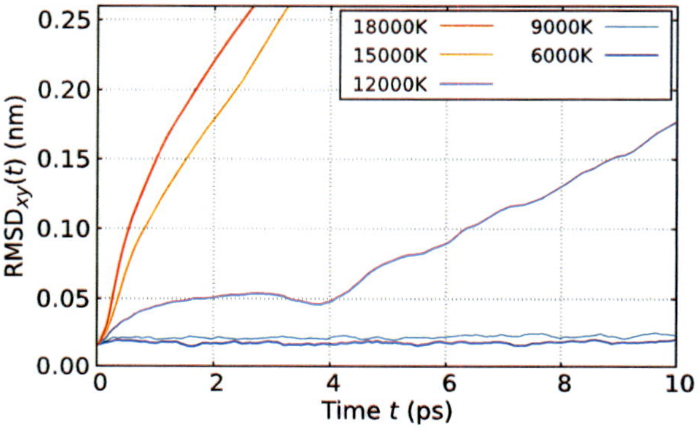

Fig. 4.63 DFT time-dependent in plane atomic root mean-square displacements RMSD$_{xy}(t)$ is shown for the thin film in z-direction at various T_e's

$T_e \leq 12000$ K, the RMSD$_z(t)$ curves shows a time-dependent oscillation, which is related to the breathing mode of the thin film. One can also recognize an additional high frequency oscillation of the RMSD$_z(t)$ curves at $T_e \leq 6000$ K and 9000 K, which has its origin in the A_{1g} phonon oscillation. For $T_e \geq 15000$, the RMSD$_z(t)$ curve increases monotonously, which indicates that the thin film expands and ablation occurs. At $T_e = 15000$ K, there is a kink in the RMSD$_z(t)$ curve located at $t = 3$ ps, since at this time the thin film breaks into two parts. This effect does not occur at the higher studied T_e's and indicates, that there still exist a weak interatomic bonding at $T_e = 15000$ K. In Fig. 4.63, we present the RMSD$_{xy}(t)$ curves for various T_e's. At $T_e \leq 9000$ K, the RMSD$_{xy}(t)$ curves increase initially weakly and stay at low values indicating that the crystal structure keeps intact. At $T_e = 12000$ K, the RMSD$_{xy}(t)$ curve first increases slightly, secondly keeps constants and then starts to increase monotonously at $t \geq 4$ ps. This indicates that the crystal structure starts to melt at $t = 4$ ps, as we already know from Fig. 4.61. For $T_e \geq 15000$ K, the RMSD$_{xy}(t)$ curves increase strongly and monotonously indicating that the crystal structure melts immediately after the increase of T_e.

4.5.8.2 Thin Film in *y*-Direction

Since the z-direction is not equivalent to the x- and y-direction, we studied additionally a thin film in y-direction ($[\bar{1}\,\bar{1}\,2]$ direction of the crystal structure). For this, we utilized again the supercell containing 384 atoms mentioned above. But now we increased the simulation volume five times in y-direction instead of increasing it in the z-direction as we did it before. In this way, we obtained a film of 1.5 nm thickness shown in Fig. 4.66. Again, we performed the FIRE algorithm to derive the optimal atomic coordinates and then we applied the Andersen thermostat to initialize the atomic coordinates and velocities at $T_i = 300$ K. We simulated the eight T_e's, namely 300 K, 3000 K, 6000 K, 9000 K, 12000 K, 15000 K, 18000 K, and 21000 K using a time step of 5 fs. We simulated the T_e's below 15000 K for 10 ps and the other ones for 2.5 ps.

Our MD simulations show that the thin film in y-directions behaves approximately in a similar way at increased T_e compared to the thin film in z-direction. In Fig. 4.64, we present the atomic root mean-square displacement in y-direction $\mathrm{RMSD}_y(t)$ analogously to Eq. (4.82) and, in Fig. 4.65, we present the in plane atomic root mean-square displacements $\mathrm{RMSD}_{xz}(t)$ analogously to Eq. (4.83) at various T_e's.

Fig. 4.64 DFT time-dependent atomic root mean-square displacements in y-direction $\mathrm{RMSD}_y(t)$ is shown for the thin film in y-direction at various T_e's

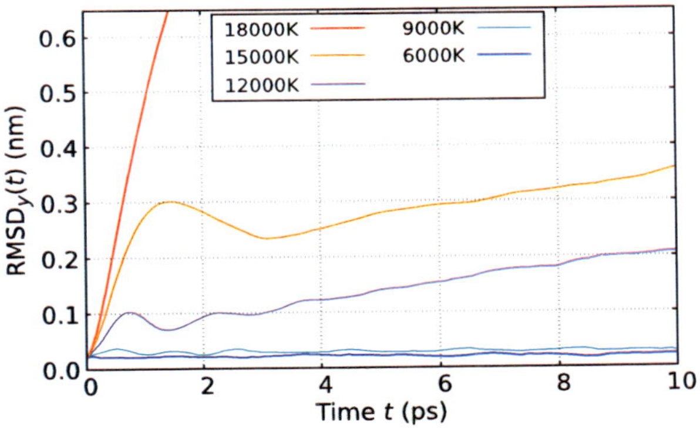

Fig. 4.65 DFT time-dependent in plane atomic root mean-square displacements $\mathrm{RMSD}_{xz}(t)$ is shown for the thin film in y-direction at various T_e's

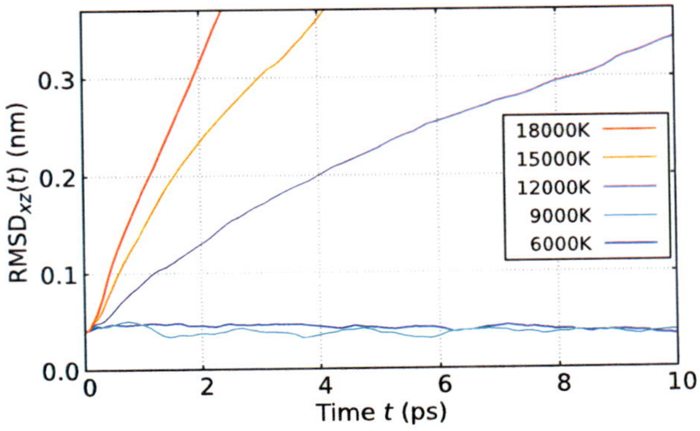

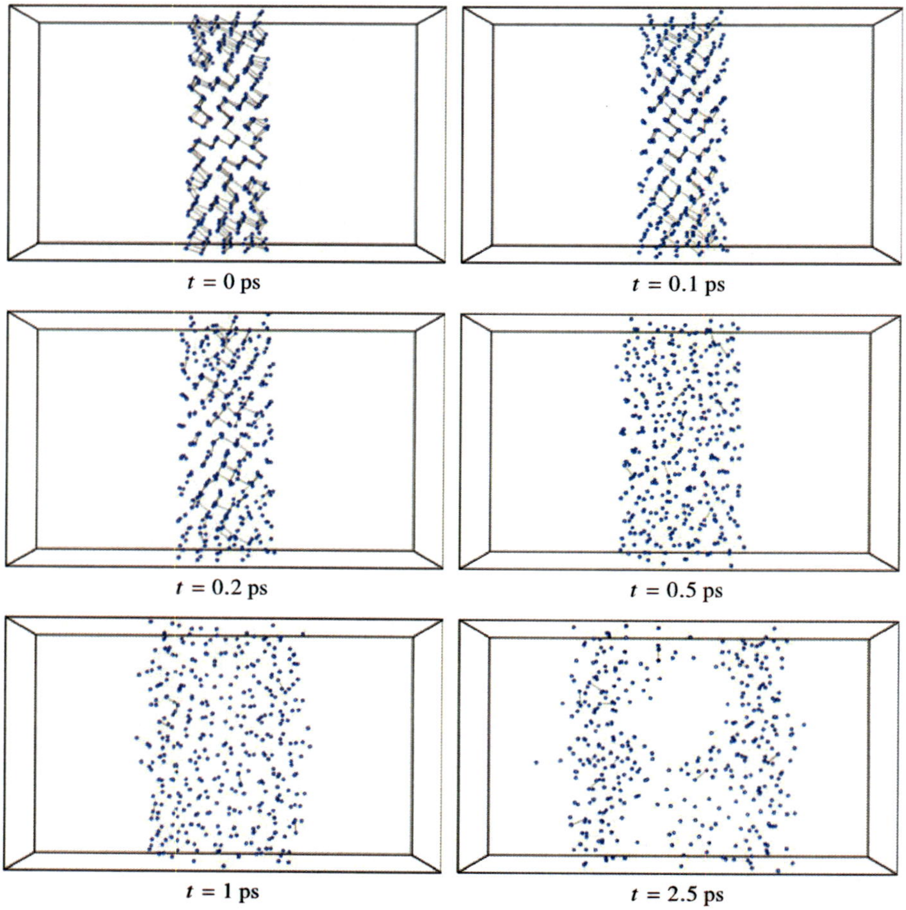

Fig. 4.66 Snapshots of the DFT-MD simulation of the thin film in y-direction containing 384 Sb atoms at $T_e = 18000$ K. Bonds between neighboring atoms are drawn

We want to note that the $\text{RMSD}_x(t)$ and $\text{RMSD}_z(t)$ curves behave in a similar way for the studied T_e's although the x- and z-direction are not equivalent.

At $T_e \leq 9000$ K, the thickness of the thin film also oscillates as a function of time, but with a smaller amplitude compared to the thin film in z-direction. During this breathing mode, the crystal structure keeps intact. At $T_e = 12000$ K, the thin film shows a few oscillations of the thickness and than expands. Here the initial expansion of the film does not prevent melting, which starts immediately after the increase of T_e as for all higher T_e's. At $T_e = 15000$ K, the thin film expands and contracts one time before it only expands. For $T_e \geq 18000$ K, it only expands, which we present, as an example, in Fig. 4.66, where snapshots of the supercell of the DFT MD simulation at $T_e = 18000$ K are shown.

4.5.9 Summary of the Effects Induced by an Increased T_e

In the past section, we studied the effects that are caused by the high-T_e-deformed PES in Sb. Now we take the time to summarize all identified relevant effects in Fig. 4.67.

We consider representatively for T_e beside 300 K the two temperatures 6000 K and 9000 K below and the two temperatures 12000 K and 15000 K above the non-thermal melting threshold. In Fig. 4.67, the non-thermal melting threshold is indicated by a vertical line and separates the data in two sections bellow and above the threshold.

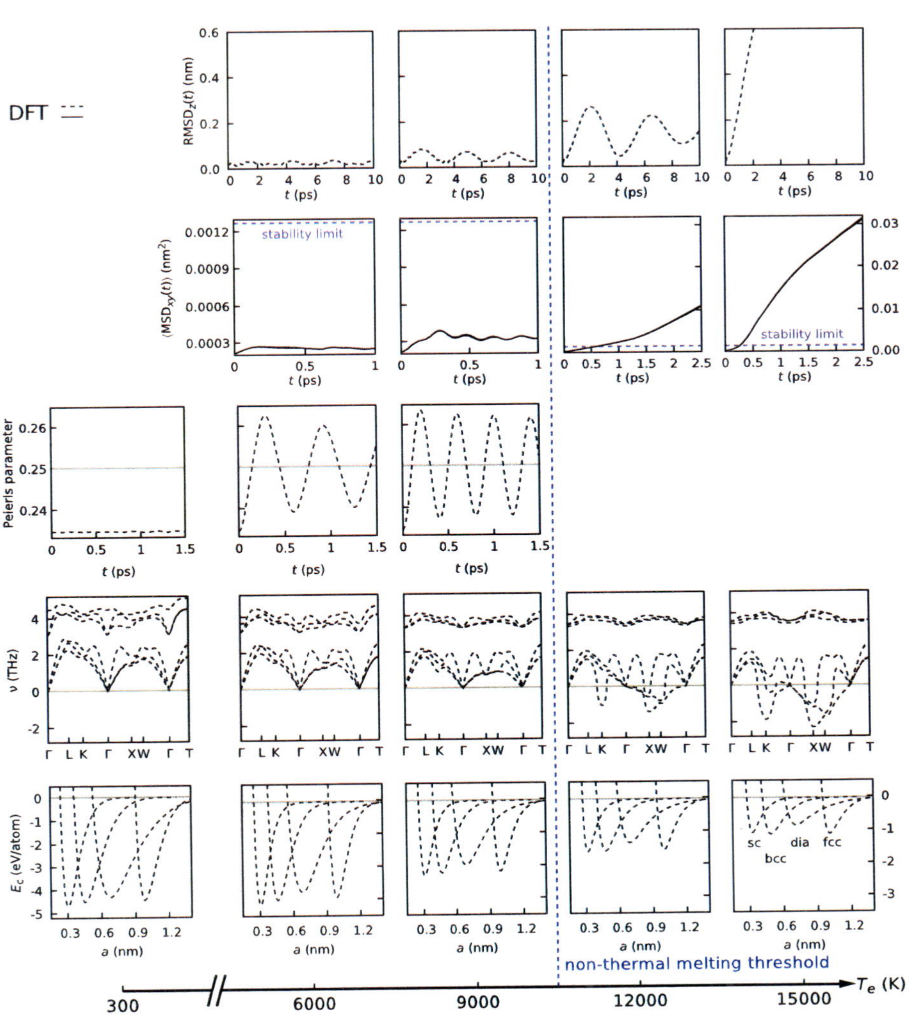

Fig. 4.67 Summary of the effects in Sb induced by the increased T_e deformed PES. Description see text

The reader should note, that the position of this line is, for presentation purposes, not located at the exact temperature, which is slightly below 12000 K. Five horizontal series of figures are presented, each one highlighting a particular quantity of interest:

1. In the top series of figures, we present the atomic root mean-square displacements in z-direction $\mathrm{RMSD}_z(t)$ of the thin-film with vacuum condition in z-direction as a function of time t. Below the non-thermal melting threshold, $\mathrm{RMSD}_z(t)$ oscillates harmonically indicating the breathing mode of the thin-film. At $T_e = 12000$ K, the oscillation of $\mathrm{RMSD}_z(t)$ is damped, since the thin-film starts to melt. At $T_e = 15000$ K, $\mathrm{RMSD}_z(t)$ simply increases rapidly as a function of time, since the thin-film expands and melts.
2. The second row shows the average in plane atomic mean-square displacements $\langle \mathrm{MSD}_{xy}(t) \rangle$ as a function of time t for the bulk. We do not present just $\langle \mathrm{MSD}(t) \rangle$, since below the non-thermal melting threshold $\langle \mathrm{MSD}_x(t) \rangle \approx \langle \mathrm{MSD}_x(t) \rangle$ differs significantly from $\langle \mathrm{MSD}_z(t) \rangle$, which behavior is dominantly given by the A_{1g} phonon oscillation. Below the non-thermal melting threshold, $\langle \mathrm{MSD}_{xy}(t) \rangle$ increases a bit after the increase of T_e and oscillates below the Lindemann stability limit indicating thermal phonon antisqueezing. Above the non-thermal melting threshold, $\langle \mathrm{MSD}_{xy}(t) \rangle$ exceeds quickly the Lindemann stability limit and shows the three stages of non-thermal melting: First the time-derivative of $\langle \mathrm{MSD}_{xy}(t) \rangle$ increases rapidly, decreases and then keeps constant, so that $\langle \mathrm{MSD}_{xy}(t) \rangle$ increases linearly with time.
3. The central row depicts the Peierls parameter as a function of time in bulk MD simulations, which is directly related to the A_{1g} phonon oscillation. We omit the corresponding diagrams above the non-thermal melting threshold, since the crystal structure melts immediately, so that the Peierls parameter is no longer defined. The increase of T_e induces a strong oscillation of the Peierls parameter.
4. The forth row shows the phonon bandstructure of the A7 structure. The optical phonon branches decrease slightly with increasing T_e. In addition, the frequency difference between the optical branches becomes smaller. The acoustic phonon branches decrease uniformly with increasing T_e. At 12000 K, parts of the acoustic phonon branches become negative.
5. The bottom series of figures shows the cohesive energies of the diamond-like, fcc, bcc and sc structures at the considered T_e's. To improve the visibility, we shifted the lattice parameter of the cohesive energy curve of the fcc structure by 0.6 nm. The increase of T_e reduces the absolute value of the cohesive energy curves and induces that the sc structure is not any more the energetically favorite structure among the four ones.

4.6 THz Emission from Coherent Phonon Oscillations in BNNTs

If a femtosecond laser-pulse excites a coherent phonon mode in matter, high amplitudes of the corresponding atomic oscillations can be induced. If the atomic bonds have a permanent electric dipole moment due to different electronegativities of the corresponding atoms, the laser induced atomic oscillations in the coherent phonon mode cause a time variation of the electrical dipole moments. The latter will emit THz radiation, since phonon frequencies in solids are located in the THz region. THz radiation is technological quite interesting, since it promises new sensing applications, like concealed weapon detection, chemical and biological agent detection, and medical diagnostics [32].

Boron nitride nanotubes (BNNTs) exhibit exceptional mechanical properties [33] and are nontoxic [34], which makes them more readily usable, especially for biological applications. Furthermore, BNNTs exhibit bonds with a permanent electric dipole moment, due to nitrogen's larger electronegativity compared to boron. A femtosecond laser-pulse excites three coherent phonon modes in thin BNNTs due to the displacive excitation mechanism. Therefore, one would expect that femtosecond-laser excited BNNTs emit THz radiation. To support this hypothesis, we simulated femtosecond-laser excited BNNTs with T_e-dependent DFT using CHIVES and analyzed the THz emission. For this, we implemented the calculation of the local electric dipole moment of the boron nitride bonds in CHIVES. We already published the obtained results in Ref. [35]. In literature, the first observation of THz emission from coherent lattice vibrations was experimentally reported in 1995 [31] on single-crystal tellurium.

4.6.1 Equilibrium Structure

The BNNT structure is created by cutting and rolling up a strip of hexagonal boron nitride (BN), which two dimensional crystal structure is generated by the grid vectors

$$\mathbf{a}_1 = d \begin{bmatrix} \sqrt{3} \\ 0 \\ 0 \end{bmatrix}, \quad \mathbf{a}_2 = d \begin{bmatrix} \frac{\sqrt{3}}{2} \\ \frac{3}{2} \\ 0 \end{bmatrix}, \tag{4.101}$$

and contains two atoms in the basis with the basis vectors

$$\mathbf{d}_1 = d \begin{bmatrix} 0 \\ 0 \\ 0 \end{bmatrix}, \quad \mathbf{d}_2 = d \begin{bmatrix} \frac{\sqrt{3}}{2} \\ -\frac{1}{2} \\ 0 \end{bmatrix}, \tag{4.102}$$

where d denotes the distance between two neighboring atoms. The boron atoms are located at basis $\mathbf{d}_1$ and the nitrogen atoms at basis $\mathbf{d}_2$. Thus, the translation vector of an atom localized in the hexagonal BN grid, which we put in the x, y-plane, is given by

$$\mathbf{T}_{n_1 n_2 \upsilon} = n_1\,\mathbf{a}_1 + n_2\,\mathbf{a}_2 + \upsilon\,\mathbf{d}_2 = d \begin{bmatrix} \sqrt{3}\,n_1 + \frac{\sqrt{3}}{2}\,n_2 + \frac{\sqrt{3}}{2}\,\upsilon \\ \frac{3}{2}\,n_2 - \frac{1}{2}\,\upsilon \\ 0 \end{bmatrix} \tag{4.103}$$

with $n_1, n_2 \in \mathbb{Z}$ and $\upsilon = 0$ corresponds to a boron and $\upsilon = 1$ to a nitrogen atom.

Definition 4.5 (*Wrapping vector* **w**) The wrapping vector $\mathbf{w} = [w_1, w_2, 0]^{\mathrm{t}}$ controls, how the strip is to be cut and which atom at the cut edge of the strip is mapped to which atom at the other cut edge when it is rolled up (see Fig. 4.68). This vector is defined with the help of two integer numbers (N_1, N_2) by

$$\mathbf{w} = N_1\,\mathbf{a}_1 + N_2\,\mathbf{a}_2 = d \begin{bmatrix} \sqrt{3}\,N_1 + \frac{\sqrt{3}}{2}\,N_2 \\ \frac{3}{2}\,N_2 \\ 0 \end{bmatrix}. \tag{4.104}$$

The diameter of the nanotube is given by the length of the wrapping vector

$$||\mathbf{w}|| = d\sqrt{3\left(N_1 + \frac{1}{2}\,N_2\right)^2 + \frac{9}{4}\,N_2^2} = \sqrt{3}\,d\,\sqrt{N_1^2 + N_2^2 + N_1\,N_2}. \tag{4.105}$$

The nanotubes are named by the two integer numbers (N_1, N_2). To construct the BNNT, we define the rotation matrix **M** that rotates from the unit vectors $\hat{\mathbf{e}}_x$, $\hat{\mathbf{e}}_y$ of the x, y-plane to the unit vectors $\hat{\mathbf{e}}_1$, $\hat{\mathbf{e}}_2$ of the strip:

$$\begin{aligned} \mathbf{M} &= \frac{1}{||\mathbf{w}||} \begin{bmatrix} w_1 & w_2 & 0 \\ -w_2 & w_1 & 0 \\ 0 & 0 & 1 \end{bmatrix} \\ &= \frac{1}{\sqrt{3}\,\sqrt{N_1^2 + N_2^2 + N_1\,N_2}} \begin{bmatrix} \sqrt{3}\,N_1 + \frac{\sqrt{3}}{2}\,N_2 & \frac{3}{2}\,N_2 & 0 \\ -\frac{3}{2}\,N_2 & \sqrt{3}\,N_1 + \frac{\sqrt{3}}{2}\,N_2 & 0 \\ 0 & 0 & 1 \end{bmatrix} \\ &= \frac{1}{2\,\sqrt{N_1^2 + N_2^2 + N_1\,N_2}} \begin{bmatrix} 2\,N_1 + N_2 & \sqrt{3}\,N_2 & 0 \\ -\sqrt{3}\,N_2 & 2\,N_1 + N_2 & 0 \\ 0 & 0 & 1 \end{bmatrix}. \end{aligned} \tag{4.106}$$

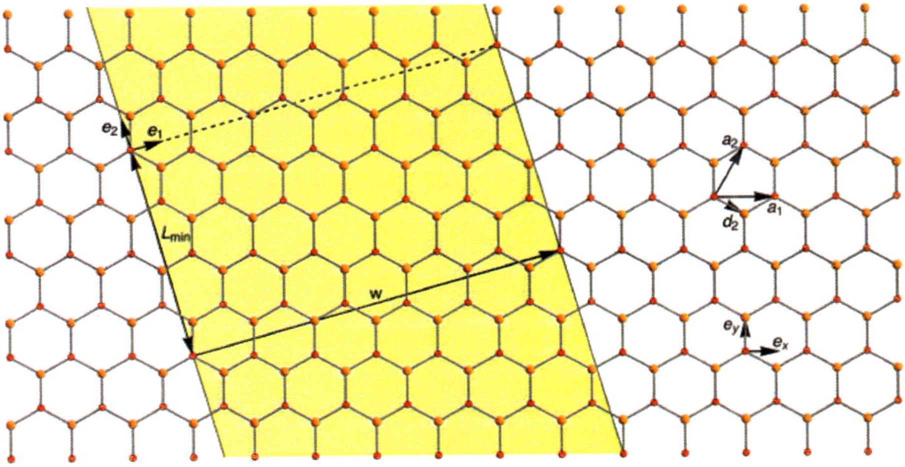

Fig. 4.68 Visualization how the stripe for rolling up the BNNT is cut off from hexagonal BN for the wrapping vector (5, 2) as an example. Boron atoms are drawn in red and nitrogen atoms in orange

The rotated translation vector $\mathbf{T}^{(\mathrm{rot})} = \left[T_1^{(\mathrm{rot})},\, T_2^{(\mathrm{rot})},\, 0\right]^{\mathrm{t}}$ of any atom in the hexagonal BN can be calculated with the help of the rotation matrix $\mathbf{M} \in \mathbb{R}^{3\times 3}$ from the translation vector $\mathbf{T}_{n_1 n_2 \upsilon}$ by

$$\begin{aligned}
&\mathbf{T}^{(\mathrm{rot})}(n_1, n_2, \upsilon) \\
&= \mathbf{M} \cdot \mathbf{T}_{n_1 n_2 \upsilon} \\
&= \frac{d}{2\sqrt{N_1^2 + N_2^2 + N_1 N_2}} \begin{bmatrix} (2N_1 + N_2)\sqrt{3}\left(n_1 + \frac{1}{2} n_2 + \frac{1}{2}\upsilon\right) + \sqrt{3} N_2 \left(\frac{3}{2} n_2 - \frac{1}{2}\upsilon\right) \\ -3N_2 \left(n_1 + \frac{1}{2} n_2 + \frac{1}{2}\upsilon\right) + (2N_1 + N_2)\left(\frac{3}{2} n_2 - \frac{1}{2}\upsilon\right) \\ 0 \end{bmatrix} \\
&= \frac{d}{2\sqrt{N_1^2 + N_2^2 + N_1 N_2}} \begin{bmatrix} \sqrt{3}\left((2N_1 + N_2) n_1 + (N_1 + 2N_2) n_2 + N_1 \upsilon\right) \\ -3N_2 n_1 + 3N_1 n_2 - (N_1 + 2N_2)\upsilon \\ 0 \end{bmatrix}.
\end{aligned} \tag{4.107}$$

If the strip is rolled up to a nanotube, $T_1^{(\mathrm{rot})}$ runs around the nanotube and $T_2^{(\mathrm{rot})}$ points in the direction of the nanotube axis. To derive the minimal periodical length L_{min} of a given BNNT with wrapping vector (N_1, N_2), we have to determine the two smallest integer numbers $(\tilde{n}_1, \tilde{n}_2) \neq (0, 0)$, so that

$$T_1^{(\mathrm{rot})}(\tilde{n}_1, \tilde{n}_2, 0) = 0.$$

We have to consider this condition, because then the boron atom characterized by $\tilde{n}_1, \tilde{n}_2$ lies on the cutout line through the zero point, and is one of the two boron atoms, which among all boron atoms of the cutout line are closest to the zero point (see Fig. 4.68). Using Eq. (4.107), we get

$$\begin{aligned} &(2\,N_1 + N_2)\,\tilde{n}_1 + (N_1 + 2\,N_2)\,\tilde{n}_2 = 0, \\ \Leftrightarrow \qquad & \tilde{n}_2 = -\,\frac{2\,N_1 + N_2}{N_1 + 2\,N_2}\,\tilde{n}_1. \end{aligned} \tag{4.108}$$

Since $\tilde{n}_2$ must be always an integer number, we have

$$\tilde{n}_1 = \frac{N_1 + 2\,N_2}{\gcd(2\,N_1 + N_2,\, N_1 + 2\,N_2)}, \tag{4.109}$$

where $\gcd(2\,N_1 + N_2, N_1 + 2\,N_2)$ denotes the greatest common divisor of the two integer numbers $2\,N_1 + N_2$ and $N_1 + 2\,N_2$. We obtain the minimal periodical length $L_{\min}$ of the nanotube, if we insert $\tilde{n}_1, \tilde{n}_2$ and $\upsilon = 0$ in $T_2^{(\mathrm{rot})}$:

$$L_{\min} = \left| T_2^{(\mathrm{rot})}(\tilde{n}_1, \tilde{n}_2, 0) \right| = \frac{3\,d\,\left| -N_2\,\tilde{n}_1 + N_1\,\tilde{n}_2 \right|}{2\sqrt{N_1^2 + N_2^2 + N_1\,N_2}}. \tag{4.110}$$

To finally set up the simulation cell for the BNNT, we utilize periodic boundary conditions and repeat N_{L} times the minimal periodical length $L_{\min}$ in the simulation cell. Now the atoms characterized by n_1, n_2, υ belong to the simulation cell, if

$$\begin{aligned} 0 \le & T_1^{(\mathrm{rot})}(n_1, n_2, 0) < \left|\left|\mathbf{w}\right|\right|, \\ 0 \le & T_2^{(\mathrm{rot})}(n_1, n_2, 0) < N_{\mathrm{L}} \cdot L_{\min} \end{aligned}$$

is fulfilled. In the above conditions, the value of υ is not relevant, because either both basis atoms belong to the simulation cell or not. We obtain further by inserting Eqs. (4.105), (4.110), and (4.107):

$$0 \le (2\,N_1 + N_2)\,n_1 + (N_1 + 2\,N_2)\,n_2 \quad < \quad 2\,\left(N_1^2 + N_2^2 + N_1\,N_2\right), \tag{4.111}$$

$$0 \le -\,N_2\,n_1 + N_1\,n_2 \quad < \quad N_{\mathrm{L}}\left|N_2\,\tilde{n}_1 + (N_1 + N_2)\,\tilde{n}_2\right|. \tag{4.112}$$

The atoms of the hexagonal BN grid, which fulfill the above conditions, are put to form a nanotube by

$$\mathbf{r}_{n_1 n_2 \upsilon} = \begin{bmatrix} \frac{L}{2} + \frac{||\mathbf{w}||}{2\pi}\,\sin\left(2\pi\,\frac{T_1^{(\mathrm{rot})}(n_1,n_2,\upsilon)}{||\mathbf{w}||}\right) \\ \frac{L}{2} + \frac{||\mathbf{w}||}{2\pi}\,\cos\left(2\pi\,\frac{T_1^{(\mathrm{rot})}(n_1,n_2,\upsilon)}{||\mathbf{w}||}\right) \\ T_2^{(\mathrm{rot})}(n_1, n_2, \upsilon) \end{bmatrix}. \tag{4.113}$$

Fig. 4.69 Chiral BNNT

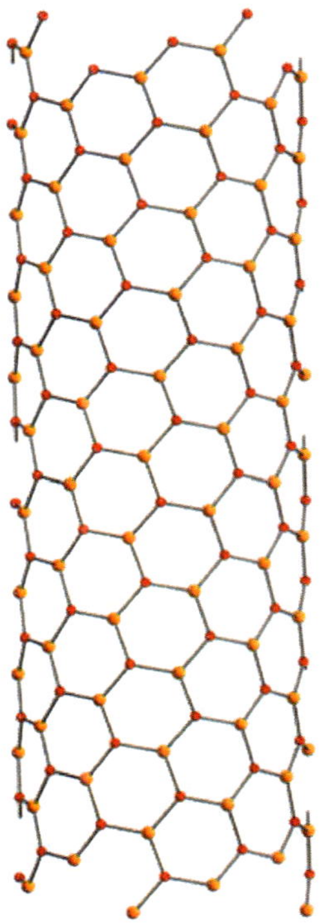

where L denotes the distance between the central axes of the nanotubes in x- and y-direction due to the periodic boundary conditions.

Definition 4.6 (*Different types of nanotubes*) Nanotubes are divided into three groups with the help of the wrapping vector (N_1, N_2):

- Nanotubes with $N_2 = 0$ are called **zigzag**,
- with $N_1 = N_2$ are called **armchair**,
- all other are called **chiral**.

We depict the different types of BNNTs in Figs. 4.69, 4.70, and 4.71. To increase visibility, we always show only half of the BNNT.

4.6.2 *Displacive Excitation of Coherent Phonons in BNNTs*

We studied the influence of a femtosecond-laser excitation on different types of BNNTs with the help of T_e-dependent DFT using CHIVES. For this, we considered

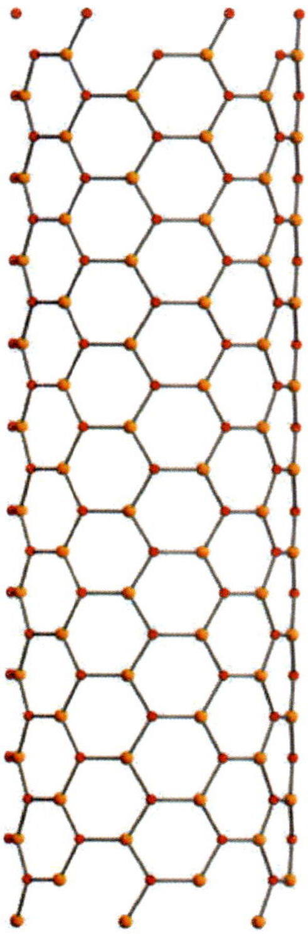

Fig. 4.70 Armchair BNNT

the (3, 3) and (5, 5) armchair, the (4, 2) and (8, 2) chiral, and the (5, 0), (6, 0), (7, 0), (8, 0), and (9, 0) zigzag BNNTs. We focused on zigzag BNNTs, because they are preferably synthesized [66–68]. In addition, the (5, 0) zigzag BNNT is theoretically the thinnest stable one [69] and the (6, 0) zigzag BNNT was the thinnest fabricated zigzag BNNT so far [70]. The thinnest so far fabricated BNNT was the (3, 3) armchair [71].

We studied infinitely long BNNTs by applying periodic boundary conditions and we used two irreducible **k**-points in the first Brillouin zone for the electrons in CHIVES. At first, we derived for each considered BNNT the equilibrium nearest neighbor distance d_{opt} at $T_{\text{e}} = 316\,\text{K}$ (1 mHa). For different values of nearest neighbor distances d, we derived the corresponding equilibrium Helmholtz free energy $F(d)$ by setting up the crystal structure of the nanotube as described in Sect. 4.6.1 and applying the FIRE algorithm [60]. Using the obtained $F(d)$ values, we fit F as a function of d as a polynomial of degree three. From this fit, we found the optimal nearest neighbor distance d_{opt} that minimizes F as a function of d. Then we set up the crystal structure as described in Sect. 4.6.1 using d_{opt} and applied the FIRE algorithm [60] to derive the equilibrium atomic coordinates for $T_{\text{e}} = 316$ K and 18900 K (60 mHa), which correspond to the electronic state before and after a

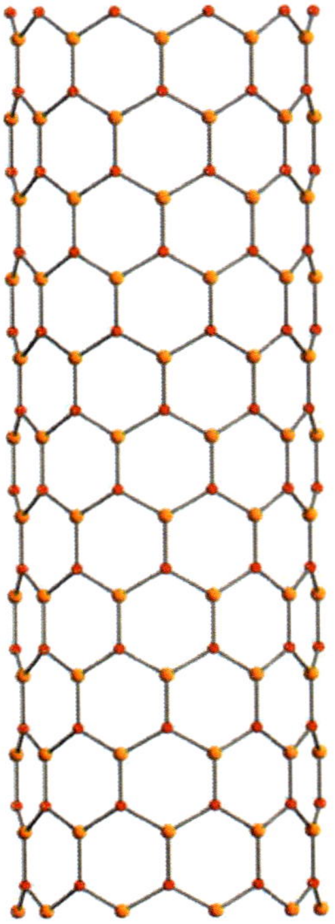

Fig. 4.71 Zigzag BNNT

typical femtosecond-laser excitation, respectively. We denote by $\mathbf{r}_i^{(\mathrm{opt})}(T_\mathrm{e}) \in \mathbb{R}^3$ the equilibrium position of atom i at an electronic temperature T_e and denote by

$$\mathbf{u}_i(T_\mathrm{e}) = \mathbf{r}_i^{(\mathrm{opt})}(T_\mathrm{e}) - \mathbf{r}_i^{(\mathrm{opt})}(316\,\mathrm{K}). \tag{4.114}$$

the T_e-induced change of the equilibrium position of atom i. Using $\mathbf{u}_i = [u_{ix}, u_{iy}, u_{iz}]^\mathrm{t}$, we calculated the mass-normalized displacements

$$U_{i\alpha}(T_\mathrm{e}) = \sqrt{m_i}\, u_{i\alpha}(T_\mathrm{e}), \tag{4.115}$$

where m_i is the mass of atom i, $\alpha = x, y, z$, and denote all of them in the vector $\mathbf{U}(T_\mathrm{e}) \in \mathbb{R}^{3\,N_\mathrm{at}}$ (c.f. Eq. (2.36)), where N_at denotes the total number of atoms in the supercell.

In addition, we calculated the dynamical matrix $\mathbf{D}(T_\mathrm{e}) \in \mathbb{R}^{3\,N_\mathrm{at} \times 3\,N_\mathrm{at}}$ at $T_\mathrm{e} = 18900\,\mathrm{K}$ by displacing separately one boron and one nitrogen atom in all room directions and by utilizing symmetry arguments see Sect. 4.1.2. By diagonalizing the dynamical matrix $\mathbf{D}(T_\mathrm{e})$ with the matrix $\mathbf{C}(T_\mathrm{e}) \in \mathbb{R}^{3\,N_\mathrm{at} \times 3\,N_\mathrm{at}}$, we obtain the phonon eigenmodes of the BNNT (see Sect. 2.2.1). With the help of the matrix $\mathbf{C}(T_\mathrm{e})$, we derived the T_e-induced displacements in direction of the phonon eigenmodes (c.f. Eq. (2.40))

$$\mathscr{U}(T_\mathrm{e}) = \mathbf{C}(T_\mathrm{e}) \cdot \mathbf{U}(T_\mathrm{e}).$$

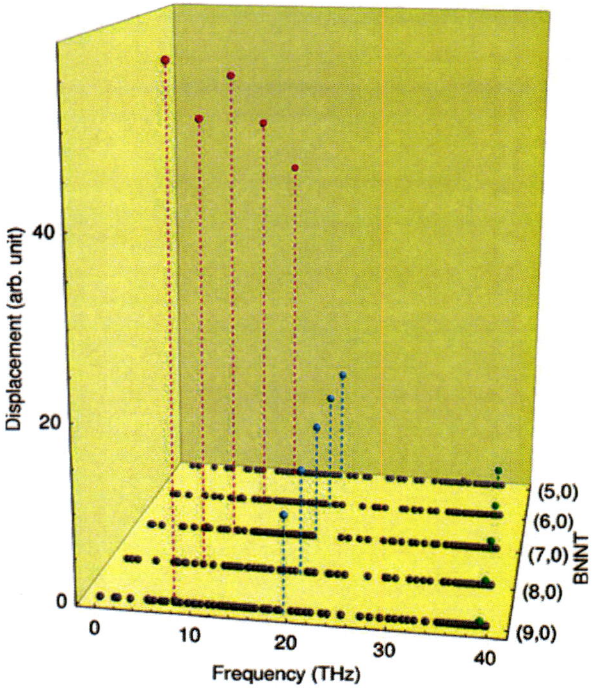

Fig. 4.72 Laser-induced displacements in direction of the phonon eigenmodes at $T_e = 18900$ K are shown for the studied zigzag BNNTs

In this way, we found out that three coherent phonon modes are displacively excited in thin zigzag BNNTs. One can clearly see this in Fig. 4.72, where the laser-excited modes are highlighted in color beside the other modes shown as black points. We identified the excited modes as the radial breathing mode, radial buckling mode and the longitudinal bond stretching mode [72]. In Figs. 4.74, 4.75, and 4.76, we present the corresponding atomic movements in these phonon modes for the (5, 0) zigzag BNNT.

In thin armchair and chiral BNNTs, only the two radial modes are excited. In Fig. 4.73, the displacements and frequencies of the excited modes are presented as a function of the diameter for the different studied BNNTs.

The femtosecond laser-pulse excited electrons from bonding to antibonding states, which causes softening of the interatomic bonding among other effects. Therefore, the atoms prefer larger distances to each other after the laser excitation, so that the equilibrium diameter of the nanotube increases, which explains the excitation of the radial breathing mode. The nitrogen atoms prefer smaller bond angles in contrast to the boron atoms [72]. Due to this, the nitrogen atoms are placed more outside and the boron atoms more inside the nanotube in the equilibrium structure, which is called buckling [73]. The laser excitation changes the bonding angle preferences of the boron and nitrogen atoms, modifies the equilibrium buckling and leads to the excitation of the radial buckling mode. Also the excitation of the longitudinal bond stretching mode has its origin in the variation of the equilibrium bond angles.

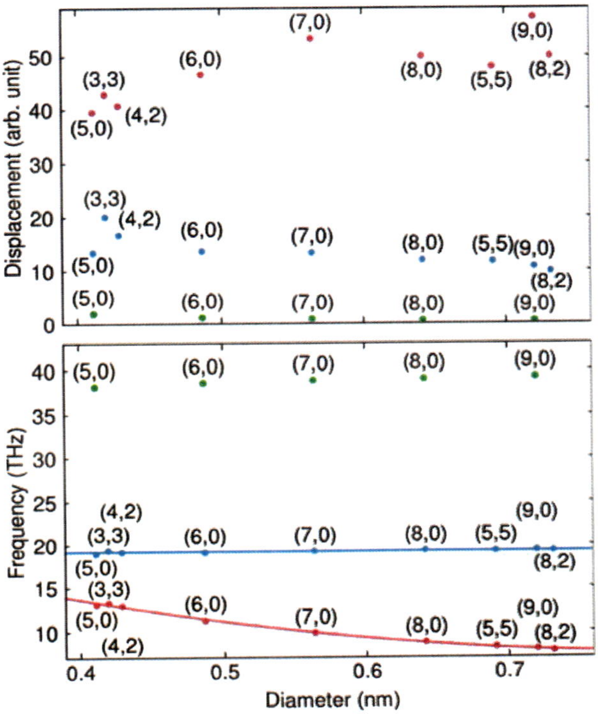

Fig. 4.73 Displacement (top) and frequency (bottom) of the three laser-excited phonon modes at $T_e = 18900$ K are shown for the studied BNNTs adapted from Fig. 3 of Ref. [35]. The lines are fits (see text)

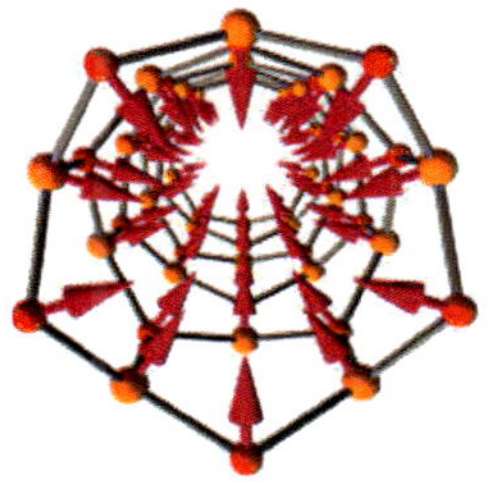

Fig. 4.74 Atomic movements in the radial breathing mode

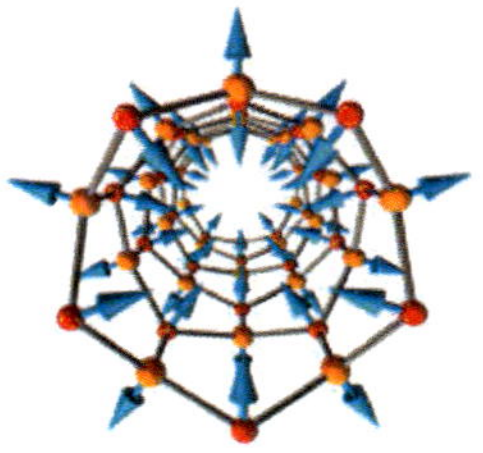

Fig. 4.75 Atomic movements in the radial buckling mode

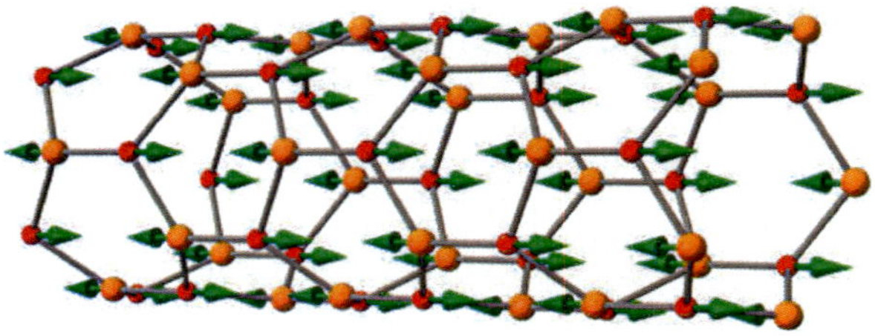

Fig. 4.76 Atomic movements in the longitudinal bond stretching mode

Thin BNNTs show a greater buckling and the buckling vanishes for BNNTs with large diameters [72]. Hence, the excitation amplitudes of the radial buckling mode and the longitudinal bond stretching mode decrease with increasing tube diameter, as one can see in the top part of Fig. 4.73. As a consequence, only the radial breathing mode can be excited in BNNTs with large diameters like in carbon nano tubes (CNTs) [74].

The frequency of the radial breathing mode decreases with increasing tube diameter $\mathbb{d}$ like $c_1 + c_2/\mathbb{d}$ with constants c_1, c_2, as one can see in the bottom part of Fig. 4.73. This behavior is also present in CNTs and was previously theoretically justified [72]. This study also shows that the frequencies of the radial buckling and longitudinal bond stretching mode weakly depend on the tube diameter. In agreement with this, we found that the frequency of the radial buckling mode is equal for all studied BNNTs indicated by the horizontal line in the bottom part of Fig. 4.73. In addition, we found that the frequency of the longitudinal bond stretching mode very weakly increases from 38.17 THz up to 39.17 THz with increasing tube diameter. Arenal et al. measured the longitudinal bond stretching mode frequency of 41.07 THz for single-walled BNNTs with large diameter using Raman spectroscopy [75]. They also measured a longitudinal stretching mode frequency of 40.92 THz for hexagonal BN, which can be interpreted as a BNNT with infinite diameter.

4.6.3 THz Radiation from Coherent Phonon Oscillations in the (5, 0) Zigzag BNNT

Now we consider to the (5, 0) zigzag BNNT, because it exhibits the largest amplitudes of the radial buckling and of the longitudinal bond stretching mode among all studied BNNTs. We utilized a supercell containing 30 boron and 30 nitrogen atoms showed in Fig. 4.77. For this cell, the minimal periodical length L_{min} is $N_{\text{L}} = 3$ times repeated. We applied periodical boundary conditions and used two irreducible **k**-points in the first Brillouin zone for the electrons.

In Fig. 4.78, the electronic charge density of the valence electrons is shown in a plane though the BNNT for the equilibrium structure at $T_{\text{e}} = 316\,\text{K}$, which we derived previously. One can clearly see that the charge density around nitrogen is significant

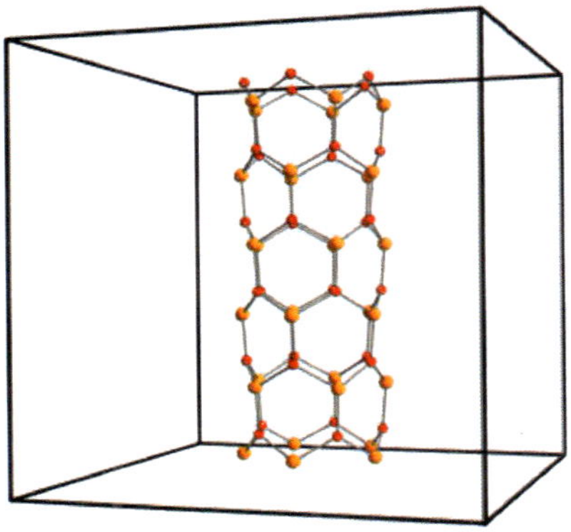

Fig. 4.77 Supercell containing 30 boron and 30 nitrogen atoms

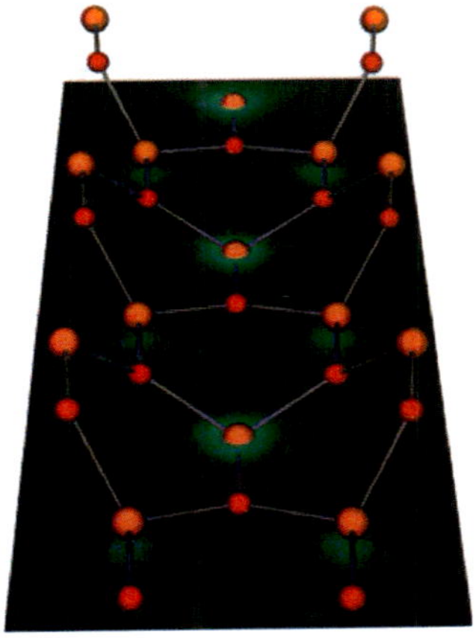

Fig. 4.78 Electronic charge density in a plane through the (5, 0) zigzag BNNT is shown for the equilibrium structure at $T_e = 316$ K adapted from Fig. 4a of Ref. [35]. For clarity, we show only the bottom half of the BNNT

higher compared with boron. This is not surprising, because the electronegativity of nitrogen is larger than that of boron. Thus, the boron nitride bonds exhibit a permanent electrical dipole moment, as initially expected.

Now we want to calculate the local dipole moment of one ring of the (5, 0) zigzag BNNT, which corresponds to the minimal periodical length L_{min} with 10 boron and 10 nitrogen atoms. We denote the indices of the ring atoms in the set I. The local dipole moment of the ring is given by

$$\mathbf{d}^{(I)} = \mathbf{d}^{(I)}_{\mathrm{ions}} + \mathbf{d}^{(I)}_{\mathrm{e}}, \tag{4.116}$$

where $\mathbf{d}^{(I)}_{\mathrm{ions}}$ is the local dipole moment caused by the effective charge of the atomic nuclei of the atomic subset I calculated by

$$\mathbf{d}^{(I)}_{\mathrm{ions}} = \sum_{i \in I} \mathrm{q}_i \, \mathbf{r}_i \tag{4.117}$$

and $\mathbf{d}_\mathrm{e}^{(I)}$ is the local dipole moment of the valence electrons of the atomic subset I calculated by

$$\mathbf{d}_\mathrm{e}^{(I)} = -e \int d^3r \; \mathbf{r} \; n_\mathrm{e}^{(I)}(\mathbf{r}). \tag{4.118}$$

Here, $n_\mathrm{e}^{(I)}(\mathbf{r})$ denotes the electronic charge density of the atomic subset I. Within T_e-dependent DFT, the total electronic charge density $n_\mathrm{e}(\mathbf{r})$ is calculated by (c.f. (2.242))

$$n_\mathrm{e}(\mathbf{r}) = \sum_m f_\mathrm{F}(\varepsilon_m, T_\mathrm{e}) \, \varphi_m^*(\mathbf{r}) \, \varphi_m(\mathbf{r}), \tag{4.119}$$

where $\varphi_m(\mathbf{r})$ are the Kohn-Sham orbitals, ε_m are the Kohn-Sham energies, and $f_\mathrm{F}(\varepsilon_m, T_\mathrm{e})$ is the Fermi distribution. The Kohn-Sham Orbitals φ_m are a linear combination of the basis functions χ_{ij} with coefficients c_{mij} (see Sect. 2.3.5)

$$\varphi_m(\mathbf{r}) = \sum_i \sum_j c_{mij} \, \chi_{ij}(\mathbf{r}), \tag{4.120}$$

where the sum over i runs over all atoms and the sum over j runs over all different types of basis functions. Since CHIVES uses atom centered basis functions (see Sect. 3.6), one can assign any basis function to a specific atom. This allows us to calculate the electronic charge density of the atomic subset I just by

$$n_\mathrm{e}^{(I)}(\mathbf{r}) = \sum_m f_\mathrm{F}(\varepsilon_m, T_\mathrm{e}) \sum_{i,i' \in I} \sum_{j,j'} c_{mij}^* \, c_{mi'j'} \, \chi_{ij}^*(\mathbf{r}) \, \chi_{i'j'}(\mathbf{r}). \tag{4.121}$$

This calculation method of the local dipole moment was developed and validated for ground state DFT in [76]. We extended this method to T_e-dependent DFT and implemented it in CHIVES.

We took the $T_\mathrm{e} = 316$ K equilibrium structure of the (5, 0) zigzag BNNT and increased T_e to the constant value of 18900 K to simulate a femtosecond-laser excitation, and performed a MD simulation with CHIVES using a time step of 1 fs. We calculated the local dipole moment $\mathbf{d}^I$ of the ring at each time step using the method described above. The local dipole moment $\mathbf{d}^I$ points in z-direction and varies significantly during time due to the large laser-induced atomic oscillations. Since only the second derivative with respect to the time is relevant for the emitted electric field $\mathbf{E}$ (3.69) and magnetic field $\mathbf{B}$ (3.66), we show in Fig. 4.79 the second time derivative of the z-component d_z of $\mathbf{d}^{(I)}$.

Since $\mathbf{d}^{(I)} = d_z \, \hat{\mathbf{e}}_z$ points in z-direction, we obtain, in spherical coordinates r, ϑ, ϕ, for the emitted magnetic field

$$\mathbf{B}(r, \vartheta, \phi, t) \overset{(3.66)}{\approx} \frac{\mu_0 \, \ddot{d}_z(t_0)}{4\pi \, \mathrm{c} \, r} \sin(\vartheta) \, \hat{\mathbf{e}}_\vartheta \tag{4.122}$$

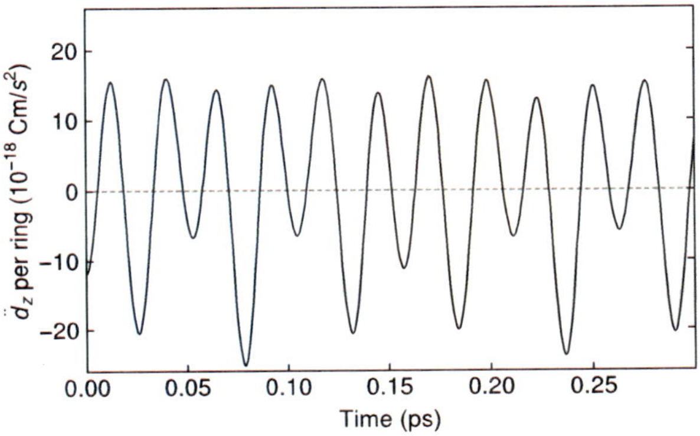

Fig. 4.79 Second time derivative of the z-component of the electric dipole moment of one ring of the (5, 0) zigzag BNNT is shown as a function of time at $T_e = 18900$ K

and for the emitted electric field

$$\mathbf{E}(r, \vartheta, \phi, t) \overset{(3.69)}{\approx} \frac{\mu_0 \, \ddot{d}_z(t_0)}{4\pi \, r} \sin(\vartheta) \, \hat{\mathbf{e}}_\phi, \tag{4.123}$$

where $t_0 \overset{(3.65)}{\approx} t - \frac{r}{c}$ denotes the retarded time. The related poynting vector is given by

$$\mathbf{S}(r, \vartheta, \phi, t) \overset{(3.76)}{=} \frac{1}{\mu_0} (\mathbf{E} \times \mathbf{B}) \approx \frac{\mu_0 \left(\ddot{d}_z(t_0)\right)^2}{16\pi^2 \, \mathbb{c} \, r^2} \sin^2(\vartheta) \, \hat{\mathbf{e}}_r. \tag{4.124}$$

With the help of the poynting vector, we obtain for the total emitted power $\mathbb{P}(t)$ at time t:

$$\begin{aligned}
\mathbb{P}(t) &= \int_{\mathsf{S}} d\mathbf{a}^{\mathrm{t}} \cdot \mathbf{S} \\
&\approx \int_0^{2\pi} d\phi \int_0^{\pi} d\vartheta \, r^2 \, \sin(\vartheta) \, \frac{\mu_0 \left(\ddot{d}_z(t_0)\right)^2}{16\pi^2 \, \mathbf{c} \, r^2} \sin^2(\vartheta) \\
&= \frac{\mu_0}{8\pi \, \mathbf{c}} \left(\ddot{d}_z(t_0)\right)^2 \underbrace{\int_0^{\pi} d\vartheta \, \sin^3(\vartheta)}_{=\frac{4}{3}} \\
&= \frac{\mu_0}{6\pi \, \mathbf{c}} \left(\ddot{d}_z(t_0)\right)^2.
\end{aligned} \tag{4.125}$$

In Fig. 4.80, we show the spectrum of the electrical field that is emitted from one ring of the (5, 0) zigzag BNNT at $T_e = 18900$ K. One can clearly see three peaks, which correspond to the three frequencies of the excited coherent phonons. At 25 THz, an additional broad and very weak peak is located, which correspond to numerical noise and can be neglected. The peak with the highest frequency corresponding to the longitudinal bond stretching mode has the highest amplitude although the

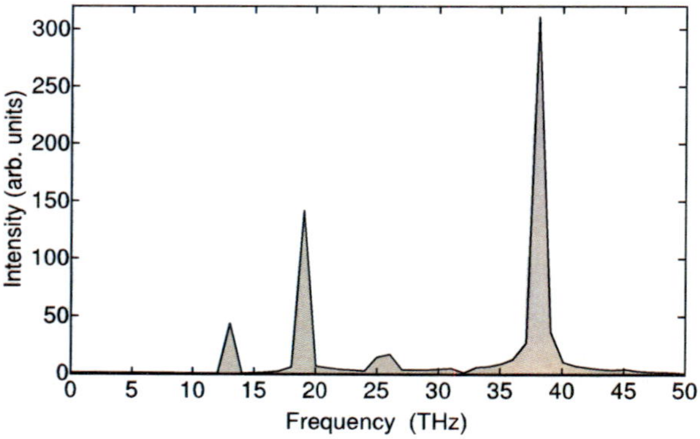

Fig. 4.80 Spectrum of the electrical field emitted from one ring of the (5, 0) zigzag BNNT is shown at $T_e = 18900$ K adapted from Fig. 4b of Ref. [35]

oscillation amplitude of this mode is the smallest compared to the other modes (see Fig. 4.73). The reason for this is that the atomic motions related to this mode emit most efficiently, because the boron and nitrogen atoms move more directly against each other than in the two other modes, as one can see in Figs. 4.74, 4.75, and 4.76. In addition, we verified, that the used simulation cell containing 60 atoms ($N_L = 3$) is big enough. For this, we repeated the calculations with a simulation cell containing 80 atoms ($N_L = 4$) and a simulation cell containing 100 atoms ($N_L = 5$) and found no differences in the result compared with the used simulation cell.

In an experimental setup, the measurement of the emitted radiation coming from an excited bundle of BNNTs may be much more suitable, since the emitted power will be higher and the fabrication of single isolated BNNTs is much more complicated. Therefore, we put (5, 0) zigzag BNNTs, each one consisting of a periodical repetition of rings, together to a closed packed structure. We calculated the emitted total average power $\langle \mathbb{P}(t) \rangle$ and obtained $4.2 \cdot 10^{-31}$ W/mm^3 for a single laser pulse excitation that increases T_e to 18900 K. To reach this T_e, the bundle of BNNTs had to absorb 2.2 J/mm^3. To derive this value, we first calculated the difference $\Delta E = E(18900\,\text{K}) - E(316\,\text{K})$ of the internal energy $E(T_e) = F(T_e) + T_e\, S_e(T_e)$ of the equilibrium structure of the (5, 0) zigzag BNNT with 60 atoms. Here, $S_e(T_e)$ denotes the electronic entropy and $F(T_e)$ the Helmholtz free energy obtained directly from T_e-dependent DFT. Then we took the nanotube density in the bundle into account. Commonly, THz pulses are created by photoconductive antennas, which reach $1.1 \cdot 10^{-6}$ W [77]. This value is, however, much bigger than the intensity obtained from the laser-excited bundle of BNNTs. But the THz radiation of the BNNTs can be shaped, since one can control the excitation of the three-laser excited coherent phonon modes by using several femtosecond laser-pulses [35].

16. W. Bludau, A. Onton, W. Heinke, J. Appl. Phys. **45**(4), 1846 (1974). https://doi.org/10.1063/1.1663501
17. V. Alex, S. Finkbeiner, J. Weber, J. Appl. Phys. **79**(9), 6943 (1996). https://doi.org/10.1063/1.362447
18. M. Beye, F. Sorgenfrei, W.F. Schlotter, W. Wurth, A. Föhlisch, Proc. Natl. Acad. Sci. **107**(39), 16772 (2010). https://doi.org/10.1073/pnas.1006499107. https://www.pnas.org/content/107/39/16772
19. M. Schultze, K. Ramasesha, C. Pemmaraju, S. Sato, D. Whitmore, A. Gandman, J.S. Prell, L.J. Borja, D. Prendergast, K. Yabana, D.M. Neumark, S.R. Leone, Science **346**(6215), 1348 (2014). https://doi.org/10.1126/science.1260311. http://science.sciencemag.org/content/346/6215/1348
20. P. Stampfli, K.H. Bennemann, Phys. Rev. B **49**, 7299 (1994). https://doi.org/10.1103/PhysRevB.49.7299
21. N. Medvedev, M. Kopecky, J. Chalupsky, L. Juha, Phys. Rev. B **99**, 100303 (2019). https://doi.org/10.1103/PhysRevB.99.100303
22. K. Ishioka, M. Kitajima, O.V. Misochko, J. Appl. Phys. **103**(12), 123505 (2008). https://doi.org/10.1063/1.2940130
23. M. Hase, K. Ushida, M. Kitajima, J. Phys. Soc. Jpn. **84**(2), 024708 (2015). https://doi.org/10.7566/JPSJ.84.024708
24. B.N. Mironov, V.O. Kompanets, S.A. Aseev, A.A. Ishchenko, O.V. Misochko, S.V. Chekalin, E.A. Ryabov, JETP Lett. **103**(8), 531 (2016). https://doi.org/10.1134/S0021364016080099
25. B.N. Mironov, V.O. Kompanets, S.A. Aseev, A.A. Ischenko, I.V. Kochikov, O.V. Misochko, S.V. Chekalin, E.A. Ryabov, J. Exp. Theor. Phys. **124**(3), 422 (2017). https://doi.org/10.1134/S1063776117020145
26. L. Waldecker, T. Vasileiadis, R. Bertoni, R. Ernstorfer, T. Zier, F.H. Valencia, M.E. Garcia, E.S. Zijlstra, Phys. Rev. B **95**, 054302 (2017). https://doi.org/10.1103/PhysRevB.95.054302
27. H.J. Beister, K. Strössner, K. Syassen, Phys. Rev. B **41**, 5535 (1990). https://doi.org/10.1103/PhysRevB.41.5535
28. E.S. Zijlstra, N. Huntemann, M.E. Garcia, New J. Phys. **10**(3), 033010 (2008). http://stacks.iop.org/1367-2630/10/i=3/a=033010
29. N. Huntemann, E.S. Zijlstra, M.E. Garcia, Appl. Phys. A **96**(1), 19 (2009). https://doi.org/10.1007/s00339-009-5175-5
30. T.N. Kolobyanina, S.S. Kabalkina, L.F. Vereshchagin, L.V. Fedina, Zh. Eksp. Teor. Fiz. **55**, 164 (1968). http://www.jetp.ac.ru/cgi-bin/dn/e_031_02_0259.pdf
31. T. Dekorsy, H. Auer, C. Waschke, H.J. Bakker, H.G. Roskos, H. Kurz, V. Wagner, P. Grosse, Phys. Rev. Lett. **74**, 738 (1995). https://doi.org/10.1103/PhysRevLett.74.738
32. K. Sakai, M. Tani, *Terahertz Optoelectronics* (Springer, 2005). https://www.springer.com/de/book/9783540200130
33. J. Garel, I. Leven, C. Zhi, K. Nagapriya, R. Popovitz-Biro, D. Golberg, Y. Bando, O. Hod, E. Joselevich, Nano Lett. **12**(12), 6347 (2012). https://doi.org/10.1021/nl303601d. http://pubs.acs.org/doi/abs/10.1021/nl303601d
34. J.C. Carrero-Sanchez, A.L. Elías, R. Mancilla, G. Arrellin, H. Terrones, J.P. Laclette, M. Terrones, Nano Lett. **6**(8), 1609 (2006). https://doi.org/10.1021/nl060548p
35. B. Bauerhenne, E.S. Zijlstra, A. Kalitsov, M.E. Garcia, ACS Appl. Nano Mater. **1**, 6932 (2018). https://doi.org/10.1021/acsanm.8b01716
36. D. Frekel, B. Smit, *Understanding Molecular Simulation* (Academic Press, 2002)
37. W.C. Swope, H.C. Andersen, P.H. Berens, K.R. Wilson, J. Chem. Phys. **76**(1), 637 (1982). https://doi.org/10.1063/1.442716
38. L. Verlet, Phys. Rev. **159**, 98 (1967). https://doi.org/10.1103/PhysRev.159.98
39. M.P. Allen, D.J. Tildesley, *Computer Simulation of Liquids* (Clarendon Press, Oxford, 1991)
40. F. Schwabl, *Statistische Mechanik*, 3rd edn. (Springer, 2006). https://www.springer.com/de/book/9783540310952
41. W. Nolting, *Grundkurs Theoretische Physik 6 Statistische Physik*, 7th edn. (Springer, 2014)
42. G.M. Fichtenholz, *Differential- und Integralrechnung 3*, 12th edn. (Verlag Harri Deutsch, 1992)

43. H.C. Andersen, J. Chem. Phys. **72**(4), 2384 (1980). https://doi.org/10.1063/1.439486
44. M.J. Wichura, J. R. Stat. Soc. Ser. C (Appl. Stat.) **37**(3), 477 (1988). http://www.jstor.org/stable/2347330
45. M. Harb, R. Ernstorfer, T. Dartigalongue, C.T. Hebeisen, R.E. Jordan, R.J.D. Miller, J. Phys. Chem. B **110**(50), 25308 (2006). https://doi.org/10.1021/jp064649n. PMID: 17165976
46. M. Harb, R. Ernstorfer, C.T. Hebeisen, G. Sciaini, W. Peng, T. Dartigalongue, M.A. Eriksson, M.G. Lagally, S.G. Kruglik, R.J.D. Miller, Phys. Rev. Lett. **100**, 155504 (2008). https://doi.org/10.1103/PhysRevLett.100.155504
47. L. Waldecker, R. Bertoni, R. Ernstorfer, J. Vorberger, Phys. Rev. X **6**, 021003 (2016). https://doi.org/10.1103/PhysRevX.6.021003
48. S.W. Epp, M. Hada, Y. Zhong, Y. Kumagai, K. Motomura, S. Mizote, T. Ono, S. Owada, D. Axford, S. Bakhtiarzadeh, H. Fukuzawa, Y. Hayashi, T. Katayama, A. Marx, H.M. Müller-Werkmeister, R.L. Owen, D.A. Sherrell, K. Tono, K. Ueda, F. Westermeier, R.J.D. Miller, Struct. Dyn. **4**(5), 054308 (2017). https://doi.org/10.1063/1.4999701
49. N.W. Ashcroft, N.D. Mermin, *Solid State Physics* (Saunders College Publishing, 1976)
50. R. Gross, A. Marx, *Festkörperphysik* (Oldenburg Wissenschaftsverlag GmbH, München, 2012)
51. D.R. Lide, *Handbook of Chemistry and Physics*, 84th edn. (CRC press, 2004)
52. B. Farid, R.W. Godby, Phys. Rev. B **43**, 14248 (1991). https://doi.org/10.1103/PhysRevB.43.14248
53. G.X. Zhang, A.M. Reilly, A. Tkatchenko, M. Scheffler, New J. Phys. **20**(6), 063020 (2018). https://doi.org/10.1088/1367-2630/aac7f0
54. L. Shokeen, P.K. Schelling, J. Appl. Phys. **109**(7), 073503 (2011). https://doi.org/10.1063/1.3554410
55. S.L. Johnson, P. Beaud, E. Vorobeva, C.J. Milne, E.D. Murray, S. Fahy, G. Ingold, Phys. Rev. Lett. **102**, 175503 (2009). https://doi.org/10.1103/PhysRevLett.102.175503
56. I. Štich, R. Car, M. Parrinello, Phys. Rev. Lett. **63**, 2240 (1989). https://doi.org/10.1103/PhysRevLett.63.2240
57. P. Blaha, K. Schwarz, G.K.H. Madsen, D. Kvasnicka, J. Luitz, R. Laskowski, F. Tranand, L.D. Marks, *WIEN2k, An Augmented Plane Wave + Local Orbitals Program for Calculating Crystal Properties* (Karlheinz Schwarz, Techn. Universität Wien, Austria, 2018). ISBN 3-9501031-1-2
58. J.P. Perdew, Int. J. Quantum Chem. **28**(S19), 497 (1985). https://doi.org/10.1002/qua.560280846. https://onlinelibrary.wiley.com/doi/abs/10.1002/qua.560280846
59. S.V. Faleev, M. van Schilfgaarde, T. Kotani, F.m.c. Léonard, M.P. Desjarlais, Phys. Rev. B **74**, 033101 (2006). https://doi.org/10.1103/PhysRevB.74.033101
60. E. Bitzek, P. Koskinen, F. Gähler, M. Moseler, P. Gumbsch, Phys. Rev. Lett. **97**, 170201 (2006). https://doi.org/10.1103/PhysRevLett.97.170201
61. E.S. Zijlstra, T. Zier, B. Bauerhenne, S. Krylow, M.E. Garcia, in *Laser Applications in Microelectronic and Optoelectronic Manufacturing (LAMOM) XXI*, vol. 9735, ed. by B. Neuenschwander, S. Roth, C.P. Grigoropoulos, T. Makimura. International Society for Optics and Photonics (SPIE, 2016), vol. 9735, pp. 99 – 104. https://doi.org/10.1117/12.2214668
62. B. Bauerhenne, E.S. Zijlstra, M.E. Garcia, Appl. Phys. A **123**(9), 608 (2017). https://doi.org/10.1007/s00339-017-1216-7
63. S. Krylow, E.S. Zijlstra, F.C. Kabeer, T. Zier, B. Bauerhenne, M.E. Garcia, Phys. Rev. Mater. **1**, 073601 (2017). https://doi.org/10.1103/PhysRevMaterials.1.073601
64. Y. Liu, R.E. Allen, Phys. Rev. B **52**, 1566 (1995). https://doi.org/10.1103/PhysRevB.52.1566
65. H.J. Zeiger, J. Vidal, T.K. Cheng, E.P. Ippen, G. Dresselhaus, M.S. Dresselhaus, Phys. Rev. B **45**, 768 (1992). https://doi.org/10.1103/PhysRevB.45.768
66. R.S. Lee, J. Gavillet, M.L. Chapelle, A. Loiseau, Phys. Rev. B **64**, 121405 (2001). https://doi.org/10.1103/PhysRevB.64.121405
67. D. Goldberg, Y. Bando, L. Bourgeois, K. Kurashima, T. Sato, Carbon **38**(14), 2017 (2000). https://doi.org/10.1016/S0008-6223(00)00058-0. http://www.sciencedirect.com/science/article/pii/S0008622300000580
68. M. Menon, D. Srivastava, Chem. Phys. Lett. **307**(5–6), 407 (1999). https://doi.org/10.1016/S0009-2614(99)00552-7. URL http://www.sciencedirect.com/science/article/pii/S0009261499005527

69. H.J. Xiang, J. Yang, J.G. Hou, Q. Zhu, Phys. Rev. B **68**, 035427 (2003). https://doi.org/10.1103/PhysRevB.68.035427
70. E. Bengu, L.D. Marks, Phys. Rev. Lett. **86**, 2385 (2001). https://doi.org/10.1103/PhysRevLett.86.2385
71. T. Xu, Y. Zhou, X. Tan, K. Yin, L. He, F. Banhart, L. Sun, Adv. Funct. Mater. **27**(19), 1603897 (2017). https://doi.org/10.1002/adfm.201603897
72. L. Wirtz, A. Rubio, Phys. Rev. B **68**, 045425 (2003). https://doi.org/10.1103/PhysRevB.68.045425
73. X. Blase, A. Rubio, S.G. Louie, M.L. Cohen, Europhys. Lett. **28**, 335 (1994). https://doi.org/10.1209/0295-5075/28/5/007. http://stacks.iop.org/0295-5075/28/i=5/a=007
74. Y.S. Lim, K.J. Yee, J.H. Kim, E.H. Haroz, J. Shaver, J. Kono, S.K. Doorn, R.H. Hauge, R.E. Smalley, Nano Lett. **6**(12), 2696 (2006). https://doi.org/10.1021/nl061599p
75. R. Arenal, A.C. Ferrari, S. Reich, L. Wirtz, J.Y. Mevellec, S. Lefrant, A. Rubio, A. Loiseau, Nano Lett. **6**(8), 1812 (2006). https://doi.org/10.1021/nl0602544
76. S. Luber, J. Chem. Phys. **141**(23), 234110 (2014). https://doi.org/10.1063/1.4903828
77. M. Tani, S. Matsuura, K. Sakai, S. Nakashima, Appl. Opt. **36**(30), 7853 (1997). https://doi.org/10.1364/AO.36.007853. http://ao.osa.org/abstract.cfm?URI=ao-36-30-7853

Chapter 5
Empirical MD Simulations of Laser-Excited Matter

Abstract Following the Born–Oppenheimer approximation, the atomic nuclei move on an effective electronic potential energy surface (PES). The ab-initio calculation of this electronic PES is only possible with quantum mechanical methods like DFT and, therefore, is quite computational expensive. If an excitation of the electrons can be neglected and they can be assumed in the ground state, which is reasonable for the simulation of many thermodynamic processes, one neglects the electrons as degree freedom and uses an interatomic potential for the description of the electronic PES or of the interatomic bonding. Simulations with interatomic potentials are less computational expensive and can be efficiently parallelized, so that simulations of up to 10^9 atoms are nowadays possible [1]. However, the commonly used interatomic potentials do not reach the precision of ab-initio calculations and neglect electronic excitations, which become relevant, if matter is irradiated by a femtosecond laser-pulse, for example. Laser-excited electrons and the related changes of the PES can be described by an increased electronic temperature T_e, which is successfully done in T_e-dependent DFT. Therefore, T_e-dependent interatomic potentials were recently introduced to take the changes in the PES due to hot electrons into account. In this chapter, interatomic potentials are introduced together with their development from experimental data and/or ab-initio calculations. Then it is reviewed how one can use a ground state interatomic potential to simulate a laser excitation of matter. The T_e-dependent interatomic potentials are presented, which are available up to now for the elements tungsten, gold, molybdenum and silicon [2–8]. The two available T_e-dependent interatomic potentials for silicon are analyzed by repeating the ab-initio calculations performed with CHIVES, which were presented in the previous chapter. It is shown, how the T_e-dependent interatomic potentials are up to now used in MD simulations to take beside the T_e-deformed PES also the electron-phonon coupling into account. Finally an universal interatomic potential parameter fitting program is presented, which can be used to fit the parameters of an interatomic potential with almost arbitrary functional form to the interatomic forces and structural cohesive energies obtained from ab-initio calculations.

© The Author(s), under exclusive license to Springer Nature Switzerland AG 2021
B. Bauerhenne, *Materials Interaction with Femtosecond Lasers*,
https://doi.org/10.1007/978-3-030-85135-4_5

5.1 Interatomic Potentials for Ground State Electrons in Solid State Physics

An interatomic potential Φ describes the energy of a system of interacting atoms. In the moment, we only deal with systems of a single type of atom. If the coordinate of atom i is denoted by $\mathbf{r}_i$ and the total number of atoms is denoted by N_{at}, the interatomic potential is just a function of all atomic coordinates

$$\Phi \equiv \Phi\big(\mathbf{r}_1, \ldots, \mathbf{r}_{N_{\mathrm{at}}}\big). \tag{5.1}$$

The electrons are ignored as a degree of freedom and are assumed to be in the ground state. Φ is not an arbitrary function of the atomic coordinates, since it should describe the energy under consideration of physical symmetries. To be able to do this, the potential Φ must fulfill the following constraints [9]:

(i) Since the potential describes interactions between atoms and does not take into account interactions from outside, it is invariant under a translation of the coordinate system:

$$\forall \mathbf{v} \in \mathbb{R}^3: \qquad \Phi\big(\mathbf{r}_1 + \mathbf{v}, \ldots, \mathbf{r}_{N_{\mathrm{at}}} + \mathbf{v}\big) = \Phi\big(\mathbf{r}_1, \ldots, \mathbf{r}_{N_{\mathrm{at}}}\big). \tag{5.2}$$

(ii) Due to the same reason as in (i), it is also invariant under a rotation or reflection of the coordinate system:

$$\forall \mathbf{M} \in \mathbb{R}^{3\times 3},\ \mathbf{M}\cdot\mathbf{M}^{\mathrm{t}} = \mathbb{1}: \qquad \Phi\big(\mathbf{M}\cdot\mathbf{r}_1, \ldots, \mathbf{M}\cdot\mathbf{r}_{N_{\mathrm{at}}}\big) = \Phi\big(\mathbf{r}_1, \ldots, \mathbf{r}_{N_{\mathrm{at}}}\big). \tag{5.3}$$

Here $\mathbf{M}$ denotes a rotation or reflection matrix and $\mathbb{1}$ the 3×3 unity matrix.

(iii) Since the numbering of the atoms is arbitrary, the potential is invariant under a permutation of the atomic numbering:

$$\forall \mathfrak{s} \in \mathfrak{S}_{N_{\mathrm{at}}}: \qquad \Phi \equiv \Phi\big(\mathbf{r}_{\mathfrak{s}(1)}, \ldots, \mathbf{r}_{\mathfrak{s}(N_{\mathrm{at}})}\big) = \Phi\big(\mathbf{r}_1, \ldots, \mathbf{r}_{N_{\mathrm{at}}}\big). \tag{5.4}$$

Here $\mathfrak{S}_{N_{\mathrm{at}}}$ denotes the set of all permutations of the numbers $1, \ldots, N_{\mathrm{at}}$.

In the most cases, the interaction between atoms becomes negligible, if the corresponding distance exceeds a certain finite limit, the so called cutoff radius $r^{(\mathrm{c})}$. One rare exception of this is ionic bonding, which has its origin in the Coulomb interaction. This decays slowly with $\propto \frac{1}{r_{ij}}$ as a function of the atomic distance r_{ij} and converges only conditionally depending on the order of summation to zero. To approximate efficiently such long range interactions, a special method called Ewald summation [10] was developed. We do not discuss this method further, since ionic bonding does not appear in systems of single atom type. Thus, if a finite cutoff radius $r^{(\mathrm{c})}$ can be assumed, the atomic interaction is local. Each atom i only interacts with the neighboring atoms that lie within distance $r^{(\mathrm{c})}$ from atom i. Consequently, only atoms in the neighborhood

$$\mathcal{N}_i = \left\{ j \in \{1, \ldots, N_{\text{at}}\},\ j \neq i, r_{ij} < r^{(c)} \right\} \tag{5.5}$$

influence the energy Φ_{loc} of atom i. Since all atoms in the system belong to the same type, the local energy Φ_{loc} is a uniform function and the energy Φ of the system is just the sum over the local energies from all contained atoms. Φ_{loc} should fulfill the properties (i) and (ii) and distances and angles between atoms are invariant under translations and rotations, such that Φ_{loc} is a function of distances and angles only rather then atomic coordinates. Let

$$r_{ij} = ||\mathbf{r}_i - \mathbf{r}_j|| \tag{5.6}$$

be the distance and

$$\hat{\mathbf{r}}_{ij} = \frac{\mathbf{r}_{ij}}{r_{ij}} \tag{5.7}$$

be the normalized distance vector between atom i and a neighbor j. From the normalized distance vectors $\hat{\mathbf{r}}_{ij}$ and $\hat{\mathbf{r}}_{ik}$, the cosine of the angle θ_{ijk} between the bond from atom i to j and i to k (see Fig. 5.2) can be directly calculated by

$$\cos(\theta_{ijk}) = \hat{\mathbf{r}}_{ij} \cdot \hat{\mathbf{r}}_{ik}. \tag{5.8}$$

Hence, using $\cos(\theta_{ijk})$ as a variable is more efficient than using the angle θ_{ijk}, for which the function arccos must be additionally applied. Consequently, Φ_{loc} should be formulated as a function of distances r_{ij} and cosine of angles $\cos(\theta_{ijk})$ between neighboring atoms:

$$\Phi = \sum_{i=1}^{N_{\text{at}}} \Phi_{\text{loc}}\Big(\big\{r_{ij},\ j \in \mathcal{N}_i\big\}, \big\{\cos(\theta_{ijk}),\ j,k \in \mathcal{N}_i, j \neq k\big\}\Big). \tag{5.9}$$

Furthermore, the function Φ_{loc} should change smoothly, if an atom ℓ enters or leaves the neighborhood $\mathcal{N}_i$ of atom i. It should be valid for any atom $\ell \notin \mathcal{N}_i$:

$$\begin{aligned} &\lim_{r_{i\ell} \nearrow r^{(c)}} \Phi_{\text{loc}}\Big(\big\{r_{ij},\ j \in \mathcal{N}_i \cup \{\ell\}\big\}, \big\{\cos(\theta_{ijk}),\ j,k \in \mathcal{N}_i \cup \{\ell\}, j \neq k\big\}\Big) \\ =&\Phi_{\text{loc}}\Big(\big\{r_{ij},\ j \in \mathcal{N}_i\big\}, \big\{\cos(\theta_{ijk}),\ j,k \in \mathcal{N}_i, j \neq k\big\}\Big). \end{aligned} \tag{5.10}$$

Since the force $\mathbf{f}_i$ on any atom i is calculated by a gradient from the potential Φ

$$\mathbf{f}_i = -\nabla_{\mathbf{r}_i} \Phi, \tag{5.11}$$

requirement (5.10) should also be postulated for the first derivatives of Φ_{loc}.

Fig. 5.1 Two-body potential Φ_2

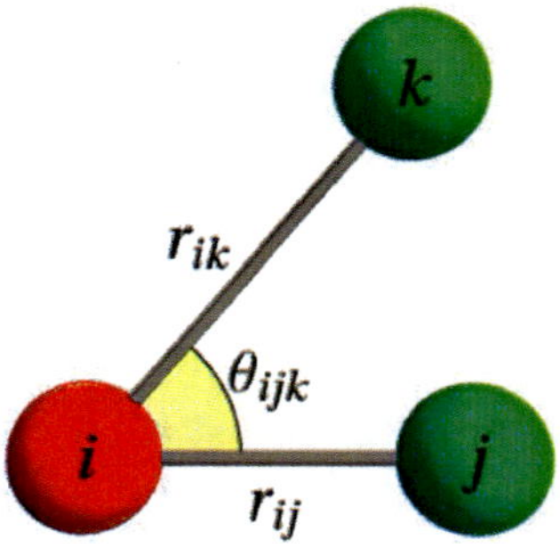

Fig. 5.2 Three-body potential Φ_3

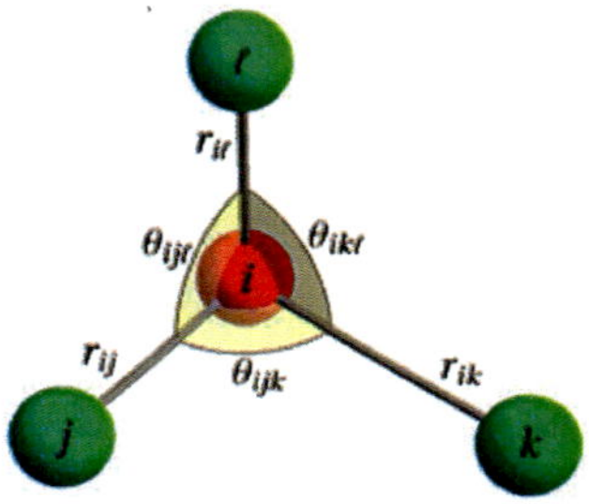

Fig. 5.3 Four-body potential Φ_4

The local energy Φ_{loc} of an atom i is constructed as a sum over different local energy contributions, which are originated from the chemical nature of atomic bonding [11]. The simplest local energy contribution is the self energy Φ_0 of atom i, which is just a constant and can be set to zero. The other ones describe a specific atomic interaction (see Figs. 5.1, 5.2, 5.3, 5.4 and 5.5) and are therefore functions of a fixed number of variables. In these functions, also further variables were introduced, which can be calculated from distances to and cosines of angles between neighboring atoms. Several of such local energy contribution functions were semi-empirically developed to describe the relevant physics of interatomic bonding in the past decades. In the following, the most commonly used are introduced.

The most obvious construction of a local energy contribution function describes the interaction of a central atom i with just one, two or more neighboring atoms:

- $\Phi_2(r_{ij})$ describes the interaction between atom i and one neighbor j, as it is illustrated in Fig. 5.1. This function is called pair or two-body potential and depends only on the distance r_{ij}, which already describes completely the relative position of the two atoms i and j.
- $\Phi_3\big(r_{ij}, r_{ik}, \cos(\theta_{ijk})\big)$ describes the interaction between atom i and two neighbors j and k, as it is illustrated in Fig. 5.2. This function is called three-body potential and depends on the distances r_{ij} and r_{ik} to and the cosine of the angle θ_{ijk} between

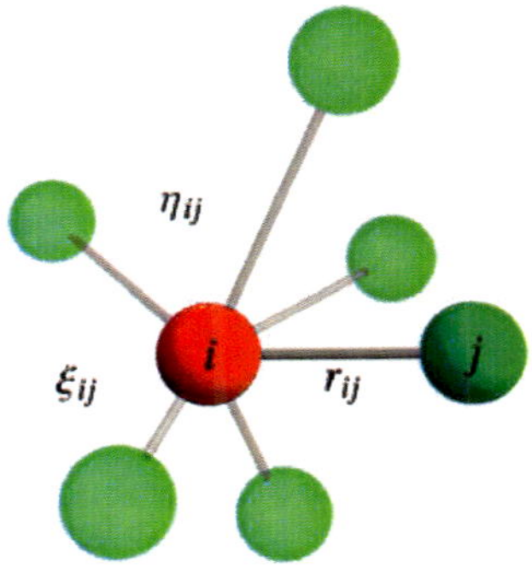

Fig. 5.4 Bond-order potential Φ_{2b}, which describes the two-body interaction under consideration of bond orders, which are constructed from the neighbors of atom i

the neighbors j and k of atom i, since these three quantities completely describe the relative positions of the three atoms i, j and k.

• $\Phi_4\big(r_{ij}, r_{ik}, r_{i\ell}, \cos(\theta_{ijk}), \cos(\theta_{ij\ell}), \cos(\theta_{ik\ell})\big)$ describes the interaction between atom i and three neighbors j, k and ℓ, as it is illustrated in Fig. 5.3. This function is called four-body potential. To describe the relative positions of the four atoms, the distance $r_{i\ell}$ and the cosine of the two angles $\theta_{ij\ell}$ and $\theta_{ik\ell}$ have to be added to the variables of the three-body potential Φ_3 described before.

In principle, the above construction can be continued straightforward to five and more neighbors. But the computational cost increases dramatically with increasing number of neighbors, so that even four-body potentials are very rarely used in computations. Three and more body potentials are introduced to describe covalent bonding, which is created by the overlap of electronic orbitals. Due to this, covalent bonds are directed and depend on the angles between atomic bonds. For example, the diamond structure, which is formed due to covalent bonds, cannot be the equilibrium structure of a pair potential, which always exhibits a closed packed structure like fcc as equilibrium structure. To stabilize the diamond structure, one has to add a three-body potential to the pair potential [12], for example. Another method for describing covalent boding is the concept of bond order, which describes the influence of other bonds connected to atom i on the bond between atom i and j. This influence of the other bonds is measured by a quantity called bond order for a given bond. The corresponding local energy contribution is constructed as a two-body potential, which also takes one or two measures for the bond order into account. One measure is used for the bonding and one for the anti-bonding part in the two-body potential.

• $\Phi_{2b}(r_{ij}, \xi_{ij}, \eta_{ij})$ describes the energy of the bond between atom i and j by using also two measures ξ_{ij}, η_{ij} of the bond order derived from the neighbors of atom i, as it is illustrated in Fig. 5.4. The two measures are constructed as

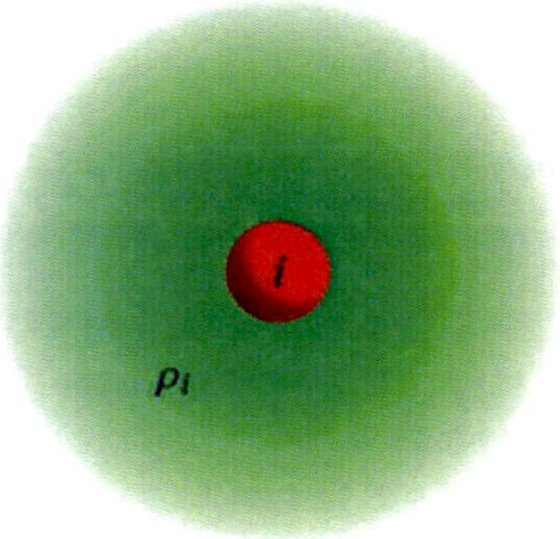

Fig. 5.5 Embedding atom potential Φ_ρ, which describes the interaction with surrounding atomic density ρ_i

$$\xi_{ij} = \sum_{\substack{k=1 \\ k \neq i, k \neq j \\ r_{ik} < r_{2b}^{(c)}}}^{N_{at}} \xi_{loc}\big(r_{ij}, r_{ik}, \cos(\theta_{ijk})\big), \tag{5.12}$$

$$\eta_{ij} = \sum_{\substack{k=1 \\ k \neq i, k \neq j \\ r_{ik} < r_{2b}^{(c)}}}^{N_{at}} \eta_{loc}\big(r_{ij}, r_{ik}, \cos(\theta_{ijk})\big) \tag{5.13}$$

by utilizing the functions ξ_{loc} and η_{loc}. These function determine how the bond from atom i to a neighboring atom k contributes to the bond order of the bond from i to j in dependence of the distances r_{ij}, r_{ik} and the cosine of the angle θ_{ijk} between the atoms i, j and k.

In metals, there exists a further bonding type due to the specifics of electronic properties of metals. The conduction and valence band overlap, so that there exists a significant amount of electrons in the conduction band already at room temperature. These electrons are able to move freely in the metal, which is also the reason for the high electric conductivity. The interaction of the electrons with this free electron gas is called metallic bonding. The corresponding local energy contribution is called embedding function and describes the interaction of an atom i with the surrounding electronic density, which is calculated from positions of the neighboring atoms.

• $\Phi_\rho(\rho_i)$ describes the interaction of atom i with the surrounding electron density from the neighboring atoms and is called embedded atom potential, as it is illustrated in Fig. 5.5. It depends only on the atom i surrounding atomic density ρ_i, which is calculated by

$$\rho_i = \sum_{\substack{j=1 \\ j \neq i \\ r_{ij} < r_\rho^{(c)}}}^{N_{at}} \rho_{loc}(r_{ij}, \chi_{ij}). \tag{5.14}$$

The function ρ_{loc} specifies how an atom j contributes to the atomic density. For metals, the function ρ_{loc} only depends on its distance r_{ij} to atom i. Baskes [13] extended the description of embedded atom potentials to describe also covalent bonding. For this, the function ρ_{loc} depends beside r_{ij} also on the bond order χ_{ij} of the bond between atom i and j. The bond order χ_{ij} is calculated by

$$\chi_{ij} = \sum_{\substack{k=1 \\ k \neq i \\ r_{ik} < r_\rho^{(c)}}}^{N_{\text{at}}} \chi_{\text{loc}}\big(r_{ij}, r_{ik}, \cos(\theta_{ijk})\big). \tag{5.15}$$

Here, the sum over k includes the atom j in contrast to ξ_{ij} and η_{ij} used to calculate Φ_{2b}.

Using the above mentioned local energy contribution functions, the potential Φ is calculated by

$$\begin{aligned}
\Phi = &\sum_{i=1}^{N_{\text{at}}} \Phi_0 + \sum_{i=1}^{N_{\text{at}}} \sum_{\substack{j=1 \\ j \neq i \\ r_{ij} < r_2^{(c)}}}^{N_{\text{at}}} \Phi_2(r_{ij}) + \sum_{i=1}^{N_{\text{at}}} \sum_{\substack{j=1 \\ j \neq i \\ r_{ij} < r_3^{(c)}}}^{N_{\text{at}}} \sum_{\substack{k=1 \\ k \neq i, k \neq j \\ r_{ik} < r_3^{(c)}}}^{N_{\text{at}}} \Phi_3\big(r_{ij}, r_{ik}, \cos(\theta_{ijk})\big) \\
&+ \sum_{i=1}^{N_{\text{at}}} \sum_{\substack{j=1 \\ j \neq i \\ r_{ij} < r_4^{(c)}}}^{N_{\text{at}}} \sum_{\substack{k=1 \\ k \neq i, k \neq j \\ r_{ik} < r_4^{(c)}}}^{N_{\text{at}}} \sum_{\substack{\ell=1 \\ \ell \neq i, \ell \neq j, \ell \neq k \\ r_{i\ell} < r_4^{(c)}}}^{N_{\text{at}}} \times \\
&\times \Phi_4\big(r_{ij}, r_{ik}, r_{i\ell}, \cos(\theta_{ijk}), \cos(\theta_{ij\ell}), \cos(\theta_{ik\ell})\big) \\
&+ \sum_{i=1}^{N_{\text{at}}} \sum_{\substack{j=1 \\ j \neq i \\ r_{ij} < r_{2b}^{(c)}}}^{N_{\text{at}}} \Phi_{2b}(r_{ij}, \xi_{ij}, \eta_{ij}) + \sum_{i=1}^{N_{\text{at}}} \Phi_\rho(\rho_i).
\end{aligned} \tag{5.16}$$

5.1.1 Classical Analytical Interatomic Potential Models

Which of the functions Φ_2, Φ_3, Φ_4, Φ_{2b}, and Φ_ρ are used and how they are defined depends on the concrete potential model. Si plays an important role in technology and has a rich phase diagram with many stable and metastable crystal structures. Thus, a decent number of analytical potential models were developed for Si within the past decades, each with particular use cases. Since there are too many analytical

Si potentials available to thoroughly review here, only the most commonly used and some of their modifications are presented.

Among the most regularly used Si potentials is the potential of Stillinger and Weber (SW) [12], since it has got a simple analytical form, allows very efficient calculations, stabilizes the diamond structure, reproduces the experimental melting temperature, and describes accurately the elastic constants. The third potential of Tersoff (T3) [14] is also noteworthy, since it also has a simple analytical form, allows very efficient calculations, stabilizes the diamond structure and, as well, describes properly the elastic constants. Several modifications of the Tersoff potential were constructed. The so called Modified Tersoff potential (MT), which was developed by Kumagai et al. [15], was used as a basis to develop a T_e-dependent interatomic potential for Si by Shokeen and Schelling [6, 7] and by Darkins et al. [8]. Since metals are very well described by embedded atom potentials [16], Baskes introduced the Modified Embedded Atom Method (MEAM) to describe Si, which becomes metallic at high pressures or after an intense laser-excitation.

In this section we present the functional form of the selected potentials in detail, since we will use them to fit a T_e-dependent interatomic potential for Si in Sect. 7.1.2.

5.1.1.1 Stillinger and Weber (SW)

The most popular potential for silicon was published by Stillinger and Weber in the beginning of 1985 [12, 17]. This potential consists of a two-body Φ_2 and a three-body interaction term Φ_3, which both share the same cutoff radius, i.e., $r^{(c)} := r_2^{(c)} = r_3^{(c)}$. A separate cutoff-function

$$f_c^{(\mathrm{SW_{orig}})}(r) = \begin{cases} e^{\frac{1}{\frac{r}{\sigma}-a}} & \frac{r}{\sigma} < a \\ 0 & \frac{r}{\sigma} \geq a \end{cases} \tag{5.17}$$

is introduced to archive that the potential decays smoothly to zero, if atomic distances reach the cutoff radius $r^{(c)}$. By construction, the cutoff-function is infinitely often continuously differentiable. In the original formulation, the potential uses the 9 parameters $\sigma, \epsilon, A, B, p, q, a, \lambda, \gamma$ and is defined as

$$\Phi_2^{(\mathrm{SW_{orig}})}(r_{ij}) = \frac{1}{2}\, f_c^{(\mathrm{SW_{orig}})}(r_{ij})\, \varepsilon\, A \left(B \left(\frac{r_{ij}}{\sigma}\right)^{-p} - \left(\frac{r_{ij}}{\sigma}\right)^{-q} \right), \tag{5.18}$$

$$\Phi_3^{(\mathrm{SW_{orig}})}\left(r_{ij}, r_{ik}, \cos(\theta_{ijk})\right) = \frac{1}{2}\, \varepsilon\, \lambda\, f_c^{(\mathrm{SW_{orig}})}(r_{ij})^{\gamma}\, f_c^{(\mathrm{SW_{orig}})}(r_{ik})^{\gamma} \left(\cos(\theta_{ijk}) + \frac{1}{3}\right)^2. \tag{5.19}$$

From the two parameters σ and a, the cutoff radius $r^{(c)}$ can be calculated by

$$r^{(c)} = \sigma\, a \tag{5.20}$$

and with the help of the cutoff radius, the cutoff function can be formulated as

$$f_c^{(SW)}(r) = \begin{cases} e^{\frac{\sigma}{r-r^{(c)}}} & r < r^{(c)} \\ 0 & r \geq r^{(c)}. \end{cases} \tag{5.21}$$

The original potential notation (5.18) and (5.19) contains the parameter products $\varepsilon\, A\, B\, \sigma^p$, $\varepsilon\, A\, \sigma^q$ and $\varepsilon\, \lambda$, which will create a numerical instable fit. To obtain a numerical stable fit, the new parameters

$$A_0 = \varepsilon\, A\, B\, \sigma^p, \tag{5.22}$$

$$B_0 = \varepsilon\, A\, \sigma^q, \tag{5.23}$$

$$\lambda_0 = \varepsilon\, \lambda \tag{5.24}$$

are defined and the potential is formulated in the 8 parameters $r^{(c)}$, σ, A_0, B_0, p, q, λ_0, γ:

$$\Phi_2^{(SW)}(r_{ij}) = \frac{1}{2}\, f_c^{(SW)}(r_{ij}) \left(A_0\, r_{ij}^{-p} - B_0\, r_{ij}^{-q}\right), \tag{5.25}$$

$$\Phi_3^{(SW)}\left(r_{ij}, r_{ik}, \cos(\theta_{ijk})\right) = \frac{1}{2}\, \lambda_0\, f_c^{(SW)}(r_{ij})^{\gamma}\, f_c^{(SW)}(r_{ik})^{\gamma} \left(\cos(\theta_{ijk}) + \frac{1}{3}\right)^2. \tag{5.26}$$

In 1989, Mistriotis et al. modified the Φ_2 and Φ_3 term of this potential and added a four-body term Φ_4 to improve the description of Si clusters [18].

5.1.1.2 1. Tersoff (T1)

At the beginning of 1986, Tersoff published a potential, which consists of a two-body interaction term with bond order Φ_{2b} [19]. Tersoff introduced the following cutoff-function

$$f_c^{(T1)}(r) = \begin{cases} 1 & r \leq r^{(1)} \\ \frac{1}{2} - \frac{1}{2}\sin\left(\frac{\pi}{2}\, \frac{2\,r - r^{(1)} - r_{2b}^{(c)}}{r_{2b}^{(c)} - r^{(1)}}\right) & r^{(1)} < r < r_{2b}^{(c)} \\ 0 & r \geq r_{2b}^{(c)}, \end{cases} \tag{5.27}$$

which is twice continuously differentiable. The potential has 10 parameters $r_{2b}^{(c)}$, $r^{(1)}$, A, B_0, λ_1, λ_2, b, n, c, d and is defined as

$$\Phi_{2b}^{(T1)}(r_{ij}, \xi_{ij}, \eta_{ij}) = \frac{1}{2} f_c^{(T1)}(r_{ij}) \left(A\, e^{-\lambda_1 r_{ij}} - B_0\, e^{-\frac{\xi_{ij}}{b}}\, e^{-\lambda_2 r_{ij}} \right), \tag{5.28}$$

$$\xi_{loc}^{(T1)}\big(r_{ij}, r_{ik}, \cos(\theta_{ijk})\big) = \left(\frac{f_c^{(T1)}(r_{ik})}{f_c^{(T1)}(r_{ij})} \right)^n \frac{e^{n\,\lambda_2\,(r_{ij}-r_{ik})}}{c + e^{-d\,\cos(\theta_{ijk})}}. \tag{5.29}$$

5.1.1.3 2. Tersoff (T2)

Tersoff modified the functional form of the two-body interaction term with bond order Φ_{2b} of his first potential in 1988 [14]. The modified potential utilizes the same cutoff-function $f_c^{(T1)}$ like the first Tersoff potential, contains the 8 parameters $r_{2b}^{(c)}$, $r^{(1)}$, A, B, λ_1, λ_2, a, n and is constructed as

$$\Phi_{2b}^{(T2)}(r_{ij}, \xi_{ij}, \eta_{ij}) = \frac{1}{2} f_c^{(T1)}(r_{ij}) \left(A\, e^{-\lambda_1 r_{ij}} - \left(1 + a^n\, \xi_{ij}^n\right)^{-\frac{1}{2n}} B\, e^{-\lambda_2 r_{ij}} \right), \tag{5.30}$$

$$\xi_{loc}^{(T2)}\big(r_{ij}, r_{ik}, \cos(\theta_{ijk})\big) = f_c^{(T1)}(r_{ik}). \tag{5.31}$$

5.1.1.4 3. Tersoff (T3)

Beside his second potential, Tersoff introduced a third potential also with a modified functional form of the two-body interaction term with bond order Φ_{2b} in the same publication [14]. This potential utilizes again the cutoff-function $f_c^{(T1)}$ of the first Tersoff potential. With the help of the 13 parameters $r_{2b}^{(c)}$, $r^{(1)}$, A, B, λ_1, λ_2, α, β, n, c, d, h, λ_3, the potential is formulated as

$$\begin{aligned}\Phi_{2b}^{(T3)}(r_{ij}, \xi_{ij}, \eta_{ij}) = &\frac{1}{2} f_c^{(T1)}(r_{ij} \Big(\left(1 + \alpha^n\, \eta_{ij}^n\right)^{-\frac{1}{2n}} A\, e^{-\lambda_1 r_{ij}} \\ &- \left(1 + \beta^n\, \xi_{ij}^n\right)^{-\frac{1}{2n}} B\, e^{-\lambda_2 r_{ij}} \Big),\end{aligned} \tag{5.32}$$

$$\eta_{loc}^{(T3)}\big(r_{ij}, r_{ik}, \cos(\theta_{ijk})\big) = f_c^{(T1)}\, e^{\lambda_3^3\,(r_{ij}-r_{ik})^3}, \tag{5.33}$$

$$\xi_{loc}^{(T3)}\big(r_{ij}, r_{ik}, \cos(\theta_{ijk})\big) = f_c^{(T1)} \left(1 + \frac{c^2}{d^2} - \frac{c^2}{d^2 + \left(h - \cos(\theta_{ijk})\right)^2} \right) e^{\lambda_3^3\,(r_{ij}-r_{ik})^3}. \tag{5.34}$$

The T3 potential is often just called Tersoff potential in literature neglecting the two lesser used other ones. Tersoff published for this potential an other parametrization with a better description of the elastic constants in the end of 1988 [20].

5.1.1.5 Dodson (D)

Dodson modified the functional form of the two-body interaction term with bond order Φ_{2b} of the first Tersoff potential in 1987 [21]. His potential utilizes the cutoff-function $f_c^{(T1)}$ of the first Tersoff potential and is constructed with the help of the 11 parameters $r_{2b}^{(c)}$, $r^{(1)}$, A, B_0, λ_1, λ_2, η, b, n, c, d as

$$\Phi_{2b}^{(D)}(r_{ij}, \xi_{ij}, \eta_{ij}) = \frac{1}{2} f_c^{(T1)}(r_{ij}) \left(A\, e^{-\lambda_1 r_{ij}} - e^{-\frac{\xi_{ij}^{\eta}}{b}} B_0\, e^{-\lambda_2 r_{ij}} \right), \tag{5.35}$$

$$\xi_{\text{loc}}^{(D)}\big(r_{ij}, r_{ik}, \cos(\theta_{ijk})\big) = \left(\frac{f_c^{(T1)}(r_{ik})}{f_c^{(T1)}(r_{ij})} \right)^n \frac{e^{n\,\lambda_2\,(r_{ij}-r_{ik})}}{c + e^{-d\,\cos(\theta_{ijk})}}. \tag{5.36}$$

5.1.1.6 Modified Tersoff (MT)

In 2007, Kumagai et al. modified the functional form of the two-body interaction term with bond order Φ_{2b} of the third Tersoff potential to improve the description of the elastic properties [15]. The new cutoff-function

$$f_c^{(MT)}(r) = \begin{cases} 1 & r \le r^{(1)} \\ \frac{1}{2} + \frac{9}{16} \cos\left(\pi \frac{r - r^{(1)}}{r_{2b}^{(c)} - r^{(1)}}\right) - \frac{1}{16} \cos\left(3\pi \frac{r - r^{(1)}}{r_{2b}^{(c)} - r^{(1)}}\right) & r^{(1)} < r < r_{2b}^{(c)} \\ 0 & r \ge r_{2b}^{(c)} \end{cases} \tag{5.37}$$

was introduced, which is three times continuously differentiable. The new potential uses the 16 parameters $r_{2b}^{(c)}$, $r^{(1)}$, A, B, λ_1, λ_2, n, δ, α, β, h, c_1, c_2, c_3, c_4, c_5 and is formulated as

$$\Phi_{2b}^{(MT)}(r_{ij}, \xi_{ij}, \eta_{ij}) = \frac{1}{2} f_c^{(MT)}(r_{ij}) \Big(A\, e^{-\lambda_1 r_{ij}} - (1 + \xi_{ij}^{\eta})^{-\delta}\, B\, e^{-\lambda_2 r_{ij}} \Big), \tag{5.38}$$

$$\begin{aligned} \xi_{\text{loc}}^{(MT)}\big(r_{ij}, r_{ik}, \cos(\theta_{ijk})\big) =& f_c^{(MT)}(r_{ik})\, e^{\alpha\,(r_{ij}-r_{ik})^{\beta}} \times \\ & \times \left(c_1 + \frac{c_2\big(h - \cos(\theta_{ijk})\big)^2}{c_3 + \big(h - \cos(\theta_{ijk})\big)^2} \left(1 + c_4\, e^{-c_5\left(h - \cos(\theta_{ijk})\right)^2} \right) \right). \end{aligned} \tag{5.39}$$

Based on this potential model, Shokeen and Schelling [6, 7, 22] developed the first and Darkins et al. [8] the second T_e-dependent interatomic potential for silicon.

5.1.1.7 Pun and Mishin (PM)

In 2017, Pun and Mishin modified the functional form of the two-body interaction term with bond order Φ_{2b} of the Modified Tersoff potential (MT) to further improve the description [23]. The cutoff function $f_c^{(MT)}$ and the bond order measure function

$\xi_{\text{loc}}^{(\text{MT})}\big(r_{ij}, r_{ik}, \cos(\theta_{ijk})\big)$ were kept and a further parameter c_0 was added in the $\Phi_{2\text{b}}$ function. The new potential uses the 17 parameters $r_{2\text{b}}^{(\text{c})}, r^{(1)}, A, B, \lambda_1, \lambda_2, n, \delta, \alpha, \beta, h, c_1, c_2, c_3, c_4, c_5, c_0$ and is formulated as

$$\Phi_{2\text{b}}^{(\text{PM})}(r_{ij}, \xi_{ij}, \eta_{ij}) = \frac{1}{2}\, f_{\text{c}}^{(\text{MT})}(r_{ij})\Big(A\, e^{-\lambda_1 r_{ij}} - (1+\xi_{ij}^n)^{-\delta}\, B\, e^{-\lambda_2 r_{ij}} + c_0\Big). \tag{5.40}$$

5.1.1.8 Modified Embedded Atom Method (MEAM)

In 1987, Baskes extended the Embedded Atom Method, which was originally developed for metals [24], to covalent bonded systems and presented a Modified Embedded Atom Method potential for silicon [13]. Two years later in 1989, he refined the Modified Embedded Atom Method potential and published optimized parameters for silicon and germanium [25]. This potential utilizes the twice continuously differentiable cutoff-function

$$f_{\text{c}}^{(\text{MEAM})}(r) = \begin{cases} 1 & r \le r^{(1)} \\ \dfrac{\left(r-r^{(\text{c})}\right)^3\left(6\,r^2+3\,r\left(r^{(\text{c})}-5\,r^{(1)}\right)+10\left(r^{(1)}\right)^2-5\,r^{(1)}\,r^{(\text{c})}+\left(r^{(\text{c})}\right)^2\right)}{\left(r^{(1)}-r^{(\text{c})}\right)^5} & r^{(1)} < r < r^{(\text{c})} \\ 0 & r \ge r^{(\text{c})} \end{cases} \tag{5.41}$$

and contains the 9 parameters $r^{(\text{c})}, r^{(1)}, r_1, E_0, B, m, \alpha, a_1, a_2$. Here the parameter r_1 denotes the equilibrium nearest neighbor distance in the diamond structure, in which silicon crystallize under ambient conditions. The potential consists of a two-body Φ_2 and a density-interaction term Φ_ρ, both with the same cutoff radius $r^{(\text{c})}$, i.e., $r_2^{(\text{c})} = r_\rho^{(\text{c})} = r^{(\text{c})}$. With the help of the radial function

$$\rho(r) = \rho_0\, f_{\text{c}}^{(\text{MEAM})}(r)\left(\frac{r}{r_1}\right)^{m_1} e^{-\alpha\,(r-r_1)}, \tag{5.42}$$

and the atomic density $\rho_{\text{dia}}(r_1)$ of the diamond structure with equilibrium nearest neighbor distance r_1, the density-interaction term is defined as

$$\Phi_\rho^{(\text{MEAM})}(\rho_i) = E_0\, \frac{\rho_i}{\rho_{\text{dia}}(r_1)}\, \ln\left(\frac{\rho_i}{\rho_{\text{dia}}(r_1)}\right), \tag{5.43}$$

$$\rho_{\text{loc}}^{(\text{MEAM})}(r_{ij}, \chi_{ij}) = \rho(r_{ij}) + \chi_{ij}, \tag{5.44}$$

$$\chi_{\text{loc}}^{(\text{MEAM})}\big(r_{ij}, r_{ik}, \cos(\theta_{ijk})\big) = \Big(a_1^2\, \cos(\theta_{ijk}) - a_2^2\, \big(1 - 3\, \cos(\theta_{ijk})^2\big)\Big)\, \rho(r_{ij})\, \rho(r_{ik}). \tag{5.45}$$

In the definition of $\Phi_\rho^{(\text{MEAM})}$, a logarithmic function is used to take into account the logarithmic dependence of the bond length on the number of bonds observed by

Pauling [26]. The remaining two-body interaction term $\Phi_2^{(\mathrm{MEAM})}$ is implicitly defined: Baskes demanded that the potential energy per atom of the diamond structure with nearest neighbor distance of r should behave like the universal energy function E_r of Rose et al. [27] in dependence of r. The latter function is defined as

$$E_r(r) = -E_0 \left(1 + \sqrt{\frac{9\,B\,\Omega_0}{r_1^2\,E_0}}\;(r - r_1)\right) e^{-\sqrt{\frac{9\,B\,\Omega_0}{r_1^2\,E_0}}\,(r-r_1)} \tag{5.46}$$

and describes the energy per atom of a crystal at uniform expansion or compression. Here r denotes the actual distance and r_1 the equilibrium distance to the first neighbors. B is the bulk modulus, E_0 is an energy constant and Ω_0 is the equilibrium atomic volume, which yields for the diamond structure at equilibrium nearest neighbor distance r_1

$$\Omega_0 = \left(\sqrt{\frac{4}{3}}\,r_1\right)^3 . \tag{5.47}$$

Inserting this atomic volume Ω_0 in (5.46) and taking the cutoff-function $f_c^{(\mathrm{MEAM})}(r)$ into account, it follows for silicon

$$E_r(r) = -E_0\, f_c^{(\mathrm{MEAM})}(r) \left(1 + \sqrt{\frac{8\sqrt{3}\,B\,r_1}{E_0}}\;(r - r_1)\right) e^{-\sqrt{\frac{8\sqrt{3}\,B\,r_1}{E_0}}\,(r-r_1)} . \tag{5.48}$$

Since the potential energy per atom of the diamond structure with nearest neighbor distance r should behave like $E_r(r)$, it is valid

$$\sum_{i=1}^{N_{\mathrm{at}}} E_r(r) = \sum_{i=1}^{N_{\mathrm{at}}} \Phi_\rho^{(\mathrm{MEAM})}\big(\rho_{\mathrm{dia}}(r)\big) + \sum_{i=1}^{N_{\mathrm{at}}} \sum_{\substack{j=1\\ j\neq i\\ r_{ij}<r^{(c)}}}^{N_{\mathrm{at}}} \phi_2^{(\mathrm{MEAM})}(r_{ij}). \tag{5.49}$$

Here $\rho_{\mathrm{dia}}(r)$ denotes the atomic density of the diamond structure with nearest neighbor distance r. For the diamond structure, $\chi_{ij} = 0$ in (5.44), since for cubic structures the sum over $\chi_{\mathrm{loc}}^{(\mathrm{MEAM})}$ equals to zero [25]. Hence, the atomic density at equilibrium nearest neighbor distance r_1 can be calculated as

$$\rho_{\mathrm{dia}}(r_1) = \sum_{s=1}^{\infty} N_s\, \rho\,(r_s)\,. \tag{5.50}$$

Here N_s denotes the number of atoms at distance r_s in the diamond structure at equilibrium. Consequently, it follows for the atomic density at arbitrary nearest neighbor distance r

$$\rho_{\text{dia}}(r) = \sum_{s=1}^{\infty} N_s \, \rho\left(\frac{r_s}{r_1} r\right). \tag{5.51}$$

Furthermore, since all atoms see the same environment in the diamond structure, one gets for any atom $i \in \{1, \ldots, N_{\text{at}}\}$ from (5.49):

$$\sum_{\substack{j=1 \\ j \neq i \\ r_{ij} < r^{(c)}}}^{N_{\text{at}}} \phi_2^{(\text{MEAM})}(r_{ij}) = E_r(r) - \Phi_\rho^{(\text{MEAM})}\big(\rho_{\text{dia}}(r)\big). \tag{5.52}$$

Taking into account the symmetry of the diamond structure, it follows

$$\sum_{s=1}^{\infty} N_s \, \Phi_2^{(\text{MEAM})}\left(\frac{r_s}{r_1} r\right) = E_r(r) - \Phi_\rho^{(\text{MEAM})}\big(\rho_{\text{dia}}(r)\big), \tag{5.53}$$

$$\Phi_2^{(\text{MEAM})}(r) + \sum_{s=2}^{\infty} \frac{N_s}{N_1} \Phi_2^{(\text{MEAM})}\left(\frac{r_s}{r_1} r\right) = \underbrace{\frac{1}{N_1}\Big(E_r(r) - \Phi_\rho^{(\text{MEAM})}\big(\rho_{\text{dia}}(r)\big)\Big)}_{=:\phi_2(r)}. \tag{5.54}$$

Since the interaction between atoms is only taken into account for distances smaller than the cutoff radius $r^{(c)}$, the sum on the left side of (5.54) is finite. For a cutoff radius $r^{(c)} < 0.539\,\text{nm}$, the sum only contains atoms up to the third nearest neighbors in the diamond structure at equilibrium. Hence, one gets

$$\Phi_2^{(\text{MEAM})}(r) = \phi_2(r) - \frac{N_2}{N_1} \Phi_2^{(\text{MEAM})}\left(\frac{r_2}{r_1} r\right) - \frac{N_3}{N_1} \Phi_2^{(\text{MEAM})}\left(\frac{r_3}{r_1} r\right). \tag{5.55}$$

It follows

$$\Phi_2^{(\text{MEAM})}\left(\frac{r_2}{r_1} r\right) = \phi_2\left(\frac{r_2}{r_1} r\right) - \frac{N_2}{N_1} \Phi_2^{(\text{MEAM})}\left(\left(\frac{r_2}{r_1}\right)^2 r\right) - \frac{N_3}{N_1} \Phi_2^{(\text{MEAM})}\left(\frac{r_3}{r_1}\frac{r_2}{r_1} r\right), \tag{5.56}$$

$$\Phi_2^{(\text{MEAM})}\left(\frac{r_3}{r_1} r\right) = \phi_2\left(\frac{r_3}{r_1} r\right) - \frac{N_2}{N_1} \Phi_2^{(\text{MEAM})}\left(\frac{r_3}{r_1}\frac{r_2}{r_1} r\right) - \frac{N_3}{N_1} \Phi_2^{(\text{MEAM})}\left(\left(\frac{r_3}{r_1}\right)^2 r\right). \tag{5.57}$$

The equilibrium nearest neighbor distance of the silicon diamond structure equals $r_1 = 0.234\,\text{nm}$. Hence it is convenient to allow only atomic distances $r \geq 0.173\,\text{nm}$ for $\Phi_2^{(\text{MEAM})}(r)$. Then, one gets

$$\Phi_2^{(\mathrm{MEAM})}\left(\frac{r_3}{r_1}\frac{r_2}{r_1}r\right)=0, \tag{5.58}$$

$$\Phi_2^{(\mathrm{MEAM})}\left(\left(\frac{r_3}{r_1}\right)^2 r\right)=0 \tag{5.59}$$

due to the cutoff radius $r^{(c)} < 0.539\,\mathrm{nm}$. It follows for (5.56) and (5.57):

$$\Phi_2^{(\mathrm{MEAM})}\left(\frac{r_2}{r_1}r\right)=\phi_2\left(\frac{r_2}{r_1}r\right)-\frac{N_2}{N_1}\Phi_2^{(\mathrm{MEAM})}\left(\left(\frac{r_2}{r_1}\right)^2 r\right), \tag{5.60}$$

$$\Phi_2^{(\mathrm{MEAM})}\left(\frac{r_3}{r_1}r\right)=\phi_2\left(\frac{r_3}{r_1}r\right). \tag{5.61}$$

Inserting (5.60) and (5.61) in (5.55) yields the explicit definition of the two-body interaction term for $0.173\,\mathrm{nm} \leq r \leq 0.539\,\mathrm{nm}$:

$$\Phi_2^{(\mathrm{MEAM})}(r) = \phi_2(r) - \frac{N_2}{N_1}\phi_2\left(\frac{r_2}{r_1}r\right) - \frac{N_3}{N_1}\phi_2\left(\frac{r_3}{r_1}r\right) + \left(\frac{N_2}{N_1}\right)^2\phi_2\left(\left(\frac{r_2}{r_1}\right)^2 r\right). \tag{5.62}$$

5.1.2 Determining of Interatomic Potential Parameters

The construction of the interatomic potential Φ requires defining the values of its parameters. One has to optimize the parameters in such a way that the corresponding interatomic potential describes deducible physical properties as accurately as possible. This optimizing is performed by a fitting procedure and the set of physical properties is called fitted data set. If the number of parameters of Φ for the same data set is increased, then the error in the description of the fitted data decreases. At the same time, however, Φ becomes less transferable, that is Φ describes not fitted physical properties less accurately. A significant error in the description of a physical property or process is called artifact. On the other hand, if one keeps the number of parameters constant and includes additional physical properties in the fitting data set, the potential becomes more transferable and the occurrence of artifacts decreases, but the error in the description of the fitted physical properties increases as well (see for this also Fig. 5.6). Therefore, a balance between number of parameters/complexity of Φ and the size of the fitted data set is paramount to find for a useful model.

In times of limited computational resources modeling was restricted to a few essential parameters and fits to a few experimental quantities like cohesive energies, elastic constants and phonon frequencies. This procedure was successful, since the functional form of the potentials were carefully constructed driven by physical arguments. The potential of Stillinger and Weber [12, 17] is a famous example of such a construction.

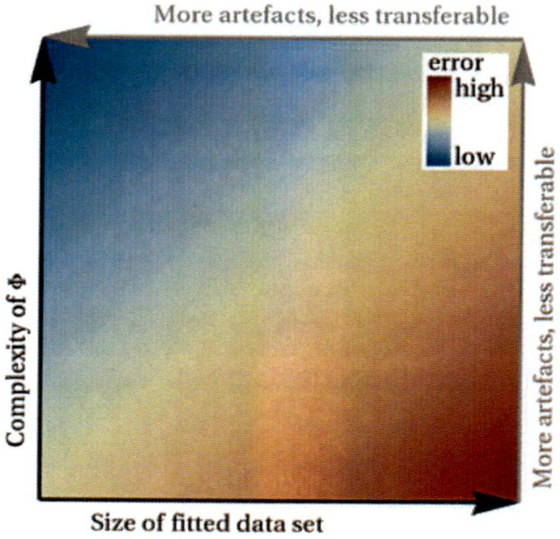

Fig. 5.6 The relationship between complexity of the potential model, the size of the fitted data set, the fit error and the occurring of artifacts in the finally obtained potential is illustrated

To increase the accuracy of the Φ description, one has to use more potential parameters and, consequently, increase the size of the fitting data set. Ercolessi et al. introduced the so called **force matching method** in 1994 [28]. They performed ab-initio calculations and MD simulations and used the obtained forces, energies and other physical quantities for the interatomic potential fitting. This allows to increase dramatically the size of the fitting data set instead of using just a few available experimental results and was one key point to derive accurate interatomic potentials.

To fit the potential parameters by using the force matching method for interatomic potentials used in solid state physics, Brommer et al. developed the open source program POTFIT [29–31], where the parameters are found by minimizing the following error function[1]

$$\mathbb{W}_{\text{err}}^{(\text{potfit})} = \sum_{i=1}^{N_{\text{at}}} \sum_{\alpha=x,y,z} \mathbb{w}_{\text{f}}^{(i)} \frac{\left(-\frac{\partial\Phi}{\partial r_{i\alpha}} - f_{i\alpha}^2\right)^2}{f_{i\alpha}^2 + \varepsilon_i} + \sum_{k=1}^{N_{\text{c}}} \mathbb{w}_{\text{A}}^{(k)} \frac{\left(A_k(\Phi) - A_k\right)^2}{A_k^2 + \varepsilon_k}, \tag{5.63}$$

which is the weighted sum ($\mathbb{w}_{\text{f}}^{(i)}$ and $\mathbb{w}_{\text{A}}^{(k)}$ are weights) of relative errors in forces and other observable like energies. $f_{i\alpha}$ denotes the ab-initio and $-\frac{\partial\Phi}{\partial r_{i\alpha}}$ the force on atom i in direction α obtained from the interatomic potential. A_k is a physical quantity, e.g. the energy, obtained from ab-initio and $A_k(\Phi)$ is the physical quantity obtained from the interatomic potential. The additional small parameters $\varepsilon_i > 0$ and $\varepsilon_k > 0$ prevent that the denominators become zero and take care, that small values $f_{i\alpha}$ of the forces or small values A_k of the fitted physical quantity are not to accurately fitted, since they are actually not known to such a precision.

In general, the search for the optimal parameters minimizing $\mathbb{W}_{\text{err}}^{(\text{potfit})}$ is quite complicated. Therefore, several distinct optimization algorithms are implemented

[1] We adapted this function to our notation.

in POTFIT: (i) the Powell algorithm [32], which is a conjugate-gradient-like optimization method that takes advantage of the particular form of $\mathbb{W}_{\rm err}^{\rm (potfit)}$ as a sum of squares, (ii) a method based on simulated annealing [33], which samples parts of the parameter space in a Monte-Carlo based fashion, (iii) a differential evolution algorithm [34], which is a variant of the genetic algorithm approach.

5.1.3 Machine Learning Interatomic Potentials

To increase the accuracy and to to be able to model properly a wide range of atomic structures with an interatomic potential, one has to fit a large amount of ab-initio data and one needs an interatomic potential model that is flexible enough. Nowadays, modern computers and efficient implementations of DFT allow to derive ab-initio forces, energies, and other properties for a huge amount of atomic structures and collecting them in a database. The functional form of the classical interatomic potentials like the Stillinger-Weber or Tersoff potential were constructed by creating physical models for describing interatomic bonding based on the bond types like covalent, metallic or ionic bonding. For this, a lot of human input is necessary, since reasonably physical approximations are indispensable. It is by no means straightforward to extend the functional form of such classical interatomic potentials to describe the available ab-initio data in the huge database more accurately. Therefore, the concept of machine learning was applied to generate an optimal functional form of the interatomic potential automatically with less human interaction from the ab-initio database [35]. This leads to the following requirements:

- The evaluation of the functional form of the machine learning potential should be fast to enable large scale MD simulations.
- One should be able to calculate analytical energy gradients for the calculation of forces or other gradient-dependent quantities.
- Minimal human interaction should be needed to develop the machine learning potential from the database and systematic improvements should be possible, if the database is increased.
- The database should only include consistent sets of electronic structure data. If experimental and theoretical or data from different levels of theory are contained in the database, this can produce inconsistencies, which could rise serious problems in the potential construction.

Physical meaningful interatomic potentials fulfill several properties, as described in the beginning of Sect. 5.1. To realize these properties in machine learning potentials, one introduces descriptors that transform the atomic coordinates to a suitable input for the machine learning algorithm. The calculation of the descriptors should be fast and they need to be differentiable with respect to the atomic positions, in order to be able to calculate analytically the gradient of the potential for the forces. According to available literature, one commonly uses atom centered symmetry functions, the bispectrum of the neighbor density, the smooth overlap of atomic positions or the Coulomb matrix as descriptor. More details can be found in the review of Behler [35]. For the machine learning method, also several different approaches have been reported [35]. The first machine learning potential was developed in 1995 by using a neural network machine learning implementation and a simple atom centered symmetry function [36]. Also kernel methods are used, where Gaußian approximation potentials are an example. Recently Bartok et al. [37] developed an Gaußian approximation potential for silicon with electrons in the ground state. For the moment, we do not proceed in discussing these machine learning approaches, since we will not use them in our simulations. The main drawback of the machine learning potentials is that they do not exhibit a functional form that is physical interpretable.

5.1.4 Performing Large Scale MD Simulations

Interatomic potentials allow MD simulations containing a large number of atoms and hence can be a crucial step towards simulation conditions closer to experiment. In a MD simulation one has to calculate the force on each atom i from the interatomic potential. For that one only has to consider the positions of the neighbors of atom i, that are located around atom i within the cutoff radius $r^{(c)}$. This requires a list that contains the neighbor atoms for each atom. To calculate this neighbor list for each atom, we must not search over all N_{at} atoms, since this procedure would scale with $O(N_{\text{at}}^2)$. Fortunately, the **link-list algorithm** was introduced [38], which is visualized in two dimensions in Fig. 5.7.

The simulation cell is divided into squares (cuboids in 3D) with lengths that are equal or slightly bigger than the cutoff radius $r^{(c)}$. Obviously, every atom is located exactly in one of these squares (cubes), which are called link cells. Within every link cell a chain of the included atoms is formed and the atom, which is the head of the chain (illustrated in red in Fig. 5.7) is stored in a so called top list. In addition, the links between the atoms, which are illustrated by arrows in Fig. 5.7, are stored in a so called link list. This list contains for each atom the atom that is the next one in the chain, if there is a next atom. To construct the top and link lists, one goes thorough the list of all atoms only ones. In this loop, one determines for the considered atom i, in which link cell it is localized, and sets this atom as the top atom for this link cell. If an atom k was previously set as the top atom, one sets the atom k as the next atom for atom i in the link list. In this way, the chains of atoms are formed from the back and the construction of the top and link lists scales as $O(N_{\text{at}})$.

Using the top and link lists, the neighbor list of all atoms i can be efficiently constructed. To calculate the neighbor list of atom i, one considers the atoms in the same link cell (shown in yellow in Figs. 5.7 and 5.8) and the atoms in half of the neighboring link cells (shown in blue in Figs. 5.7 and 5.8). One only needs to consider half of the neighboring link cells, because if atom j is a neighbor of atom i, then atom i is a neighbor of atom j. Therefore, if we notice that atom j is a neighbor of atom i, then we set additionally atom i as a neighbor of atom j. By doing this,

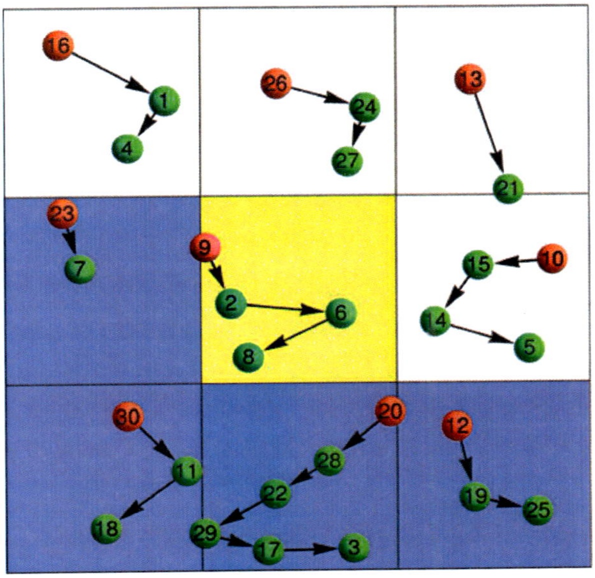

Fig. 5.7 The usage of link cells, the top and link lists are shown in two dimensions. Description see text

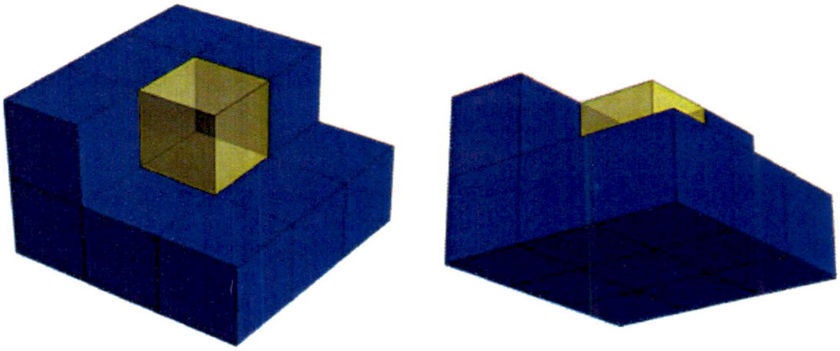

Fig. 5.8 In three dimensions, the neighboring link cells, which are used to calculate the neighbor list of atoms contained in the yellow link cell, are shown in blue

one only needs to consider half of the neighboring link cells, since one calculates the neighbor list for each atom i. Furthermore, to consider the atoms of a link cell, one just starts with the top atom from the top list and goes through the chain of atoms using the link list. In this way, the calculation of the neighbor list for each atom scales just with $O(N_{\text{at}})$.

Large scale MD simulations using an interatomic potential can be dramatically speed up by using a massive parallelization of the calculation. For this, the message passing interface (MPI) is commonly used, which allows to run a program on thousands of processors distributed on many computers. The simulation cell is disjointly divided into MPI cells (see Fig. 5.9), which each one is treated by one processor.

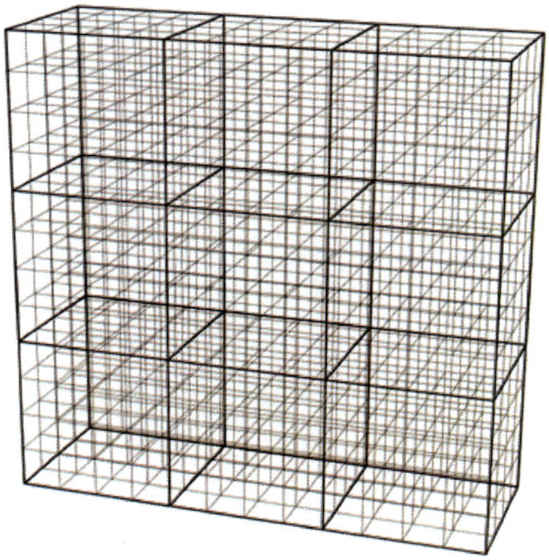

Fig. 5.9 The simulation volume is disjointly divided into MPI cells (big cells), which are disjointly divided into link cells (small cells)

Each MPI cell is disjointly divided into link cells, so that the link list algorithm can be applied to calculate efficiently the atomic neighbor lists. The most time consuming part in MD simulations are the calculations of the interatomic forces. Since one only needs the neighbors within the cutoff distance $r^{(c)}$ of atom i to calculate the force on atom i from the interatomic potential, one needs for the force calculation of one MPI cell only information from the 26 neighboring MPI cells about the atoms that are located in the adjacent link cells of the considered MPI cell. Therefore, the calculation of the forces in one MPI cell is independently from all other MPI cells except for the 26 direct neighbors, from which a few information is needed about the atoms in the adjacent link cells. This is the key point, why parallelization works, and opens the way for the simulation of a huge number of atoms, since only the number of available processors limits the system size.

5.2 Simulation of Laser Excitation via Two Temperatures and Velocity Scaling

If an intense femtosecond laser-pulse is irradiated on matter, a significant fraction of electrons can get excited and instantly reaching a high temperature while the ions remain initially cold. In order to describe this nonequilibrium state with hot electrons and cold ions, Anisimov et al. [39] introduced in 1974 the **two-temperature model** (TTM), in which they associate an electronic temperature T_{i} to the electrons and an ionic temperature T_{i} to the ions. They defined these temperatures locally by dividing the simulation cell into small sub cells, each one with individual T_{e} and T_{i}. They treated the electrons and ions at the continuum level and described the time evolution of T_{e} and T_{i} by the following two coupled differential equations using energy balance:

$$C_e(T_e)\,\frac{\partial T_e}{\partial t} = \nabla\cdot\Big(K_e(T_e)\,\nabla T_e\Big) - G_{ep}\,(T_e - T_i) + \frac{\partial E_{L_{abs}}(t)}{\partial t}, \tag{5.64}$$

$$C_i(T_i)\,\frac{\partial T_i}{\partial t} = \nabla\cdot\Big(K_i(T_i)\,\nabla T_i\Big) + G_{ep}\,(T_e - T_i). \tag{5.65}$$

$C_e(T_e)$ denotes the electronic and $C_i(T_i)$ denotes the ionic specific heat. $K_e(T_e)$ is the electronic and $K_i(T_i)$ is the phonon head conductivity. In Eq. (5.64), the first term on the right hand side takes the electronic heat flow into the neighboring sub cells into account, the second term describes the energy that is moved to the phonons due to the electron-phonon coupling and the third term describes the energy absorption rate from the laser field at time t. G_{ep} denotes the electron-phonon coupling constant and $E_{L_{abs}}(t)$ the energy, that is deposited by the laser in the electronic system up to time t. In Eq. (5.65), the first term on the right hand side takes the phonon related heat flow into the neighboring sub cells into account, which is commonly omitted, since the phonon heat conductivity is significantly smaller then the electronic heat conductivity. One can solve the system of differential equations (5.64) and (5.65) using a finite difference method and obtains the spatial and time evolution of the electronic and ionic temperatures. The TTM allows to simulate excited electrons at the continuum level. However, TTM cannot describe the physics of ultrafast phase transitions and damages occurring under highly nonequilibrium conditions induced by femtosecond laser-pulses. For this, one has to take explicitly the atomic movements into account.

Thus, Ivanov and Zhigilei introduced a combined atomistic-continuum modeling method named TTM-MD in 2003 [40]. They take explicitly the atoms into account by using an interatomic potential developed for electrons in the ground state. To consider excited electrons, they associate a temperature T_e to the electrons and treat them at the continuum level, so that the electronic energy $E_e(T_e)$ is just a function of T_e, from which one gets the electronic specific heat by

$$C_e(T_e) = \frac{\partial E_e(T_e)}{\partial T_e}. \tag{5.66}$$

They also associate a temperature T_i to the ions from their velocities by using Eq. (4.16).

In contrast to DFT calculations, one can use quite large simulation cells with one length being 100 nm or more in MD simulations with classical interatomic potentials. Therefore, one can define local electronic and ionic temperatures to take into account the spatial laser pulse profile and the laser intensity decay during material penetration. Practically, one associates to each link cell (or a set of neighboring link cells) a separate electronic and ionic temperature. To determine the local ionic temperature, only the atoms of the corresponding link cell are considered in Eq. (4.16).

Furthermore, Ivanov and Zhigilei distinguish between the thermal velocities $\mathbf{v}_i^{(T)}$ of the atoms and the collective velocity $\mathbf{v}^{(c)}$ of the atoms in each link cell. They calculate the collective velocity by

$$\mathbf{v}^{(c)} = \frac{\sum_i m_i \, \mathbf{v}_i}{\sum_i m_i}, \tag{5.67}$$

where the sum runs over the atoms of the link cell and m_i denotes the mass of atom i. They calculate the thermal velocity by

$$\mathbf{v}_i^{(T)} = \mathbf{v}_i - \mathbf{v}^{(c)}. \tag{5.68}$$

Using the thermal velocities $\mathbf{v}_i^{(T)}$, they define the local thermal kinetic energy

$$E_{\text{kin}}^{(T)} = \frac{1}{2} \sum_i m_i \left(\mathbf{v}_i^{(T)}\right)^2 \tag{5.69}$$

and the local ionic temperature T_i from Eq. (4.16) for each link cell. Furthermore, they take the electronic heat flow between the different link cells into account, which occurs due to the different local electronic temperatures. For this, they utilize the electronic heat conductivity $K_\text{e}(T_\text{e})$, which is a function of the electronic temperature T_e. Finally, they control the change of the electronic temperature T_e in any link cell by using Eq. (5.64) from the two-temperature model. To take the energy flow, which is related to the electron-phonon coupling, to the phonons into account, they modified the equation of motion for the ions to

$$\boxed{m_i \, \frac{d\mathbf{v}_i}{dt} = -\nabla_{\mathbf{r}_i} \Phi + \xi \, m_i \, \mathbf{v}_i^{(T)}.} \tag{5.70}$$

The first term on the right hand side describes the forces related to interatomic bonding and the second term describes the acceleration of the ions due to the electron-phonon coupling. Here, an atom is accelerated in the direction of its thermal velocity. They obtained the constant ξ, which controls the acceleration strength, from the energy conservation: The thermal kinetic energy of the ions is exactly increased by the energy contribution $G_\text{ep}\,(T_\text{e} - T_\text{i})$ that the electrons lost due to the electron-phonon coupling. This leads to

$$\xi = \frac{G_\text{ep}\,(T_\text{e} - T_\text{i})}{2\,E_\text{kin}^{(T)}}. \tag{5.71}$$

The numerical solution of the electronic heat diffusion between the different link cells needs a very small time step Δt_D, which is much smaller than the time step Δt used to solve the equations of motions for the ions. Therefore, Ivanov and Zhigilei used two time steps in the simulation. They solve the ionic equations of motions using the time step Δt. In one time step Δt, they perform $N_\text{D} \in \mathbb{N}$ time steps Δt_D to solve the equations for the electronic heat diffusion, so that the local electronic temperatures change and the ionic positions and the local ionic temperatures keep constant during the time steps Δt_{D}. Therefore, they have

$$\Delta t = N_{\mathrm{D}}\,\Delta t_{\mathrm{D}}, \tag{5.72}$$

and they use

$$\xi = \frac{\frac{1}{N_{\mathrm{D}}}\sum_{k=0}^{N_{\mathrm{D}}-1} G_{\mathrm{ep}}\left(T_{\mathrm{e}}(t + k\,\Delta t_{\mathrm{D}}) - T_{\mathrm{i}}(t)\right)}{2\,E_{\mathrm{kin}}^{(\mathrm{T})}(t)}. \tag{5.73}$$

to perform one time step Δt for the ions.

5.3 Simulation of Laser Excitation via Bond-Softening in the Tersoff Potential

A femtosecond laser-pulse potentially excites many electrons from bonding to anti-bonding states, which softens the interatomic bonding. Therefore, Lee and Park [41] recently introduced the parameter $\alpha_{ij}^{(\mathrm{att})}$ that scales the bonding part in the third Tersoff potential (T3) (c.f. Eq. (5.32)):

$$\Phi_{2\mathrm{b}}^{(\mathrm{Lee})}(r_{ij}, \xi_{ij}, \eta_{ij}) = \frac{1}{2}\, f_{\mathrm{c}}^{(\mathrm{T1})}(r_{ij}) \Bigg(\underbrace{\left(1 + \alpha^{n}\, \eta_{ij}^{n}\right)^{-\frac{1}{2n}} A\, e^{-\lambda_1 r_{ij}}}_{\text{antibonding}} - \alpha_{ij}^{(\mathrm{att})} \underbrace{\left(1 + \beta^{n}\, \xi_{ij}^{n}\right)^{-\frac{1}{2n}} B\, e^{-\lambda_2 r_{ij}}}_{\text{bonding}} \Bigg). \tag{5.74}$$

They set

$$\alpha_{ij}^{(\mathrm{att})} = \begin{cases} 1 & \text{for normal bonds} \\ 0.5 & \text{for broken bonds} \end{cases} \tag{5.75}$$

and kept, beside the parameter $\alpha_{ij}^{(\mathrm{att})}$, the original T3 potential from Ref. [20]. To simulate the effects of a femtosecond-laser excitation occurring within the first picosecond, they divided the simulation volume into small subcells and determined, for each subcell, the number of broken bonds

$$N_{\mathrm{broken}} = \left\lfloor \frac{E_{\mathrm{L_{abs}}}(t)}{\hbar\,\omega} \right\rfloor. \tag{5.76}$$

from the laser-absorbed energy $E_{\mathrm{L_{abs}}}(t)$ using the one-photon energy $\hbar\,\omega = 2.87\,\mathrm{eV}$. Here, $\lfloor x \rfloor$ denotes the largest integer number that is smaller or equal than x. Such a simulation setup allowed Lee and Park to simulate the non-thermal melting of silicon [41]. They took into account, that subcells close to the surface absorb more laser energy than subcells further away from the surface. Due to this, there are more broken bonds close to the surface, since the electrons absorb more energy. However, this simulation setup is not based on energy conservation, i.e., the potential energy is not increased by the amount of energy that is absorbed from the laser. No electronic temperature is related to the electrons. Therefore, a local electronic heat flow and the spatial relaxation of the electronic temperature cannot be simulated. In addition, the electron-phonon coupling was not taken into account and it would be very hard to implement this effect, since there are neither ionic and electronic temperatures nor an energy conservation.

5.4 T_e-Dependent Interatomic Potentials

A femtosecond laser-pulse excites mainly the electrons, which thermalize to a high electronic temperature T_e due to electron-electron collisions within tens of femtoseconds (see Sect. 3.4.2). This state with electrons at high T_e and quite cold ions exists for several ps (see Sect. 3.4.3). The PES generated by the hot electrons differs significantly from the PES generated from electrons in the ground state and is well described by T_e-dependent DFT. Therefore, in 2008, Khakshouri et al. introduced the first T_e-dependent interatomic potential that depend beside on the atomic coordinates also on T_e to take the significant changes in the PES due to the hot electrons into account [2]. This first T_e-dependent interatomic potential was developed for the metal tungsten. In the following years, five further T_e-dependent interatomic potentials were developed. Now we briefly introduce the up to now available T_e-dependent interatomic potentials in the literature:

- In 2008 for tungsten, Khakshouri et al. fitted the potential parameters to the Helmholtz free energies of the ideal bulk bcc structure obtained from T_e-dependent DFT for different lattice parameters at various T_e's [2]. They also provided a second parameterization, where they fitted the pressure, which they obtained from T_e-dependent DFT calculations of the ideal bulk bcc structure with $T_e = 300\,\text{K}$ equilibrium lattice parameter at various T_e's.
- In 2015, Murphy et al. developed an alternative T_e-dependent interatomic potential for tungsten [3]. They fitted the parameters of an extended embedded atom potential to the T_e-dependent DFT Helmholtz free energies of the ideal bulk bcc structure for different lattice parameters at various T_e's.
- In 2012, Norman et al. developed a T_e-dependent interatomic potential for the metal gold [4]. They fitted the parameters of an embedded atom potential to the T_e-dependent DFT Helmholtz free energies and forces of several reference structures of crystal and molten material at various T_e's.
- In 2012, Moriarty et al. developed a T_e-dependent interatomic potential for the transition metal molybdenum with the help of generalized pseudopotential theory [5].

- In 2010, Shokeen and Schelling developed a T_e-dependent interatomic potential for the semiconductor silicon [6, 7]. They fitted the parameters of the Modified Tersoff (MT) potential to the Helmholtz free cohesive energies obtained from T_e-dependent DFT of several ideal bulk crystal structures at various lattice parameters and T_e's.
- In 2018, Darkins et al. developed a second T_e-dependent interatomic potential for silicon [8]. In contrast to Shokeen and Schelling, they fitted the parameters of a modified version of the Modified Tersoff (MT) potential to the T_e-dependent DFT cohesive energies of the diamond structure at various lattice parameters and T_e's and to the Bragg peak decay measured by Harb et al. [42, 43].

We intensively studied the effects that are caused by electrons at increased T_e in silicon with the help of T_e-dependent DFT (see Sect. 4.4). Therefore, we will analyze now the performance of the two available T_e-dependent interatomic potentials for silicon in describing the increased T_e-induced effects obtained from T_e-dependent DFT. We already published the obtained results in Ref. [44].

5.4.1 Si Potential of Shokeen and Schelling

5.4.1.1 Construction

Shokeen and Schelling employed the Modified Tersoff (MT) potential (see Eqs. (5.37), (5.38) and (5.39)). They increased the original cutoff radius from $r_{2b}^{(c)} = 0.33\,\mathrm{nm}$ to $r_{2b}^{(c)} = 0.34\,\mathrm{nm}$ and the parameter $r^{(1)}$ from $r^{(1)} = 0.27\,\mathrm{nm}$ to $r^{(1)} = 0.31\,\mathrm{nm}$, but kept the 6 potential parameters $\lambda_1, \eta, \beta, h, c_3, c_5$. The remaining 8 potential parameters $A, B, \lambda_2, \delta, \alpha, c_1, c_2, c_4$ were set T_e-dependent. To do so, they were fitted at a given T_e to the Helmholtz free cohesive energies (4.75) of the diamond-like, fcc, bcc and sc structures at seven different lattice parameters near equilibrium. The cohesive energies were derived from T_e-dependent DFT calculations in the LDA approximation using the free available code ABINIT.[2] In total, 28 cohesive energies were fitted at 48 different T_e's ranging from 2321 K to 29592 K in increments of 580 K.[3] Furthermore, all cohesive energies were shifted to get the experimental cohesive energy of 4.63 eV for the diamond-like structure at equilibrium lattice parameter and $T_e = 0\,\mathrm{K}$. At $T_e = 0\,\mathrm{K}$, the potential coefficients of Kumagai et al. [15] were used except for $r_{2b}^{(c)} = 0.34\,\mathrm{nm}$, $r^{(1)} = 0.31\,\mathrm{nm}$, and $\alpha = 1.9$. The parameter α was set in such a way to maintain the experimental melting temperature of $T_m = 1687\,\mathrm{K}$ [45] at the increased $r_{2b}^{(c)}$ and $r^{(1)}$ values, since the potential of Kumagai et al. [15] already exhibits the experimental melting temperature. Also the T_e-dependent Helmholtz Free energy F_0 of an isolated Si atom was derived from T_e-dependent DFT and added to the potential.

---

[2] www.abinit.org.

[3] The eV unit was used for T_e in the calculations: 1 eV = 11604.668902282167 K.

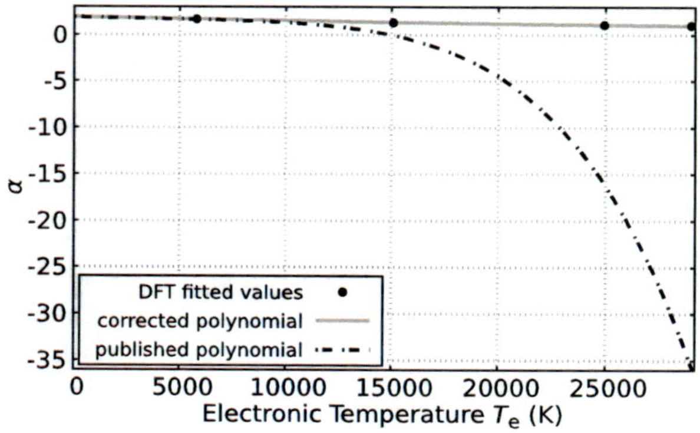

Fig. 5.10 For potential parameter α, the corresponding published and corrected polynomial is shown together with the supporting points adapted from Fig. 5 of Ref. [44]

In a first step, Shokeen and Schelling published F_0 and the 8 fitted potential coefficients at 4 selected T_e's in Ref. [6]. In a second step to finally get a smooth T_e-dependent interatomic potential, F_0 and all 8 T_e-dependent potential coefficients were approximated by a polynomial of degree 6 using the corresponding DFT fitted values as supporting points. This smooth T_e-dependent interatomic potential was published in Ref. [7] and will be called Shokeen–Schelling potential in the further study. Unfortunately, there is a miss print in the tabulation of the polynomial for the coefficient α in TABLE I of Ref. [7]: $a_5 = 0.4140$ is written, but it should be $a_5 = 0.0414$. To verify this, the in Ref. [7] published and the corrected polynomial is shown for α together with the corresponding DFT fitted values from Ref. [6] in Fig. 5.10. The corrected polynomial fits perfectly to the DFT fitted values, which act as supporting points, while the published one deviates dramatically at T_e's above 15000 K. Hence, the corrected polynomial is used in the further study.

5.4.1.2 Accuracy in the Description of Si at Increased T_e

We repeated the calculations and MD simulations presented in Sect. 4.4 at constant T_e with the Shokeen–Schelling potential. To perform the corresponding MD simulations, we used the DFT initializations. In Fig. 5.11, we present the summary of the comparison between the Shokeen–Schelling potential and DFT results (c.f. Fig. 4.30).

Since Shokeen and Schelling shifted the cohesive energies curves obtained from DFT before fitting, the cohesive energy curves do not match to the DFT curves. Therefore, we skip the DFT curves in the bottom series of figures in Fig. 5.11. In addition, the absolute values of the cohesive energies become bigger for increasing T_e in contrast to DFT.

The optical phonon modes of the diamond-like structure are well described at increased T_e's. The acoustic phonon modes are well described at $T_e = 12631$ K and show a softening compared to $T_e = 316$ K. But, parts of them become negative at $T_e = 15789$ K in contrast to DFT, where negative acoustic phonon frequencies only occur

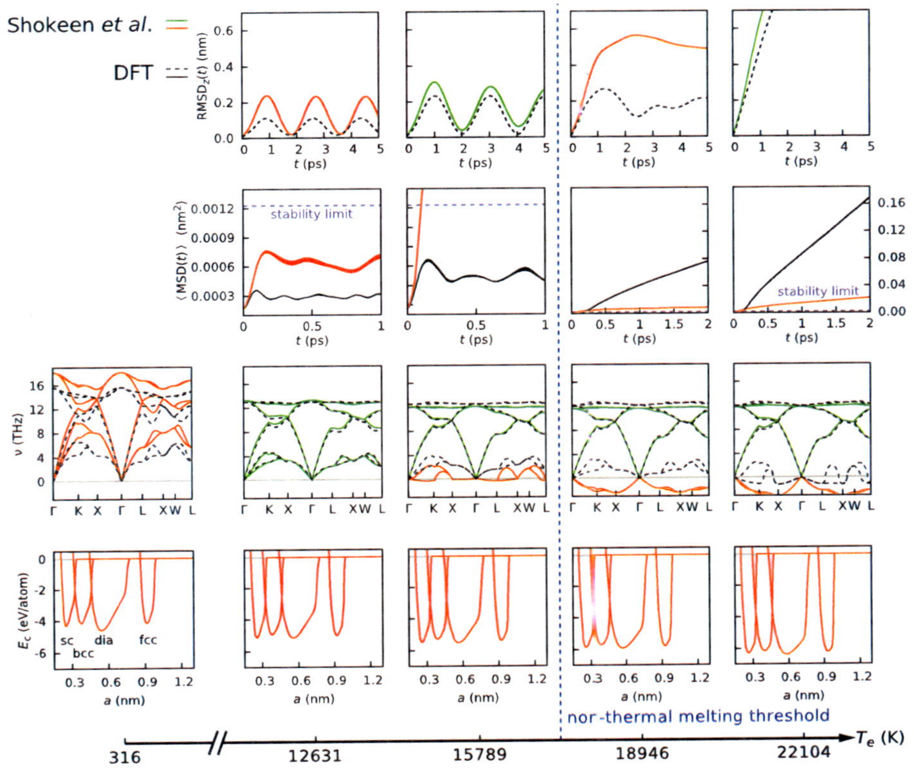

Fig. 5.11 Summary of the performance of the Shokeen–Schelling potential in describing the effects in Si induced by the increased T_e deformed PES. Description see text

at $T_e \geq 22104$ K. At $T_e = 18946$ K, all acoustic branches are negative in contrast to DFT, where all phonon frequencies are still positive.

In bulk MD simulations, the Shokeen–Schelling potential exhibits thermal phonon antisqueezing at $T_e = 12631$ K. But the averaged atomic mean square displacements $\langle \text{MSD}(t) \rangle$ become twice as big compared to DFT. At $T_e = 15789$ K, the bulk crystal lattice already melts and no phonon squeezing occurs in contrast to DFT. At $T_e = 18946$ K, the non-thermal melting occurs in four stages contrary to the three stages found from DFT. After the super and the fractional diffusive stage, the normally diffusive stage splits into two areas with different diffusion coefficients D:

0.63 ps after the laser excitation, the $\langle \text{MSD}(t) \rangle$ increases linearly in time with diffusion coefficient

$$D = 0.0002479 \pm 0.0000002 \text{ (nm}^2/\text{ps)}.$$

Then at 1.86 ps, the diffusion coefficient suddenly changes to

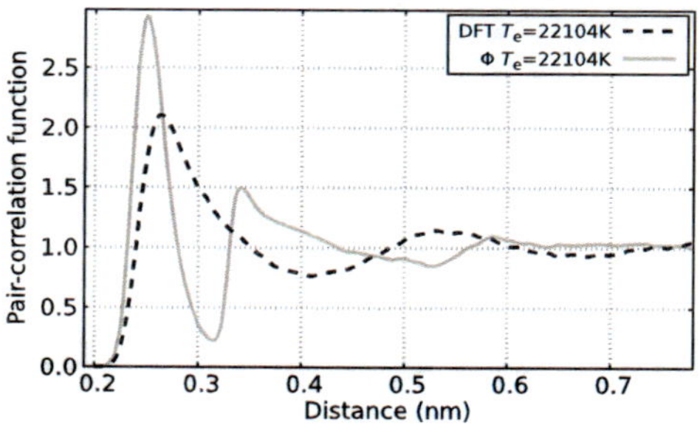

Fig. 5.12 The pair-correlation function at $T_e = 22104\,\text{K}$ is shown 1 ps after the laser excitation obtained from the Shokeen–Schelling potential Φ and from DFT

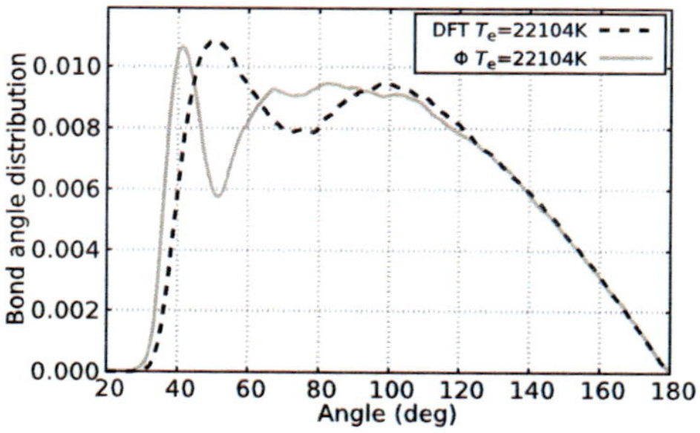

Fig. 5.13 The bond angle distribution for distances up to 0.42 nm at $T_e = 22104\,\text{K}$ is shown 1 ps after the laser excitation obtained from the Shokeen–Schelling potential Φ and from DFT

$$D = 0.0001558 \pm 0.0000001\ (\text{nm}^2/\text{ps}).$$

This artifact of two diffusion coefficients D vanishes at higher T_e's, so that only above $T_e = 18946\,\text{K}$, the Shokeen–Schelling potential exhibits the three melting stages discovered in DFT. Furthermore, at all studied T_e's above the DFT non-thermal melting threshold, the averaged atomic mean-square displacement $\langle \text{MSD}(t) \rangle$ is always much smaller compared to DFT. This indicates that the non-thermal melting behaves significantly to slow when described by means of the Shokeen–Schelling potential. Consequently, the pair-correlation function and the bond angle distribution is poorly described by the Schokeen–Schelling potential after the laser excitation, which is shown in Figs. 5.12 and 5.13 at $T_e = 22104\,\text{K}$, as an example.

The MD simulations of Si in thin-film geometry at increased T_e are in qualitative agreement with DFT. At $T_e \leq 15789\,\text{K}$, the thin film does not melt and it's thickness oscillates periodically in time. At $T_e = 18946\,\text{K}$, the thin film starts to melt. Here, the thickness firstly increases followed by a very weak decrease. At higher T_e's, the thin film just melts and expands. This behavior of the thin film at the various T_e's can be clearly seen from the atomic root mean-square displacements in z-direction $\text{RMSD}_z(t)$, which are presented in the top series of figures in Fig. 5.11.

5.4.1.3 Analysis of Electronic Absorbed Energy and Electronic Specific Heat

The T_e-dependent interatomic potential Φ corresponds to the Helmholtz free energy of the electrons. Therefore, using thermodynamic relations, the electronic internal energy is given by

$$E_e = \Phi - T_e \frac{\partial \Phi}{\partial T_e} \tag{5.77}$$

and the electronic specific heat is given by

$$C_e = -T_e \frac{\partial^2 \Phi}{\partial T_e^2}. \tag{5.78}$$

To analyze E_e and C_e, both quantities were calculated for the ideal diamond-like structure as a function of T_e in the T_e-validity range of the Shokeen–Schelling potential. The resulting curves are presented in Figs. 5.14 and 5.15 together with the corresponding curves obtained from T_e-dependent DFT using CHIVES. Unfortunately, E_e and C_e obtained from the potential are nonphysical. Both quantities should be at least non negative and E_e should increase monotonously as a function T_e, but these conditions are violated.

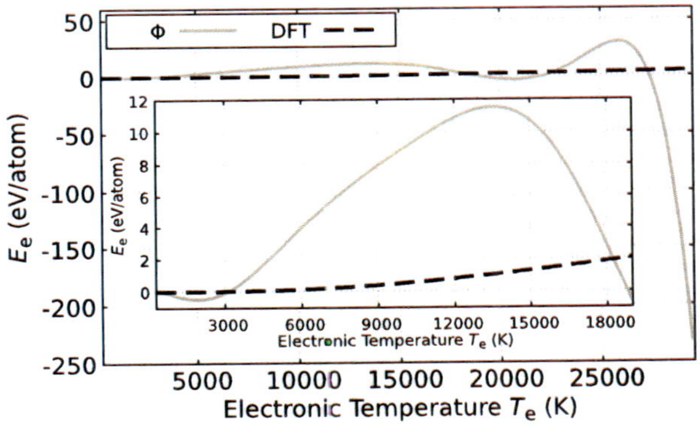

Fig. 5.14 Electronic internal energy E_e of the ideal diamond-like structure as a function of T_e for DFT and the Shokeen–Schelling potential Φ

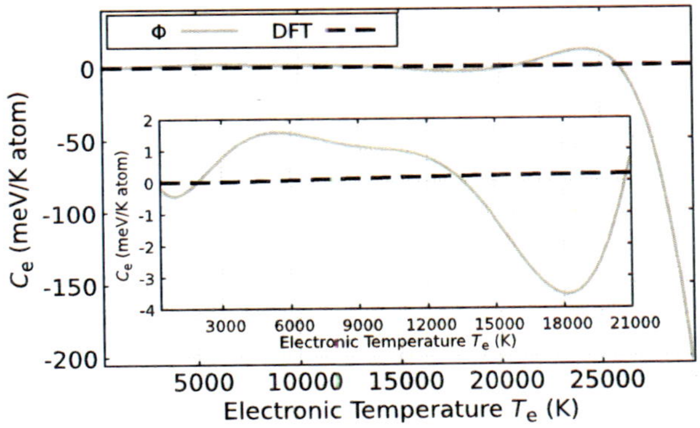

Fig. 5.15 Electronic specific heat C_e of the ideal diamond-like structure as a function of T_e for DFT and the Shokeen–Schelling potential Φ

5.4.2 *Si Potential of Darkins et al.*

5.4.2.1 Construction

Darkins et al. [8] also used the Modified Tersoff (MT) potential (see Eqs. (5.37), (5.38) and (5.39)). But in contrast to the Shokeen–Schelling potential, Darkins et al. modified the $\xi_{\rm loc}$-function from Eq. (5.39) with the help of the additional parameter λ and the additional function Λ:

$$\xi_{\rm loc}\big(r_{ij}, r_{ik}, \cos(\theta_{ijk})\big) = f_{\rm c}^{\rm (MT)}(r_{ik})\, e^{\alpha\,(r_{ij}-r_{ik})^{\beta}} \times \\ \times \Big(g\big(\cos(\theta_{ijk})\big) - \Lambda\Big(g\big(\cos(\theta_{ijk})\big) - g(-1/3), \lambda\Big)\Big), \tag{5.79}$$

$$g\big(\cos(\theta_{ijk})\big) = c_1 + \frac{c_2\,\big(h-\cos(\theta_{ijk})\big)^2}{c_3 + \big(h-\cos(\theta_{ijk})\big)^2}\left(1 + c_4\, e^{-c_5\,\left(h-\cos(\theta_{ijk})\right)^2}\right), \tag{5.80}$$

$$\Lambda(x, y) = \mathrm{sgn}(x)\begin{cases} |x| & |x| \le \frac{y}{\sqrt{2}} \\ \sqrt{y^2 - \big(\sqrt{2}\,y - |x|\big)^2} & \frac{y}{\sqrt{2}} < x < \sqrt{2}\,y \\ y & x \ge \sqrt{2}\,y. \end{cases} \tag{5.81}$$

Here, the function $g\big(\cos(\theta_{ijk})\big)$ corresponds to the original $\cos(\theta_{ijk})$-dependence of the $\xi_{\rm loc}$-function in Eq. (5.39) from the Modified Tersoff potential.

Darkins et al. set $r_{\rm 2b}^{\rm (c)} = 0.34\,$nm and $r^{(1)} = 0.31\,$nm similar to Shokeen and Schelling, chose $\alpha = 1.9$ and kept the 12 parameters $\lambda_1, \lambda_2, \eta, \delta, \alpha, \beta, h, c_1, c_2, c_3, c_4, c_5$ from the Modified Tersoff potential. Only the parameters A, B, λ depend on $T_{\rm e}$. For this, A, B were fitted at a given $T_{\rm e}$ to the internal cohesive energy of the diamond structure at various lattice parameters near equilibrium. To get the internal cohesive energy, the internal energy of the diamond structure was reduced by the internal energy of the contained atoms at isolated state. The internal energies were derived from DFT using the code VASP.[4] To define the remaining parameter λ, the measurements of Harb et al. [42, 43] were considered, in which the femtosecond laser excitation of a free standing 50 nm Si film was measured using femtosecond electron diffraction. To consider these measurements, MD simulations of the laser-excited 50 nm Si film were performed with the potential including the electron-phonon coupling by applying a Langevin formalism with a damping parameter γ. To determine this damping parameter γ, $\lambda = 0$ was assumed below the damage threshold. Then, the damping parameter γ was optimized to get a similar evolution of the ionic temperature $T_{\rm i}$ compared to the corresponding experimental data of Ref. [42] at a laser fluence of 5.6 mJ/cm^2 below the damage threshold. From this, $\lambda = 1\,\frac{\rm g}{\rm mol\ ps}$ was found and uniformly used for all further MD simulations. Using this λ value, the damage

[4] https://www.vasp.at.

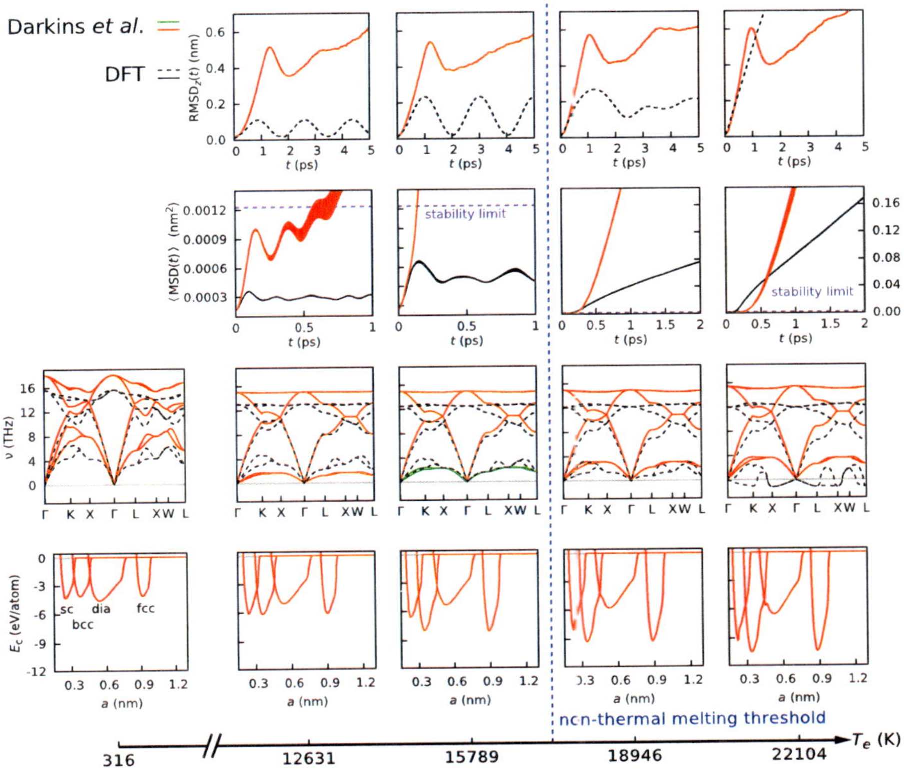

Fig. 5.16 Summary of the performance of the potential of Darkins et al. in describing the effects in Si induced by the increased T_e deformed PES. Description see text

threshold was located at $T_e = 5222$ K in the MD simulations. Finally, to define λ above the damage threshold, it was set to get a similar decay of the Bragg peaks from the potential MD simulations with respect to the experimental data of Ref. [43]. Darkins et al. tabulated the obtained values of A, B and λ at various T_e's in Ref. [8]. Since there is no smooth T_e-dependent function provided for these parameters, the electronic internal energy and the electronic specific heat cannot be directly calculated from the potential by means of T_e derivatives using thermodynamic relations.

5.4.2.2 Accuracy in the Description of Si at Increased T_e

Again, we repeated the calculations and MD simulations presented in Sect. 4.4 at constant T_e with the potential of Darkins et al. and present the summary of the comparison between the potential and DFT results in Fig. 5.16.

Since Darkins et al. also shifted the cohesive energies obtained from DFT before fitting, the cohesive energy curves do not match the DFT curves. Therefore, we skip

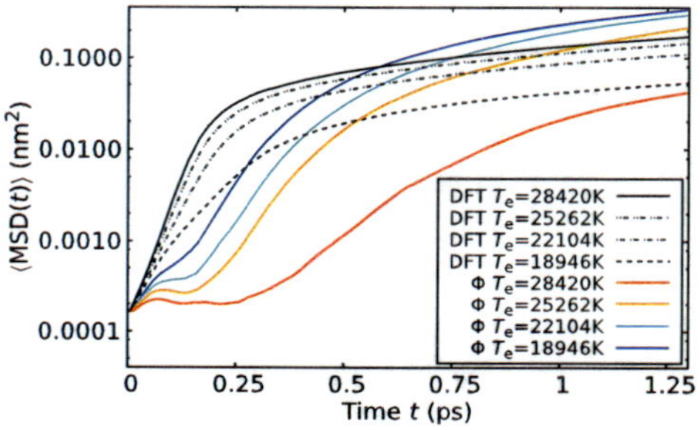

Fig. 5.17 Averaged mean-square displacement $\langle \mathrm{MSD}(t) \rangle$ of bulk Si is shown for DFT and the potential Φ of Darkins et al. at the studied T_e's above the non-thermal melting threshold

the corresponding DFT curves in the bottom series of Fig. 5.16. In addition, the absolute values of the cohesive energy curves become bigger for increasing T_e in contrast to DFT.

For the potential of Darkins et al., all phonon frequencies of the diamond-like structure show a softening, if T_e is raised from $T_e = 316$ K up to $T_e = 12631$ K, as one can see in Fig. 5.16. This indicates that phonon squeezing may be observable at $T_e = 12631$ K. But at $T_e = 12631$ K and $T_e = 15789$ K, the bulk crystal structure already melts in contrast to DFT, as it can be clearly seen from the time-dependent averaged atomic mean-square displacements $\langle \mathrm{MSD}(t) \rangle$ in Fig. 5.16, and phonon squeezing cannot be observed.

At higher T_e's, the non-thermal melting in bulk silicon obtained from the potential of Darkins et al. occurs in four stages contrary to the three stages obtained from DFT. In detail, the derivative of the atomic mean-square displacements increases, decreases, increases and keeps then constant as a function of time (see Fig. 5.17). Beside this, the atomic mean square-displacements are smaller for higher T_e's, or, in other words, the non-thermal melting becomes slower for higher T_e's. One explanation of this may be that atomic bonding in the potential of Darkins et al. become stronger for increasing T_e's. This is indicated from the phonon bandstructure, since all phonon frequencies exhibit a hardening and, consequently, the acoustic ones become never negative, if T_e is further increased from $T_e = 12631$ K, as it can be seen in Fig. 5.16. This is again a direct contradiction to DFT, where an increasing T_e causes phonon softening and, consequently, weakens atomic bonds.

The thin-film shrinks and starts to melt at $T_e \geq 12631$ K in contradiction to DFT. This behavior is illustrated in Fig. 5.18 at $T_e = 12631$ K, where the averaged z-positions of the crystal planes in the thin film are shown as a function of time. Consequently, the atomic root mean-square displacements in z-direction $\mathrm{RMSD}_z(t)$ obtained from the potential of Darkins et al. differ significantly from DFT, as it can be seen in Fig. 5.16.

In Summary, the potential of Darkins et al. does not even give a qualitative description of the effects in Si caused by the increased T_e in contrast to the Shokeen–Schelling potential. The reason of this may be the special method of fitting the potential parameters: These were fitted to DFT and simultaneously to experimental data by also including the effect of electron-phonon coupling. This procedure may be critical due

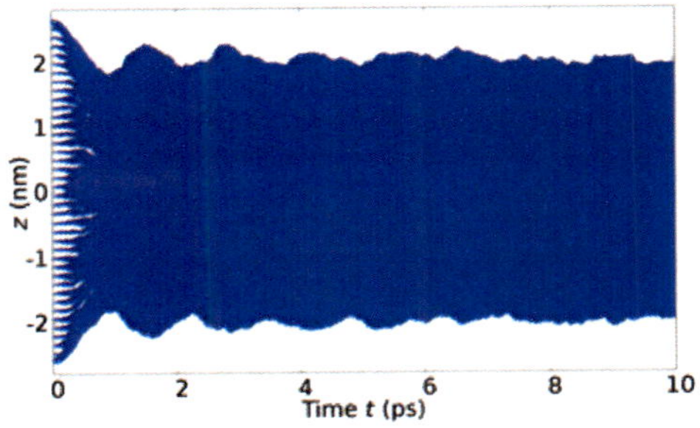

Fig. 5.18 Averaged z-position of the crystal planes with respect to the film center is shown for the potential Φ of Darkins et al. as a function of time at $T_e = 12631\,\text{K}$. The line width corresponds to the standard deviations of the atomic z-coordinates in each plane

to two reasons: On the one hand, the task of the potential is to describe the PES or, in other words, the effective ion interactions at a given T_e. But the PES or the ion interactions are not influenced by the electron-phonon coupling. Hence, electron-phonon coupling should not be included in the fitting procedure. On the other hand, it seems that the experimental data contradicts the used DFT results: All potential parameters, except of λ, were firstly obtained from DFT. Then, the remaining potential parameter λ was defined by comparing the Bragg peak decay derived from MD simulations using the potential and electron-phonon coupling with experimental data. It seems, that the Bragg peak decay or the melting speed obtained from the potential MD simulations with the electron-phonon coupling is to fast with respect to the experimental data. Consequently, the non-thermal melting obtained from the potential must be slowed down by λ, so that the results match to the experiment. This produces slower non-thermal melting at higher T_e's in contrast to DFT.

5.4.3 MD Simulations with a T_e-Dependent Interatomic Potential

Murphy et al. [3], Norman et al. [4], Darkins et al. [8] and Shokeen and Schelling [22] presented MD simulations with their individual T_e-dependent interatomic potential including also the electron-phonon coupling. They all except Darkins et al. and Shokeen and Schelling use Eq. (5.64)

$$C_e(T_e)\,\frac{\partial T_e}{\partial t} = \nabla\cdot\Big(K_e(T_e)\,\nabla T_e\Big) - G_{ep}\,(T_e - T_i) + \frac{\partial E_{L_{abs}}(t)}{\partial t}.$$

of Ivanov and Zhigilei [40] to control T_e. They all calculated the ionic temperature T_i from the ionic velocities $\mathbf{v}_i$ via Eq. (4.16) similar to Ivanov and Zhigilei. To simulate the effects of the increased T_e deformed PES and the effects of the electron-phonon coupling on the ions, they all use the following ionic equation of motion:

$$m_i\,\frac{d\mathbf{v}_i}{dt} = -\nabla_{\mathbf{r}_i}\Phi - \gamma\,\mathbf{v}_i + \mathbf{f}_i^{(s)}. \tag{5.82}$$

The first term on the right hand side is to the force on atom i obtained from the T_e-dependent interatomic potential Φ to take the T_e-induced changes in the PES into account. The second and third term are present to take the electron-phonon coupling into account. The second term describes an electronic friction on the ions with constant γ that controls the timescale of the ionic and electronic temperature relaxation. The third term is a stochastic force that is set to induce a relaxation of the ionic and electronic temperatures to the same value. This stochastic force follows a Gaussian distribution, which ensemble average $\langle . \rangle$ obeys $\left\langle \mathbf{f}_{\mathrm{i}}^{\,(\mathrm{s})} \right\rangle = \mathbf{0}$ and where $\left\langle \mathbf{f}_{\mathrm{i}}^{\,(\mathrm{s})\mathrm{t}} \cdot \mathbf{f}_{\mathrm{i}}^{\,(\mathrm{s})} \right\rangle$ is set in such a way, that the ionic and electronic temperatures relax to the same value. Norman et al. added in Eq. (5.82) an extra term, which describes the force of entrainment of the ideal gas of delocalized electrons in gold on the ions. Darkins et al. and Shokeen and Schelling did not consider the electron-phonon coupling constant G_{ep}, therefore they use

$$\frac{\partial E_{\mathrm{e}}(T_{\mathrm{e}})}{\partial t} = \nabla \cdot \Big(K_{\mathrm{e}}(T_{\mathrm{e}})\, \nabla T_{\mathrm{e}} \Big) - \sum_i \mathbf{v}_i^{\mathrm{t}} \cdot \Big(\mathbf{f}_i^{\,(\mathrm{s})} - \gamma\, \mathbf{v}_i \Big) + \frac{\partial E_{\mathrm{L_{abs}}}(t)}{\partial t} \tag{5.83}$$

to control T_e instead of Eq. (5.64). Darkins et al. did not have the first term, since they used one global T_e instead of local T_e's in the simulation cell. In Eq. (5.83), the energy loss of the electrons due to the electron-phonon coupling is described by $-\sum_i \left(\mathbf{f}_i^{\,(\mathrm{s})\mathrm{t}} - \gamma\, \mathbf{v}_i^{\mathrm{t}} \right) \cdot \mathbf{v}_i$ instead of $-G_{\mathrm{ep}}\,(T_e - T_{\mathrm{i}})$. The sum runs over all atoms i that are included in the sub cell with local electronic temperature T_e. $E_e(T_e)$ denotes the internal energy of the electrons in the sub cell, which Shokeen and Schelling calculated from their T_e-dependent interatomic potential Φ by

$$E_{\mathrm{e}}(T_{\mathrm{e}}) = \Phi(T_{\mathrm{e}}) - T_{\mathrm{e}}\, \frac{\partial \Phi(T_{\mathrm{e}})}{\partial T_{\mathrm{e}}} - \Phi(T_{\mathrm{e}} = 0\,\mathrm{K}). \tag{5.84}$$

Using Eq. (5.83), Shokeen and Schelling determined the change $E_e(t + \Delta t) - E_e(t)$ of the electronic energy and derived the new electronic temperature $T_e(t + \Delta t)$ of the next time step by

$$T_{\mathrm{e}}(t + \Delta t) = \sqrt{T_{\mathrm{e}}(t)^2 + \frac{2\, T_{\mathrm{e}}(t) \Big(E_{\mathrm{e}}(t + \Delta t) - E_{\mathrm{e}}(t) \Big)}{C_{\mathrm{e}}\big(T_{\mathrm{e}}(t)\big)}}. \tag{5.85}$$

Here, C_e denotes the electronic specific heat, which Shokeen and Schelling did not calculate from their T_e-dependent interatomic potential Φ using thermodynamic relations by $C_e = -T_e \frac{\partial^2 \Phi}{\partial T_e^2}$, since the obtained electronic specific heat would behave in a non physical way (see Fig. 5.15). Therefore, Shokeen and Schelling fitted $C_e(T_e)$ as a polynomial in T_e using results obtained from T_e-dependent DFT for Si in the ideal diamond-like structure.

In Summary, all simulation methods take into account, that a specific amount of energy is removed from the electrons due to the electron-phonon coupling. But all except for Darkins et al. did not simulate the energy gain or loss of the ions due to electron-phonon coupling based on energy conservation, so that the removed energy from the electrons is transferred to the ions. They all, except for Darkins et al., manipulate the ionic motion just with an adapted Langevin thermostat for ensuring that the electronic and ionic temperature relax to the same value. Therefore, Darkins et al. [8] developed a MD simulation formalism based on energy conservation. For this, they claimed that the T_e-dependent interatomic potential should describe the internal energy instead of the Helmholtz free energy of the electrons, which was the case for all other presented T_e-dependent interatomic potentials. This allows them to formulate an energy conservation using the following total energy:

$$E_{\text{tot}} = \sum_i \frac{1}{2} m_i v_i^2 + \Phi + \int_0^{T_e} dT_e' \, w\left(\frac{\partial \Phi}{\partial T_e}, \varepsilon\right) \tag{5.86}$$

Here, the sum over i runs over all atoms in the simulation cell and the integral over the function $w\left(\frac{\partial \Phi}{\partial T_e}, \varepsilon\right)$ takes care that the electronic specific heat

$$C_e = \frac{\partial E_{\text{tot}}}{\partial T_e} = \frac{\partial \Phi}{\partial T_e} + w\left(\frac{\partial \Phi}{\partial T_e}, \varepsilon\right) \tag{5.87}$$

is always bigger or equal than a given constant $\varepsilon > 0$. To formulate the energy conservation, the electronic specific heat must be calculated from the T_e-dependent interatomic potential via thermodynamic relations. To prevent negative values or values to close to zero of the electronic specific heat derived from the T_e-dependent interatomic potential, Darkins et al. introduced the integral term in Eq. (5.86) as an on-the-fly correction of the T_e-dependent interatomic potential during the MD simulation. But this correction introduces further artificial forces on the ions. To derive the equation for controlling T_e, Darkins et al. consider the total time derivative of the total energy:

$$\frac{dE_{\text{tot}}}{dt} = \sum_i m_i \mathbf{v}_i^{\text{t}} \cdot \frac{d\mathbf{v}_i}{dt} + \sum_i \mathbf{v}_i^{\text{t}} \cdot \nabla_{\mathbf{r}_i} \Phi + \underbrace{\frac{\partial E_{\text{tot}}}{\partial T_e}}_{=C_e} \frac{dT_e}{dt}. \tag{5.88}$$

The change of the total energy corresponds to the energy absorption rate from the laser field, i.e., $\frac{dE_{\text{tot}}}{dt} = \frac{\partial E_{\text{L}_{\text{abs}}}(t)}{\partial t}$. Since Darkins et al. also use Eq. (5.82) as the equation of motion for the ions, they obtained by inserting the latter for $m_i \frac{d\mathbf{v}_i}{dt}$ in Eq. (5.88):

$$\frac{\partial E_{\text{L}_{\text{abs}}}(t)}{\partial t} = \sum_i \mathbf{v}_i^{\text{t}} \cdot \left(-\nabla_{\mathbf{r}_i} \Phi - \gamma \, \mathbf{v}_i + \mathbf{f}_i^{(\text{s})}\right) + \sum_i \mathbf{v}_i^{\text{t}} \cdot \nabla_{\mathbf{r}_i} \Phi + C_e \frac{dT_e}{dt}. \tag{5.89}$$

By rearranging they get the final equation

$$C_e \frac{dT_e}{dt} = -\sum_i \mathbf{v}_i^t \cdot \left(\mathbf{f}_i^{(s)} - \gamma\, \mathbf{v}_i\right) + \frac{\partial E_{L_{abs}}(t)}{\partial t}. \tag{5.90}$$

Since $C_e \frac{dT_e}{dt}$ can be interpreted as the change of the electronic internal energy $\frac{\partial E_e(T_e)}{\partial t}$, the above equation is equivalent to Eq. (5.83) used by Shokeen and Schelling to control T_e. Only the term, which considers the local heat flow is missing, since Darkins et al. only consider one global T_e for the whole simulation cell. The main difference is that Darkins et al. based their method on energy conservation in contrast to the others. Therefore, they calculated C_e via Eq. (5.87) from their T_e-dependent interatomic potential instead of using a precomputed C_e.

The simulation method of Darkins et al. is based on the fact that the T_e-dependent interatomic potential should describe the internal energy of the electrons. However, this is not sufficient. If the electronic state is described by an electronic temperature T_e, the thermodynamically correct quantity to consider for the energy is the Helmholtz free energy. T_e-dependent DFT is only working, since the equilibrium density at electronic temperature T_e minimizes the Helmholtz free energy. If one sets a Fermi distribution of the occupation numbers of the electronic orbitals in DFT, then one has to consider the Helmholtz free energy. If one were to consider the internal energy of the electrons, the energy functional would not be variational [46]. In addition, in MD simulations with electrons at temperature T_e, only the Helmholtz free energy is conserved [47]. Therefore, the simulation method of Darkins et al. takes the forces on the ions wrongly into account, since their T_e-dependent interatomic potential describes the internal and not the Helmholtz free energy of the electrons. Beside this, one can expect additional artificial forces due to the on-the-fly C_e correction, which should be avoided by constructing a T_e-dependent interatomic potential that delivers a physical meaningful C_e.

5.5 Universal Interatomic Potential Parameter Fitting Program

Since POTFIT can only fit the parameters of a few fixed analytical interatomic potential models and we do not like the presence of the additional parameters ε_i and ε_k in the fit error function $\mathbb{W}_{\text{err}}^{(\text{potfit})}$ (5.63), I developed an own interatomic potential parameter fitting program. This code allows to fit the parameters of a huge amount of different analytical interatomic potential models, since one can change quite easily the used analytical form of the interatomic potential. Later, this code is used to fit the parameters of the potential models MT, T1, T2, T3, D, PM, SW and MEAM to develop potentials for Si (see Sect. 7.1.2), which is impossible with POTFIT.

The interatomic potential parameters are fitted at a given fixed electronic temperatures T_e. For this, at the considered T_e, atomic forces and Helmholtz free structural cohesive energies of several reference T_e-dependent DFT simulations s_ℓ are used.

5.5.1 Construction of Fit Error Function

To measure the deviation between the atomic forces $\mathbf{f}_i^{(s_\ell)}(T_e, t_k)$ obtained from DFT and the atomic forces $-\nabla_{\mathbf{r}_i^{(s_\ell)}}\Phi\left(\left\{\mathbf{r}_j^{(s_\ell)}(T_e, t_k)\right\}, T_e\right)$ obtained from the T_e-dependent interatomic potential Φ for a specific reference simulation s_ℓ at a given T_e, we define the normalized root-mean square deviation in atomic forces

$$f_{\mathrm{err}}^{(s_\ell)}(T_e) = \sqrt{\frac{\sum_{k=1}^{N_t^{(s_\ell)}} \sum_{i=1}^{N_{\mathrm{at}}^{(s_\ell)}} \left| -\nabla_{\mathbf{r}_i^{(s_\ell)}}\Phi\left(\left\{\mathbf{r}_j^{(s_\ell)}(T_e, t_k)\right\}, T_e\right) - \mathbf{f}_i^{(s_\ell)}(T_e, t_k)\right|^2}{\sum_{k=1}^{N_t^{(s_\ell)}} \sum_{i=1}^{N_{\mathrm{at}}^{(s_\ell)}} \left|\mathbf{f}_i^{(s_\ell)}(T_e, t_k)\right|^2}}. \tag{5.91}$$

To measure analogously the deviation between Helmholtz free structural cohesive energies $E_c^{(s_\ell)}(T_e, t_k)$ obtained from DFT and Helmholtz free structural cohesive energies

$$\Phi\left(\left\{\mathbf{r}_j^{(s_\ell)}(T_e, t_k)\right\}, T_e\right) - \sum_{i=1}^{N_{\mathrm{at}}^{(s_\ell)}} \Phi_0(T_e)$$

obtained from the potential Φ for a specific reference simulation s_ℓ at given T_e, we define the normalized root-mean square deviation in Helmholtz free structural cohesive energies

$$E_{\mathrm{err}}^{(s_\ell)}(T_e) = \sqrt{\frac{\sum_{k=1}^{N_t^{(s_\ell)}} \left(\Phi\left(\left\{\mathbf{r}_j^{(s_\ell)}(T_e, t_k)\right\}, T_e\right) - \sum_{i=1}^{N_{\mathrm{at}}^{(s_\ell)}} \Phi_0(T_e) - E_c^{(s_\ell)}(T_e, t_k)\right)^2}{\sum_{k=1}^{N_t^{(s_\ell)}} \left(E_c^{(s_\ell)}(T_e, t_k)\right)^2}}. \tag{5.92}$$

Both deviations are unitless by definition. They are also normalized, since the constant potential

$$\Phi(T_e) \equiv \sum_{i=1}^{N_{\mathrm{at}}} \Phi_0(T_e), \tag{5.93}$$

which only takes the Helmholtz free energies $\Phi_0(T_e)$ of the isolated atoms into account and, therefore, always cause zero atomic forces, leads to

$$f_{\mathrm{err}}^{(s_\ell)}(T_\mathrm{e}) = 1, \qquad E_{\mathrm{err}}^{(s_\ell)}(T_\mathrm{e}) = 1.$$

Due to this, $f_{\mathrm{err}}^{(s_\ell)}(T_\mathrm{e})$ can be interpreted as the relative error in the atomic forces and $E_{\mathrm{err}}^{(s_\ell)}(T_\mathrm{e})$ as the relative error in the Helmholtz free structural cohesive energies. For a reasonable reference simulation s_ℓ, both relative errors are well defined, since the occurring denominators in Eqs. (5.91) and (5.92) are positive. In detail, if all reference structural cohesive energies $E^{(s_\ell)}(T_\mathrm{e}, t_k)$ are zero, the corresponding reference simulation s_ℓ consists only of structures, in which the atoms do not interact with each other. Hence, this run contains no information about interactions and should not be used in $E_{\mathrm{err}}^{(s_\ell)}(T_\mathrm{e})$. Furthermore, if all reference atomic forces $\mathbf{f}_i^{(s_\ell)}(T_\mathrm{e}, t_k)$ are zero, the corresponding reference simulation contains structures, in which atoms are all at equilibrium positions or do not interact with each other. As previous mentioned, structures of the latter case should not be fitted. Moreover, if the zero forces of a structure are caused by crystal symmetry, any potential produces zero forces and, consequently, fitting of atomic forces from such a structure should not be done. Only if the zero forces are not linked to crystal symmetry, the atomic forces of the corresponding structure contain further information and can be used for fitting. If a run only contains such structures, a structure with non zero forces must be added to properly define $f_{\mathrm{err}}^{(s_\ell)}(T_\mathrm{e})$.

At a given T_e, we search for the corresponding optimal potential parameters that exhibit as small as possible relative errors in atomic forces $f_{\mathrm{err}}^{(s_\ell)}(T_\mathrm{e})$ and Helmholtz free cohesive structural energies $E_{\mathrm{err}}^{(s_\ell)}(T_\mathrm{e})$. We determine these optimal potential parameters by minimization of the fit error function $\mathbb{W}_{\mathrm{err}}(T_\mathrm{e})$. By utilizing the fit weights $\mathbb{w}_{\mathrm{f}}^{(s_\ell)}(T_\mathrm{e}) \geq 0$, $\mathbb{w}_{\mathrm{E}}^{(s_\ell)}(T_\mathrm{e}) \geq 0$, which obey the sum rule

$$\sum_{s_\ell \in \mathcal{M}_\mathrm{s}(T_\mathrm{e})} \mathbb{w}_{\mathrm{f}}^{(s_\ell)}(T_\mathrm{e}) + \sum_{s_\ell \in \mathcal{M}_\mathrm{s}(T_\mathrm{e})} \mathbb{w}_{\mathrm{E}}^{(s_\ell)}(T_\mathrm{e}) = 1. \tag{5.94}$$

we define this fit error function $\mathbb{W}_{\mathrm{err}}(T_\mathrm{e})$ as the weighted sum of the relative errors in atomic forces and Helmholtz free cohesive structural energies of the different available reference simulations s_ℓ at the considered T_e:

$$
\begin{aligned}
\mathbb{W}_{\text{err}}(T_{\text{e}}) &= \sum_{s_\ell \in \mathcal{M}_{\text{s}}(T_{\text{e}})} \mathbb{W}_{\text{f}}^{(s_\ell)}(T_{\text{e}}) \left(f_{\text{err}}^{(s_\ell)}(T_{\text{e}}) \right)^2 + \sum_{s_\ell \in \mathcal{M}_{\text{s}}(T_{\text{e}})} \mathbb{W}_{\text{E}}^{(s_\ell)}(T_{\text{e}}) \left(E_{\text{err}}^{(s_\ell)}(T_{\text{e}}) \right)^2 \\
&= \sum_{s_\ell \in \mathcal{M}_{\text{s}}(T_{\text{e}})} \mathbb{W}_{\text{f}}^{(s_\ell)}(T_{\text{e}}) \times \\
&\quad \times \frac{\sum_{k=1}^{N_t^{(s_\ell)}} \sum_{i=1}^{N_{\text{at}}^{(s_\ell)}} \left| -\nabla_{\mathbf{r}_i^{(s_\ell)}} \Phi\left(\left\{ \mathbf{r}_j^{(s_\ell)}(T_{\text{e}}, t_k) \right\}, T_{\text{e}} \right) - \mathbf{f}_i^{(s_\ell)}(T_{\text{e}}, t_k) \right|^2}{\sum_{k=1}^{N_t^{(s_\ell)}} \sum_{i=1}^{N_{\text{at}}^{(s_\ell)}} \left| \mathbf{f}_i^{(s_\ell)}(T_{\text{e}}, t_k) \right|^2} \\
&\quad + \sum_{s_\ell \in \mathcal{M}_{\text{s}}(T_{\text{e}})} \mathbb{W}_{\text{E}}^{(s_\ell)}(T_{\text{e}}) \times \\
&\quad \times \frac{\sum_{k=1}^{N_t^{(s_\ell)}} \left(\Phi\left(\left\{ \mathbf{r}_j^{(s_\ell)}(T_{\text{e}}, t_k) \right\}, T_{\text{e}} \right) - \sum_{i=1}^{N_{\text{at}}^{(s_\ell)}} \Phi_0(T_{\text{e}}) - E_{\text{c}}^{(s_\ell)}(T_{\text{e}}, t_k) \right)^2}{\sum_{k=1}^{N_t^{(s_\ell)}} \left(E_{\text{c}}^{(s_\ell)}(T_{\text{e}}, t_k) \right)^2} .
\end{aligned}
\tag{5.95}
$$

Here the set $\mathcal{M}_{\text{s}}(T_{\text{e}})$ contains the reference simulations at the considered T_{e}. Since the weights obey the sum rule (5.94) at each T_{e}, the constant potential (5.93) leads to the fit error $\mathbb{W}_{\text{err}}(T_{\text{e}}) = 1$, by construction. This normalization is not present in the error function of Ercolessi and Adams [28], whereas it is present in the error function of POTFIT [30], if the corresponding weights are selected accordingly. But the POTFIT error function introduces, beside the fitting weights, additional fitting parameters ε_i and ε_k, which influence the error and are avoided in the definition of $\mathbb{W}_{\text{err}}(T_{\text{e}})$.

5.5.2 *General Definition of the Analytical Form of the Interatomic Potential*

To be very flexible, the local interaction energy contribution functions (see Eq. (5.16))

$$
\Phi = \sum_{i=1}^{N_{\text{at}}} \Phi_0 + \sum_{i=1}^{N_{\text{at}}} \sum_{\substack{j=1 \\ j \neq i \\ r_{ij} < r_{2\text{b}}^{(\text{c})}}}^{N_{\text{at}}} \Phi_{2\text{b}}(r_{ij}, \xi_{ij}, \eta_{ij}) + \sum_{i=1}^{N_{\text{at}}} \Phi_\rho(\rho_i) \tag{5.96}
$$

were implemented in the fit program using

$$\xi_{ij} = \sum_{\substack{k=1 \\ k \neq i, k \neq j \\ r_{ik} < r_{2b}^{(c)}}}^{N_{at}} \xi_{loc}\big(r_{ij}, r_{ik}, \cos(\theta_{ijk})\big), \tag{5.97}$$

$$\eta_{ij} = \sum_{\substack{k=1 \\ k \neq i, k \neq j \\ r_{ik} < r_{2b}^{(c)}}}^{N_{at}} \eta_{loc}\big(r_{ij}, r_{ik}, \cos(\theta_{ijk})\big), \tag{5.98}$$

$$\rho_i = \sum_{\substack{j=1 \\ j \neq i \\ r_{ik} < r_{\rho}^{(c)}}}^{N_{at}} \rho_{loc}(r_{ij}, \chi_{ij}), \tag{5.99}$$

$$\chi_{ij} = \sum_{\substack{k=1 \\ k \neq i \\ r_{ik} < r_{\rho}^{(c)}}}^{N_{at}} \chi_{loc}\big(r_{ij}, r_{ik}, \cos(\theta_{ijk})\big). \tag{5.100}$$

Although the interaction terms $\Phi_2(r_{ij})$ and $\Phi_3\big(r_{ij}, r_{ik}, \cos(\theta_{ijk})\big)$ are missing in Eq. (5.96), a potential using these terms like SW can be treated by using $\Phi_{2b}(r_{ij}, \xi_{ij}, \eta_{ij})$ with the trick

$$\Phi_{2b}(r_{ij}, \xi_{ij}, \eta_{ij}) = \Phi_2(r_{ij}) + \xi_{ij}, \qquad \xi_{loc}\big(r_{ij}, r_{ik}, \cos(\theta_{ijk})\big) = \Phi_3\big(r_{ij}, r_{ik}, \cos(\theta_{ijk})\big). \tag{5.101}$$

5.5.3 *Analytical Expressions for the Interatomic Potential Parameter Derivatives*

In order to derive the electronic specific heat of the electrons from a T_e-dependent interatomic potential via thermodynamic relations, one has to consider the Helmholtz free energy of an isolated atom $\Phi_0(T_e)$ as a function of T_e. $\Phi_0(T_e)$ can be directly obtained from DFT calculations of an isolated atom. The remaining parameters of the potential are derived from the fit error function $\mathbb{W}_{err}(T_e)$ (5.95). For this, the force $\mathcal{F}_{n\gamma}$ obtained from the potential Φ on atom $n \in \{1, \ldots, N_{at}\}$ in direction $\gamma \in \{x, y, z\}$ is calculated with the help of

$$\mathcal{F}_{n\gamma} = -\frac{\partial \Phi}{\partial r_{n\gamma}} = -\sum_{i=1}^{N_{at}} \sum_{\substack{j=1 \\ j \neq i \\ r_{ij} < r_{2b}^{(c)}}}^{N_{at}} \frac{\partial \Phi_{2b}}{\partial r_{n\gamma}} - \sum_{i=1}^{N_{at}} \frac{\partial \Phi_\rho}{\partial \rho_i} \frac{\partial \rho_i}{\partial r_{n\gamma}} \tag{5.102}$$

by using

$$\frac{\partial \Phi_{2b}}{\partial r_{n\gamma}} = \frac{\partial \Phi_{2b}}{\partial r_{ij}} \frac{dr_{ij}}{dr_{n\gamma}} + \frac{\partial \Phi_{2b}}{\partial \xi_{ij}} \frac{\partial \xi_{ij}}{\partial r_{n\gamma}} + \frac{\partial \Phi_{2b}}{\partial \eta_{ij}} \frac{\partial \eta_{ij}}{\partial r_{n\gamma}}, \tag{5.103}$$

$$\frac{\partial \xi_{ij}}{\partial r_{n\gamma}} = \sum_{\substack{k=1 \\ k \neq i, k \neq j \\ r_{ik} < r_{2b}^{(c)}}}^{N_{at}} \left(\frac{\partial \xi_{loc}}{\partial r_{ij}} \frac{dr_{ij}}{dr_{n\gamma}} + \frac{\partial \xi_{loc}}{\partial r_{ik}} \frac{dr_{ik}}{dr_{n\gamma}} + \frac{\partial \xi_{loc}}{d\cos(\theta_{ijk})} \frac{d\cos(\theta_{ijk})}{dr_{n\gamma}} \right), \tag{5.104}$$

$$\frac{\partial \eta_{ij}}{\partial r_{n\gamma}} = \sum_{\substack{k=1 \\ k \neq i, k \neq j \\ r_{ik} < r_{2b}^{(c)}}}^{N_{at}} \left(\frac{\partial \eta_{loc}}{\partial r_{ij}} \frac{dr_{ij}}{dr_{n\gamma}} + \frac{\partial \eta_{loc}}{\partial r_{ik}} \frac{dr_{ik}}{dr_{n\gamma}} + \frac{\partial \eta_{loc}}{d\cos(\theta_{ijk})} \frac{d\cos(\theta_{ijk})}{dr_{n\gamma}} \right) \tag{5.105}$$

and

$$\frac{\partial \rho_i}{\partial r_{n\gamma}} = \sum_{\substack{j=1 \\ j \neq i \\ r_{ik} < r_{\rho}^{(c)}}}^{N_{at}} \left(\frac{\partial \rho_{loc}}{\partial r_{ij}} \frac{dr_{ij}}{dr_{n\gamma}} + \frac{\partial \rho_{loc}}{\partial \chi_{ij}} \frac{\partial \chi_{ij}}{\partial r_{n\gamma}} \right), \tag{5.106}$$

$$\frac{\partial \chi_{ij}}{\partial r_{n\gamma}} = \sum_{\substack{k=1 \\ k \neq i \\ r_{ik} < r_{\rho}^{(c)}}}^{N_{at}} \left(\frac{\partial \chi_{loc}}{\partial r_{ij}} \frac{dr_{ij}}{dr_{n\gamma}} + \frac{\partial \chi_{loc}}{\partial r_{ik}} \frac{dr_{ik}}{dr_{n\gamma}} + \frac{\partial \chi_{loc}}{d\cos(\theta_{ijk})} \frac{d\cos(\theta_{ijk})}{dr_{n\gamma}} \right). \tag{5.107}$$

and

$$\frac{dr_{ij}}{dr_{n\gamma}} = \begin{cases} \hat{\mathbf{r}}_{ij\gamma} & n = i \\ -\hat{\mathbf{r}}_{ij\gamma} & n = j \\ 0 & n \neq i, n \neq j \end{cases} \tag{5.108}$$

$$\frac{d\cos(\theta_{ijk})}{dr_{n\gamma}} = \begin{cases} \frac{\hat{\mathbf{r}}_{ij\gamma}}{r_{ik}} + \frac{\hat{\mathbf{r}}_{ik\gamma}}{r_{ij}} - \cos(\theta_{ijk}) \left(\frac{\hat{\mathbf{r}}_{ij\gamma}}{r_{ij}} + \frac{\hat{\mathbf{r}}_{ik\gamma}}{r_{ik}} \right) & n = i \\ -\frac{\hat{\mathbf{r}}_{ik\gamma}}{r_{ij}} + \cos(\theta_{ijk}) \frac{\hat{\mathbf{r}}_{ij\gamma}}{r_{ij}} & n = j \\ -\frac{\hat{\mathbf{r}}_{ij\gamma}}{r_{ik}} + \cos(\theta_{ijk}) \frac{\hat{\mathbf{r}}_{ik\gamma}}{r_{ik}} & n = k \\ 0 & n \neq i, n \neq j, n \neq k. \end{cases} \tag{5.109}$$

Let be

$$\mathcal{F}_{n\gamma}^{(s_\ell)}(T_e, t_k) = \frac{\partial \Phi\left(\left\{\mathbf{r}_j^{(s_\ell)}(T_e, t_k)\right\}, T_e\right)}{\partial r_{n\gamma}} \tag{5.110}$$

the force acting on atom n in direction γ and

$$\Phi^{(s_\ell)}(T_e, t_k) = \Phi\Big(\big\{\mathbf{r}_j^{(s_\ell)}(T_e, t_k)\big\}, T_e\Big) - \sum_{i=1}^{N_{at}^{(s_\ell)}} \Phi_0(T_e) \tag{5.111}$$

the cohesive structural energy obtained from the potential for the time step t_k of the reference simulation s_ℓ at a given electronic temperature T_e. Then, if we define

$$f_{sum}^{(s_\ell)}(T_e) = \sum_{k=1}^{N_t^{(s_\ell)}} \sum_{i=1}^{N_{at}^{(s_\ell)}} \left|\mathbf{f}_i^{(s_\ell)}(T_e, t_k)\right|^2, \tag{5.112}$$

$$E_{sum}^{(s_\ell)}(T_e) = \sum_{k=1}^{N_t^{(s_\ell)}} \Big(E_c^{(s_\ell)}(T_e, t_k)\Big)^2, \tag{5.113}$$

the fit error function (5.95) can be shortly written as

$$\begin{aligned}\mathbb{W}_{err}(T_e) = &\sum_{s_\ell \in \mathcal{M}_s(T_e)} \frac{\mathbb{W}_f^{(s_\ell)}(T_e)}{f_{sum}^{(s_\ell)}(T_e)} \sum_{k=1}^{N_t^{(s_\ell)}} \sum_{n=1}^{N_{at}^{(s_\ell)}} \sum_{\gamma=x,y,z} \Big(\mathcal{F}_{n\gamma}^{(s_\ell)}(T_e, t_k) - f_{i\gamma}^{(s_\ell)}(T_e, t_k)\Big)^2 \\ &+ \sum_{s_\ell \in \mathcal{M}_s(T_e)} \frac{\mathbb{W}_E^{(s_\ell)}(T_e)}{E_{sum}^{(s_\ell)}(T_e)} \sum_{k=1}^{N_t^{(s_\ell)}} \Big(\Phi^{(s_\ell)}(T_e, t_k) - E_c^{(s_\ell)}(T_e, t_k)\Big)^2. \end{aligned} \tag{5.114}$$

To determine potential parameters that minimize the fit error function $\mathbb{W}_{err}(T_e)$, the derivative of $\mathbb{W}_{err}(T_e)$ with respect to any potential parameter $P \neq \Phi_0$ must be calculated, which is done as followed:

$$\begin{aligned}\frac{\partial \mathbb{W}_{err}(T_e)}{\partial P} = &\sum_{s_\ell \in \mathcal{M}_s(T_e)} 2\,\frac{\mathbb{W}_f^{(s_\ell)}(T_e)}{f_{sum}^{(s_\ell)}(T_e)} \sum_{k=1}^{N_t^{(s_\ell)}} \sum_{n=1}^{N_{at}^{(s_\ell)}} \sum_{\gamma=x,y,z} \Big(\mathcal{F}_{n\gamma}^{(s_\ell)}(T_e, t_k) - f_{i\gamma}^{(s_\ell)}(T_e, t_k)\Big) \times \\ &\times \frac{\partial \mathcal{F}_{n\gamma}^{(s_\ell)}(T_e, t_k)}{\partial P} \\ &+ \sum_{s_\ell \in \mathcal{M}_s(T_e)} 2\,\frac{\mathbb{W}_E^{(s_\ell)}(T_e)}{E_{sum}^{(s_\ell)}(T_e)} \sum_{k=1}^{N_t^{(s_\ell)}} \Big(\Phi^{(s_\ell)}(T_e, t_k) - E^{(s_\ell)}(T_e, t_k)\Big) \frac{\partial \Phi^{(s_\ell)}(T_e, t_k)}{\partial P}. \end{aligned} \tag{5.115}$$

This is calculated with the help of (the argument (T_e, t_k) is omitted for clarity)

$$\frac{\partial \mathcal{F}_{n\gamma}^{(s_\ell)}}{\partial P} = -\sum_{i=1}^{N_{\rm at}} \sum_{\substack{j=1 \\ j\neq i \\ r_{ij}<r_{\rm 2b}^{(c)}}}^{N_{\rm at}} \frac{\partial^2 \Phi_{\rm 2b}}{\partial r_{n\gamma}\,\partial P}$$

$$-\sum_{i=1}^{N_{\rm at}} \left(\left(\frac{\partial^2 \Phi_\rho}{\partial \rho_i\,\partial P} + \frac{\partial^2 \Phi_\rho}{\partial \rho_i^2} \frac{\partial \rho_i}{\partial P} \right) \frac{\partial \rho_i}{\partial r_{n\gamma}} + \frac{\partial \Phi_\rho}{\partial \rho_i} \frac{\partial^2 \rho_i}{\partial r_{n\gamma}\,\partial P} \right), \tag{5.116}$$

$$\frac{\partial \Phi^{(s_\ell)}}{\partial P} = \sum_{i=1}^{N_{\rm at}} \sum_{\substack{j=1 \\ j\neq i \\ r_{ij}<r_{\rm 2b}^{(c)}}}^{N_{\rm at}} \left(\frac{\partial \Phi_{\rm 2b}}{\partial P} + \frac{\partial \Phi_{\rm 2b}}{\partial \xi_{ij}} \frac{\partial \xi_{ij}}{\partial P} + \frac{\partial \Phi_{\rm 2b}}{\partial r_{ij}} \frac{\partial \eta_{ij}}{\partial P} \right)$$

$$+\sum_{i=1}^{N_{\rm at}} \left(\frac{\partial \Phi_\rho}{\partial P} + \frac{\partial \Phi_\rho}{\partial \rho_i} \frac{\partial \rho_i}{\partial P} \right) \tag{5.117}$$

by using

$$\frac{\partial^2 \Phi_{\rm 2b}}{\partial r_{n\gamma}\,\partial P} = \left(\frac{\partial^2 \Phi_{\rm 2b}}{\partial r_{ij}\,\partial P} + \frac{\partial^2 \Phi_{\rm 2b}}{\partial r_{ij}\,\partial \xi_{ij}} \frac{\partial \xi_{ij}}{\partial P} + \frac{\partial^2 \Phi_{\rm 2b}}{\partial r_{ij}\,\partial \eta_{ij}} \frac{\partial \eta_{ij}}{\partial P} \right) \frac{dr_{ij}}{dr_{n\gamma}}$$

$$+ \left(\frac{\partial^2 \Phi_{\rm 2b}}{\partial \xi_{ij}\,\partial P} + \frac{\partial^2 \Phi_{\rm 2b}}{\partial \xi_{ij}^2} \frac{\partial \xi_{ij}}{\partial P} + \frac{\partial^2 \Phi_{\rm 2b}}{\partial \xi_{ij}\,\partial \eta_{ij}} \frac{\partial \eta_{ij}}{\partial P} \right) \frac{\partial \xi_{ij}}{\partial r_{n\gamma}} + \frac{\partial \Phi_{\rm 2b}}{\partial \xi_{ij}} \frac{\partial^2 \xi_{ij}}{\partial r_{n\gamma}\,\partial P}$$

$$+ \left(\frac{\partial^2 \Phi_{\rm 2b}}{\partial \eta_{ij}\,\partial P} + \frac{\partial^2 \Phi_{\rm 2b}}{\partial \eta_{ij}^2} \frac{\partial \eta_{ij}}{\partial P} + \frac{\partial^2 \Phi_{\rm 2b}}{\partial \xi_{ij}\,\partial \eta_{ij}} \frac{\partial \xi_{ij}}{\partial P} \right) \frac{\partial \eta_{ij}}{\partial r_{n\gamma}} + \frac{\partial \Phi_{\rm 2b}}{\partial \eta_{ij}} \frac{\partial^2 \eta_{ij}}{\partial r_{n\gamma}\,\partial P}, \tag{5.118}$$

$$\frac{\partial \xi_{ij}}{\partial P} = \sum_{\substack{k=1 \\ k\neq i, k\neq j \\ r_{ik}<r_{\rm 2b}^{(c)}}}^{N_{\rm at}} \frac{\partial \xi_{\rm loc}}{\partial P}, \tag{5.119}$$

$$\frac{\partial \eta_{ij}}{\partial P} = \sum_{\substack{k=1 \\ k\neq i, k\neq j \\ r_{ik}<r_{\rm 2b}^{(c)}}}^{N_{\rm at}} \frac{\partial \eta_{\rm loc}}{\partial P}, \tag{5.120}$$

$$\frac{\partial^2 \xi_{ij}}{\partial r_{n\gamma}\,\partial P} = \sum_{\substack{k=1 \\ k\neq i, k\neq j \\ r_{ik}<r_{\rm 2b}^{(c)}}}^{N_{\rm at}} \left(\frac{\partial^2 \xi_{\rm loc}}{\partial r_{ij}\,\partial P} \frac{dr_{ij}}{dr_{n\gamma}} + \frac{\partial^2 \xi_{\rm loc}}{\partial r_{ik}\,\partial P} \frac{dr_{ik}}{dr_{n\gamma}} + \frac{\partial^2 \xi_{\rm loc}}{d\cos(\theta_{ijk})\,\partial P} \frac{d\cos(\theta_{ijk})}{dr_{n\gamma}} \right), \tag{5.121}$$

$$\frac{\partial^2 \eta_{ij}}{\partial r_{n\gamma}\,\partial P} = \sum_{\substack{k=1 \\ k\neq i, k\neq j \\ r_{ik}<r_{\rm 2b}^{(c)}}}^{N_{\rm at}} \left(\frac{\partial^2 \eta_{\rm loc}}{\partial r_{ij}\,\partial P} \frac{dr_{ij}}{dr_{n\gamma}} + \frac{\partial^2 \eta_{\rm loc}}{\partial r_{ik}\,\partial P} \frac{dr_{ik}}{dr_{n\gamma}} + \frac{\partial^2 \eta_{\rm loc}}{d\cos(\theta_{ijk})\,\partial P} \frac{d\cos(\theta_{ijk})}{dr_{n\gamma}} \right) \tag{5.122}$$

and

$$\frac{\partial \rho_i}{\partial P} = \sum_{\substack{j=1 \\ j \neq i \\ r_{ik} < r_\rho^{(c)}}}^{N_{at}} \left(\frac{\partial \rho_{loc}}{\partial P} + \frac{\partial \rho_{loc}}{\partial \chi_{ij}} \frac{\partial \chi_{ij}}{\partial P} \right), \tag{5.123}$$

$$\frac{\partial \chi_{ij}}{\partial P} = \sum_{\substack{k=1 \\ k \neq i \\ r_{ik} < r_\rho^{(c)}}}^{N_{at}} \frac{\partial \chi_{loc}}{\partial P}, \tag{5.124}$$

$$\frac{\partial^2 \rho_i}{\partial r_{n\gamma}\, \partial P} = \sum_{\substack{j=1 \\ j \neq i \\ r_{ik} < r_\rho^{(c)}}}^{N_{at}} \left(\left(\frac{\partial^2 \rho_{loc}}{\partial r_{ij}\, \partial P} + \frac{\partial^2 \rho_{loc}}{\partial r_{ij}\, \partial \chi_{ij}} \frac{\partial \chi_{ij}}{\partial P} \right) \frac{dr_{ij}}{dr_{n\gamma}} \right.$$
$$\left. + \left(\frac{\partial^2 \rho_{loc}}{\partial \chi_{ij}\, \partial P} + \frac{\partial^2 \rho_{loc}}{\partial \chi_{ij}^2} \frac{\partial \chi_{ij}}{\partial P} \right) \frac{\partial \chi_{ij}}{\partial r_{n\gamma}} + \frac{\partial \rho_{loc}}{\partial \chi_{ij}} \frac{\partial^2 \chi_{ij}}{\partial r_{n\gamma}\, \partial P} \right), \tag{5.125}$$

$$\frac{\partial \chi_{ij}}{\partial P} = \sum_{\substack{k=1 \\ k \neq i \\ r_{ik} < r_\rho^{(c)}}}^{N_{at}} \frac{\partial \chi_{loc}}{\partial P}, \tag{5.126}$$

$$\frac{\partial^2 \chi_{ij}}{\partial r_{n\gamma}\, \partial P} = \sum_{\substack{k=1 \\ k \neq i \\ r_{ik} < r_\rho^{(c)}}}^{N_{at}} \left(\frac{\partial^2 \chi_{loc}}{\partial r_{ij}\, \partial P} \frac{dr_{ij}}{dr_{n\gamma}} + \frac{\partial^2 \chi_{loc}}{\partial r_{ik}\, \partial P} \frac{dr_{ik}}{dr_{n\gamma}} + \frac{\partial^2 \chi_{loc}}{d\cos(\theta_{ijk})\, \partial P} \frac{d\cos(\theta_{ijk})}{dr_n^{(\gamma)}} \right). \tag{5.127}$$

5.5.4 Efficient and Parallelized Implementation in Fortran

The fit program was written in the FORTRAN language. To be as flexible as possible, an additional MATHEMATICA[5] program was developed, which calculates the above mentioned derivatives for any choice of the functions $\Phi_{2b}(r_{ij}, \xi_{ij}, \eta_{ij})$, $\Phi_\rho(\rho_i)$, $\xi_{loc}\big(r_{ij}, r_{ik}, \cos(\theta_{ijk})\big)$, $\eta_{loc}\big(r_{ij}, r_{ik}, \cos(\theta_{ijk})\big)$, $\rho_{loc}\big(r_{ij}, \cos(\theta_{ijk})\big)$, $\chi_{loc}\big(r_{ij}, r_{ik}, \cos(\theta_{ijk})\big)$. Then the MATHEMATICA program transforms each needed expression to Fortran syntax and inserts it into the Fortran fit program. The fit program allows to perform a local optimization search for the potential parameters with the BFGS [48] or FIRE [49] algorithm. Here the program benefits from the usage of the ana-

[5] http://www.wolfram.com/mathematica.

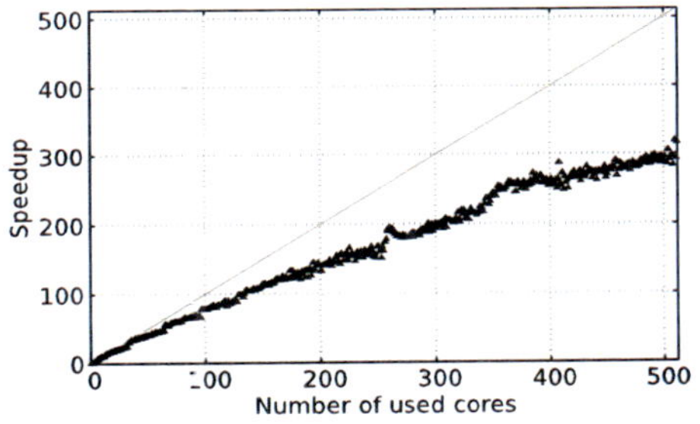

Fig. 5.19 Fitting time speedup is shown as a function of the number of used cores. The line indicates the ideal proportionality of 1

lytical derivatives of $\mathbb{W}_{\text{err}}(T_e)$ with respect to the potential parameters. In POTFIT, these analytical derivatives are not known and must be numerically calculated, which reduces numerical precision and increases computation time. In addition, POTIFT can only handle a fixed number of implemented analytical potential models, whereas a change of the potential model is not possible on the fly.

During the fitting process, the used reference atomic configurations stay fixed. For any atom, its neighbors and the distances to them remain constant. Hence, at the beginning, the fit program calculates once for each atom its neighbor list, which contains, beside the atom number of the neighbor, also the distance, the inverse distance and the normalized distance vector to the neighbor. Then this neighbor list is used in all fitting steps.

To further speed up the fitting process, the fit program was parallelized using the messing passing interface (MPI). In detail, the different time steps t_k of the reference simulations s_ℓ were distributed between the involved cores, since energy, forces, and derivatives of $\mathbb{W}_{\text{err}}$ with respect of P can be independently calculated for the different steps t_k. To analyze the efficiency of the parallelization, we performed a benchmark test on InfiniBand-connected computers, each one containing two AMD Opteron 6276 processors with 16 cores. For this, we measured the fitting time of the MT potential parameters at $T_e = 316\,\text{K}$ for various numbers of used cores. The resulting speedup with respect to the serial program version is shown in Fig. 5.19. The fit program reaches approximately the ideal speedup for number of cores up to 50. Then the speedup starts to increases less and less with the number of used cores, since communication time and/or load balance problems may become more and more noticeable. This lesser increase of the speedup may occur at bigger number of cores, if more time steps t_k are used in total for the fitting.

5.6 Summary

A ground state interatomic potential is a function of the atomic coordinates and describes the PES of the electrons in the ground state. Since the electrons are neglected as degree of freedom, calculations with interatomic potentials scale with $O\left(N_{\text{at}}\right)$ in contrast to DFT calculations that scale with $O\left(N_{\text{at}}^3\right)$ due to the matrix diagonalization needed to solve the electronic part of the problem. Here N_{at} denotes the total number of atoms. Based on physical arguments, many analytical interatomic potential models were semi-empirically developed. These interatomic potential models contain parameters that are fit to DFT calculations and/or experimental results. With the help of large databases of ab-initio structural data, machine learning potentials were automatically developed with minimal human interaction. Such potentials allow to reach high accuracy in describing many different atomic structures with electrons being in the ground state. However, they exhibit quite complicated functional forms, which cannot be physically interpreted. Ground state interatomic potentials cannot describe laser-excited electrons. If one neglects the changes in the PES due to excited electrons, one can describe the laser excitation of a solid just with a ground state interatomic potential. For this, the electrons are treated in a continuum and the effect of the laser-excited electrons on the nuclei is considered by the electron-phonon coupling.

To take also the changes of the PES due to the excited electrons into account, T_e-dependent interatomic potentials were recently introduced. Such interatomic potentials are up to now available for tungsten, gold, molybdenum and silicon. We analyzed the two available T_e-dependent interatomic potentials for silicon. The one developed by Shokeen and Schelling provides a rough description of the increased T_e-induced effects in silicon. The one developed by Darkins et al. cannot describe the increased T_e-induced effects in silicon and shows even an nonphysical behavior. Reason for this may be the fitting of the corresponding interatomic potential parameters to DFT and experimental results at the same time. Also an universal interatomic potential fitting program is presented, that allows to fit the parameters of a T_e-dependent interatomic potential with almost arbitrary analytical functional form to cohesive energies and forces.

Up to now available MD simulation setups with T_e-dependent interatomic potentials do not contain the description of the electron-phonon coupling in a convenient way. All methods except the one of Darkins et al. include the electron-phonon coupling not using energy conservation but rather using a Langevin thermostat, which is adjusted, so that the electronic and ionic temperature relax to the same value on a given timescale. Darkins et al. developed a method based on energy conservation, which is more convenient. However, their method is based on the nonphysical assumption that the T_e-dependent interatomic potential should describe the internal and not the Helmholtz free energy of the electrons.

References

1. Y. Shibuta, S. Sakane, E. Miyoshi, S. Okita, T. Takaki, M. Ohno, Nature Commun. **8**, 10 (2017). https://doi.org/10.1038/s41467-017-00017-5
2. S. Khakshouri, D. Alfè, D.M. Duffy, Phys. Rev. B **78**, 224304 (2008). https://doi.org/10.1103/PhysRevB.78.224304. http://link.aps.org/doi/10.1103/PhysRevB.78.224304
3. S.T. Murphy, S.L. Daraszewicz, Y. Giret, M. Watkins, A.L. Shluger, K. Tanimura, D.M. Duffy, Phys. Rev. B **92**, 134110 (2015). https://doi.org/10.1103/PhysRevB.92.134110. http://link.aps.

org/doi/10.1103/PhysRevB.92.134110
4. G.E. Norman, S.V. Starikov, V.V. Stegailov, J. Exp. Theor. Phys. **114**(5), 792 (2012). https://doi.org/10.1134/S1063776112040115
5. J.A. Moriarty, R.Q. Hood, L.H. Yang, Phys. Rev. Lett. **108**, 036401 (2012). https://doi.org/10.1103/PhysRevLett.108.036401. http://link.aps.org/doi/10.1103/PhysRevLett.108.036401
6. L. Shokeen, P.K. Schelling, Appl. Phys. Lett. **97**(15), 151907 (2010). https://doi.org/10.1063/1.3499296
7. L. Shokeen, P.K. Schelling, J. Appl. Phys. **109**(7), 073503 (2011). https://doi.org/10.1063/1.3554410
8. R. Darkins, P.W. Ma, S.T. Murphy, D.M. Duffy, Phys. Rev. B **98**, 024304 (2018). https://doi.org/10.1103/PhysRevB.98.024304. https://link.aps.org/doi/10.1103/PhysRevB.98.024304
9. A.V. Shapeev, Multiscale Modeling Simul. **14**(3), 1153 (2016). https://doi.org/10.1137/15M1054183
10. P.P. Ewald, Annalen der Physik **369**(3), 253 (1921). https://doi.org/10.1002/andp.19213690304. https://onlinelibrary.wiley.com/doi/abs/10.1002/andp.19213690304
11. G. Czycholl, *Theoretische Festkörper Physik Band 1*, 4th edn. (Springer, Berlin, 2016). https://www.springer.com/de/book/9783662471401
12. F.H. Stillinger, T.A. Weber, Phys. Rev. B **31**, 5262 (1985). https://doi.org/10.1103/PhysRevB.31.5262. http://link.aps.org/doi/10.1103/PhysRevB.31.5262
13. M.I. Baskes, Phys. Rev. Lett. **59**, 2666 (1987). https://doi.org/10.1103/PhysRevLett.59.2666. http://link.aps.org/doi/10.1103/PhysRevLett.59.2666
14. J. Tersoff, Phys. Rev. B **37**, 6991 (1988). https://doi.org/10.1103/PhysRevB.37.6991. http://link.aps.org/doi/10.1103/PhysRevB.37.6991
15. T. Kumagai, S. Izumi, S. Hara, S. Sakai, Comput. Mater. Sci. **39**(2), 457 (2007). https://doi.org/10.1016/j.commatsci.2006.07.013. http://www.sciencedirect.com/science/article/pii/S0927025606002254
16. V.V. Zhakhovskii, N.A. Inogamov, Y.V. Petrov, S.I. Ashitkov, K. Nishihara, Appl. Surface Sci. **255**(24), 9592 (2009). https://doi.org/10.1016/j.apsusc.2009.04.082. http://www.sciencedirect.com/science/article/pii/S0169433209004346
17. F.H. Stillinger, T.A. Weber, Phys. Rev. B **33**, 1451 (1986). https://doi.org/10.1103/PhysRevB.33.1451. https://link.aps.org/doi/10.1103/PhysRevB.33.1451
18. A.D. Mistriotis, N. Flytzanis, S.C. Farantos, Phys. Rev. B **39**, 1212 (1989). https://doi.org/10.1103/PhysRevB.39.1212. https://link.aps.org/doi/10.1103/PhysRevB.39.1212
19. J. Tersoff, Phys. Rev. Lett. **56**, 632 (1986). https://doi.org/10.1103/PhysRevLett.56.632. http://link.aps.org/doi/10.1103/PhysRevLett.56.632
20. J. Tersoff, Phys. Rev. B **38**, 9902 (1988). https://doi.org/10.1103/PhysRevB.38.9902. http://link.aps.org/doi/10.1103/PhysRevB.38.9902
21. B.W. Dodson, Phys. Rev. B **35**, 2795 (1987). https://doi.org/10.1103/PhysRevB.35.2795. http://link.aps.org/doi/10.1103/PhysRevB.35.2795
22. L. Shokeen, P.K. Schelling, Comput. Mater. Sci. **67**, 316 (2013). http://dx.doi.org/10.1016/j.commatsci.2012.07.042. http://www.sciencedirect.com/science/article/pii/S0927025612004910
23. G.P.P. Pun, Y. Mishin, Phys. Rev. B **95**, 224103 (2017). https://doi.org/10.1103/PhysRevB.95.224103. https://link.aps.org/doi/10.1103/PhysRevB.95.224103
24. M.S. Daw, M.I. Baskes, Phys. Rev. Lett. **50**, 1285 (1983). https://doi.org/10.1103/PhysRevLett.50.1285. https://link.aps.org/doi/10.1103/PhysRevLett.50.1285
25. M.I. Baskes, J.S. Nelson, A.F. Wright, Phys. Rev. B **40**, 6085 (1989). https://doi.org/10.1103/PhysRevB.40.6085. http://link.aps.org/doi/10.1103/PhysRevB.40.6085
26. L. Pauling, *The Nature of the Chemical Bond - An Introduction to Modern Structural Chemistry*, 3rd edn. (Cornell University Press, Ithaka, New York, 1960). http://www.cornellpress.cornell.edu/book/?GCOI=80140100962220
27. J.H. Rose, J.R. Smith, F. Guinea, J. Ferrante, Phys. Rev. B **29**, 2963 (1984). https://doi.org/10.1103/PhysRevB.29.2963. http://link.aps.org/doi/10.1103/PhysRevB.29.2963

28. F. Ercolessi, J.B. Adams, Europhys. Lett. **26**(8), 583 (1994). https://doi.org/10.1209/0295-5075/26/8/005. http://stacks.iop.org/0295-5075/26/i=8/a=005
29. P. Brommer, F. Gähler, Philos. Mag. **86**(6–8), 753 (2006). https://doi.org/10.1080/14786430500333349
30. P. Brommer, F. Gähler, Modell. Simul. Mater. Sci. Eng. **15**(3), 295 (2007). http://stacks.iop.org/0965-0393/15/i=3/a=008
31. P. Brommer, A. Kiselev, D. Schopf, P. Beck, J. Roth, H.R. Trebin, Modelling Simul. Mater. Sci. Eng. **23**(7), 074002 (2015). http://stacks.iop.org/0965-0393/23/i=7/a=074002
32. M.J.D. Powell, Comput. J. **7**, 303 (1965). https://academic.oup.com/comjnl/article/7/4/303/354228
33. A. Corana, M. Marchesi, C. Martini, S. Ridella, ACM Trans. Math. Softw. **13**(3), 262 (1987). https://doi.org/10.1145/29380.29864
34. R. Storn, K. Price, J. Global Opt. **11**, 341 (1997). https://doi.org/10.1023/A:1008202821328
35. J. Behler, J. Chem. Phys. **145**(17), 170901 (2016). https://doi.org/10.1063/1.4966192
36. T.B. Blank, S.D. Brown, A.W. Calhoun, D.J. Doren, J. Chem. Phys. **103**(10), 4129 (1995). https://doi.org/10.1063/1.469597
37. A.P. Bartók, J. Kermode, N. Bernstein, G. Csányi, Phys. Rev. X **8**, 041048 (2018). https://doi.org/10.1103/PhysRevX.8.041048. https://link.aps.org/doi/10.1103/PhysRevX.8.041048
38. D. Frekel, B. Smit, *Understanding Molecular Simulation* (Academic, Cambridge, 2002)
39. S.I. Anisimov, B.L. Kapeliovich, T.L. Perel'man, JETP **39**(2), 375 (1974). http://www.jetp.ac.ru/cgi-bin/e/index/e/39/2/p375?a=list
40. D.S. Ivanov, L.V. Zhigilei, Phys. Rev. B **68**, 064114 (2003). https://doi.org/10.1103/PhysRevB.68.064114. https://link.aps.org/doi/10.1103/PhysRevB.68.064114
41. B.S. Lee, S. Park, J. Appl. Phys. **126**(4), 045109 (2019). https://doi.org/10.1063/1.5096013
42. M. Harb, R. Ernstorfer, T. Dartigalongue, C.T. Hebeisen, R.E. Jordan, R.J.D. Miller, J. Phys. Chem. B **110**(50), 25308 (2006). https://doi.org/10.1021/jp064649n. PMID: 17165976
43. M. Harb, R. Ernstorfer, C.T. Hebeisen, G. Sciaini, W. Peng, T. Dartigalongue, M.A. Eriksson, M.G. Lagally, S.G. Kruglik, R.J.D. Miller, Phys. Rev. Lett. **100**, 155504 (2008). https://doi.org/10.1103/PhysRevLett.100.155504. https://link.aps.org/doi/10.1103/PhysRevLett.100.155504
44. B. Bauerhenne, M.E. Garcia, Eur. Phys. J. Special Topics **227**(14), 1615 (2019). https://doi.org/10.1140/epjst/e2019-800181-3
45. K. Yamaguchi, K. Itagaki, J. Thermal Anal. Calorimetry **69**(3), 1059 (2002). https://doi.org/10.1023/A:1020609517891
46. M. Weinert, J.W. Davenport, Phys. Rev. B **45**, 13709 (1992). https://doi.org/10.1103/PhysRevB.45.13709. https://link.aps.org/doi/10.1103/PhysRevB.45.13709
47. R.M. Wentzcovitch, J.L. Martins, P.B. Allen, Phys. Rev. B **45**, 11372 (1992). https://doi.org/10.1103/PhysRevB.45.11372. https://link.aps.org/doi/10.1103/PhysRevB.45.11372
48. D.F. Shanno, Math. Comp. **24**, 647 (1970). https://doi.org/10.1090/S0025-5718-1970-0274029-X
49. E. Bitzek, P. Koskinen, F. Gähler, M. Moseler, P. Gumbsch, Phys. Rev. Lett. **97**, 170201 (2006). https://doi.org/10.1103/PhysRevLett.97.170201. https://link.aps.org/doi/10.1103/PhysRevLett.97.170201

Chapter 6
Ab-Initio Theory Considering Excited Potential Energy Surface and e^--Phonon Coupling

Abstract An intense femtosecond laser-pulse can be tailored to excite the electrons of matter to high electronic temperatures T_e, whereas the ions remain mostly unaffected for the duration of the pulse and for some time afterwards. Within the Born-Oppenheimer approximation, the potential energy surface (PES), on which the ions move, changes significantly due to the increased T_e. The femtosecond-laser induced transition state with hot electrons and cold ions exists for one to several picoseconds and causes many ultrafast phenomena. Equilibration of the two subsystems is driven by the electron-phonon coupling (EPC): Incoherent electron-phonon collisions induce an energy transfer from the electrons to the ions, so that T_e decreases and the ionic temperature T_i increases until both are equal. However, a physically satisfactory simulation including both, the EPC and the excited PES was so far not possible. The reason for this is that the excited PES corresponds to the potential part of the internal energy of the ions and to the Helmholtz free energy of the electrons, since the electrons are modeled by a temperature T_e. However, the energy conservation can only be formulated with the internal energy and the energy conservation is crucial to simulate properly the transference of energy between the hot electrons and the ions induced by the EPC. One finds commonly in literature only the simulation of the PES at increased T_e without EPC, so that T_e remains constant (see Chap. 4), or the simulation of the EPC with PES at $T_e = 0$ K (see Sect. 5.2). In addition, there are approaches that simulate the PES at increased T_e along with the EPC without energy conservation (see Sect. 5.4.3). However, these approaches are inconvenient, since the energy taken from the hot electrons due to the EPC is not entirely transferred to the ions. Yet another approach was recently developed that simulates the PES at increased T_e with the EPC based on energy conservation but using non-physical assumptions (see Sect. 5.4.3). T_e-dependent DFT allows for precise calculations of the PES at increased T_e. Such calculations come with high computational cost and are restricted to several hundreds of atoms only. Therefore, simulations with hundreds of millions of atoms, which are necessary for understanding many relevant laser-induced processes like laser material processing, can only be done with T_e-dependent interatomic potentials, which, up till now, only describe roughly the excited PES compared to T_e-dependent DFT. In Sect. 5.4 the available T_e-dependent interatomic potentials for predicting the ultrafast behavior of Si were analyzed. In this chapter, a theory is introduced that leads to a MD simulation method, which is

© The Author(s), under exclusive license to Springer Nature Switzerland AG 2021
B. Bauerhenne, *Materials Interaction with Femtosecond Lasers*,
https://doi.org/10.1007/978-3-030-85135-4_6

designed to simulate the effects of the excited PES including the EPC and at the same time, fulfills full conversation of the energy. The theory provides the construction of a suitable energy function that is conserved. Based on the theory, a simulation method is derived for one global electronic and ionic temperature in the simulation cell, which can be especially used within T_e-dependent DFT calculations. Then the theory is extended to larger simulation cells containing sub cells with local electronic and ionic temperatures. This allows to derive a simulation method accounting for a spatial distribution of the laser radiation and local electronic heat flow. For such large simulation cells, a T_e-dependent interatomic potential is required. At the end of the chapter, a method for developing accurate T_e-dependent interatomic potentials from T_e-dependent DFT calculations is presented. These T_e-dependent interatomic potentials use an easy and physically interpretable analytical functional form, which can be optimally adjusted to describe the ab-initio data most accurately and efficiently. In addition, the interatomic potential parameters can always be optimally fitted and the interatomic potential can be efficiently evaluated.

6.1 Usage of Global Temperatures in the Simulation Cell

We consider a system of N_{at} atoms within a simulation cell with the simplification that all atoms are of the same type and mass m. To simulate laser-excited electrons, we separate the atoms into ions and electrons. The ions are described by their positions $\mathbf{r}_1, \ldots, \mathbf{r}_{N_{at}}$ and velocities $\mathbf{v}_1, \ldots, \mathbf{v}_{N_{at}}$. We group together all the ionic positions into the vector

$$\mathcal{R} = \begin{bmatrix} \mathbf{r}_1 \\ \vdots \\ \mathbf{r}_{N_{at}} \end{bmatrix} \in \mathbb{R}^{3N_{at}} \tag{6.1}$$

and all ionic velocities into the vector

$$\mathcal{V} = \begin{bmatrix} \mathbf{v}_1 \\ \vdots \\ \mathbf{v}_{N_{at}} \end{bmatrix} \in \mathbb{R}^{3N_{at}}. \tag{6.2}$$

We describe the state of the electronic system by a global temperature T_e within the simulation cell. The interaction energy of this system with ions at positions $\mathbf{r}_1, \ldots, \mathbf{r}_{N_{at}}$, and electrons at temperature T_e is given by $\Phi(\mathbf{r}_1, \ldots, \mathbf{r}_{N_{at}}, T_e)$, which can be obtained from T_e-dependent DFT or from a T_e-dependent interatomic potential. We write all atomic forces obtained from Φ in the vector

$$\mathcal{F} = \begin{bmatrix} -\nabla_{\mathbf{r}_1} \Phi \\ \vdots \\ -\nabla_{\mathbf{r}_{N_{at}}} \Phi \end{bmatrix} \in \mathbb{R}^{3\,N_{at}}. \tag{6.3}$$

Since the state of the electrons is controlled by setting the temperature T_e, the electrons are thermodynamically described in the canonical ensemble and exchange energy with a heat bath (see Fig. 6.1). This heat bath sets formally the temperature of the electrons and is so large, that the influence on the heat bath due to the energy exchange with the electrons is negligible. Consequently, Φ is the Helmholtz free energy of the electrons and both the electronic entropy, S_e, and the electronic heat capacity, C_e, can be calculated from Φ using thermodynamic relations:

$$S_e = -\frac{\partial \Phi}{\partial T_e}, \tag{6.4}$$

$$C_e \overset{(5.78)}{=} -T_e \frac{\partial^2 \Phi}{\partial T_e^2}.$$

In contrast to this, the ions are thermodynamically described in the microcanonical ensemble, because their state is defined by their positions and velocities. Using the kinetic energy or the velocities of the ions, we can assign a global ionic temperature T_i within the whole simulation cell to the ions. Or in general, we can assign several different global ionic temperatures, each one for a given set of phonon modes. For this, we use the polarization vectors $\mathbf{e}^{(1)}, \ldots, \mathbf{e}^{(3N_{at})} \in \mathbb{R}^{3N_{at}}$ from Eq. (2.39) of the different phonon modes, which are derived by diagonalizing the dynamical matrix $\mathbf{D}$. The polarization vectors are orthonormal, i.e.,

$$\forall\, i, j \in \{1, \ldots, 3N_{at}\}: \qquad \left(\mathbf{e}^{(i)}\right)^t \cdot \mathbf{e}^{(j)} = \delta_{ij}, \tag{6.5}$$

and they form a complete basis set of $\mathbb{R}^{3N_{at}}$, i.e.,

$$\sum_{j=1}^{3N_{at}} \mathbf{e}^{(j)} \cdot \left(\mathbf{e}^{(j)}\right)^t = \mathbb{1}. \tag{6.6}$$

$\mathbb{1}$ is the $3N_{at} \times 3N_{at}$ unity matrix. We divide the phonon modes into $N_{\mathcal{M}}$ disjoint subsets $\mathcal{M}_k$:

$$\bigcup_{k=1}^{N_{\mathcal{M}}} \mathcal{M}_k = \{1, \ldots, 3N_{at}\}, \tag{6.7}$$

$$\forall\, k, \ell \in \{1, \ldots, N_{\mathcal{M}}\},\; k \neq \ell: \qquad \mathcal{M}_k \cap \mathcal{M}_\ell = \{\}. \tag{6.8}$$

Each phonon mode is labeled by an integer index and the set $\mathcal{M}_k$ contains the indices of the corresponding phonon modes. We denote by $|\mathcal{M}_k|$ the number of corresponding phonon modes in set $\mathcal{M}_k$. For each set $\mathcal{M}_k$, we define the corresponding projection operator $\mathbf{P}_{\mathcal{M}_k} \in \mathbb{R}^{3N_{at} \times 3N_{at}}$ by

$$\mathbf{P}_{\mathcal{M}_k} = \sum_{j \in \mathcal{M}_k} \mathbf{e}^{(j)} \cdot \left(\mathbf{e}^{(j)}\right)^t. \tag{6.9}$$

These projection operators fulfill the following rules:

$$\forall\, k \in \{1,\ldots,N_{\mathcal{M}}\}: \quad \mathbf{P}^{\mathrm{t}}_{\mathcal{M}_k} = \mathbf{P}_{\mathcal{M}_k}, \tag{6.10}$$

$$\forall\, k \in \{1,\ldots,N_{\mathcal{M}}\}: \quad \mathbf{P}_{\mathcal{M}_k} \cdot \mathbf{P}_{\mathcal{M}_k} = \mathbf{P}_{\mathcal{M}_k}, \tag{6.11}$$

$$\forall\, k,\ell \in \{1,\ldots,N_{\mathcal{M}}\},\ k \neq \ell: \quad \mathbf{P}_{\mathcal{M}_k} \cdot \mathbf{P}_{\mathcal{M}_\ell} = \mathbb{0}, \tag{6.12}$$

$$\sum_{k=1}^{N_{\mathcal{M}}} \mathbf{P}_{\mathcal{M}_k} = \mathbb{1}. \tag{6.13}$$

$\mathbb{0}$ is the $3\,N_{\mathrm{at}} \times 3\,N_{\mathrm{at}}$ zero matrix. $\mathbf{P}_{\mathcal{M}_k} \cdot \mathcal{V}$ projects the atomic velocities $\mathcal{V}$ onto the directions of the phonon modes of set $\mathcal{M}_k$. From the velocities, the total kinetic energy of the ions is calculated by

$$E_{\mathrm{kin}} = \frac{1}{2} \sum_{j=1}^{N_{\mathrm{at}}} m\, \mathbf{v}_j^2 = \frac{m}{2}\, \mathcal{V}^{\mathrm{t}} \cdot \mathcal{V}. \tag{6.14}$$

Using the projection $\mathbf{P}_{\mathcal{M}_k}$, we define the kinetic energy in the phonon modes of set $\mathcal{M}_k$ as

$$E_{\mathrm{kin}_{\mathcal{M}_k}} = \frac{m}{2} \left(\mathbf{P}_{\mathcal{M}_k} \cdot \mathcal{V}\right)^{\mathrm{t}} \cdot \mathbf{P}_{\mathcal{M}_k} \cdot \mathcal{V} = \frac{m}{2}\, \mathcal{V}^{\mathrm{t}} \cdot \mathbf{P}_{\mathcal{M}_k} \cdot \mathcal{V}. \tag{6.15}$$

By construction, the total kinetic energy can be calculated by

$$E_{\mathrm{kin}} = \sum_{k=1}^{N_{\mathcal{M}}} E_{\mathrm{kin}_{\mathcal{M}_k}} = \sum_{k=1}^{N_{\mathcal{M}}} \frac{m}{2}\, \mathcal{V}^{\mathrm{t}} \cdot \mathbf{P}_{\mathcal{M}_k} \cdot \mathcal{V}. \tag{6.16}$$

Using the assumption of thermodynamical equilibrium, we assign an ionic temperature $T_{\mathrm{i}_{\mathcal{M}_k}}$ to the phonon modes of set $\mathcal{M}_k$ from the corresponding kinetic energy $E_{\mathrm{kin}_{\mathcal{M}_k}}$ by (c.f. (4.16))

$$T_{\mathrm{i}_{\mathcal{M}_k}} = \frac{2\, E_{\mathrm{kin}_{\mathcal{M}_k}}}{|\mathcal{M}_k|\, k_{\mathrm{B}}} = \frac{m\, \mathcal{V}^{\mathrm{t}} \cdot \mathbf{P}_{\mathcal{M}_k} \cdot \mathcal{V}}{|\mathcal{M}_k|\, k_{\mathrm{B}}}. \tag{6.17}$$

k_{B} denotes the Boltzmann constant.

Now we consider how the energy of the system changes with time. If T_{e} is constant, one has

$$\Phi\big|_{T_{\mathrm{e}}} + E_{\mathrm{kin}} = \mathrm{const.}, \tag{6.18}$$

since the ions are treated in the microcanonical ensemble and no energy is needed to change T_{e}. This form of the energy conservation principle is also used in T_{e}-dependent DFT MD simulations: The PES of ions at positions $\mathbf{r}_1,\ldots,\mathbf{r}_{N_{\mathrm{at}}}$ and electrons at temperature T_{e} is given by the Helmholtz free energy F, which is calculated by $F = E - T_{\mathrm{e}}\, S_{\mathrm{e}}$, where E denotes the potential part of the internal energy of ions and the internal energy of the electrons [1–3]. F is used in MD simulations with T_e-dependent DFT at constant T_e to formulate the energy conservation similar to Eq. (6.18) and to calculate the forces on the atoms [1–5].

If T_e is not constant, the energy conservation cannot be formulated as simply as in Eq. (6.18). To formulate the general energy conservation, we must transform Φ to the internal energy of the electrons. If the positions $\mathbf{r}_1, \ldots, \mathbf{r}_{N_{at}}$ of the ions are kept constant, we can easily do this transformation by

$$E\Big|_{\mathbf{r}_1,\ldots,\mathbf{r}_{N_{at}}} = \Phi + T_e\, S_e \overset{(6.4)}{=} \Phi - T_e \frac{\partial \Phi}{\partial T_e}. \tag{6.19}$$

But this transformation would not be correct when the ions are moving, since Φ already denotes the potential part of the internal energy of the ions and only the electronic contribution must be transformed to the internal energy. In order to achieve this transformation, we first consider the change in time of the potential part of the internal energy of ions and the internal energy of the electrons and derive the transformation from Φ:

$$\frac{dE}{dt} = \frac{d\Phi}{dt} - \underbrace{\frac{d}{dt}\left(T_e \frac{\partial \Phi}{\partial T_e}\right)}_{\text{only } e^-}.$$

The second term needs to only take the electronic degree of freedom T_e into account. This derivative differs from the complete one in that we have to drop the term $T_e \sum_{j=1}^{N_{at}} \left(\nabla_{\mathbf{r}_j} \frac{\partial \Phi}{\partial T_e}\right)^t \cdot \frac{d\mathbf{r}_j}{dt}$, that appears in the total derivative because of the dependency of $\frac{\partial \Phi}{\partial T_e}$ on the ionic coordinates $\mathbf{r}_1(t), \ldots, \mathbf{r}_{N_{at}}(t)$:

$$\begin{aligned} \frac{dE}{dt} &= \frac{d\Phi}{dt} - \frac{dT_e}{dt}\frac{\partial \Phi}{\partial T_e} - \underbrace{T_e \sum_{j=1}^{N_{at}} \left(\nabla_{\mathbf{r}_j} \frac{\partial \Phi}{\partial T_e}\right)^t \cdot \frac{d\mathbf{r}_j}{dt}}_{\text{ignored}} - T_e \frac{\partial^2 \Phi}{\partial T_e^2}\frac{dT_e}{dt} \\ &= \frac{d\Phi}{dt} - \frac{dT_e}{dt}\frac{\partial \Phi}{\partial T_e} - T_e \frac{\partial^2 \Phi}{\partial T_e^2}\frac{dT_e}{dt}. \end{aligned} \tag{6.20}$$

In order to get finally $E(t_\ell)$ the potential part of the internal energy of the ions and the internal energy of the electrons at time t_ℓ in a MD simulation starting at time t_0, we integrate Eq. (6.20) over time from t_0 to t_ℓ:

$$E(t_\ell) = \int_{t_0}^{t_\ell} dt \left(\frac{d\Phi}{dt} - \frac{dT_e}{dt} \frac{\partial \Phi}{\partial T_e} - T_e \frac{\partial^2 \Phi}{\partial T_e^2} \frac{dT_e}{dt} \right)$$

$$= \Phi(t_\ell) - \Phi(t_0) - \int_{t_0}^{t_\ell} dt \left(\frac{\partial \Phi}{\partial T_e} \frac{dT_e}{dt} + T_e \frac{\partial^2 \Phi}{\partial T_e^2} \frac{dT_e}{dt} \right).$$

We write $\Phi(t_\ell)$ instead of $\Phi\Big(\mathbf{r}_1(t_\ell), \ldots, \mathbf{r}_{N_{at}}(t_\ell), T_e(t_\ell)\Big)$ for readability. Since the energy is only defined up to a constant, we can omit $\Phi(t_0)$ and get finally

$$\boxed{E(t_\ell) = \Phi(t_\ell) - \int_{t_0}^{t_\ell} dt \left(\frac{\partial \Phi}{\partial T_e} + T_e \frac{\partial^2 \Phi}{\partial T_e^2} \right) \frac{dT_e}{dt}.} \tag{6.21}$$

We can use the thermodynamical relations (6.4) for the electronic entropy S_e and (5.78) for the electronic heat capacity C_e to express $E(t_\ell)$:

$$E(t_\ell) = \Phi(t_\ell) + \int_{t_0}^{t_\ell} dt\, (S_e + C_e) \frac{dT_e}{dt}. \tag{6.22}$$

Using $E(t_\ell)$, we are able to formulate the energy conservation for moving ions and changing T_e. With a none constant T_e, we can also take into account that the electrons can absorb energy from a laser field. By defining $E_{L_{abs}}(t_\ell)$ as the total energy absorbed from the laser up to time t_ℓ, we can write the energy conservation as

$$E(t_\ell) + E_{kin} = E_{L_{abs}}(t_\ell) + \text{const}. \tag{6.23}$$

If T_e keeps constant, $E(t_\ell)$ transforms to $\Phi(t_\ell)$, since the integral term in Eq. (6.21) vanishes due to $\frac{dT_e}{dt} = 0$. Consequently, for constant T_e and no energy absorption from the laser $E_{L_{abs}} \equiv 0$, our energy conservation in Eq. (6.23) transforms to the energy conservation of Eq. (6.18), which is commonly used in T_e-dependent DFT MD simulations [4–6] at constant T_e. Next we derive the equations of motion: For this, we calculate the time derivative of the total kinetic energy of the ions

$$\frac{dE_{kin}}{dt} = m \sum_{j=1}^{N_{at}} \mathbf{v}_j^t \cdot \frac{d\mathbf{v}_j}{dt} = m\, \mathcal{V}^t \cdot \frac{d\mathcal{V}}{dt} \tag{6.24}$$

and the time derivative of E from Eq. (6.21)

Fig. 6.1 Energy flows during a MD simulation

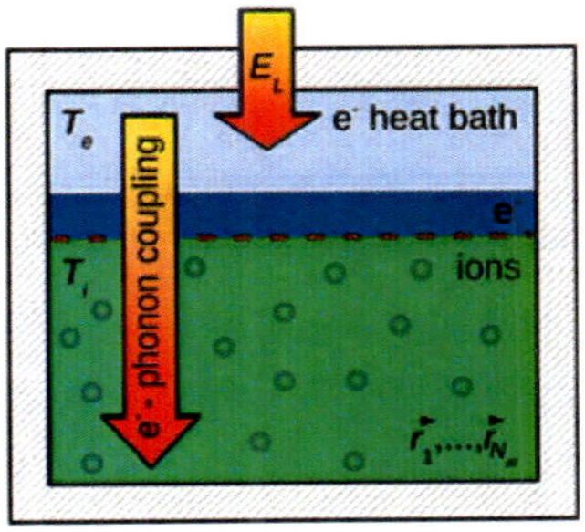

$$\begin{aligned}\frac{dE}{dt} &= \sum_{j=1}^{N_{\text{at}}} \left(\nabla_{\mathbf{r}_j}\Phi\right)^{\text{t}} \cdot \frac{d\mathbf{r}_j}{dt} + \frac{\partial \Phi}{\partial T_{\text{e}}}\frac{dT_{\text{e}}}{dt} - \left(\frac{\partial \Phi}{\partial T_{\text{e}}} + T_{\text{e}}\frac{\partial^2 \Phi}{\partial T_{\text{e}}^2}\right)\frac{dT_{\text{e}}}{dt} \\ &= \sum_{j=1}^{N_{\text{at}}} \mathbf{v}_j^{\text{t}} \cdot \nabla_{\mathbf{r}_j}\Phi - T_{\text{e}}\frac{\partial^2 \Phi}{\partial T_{\text{e}}^2}\frac{dT_{\text{e}}}{dt} \\ &\overset{(5.78)}{=} -\mathcal{V}^{\text{t}} \cdot \mathcal{F} + C_{\text{e}}\frac{dT_{\text{e}}}{dt}. \end{aligned} \tag{6.25}$$

We interpret the second term as the time derivative of the internal energy of the electrons

$$\frac{dE_{\text{e}}}{dt} = C_{\text{e}}\frac{dT_{\text{e}}}{dt}. \tag{6.26}$$

Since we neglect the local electronic heat flow by using a global ionic and electronic temperature, we consider only two processes that change the internal energy E_{e} of the electrons (see Fig. 6.1):

- The electrons can interchange energy with the ions due to electron-phonon coupling. We denote the total energy that the electrons interchange with the ions up to time t_ℓ by $E_{\text{ep}}(t_\ell)$. Since the electrons couple with different strengths to different phonon modes of the ions, we use different electron-phonon coupling constants $G_{\text{ep}_{\mathcal{M}_k}}$. The constant $G_{\text{ep}_{\mathcal{M}_k}}$ indicates how big the energy flow per phonon mode is from the phonons of set $\mathcal{M}_k$ to the electrons depending on the temperature difference $T_{\text{e}} - T_{\text{i}_{\mathcal{M}_k}}$. Hence, we obtain for the total energy flow from the ions to the electrons
$$\frac{dE_{\text{ep}}}{dt} = -\sum_{k=1}^{N_{\mathcal{M}}} |\mathcal{M}_k|\, G_{\text{ep}_{\mathcal{M}_k}}\left(T_{\text{e}} - T_{\text{i}_{\mathcal{M}_k}}\right). \tag{6.27}$$
- The electrons can absorb energy from a laser field. We denote the total absorbed energy flow from the laser by $\frac{dE_{\text{L-abs}}}{dt}$. Since the electrons are much lighter than the ions, we assume that only the electrons absorb the energy of the laser.

Hence, the time derivative of the internal energy of the electrons is given by

$$\begin{aligned}\frac{dE_{\mathrm{e}}}{dt} =& \frac{dE_{\mathrm{ep}}}{dt} + \frac{dE_{\mathrm{L_{abs}}}}{dt}\\ =& -\sum_{k=1}^{N_{\mathcal{M}}} |\mathcal{M}_k|\, G_{\mathrm{ep}_{\mathcal{M}_k}} \left(T_{\mathrm{e}} - T_{\mathrm{i}_{\mathcal{M}_k}}\right) + \frac{dE_{\mathrm{L_{abs}}}}{dt}.\end{aligned} \tag{6.28}$$

Equating with Eq. (6.26) gives a differential equation to calculate the time change of T_{e}:

$$\boxed{C_{\mathrm{e}} \frac{dT_{\mathrm{e}}}{dt} = -\sum_{k=1}^{N_{\mathcal{M}}} |\mathcal{M}_k|\, G_{\mathrm{ep}_{\mathcal{M}_k}} \left(T_{\mathrm{e}} - T_{\mathrm{i}_{\mathcal{M}_k}}\right) + \frac{dE_{\mathrm{L_{abs}}}}{dt}.} \tag{6.29}$$

Now, we derive the equations of motions for the ions. For this, we calculate the time derivative of the energy conservation (6.23) and insert Eqs. (6.24) and (6.25):

$$\begin{aligned} & \frac{dE}{dt} + \frac{dE_{\mathrm{kin}}}{dt} =\frac{dE_{\mathrm{L_{abs}}}}{dt},\\ \Leftrightarrow\quad & -\mathcal{V}^{\mathrm{t}} \cdot \mathcal{F} + C_{\mathrm{e}} \frac{dT_{\mathrm{e}}}{dt} + m\, \mathcal{V}^{\mathrm{t}} \cdot \frac{d\mathcal{V}}{dt} =\frac{dE_{\mathrm{L_{abs}}}}{dt},\\ \Leftrightarrow\quad & m\, \mathcal{V}^{\mathrm{t}} \cdot \frac{d\mathcal{V}}{dt} - \mathcal{V}^{\mathrm{t}} \cdot \mathcal{F} + C_{\mathrm{e}} \frac{dT_{\mathrm{e}}}{dt} - \frac{dE_{\mathrm{L_{abs}}}}{dt} =0.\end{aligned}$$

Inserting (6.29) for $C_{\mathrm{e}} \frac{dT_{\mathrm{e}}}{dt}$ yields

$$\begin{aligned} & m\, \mathcal{V}^{\mathrm{t}} \cdot \frac{d\mathcal{V}}{dt} - \mathcal{V}^{\mathrm{t}} \cdot \mathcal{F} - \sum_{k=1}^{N_{\mathcal{M}}} |\mathcal{M}_k|\, G_{\mathrm{ep}_{\mathcal{M}_k}} \left(T_{\mathrm{e}} - T_{\mathrm{i}_{\mathcal{M}_k}}\right) =0,\\ \Leftrightarrow\quad & m\, \mathcal{V}^{\mathrm{t}} \cdot \frac{d\mathcal{V}}{dt} - \mathcal{V}^{\mathrm{t}} \cdot \mathcal{F} - \sum_{k=1}^{N_{\mathcal{M}}} \underbrace{\frac{m\, \mathcal{V}^{\mathrm{t}} \cdot \mathbf{P}_{\mathcal{M}_k} \cdot \mathcal{V}}{2\, E_{\mathrm{kin}_{\mathcal{M}_k}}}}_{=1} |\mathcal{M}_k|\, G_{\mathrm{ep}_{\mathcal{M}_k}} \left(T_{\mathrm{e}} - T_{\mathrm{i}_{\mathcal{M}_k}}\right) =0,\\ \Leftrightarrow\quad & m\, \mathcal{V}^{\mathrm{t}} \cdot \frac{d\mathcal{V}}{dt} - \mathcal{V}^{\mathrm{t}} \cdot \mathcal{F} - \mathcal{V}^{\mathrm{t}} \cdot \sum_{k=1}^{N_{\mathcal{M}}} \frac{|\mathcal{M}_k|\, G_{\mathrm{ep}_{\mathcal{M}_k}} \left(T_{\mathrm{e}} - T_{\mathrm{i}_{\mathcal{M}_k}}\right)}{2\, E_{\mathrm{kin}_{\mathcal{M}_k}}}\, m\, \mathbf{P}_{\mathcal{M}_k} \cdot \mathcal{V} =0,\\ \Leftrightarrow\quad & \mathcal{V}^{\mathrm{t}} \cdot \left(m \frac{d\mathcal{V}}{dt} - \mathcal{F} - \sum_{k=1}^{N_{\mathcal{M}}} \frac{|\mathcal{M}_k|\, G_{\mathrm{ep}_{\mathcal{M}_k}} \left(T_{\mathrm{e}} - T_{\mathrm{i}_{\mathcal{M}_k}}\right)}{2\, E_{\mathrm{kin}_{\mathcal{M}_k}}}\, m\, \mathbf{P}_{\mathcal{M}_k} \cdot \mathcal{V} \right) =0.\end{aligned} \tag{6.30}$$

The above equation is valid, if

$$m\,\frac{d\mathcal{V}}{dt} = \mathcal{F} + \sum_{k=1}^{N_{\mathcal{M}}} \frac{|\mathcal{M}_k|\,G_{\mathrm{ep}_{\mathcal{M}_k}}\left(T_{\mathrm{e}} - T_{\mathrm{i}_{\mathcal{M}_k}}\right)}{2\,E_{\mathrm{kin}_{\mathcal{M}_k}}}\,m\,\mathbf{P}_{\mathcal{M}_k}\cdot\mathcal{V} \tag{6.31}$$

is fulfilled. This is the ionic equation of motion we are looking for. The first term of the right hand side is the force coming from the potential energy surface at T_{e} and the second term is the force coming from the electron-phonon coupling. Defining

$$\xi_{\mathcal{M}_k} = \frac{|\mathcal{M}_k|\,G_{\mathrm{ep}_{\mathcal{M}_k}}\left(T_{\mathrm{e}} - T_{\mathrm{i}_{\mathcal{M}_k}}\right)}{2\,E_{\mathrm{kin}_{\mathcal{M}_k}}}, \tag{6.32}$$

we can write the total ionic force vector as

$$\mathcal{F}_{\mathrm{tot}} = \mathcal{F} + \sum_{k=1}^{N_{\mathcal{M}}} \xi_{\mathcal{M}_k}\,m\,\mathbf{P}_{\mathcal{M}_k}\cdot\mathcal{V}. \tag{6.33}$$

Finally we summarize our MD simulation procedure in the next theorem:

Theorem 6.1 (MD simulations with excited PES and EPC with global temperatures) *Let us consider a simulation cell containing N_{at} atoms of identical type with mass m. Let us divide the phonon modes of the simulation cell into $N_{\mathcal{M}}$ different sets $\mathcal{M}_k$ and let us associate to each set an ionic temperature $T_{i_{\mathcal{M}_k}}$ by Eq.* (6.17)

$$T_{i_{\mathcal{M}_k}} = \frac{2\,E_{kin_{\mathcal{M}_k}}}{|\mathcal{M}_k|\,k_B} = \frac{m\,\mathcal{V}^t\cdot\mathbf{P}_{\mathcal{M}_k}\cdot\mathcal{V}}{|\mathcal{M}_k|\,k_B}.$$

Here, $E_{kin_{\mathcal{M}_k}}$ denotes the kinetic energy of the $|\mathcal{M}_k|$ phonon modes of set $\mathcal{M}_k$, k_B the Boltzmann constant, $\mathcal{V}$ the vector of all velocities of the atoms, and $\mathbf{P}_{\mathcal{M}_k}$ the matrix that projects onto the phonon modes of set $\mathcal{M}_k$. Let us also associate a temperature T_e to the electrons in the simulation cell. Then, we obtain the total internal energy of the ions and electrons at time t_ℓ by Eq. (6.21)

$$E(t_\ell) = \Phi(t_\ell) - \int_{t_0}^{t_\ell} dt\left(\frac{\partial\Phi}{\partial T_e} + T_e\,\frac{\partial^2\Phi}{\partial T_e^2}\right)\frac{dT_e}{dt}$$

with starting time t_0. Φ denotes the Helmholtz free energy of the electrons obtained from T_e-dependent DFT or a T_e-dependent interatomic potential. The time propagation of T_e is given by Eq. (6.29)

$$C_e \frac{dT_e}{dt} = -\sum_{k=1}^{N_M} |\mathcal{M}_k| \, G_{ep_{\mathcal{M}_k}} \left(T_e - T_{i_{\mathcal{M}_k}}\right) + \frac{dE_{L_{abs}}}{dt}.$$

Here, $C_e = -T_e \frac{\partial^2 \Phi}{\partial T_e^2}$ denotes the electronic specific heat capacity, $G_{ep_{\mathcal{M}_k}}$ the electron-phonon coupling constant for the phonon modes of set $\mathcal{M}_k$, and $\frac{dE_{L_{abs}}}{dt}$ the energy absorption rate from the laser field. From energy conservation, we obtain the equation of motion for the ions by Eq. (6.31)

$$m \frac{d\mathcal{V}}{dt} = \mathcal{F} + \sum_{k=1}^{N_M} \frac{|\mathcal{M}_k| \, G_{ep_{\mathcal{M}_k}} \left(T_e - T_{i_{\mathcal{M}_k}}\right)}{2 \, E_{kin_{\mathcal{M}_k}}} \, m \, \mathbf{P}_{\mathcal{M}_k} \cdot \mathcal{V}.$$

$\mathcal{F}$ denotes the vector of all atomic forces obtained from Φ.

6.1.1 Implementation in the Velocity Verlet Algorithm

In any MD simulation, we start from initial conditions $T_e(t_0)$, $\mathcal{R}(t_0)$, $\mathcal{V}(t_0)$, choose a time increment $\Delta t > 0$ and consider discrete times $t_\ell = t_0 + \ell \, \Delta t$, $\ell \in \mathbb{N}$, at which we want to calculate $T_e(t_\ell)$, $\mathcal{R}(t_\ell)$, $\mathcal{V}(t_\ell)$. For this, we have to integrate numerically the coupled differential equations of the ionic motions

$$m \frac{d\mathcal{V}}{dt} = \mathcal{F}_{\text{tot}}$$

and the differential equation for T_e

$$C_e \frac{dT_e}{dt} = \frac{dE_e}{dt}.$$

To integrate the ionic equations of motion, we choose the Velocity Verlet Algorithm, which calculates $\mathcal{R}$, $\mathcal{V}$, $\mathcal{F}$ at the next time step $t_\ell + \Delta t$ from the corresponding values of the previous time step t_ℓ via (c.f. Eqs. (4.4) and (4.5))

$$\mathcal{R}(t_{\ell+1}) = \mathcal{R}(t_\ell) + \Delta t \; \mathcal{V}(t_\ell) + \frac{\Delta t^2}{2m} \mathcal{F}_{\text{tot}}(t_\ell), \tag{6.34}$$

$$\mathcal{V}(t_{\ell+1}) = \mathcal{V}(t_\ell) + \frac{\Delta t}{2m} \Big(\mathcal{F}_{\text{tot}}(t_\ell) + \mathcal{F}_{\text{tot}}(t_{\ell+1}) \Big). \tag{6.35}$$

Using $\mathcal{R}(t_0)$, we can calculate $\mathcal{F}(t_0)$, $\Phi(t_0)$, $S_e(t_0)$ and $C_e(t_0)$. We also know the phonon mode projection operators $P_{\mathcal{M}_k}$, which we assume to be time-independent. With the help of these projection operators $P_{\mathcal{M}_k}$ and $\mathcal{V}(t_0)$, we are able to calculate the kinetic energies $E_{\mathrm{kin}_{\mathcal{M}_k}}(t_0)$, the temperatures $T_{\mathrm{i}_{\mathcal{M}_k}}(t_0)$, and the terms $\xi_{\mathcal{M}_k}(t_0)$ of the different phonon mode sets $\mathcal{M}_k$. Consequently, we are able to calculate the total force vector $\mathcal{F}_{\mathrm{tot}}(t_0)$ at the starting time t_0 immediately from the initial conditions using Eq. (6.33). From $\mathcal{R}(t_0)$, $\mathcal{V}(t_0)$ and $\mathcal{F}_{\mathrm{tot}}(t_0)$, we are able to calculate $\mathcal{R}(t_1)$ from Eq. (6.34).

However, now we cannot calculate directly $\mathcal{V}(t_1)$ from Eq. (6.35), since we need for this $\mathcal{F}_{\mathrm{tot}}(t_1)$, which can only be calculated from $\mathcal{V}(t_1)$ using Eq. (6.33). If the electron-phonon coupling is neglected, $\mathcal{F}_{\mathrm{tot}}(t_1)$ can be directly calculated from $\mathcal{R}(t_1)$ using Φ because of $\mathcal{F}_{\mathrm{tot}}(t_1) = \mathcal{F}(t_1)$ and $\mathcal{V}(t_1)$ can be just calculated from Eq. (6.35).

In the general case the electron-phonon coupling cannot be neglected, and we have to modify the procedure as follows: We consider a time step $t_\ell \geq t_0$ at which all quantities are known and want to calculate all quantities at time step $t_{\ell+1}$. At first, we calculate $\mathcal{R}(t_{\ell+1})$ from Eq. (6.34). To derive $T_{\mathrm{e}}(t_{\ell+1})$, we define further

$$\Delta E_{\mathrm{L_{abs}}}(t_\ell) = \int_{t_\ell}^{t_{\ell+1}} dt\, \frac{dE_{\mathrm{L_{abs}}}(t)}{dt} = E_{\mathrm{L_{abs}}}(t_{\ell+1}) - E_{\mathrm{L_{abs}}}(t_\ell) \tag{6.36}$$

as the energy that is absorbed by the electrons from the laser at time step t_ℓ and $\Delta E_{\mathrm{ep}}(t_\ell)$ as the total energy that is transferred to the electrons from the ions due to electron-phonon coupling at time step t_ℓ. We calculate numerically $\Delta E_{\mathrm{ep}}(t_\ell)$ by

$$\Delta E_{\mathrm{ep}}(t_\ell) \overset{(6.27)}{=} -\sum_{k=1}^{N_{\mathcal{M}}} |\mathcal{M}_k|\, G_{\mathrm{ep}_{\mathcal{M}_k}}(t_\ell) \left(T_{\mathrm{e}}(t_\ell) - T_{\mathrm{i}_{\mathcal{M}_k}}(t_\ell)\right) \Delta t. \tag{6.37}$$

From the total change of the electronic energy at time step t_ℓ

$$\Delta E_{\mathrm{e}}(t_\ell) = \Delta E_{\mathrm{ep}}(t_\ell) + \Delta E_{\mathrm{L_{abs}}}(t_\ell), \tag{6.38}$$

we can calculate numerically the related change of T_{e} for $C_{\mathrm{e}}(t_\ell) > 0$ by

$$\Delta T_{\mathrm{e}}(t_\ell) \overset{(6.26)}{=} \frac{\Delta E_{\mathrm{e}}(t_\ell)}{C_{\mathrm{e}}(t_\ell)} \overset{(6.38)}{=} \frac{\Delta E_{\mathrm{ep}}(t_\ell) + \Delta E_{\mathrm{L_{abs}}}(t_\ell)}{C_{\mathrm{e}}(t_\ell)}. \tag{6.39}$$

From $\Delta T_{\mathrm{e}}(t_\ell)$, we obtain $T_{\mathrm{e}}(t_{\ell+1})$ just by

$$T_{\mathrm{e}}(t_{\ell+1}) = T_{\mathrm{e}}(t_\ell) + \Delta T_{\mathrm{e}}(t_\ell). \tag{6.40}$$

For $C_{\mathrm{e}}(t_\ell) = 0$, we assume formally the ions as fixed and assign the change of the internal energy that is caused by varying T_{e} to the electrons

$$
\begin{aligned}
\Delta E_{\mathrm{e}}(t_\ell) &= \Delta E(t_\ell) \\
&\overset{(6.19)}{=} \Phi\left(T_{\mathrm{e}}(t_{\ell+1}), \mathcal{R}(t_\ell)\right) - T_{\mathrm{e}}(t_{\ell+1}) \frac{d\Phi\left(T_{\mathrm{e}}(t_{\ell+1}), \mathcal{R}(t_\ell)\right)}{dT_{\mathrm{e}}} \\
&\quad - \Phi\Big(T_{\mathrm{e}}(t_\ell), \mathcal{R}(t_\ell)\Big) + T_{\mathrm{e}}(t_\ell) \frac{d\Phi\Big(T_{\mathrm{e}}(t_\ell), \mathcal{R}(t_\ell)\Big)}{dT_{\mathrm{e}}}. \qquad (6.41)
\end{aligned}
$$

Now we solve numerically the equation above for $T_{\mathrm{e}}(t_{\ell+1})$. From $T_{\mathrm{e}}(t_{\ell+1})$ and $\mathcal{R}(t_{\ell+1})$, we can determine the quantities $\mathcal{F}(t_{\ell+1})$, $S_{\mathrm{e}}(t_{\ell+1})$ and $C_{\mathrm{e}}(t_{\ell+1})$. It is convenient to calculate also $G_{\mathrm{ep}_{\mathcal{M}_k}}(t_{\ell+1})$ at this point. In the most general case, we have

$$
G_{\mathrm{ep}_{\mathcal{M}_k}}(t_{\ell+1}) \equiv G_{\mathrm{ep}_{\mathcal{M}_k}}\left(T_{\mathrm{e}}(t_{\ell+1}), \mathcal{R}(t_{\ell+1}), \mathcal{V}(t_{\ell+1})\right).
$$

Since $G_{\mathrm{ep}_{\mathcal{M}_k}}$ is a nontrivial function of $\mathcal{V}(t_{\ell+1})$ and we do not know $\mathcal{V}(t_{\ell+1})$ at this moment, we use $\mathcal{V}(t_\ell)$ to calculate $G_{\mathrm{ep}_{\mathcal{M}_k}}(t_{\ell+1})$:

$$
G_{\mathrm{ep}_{\mathcal{M}_k}}(t_{\ell+1}) \approx G_{\mathrm{ep}_{\mathcal{M}_k}}\left(T_{\mathrm{e}}(t_{\ell+1}), \mathcal{R}(t_{\ell+1}), \mathcal{V}(t_\ell)\right). \qquad (6.42)
$$

Furthermore, we get for $\mathcal{V}(t_{\ell+1})$ by inserting Eq. (6.33) for $\mathcal{F}_{\mathrm{tot}}(t_{\ell+1})$:

$$
\begin{aligned}
\mathcal{V}(t_{\ell+1}) &\overset{(6.35)}{=} \mathcal{V}(t_\ell) + \frac{\Delta t}{2m}\Big(\mathcal{F}_{\mathrm{tot}}(t_\ell) + \mathcal{F}_{\mathrm{tot}}(t_{\ell+1})\Big) \\
&\overset{(6.33)}{=} \mathcal{V}(t_\ell) + \frac{\Delta t}{2m}\Big(\mathcal{F}_{\mathrm{tot}}(t_\ell) + \mathcal{F}(t_{\ell+1})\Big) + \frac{\Delta t}{2} \sum_{k=1}^{N_{\mathcal{M}}} \xi_{\mathcal{M}_k}(t_{\ell+1})\, \mathbf{P}_{\mathcal{M}_k} \cdot \mathcal{V}(t_{\ell+1}).
\end{aligned}
$$

We define

$$
\mathcal{W}(t_{\ell+1}) := \mathcal{V}(t_\ell) + \frac{\Delta t}{2m}\Big(\mathcal{F}_{\mathrm{tot}}(t_\ell) + \mathcal{F}(t_{\ell+1})\Big), \qquad (6.43)
$$

which we can calculate, since $\mathcal{F}(t_{\ell+1})$ can be determined from the already known $\mathcal{R}(t_{\ell+1})$ and $T_{\mathrm{e}}(t_{\ell+1})$ using Φ. $\mathcal{W}(t_{\ell+1})$ corresponds to the velocity vector at time $t_{\ell+1}$, if the influence of the electron-phonon coupling is neglected at the time step $t_{\ell+1}$. We obtain the following from the properties of the phonon mode projection operators

$$
\begin{aligned}
\mathcal{V}(t_{\ell+1}) &= \mathcal{W}(t_{\ell+1}) + \frac{\Delta t}{2} \sum_{k=1}^{N_{\mathcal{M}}} \xi_{\mathcal{M}_k}(t_{\ell+1})\, \mathbf{P}_{\mathcal{M}_k} \cdot \mathcal{V}(t_{\ell+1}), \\
\Leftrightarrow \quad \underbrace{\sum_{k=1}^{N_{\mathcal{M}}} \mathbf{P}_{\mathcal{M}_k}}_{=\mathbb{1}} \cdot \mathcal{V}(t_{\ell+1}) &= \underbrace{\sum_{k=1}^{N_{\mathcal{M}}} \mathbf{P}_{\mathcal{M}_k}}_{=\mathbb{1}} \cdot \mathcal{W}(t_{\ell+1}) + \frac{\Delta t}{2} \sum_{k=1}^{N_{\mathcal{M}}} \xi_{\mathcal{M}_k}(t_{\ell+1})\, \mathbf{P}_{\mathcal{M}_k} \cdot \mathcal{V}(t_{\ell+1}).
\end{aligned}
$$

Multiplying on the left side by $\mathbf{P}_{\mathcal{M}_i}$, where i is arbitrary from $\{1, \ldots, N_{\mathcal{M}}\}$, yields

$$\begin{aligned} \mathbf{P}_{\mathcal{M}_i} \cdot \mathcal{V}(t_{\ell+1}) &= \mathbf{P}_{\mathcal{M}_i} \cdot \mathcal{W}(t_{\ell+1}) + \frac{\Delta t}{2}\, \xi_{\mathcal{M}_i}(t_{\ell+1})\, \mathbf{P}_{\mathcal{M}_i} \cdot \mathcal{V}(t_{\ell+1}), \\ \Leftrightarrow \quad \mathbf{P}_{\mathcal{M}_i} \cdot \mathcal{W}(t_{\ell+1}) &= \left(1 - \frac{\Delta t}{2}\, \xi_{\mathcal{M}_i}(t_{\ell+1})\right) \mathbf{P}_{\mathcal{M}_i} \cdot \mathcal{V}(t_{\ell+1}). \end{aligned}$$

Since i was chosen arbitrarily, the above equation is valid for all $i \in \{1, \ldots, N_{\mathcal{M}}\}$. For a consistent notation, we change i back to k and obtain $\forall\, k \in \{1, \ldots, N_{\mathcal{M}}\}$:

$$\mathbf{P}_{\mathcal{M}_k} \cdot \mathcal{V}(t_{\ell+1}) = \frac{1}{1 - \frac{\Delta t}{2}\, \xi_{\mathcal{M}_k}(t_{\ell+1})}\, \mathbf{P}_{\mathcal{M}_k} \cdot \mathcal{W}(t_{\ell+1}). \tag{6.44}$$

$\mathbf{P}_{\mathcal{M}_k} \cdot \mathcal{V}(t_{\ell+1})$ could be calculated from this equation if we knew $\xi_{\mathcal{M}_k}(t_{\ell+1})$, since we already determined $\mathcal{W}(t_{\ell+1})$. To derive $\xi_{\mathcal{M}_k}(t_{\ell+1})$, we consider the kinetic energy $E_{\mathrm{kin}_{\mathcal{M}_k}}(t_{\ell+1})$ of the phonon mode set $\mathcal{M}_k$:

$$\begin{aligned} E_{\mathrm{kin}_{\mathcal{M}_k}}(t_{\ell+1}) &\overset{(6.15)}{=} \frac{m}{2} \left(\mathbf{P}_{\mathcal{M}_k} \cdot \mathcal{V}(t_{\ell+1})\right)^{\mathrm{t}} \cdot \mathbf{P}_{\mathcal{M}_k} \cdot \mathcal{V}(t_{\ell+1}) \\ &\overset{(6.44)}{=} \frac{1}{\left(1 - \frac{\Delta t}{2}\, \xi_{\mathcal{M}_i}(t_{\ell+1})\right)^2} \frac{m}{2} \left(\mathbf{P}_{\mathcal{M}_k} \cdot \mathcal{W}(t_{\ell+1})\right)^{\mathrm{t}} \cdot \mathbf{P}_{\mathcal{M}_k} \cdot \mathcal{W}(t_{\ell+1}). \end{aligned}$$

We define

$$\mathcal{H}_{\mathcal{M}_k}(t_{\ell+1}) := \frac{m}{2} \left(\mathbf{P}_{\mathcal{M}_k} \cdot \mathcal{W}(t_{\ell+1})\right)^{\mathrm{t}} \cdot \mathbf{P}_{\mathcal{M}_k} \cdot \mathcal{W}(t_{\ell+1}), \tag{6.45}$$

which we can calculate directly from the determined $\mathcal{W}(t_{\ell+1})$. $\mathcal{H}_{\mathcal{M}_k}(t_{\ell+1})$ corresponds to the kinetic energy of the phonon mode set $\mathcal{M}_k$ at time step $t_{\ell+1}$, if the influence of the electron-phonon coupling is neglected at the time step $t_{\ell+1}$. We get

$$E_{\mathrm{kin}_{\mathcal{M}_k}}(t_{\ell+1}) = \frac{\mathcal{H}_{\mathcal{M}_k}(t_{\ell+1})}{1 - \Delta t\, \xi_{\mathcal{M}_\ell}(t_{\ell+1}) + \frac{\Delta t^2}{4}\, \xi_{\mathcal{M}_\ell}(t_{\ell+1})^2} \tag{6.46}$$

and obtain for the parameter $\xi_{\mathcal{M}_k}$ at time step $t_{\ell+1}$:

$$\begin{aligned} \xi_{\mathcal{M}_k}(t_{\ell+1}) &\overset{(6.32)}{=} \frac{|\mathcal{M}_k|\, G_{\mathrm{ep}_{\mathcal{M}_k}}(t_{\ell+1}) \left(T_{\mathrm{e}}(t_{\ell+1}) - T_{\mathrm{i}_{\mathcal{M}_k}}(t_{\ell+1})\right)}{2\, E_{\mathrm{kin}_{\mathcal{M}_k}}(t_{\ell+1})} \\ &\overset{(6.17)}{=} \frac{|\mathcal{M}_k|\, G_{\mathrm{ep}_{\mathcal{M}_k}}(t_{\ell+1}) \left(T_{\mathrm{e}}(t_{\ell+1}) - \frac{2\, E_{\mathrm{kin}_{\mathcal{M}_k}}(t_{\ell+1})}{|\mathcal{M}_k|\, k_{\mathrm{B}}}\right)}{2\, E_{\mathrm{kin}_{\mathcal{M}_k}}(t_{\ell+1})} \\ &= \frac{|\mathcal{M}_k|\, G_{\mathrm{ep}_{\mathcal{M}_k}}(t_{\ell+1})\, T_{\mathrm{e}}(t_{\ell+1})}{2\, E_{\mathrm{kin}_{\mathcal{M}_k}}(t_{\ell+1})} - \frac{G_{\mathrm{ep}_{\mathcal{M}_k}}(t_{\ell+1})}{k_{\mathrm{B}}} \end{aligned}$$

$$\overset{(6.46)}{=} \frac{|\mathcal{M}_k|\, G_{\mathrm{ep}_{\mathcal{M}_k}}(t_{\ell+1})\, T_{\mathrm{e}}(t_{\ell+1})}{2\, \mathcal{H}_{\mathcal{M}_k}(t_{\ell+1})} \left(1 - \Delta t\, \xi_{\mathcal{M}_\ell}(t_{\ell+1}) + \frac{\Delta t^2}{4}\, \xi_{\mathcal{M}_\ell}(t_{\ell+1})^2\right) - \frac{G_{\mathrm{ep}_{\mathcal{M}_k}}(t_{\ell+1})}{k_{\mathrm{B}}}. \tag{6.47}$$

Since all quantities are considered at time step $t_{\ell+1}$, we omit the time argument $(t_{\ell+1})$ in the following for brevity. Now we solve this quadratic equation for $\xi_{\mathcal{M}_k}$:

$$\begin{aligned}
0 =& \frac{|\mathcal{M}_k|\, G_{\mathrm{ep}_{\mathcal{M}_k}}\, T_{\mathrm{e}}\, \Delta t^2}{8\, \mathcal{H}_{\mathcal{M}_k}}\, \xi_{\mathcal{M}_k}^2 - \left(1 + \frac{|\mathcal{M}_k|\, G_{\mathrm{ep}_{\mathcal{M}_k}}\, T_{\mathrm{e}}\, \Delta t}{2\, \mathcal{H}_{\mathcal{M}_k}}\right) \xi_{\mathcal{M}_k} \\
&+ \frac{|\mathcal{M}_k|\, G_{\mathrm{ep}_{\mathcal{M}_k}}\, T_{\mathrm{e}}}{2\, \mathcal{H}_{\mathcal{M}_k}} - \frac{G_{\mathrm{ep}_{\mathcal{M}_k}}}{k_{\mathrm{B}}}, \\
\Leftrightarrow \quad 0 =& \xi_{\mathcal{M}_k}^2 - \left(\frac{8\, \mathcal{H}_{\mathcal{M}_k}}{|\mathcal{M}_k|\, G_{\mathrm{ep}_{\mathcal{M}_k}}\, T_{\mathrm{e}}\, \Delta t^2} + \frac{4}{\Delta t}\right) \xi_{\mathcal{M}_k} \\
&+ \frac{4}{\Delta t^2} - \frac{8\, \mathcal{H}_{\mathcal{M}_k}}{k_{\mathrm{B}}\, |\mathcal{M}_k|\, G_{\mathrm{ep}_{\mathcal{M}_k}}\, T_{\mathrm{e}}\, \Delta t^2}, \\
\Rightarrow \quad \xi_{\mathcal{M}_k} =& \frac{4\, \mathcal{H}_{\mathcal{M}_k}}{|\mathcal{M}_k|\, G_{\mathrm{ep}_{\mathcal{M}_k}}\, T_{\mathrm{e}}\, \Delta t^2} + \frac{2}{\Delta t} \\
&- \sqrt{\left(\frac{4\, \mathcal{H}_{\mathcal{M}_k}}{|\mathcal{M}_k|\, G_{\mathrm{ep}_{\mathcal{M}_k}}\, T_{\mathrm{e}}\, \Delta t^2} + \frac{2}{\Delta t}\right)^2 + \frac{8\, \mathcal{H}_{\mathcal{M}_k}}{k_{\mathrm{B}}\, |\mathcal{M}_k|\, G_{\mathrm{ep}_{\mathcal{M}_k}}\, T_{\mathrm{e}}\, \Delta t^2} - \frac{4}{\Delta t^2}}.
\end{aligned} \tag{6.48}$$

$\xi_{\mathcal{M}_k}$ increases or decreases the velocity of ions in the direction of the velocities of the phonon modes from set $\mathcal{M}_k$. Since the velocities enter the kinetic energy quadratically, the sign of the velocity is not relevant for the energy conservation. Hence, there are two mathematical solutions for $\xi_{\mathcal{M}_k}$. The first solution changes the velocity slightly and thus corresponds to a small $\xi_{\mathcal{M}_k}$. That is the solution of Eq. (6.48). The second solution changes the sign of the velocity or changes the direction of movement and thus corresponds to a large absolute value of $\xi_{\mathcal{M}_k}$. This unphysical solution is given by the "+" solution of the quadratic equation. To see that the "+" solution is one with bigger magnitude, we let a be the first two summands and b be the square root in Eq. (6.48), i.e., $\xi_{\mathcal{M}_k} \overset{(6.48)}{=} a \pm b$. We obtain $a \geq 0$ by construction and $b \geq 0$, if the square root has a real solution. We get finally from the triangle inequality: $|a - b| \leq |a| + |b| = a + b = |a + b|$.

Equation (6.48) is not valid for $G_{\mathrm{ep}_{\mathcal{M}_k}} = 0$. In this case, we obtain $\xi_{\mathcal{M}_k} = 0$ from the definition of $\xi_{\mathcal{M}_k}$ in Eq. (6.32). Equation (6.48) is also not valid for $T_{\mathrm{e}} = 0$. Here, we get $\xi_{\mathcal{M}_k} = -\frac{G_{\mathrm{ep}_{\mathcal{M}_k}}}{k_{\mathrm{B}}}$ from Eq. (6.47).

To calculate $E(t_{\ell+1})$ from Eq. (6.22), we define

$$I(t_\ell) := \int_{t_0}^{t_\ell} dt\,(S_e + C_e)\,\frac{dT_e}{dt}. \tag{6.49}$$

Thus, we get $I(t_0) = 0$ and $U(t_{\ell+1}) = \Phi(t_{\ell+1}) + I(t_{\ell+1})$. To calculate numerically $I(t_{\ell+1})$, we use $I(t_\ell)$ and approximate the remaining integral from t_ℓ to $t_{\ell+1}$ in (6.49) by using the trapezoidal rule:

$$I(t_{\ell+1}) = I(t_\ell) + \frac{1}{2}\Big(\big(S_e(t_\ell) + C_e(t_\ell)\big)\,\Delta T_e(t_\ell) + \big(S_e(t_{\ell+1}) + C_e(t_{\ell+1})\big)\,\Delta T_e(t_{\ell+1})\Big). \tag{6.50}$$

Now we have calculated all quantities at time step $t_{\ell+1}$. We provide a summary of the calculation setup in Sect. A.3 of the appendix.

6.1.2 Remarks

- As already mentioned, our modeling of the electron-phonon coupling can also be implemented in T_e-dependent DFT, so that Φ corresponds to the Helmholtz free energy F of the electrons.
- If no energy is absorbed from the laser field, i.e., $E_{L_{abs}} \equiv 0$, and the electron-phonon coupling is neglected, i.e., $G_{ep_{\mathcal{M}_k}} = 0$ for all $k \in \{1, \ldots, N_{\mathcal{M}}\}$, the electronic temperature T_e keeps constant due to Eq. (6.29). Then, the potential energy $E(t_\ell)$ of the ions and the electrons at T_e reduces to $\Phi(t_\ell)$ due to Eq. (6.21), the energy conservation (6.23) becomes $\Phi + E_{kin} = \text{const.}$ and the equation of motion for the ions (6.31) reduces to $m\,\frac{d\mathcal{V}}{dt} = \mathcal{F}$. This corresponds just to the MD simulation setup commonly used in T_e-dependent DFT.
- This modeling predicts perfect conservation of energy. Hence, the presented numerical implementation in the Velocity Verlet algorithm must not show any drift or fluctuation of the total energy $E + E_{kin} - E_{L_{abs}}$ during time propagation in the limit of $\Delta t \to 0$. This feature can be used to check the numerical implementation of the algorithm in the program.
- If a phonon mode set $\mathcal{M}_k$ contains only a small number of different modes, the corresponding ionic temperature $T_{i_{\mathcal{M}_k}}$ will show significant fluctuations in time. Especially, if the set just contains one single phonon mode, the corresponding ionic temperature may be ill defined due to this fluctuations. Hence, every phonon mode set $\mathcal{M}_k$ should contain enough modes.
- If the lattice melts due to the laser excitation, the symmetry of the structure breaks. Therefore, in such cases, the consideration of different electron-phonon coupling constants $G_{ep_{\mathcal{M}_k}}$ for specific phonon mode sets $\mathcal{M}_k$ may be non-physical, so that only one coupling constant should be used instead.

- Laser pulses with a Gaussian-shaped time profile are commonly used in experiments. Such pulses are characterized by the FWHM (full width at half maximum) time width τ. Let $E_{\mathrm{L_{tot}}}$ be the total laser-absorbed energy of the pulse. Then the total laser-absorbed energy rate at time t_ℓ is given by

$$\frac{dE_{\mathrm{L_{abs}}}(t_\ell)}{dt} = \frac{E_{\mathrm{L_{tot}}}}{\tau}\sqrt{\frac{\log(16)}{\pi}}\,\exp\left(-\frac{(t_\ell - 2\,\tau)^2}{\tau^2}\,\log(16)\right). \tag{6.51}$$

Here, the MD simulation starts at $t_0 = 0$ and the maximal energy absorption rate is reached at time $2\,\tau$. 99.99975% of $E_{\mathrm{L_{tot}}}$ is absorbed during $t = 0$ and $t = 4\,\tau$. Using the Gauss error function

$$\mathrm{erf}(x) = \frac{2}{\sqrt{\pi}}\int_0^x dt\, \mathrm{e}^{-t^2} \tag{6.52}$$

The total laser-absorbed energy up to time t_ℓ can be analytically calculated by:

$$\begin{aligned} E_{\mathrm{L_{abs}}}(t_\ell) &= \int_0^{t_\ell} dt\,\frac{dE_{\mathrm{L_{abs}}}(t)}{dt} \\ &= \frac{E_{\mathrm{L_{tot}}}}{2}\left(\mathrm{erf}\left(\sqrt{\log(65536)}\right) + \mathrm{erf}\left(\frac{t_\ell - 2\,\tau}{\tau}\sqrt{\log(16)}\right)\right). \end{aligned} \tag{6.53}$$

Analogously, the total laser-absorbed energy at time step t_ℓ is calculated by

$$\begin{aligned} \Delta E_{\mathrm{L_{abs}}}(t_\ell) &= \int_{t_\ell}^{t_{\ell+1}} dt\,\frac{dE_{\mathrm{L_{abs}}}(t)}{dt} \\ &= \frac{E_{\mathrm{L_{tot}}}}{2}\left(-\mathrm{erf}\left(\frac{t_\ell - 2\,\tau}{\tau}\sqrt{\log(16)}\right) + \mathrm{erf}\left(\frac{t_{\ell+1} - 2\,\tau}{\tau}\sqrt{\log(16)}\right)\right). \end{aligned} \tag{6.54}$$

- If a T_{e}-dependent interatomic potential Φ is used in the MD simulation, Φ must exhibit a physical specific electronic heat

$$C_{\mathrm{e}} = -T_{\mathrm{e}}\,\frac{\partial^2\Phi}{\partial T_{\mathrm{e}}^2}.$$

The minimal requirement for a meaningful MD simulation is $C_{\mathrm{e}} \geq 0$.

- If one uses a T_{e}-dependent interatomic potential in the MD simulation, the force $-\nabla_{\mathbf{r}_i}\Phi$ on any atom i can be calculated only from the knowledge of the positions of the neighbors j of atom i within the cutoff radius $r^{(\mathrm{c})}$ of Φ. This is used to parallelize

MD simulations with an interatomic potential by separating the simulation cell in small sub cells, which can be treated independently of each other except for a small information exchange between neighboring cells (see Sect. 5.1.4). Using this parallelization, MD simulations of hundreds of millions of atoms are possible in a reasonable time. This parallelization may be impossible, if the projection on the phonon mode sets $\mathcal{M}_k$ is used to describe the electron-phonon coupling: In any phonon mode, all atoms of the structure show a collective motion. Hence, to calculate any component of

$$\mathcal{F}_{\text{tot}} = \mathcal{F} + \sum_{k=1}^{N_{\mathcal{M}}} \xi_{\mathcal{M}_k} \, m \, \mathbf{P}_{\mathcal{N}_k} \cdot \mathcal{V},$$

we need always the knowledge of all atomic velocities. The total force on any atom i can only be calculated, if the velocities of all other atoms are known to perform the necessary projection $\mathbf{P}_{\mathcal{M}_k} \cdot \mathcal{V}$. One solution to this problem may be a local definition of the projection operators $\mathbf{P}_{\mathcal{M}_k}$. The projection on the different phonon modes should only take explicitly the movement of neighboring atoms into account to be able to perform the common parallelization for speeding up. This is no longer a problem, when one uses local varying T_e's and T_i's in the simulation cell, which is described in the next section.

6.2 Usage of Local Temperatures in the Simulation Cell

In order to be able to simulate a spatial laser profile or the exponential decay of the laser intensity as the pulse penetrates the sample, we introduce local electronic and ionic temperatures. For this, we divide uniformly the complete simulation cell with cell lengths c_x, c_y, c_z into disjunct sub cells, as it is illustrated in Fig. 6.2. Such a simulation setup makes sense only for large simulation cells that are only treatable with T_e-dependent interatomic potentials. Therefore, it is most convenient that the sub cells correspond to the link cells used for the efficient calculation of the atomic neighbor lists (see Sect. 5.1.4). Next, we perform the calculation procedure, that is described in Sect. 6.1 for the total simulation cell, for each sub cell individually, but with some modifications, which we will discuss next.

Let Δx, Δy and Δz be the cell lengths of the sub cells, so that we have

$$N_x = \frac{c_x}{\Delta x}, \qquad N_y = \frac{c_y}{\Delta y}, \qquad N_z = \frac{c_z}{\Delta z} \tag{6.55}$$

sub cells in the x, y, and z-direction, respectively. We number the sub cells using the index vector $\mathbf{i} = [i_x, i_y, i_z]^t$. The sub cell with index $\mathbf{i}$ describes the volume

$$\left[i_x \, \Delta x, (i_x + 1) \, \Delta x\right) \times \left[i_y \, \Delta y, (i_y + 1) \, \Delta y\right) \times \left[i_z \, \Delta z, (i_z + 1) \, \Delta z\right) \tag{6.56}$$

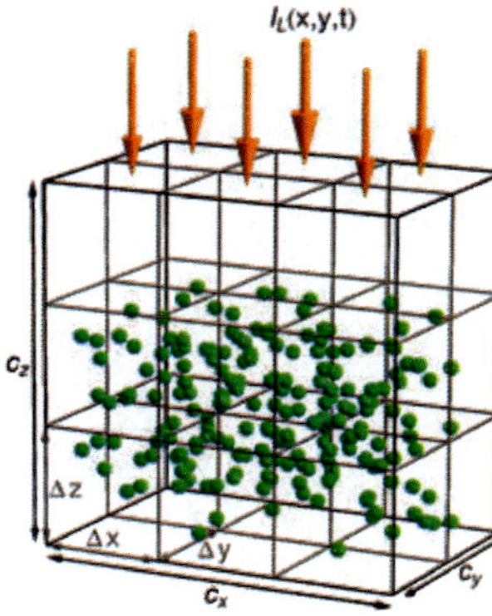

Fig. 6.2 Partition of total simulation cell into sub cells is illustrated

and contains $N_{\text{at}}^{(\mathbf{i})}$ atoms, to which an electronic temperature $T_{\text{e}}^{(\mathbf{i})}$ is assigned. Hence, any atom in sub cell **i** feels a $T_{\text{e}}^{(\mathbf{i})}$ temperature. N_{at} is still the total number of atoms in the total simulation cell. We define

$$S_{\text{e}}^{(\mathbf{i})} = -\frac{\partial \Phi}{\partial T_{\text{e}}^{(\mathbf{i})}} \tag{6.57}$$

as the local electronic entropy of sub cell **i** and denote

$$\mathbf{I} = \left\{ \mathbf{i} \middle| 0 \leq i_x \leq N_x - 1,\ 0 \leq i_y \leq N_y - 1,\ 0 \leq i_z \leq N_z - 1 \right\} \tag{6.58}$$

as the set of all sub cells in the total simulation cell. To derive $E(t_\ell)$ the total potential part of the internal energy of the ions and the total internal energy of the electrons for local ionic and electronic temperatures at time t_ℓ, we start by performing again the transformation of the electrons from the canonical to the microcanonical ensemble similar to how we got Eq. (6.21) using a global T_{e}:

$$\begin{aligned} E(t_\ell) &= \Phi(t_\ell) + \sum_{\mathbf{i}\in\mathbf{I}} \underbrace{T_{\text{e}}^{(\mathbf{i})} S_{\text{e}}^{(\mathbf{i})}}_{\text{only e}^-} \\ &\overset{(6.57)}{=} \Phi(t_\ell) - \sum_{\mathbf{i}\in\mathbf{I}} \underbrace{T_{\text{e}}^{(\mathbf{i})} \frac{\partial \Phi}{\partial T_{\text{e}}^{(\mathbf{i})}}}_{\text{only e}^-} \\ &= \Phi(t_\ell) - \sum_{\mathbf{i}\in\mathbf{I}} \int_{t_0}^{t_\ell} dt\, \frac{d}{dt} \underbrace{\left(T_{\text{e}}^{(\mathbf{i})} \frac{\partial \Phi}{\partial T_{\text{e}}^{(\mathbf{i})}} \right)}_{\text{only e}^-} \end{aligned}$$

$$
\begin{aligned}
&= \Phi(t_\ell) - \sum_{\mathbf{i}\in\mathbf{I}} \int_{t_0}^{t_\ell} dt \left(\frac{dT_\mathrm{e}^{(\mathbf{i})}}{dt} \frac{\partial \Phi}{\partial T_\mathrm{e}^{(\mathbf{i})}} + T_\mathrm{e}^{(\mathbf{i})} \underbrace{\sum_{j=1}^{N_\mathrm{at}} \left(\nabla_{\mathbf{r}_j} \frac{\partial \Phi}{\partial T_\mathrm{e}^{(\mathbf{i})}} \right)^\mathrm{t} \cdot \frac{d\mathbf{r}_j}{dt}}_{\text{ignored}} \right. \\
&\qquad \left. + T_\mathrm{e}^{(\mathbf{i})} \frac{\partial^2 \Phi}{\partial \left(T_\mathrm{e}^{(\mathbf{i})}\right)^2} \frac{dT_\mathrm{e}^{(\mathbf{i})}}{dt} \right) \\
&= \Phi(t_\ell) + \sum_{\mathbf{i}\in\mathbf{I}} \int_{t_0}^{t_\ell} dt \left(- \frac{\partial \Phi}{\partial T_\mathrm{e}^{(\mathbf{i})}} - T_\mathrm{e}^{(\mathbf{i})} \frac{\partial^2 \Phi}{\partial \left(T_\mathrm{e}^{(\mathbf{i})}\right)^2} \right) \frac{dT_\mathrm{e}^{(\mathbf{i})}}{dt} \\
&\overset{(6.57)}{=} \Phi(t_\ell) + \sum_{\mathbf{i}\in\mathbf{I}} \int_{t_0}^{t_\ell} dt \left(S_\mathrm{e}^{(\mathbf{i})} - T_\mathrm{e}^{(\mathbf{i})} \frac{\partial^2 \Phi}{\partial \left(T_\mathrm{e}^{(\mathbf{i})}\right)^2} \right) \frac{dT_\mathrm{e}^{(\mathbf{i})}}{dt}.
\end{aligned}
$$

To finally derive $E(t_\ell)$, we have to modify the equation above. By comparing with Eq. (6.22), which is valid for a global ionic and electronic temperature, it is tempting to define

$$
-T_\mathrm{e}^{(\mathbf{i})} \frac{\partial^2 \Phi}{\partial \left(T_\mathrm{e}^{(\mathbf{i})}\right)^2}
$$

as the local electronic heat capacity $C_\mathrm{e}^{(\mathbf{i})}$ of sub cell $\mathbf{i}$. But this is a bad definition, since the electrons are in fact quantum mechanical objects, which are described by a delocalized wave function. Hence, the electrons spread across the whole simulation cell instead of being confined within each small sub cell $\mathbf{i}$, so that the total electronic heat capacity

$$
C_\mathrm{e} = - \sum_{\mathbf{i}\in\mathbf{I}} T_\mathrm{e}^{(\mathbf{i})} \frac{\partial^2 \Phi}{\partial \left(T_\mathrm{e}^{(\mathbf{i})}\right)^2} \tag{6.59}
$$

is well defined. The local electronic heat capacity $C_\mathrm{e}^{(\mathbf{i})}$ of sub cell $\mathbf{i}$ is just the fraction $N_\mathrm{at}^{(\mathbf{i})}/N_\mathrm{at}$ of the total heat capacity:

$$
\begin{aligned}
C_\mathrm{e}^{(\mathbf{i})} &= \frac{N_\mathrm{at}^{(\mathbf{i})}}{N_\mathrm{at}} C_\mathrm{e} \\
&\overset{(6.59)}{=} - \frac{N_\mathrm{at}^{(\mathbf{i})}}{N_\mathrm{at}} \sum_{\mathbf{i}\in\mathbf{I}} T_\mathrm{e}^{(\mathbf{i})} \frac{\partial^2 \Phi}{\partial \left(T_\mathrm{e}^{(\mathbf{i})}\right)^2}.
\end{aligned} \tag{6.60}
$$

Using this definition, we define the potential part of the internal energy of the ions and the internal energy of the electrons at time t_ℓ to

$$\boxed{E(t_\ell) = \Phi(t_\ell) + \sum_{\mathbf{i}\in\mathbf{I}} \int_{t_0}^{t_\ell} dt \left(-\frac{\partial \Phi}{\partial T_{\mathrm{e}}^{(\mathbf{i})}} - \frac{N_{\mathrm{at}}^{(\mathbf{i})}}{N_{\mathrm{at}}} \left(\sum_{\mathbf{i}'\in\mathbf{I}} T_{\mathrm{e}}^{(\mathbf{i}')} \frac{\partial^2 \Phi}{\partial \left(T_{\mathrm{e}}^{(\mathbf{i}')}\right)^2} \right) \right) \frac{dT_{\mathrm{e}}^{(\mathbf{i})}}{dt}.} \tag{6.61}$$

Inserting the definition (6.57) of the local electronic entropy $S_{\mathrm{e}}^{(\mathbf{i})}$ and the definition (6.60) of the local electronic heat capacity $C_{\mathrm{e}}^{(\mathbf{i})}$ of sub cell $\mathbf{i}$, we obtain

$$E(t_\ell) = \Phi(t_\ell) + \sum_{\mathbf{i}\in\mathbf{I}} \int_{t_0}^{t_\ell} dt \left(S_{\mathrm{e}}^{(\mathbf{i})} + \frac{N_{\mathrm{at}}^{(\mathbf{i})}}{N_{\mathrm{at}}} C_{\mathrm{e}} \right) \frac{dT_{\mathrm{e}}^{(\mathbf{i})}}{dt}. \tag{6.62}$$

We designate formally the coordinates, velocities and Φ-forces of the atoms contained in sub cell $\mathbf{i}$ with $\mathcal{R}^{(\mathbf{i})}, \mathcal{V}^{(\mathbf{i})}, \mathcal{F}^{(\mathbf{i})} \in \mathbb{R}^{3\,N_{\mathrm{at}}^{(\mathbf{i})}}$, respectively. Furthermore, we define local phonon mode subset projection operators $\mathbf{P}^{(\mathbf{i})}_{\mathcal{M}_k^{(\mathbf{i})}} \in \mathbb{R}^{3\,N_{\mathrm{at}}^{(\mathbf{i})}} \times \mathbb{R}^{3\,N_{\mathrm{at}}^{(\mathbf{i})}}$ for each sub cell $\mathbf{i}$. Using $\mathbf{P}^{(\mathbf{i})}_{\mathcal{M}_k^{(\mathbf{i})}}$ and $\mathcal{V}^{(\mathbf{i})}$, we calculate the local kinetic energies $E^{(\mathbf{i})}_{\mathrm{kin}_{\mathcal{M}_k^{(\mathbf{i})}}}$ and the local ionic temperatures $T^{(\mathbf{i})}_{\mathrm{i}_{\mathcal{M}_k^{(\mathbf{i})}}}$ of the different phonon mode subsets $\mathcal{M}_k^{(\mathbf{i})}$ in sub cell $\mathbf{i}$:

$$E^{(\mathbf{i})}_{\mathrm{kin}_{\mathcal{M}_k^{(\mathbf{i})}}} = \frac{m}{2} \left(\mathcal{V}^{(\mathbf{i})}\right)^{\mathrm{t}} \cdot \mathbf{P}_{\mathcal{M}_k^{(\mathbf{i})}} \cdot \mathcal{V}^{(\mathbf{i})}, \tag{6.63}$$

$$T^{(\mathbf{i})}_{\mathrm{i}_{\mathcal{M}_k^{(\mathbf{i})}}} = \frac{2\, E^{(\mathbf{i})}_{\mathrm{kin}_{\mathcal{M}_k^{(\mathbf{i})}}}}{\left|\mathcal{M}_k^{(\mathbf{i})}\right| k_{\mathrm{B}}} = \frac{m \left(\mathcal{V}^{(\mathbf{i})}\right)^{\mathrm{t}} \cdot \mathbf{P}_{\mathcal{M}_k^{(\mathbf{i})}} \cdot \mathcal{V}^{(\mathbf{i})}}{\left|\mathcal{M}_k^{(\mathbf{i})}\right| k_{\mathrm{B}}}. \tag{6.64}$$

We get for the time derivative of the total kinetic energy

$$\frac{dE_{\mathrm{kin}}}{dt} = m \sum_{\mathbf{i}\in\mathbf{I}} \left(\mathcal{V}^{(\mathbf{i})}\right)^{\mathrm{t}} \cdot \frac{d\mathcal{V}^{(\mathbf{i})}}{dt} \tag{6.65}$$

and for the time derivative of the energy E from Eq. (6.61) and inserting Eq. (6.59)

$$\begin{aligned}
\frac{dE}{dt} &= \sum_{\mathbf{i}\in\mathbf{I}} \left(\sum_{j=1}^{N_{\mathrm{at}}^{(\mathbf{i})}} \left(\nabla_{\mathbf{r}_j^{(\mathbf{i})}} \Phi\right)^{\mathrm{t}} \cdot \frac{d\mathbf{r}_j^{(\mathbf{i})}}{dt} + \frac{\partial \Phi}{\partial T_{\mathrm{e}}^{(\mathbf{i})}} \frac{dT_{\mathrm{e}}^{(\mathbf{i})}}{dt} + \left(-\frac{\partial \Phi}{\partial T_{\mathrm{e}}^{(\mathbf{i})}} + \frac{N_{\mathrm{at}}^{(\mathbf{i})}}{N_{\mathrm{at}}} C_{\mathrm{e}} \right) \frac{dT_{\mathrm{e}}^{(\mathbf{i})}}{dt} \right) \\
&= \sum_{\mathbf{i}\in\mathbf{I}} \left(\sum_{j=1}^{N_{\mathrm{at}}^{(\mathbf{i})}} \left(\mathbf{v}_j^{(\mathbf{i})}\right)^{\mathrm{t}} \cdot \nabla_{\mathbf{r}_j^{(\mathbf{i})}} \Phi + \frac{N_{\mathrm{at}}^{(\mathbf{i})}}{N_{\mathrm{at}}} C_{\mathrm{e}} \frac{dT_{\mathrm{e}}^{(\mathbf{i})}}{dt} \right) \\
&= \sum_{\mathbf{i}\in\mathbf{I}} \left(-\left(\mathcal{V}^{(\mathbf{i})}\right)^{\mathrm{t}} \cdot \mathcal{F}^{(\mathbf{i})} + \frac{N_{\mathrm{at}}^{(\mathbf{i})}}{N_{\mathrm{at}}} C_{\mathrm{e}} \frac{dT_{\mathrm{e}}^{(\mathbf{i})}}{dt} \right).
\end{aligned} \tag{6.66}$$

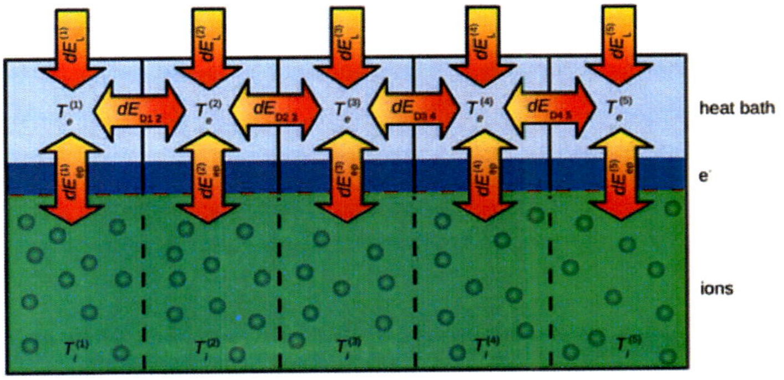

Fig. 6.3 Energy flows during a MD simulation for a one-dimensional simulation cell partitioned into sub cells

From the equation above, we define

$$\frac{dE_{\mathrm{e}}^{(\mathbf{i})}}{dt} = \frac{N_{\mathrm{at}}^{(\mathbf{i})}}{N_{\mathrm{at}}} C_{\mathrm{e}} \frac{dT_{\mathrm{e}}^{(\mathbf{i})}}{dt} \tag{6.67}$$

as the time derivative of the local internal energy of the electrons of sub cell **i**. Three contributions are responsible for the energy change $\frac{dE_{\mathrm{e}}^{(\mathbf{i})}}{dt}$ of the electrons in sub cell **i** (see Fig. 6.3):

- The energy transfer rate to the ions due to the electron-phonon coupling, $\frac{dE_{\mathrm{ep}}^{(\mathbf{i})}}{dt}$
- The energy absorption rate from the laser, $\frac{dE_{\mathrm{L_{abs}}}^{(\mathbf{i})}}{dt}$
- The energy transfer rate, $\frac{dE_{\mathrm{D}}^{(\mathbf{i})}}{dt}$, to the neighboring sub cells due to heat conductivity caused by different electronic temperatures

Therefore, we have

$$\frac{dE_{\mathrm{e}}^{(\mathbf{i})}}{dt} = \frac{dE_{\mathrm{ep}}^{(\mathbf{i})}}{dt} + \frac{dE_{\mathrm{L_{abs}}}^{(\mathbf{i})}}{dt} + \frac{dE_{\mathrm{D}}^{(\mathbf{i})}}{dt}. \tag{6.68}$$

Using the local electronic temperature $T_{\mathrm{e}}^{(\mathbf{i})}(t_\ell)$, the local ionic temperatures $T_{\mathrm{i}_{\mathcal{M}_k^{(\mathbf{i})}}}^{(\mathbf{i})}(t_\ell)$ and the atomic coordinates $\mathcal{R}^{(\mathbf{i})}(t_\ell)$ of the sub cell **i**, we define a local electron-phonon coupling constant $G_{\mathrm{ep}_{\mathcal{M}_k^{(\mathbf{i})}}}^{(\mathbf{i})}(t_\ell)$ for each phonon mode subset. Analogously to the case with a global electronic and ionic temperature described in Eq. (6.26), we define individually the energy transfer rate due to electron phonon coupling for each sub cell **i** by

$$\frac{dE_{\mathrm{ep}}^{(\mathbf{i})}}{dt} = -\sum_{k=1}^{N_{\mathcal{M}}} \left|\mathcal{M}_k^{(\mathbf{i})}\right| G_{\mathrm{ep}_{\mathcal{M}_k^{(\mathbf{i})}}}^{(\mathbf{i})} \left(T_{\mathrm{e}}^{(\mathbf{i})} - T_{\mathrm{i}_{\mathcal{M}_k^{(\mathbf{i})}}}^{(\mathbf{i})}\right). \tag{6.69}$$

Inserting this in Eq. (6.68) and equating with Eq. (6.67) yields

$$\boxed{\frac{N_{\mathrm{at}}^{(\mathbf{i})}}{N_{\mathrm{at}}} C_{\mathrm{e}} \frac{dT_{\mathrm{e}}^{(\mathbf{i})}}{dt} = -\sum_{k=1}^{N_{\mathcal{M}}} \left|\mathcal{M}_k^{(\mathbf{i})}\right| G_{\mathrm{ep}_{\mathcal{M}_k^{(\mathbf{i})}}}^{(\mathbf{i})} \left(T_{\mathrm{e}}^{(\mathbf{i})} - T_{\mathrm{i}_{\mathcal{M}_k^{(\mathbf{i})}}}^{(\mathbf{i})}\right) + \frac{dE_{\mathrm{L_{abs}}}^{(\mathbf{i})}}{dt} + \frac{dE_{\mathrm{D}}^{(\mathbf{i})}}{dt},} \tag{6.70}$$

which is the differential equation for the the local electronic temperature $T_{\mathrm{e}}^{(\mathbf{i})}$ in time. Now we will derive the ionic equations of motions. Summing up the laser absorption energy rate of all sub cells $\mathbf{i}$ yields the total laser absorption energy rate of the complete simulation cell

$$\sum_{\mathbf{i}\in\mathbf{I}} \frac{dE_{\mathrm{L_{abs}}}^{(\mathbf{i})}}{dt} = \frac{dE_{\mathrm{L_{abs}}}}{dt} \tag{6.71}$$

and summing up the energy transfer due to heat conductivity of all cells $\mathbf{i}$ yields the total heat conductivity rate in or out of the complete simulation cell

$$\sum_{\mathbf{i}\in\mathbf{I}} \frac{dE_{\mathrm{D}}^{(\mathbf{i})}}{dt} = \frac{dE_{\mathrm{D}}}{dt}. \tag{6.72}$$

We denote the total energy absorbed from the laser up to time t_ℓ by $E_{\mathrm{L_{abs}}}(t_\ell)$ and the total energy, which was transferred in or out of the complete simulation cell due to heat conductivity up to time t_ℓ, by $E_{\mathrm{D}}(t_\ell)$. Using the latter definitions, we formulate the energy conservation of the complete simulation cell by

$$E(t_\ell) + E_{\mathrm{kin}} = E_{\mathrm{L_{abs}}}(t_\ell) + E_{\mathrm{D}}(t_\ell) + \mathrm{const.} \tag{6.73}$$

We obtain from the time derivative of the energy conservation and by inserting Eqs. (6.66) and (6.65) the following:

$$\begin{aligned}
&\frac{dE}{dt} + \frac{dE_{\mathrm{kin}}}{dt} = \frac{dE_{\mathrm{L_{abs}}}}{dt} + \frac{dE_{\mathrm{D}}}{dt},\\
\Leftrightarrow\quad &\sum_{\mathbf{i}\in\mathbf{I}} \left(-\left(\mathcal{V}^{(\mathbf{i})}\right)^{\mathrm{t}} \cdot \mathcal{F}^{(\mathbf{i})} + \frac{N_{\mathrm{at}}^{(\mathbf{i})}}{N_{\mathrm{at}}} C_{\mathrm{e}} \frac{dT_{\mathrm{e}}^{(\mathbf{i})}}{dt} + m\left(\mathcal{V}^{(\mathbf{i})}\right)^{\mathrm{t}} \cdot \frac{d\mathcal{V}^{(\mathbf{i})}}{dt}\right) = \frac{dE_{\mathrm{L_{abs}}}}{dt} + \frac{dE_{\mathrm{D}}}{dt},\\
\Leftrightarrow\quad &\sum_{\mathbf{i}\in\mathbf{I}} \left(m\left(\mathcal{V}^{(\mathbf{i})}\right)^{\mathrm{t}} \cdot \frac{d\mathcal{V}^{(\mathbf{i})}}{dt} - \left(\mathcal{V}^{(\mathbf{i})}\right)^{\mathrm{t}} \cdot \mathcal{F}^{(\mathbf{i})} + \frac{N_{\mathrm{at}}^{(\mathbf{i})}}{N_{\mathrm{at}}} C_{\mathrm{e}} \frac{dT_{\mathrm{e}}^{(\mathbf{i})}}{dt}\right) = \frac{dE_{\mathrm{L_{abs}}}}{dt} + \frac{dE_{\mathrm{D}}}{dt}.
\end{aligned}$$

In addition, if we insert Eq. (6.70) for $\frac{N_{\mathrm{at}}^{(\mathbf{i})}}{N_{\mathrm{at}}} C_{\mathrm{e}} \frac{dT_{\mathrm{e}}^{(\mathbf{i})}}{dt}$, the following applies:

$$\sum_{\mathbf{i}\in\mathbf{I}} \left(m\left(\mathcal{V}^{(\mathbf{i})}\right)^{\mathrm{t}} \cdot \frac{d\mathcal{V}^{(\mathbf{i})}}{dt} - \left(\mathcal{V}^{(\mathbf{i})}\right)^{\mathrm{t}} \cdot \mathcal{F}^{(\mathbf{i})} - \sum_{k=1}^{N_{\mathcal{M}}} \left|\mathcal{M}_k^{(\mathbf{i})}\right| G_{\mathrm{ep}_{\mathcal{M}_k^{(\mathbf{i})}}}^{(\mathbf{i})} \left(T_{\mathrm{e}}^{(\mathbf{i})} - T_{\mathrm{i}_{\mathcal{M}_k^{(\mathbf{i})}}}^{(\mathbf{i})}\right)\right) = 0.$$

Since each atom only belongs to one sub cell and the above equation must be generally valid, it follows for any $\mathbf{i} \in \mathbf{I}$:

$$m\left(\mathcal{V}^{(\mathbf{i})}\right)^{\mathrm{t}} \cdot \frac{d\mathcal{V}^{(\mathbf{i})}}{dt} - \left(\mathcal{V}^{(\mathbf{i})}\right)^{\mathrm{t}} \cdot \mathcal{F}^{(\mathbf{i})} - \sum_{k=1}^{N_{\mathcal{M}}} \left|\mathcal{M}_k^{(\mathbf{i})}\right| G^{(\mathbf{i})}_{\mathrm{ep}_{\mathcal{M}_k^{(\mathbf{i})}}} \left(T_{\mathrm{e}}^{(\mathbf{i})} - T^{(\mathbf{i})}_{\mathrm{i}_{\mathcal{M}_k^{(\mathbf{i})}}}\right) = 0,$$

$$\Leftrightarrow \quad m\left(\mathcal{V}^{(\mathbf{i})}\right)^{\mathrm{t}} \cdot \frac{d\mathcal{V}^{(\mathbf{i})}}{dt} - \left(\mathcal{V}^{(\mathbf{i})}\right)^{\mathrm{t}} \cdot \mathcal{F}^{(\mathbf{i})} - \sum_{k=1}^{N_{\mathcal{M}}} \underbrace{\frac{m\left(\mathcal{V}^{(\mathbf{i})}\right)^{\mathrm{t}} \cdot \mathbf{P}^{(\mathbf{i})}_{\mathcal{M}_k^{(\mathbf{i})}} \cdot \mathcal{V}^{(\mathbf{i})}}{2\,E^{(\mathbf{i})}_{\mathrm{kin}_{\mathcal{M}_k^{(\mathbf{i})}}}}}_{=1} \left|\mathcal{M}_k^{(\mathbf{i})}\right| G^{(\mathbf{i})}_{\mathrm{ep}_{\mathcal{M}_k^{(\mathbf{i})}}} \left(T_{\mathrm{e}}^{(\mathbf{i})} - T^{(\mathbf{i})}_{\mathrm{i}_{\mathcal{M}_k^{(\mathbf{i})}}}\right) = 0,$$

$$\Leftrightarrow \quad m\left(\mathcal{V}^{(\mathbf{i})}\right)^{\mathrm{t}} \cdot \frac{d\mathcal{V}^{(\mathbf{i})}}{dt} - \left(\mathcal{V}^{(\mathbf{i})}\right)^{\mathrm{t}} \cdot \mathcal{F}^{(\mathbf{i})} - \left(\mathcal{V}^{(\mathbf{i})}\right)^{\mathrm{t}} \cdot \sum_{k=1}^{N_{\mathcal{M}}} \frac{\left|\mathcal{M}_k^{(\mathbf{i})}\right| G^{(\mathbf{i})}_{\mathrm{ep}_{\mathcal{M}_k^{(\mathbf{i})}}} \left(T_{\mathrm{e}}^{(\mathbf{i})} - T^{(\mathbf{i})}_{\mathrm{i}_{\mathcal{M}_k^{(\mathbf{i})}}}\right)}{2\,E^{(\mathbf{i})}_{\mathrm{kin}_{\mathcal{M}_k^{(\mathbf{i})}}}} m\,\mathbf{P}^{(\mathbf{i})}_{\mathcal{M}_k^{(\mathbf{i})}} \cdot \mathcal{V}^{(\mathbf{i})} = 0,$$

$$\Leftrightarrow \quad \left(\mathcal{V}^{(\mathbf{i})}\right)^{\mathrm{t}} \cdot \left(m\frac{d\mathcal{V}^{(\mathbf{i})}}{dt} - \mathcal{F}^{(\mathbf{i})} - \sum_{k=1}^{N_{\mathcal{M}}} \frac{\left|\mathcal{M}_k^{(\mathbf{i})}\right| G^{(\mathbf{i})}_{\mathrm{ep}_{\mathcal{M}_k^{(\mathbf{i})}}} \left(T_{\mathrm{e}}^{(\mathbf{i})} - T^{(\mathbf{i})}_{\mathrm{i}_{\mathcal{M}_k^{(\mathbf{i})}}}\right)}{2\,E^{(\mathbf{i})}_{\mathrm{kin}_{\mathcal{M}_k^{(\mathbf{i})}}}} m\,\mathbf{P}^{(\mathbf{i})}_{\mathcal{M}_k^{(\mathbf{i})}} \cdot \mathcal{V}^{(\mathbf{i})} \right) = 0.$$

The last equation is generally valid, if the term in the second bracket is zero:

$$\boxed{m\frac{d\mathcal{V}^{(\mathbf{i})}}{dt} = \mathcal{F}^{(\mathbf{i})} + \sum_{k=1}^{N_{\mathcal{M}}} \frac{\left|\mathcal{M}_k^{(\mathbf{i})}\right| G^{(\mathbf{i})}_{\mathrm{ep}_{\mathcal{M}_k^{(\mathbf{i})}}} \left(T_{\mathrm{e}}^{(\mathbf{i})} - T^{(\mathbf{i})}_{\mathrm{i}_{\mathcal{M}_k^{(\mathbf{i})}}}\right)}{2\,E^{(\mathbf{i})}_{\mathrm{kin}_{\mathcal{M}_k^{(\mathbf{i})}}}} m\,\mathbf{P}^{(\mathbf{i})}_{\mathcal{M}_k^{(\mathbf{i})}} \cdot \mathcal{V}^{(\mathbf{i})}.} \tag{6.74}$$

This is the ionic equation of motion for sub cell $\mathbf{i}$. Hence, the total atomic force vector of sub cell $\mathbf{i}$ at time t_ℓ is given by

$$\mathcal{F}^{(\mathbf{i})}_{\mathrm{tot}}(t_\ell) = \mathcal{F}^{(\mathbf{i})}(t_\ell) + \sum_{k=1}^{N_{\mathcal{M}}} \xi^{(\mathbf{i})}_{\mathcal{M}_k^{(\mathbf{i})}}(t_\ell)\, m\,\mathbf{P}^{(\mathbf{i})}_{\mathcal{M}_k^{(\mathbf{i})}} \cdot \mathcal{V}^{(\mathbf{i})}(t_\ell) \tag{6.75}$$

using the local defined parameter

$$\xi^{(\mathbf{i})}_{\mathcal{M}_k^{(\mathbf{i})}}(t_\ell) = \frac{\left|\mathcal{M}_k^{(\mathbf{i})}\right| G^{(\mathbf{i})}_{\mathrm{ep}_{\mathcal{M}_k^{(\mathbf{i})}}}(t_\ell) \left(T_{\mathrm{e}}^{(\mathbf{i})}(t_\ell) - T^{(\mathbf{i})}_{\mathrm{i}_{\mathcal{M}_k^{(\mathbf{i})}}}(t_\ell)\right)}{2\,E^{(\mathbf{i})}_{\mathrm{kin}_{\mathcal{M}_k^{(\mathbf{i})}}}(t_\ell)}. \tag{6.76}$$

In the Velocity Verlet Algorithm, this parameter can be calculated from Eq. (6.48), if all involved quantities are considered locally for sub cell $\mathbf{i}$. Finally, we summarize our extended MD simulation procedure in the next theorem:

Theorem 6.2 (MD simulations with excited PES and EPC with local temperatures) *Let us consider a simulation cell with N_{at} atoms of identical type with mass m. Let us divide the simulation cell uniformly into sub cells, which are labeled by the index* $\mathbf{i}$. $\mathbf{I}$ *denotes the set of all sub cell indices* $\mathbf{i}$. *Let us divide the phonon modes of a sub cell* $\mathbf{i}$ *into $N_{\mathcal{M}}$ different sets $\mathcal{M}_k^{(\mathbf{i})}$, each one containing $\left|\mathcal{M}_k^{(\mathbf{i})}\right|$ phonon modes. For each sub cell* $\mathbf{i}$, *let us define a local ionic temperature of the phonon mode subset $\mathcal{M}_k^{(\mathbf{i})}$ by Eq.* (6.64)

$$T_{i_{\mathcal{M}_k^{(\mathbf{i})}}}^{(\mathbf{i})} = \frac{2\,E_{kin_{\mathcal{M}_k^{(\mathbf{i})}}}^{(\mathbf{i})}}{\left|\mathcal{M}_k^{(\mathbf{i})}\right|\,k_B} = \frac{m\,\left(\mathcal{V}^{(\mathbf{i})}\right)^t \cdot \mathbf{P}_{\mathcal{M}_k^{(\mathbf{i})}}^{(\mathbf{i})} \cdot \mathcal{V}^{(\mathbf{i})}}{\left|\mathcal{M}_k^{(\mathbf{i})}\right|\,k_B}.$$

Here, $E_{kin_{\mathcal{M}_k^{(\mathbf{i})}}}^{(\mathbf{i})}$ denotes the kinetic energy of the phonon modes of set $\mathcal{M}_k^{(\mathbf{i})}$ in sub cell $\mathbf{i}$, *k_B the Boltzmann constant, $\mathcal{V}^{(\mathbf{i})}$ the vector of the velocities of the atoms in sub cell* $\mathbf{i}$, *and $\mathbf{P}_{\mathcal{M}_k^{(\mathbf{i})}}^{(\mathbf{i})}$ the matrix that projects onto the phonon modes of set $\mathcal{M}_k^{(\mathbf{i})}$. Let us associate to each sub cell* $\mathbf{i}$ *a local electronic temperature $T_e^{(\mathbf{i})}$ and a local electronic specific heat by Eq.* (6.60)

$$C_e^{(\mathbf{i})} = -\frac{N_{at}^{(\mathbf{i})}}{N_{at}} \sum_{\mathbf{i} \in \mathbf{I}} T_e^{(\mathbf{i})} \frac{\partial^2 \Phi}{\partial \left(T_e^{(\mathbf{i})}\right)^2}.$$

Here, Φ denotes the T_e-dependent interatomic potential and $N_{at}^{(\mathbf{i})}$ the total number of atoms within sub cell $\mathbf{i}$. *Then, we obtain the total potential part of the internal energy of the ions and the total internal energy of the electrons in the simulation cell at time t_ℓ by Eq.* (6.61)

$$E(t_\ell) = \Phi(t_\ell) + \sum_{\mathbf{i} \in \mathbf{I}} \int_{t_0}^{t_\ell} dt \left(-\frac{\partial \Phi}{\partial T_e^{(\mathbf{i})}} - \frac{N_{at}^{(\mathbf{i})}}{N_{at}} \left(\sum_{\mathbf{i}' \in \mathbf{I}} T_e^{(\mathbf{i}')} \frac{\partial^2 \Phi}{\partial \left(T_e^{(\mathbf{i}')}\right)^2} \right) \right) \frac{dT_e^{(\mathbf{i})}}{dt}.$$

The time propagation of $T_e^{(\mathbf{i})}$ is given by Eq. (6.70)

$$C_e^{(\mathbf{i})} \frac{dT_e^{(\mathbf{i})}}{dt} = -\sum_{k=1}^{N_{\mathcal{M}}} \left|\mathcal{M}_k^{(\mathbf{i})}\right| G_{ep_{\mathcal{M}_k^{(\mathbf{i})}}}^{(\mathbf{i})} \left(T_e^{(\mathbf{i})} - T_{i_{\mathcal{M}_k^{(\mathbf{i})}}}^{(\mathbf{i})} \right) + \frac{dE_{L_{abs}}^{(\mathbf{i})}}{dt} + \frac{dE_D^{(\mathbf{i})}}{dt}.$$

$G^{(\mathbf{i})}_{ep_{\mathcal{M}_k^{(\mathbf{i})}}}$ denotes the electron-phonon coupling constant for the phonon modes of set $\mathcal{M}_k^{(\mathbf{i})}$ in sub cell **i**, *$\frac{dE^{(\mathbf{i})}_{L_{abs}}}{dt}$ the energy absorption rate from the laser field in sub cell* **i**, *and $\frac{dE^{(\mathbf{i})}_D}{dt}$ the energy transfer rate from the neighboring sub cells to sub cell* **i** *due to heat diffusion related to the different local electronic temperatures. From energy conservation, we obtain the equation of motion for the ions of sub cell* **i** *by Eq. (6.74)*

$$m\,\frac{d\mathcal{V}^{(\mathbf{i})}}{dt} = \mathcal{F}^{(\mathbf{i})} + \sum_{k=1}^{N_{\mathcal{M}}} \frac{\left|\mathcal{M}_k^{(\mathbf{i})}\right| G^{(\mathbf{i})}_{ep_{\mathcal{M}_k^{(\mathbf{i})}}} \left(T_e^{(\mathbf{i})} - T^{(\mathbf{i})}_{i_{\mathcal{M}_k^{(\mathbf{i})}}}\right)}{2\,E^{(\mathbf{i})}_{kin_{\mathcal{M}_k^{(\mathbf{i})}}}}\, m\,\mathbf{P}^{(\mathbf{i})}_{\mathcal{M}_k^{(\mathbf{i})}} \cdot \mathcal{V}^{(\mathbf{i})}.$$

$\mathcal{F}^{(\mathbf{i})}$ denotes the vector of the forces obtained from Φ for all atoms in sub cell **i**.

6.2.1 Numerical Implementation

To obtain concretely the local electronic temperature $T_\mathrm{e}^{(\mathbf{i})}$ at the considered time step t_ℓ, we derive the total change of the internal energy of the electrons in sub cell **i** during the time step t_ℓ by

$$\Delta E_\mathrm{e}^{(\mathbf{i})}(t_\ell) = \Delta E_\mathrm{ep}^{(\mathbf{i})}(t_\ell) + \Delta E^{(\mathbf{i})}_{\mathrm{L_{abs}}}(t_\ell) + \Delta E_\mathrm{D}^{(\mathbf{i})}(t_\ell) \tag{6.77}$$

and, from this, we get the final change of the local electronic temperature $T_\mathrm{e}^{(\mathbf{i})}$ at time step t_ℓ for $C_\mathrm{e}(t_\ell) > 0$ by

$$\Delta T_\mathrm{e}^{(\mathbf{i})}(t_\ell) \overset{(6.67)}{=} \frac{N_\mathrm{at}}{N_\mathrm{at}^{(\mathbf{i})}} \frac{\Delta E_\mathrm{e}^{(\mathbf{i})}(t_\ell)}{C_\mathrm{e}(t_\ell)}. \tag{6.78}$$

For $C_\mathrm{e}(t_\ell) = 0$, we use again Eq. (6.41), but we consider all involved quantities locally for sub cell **i**. If the sub cell **i** contains no atoms, i.e., $N_\mathrm{at}^{(\mathbf{i})} = 0$, the corresponding local electronic temperature $T_\mathrm{e}^{(\mathbf{i})}$ is not defined. Now we focus on the calculation of $\Delta E_\mathrm{ep}^{(\mathbf{i})}(t_\ell)$, $\Delta E^{(\mathbf{i})}_{\mathrm{L_{abs}}}(t_\ell)$, and $\Delta E_\mathrm{D}^{(\mathbf{i})}(t_\ell)$. The amount of energy $\Delta E_\mathrm{ep}^{(\mathbf{i})}(t_\ell)$ that is transfered from the ions to the electrons due to electron-phonon coupling in sub cell **i** at time t_ℓ is given by

$$\Delta E_\mathrm{ep}^{(\mathbf{i})}(t_\ell) = -\sum_{k=1}^{N_{\mathcal{M}}} \left|\mathcal{M}_k^{(\mathbf{i})}\right| G^{(\mathbf{i})}_{\mathrm{ep}_{\mathcal{M}_k^{(\mathbf{i})}}}(t_\ell) \left(T_\mathrm{e}^{(\mathbf{i})}(t_\ell) - T^{(\mathbf{i})}_{\mathrm{i}_{\mathcal{M}_k^{(\mathbf{i})}}}(t_\ell)\right) \Delta t. \tag{6.79}$$

To derive the amount of energy $\Delta E_{\mathrm{L_{abs}}}^{(\mathbf{i})}(t_\ell)$, that is absorbed from the laser in sub cell $\mathbf{i}$ at time step t_ℓ, we take into account that the simulation cell is irradiated with the spatial incident laser intensity $I_{\mathrm{inc}}(x, y, t)$ from the top. From the spatial incident laser intensity and the reflectivity $\mathfrak{R}$, we can derive the total laser energy that is penetrating into the total simulation volume from the time t_0 to t_ℓ as

$$E_{\mathrm{L}}(t_\ell) = \int_0^{c_x} dx \int_0^{c_y} dy \int_{t_0}^{t_\ell} dt\, (1 - \mathfrak{R})\, I_{\mathrm{inc}}(x, y, t). \tag{6.80}$$

The total absorbed laser fluence $I_{\mathrm{L_{tot}}}(x, y)$ at the surface is defined as

$$I_{\mathrm{L_{tot}}}(x, y) = \int_{-\infty}^{\infty} dt\, (1 - \mathfrak{R})\, I_{\mathrm{inc}}(x, y, t). \tag{6.81}$$

The quantity commonly measured in the experiments is the total incident laser fluence:

$$I_{\mathrm{inc_{tot}}}(x, y) = \int_{-\infty}^{\infty} dt\, I_{\mathrm{inc}}(x, y, t). \tag{6.82}$$

With the help of the optical absorption coefficient α_{abs} and the Lambert–Beer law, we can calculate the total laser energy that is absorbed in the total simulation volume from time t_0 to t_ℓ as

$$\begin{aligned} E_{\mathrm{L_{abs}}}(t_\ell) &= \left(1 - e^{-\alpha_{\mathrm{abs}}\, c_z}\right) E_{\mathrm{L}}(t_\ell) \\ &= \left(1 - e^{-\alpha_{\mathrm{abs}}\, c_z}\right) \int_0^{c_x} dx \int_0^{c_y} dy \int_{t_0}^{t_\ell} dt\, (1 - \mathfrak{R})\, I_{\mathrm{inc}}(x, y, t). \end{aligned} \tag{6.83}$$

We define $\Delta E_{\mathrm{L}}^{(\mathbf{i})}(t_\ell)$ as the laser energy that is penetrating into the sub cell $\mathbf{i}$ during time step t_ℓ. We can calculate directly the laser energy that is penetrating into the top sub cells $\mathbf{i}_{\mathrm{top}} = [i_x, i_y, c_z]^{\mathrm{t}}$ during time step t_ℓ from the spatial incident laser intensity $I_{\mathrm{inc}}(x, y, t)$ as follows

$$\Delta E_{\mathrm{L}}^{(\mathbf{i}_{\mathrm{top}})}(t_\ell) = \int_{i_x \Delta x}^{(i_x+1)\Delta x} dx \int_{i_y \Delta y}^{(i_y+1)\Delta y} dy \int_{t_\ell}^{t_{\ell+1}} dt\, (1 - \mathfrak{R})\, I_{\mathrm{inc}}(x, y, t). \tag{6.84}$$

We define $N_{\mathrm{at}}^{(\mathrm{ref})}$ as the number of atoms in a sub cell under ambient conditions. Using again the Lambert–Beer law, we obtain for the energy that is absorbed from the laser:

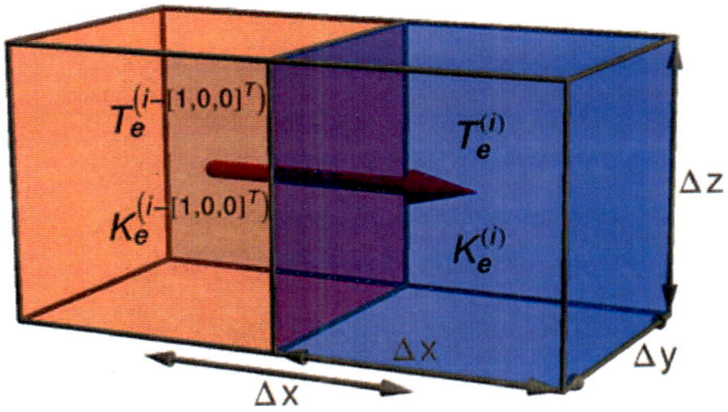

Fig. 6.4 Electronic heat diffusion between two neighboring sub cells with different electronic temperatures is illustrated

$$\Delta E^{(\mathbf{i})}_{\mathrm{L_{abs}}}(t_\ell) = \left(1 - e^{-\alpha_{\mathrm{abs}} \frac{N^{(\mathbf{i})}_{\mathrm{at}}}{N^{(\mathrm{ref})}_{\mathrm{at}}} \Delta z}\right) \Delta E^{(\mathbf{i})}_{\mathrm{L}}(t_\ell). \tag{6.85}$$

Here we put the fraction $N^{(\mathbf{i})}_{\mathrm{at}}/N^{(\mathrm{ref})}_{\mathrm{at}}$, since the optical absorption coefficient yields the value α_{abs} only for the atomic density at ambient conditions. The absorption coefficient related to a sub cell **i** decreases with decreasing number of contained atoms $N^{(\mathbf{i})}_{\mathrm{at}}$. In particular, if there are no atoms, the absorption coefficient will be zero and there will be no absorption. Because the laser energy flows in the negative z-direction, we get from the energy conservation:

$$\Delta E^{(\mathbf{i}-[0,0,1]^{\mathrm{t}})}_{\mathrm{L}}(t_\ell) = \Delta E^{(\mathbf{i})}_{\mathrm{L}}(t_\ell) - \Delta E^{(\mathbf{i})}_{\mathrm{L_{abs}}}(t_\ell). \tag{6.86}$$

Using Eqs. (6.86) and (6.85), we can calculate successively the laser energy that is penetrating into and that is absorbed by each sub cell **i** from the top to the bottom.

Since the sub cells exhibit different electronic temperatures, we also have to consider the electronic heat diffusion between the sub cells. We denote the electronic heat conductivity of a sub cell **i** at time step t_ℓ by $K^{(\mathbf{i})}_{\mathrm{e}}(t_\ell)$ and assume that it is a function of the electronic and ionic temperatures of the sub cell:

$$K^{(\mathbf{i})}_{\mathrm{e}}(t_\ell) \equiv K_{\mathrm{e}}\left(T^{(\mathbf{i})}_{\mathrm{e}}(t_\ell), T^{(\mathbf{i})}_{\mathrm{i}_{M_1}}(t_\ell), \ldots, T^{(\mathbf{i})}_{\mathrm{i}_{M_{N_M}}}(t_\ell)\right). \tag{6.87}$$

In general, the electronic heat conductivity may depend also on the atomic coordinates. We calculate the electronic heat diffusion from sub cell $\mathbf{i} - [0, 0, 1]^{\mathrm{t}}$ into sub cell **i** during time step t_ℓ, which is illustrated in Fig. 6.4, by

$$\Delta E^{(\mathbf{i},\mathbf{i}-[1,0,0]^{\mathrm{t}})}_{\mathrm{D}}(t_\ell) = \frac{K^{(\mathbf{i}-[1,0,0]^{\mathrm{t}})}_{\mathrm{e}}(t_\ell) + K^{(\mathbf{i})}_{\mathrm{e}}(t_\ell)}{2} \, \frac{T^{(\mathbf{i}-[1,0,0]^{\mathrm{t}})}_{\mathrm{e}}(t_\ell) - T^{(\mathbf{i})}_{\mathrm{e}}(t_\ell)}{\Delta x} \Delta y \, \Delta z \, \Delta t. \tag{6.88}$$

We average the electronic heat conductivities of the two involved sub cells, since the diffusion propagates formally from the middle of the one sub cell to the middle of the other sub cell. Furthermore, we do not consider a sub cell for electronic heat diffusion if it contains less atoms than $\frac{1}{10}\,N_{\mathrm{at}}^{(\mathrm{ref})}$. To get finally the total electronic heat diffusion $\Delta E_{\mathrm{D}}^{(\mathbf{i})}(t_\ell)$ of sub cell $\mathbf{i}$, we have just to take the heat diffusion of the corresponding six neighbors into account:

$$\begin{aligned}
\Delta E_{\mathrm{D}}^{(\mathbf{i})}(t_\ell) = \;& \Delta E_{\mathrm{D}}^{(\mathbf{i},\mathbf{i}+[1,0,0]')}(t_\ell) + \Delta E_{\mathrm{D}}^{(\mathbf{i},\mathbf{i}-[1,0,0]')}(t_\ell) \\
&+ \Delta E_{\mathrm{D}}^{(\mathbf{i},\mathbf{i}+[0,1,0]')}(t_\ell) + \Delta E_{\mathrm{D}}^{(\mathbf{i},\mathbf{i}-[0,1,0]')}(t_\ell) \\
&+ \Delta E_{\mathrm{D}}^{(\mathbf{i},\mathbf{i}+[0,0,1]')}(t_\ell) + \Delta E_{\mathrm{D}}^{(\mathbf{i},\mathbf{i}-[0,0,1]')}(t_\ell) \\
=\;& \frac{K_{\mathrm{e}}^{(\mathbf{i}+[1,0,0]')}(t_\ell) + K_{\mathrm{e}}^{(\mathbf{i})}(t_\ell)}{2}\,\frac{T_{\mathrm{e}}^{(\mathbf{i}+[1,0,0]')}(t_\ell) - T_{\mathrm{e}}^{(\mathbf{i})}(t_\ell)}{\Delta x}\,\Delta y\,\Delta z\,\Delta t \\
&+ \frac{K_{\mathrm{e}}^{(\mathbf{i}-[1,0,0]')}(t_\ell) + K_{\mathrm{e}}^{(\mathbf{i})}(t_\ell)}{2}\,\frac{T_{\mathrm{e}}^{(\mathbf{i}-[1,0,0]')}(t_\ell) - T_{\mathrm{e}}^{(\mathbf{i})}(t_\ell)}{\Delta x}\,\Delta y\,\Delta z\,\Delta t \\
&+ \frac{K_{\mathrm{e}}^{(\mathbf{i}+[0,1,0]')}(t_\ell) + K_{\mathrm{e}}^{(\mathbf{i})}(t_\ell)}{2}\,\frac{T_{\mathrm{e}}^{(\mathbf{i}+[0,1,0]')}(t_\ell) - T_{\mathrm{e}}^{(\mathbf{i})}(t_\ell)}{\Delta y}\,\Delta x\,\Delta z\,\Delta t \\
&+ \frac{K_{\mathrm{e}}^{(\mathbf{i}-[0,1,0]')}(t_\ell) + K_{\mathrm{e}}^{(\mathbf{i})}(t_\ell)}{2}\,\frac{T_{\mathrm{e}}^{(\mathbf{i}-[0,1,0]')}(t_\ell) - T_{\mathrm{e}}^{(\mathbf{i})}(t_\ell)}{\Delta y}\,\Delta x\,\Delta z\,\Delta t \\
&+ \frac{K_{\mathrm{e}}^{(\mathbf{i}+[0,0,1]')}(t_\ell) + K_{\mathrm{e}}^{(\mathbf{i})}(t_\ell)}{2}\,\frac{T_{\mathrm{e}}^{(\mathbf{i}+[0,0,1]')}(t_\ell) - T_{\mathrm{e}}^{(\mathbf{i})}(t_\ell)}{\Delta z}\,\Delta x\,\Delta y\,\Delta t \\
&+ \frac{K_{\mathrm{e}}^{(\mathbf{i}-[0,0,1]')}(t_\ell) + K_{\mathrm{e}}^{(\mathbf{i})}(t_\ell)}{2}\,\frac{T_{\mathrm{e}}^{(\mathbf{i}-[0,0,1]')}(t_\ell) - T_{\mathrm{e}}^{(\mathbf{i})}(t_\ell)}{\Delta z}\,\Delta x\,\Delta y\,\Delta t.
\end{aligned} \tag{6.89}$$

The above equation must be modified accordingly for the sub cells located at the border of the simulation cell. This can be done straight forward for directions with periodic boundary conditions. In directions with open boundary conditions, we consider two cases: There is no electronic heat diffusion out of the simulation cell in this direction or there is a heat diffusion to a heat bath outside, which exhibits a constant electronic temperature.

Unfortunately, the finite difference scheme (6.89) for describing the head diffusion between the sub cells is numerically stable, only if the time step Δt_{D} fulfills the von Neumann stability criterion [7]

$$\Delta t_{\mathrm{D}} \leq \frac{1}{2}\,\frac{1}{\frac{1}{\Delta x^2} + \frac{1}{\Delta y^2} + \frac{1}{\Delta z^2}}\,\min_{\mathbf{i}}\,\frac{\rho_{\mathrm{at}}\,C_{\mathrm{e}}^{(\mathbf{i})}}{N_{\mathrm{at}}^{(\mathbf{i})}\,K_{\mathrm{e}}^{(\mathbf{i})}}. \tag{6.90}$$

Here, Δx^2 denotes $\Delta x\,\Delta x$ and ρ_{at} describes how many atoms per volume are present under ambient conditions. If the electronic heat conductivity is only taken into account in z-direction and there is no sub cell partitioning along the x- and y-direction, which corresponds to the same laser intensity in x- and y-direction $I_L(x, y, t) \equiv I_L(t)$, the von Neumann stability criterion for the time step reads

$$\Delta t_D \leq \frac{1}{2} \Delta z^2 \min_{\mathbf{i}} \frac{\rho_{at} C_e^{(\mathbf{i})}}{N_{at}^{(\mathbf{i})} K_e^{(\mathbf{i})}}. \tag{6.91}$$

Since Δt_D is typically much smaller than the MD time step Δt used to integrate the equations of atomic motions, we introduce a separate time step Δt_D to calculate $\Delta E_D^{(\mathbf{i})}(t_\ell)$ and $\Delta E_{L_{abs}}^{(\mathbf{i})}(t_\ell)$ and changing $T_e^{(\mathbf{i})}$. Notice that the MD time step Δt should correspond to an integer multiple of the Δt_D:

$$\Delta t = N_D \, \Delta t_D. \tag{6.92}$$

Furthermore, the total energy that is transferred from the ions to the electrons in sub cell $\mathbf{i}$ is accumulated for the N_D steps:

$$\Delta E_{ep}^{(\mathbf{i})}(t_\ell) = - \sum_{j=0}^{N_D-1} \sum_{k=1}^{N_\mathcal{M}} \left| \mathcal{M}_k^{(\mathbf{i})} \right| G_{ep_{\mathcal{M}_k^{(\mathbf{i})}}}^{(\mathbf{i})}(t_\ell + j \, \Delta t_D) \left(T_e^{(\mathbf{i})}(t_\ell + j \, \Delta t_D) - T_{\mathbf{i}_{\mathcal{M}_k^{(\mathbf{i})}}}^{(\mathbf{i})}(t_\ell) \right) \Delta t_D, \tag{6.93}$$

$$\xi_{\mathcal{M}_k^{(\mathbf{i})}}^{(\mathbf{i})}(t_\ell) = \frac{1}{N_D} \sum_{j=0}^{N_D-1} \frac{\left| \mathcal{M}_k^{(\mathbf{i})} \right| G_{ep_{\mathcal{M}_k^{(\mathbf{i})}}}^{(\mathbf{i})}(t_\ell + j \, \Delta t_D) \left(T_e^{(\mathbf{i})}(t_\ell + j \, \Delta t_D) - T_{\mathbf{i}_{\mathcal{M}_k^{(\mathbf{i})}}}^{(\mathbf{i})}(t_\ell) \right)}{2 \, E_{kin_{\mathcal{M}_k^{(\mathbf{i})}}}^{(\mathbf{i})}(t_\ell)}. \tag{6.94}$$

6.2.2 *Remarks*

- Because the simulation cell is divided into sub cells with individual local electronic temperatures and atoms can move between the different sub cells, there are fluctuations in the total energy, which do not vanish in the limit $\Delta t \to 0$.
- Our simulation setup reduces to the simulation method of Ivanov and Zhigilei described in Sect. 5.2, if the effects of the excited PES are neglected, i.e., $\frac{\partial \Phi}{\partial T_e^{(\mathbf{i})}} = 0$ for all $\mathbf{i} \in \mathbf{I}$, and one electron-phonon coupling constant $G_{ep}^{(\mathbf{i})}$ is used for all phonon modes. In this case, one has to define the internal energy of the electrons of sub cell $\mathbf{i}$ from

$$E_e^{(\mathbf{i})}(T_e) = N_{at}^{(\mathbf{i})} \int_0^{T_e} dT_e' \, C_e(T_e') \tag{6.95}$$

by using the electronic specific heat per atom $C_e(T_e)$, which is a given function of the electronic temperature T_e. Now the potential energy of the ions and the electrons (6.61) reduces to

$$E(t_\ell) = \Phi(t_\ell) + \sum_{\mathbf{i}\in\mathbf{I}} E_e^{(\mathbf{i})}(T_e), \tag{6.96}$$

since $\frac{\partial\Phi}{\partial T_e^{(\mathbf{i})}} = 0$ for all $\mathbf{i} \in \mathbf{I}$ and one has to add $E_e^{(\mathbf{i})}(T_e)$ from Eq. (6.95), because the ground state potential Φ does not contain any information about the electronic energy. In Eq. (6.70), one uses the given function $C_e(T_e)$ and only one electron-phonon coupling constant $G_{ep}^{(\mathbf{i})}$ and one ionic temperature $T_i^{(\mathbf{i})}$ for all phonon modes:

$$\frac{N_{at}^{(\mathbf{i})}}{N_{at}} C_e(T_e) \frac{dT_e^{(\mathbf{i})}}{dt} = -G_{ep}^{(\mathbf{i})} \left(T_e^{(\mathbf{i})} - T_i^{(\mathbf{i})}\right) + \frac{dE_{L_{abs}}^{(\mathbf{i})}}{dt} + \frac{dE_D^{(\mathbf{i})}}{dt}. \tag{6.97}$$

This equation is equivalent to Eq. (5.64) used by Ivanov and Zhigilei. The ionic equation of motion (6.74) reduces to

$$m \frac{d\mathcal{V}^{(\mathbf{i})}}{dt} = \mathcal{F}^{(\mathbf{i})} + \frac{G_{ep}^{(\mathbf{i})} \left(T_e^{(\mathbf{i})} - T_i^{(\mathbf{i})}\right)}{2\,E_{kin}^{(\mathbf{i})}} m\,\mathcal{V}^{(\mathbf{i})}, \tag{6.98}$$

which is equivalent to Eqs. (5.70) and (5.71) used by by Ivanov and Zhigilei. The only difference is that we do not introduce thermal motions of the ions within the subcells. But we could include this in our approach, if necessary.

- As suggested in Ref. [8], one can to some extend include the effect of ballistic electrons by increasing the penetration depth of the laser field. For this, one uses the modified absorption coefficient $\alpha_{abs} = \frac{1}{L_o + L_b}$, where L_o is the optical penetration depth and L_b is the ballistic range.

6.3 Polynomial T_e-Dependent Interatomic Potential Model

To perform meaningful large scale MD simulations of a femtosecond-laser excitation, one needs a T_e-dependent interatomic potential that describes accurately the PES of electrons at temperature T_e. This PES can be obtained ab-initio from T_e-dependent DFT. Beside this, the T_e-dependent interatomic potential must provide a physical electronic specific heat C_e, so that our simulation method based on energy conservation can be applied. However, the T_e-dependent interatomic potentials available up to now for Si provide a rough or even nonphysical description of the PES at increased T_e and do not provide a meaningful C_e (see Sect. 5.4). Reason for this may be the small ab-initio data set used for fitting and the used functional forms, which were developed for electrons in the ground state.

To obtain a T_e-dependent interatomic potential with the desired properties one needs, in principle, two things (see Sect. 5.1.2):

1. One should fit a large data set of ab-initio simulations of various atomic structures to obtain enough information about the PES. One must take into account that, ideally, the interatomic potential should reproduce forces and energies over any pathway on the ab-initio PES. In practice, we recommend to perform reference ab-initio simulations on a small cell that is treatable with T_e-dependent DFT and require that the interatomic potential matches forces and cohesive energies at every simulation step t. For the reference simulations, we advise to use **thin-film geometry** to take both bulk and surface effects into account. Furthermore, the construction of a reliable interatomic potential needs a very good sampling of the available phase space within the reference simulations. In case of laser excitation, the reference simulations should include extreme fluctuations of local atomic configurations. For this reason, we recommend to perform various sets $\mathcal{M}_s$ of reference ab-initio simulations s_ℓ, so that expansion, compression, and melting of the material is included. It is convenient to derive such a set $\mathcal{M}_s(T_e)$ of reference ab-initio simulations s_ℓ at a constant T_e and to calculate such sets for several T_e's starting from 300 K up to a value, where ablation occurs.
2. One should use a flexible functional form for the interatomic potential that can be easily adapted to describe the fitted ab-initio data accurately and efficiently.

The development of a meaningful and flexible functional form is quite challenging: On the one hand, the analytical functional forms of the classical interatomic potentials, which we introduced in Sect. 5.1.1, were developed for electrons in the ground state. It is not straight forward to extend these functional forms to increase accuracy and/or to describe excited electrons. On the other hand, the functional forms of machine learning potentials available up to now, which we briefly introduced in Sect. 5.1.3, are quite complicated and do not allow any physical interpretation, which is one great benefit of the classical interatomic potentials. Beside this, it is very complicated to fit the parameters of the classical and the machine learning interatomic potentials, since the fit error $\mathbb{W}_{err}$ (see Sect. 5.5) is a complicated function of these parameters and exhibits in general a lot of local minima.

6.3.1 Polynomial Functional Form

Therefore, we introduce the analytical functional form described below for T_e-dependent interatomic potentials, which has got the following unique properties:

- It fulfills the necessary properties for a meaningful interatomic potential discussed in Sect. 5.1.
- It is simple and can be physically interpreted.
- It is very flexible and can be always **optimally adjusted** to describe accurately and efficiently the fitted ab-initio data.

- One can always find the optimal parameters that correspond to the **global minimum** of the fit error $\mathbb{W}_{\text{err}}$.

To achieve this, we perform the following construction: To get a physical appealing analytical functional form and to fulfill the necessary properties for a meaningful interatomic potential, we construct our T_e-dependent interatomic potential as a sum over a one-body Φ_0, two-body Φ_2, three-body Φ_3, four-body Φ_4 and an embedded atom potential term Φ_ρ:

$$
\begin{aligned}
\Phi = &\sum_{i=1}^{N_{\text{at}}} \Phi_0(T_e) + \sum_{i=1}^{N_{\text{at}}} \sum_{\substack{j=1 \\ j \neq i \\ r_{ij} < r_2^{(c)}}}^{N_{\text{at}}} \Phi_2(r_{ij}, T_e) \\
&+ \sum_{i=1}^{N_{\text{at}}} \sum_{\substack{j=1 \\ j \neq i \\ r_{ij} < r_3^{(c)}}}^{N_{\text{at}}} \sum_{\substack{k=1 \\ k \neq i, k \neq j \\ r_{ik} < r_3^{(c)}}}^{N_{\text{at}}} \Phi_3\Big(r_{ij}, r_{ik}, \cos(\theta_{ijk}), T_e\Big) \\
&+ \sum_{i=1}^{N_{\text{at}}} \sum_{\substack{j=1 \\ j \neq i \\ r_{ij} < r_4^{(c)}}}^{N_{\text{at}}} \sum_{\substack{k=1 \\ k \neq i, k \neq j \\ r_{ik} < r_4^{(c)}}}^{N_{\text{at}}} \sum_{\substack{\ell=1 \\ \ell \neq i, \ell \neq j, \ell \neq k \\ r_{i\ell} < r_4^{(c)}}}^{N_{\text{at}}} \times \\
&\times \Phi_4\Big(r_{ij}, r_{ik}, r_{i\ell}, \cos(\theta_{ijk}), \cos(\theta_{ij\ell}), \cos(\theta_{ik\ell}), T_e\Big) \\
&+ \sum_{i=1}^{N_{\text{at}}} \Phi_\rho\left(\rho_i^{(2)}, \rho_i^{(3)}, \ldots, \rho_i^{(N_\rho^{(r)})}, T_e\right).
\end{aligned}
\tag{6.99}
$$

Here, $\rho_i^{(2)}, \rho_i^{(3)}, \ldots, \rho_i^{(N_\rho^{(r)})}$ are different measures for the atomic density surrounding atom i discussed below and $r_2^{(c)}, r_3^{(c)}, r_4^{(c)}, r_\rho^{(c)}$ are the individual cutoff radii of the different terms. The one-body term $\Phi_0(T_e)$ describes the Helmholtz free energy of an isolated atom and must be included to be able to calculate C_e from Eq. (5.78) using thermodynamic relations. $\Phi_0(T_e)$ can be directly obtained from T_e-dependent DFT calculations of a single atom. We include the two-body term Φ_2, the three-body term Φ_3, and the four-body term Φ_4 to describe covalent bonding. The Stillinger–Weber potential [9] utilizes a two- and a three-body term and describes successfully Si at the ground state, where covalent bonds are responsible for the tetragonal bonding geometry. We add the embedded atom term Φ_ρ to describe metallic bonding. Embedding functions were already successfully applied to describe metals [10, 11].

To obtain an analytical functional form that is very flexible and simple enough to allow the optimization procedure to find the global minimum of the fit error $\mathbb{W}_{\text{err}}$ in the parameter space, we expand the terms Φ_2, Φ_3, Φ_4 and Φ_ρ into **polynomials**, which can, in principle, reproduce any physically reasonable function at least for a finite set of arguments. To achieve high numerical stability, the polynomials must

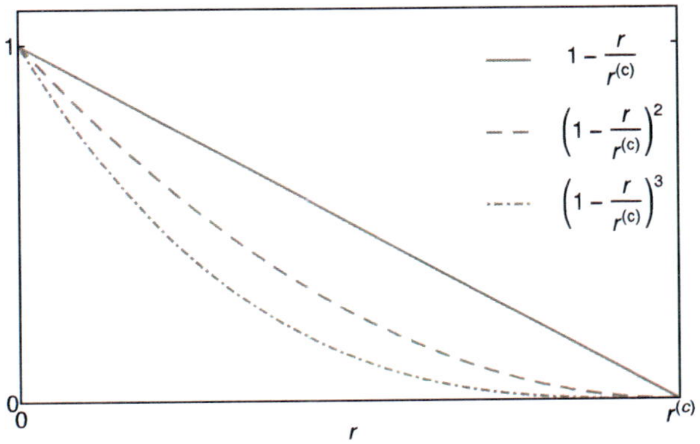

Fig. 6.5 Used term for distances r in the polynomials

be functions of variables lying in the interval $[-1, 1]$. Hence, we use $\cos(\theta)$ with θ being a bond angle as a variable for the polynomials. The interatomic distance r_{ij} is located in the interval $\left(0, r^{(c)}\right)$ and, therefore, cannot be used as a variable for the polynomial. To take also into account that the atomic interaction should decrease to zero at distances reaching the cutoff radii, we use $1 - r/r^{(c)}$ with r being an interatomic distance as a variable for the polynomials (see Fig. 6.5). We start the powers of the latter from degree two to let the interatomic potential and its first derivatives continuously decreasing to zero as distances reach the cutoff radii. Due to this, we construct Φ_2, Φ_3 and Φ_4 as

$$\Phi_2 = \sum_{q=2}^{N_2^{(r)}} c_2^{(q)} \left(1 - \frac{r_{ij}}{r_2^{(c)}}\right)^q, \tag{6.100}$$

$$\Phi_3 = \sum_{q_1=2}^{N_3^{(r)}} \sum_{q_2=q_1}^{N_3^{(r)}} \sum_{q_3=0}^{N_3^{(\theta)}} c_3^{(q_1 q_2 q_3)} \left(1 - \frac{r_{ij}}{r_3^{(c)}}\right)^{q_1} \left(1 - \frac{r_{ik}}{r_3^{(c)}}\right)^{q_2} \left(\cos(\theta_{ijk})\right)^{q_3}, \tag{6.101}$$

$$\Phi_4 = \sum_{q_1=2}^{N_4^{(r)}} \sum_{q_2=q_1}^{N_4^{(r)}} \sum_{q_3=q_2}^{N_4^{(r)}} \sum_{\substack{(q_4, q_5, q_6) \in \\ \mathcal{M}_{N_4^{(\theta)}}(q_1, q_2, q_3)}} c_4^{(q_1 q_2 q_3 q_4 q_5 q_6)} \left(1 - \frac{r_{ij}}{r_4^{(c)}}\right)^{q_1} \left(1 - \frac{r_{ik}}{r_4^{(c)}}\right)^{q_2} \times$$
$$\times \left(1 - \frac{r_{i\ell}}{r_4^{(c)}}\right)^{q_3} \left(\cos(\theta_{ik\ell})\right)^{q_4} \left(\cos(\theta_{ij\ell})\right)^{q_5} \left(\cos(\theta_{ijk})\right)^{q_6}. \tag{6.102}$$

Here, the set $\mathcal{M}_{N_4^{(\theta)}}(q_1, q_2, q_3)$ contains the allowed powers (q_4, q_5, q_6) of the bond angles as a function of $N_4^{(\theta)}$ and the distance powers (q_1, q_2, q_3). $\mathcal{M}_{N_4^{(\theta)}}(q_1, q_2, q_3)$ is constructed as

$$\mathcal{M}_{N_4^{(\theta)}}(q_1, q_2, q_3) = \Big\{ (q_4, q_5, q_6) \,\Big|\, q_4, q_5, q_6 \in \big\{0, 1, \ldots, N_4^{(\theta)}\big\}, \\ \big(q_1 < q_2 \wedge q_2 < q_3\big) \\ \vee \big(q_1 = q_2 \wedge q_2 < q_3 \wedge q_4 \leq q_5\big) \\ \vee \big(q_1 < q_2 \wedge q_2 = q_3 \wedge q_5 \leq q_6\big) \\ \vee \big(q_1 = q_2 \wedge q_2 = q_3 \wedge q_4 \leq q_5 \wedge q_5 \leq q_6\big) \Big\}. \tag{6.103}$$

This condition must be set for achieving that the occurring coefficients $c_4^{(q_1\, q_2\, q_3\, q_4\, q_5\, q_6)}$ are linearly independent. For $q_1 = 2, 3, \ldots, N_\rho^{(r)}$, we construct the measures for the atomic density surrounding atom i as

$$\rho_i^{(q_1)} = \sum_{\substack{j = 1 \\ j \neq i \\ r_{ij} < r_\rho^{(c)}}}^{N_{\text{at}}} \left(1 - \frac{r_{ij}}{r_\rho^{(c)}}\right)^{q_1} . \tag{6.104}$$

We obtain $\rho_i^{(q_1)} \in [0, \infty)$ by definition. To construct Φ_ρ, we take into account the following: If all $\rho_i^{(q_1)}$ are zero, then Φ_ρ should be also zero. Hence, to construct Φ_ρ as simple as possible, we use powers of $\rho/(1+\rho)$, starting from degree one, for expanding the atomic density ρ, since $\rho/(1+\rho)$ is zero for $\rho = 0$ and converges to one for $\rho \to \infty$:

$$\Phi_\rho = \sum_{q_1=2}^{N_\rho^{(r)}} \sum_{q_2=1}^{N_\rho^{(\rho)}} c_\rho^{(q_1\, q_2)} \left(\frac{\rho_i^{(q_1)}}{1 + \rho_i^{(q_1)}}\right)^{q_2} . \tag{6.105}$$

We want to note that we also choose the term $\rho/(1+\rho)$, since it behaves for physically relevant densities ρ in a similar way to the term $\ln(\rho/\rho_0)$ used in literature (see Sect. 5.1.1). In addition, we use several measures for the atomic density surrounding atom i to increase flexibility and to compensate that the functional form (6.104) of $\rho_i^{(q_1)}$ is fixed. By this construction, the coefficients $\{c_2^{(q)}\}$, $\{c_3^{(q_1\, q_2\, q_3)}\}$, $\{c_4^{(q_1\, q_2\, q_3\, q_4\, q_5\, q_6)}\}$, $\{c_\rho^{(q_1\, q_2)}\}$ occur linearly in the interatomic potential Φ and are linearly independent of each other.

We want to note that the usage of polynomials for the construction of interatomic potentials is not a new approach. Bowman introduced a special class of interatomic potentials for describing accurately molecules with a few atoms [12–14]. But his method cannot be applied to develop interatomic potentials describing solids.

6.3.2 Fitting of Coefficients

To fit the interatomic potential coefficients, we have to minimize the error of the interatomic potential in describing the ab-initio forces $\mathbf{f}_i^{(s_\ell)}$ and cohesive energies $E_c^{(s_\ell)}$ of the reference simulations s_ℓ contained in the set $\mathcal{M}_s(T_e)$ at a given T_e. For this, we utilize our fit error $\mathbb{W}_{err}(T_e)$ described in detail in Sect. 5.5.1 (see Eq. (5.95)):

$$\begin{aligned}\mathbb{W}_{err}(T_e) = &\sum_{s_\ell \in \mathcal{M}_s(T_e)} \frac{\mathbb{w}_f^{(s_\ell)}(T_e)}{\sum_{k=1}^{N_t^{(s_\ell)}} \sum_{i=1}^{N_{at}^{(s_\ell)}} \left|\mathbf{f}_i^{(s_\ell)}(T_e, t_k)\right|^2} \times \\ &\times \sum_{k=1}^{N_t^{(s_\ell)}} \sum_{i=1}^{N_{at}^{(s_\ell)}} \left| -\nabla_{\mathbf{r}_i^{(s_\ell)}} \Phi\left(\left\{\mathbf{r}_j^{(s_\ell)}(T_e, t_k)\right\}, T_e\right) - \mathbf{f}_i^{(s_\ell)}(T_e, t_k)\right|^2 \\ &+ \sum_{s_\ell \in \mathcal{M}_s(T_e)} \frac{\mathbb{w}_E^{(s_\ell)}(T_e)}{\sum_{k=1}^{N_t^{(s_\ell)}} \left(E_c^{(s_\ell)}(T_e, t_k)\right)^2} \times \\ &\times \sum_{k=1}^{N_t^{(s_\ell)}} \left(\Phi\left(\left\{\mathbf{r}_j^{(s_\ell)}(T_e, t_k)\right\}, T_e\right) - \sum_{i=1}^{N_{at}^{(s_\ell)}} \Phi_0(T_e) - E_c^{(s_\ell)}(T_e, t_k)\right)^2 .\end{aligned}$$

The N_C coefficients $\left\{c_2^{(q)}\right\}$, $\left\{c_3^{(q_1 q_2 q_3)}\right\}$, $\left\{c_4^{(q_1 q_2 q_3 q_4 q_5 q_6)}\right\}$, $\left\{c_\rho^{(q_1 q_2)}\right\}$ occur linearly in our interatomic potential. Therefore, if we denote the coefficients by $\alpha_1, \ldots, \alpha_{N_C}$ to simplify the notation, the fit error $\mathbb{W}_{err}(T_e)$ can be written as

$$\mathbb{W}_{err}(T_e) = \left|\sum_{k=1}^{N_C} \alpha_k \mathbf{v}_k - \mathbf{V}\right|^2 . \tag{6.106}$$

The vector $\mathbf{V} \in \mathbb{R}^{N_d}$ contains the ab-initio forces and cohesive energies multiplied with the corresponding weights, where the dimension N_d corresponds to the total number of ab-initio forces and cohesive energies. The vector $\mathbf{v}_k \in \mathbb{R}^{N_d}$ contains the forces and cohesive energies of the interatomic potential multiplied with the corresponding weights, if $\alpha_k = 1$ and all other coefficients are zero. We denote all coefficients in the vector $\mathbf{a} = [\alpha_1, \ldots, \alpha_{N_C}]^t$. The optimal coefficients that minimize $\mathbb{W}_{err}(T_e)$ can be easily found by solving a system of linear equations (see Sect. A.2 in the appendix). This system of linear equations must be solved in a numerical stable way, since N_d is quite large in the range of millions due to the fact that a large number of forces and cohesive energies should be fit for an accurate interatomic potential. For this, we define the matrix

$$\mathbf{M} = \left[\mathbf{v}_1, \ldots, \mathbf{v}_{N_C}\right] \in \mathbb{R}^{N_d \times N_C}, \tag{6.107}$$

which column vectors correspond to the vectors $\mathbf{v}_k$. Then, the fit error $\mathbb{W}_{\text{err}}(T_e)$ reads

$$\mathbb{W}_{\text{err}}(T_e) = \left|\mathbf{M}\cdot\mathbf{a}-\mathbf{V}\right|^2. \tag{6.108}$$

We can transform the matrix $\mathbf{M}$ without changing the fit error $\mathbb{W}_{\text{err}}(T_e)$ by applying a set of Givens rotations $\mathbf{G}^{(\text{all})} \in \mathbb{R}^{N_d\times N_C}$. The reason for this is that the norm of any vector does not change, if it is transformed with Givens rotations (more details can be found in Sect. A.2 in the appendix) Thus, we get from Eq. (6.108):

$$\mathbb{W}_{\text{err}}(T_e) = \left|\mathbf{M}\cdot\mathbf{a}-\mathbf{V}\right|^2 = \left|\mathbf{G}^{(\text{all})}\cdot\mathbf{M}\cdot\mathbf{a}-\mathbf{G}^{(\text{all})}\cdot\mathbf{V}\right|^2.$$

For short, we define $\mathbf{M}'' = \mathbf{G}^{(\text{all})}\cdot\mathbf{M}$ and $\mathbf{V}'' = \mathbf{G}^{(\text{all})}\cdot\mathbf{V}$. Now we can choose the set of Givens rotations $\mathbf{G}^{(\text{all})} \in \mathbb{R}^{N_d\times N_C}$ in such a way that the matrix $\mathbf{M}''$ has tridiagonal form:

$$\mathbb{W}_{\text{err}}(T_e) = \left|\mathbf{M}''\cdot\mathbf{a}-\mathbf{V}''\right|^2 = \left\| \begin{bmatrix} M''_{11} & \cdots & * \\ 0 & \ddots & \vdots \\ \vdots & \ddots & M''_{N_C N_C} \\ \vdots & & 0 \\ \vdots & & \vdots \\ 0 & \cdots & 0 \end{bmatrix} \cdot \begin{bmatrix} \alpha_1 \\ \vdots \\ \alpha_{N_C} \end{bmatrix} - \begin{bmatrix} V''_1 \\ \vdots \\ V''_{N_C} \\ V''_{N_C+1} \\ \vdots \\ V''_{N_d} \end{bmatrix} \right\|^2. \tag{6.109}$$

We remember that we search for the optimal coefficients $\alpha_1, \ldots, \alpha_{N_C}$ that minimize the fit error $\mathbb{W}_{\text{err}}(T_e)$. Due to the tridiagonal form of the matrix $\mathbf{M}''$, the optimal coefficients are the solution of the linear system of equations

$$\begin{aligned} M''_{11}\alpha_1 + M''_{12}\alpha_2 + \cdots + M''_{1N_C}\alpha_{N_C} &= V''_1, \\ M''_{22}\alpha_2 + \cdots + M''_{2N_C}\alpha_{N_C} &= V''_2, \\ \ddots \qquad \vdots \qquad &\quad \vdots \\ M''_{N_C N_C}\alpha_{N_C} &= V''_{N_C}, \end{aligned} \tag{6.110}$$

and the minimum of the fit error is given by

$$\mathbb{W}_{\text{err}}(T_e) = \sum_{i=N_C+1}^{N_d} \left(V''_i\right)^2, \tag{6.111}$$

since the solution of Eq. (6.110) sets the first N_C elements of the vector $\mathbf{M}''\cdot\mathbf{a}-\mathbf{V}''$ to zero and the remaining elements are not influenced by the coefficients $\alpha_1, \ldots, \alpha_{N_C}$. In addition, Eq. (6.110) can be easily solved from the bottom up to the top. If a diagonal element M''_{jj} is equal to zero, the used coefficients are linearly dependent and the coefficient α_j cannot be chosen to set the jth element of the vector $\mathbf{M}'' \cdot \mathbf{a} - \mathbf{V}''$ to zero. In such a case, we add $\left(V_j''\right)^2$ to the sum in Eq. (6.111) and set the coefficient α_j to zero.

For given cutoff radii $r_2^{(c)}, r_3^{(c)}, r_4^{(c)}, r_\rho^{(c)}$ and polynomial-degrees $N_2^{(r)}, N_3^{(r)}, N_3^{(\theta)}, N_4^{(r)}, N_4^{(\theta)}, N_\rho^{(r)}, N_\rho^{(\rho)}$, the above mentioned method allows to calculate the minimum of the fit error and the corresponding optimal coefficients $\left\{c_2^{(q)}\right\}, \left\{c_3^{(q_1\, q_2\, q_3)}\right\}, \left\{c_4^{(q_1\, q_2\, q_3\, q_4\, q_5\, q_6)}\right\}, \left\{c_\rho^{(q_1\, q_2)}\right\}$ in a numerical stable way. For this, we have to perform Givens rotations on the matrix $\mathbf{M}$ to transform it to the matrix $\mathbf{M}''$ with triangular form. Transforming also the vector $\mathbf{V}$ with these Givens rotations, we can derive the minimal fit error from Eq. (6.111) by summing up the squares of the elements of the transformed vector $\mathbf{V}''$ with indices $N_C + 1, N_C + 2, \ldots, N_d$.

Moreover, after we have performed the Givens rotations to get $\mathbf{M}''$ and $\mathbf{V}''$ for the given polynomial-degree combination, we can efficiently calculate the minimum of the fit error for all existing polynomial-degree combinations with lower degrees than the initial ones at the given cutoff radii $r_2^{(c)}, r_3^{(c)}, r_4^{(c)}, r_\rho^{(c)}$: For this, we consider the $N_C \times N_C$ sub matrix $\mathbf{M}^{(\text{sub})}$ containing the part with the triangle with non-zero elements of $\mathbf{M}''$, the vector $\mathbf{V}^{(\text{sub})} \in \mathbb{R}^{N_C}$ containing the first N_C elements of $\mathbf{V}''$ and the fit error $\mathbb{W}_{\text{err}}^{(\text{all})}$ calculated from Eq. (6.111) of the initial degree combination. To derive the fit error of any degree combination with lower degrees, i.e., the new total number of coefficients $N_C' < N_C$ becomes smaller, we just skip the corresponding columns in the sub matrix $\mathbf{M}^{(\text{sub})}$ and transform it again with Givens rotations to the triangular form. We also transform $\mathbf{V}^{(\text{sub})}$ with the Givens rotations and obtain the fit error of the considered degree combination by adding to $\mathbb{W}_{\text{err}}^{(\text{all})}$ the squares of the elements of the transformed vector $\mathbf{V}^{(\text{sub})}$ with indices $N_C' + 1, N_C' + 2, \ldots, N_C$. This procedure is quite efficient, since $N_C << N_d$ and, after dropping the corresponding columns in $\mathbf{M}^{(\text{sub})}$, there are still many zeros below the diagonal, because $\mathbf{M}^{(\text{sub})}$ was tridiagonal. Beside this, if only the k columns $N_C - k + 1, N_C - k + 2, \ldots, N_C - 1, N_C$ with $1 \leq k < N_C$ are dropped from $\mathbf{M}^{(\text{sub})}$, the obtained matrix is still tridiagonal and no Givens rotations must be performed. This happens quite often during a degree reduction, since we order the polynomial powers in an ascending order.

We summarize what we established so far: Using the above mentioned method, we are able to calculate efficiently the minimum of the fit error for all existing polynomial-degree combinations up to upper degree limits for given values of the cutoff radii $r_2^{(c)}, r_3^{(c)}, r_4^{(c)}, r_\rho^{(c)}$. To derive the global minimum of the fit error, we also have to optimize the cutoff radii. However, the cutoff radii do not appear linearly in our interatomic potential, so that their optimal values cannot be found directly by a least square fit. Fortunately, in covalent and metallic bonded systems, the interaction between atoms can be neglected beyond a given distance. This can be clearly seen in Fig. 4.5 for Si and in Fig. 4.35 for Sb, where the cohesive energy of several bulk structures is shown as a function of the lattice parameter. Consequently, we can assume an upper limit for the cutoff radii, which specific value we may obtain from the ab-initio calculations of the cohesive energies as a function of the lattice parameter. We can also set the nearest neighbor distance in the equilibrium structure as a lower limit of the cutoff radii, since smaller cutoff radii do not make sense. Thus, physical reasonable cutoff radii lie in a finite interval. Moreover, the fit error $\mathbb{W}_{\text{err}}$ depends continuously on the cutoff radii, since the interatomic potential and its first derivatives needed for the atomic forces converge to zero as distances reach the cutoff radii. Due to this, the fit error $\mathbb{W}_{\text{err}}$ only changes slightly, if the cutoff radii are slightly changed. Consequently, if we check discretely all cutoff radii lying in the previously described finite interval, we find surely the optimal cutoff radii. We should only beware that the used increment for the cutoff-radii is not too big.

To perform the above mentioned cutoff radii check efficiently, we used the following method: We set up the matrix $\mathbf{M}$ for all discrete values of the cutoff radii within the given interval and for all polynomial terms up to given upper degree limits. Then we perform Givens rotations to derive $\mathbf{M}''$ with tridiagonal form and the corresponding vector $\mathbf{V}''$. We calculate the fit error $\mathbb{W}_{\text{err}}^{(\text{all})}$ from Eq. (6.111) for the usage of all cutoff radii at the same time. Then we extract from $\mathbf{M}''$ the columns belonging to a specific cutoff radii combination and derive with the above mentioned method the minimal fit error for all polynomial-degree combinations up to the upper degree limits for this cutoff radii combination. By doing this for all possible cutoff radii combinations, we are able to calculate the global minimum of the fit error and the related optimal cutoff radii for each polynomial-degree combination up to the upper degree limits.

I implemented the FORTRAN program `polypot1_create_LGS`, that first creates the matrix $\mathbf{M}$ and the corresponding vector $\mathbf{V}$, secondly performs the Givens rotations to derive the matrix $\mathbf{M}''$ and the vector $\mathbf{V}''$, thirdly calculates $\mathbb{W}_{\text{err}}^{(\text{all})}$ from Eq. (6.111) and finally writes $\mathbb{W}_{\text{err}}^{(\text{all})}$, only the non-zero elements of the triangle in $\mathbf{M}''$, and the first N_{C} elements of $\mathbf{V}''$ to a file. We implemented the creation of $\mathbf{M}$ using a MPI parallelization over the time steps of the reference simulations similar to the universal fitting program described in Sect. 5.5.4. Also the transformation with Givens rotations is MPI parallelized. In addition, we speed up this program by calculating for each atom its neighbor list at the beginning, which contains, beside the atom number of the neighbor, also the distance, the inverse distance and the normalized distance vector to the neighbor. This usage of a neighbor list is similar to our universal fitting program described in Sect. 5.5.4.

I implemented a second FORTRAN program `polypot1_analyse_LGS`, that firstly loads the data from the previous mentioned file, secondly derives for all cutoff radii combinations the corresponding minimal error for all polynomial-degree combinations up to the upper degree limits, thirdly derives for all polynomial-degree combinations the corresponding optimal cutoff radii and the global minimum of the fit error, and finally writes the latter to a text file. This program is also MPI parallelized. Here, the checking of the different cutoff radii combinations is distributed between different processors, since these calculations are independent of each other.

Using the two FORTRAN programs described above, we can derive the global minimum of the fit error for all reasonable polynomial-degree combinations. It is not necessary to check for this an infinite number of polynomial-degree combinations, since the polynomial expansion of Eqs. (6.100)–(6.105) converges to the ab-initio determined PES relatively quickly with increasing degrees, so that the inclusion of high degrees does not bring significant reduction of the fit error. We can see this for example in Fig. 7.4 of the next chapter, where we apply our method for Si. Consequently, high degrees can be excluded and we only need to check all polynomial-degree combinations up to upper degree limits. These upper degree limits are not known a priori. Therefore, we initially set arbitrary upper degree limits and determined the minimal error for all polynomial-degree combinations up to these limits using the two FORTRAN programs described. From the corresponding calculations, we can check whether the initially chosen upper degree limits were high enough and can eventually increase them. Finally, we obtain a long list containing the global minimum of the fit error and the corresponding optimal cutoff radii for all reasonable polynomial-degree combinations.

6.3.3 Optimal Polynomial-Degree Combination Selection Procedure

Our goal is to describe the ab-initio data most accurately and efficiently with the interatomic potential. To do so, we have to adapt its functional form. The complexity of the different terms Φ_2, Φ_3, Φ_4, and Φ_ρ is controlled by the polynomial-degrees $N_2^{(r)}, N_3^{(r)}, N_3^{(\theta)}, N_4^{(r)}, N_4^{(\theta)}, N_\rho^{(r)}, N_\rho^{(\rho)}$. Concretely, we have to select the polynomial-degrees in such a way that the interatomic potential has got a minimal total number of coefficients N_C to be computational efficient and that the fit error is as small as possible. In addition, the same polynomial-degree combination should be used for all T_e's, since, finally, the potential parameters should be a smooth function of T_e, which allows to calculate C_e from Eq. (5.78) by a second derivative. Therefore, we use the average of $\mathbb{W}_{err}(T_e)$ over several T_e's denoted as $\langle \mathbb{W}_{err} \rangle$ for our further considerations.

To select the polynomial-degree combination of the final interatomic potential showing physical behavior, having minimal $\langle \mathbb{W}_{err} \rangle$ and reasonably low computational cost, we first extracted a relatively small subset $\mathcal{S}_{best}$ of reliable polynomial-degree combinations that exhibit low $\langle \mathbb{W}_{err} \rangle$ combined with a relatively small N_C. To construct $\mathcal{S}_{best}$, we applied an iterative procedure, that selects a series $\Phi^{(1)}, \Phi^{(2)}, \ldots$ of polynomial-degree combinations with $N_C^{(1)} < N_C^{(2)} < \cdots$ and $\langle \mathbb{W}_{err} \rangle^{(1)} > \langle \mathbb{W}_{err} \rangle^{(2)} > \cdots$. We start our procedure by setting $\Phi^{(0)} = \sum_i \Phi_0(T_e)$ as the constant potential with $\langle \mathbb{W}_{err} \rangle^{(0)} = 1$ and $N_C^{(0)} = 0$. On each further step $k = 1, 2, \ldots,$ we select $\Phi^{(k)}$ such that it maximizes the error reduction per added coefficients

$$\frac{\Delta \langle \mathbb{W}_{err} \rangle}{\Delta N_C} = \frac{\langle \mathbb{W}_{err} \rangle^{(k-1)} - \langle \mathbb{W}_{err} \rangle}{N_C - N_C^{(k-1)}} \tag{6.112}$$

among all polynomial-degree combinations with $N_C > N_C^{(k-1)}$. We continue the iterations until step k for which $\Phi^{(k)}$ reaches the upper degree limits. $\mathcal{S}_{best}$ contains not only the elements of the series $\Phi^{(1)}, \Phi^{(2)}, \ldots$ but also all polynomial-degree combinations that exhibit a 10% lower value of $\Delta \langle \mathbb{W}_{err} \rangle / \Delta N_C$ than $\Phi^{(k)}$. In this way, we developed a self-learning method that constructs the small subset $\mathcal{S}_{best}$ containing optimally adjusted polynomial-degree combinations. The small number of polynomial-degree combinations in $\mathcal{S}_{best}$ allows to perform a manual analysis of the physical properties of the corresponding interatomic potentials to select the final one. I implemented a FORTRAN program that first calculates the average fit error $\langle \mathbb{W}_{\text{err}} \rangle$ from the text files containing the list of global fit errors for the physical reasonable degree combinations at a given T_{e}, and secondly performs the iterative procedure to construct $\mathcal{S}_{\text{best}}$.

6.3.4 Easy Evaluation via Power Lists

To calculate the energy and forces from our interatomic potential, one has to determine the occurring powers in the polynomials. This can be efficiently done by using lists that store the powers in ascending order. Such a list can be created just by using multiplications. For this, one sets the original term as the first element. The square of the term is the second element, which is derived by multiplying the first element with the term. Then, any further needed element with higher power is derived by multiplying the previously calculated element by the term. Finally, using these power lists, one can determine the energy and the forces of our interatomic potential without direct calculations of powers. Moreover, if the interatomic distance vectors, distances and cosines of bond angles are calculated, something one has to do for any interatomic potential, only multiplications and additions, except for the $N_\rho^{(\rho)}$ divisions to derive the terms $\rho_i^{(q_1)}/\left(1+\rho_i^{(q_1)}\right)$, are needed to evaluate our interatomic potential and to derive the atomic forces using the power lists. The constants $-1/r_2^{(\text{c})}$, $-1/r_3^{(\text{c})}$, $-1/r_4^{(\text{c})}$, $-1/r_\rho^{(\text{c})}$ need to be calculated only once at the beginning. This procedure is easy to implement and much faster than the calculation of the occurring powers on the fly.

However, the most time consuming part are the calculations of the four-body and three-body terms. Therefore, to speed up the calculations, we derived a method to calculate the energies and forces of the four-body and three-body terms as for a two-body term with the help of the spherical harmonics.

6.3.5 Efficient Evaluation of the Three-Body Term

We write the total three-body term as

$$\Phi_3^{(\text{tot})} = \sum_{i=1}^{N_{\text{at}}} \sum_{\substack{j=1 \\ j \neq i \\ r_{ij} < r_3^{(\text{c})}}}^{N_{\text{at}}} \sum_{\substack{k=1 \\ k \neq i, k \neq j \\ r_{ik} < r_3^{(\text{c})}}}^{N_{\text{at}}} \Phi_3\left(r_{ij}, r_{ik}, \cos(\theta_{ijk}), T_{\text{e}}\right). \tag{6.113}$$

Using the Legendre polynomials $\wp_{q_3}(x)$, we can cast the three-body term in the form as

$$\Phi_3 = \sum_{q_1=2}^{N_3^{(r)}} \sum_{q_2=q_1}^{N_3^{(r)}} \sum_{q_3=0}^{N_3^{(\theta)}} \tilde{c}_3^{(q_1\, q_2\, q_3)} \left(1 - \frac{r_{ij}}{r_3^{(c)}}\right)^{q_1} \left(1 - \frac{r_{ik}}{r_3^{(c)}}\right)^{q_2} \wp_{q_3}\Big(\cos(\theta_{ijk})\Big) \quad (6.114)$$

with adapted coefficients $\tilde{c}_3^{(q_1\, q_2\, q_3)}$ compared to Eq. (6.101). Here $\wp_{q_3}\Big(\cos(\theta_{ijk})\Big)$ denotes the q_3th Legendre polynomial in $\cos(\theta_{ijk})$. We have

$$\Phi_3^{(\mathrm{tot})} = \underbrace{\sum_{i=1}^{N_{\mathrm{at}}} \sum_{\substack{j=1 \\ j \neq i \\ r_{ij} < r_3^{(c)}}}^{N_{\mathrm{at}}} \sum_{\substack{k=1 \\ k \neq i \\ r_{ik} < r_3^{(c)}}}^{N_{\mathrm{at}}} \Phi_3\Big(r_{ij}, r_{ik}, \cos(\theta_{ijk}), T_e\Big)}_{=:\tilde{\Phi}_3^{(\mathrm{tot})}} - \sum_{i=1}^{N_{\mathrm{at}}} \sum_{\substack{j=1 \\ j \neq i \\ r_{ij} < r_3^{(c)}}}^{N_{\mathrm{at}}} \Phi_3\Big(r_{ij}, r_{ij}, \underbrace{\cos(\theta_{ijj})}_{=1}, T_e\Big). \quad (6.115)$$

Due to $\theta_{ijj} = 0$, we obtain $\cos(\theta_{ijj}) = 1$ and the last term transforms to a two-body term:

$$\Phi_3(r_{ij}, r_{ij}, 1, T_e) = \sum_{q_1=2}^{N_3^{(r)}} \sum_{q_2=q_1}^{N_3^{(r)}} \left(1 - \frac{r_{ij}}{r_3^{(c)}}\right)^{q_1+q_2} \sum_{q_3=0}^{N_3^{(\theta)}} \tilde{c}_3^{(q_1\, q_2\, q_3)}. \quad (6.116)$$

If we define proper coefficients $\tilde{c}_3^{(q_1)}$, we can transform the above two-body term to

$$\Phi_3(r_{ij}, r_{ij}, 1, T_e) = \sum_{q_1=4}^{2\,N_3^{(r)}} \tilde{c}_3^{(q_1)} \left(1 - \frac{r_{ij}}{r_3^{(c)}}\right)^{q_1}. \quad (6.117)$$

Mathematical Theorem 6.1 (Addition theorem for the spherical harmonics) *We consider two arbitrary unit vectors $\hat{\mathbf{r}}_1, \hat{\mathbf{r}}_2 \in \mathbb{R}^3$. Let $Y_{\ell m}(\hat{\mathbf{r}})$ be a spherical harmonics and $\wp_\ell(x)$ be the ℓth Legendre polynomial. Then one obtains*

$$\wp_\ell\left(\hat{\mathbf{r}}_1^t \cdot \hat{\mathbf{r}}_2\right) = \frac{4\pi}{2\,\ell + 1} \sum_{m=-\ell}^{\ell} Y_{\ell m}\left(\hat{\mathbf{r}}_1\right) Y_{\ell m}^*\left(\hat{\mathbf{r}}_2\right). \quad (6.118)$$

Here, $Y_{\ell m}^*\left(\hat{\mathbf{r}}_2\right)$ denotes the complex conjugated of $Y_{\ell m}\left(\hat{\mathbf{r}}_2\right)$.

The proof can be found in Ref. [15]. To use this theorem, we consider the two unit vectors $\hat{\mathbf{r}}_{ij}, \hat{\mathbf{r}}_{ik} \in \mathbb{R}^3$ and calculate the corresponding spherical coordinate angles $\vartheta_{ij}, \phi_{ij}$ and $\vartheta_{ik}, \phi_{ik}$ from

$$\vartheta_{ij} = \arccos\left(\frac{r_{ijz}}{r_{ij}}\right), \quad \vartheta_{ik} = \arccos\left(\frac{r_{ikz}}{r_{ik}}\right), \tag{6.119}$$

$$\phi_{ij} = \arg(\mathbf{r}_{ij}), \qquad \phi_{ik} = \arg(\mathbf{r}_{ik}). \tag{6.120}$$

Here

$$\arg(\mathbf{r}) = \begin{cases} \arctan\left(\frac{r_y}{r_x}\right) & r_x > 0 \\ \arctan\left(\frac{r_y}{r_x}\right) + \pi & r_x < 0 \wedge r_y \geq 0 \\ \arctan\left(\frac{r_y}{r_x}\right) - \pi & r_x < 0 \wedge r_y < 0 \\ \frac{\pi}{2} & r_x = 0 \wedge r_y > 0 \\ -\frac{\pi}{2} & r_x = 0 \wedge r_y < 0 \\ 0 & r_x = 0 \wedge r_y = 0 \end{cases} \tag{6.121}$$

denotes the argument function. Now we can write the spherical harmonics as a function of the spherical coordinate angles:

$$Y_{\ell m}(\hat{\mathbf{r}}_{ij}) \equiv Y_{\ell m}(\vartheta_{ij}, \phi_{ij}), \qquad Y_{\ell m}(\hat{\mathbf{r}}_{ik}) \equiv Y_{\ell m}(\vartheta_{ik}, \phi_{ik}).$$

Since we have $\cos(\theta_{ijk}) = \hat{\mathbf{r}}_{ij}^{\mathrm{t}} \cdot \hat{\mathbf{r}}_{ik}$, we get from Eq. (6.118):

$$\wp_{q_3}\Big(\cos(\theta_{ijk})\Big) = \frac{4\pi}{2\,q_3 + 1} \sum_{q_4=-q_3}^{q_3} Y_{q_3 q_4}(\vartheta_{ij}, \phi_{ij})\, Y^*_{q_3 q_4}(\vartheta_{ik}, \phi_{ik}). \tag{6.122}$$

Using this, we obtain

$$\tilde{\Phi}_3^{(\mathrm{tot})} = \sum_{i=1}^{N_{\mathrm{at}}} \sum_{\substack{j=1 \\ j \neq i \\ r_{ij} < r_3^{(\mathrm{c})}}}^{N_{\mathrm{at}}} \sum_{\substack{k=1 \\ k \neq i \\ r_{ik} < r_3^{(\mathrm{c})}}}^{N_{\mathrm{at}}} \Phi_3\Big(r_{ij}, r_{ik}, \cos(\theta_{ijk}), T_{\mathrm{e}}\Big)$$

$$\stackrel{(6.114)}{=} \sum_{i=1}^{N_{\mathrm{at}}} \sum_{\substack{j=1 \\ j \neq i \\ r_{ij} < r_3^{(\mathrm{c})}}}^{N_{\mathrm{at}}} \sum_{\substack{k=1 \\ k \neq i \\ r_{ik} < r_3^{(\mathrm{c})}}}^{N_{\mathrm{at}}} \sum_{q_1=2}^{N_3^{(r)}} \sum_{q_2=q_1}^{N_3^{(r)}} \sum_{q_3=0}^{N_3^{(\theta)}} \tilde{c}_3^{(q_1\, q_2\, q_3)} \left(1 - \frac{r_{ij}}{r_3^{(\mathrm{c})}}\right)^{q_1} \left(1 - \frac{r_{ik}}{r_3^{(\mathrm{c})}}\right)^{q_2} \times$$

$$\times\ \wp_{q_3}\Big(\cos(\theta_{ijk})\Big)$$

$$\overset{(6.118)}{=} \sum_{i=1}^{N_{\rm at}} \sum_{\substack{j=1\\ j\neq i\\ r_{ij}<r_3^{(c)}}}^{N_{\rm at}} \sum_{\substack{k=1\\ k\neq i\\ r_{ik}<r_3^{(c)}}}^{N_{\rm at}} \sum_{q_1=2}^{N_3^{(r)}} \sum_{q_2=q_1}^{N_3^{(r)}} \sum_{q_3=0}^{N_3^{(\theta)}} \tilde{c}_3^{(q_1\,q_2\,q_3)} \left(1-\frac{r_{ij}}{r_3^{(c)}}\right)^{q_1} \left(1-\frac{r_{ik}}{r_3^{(c)}}\right)^{q_2} \times$$

$$\times\ \frac{4\pi}{2\,q_3+1} \sum_{q_4=-q_3}^{q_3} Y_{q_3q_4}(\vartheta_{ij},\phi_{ij})\, Y^*_{q_3q_4}(\vartheta_{ik},\phi_{ik})$$

$$= \sum_{i=1}^{N_{\rm at}} \sum_{q_1=2}^{N_3^{(r)}} \sum_{q_2=q_1}^{N_3^{(r)}} \sum_{q_3=0}^{N_3^{(\theta)}} \tilde{c}_3^{(q_1\,q_2\,q_3)} \frac{4\pi}{2\,q_3+1} \times$$

$$\times \sum_{q_4=-q_3}^{q_3} \sum_{\substack{j=1\\ j\neq i\\ r_{ij}<r_3^{(c)}}}^{N_{\rm at}} \sum_{\substack{k=1\\ k\neq i\\ r_{ik}<r_3^{(c)}}}^{N_{\rm at}} \left(1-\frac{r_{ij}}{r_3^{(c)}}\right)^{q_1} Y_{q_3q_4}(\vartheta_{ij},\phi_{ij}) \left(1-\frac{r_{ik}}{r_3^{(c)}}\right)^{q_2} \times$$

$$\times\ Y^*_{q_3q_4}(\vartheta_{ik},\phi_{ik})$$

$$= \sum_{i=1}^{N_{\rm at}} \sum_{q_1=2}^{N_3^{(r)}} \sum_{q_2=q_1}^{N_3^{(r)}} \sum_{q_3=0}^{N_3^{(\theta)}} \tilde{c}_3^{(q_1\,q_2\,q_3)} \frac{4\pi}{2\,q_3+1} \times$$

$$\times \sum_{q_4=-q_3}^{q_3} \left(\sum_{\substack{j=1\\ j\neq i\\ r_{ij}<r_3^{(c)}}}^{N_{\rm at}} \left(1-\frac{r_{ij}}{r_3^{(c)}}\right)^{q_1} Y_{q_3q_4}(\vartheta_{ij},\phi_{ij}) \right) \times$$

$$\times \left(\sum_{\substack{k=1\\ k\neq i\\ r_{ik}<r_3^{(c)}}}^{N_{\rm at}} \left(1-\frac{r_{ik}}{r_3^{(c)}}\right)^{q_2} Y^*_{q_3q_4}(\vartheta_{ik},\phi_{ik}) \right)$$

and finally

$$\boxed{\tilde{\Phi}_3^{(\mathrm{tot})} = \sum_{i=1}^{N_{\mathrm{at}}} \sum_{q_1=2}^{N_3^{(r)}} \sum_{q_2=q_1}^{N_3^{(r)}} \sum_{q_3=0}^{N_3^{(\theta)}} \tilde{c}_3^{(q_1\,q_2\,q_3)} \frac{4\pi}{2\,q_3+1} \sum_{q_4=-q_3}^{q_3} \chi_{q_1 q_3 q_4}^{(3i)} \left(\chi_{q_2 q_3 q_4}^{(3i)}\right)^{*}.} \tag{6.123}$$

The above equations allows us to calculate the entire three-body term only as a sum over atoms i and j. For this, we have to derive previously the terms

$$\boxed{\chi_{q_1 q_3 q_4}^{(3i)} = \sum_{\substack{j=1 \\ j \neq i \\ r_{ij} < r_3^{(c)}}}^{N_{\mathrm{at}}} \left(1 - \frac{r_{ij}}{r_3^{(c)}}\right)^{q_1} Y_{q_3 q_4}(\vartheta_{ij}, \phi_{ij})} \tag{6.124}$$

for atom i as a sum over the neighboring atoms j. To calculate the atomic forces, we apply the gradient $\nabla_{\mathbf{r}_k}$ on Eq. (6.123):

$$\begin{aligned} \nabla_{\mathbf{r}_k} \tilde{\Phi}_3^{(\mathrm{tot})} &= \sum_{i=1}^{N_{\mathrm{at}}} \sum_{q_1=2}^{N_3^{(r)}} \sum_{q_2=}^{N_3^{(r)}} \sum_{q_3=0}^{N_3^{(\theta)}} \tilde{c}_3^{(q_1\,q_2\,q_3)} \frac{4\pi}{2\,q_3+1} \times \\ &\times \sum_{q_4=-q_3}^{q_3} \left(\left(\chi_{q_2 q_3 q_4}^{(3i)}\right)^{*} \nabla_{\mathbf{r}_k} \chi_{q_1 q_3 q_4}^{(3i)} + \chi_{q_1 q_3 q_4}^{(3i)} \nabla_{\mathbf{r}_k} \left(\chi_{q_2 q_3 q_4}^{(3i)}\right)^{*} \right). \end{aligned} \tag{6.125}$$

In spherical coordinates, the gradient denotes [16]

$$\nabla E = \hat{\mathbf{e}}_r \frac{\partial E}{\partial r} + \hat{\mathbf{e}}_\vartheta \frac{1}{r} \frac{\partial E}{\partial \vartheta} + \hat{\mathbf{e}}_\phi \frac{1}{r \sin(\vartheta)} \frac{\partial E}{\partial \phi}. \tag{6.126}$$

In Cartesian coordinates, the unit vectors of the spherical coordinates are given by [16]

$$\hat{\mathbf{e}}_{r_{ij}} = \begin{bmatrix} \cos(\phi_{ij}) \sin(\vartheta_{ij}) \\ \sin(\phi_{ij}) \sin(\vartheta_{ij}) \\ \cos(\vartheta_{ij}) \end{bmatrix} = \hat{\mathbf{r}}_{ij}, \tag{6.127}$$

$$\hat{\mathbf{e}}_{\vartheta_{ij}} = \begin{bmatrix} \cos(\phi_{ij}) \cos(\vartheta_{ij}) \\ \sin(\phi_{ij}) \cos(\vartheta_{ij}) \\ -\sin(\vartheta_{ij}) \end{bmatrix}, \tag{6.128}$$

$$\hat{\mathbf{e}}_{\phi_{ij}} = \begin{bmatrix} -\sin(\phi_{ij}) \\ \cos(\phi_{ij}) \\ 0 \end{bmatrix}. \tag{6.129}$$

Thus, we obtain for atom i

$$\nabla_{\mathbf{r}_i} \chi^{(3i)}_{q_1 q_3 q_4} = \sum_{\substack{j=1 \\ j \neq i \\ r_{ij} < r_3^{(c)}}}^{N_{at}} \nabla_{\mathbf{r}_{ij}} \left(\left(1 - \frac{r_{ij}}{r_3^{(c)}} \right)^{q_1} Y_{q_3 q_4}(\vartheta_{ij}, \phi_{ij}) \right) \tag{6.130}$$

and for atom j

$$\nabla_{\mathbf{r}_j} \chi^{(3i)}_{q_1 q_3 q_4} = -\nabla_{\mathbf{r}_{ij}} \left(\left(1 - \frac{r_{ij}}{r_3^{(c)}} \right)^{q_1} Y_{q_3 q_4}(\vartheta_{ij}, \phi_{ij}) \right) \tag{6.131}$$

with

$$\begin{aligned} &\nabla_{\mathbf{r}_{ij}} \left(\left(1 - \frac{r_{ij}}{r_3^{(c)}} \right)^{q_1} Y_{q_3 q_4}(\vartheta_{ij}, \phi_{ij}) \right) \\ &= -\hat{\mathbf{r}}_{ij} \frac{q_1 \left(1 - \frac{r_{ij}}{r_3^{(c)}}\right)^{q_1 - 1}}{r_3^{(c)}} Y_{q_3 q_4}(\vartheta_{ij}, \phi_{ij}) + \hat{\mathbf{e}}_{\vartheta_{ij}} \frac{\left(1 - \frac{r_{ij}}{r_3^{(c)}}\right)^{q_1}}{r_{ij}} \frac{\partial Y_{q_3 q_4}(\vartheta_{ij}, \phi_{ij})}{\partial \vartheta_{ij}} \\ &+ \hat{\mathbf{e}}_{\phi_{ij}} \frac{\left(1 - \frac{r_{ij}}{r_3^{(c)}}\right)^{q_1}}{r_{ij} \sin(\vartheta_{ij})} \frac{\partial Y_{q_3 q_4}(\vartheta_{ij}, \phi_{ij})}{\partial \phi_{ij}}. \end{aligned} \tag{6.132}$$

The derivatives of the spherical harmonics are given by[1]

$$\begin{aligned} \frac{\partial Y_{q_3 q_4}(\vartheta_{ij}, \phi_{ij})}{\partial \vartheta_{ij}} =& q_4 \frac{\cos(\vartheta_{ij})}{\sin(\vartheta_{ij})} Y_{q_3 q_4}(\vartheta_{ij}, \phi_{ij}) + e^{-i\phi} \times \\ & \times \sqrt{(q_3 - q_4)(q_3 + q_4 + 1)}\, Y_{q_3 q_4 + 1}(\vartheta_{ij}, \phi_{ij}) \end{aligned} \tag{6.133}$$

and

$$\frac{\partial Y_{q_3 q_4}(\vartheta_{ij}, \phi_{ij})}{\partial \phi_{ij}} = \mathrm{i}\, q_4\, Y_{q_3 q_4}(\vartheta_{ij}, \phi_{ij}). \tag{6.134}$$

Here, one should take into account that $Y_{q_3 q_4}(\vartheta_{ij}, \phi_{ij}) = 0$ for $q_4 > q_3$ and $q_4 < -q_3$. To numerate the coefficients of the spherical harmonics for an efficient programming, we utilize the following function:

$$\mathcal{N}(q_3, q_4) = (q_3 + 1)^2 - q_3 + q_4 = q_3\, q_3 + q_3 + q_4 + 1, \tag{6.135}$$

since $q_3 = 0, 1, \ldots$, and, for each value of q_3, one has the $2\, q_3 + 1$ values $-q_3, -q_3 + 1, \ldots, q_3$ for q_4. The above method allows us to calculate the energy and forces of the three-body term like for a two-body term.

---

[1] The equations were generated by MATHEMATICA.

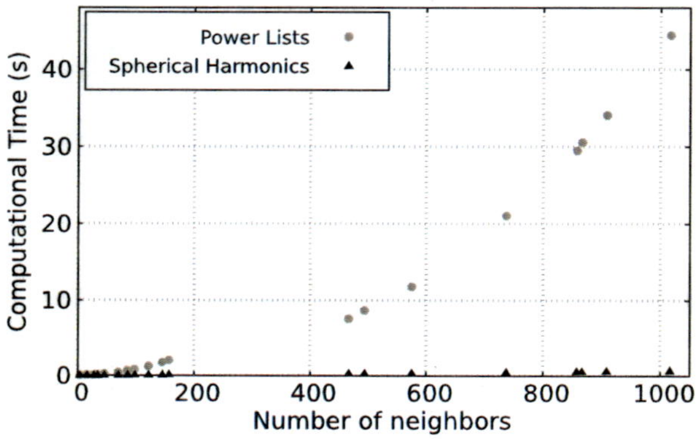

Fig. 6.6 Computational time to calculate the three-body term of the Si interatomic potential for the different calculation methods

We implemented this method and performed a benchmark test for our Si interatomic potential, which we developed in Sect. 7.1.3 of the next chapter. For this, we considered a simulation cell containing 1000 atoms forming the ideal diamond-like structure. This cell consists of $5 \times 5 \times 5$ primitive cells. We utilized periodic boundary conditions and varied the lattice parameter. We calculated the energy and forces from the three-body term on a Intel I5-8250U processor and did not use any parallelization.

In Fig. 6.6, we present the computational time needed to calculate the three-body term as a function of the number of neighbors, which must be taken into account to evaluate the interatomic potential, for the method using spherical harmonics and the easy method using power lists described in Sect. 6.3.4. One can clearly see that the computational time grows quadratically as a function of the neighbor number for the power lists in contrast to the spherical harmonics, which only grows linearly, as expected. For a small number of neighbors, both methods need the same computational time. At greater numbers of neighbors, the method using spherical harmonics provides a large speed up. For Si under ambient conditions, one needs to consider 16 neighbors to evaluate the three-body term. However, at such a small number of neighbors, the spherical harmonics method provides just a moderate speedup.

This method was originally developed by Biswas and Hamann in 1985 [17] for three-body terms with a symmetric distance dependence:

$$\Phi_3\Big(r_{ij}, r_{ik}, \cos(\theta_{ijk})\Big) = \Phi_3\Big(r_{ik}, r_{ij}, \cos(\theta_{ijk})\Big).$$

We extended this method, so that it can be applied to our non-symmetric three-body term. We even went one step further and extended the method to four-body terms.

6.3.6 Efficient Evaluation of the Four-Body Term

We denote the total four-body term as

$$\Phi_4^{(\text{tot})} = \sum_{i=1}^{N_{\text{at}}} \sum_{\substack{j=1 \\ j \neq i \\ r_{ij} < r_4^{(c)}}}^{N_{\text{at}}} \sum_{\substack{k=1 \\ k \neq i, k \neq j \\ r_{ik} < r_4^{(c)}}}^{N_{\text{at}}} \sum_{\substack{\ell=1 \\ \ell \neq i, \ell \neq j, \ell \neq k \\ r_{i\ell} < r_4^{(c)}}}^{N_{\text{at}}} \times$$

$$\times \Phi_4\Big(r_{ij}, r_{ik}, r_{i\ell}, \cos(\theta_{ijk}), \cos(\theta_{ij\ell}), \cos(\theta_{ik\ell}), T_e\Big). \tag{6.136}$$

Using the Legendre polynomials, we can denote the four-body term as

$$\Phi_4 = \sum_{q_1=2}^{N_4^{(r)}} \sum_{q_2=q_1}^{N_4^{(r)}} \sum_{q_3=q_2}^{N_4^{(r)}} \sum_{\substack{(q_4, q_5, q_6) \in \\ \mathcal{M}_{N_4^{(\theta)}}(q_1, q_2, q_3)}} \tilde{c}_4^{(q_1 q_2 q_3 q_4 q_5 q_6)} \left(1 - \frac{r_{ij}}{r_4^{(c)}}\right)^{q_1} \left(1 - \frac{r_{ik}}{r_4^{(c)}}\right)^{q_2} \times$$

$$\times \left(1 - \frac{r_{i\ell}}{r_4^{(c)}}\right)^{q_3} \wp_{q_4}\Big(\cos(\theta_{ijk})\Big) \wp_{q_5}\Big(\cos(\theta_{ij\ell})\Big) \wp_{q_6}\Big(\cos(\theta_{ik\ell})\Big). \tag{6.137}$$

with adapted coefficients $\tilde{c}_4^{(q_1 q_2 q_3 q_4 q_5 q_6)}$ compared to Eq. (6.102). $\wp_{q_4}\Big(\cos(\theta_{ijk})\Big)$ denotes the q_4th Legendre polynomial in $\cos(\theta_{ijk})$. Using $\theta_{ikj} = \theta_{ijk}$, we have:

$$\Phi_4^{(\text{tot})} = \sum_{i=1}^{N_{\text{at}}} \sum_{\substack{j=1 \\ j \neq i \\ r_{ij} < r_4^{(c)}}}^{N_{\text{at}}} \sum_{\substack{k=1 \\ k \neq i, k \neq j \\ r_{ik} < r_4^{(c)}}}^{N_{\text{at}}} \sum_{\substack{\ell=1 \\ \ell \neq i \\ r_{i\ell} < r_4^{(c)}}}^{N_{\text{at}}} \Phi_4\Big(r_{ij}, r_{ik}, r_{i\ell}, \cos(\theta_{ijk}), \cos(\theta_{ij\ell}), \cos(\theta_{ik\ell}), T_e\Big)$$

$$- \sum_{i=1}^{N_{\text{at}}} \sum_{\substack{j=1 \\ j \neq i \\ r_{ij} < r_4^{(c)}}}^{N_{\text{at}}} \sum_{\substack{k=1 \\ k \neq i, k \neq j \\ r_{ik} < r_4^{(c)}}}^{N_{\text{at}}} \Phi_4\Big(r_{ij}, r_{ik}, r_{ik}, \cos(\theta_{ijk}), \cos(\theta_{ijk}), \cos(\theta_{ikk}), T_e\Big)$$

$$- \sum_{i=1}^{N_{\text{at}}} \sum_{\substack{j=1 \\ j \neq i \\ r_{ij} < r_4^{(c)}}}^{N_{\text{at}}} \sum_{\substack{k=1 \\ k \neq i, k \neq j \\ r_{ik} < r_4^{(c)}}}^{N_{\text{at}}} \Phi_4\Big(r_{ij}, r_{ik}, r_{ij}, \cos(\theta_{ijk}), \cos(\theta_{ijj}), \cos(\theta_{ijk}), T_e\Big)$$

$$= \sum_{i=1}^{N_{\text{at}}} \sum_{\substack{j=1 \\ j \neq i \\ r_{ij} < r_4^{(c)}}}^{N_{\text{at}}} \sum_{\substack{k=1 \\ k \neq i \\ r_{ik} < r_4^{(c)}}}^{N_{\text{at}}} \sum_{\substack{\ell=1 \\ \ell \neq i \\ r_{i\ell} < r_4^{(c)}}}^{N_{\text{at}}} \Phi_4\Big(r_{ij}, r_{ik}, r_{i\ell}, \cos(\theta_{ijk}), \cos(\theta_{ij\ell}), \cos(\theta_{ik\ell}), T_e\Big)$$

$$
\begin{aligned}
&-2\sum_{i=1}^{N_{\mathrm{at}}}\sum_{\substack{j=1\\ j\neq i\\ r_{ij}<r_4^{(\mathrm{c})}}}^{N_{\mathrm{at}}} \Phi_4\Big(r_{ij}, r_{ij}, r_{ij}, \cos(\theta_{ijj}), \cos(\theta_{ijj}), \cos(\theta_{ijj}), T_{\mathrm{e}}\Big)\\
&-\sum_{i=1}^{N_{\mathrm{at}}}\sum_{\substack{j=1\\ j\neq i\\ r_{ij}<r_4^{(\mathrm{c})}}}^{N_{\mathrm{at}}}\sum_{\substack{k=1\\ k\neq i\\ r_{ik}<r_4^{(\mathrm{c})}}}^{N_{\mathrm{at}}} \Phi_4\Big(r_{ij}, r_{ik}, r_{ik}, \cos(\theta_{ijk}), \cos(\theta_{ijk}), \cos(\theta_{ikk}), T_{\mathrm{e}}\Big)\\
&-\sum_{i=1}^{N_{\mathrm{at}}}\sum_{\substack{j=1\\ j\neq i\\ r_{ij}<r_4^{(\mathrm{c})}}}^{N_{\mathrm{at}}}\sum_{\substack{k=1\\ k\neq i\\ r_{ik}<r_4^{(\mathrm{c})}}}^{N_{\mathrm{at}}} \Phi_4\Big(r_{ij}, r_{ik}, r_{ij}, \cos(\theta_{ijk}), \cos(\theta_{ijj}), \cos(\theta_{ijk}), T_{\mathrm{e}}\Big)\\
&-\sum_{i=1}^{N_{\mathrm{at}}}\sum_{\substack{j=1\\ j\neq i\\ r_{ij}<r_4^{(\mathrm{c})}}}^{N_{\mathrm{at}}}\sum_{\substack{k=1\\ k\neq i\\ r_{ik}<r_4^{(\mathrm{c})}}}^{N_{\mathrm{at}}} \Phi_4\Big(r_{ij}, r_{ij}, r_{ik}, \cos(\theta_{ijj}), \cos(\theta_{ijk}), \cos(\theta_{ijk}), T_{\mathrm{e}}\Big)
\end{aligned}
$$

$$
\begin{aligned}
&=\underbrace{\sum_{i=1}^{N_{\mathrm{at}}}\sum_{\substack{j=1\\ j\neq i\\ r_{ij}<r_4^{(\mathrm{c})}}}^{N_{\mathrm{at}}}\sum_{\substack{k=1\\ k\neq i\\ r_{ik}<r_4^{(\mathrm{c})}}}^{N_{\mathrm{at}}}\sum_{\substack{\ell=1\\ \ell\neq i\\ r_{i\ell}<r_4^{(\mathrm{c})}}}^{N_{\mathrm{at}}} \Phi_4\Big(r_{ij}, r_{ik}, r_{i\ell}, \cos(\theta_{ijk}), \cos(\theta_{ij\ell}), \cos(\theta_{ik\ell}), T_{\mathrm{e}}\Big)}_{=:\tilde{\Phi}_4^{(\mathrm{tot})}}\\
&-2\sum_{i=1}^{N_{\mathrm{at}}}\sum_{\substack{j=1\\ j\neq i\\ r_{ij}<r_4^{(\mathrm{c})}}}^{N_{\mathrm{at}}} \Phi_4\Big(r_{ij}, r_{ij}, r_{ij}, \underbrace{\cos(\theta_{ijj})}_{=1}, \underbrace{\cos(\theta_{ijj})}_{=1}, \underbrace{\cos(\theta_{ijj})}_{=1}, T_{\mathrm{e}}\Big)\\
&-\sum_{i=1}^{N_{\mathrm{at}}}\sum_{\substack{j=1\\ j\neq i\\ r_{ij}<r_4^{(\mathrm{c})}}}^{N_{\mathrm{at}}}\sum_{\substack{k=1\\ k\neq i\\ r_{ik}<r_4^{(\mathrm{c})}}}^{N_{\mathrm{at}}} \Phi_4\Big(r_{ij}, r_{ik}, r_{ik}, \cos(\theta_{ijk}), \cos(\theta_{ijk}), \underbrace{\cos(\theta_{ikk})}_{=1}, T_{\mathrm{e}}\Big)\\
&-\sum_{i=1}^{N_{\mathrm{at}}}\sum_{\substack{j=1\\ j\neq i\\ r_{ij}<r_4^{(\mathrm{c})}}}^{N_{\mathrm{at}}}\sum_{\substack{k=1\\ k\neq i\\ r_{ik}<r_4^{(\mathrm{c})}}}^{N_{\mathrm{at}}} \Phi_4\Big(r_{ik}, r_{ij}, r_{ik}, \cos(\theta_{ijk}), \underbrace{\cos(\theta_{ikk})}_{=1}, \cos(\theta_{ijk}), T_{\mathrm{e}}\Big)
\end{aligned}
$$

$$-\sum_{i=1}^{N_{at}} \sum_{\substack{j=1 \\ j \neq i \\ r_{ij} < r_4^{(c)}}}^{N_{at}} \sum_{\substack{k=1 \\ k \neq i \\ r_{ik} < r_4^{(c)}}}^{N_{at}} \Phi_4\Big(r_{ik}, r_{ik}, r_{ij}, \underbrace{\cos(\theta_{ikk})}_{=1}, \cos(\theta_{ijk}), \cos(\theta_{ijk}), T_e\Big) \tag{6.138}$$

It follows from $\theta_{ikk} = 0$ that $\cos(\theta_{ikk}) = 1$ and $\wp_{q_z}\Big(\cos(\theta_{ikk})\Big) = 1$ for all Legendre Polyonomials. The same holds for $\theta_{ijj} = 0$. Therefore, we obtain

$$\Phi_4(r_{ij}, r_{ij}, r_{ij}, 1, 1, 1, T_e) = \sum_{q_1=2}^{N_4^{(r)}} \sum_{q_2=q_1}^{N_4^{(r)}} \sum_{q_3=q_2}^{N_4^{(r)}} \left(1 - \frac{r_{ij}}{r_4^{(c)}}\right)^{q_1+q_2+q_3} \times$$
$$\times \sum_{\substack{(q_4, q_5, q_6) \in \\ \mathcal{M}_{N_4^{(\theta)}}(q_1, q_2, q_3)}} \tilde{c}_4^{(q_1 q_2 q_3 q_4 q_5 q_6)}. \tag{6.139}$$

If we define proper coefficients $\tilde{c}_4^{(q_1)}$, we can transform the above two-body term to

$$\Phi_4(r_{ij}, r_{ij}, r_{ij}, 1, 1, 1, T_e) = \sum_{q_1=6}^{3N_4^{(r)}} \tilde{c}_4^{(q_1)} \left(1 - \frac{r_{ij}}{r_4^{(c)}}\right)^{q_1}. \tag{6.140}$$

Furthermore, we get

$$\Phi_4\Big(r_{ij}, r_{ik}, r_{ik}, \cos(\theta_{ijk}), \cos(\theta_{ijk}), 1, T_e\Big)$$
$$= \sum_{q_1=2}^{N_4^{(r)}} \sum_{q_2=q_1}^{N_4^{(r)}} \sum_{q_3=q_2}^{N_4^{(r)}} \left(1 - \frac{r_{ij}}{r_4^{(c)}}\right)^{q_1} \left(1 - \frac{r_{ik}}{r_4^{(c)}}\right)^{q_2-q_3} \times$$
$$\times \sum_{\substack{(q_4, q_5, q_6) \in \\ \mathcal{M}_{N_4^{(\theta)}}(q_1, q_2, q_3)}} \tilde{c}_4^{(q_1 q_2 q_3 q_4 q_5 q_6)} \wp_{q_4}\Big(\cos(\theta_{ijk})\Big) \wp_{q_5}\Big(\cos(\theta_{ijk})\Big), \tag{6.141}$$

$$\Phi_4\Big(r_{ik}, r_{ij}, r_{ik}, \cos(\theta_{ijk}), 1, \cos(\theta_{ijk}), T_e\Big)$$
$$= \sum_{q_1=2}^{N_4^{(r)}} \sum_{q_2=q_1}^{N_4^{(r)}} \sum_{q_3=q_2}^{N_4^{(r)}} \left(1 - \frac{r_{ij}}{r_4^{(c)}}\right)^{q_2} \left(1 - \frac{r_{ik}}{r_4^{(c)}}\right)^{q_1+q_3} \times$$
$$\times \sum_{\substack{(q_4, q_5, q_6) \in \\ \mathcal{M}_{N_4^{(\theta)}}(q_1, q_2, q_3)}} \tilde{c}_4^{(q_1 q_2 q_3 q_4 q_5 q_6)} \wp_{q_4}\Big(\cos(\theta_{ijk})\Big) \wp_{q_6}\Big(\cos(\theta_{ijk})\Big), \tag{6.142}$$

$$\Phi_4\Big(r_{ik}, r_{ik}, r_{ij}, 1, \cos(\theta_{ijk}), \cos(\theta_{ijk}), T_e\Big)$$

$$= \sum_{q_1=2}^{N_4^{(r)}} \sum_{q_2=q_1}^{N_4^{(r)}} \sum_{q_3=q_2}^{N_4^{(r)}} \left(1 - \frac{r_{ij}}{r_4^{(c)}}\right)^{q_3} \left(1 - \frac{r_{ik}}{r_4^{(c)}}\right)^{q_1+q_2} \times$$
$$\times \sum_{\substack{(q_4,\, q_5,\, q_6) \in \\ \mathcal{M}_{N_4^{(\theta)}}(q_1,\, q_2,\, q_3)}} \tilde{c}_4^{(q_1\, q_2\, q_3\, q_4\, q_5\, q_6)} \wp_{q_5}\Big(\cos(\theta_{ijk})\Big) \wp_{q_6}\Big(\cos(\theta_{ijk})\Big). \tag{6.143}$$

If we define proper coefficients $\tilde{c}_4^{(q_1\, q_2\, q_3)}$, we can transform the above three-body terms to

$$\Phi_4\Big(r_{ij}, r_{ik}, r_{ik}, \cos(\theta_{ijj}), \cos(\theta_{ijk}), 1, T_e\Big) + \Phi_4\Big(r_{ik}, r_{ij}, r_{ik}, \cos(\theta_{ijk}), 1, \cos(\theta_{ijj}), T_e\Big)$$
$$+ \Phi_4\Big(r_{ik}, r_{ik}, r_{ij}, 1, \cos(\theta_{ijk}), \cos(\theta_{ijj}), T_e\Big)$$

$$= \sum_{q_1=2}^{2\,N_4^{(r)}} \sum_{q_2=q_1}^{2\,N_4^{(r)}} \sum_{q_3=0}^{2\,N_4^{(\theta)}} \tilde{c}_4^{(q_1\, q_2\, q_3)} \left(1 - \frac{r_{ij}}{r_4^{(c)}}\right)^{q_1} \left(1 - \frac{r_{ik}}{r_4^{(c)}}\right)^{q_2} \wp_{q_3}\Big(\cos(\theta_{ijk})\Big). \tag{6.144}$$

This three-body term can be evaluated like a two-body term using the method described in Sect. 6.3.5. We obtain further from the addition theorem Eq. (6.118) of the spherical harmonics

$$\tilde{\Phi}_4^{(\text{tot})} = \sum_{i=1}^{N_{\text{at}}} \sum_{\substack{j=1 \\ j \neq i \\ r_{ij} < r_4^{(c)}}}^{N_{\text{at}}} \sum_{\substack{k=1 \\ k \neq i \\ r_{ik} < r_4^{(c)}}}^{N_{\text{at}}} \sum_{\substack{\ell=1 \\ \ell \neq i \\ r_{i\ell} < r_4^{(c)}}}^{N_{\text{at}}} \sum_{q_1=2}^{N_4^{(r)}} \sum_{q_2=q_1}^{N_4^{(r)}} \sum_{q_3=q_2}^{N_4^{(r)}} \sum_{\substack{(q_4,\, q_5,\, q_6) \in \\ \mathcal{M}_{N_4^{(\theta)}}(q_1,\, q_2,\, q_3)}} \tilde{c}_4^{(q_1\, q_2\, q_3\, q_4\, q_5\, q_6)} \times$$
$$\times \left(1 - \frac{r_{ij}}{r_4^{(c)}}\right)^{q_1} \left(1 - \frac{r_{ik}}{r_4^{(c)}}\right)^{q_2} \left(1 - \frac{r_{i\ell}}{r_4^{(c)}}\right)^{q_3} \times$$
$$\times \wp_{q_4}\Big(\cos(\theta_{ijk})\Big) \wp_{q_5}\Big(\cos(\theta_{ij\ell})\Big) \wp_{q_6}\Big(\cos(\theta_{ik\ell})\Big)$$

$$= \sum_{i=1}^{N_{\text{at}}} \sum_{\substack{j=1 \\ j \neq i \\ r_{ij} < r_4^{(c)}}}^{N_{\text{at}}} \sum_{\substack{k=1 \\ k \neq i \\ r_{ik} < r_4^{(c)}}}^{N_{\text{at}}} \sum_{\substack{\ell=1 \\ \ell \neq i \\ r_{i\ell} < r_4^{(c)}}}^{N_{\text{at}}} \sum_{q_1=2}^{N_4^{(r)}} \sum_{q_2=q_1}^{N_4^{(r)}} \sum_{q_3=q_2}^{N_4^{(r)}} \sum_{\substack{(q_4,\, q_5,\, q_6) \in \\ \mathcal{M}_{N_4^{(\theta)}}(q_1,\, q_2,\, q_3)}} \tilde{c}_4^{(q_1\, q_2\, q_3\, q_4\, q_5\, q_6)} \times$$
$$\times \left(1 - \frac{r_{ij}}{r_4^{(c)}}\right)^{q_1} \left(1 - \frac{r_{ik}}{r_4^{(c)}}\right)^{q_2} \left(1 - \frac{r_{i\ell}}{r_4^{(c)}}\right)^{q_3} \times$$
$$\times \left(\frac{4\pi}{2\,q_4 + 1} \sum_{q_7=-q_4}^{q_4} Y_{q_4 q_7}(\vartheta_{ij}, \varphi_{ij})\, Y^*_{q_4 q_7}(\vartheta_{ik}, \varphi_{ik})\right) \times$$
$$\times \left(\frac{4\pi}{2\,q_5 + 1} \sum_{q_8=-q_5}^{q_5} Y_{q_5 q_8}(\vartheta_{ij}, \varphi_{ij})\, Y^*_{q_5 q_8}(\vartheta_{i\ell}, \varphi_{i\ell})\right) \times$$

$$\times \left(\frac{4\pi}{2\,q_6+1} \sum_{q_9=-q_6}^{q_6} Y_{q_6 q_9}(\vartheta_{ik},\varphi_{ik})\, Y^*_{q_6 q_9}(\vartheta_{i\ell},\varphi_{i\ell}) \right)$$

$$= \sum_{i=1}^{N_{\mathrm{at}}} \sum_{q_1=2}^{N_4^{(r)}} \sum_{q_2=q_1}^{N_4^{(r)}} \sum_{q_3=q_2}^{N_4^{(r)}} \sum_{\substack{(q_4,q_5,q_6)\in \\ \mathcal{M}_{N_4^{(\theta)}}(q_1,q_2,q_3)}} \frac{\tilde{c}_4^{\,(q_1\,q_2\,q_3\,q_4\,q_5\,q_6)}\, 64\,\pi^3}{(2\,q_4+1)\,(2\,q_5+1)\,(2\,q_6+1)} \times$$

$$\times \sum_{q_7=-q_4}^{q_4} \sum_{q_8=-q_5}^{q_5} \sum_{q_9=-q_6}^{q_6} \sum_{\substack{j=1\\ j\neq i\\ r_{ij}<r_4^{(c)}}}^{N_{\mathrm{at}}} \sum_{\substack{k=1\\ k\neq i\\ r_{ik}<r_4^{(c)}}}^{N_{\mathrm{at}}} \sum_{\substack{\ell=1\\ \ell\neq i\\ r_{i\ell}<r_4^{(c)}}}^{N_{\mathrm{at}}} \times$$

$$\times \left(1-\frac{r_{ij}}{r_4^{(c)}}\right)^{q_1} Y_{q_4 q_7}(\vartheta_{ij},\varphi_{ij})\, Y_{q_5 q_8}(\vartheta_{ij},\varphi_{ij}) \times$$

$$\times \left(1-\frac{r_{ik}}{r_4^{(c)}}\right)^{q_2} Y^*_{q_4 q_7}(\vartheta_{ik},\varphi_{ik})\, Y_{q_6 q_9}(\vartheta_{ik},\varphi_{ik}) \times$$

$$\times \left(1-\frac{r_{i\ell}}{r_4^{(c)}}\right)^{q_3} Y^*_{q_5 q_8}(\vartheta_{i\ell},\varphi_{i\ell})\, Y^*_{q_6 q_9}(\vartheta_{i\ell},\varphi_{i\ell})$$

$$= \sum_{i=1}^{N_{\mathrm{at}}} \sum_{q_1=2}^{N_4^{(r)}} \sum_{q_2=q_1}^{N_4^{(r)}} \sum_{q_3=q_2}^{N_4^{(r)}} \sum_{\substack{(q_4,q_5,q_6)\in \\ \mathcal{M}_{N_4^{(\theta)}}(q_1,q_2,q_3)}} \frac{\tilde{c}_4^{\,(q_1\,q_2\,q_3\,q_4\,q_5\,q_6)}\, 64\,\pi^3}{(2\,q_4+1)\,(2\,q_5+1)\,(2\,q_6+1)} \times$$

$$\times \sum_{q_7=-q_4}^{q_4} \sum_{q_8=-q_5}^{q_5} \sum_{q_9=-q_6}^{q_6} \left(\sum_{\substack{j=1\\ j\neq i\\ r_{ij}<r_4^{(c)}}}^{N_{\mathrm{at}}} \left(1-\frac{r_{ij}}{r_4^{(c)}}\right)^{q_1} Y_{q_4 q_7}(\vartheta_{ij},\varphi_{ij})\, Y_{q_5 q_8}(\vartheta_{ij},\varphi_{ij}) \right) \times$$

$$\times \left(\sum_{\substack{k=1\\ k\neq i\\ r_{ik}<r_4^{(c)}}}^{N_{\mathrm{at}}} \left(1-\frac{r_{ik}}{r_4^{(c)}}\right)^{q_2} Y^*_{q_4 q_7}(\vartheta_{ik},\varphi_{ik})\, Y_{q_6 q_9}(\vartheta_{ik},\varphi_{ik}) \right) \times$$

$$\times \left(\sum_{\substack{\ell=1\\ \ell\neq i\\ r_{i\ell}<r_4^{(c)}}}^{N_{\mathrm{at}}} \left(1-\frac{r_{i\ell}}{r_4^{(c)}}\right)^{q_3} Y^*_{q_5 q_8}(\vartheta_{i\ell},\varphi_{i\ell})\, Y^*_{q_6 q_9}(\vartheta_{i\ell},\varphi_{i\ell}) \right)$$

and finally

$$
\begin{aligned}
\tilde{\Phi}_4^{(\mathrm{tot})} = & \sum_{i=1}^{N_{\mathrm{at}}} \sum_{q_1=2}^{N_4^{(r)}} \sum_{q_2=q_1}^{N_4^{(r)}} \sum_{q_3=q_2}^{N_4^{(r)}} \sum_{\substack{(q_4,q_5,q_6) \in \\ \mathcal{M}_{N_4^{(\theta)}}(q_1,q_2,q_3)}} \times \\
& \times \frac{\tilde{c}_4^{(q_1\, q_2\, q_3\, q_4\, q_5\, q_6)}\, 64\,\pi^3}{(2\,q_4+1)\,(2\,q_5+1)\,(2\,q_6+1)} \times \\
& \times \sum_{q_7=-q_4}^{q_4} \sum_{q_8=-q_5}^{q_5} \sum_{q_9=-q_6}^{q_6} \times \\
& \times \left(\chi^{(4i)}_{q_1 q_4 q_7 q_5 q_8}\, \kappa^{(4i)}_{q_2 q_4 q_7 q_6 q_9}\, \left(\chi^{(4i)}_{q_3 q_5 q_8 q_6 q_9}\right)^* \right)
\end{aligned}
\tag{6.145}
$$

using

$$
\chi^{(4i)}_{q_1 q_4 q_7 q_5 q_8} = \sum_{\substack{j=1 \\ j \neq i \\ r_{ij} < r_4^{(\mathrm{c})}}}^{N_{\mathrm{at}}} \left(1 - \frac{r_{ij}}{r_4^{(\mathrm{c})}}\right)^{q_1} Y_{q_4 q_7}(\vartheta_{ij}, \varphi_{ij})\, Y_{q_5 q_8}(\vartheta_{ij}, \varphi_{ij})
\tag{6.146}
$$

and

$$
\kappa^{(4i)}_{q_2 q_4 q_7 q_6 q_9} = \sum_{\substack{j=1 \\ j \neq i \\ r_{ij} < r_4^{(\mathrm{c})}}}^{N_{\mathrm{at}}} \left(1 - \frac{r_{ij}}{r_4^{(\mathrm{c})}}\right)^{q_2} Y^*_{q_4 q_7}(\vartheta_{ij}, \varphi_{ij})\, Y_{q_6 q_9}(\vartheta_{ij}, \varphi_{ij}).
\tag{6.147}
$$

The occurring terms $\chi^{(4i)}_{q_1 q_4 q_7 q_5 q_8}$, $\kappa^{(4i)}_{q_2 q_4 q_7 q_6 q_9}$, $\left(\chi^{(4i)}_{q_3 q_5 q_8 q_6 q_9}\right)^*$ are previously calculated for each atom i from the sum over the neighboring atoms j. To calculate the force on atom k, we just apply the gradient $\nabla_{\mathbf{r}_k}$ to Eq. (6.145):

$$
\begin{aligned}
\nabla_{\mathbf{r}_k} \tilde{\Phi}_4 = & \sum_{i=1}^{N_{\mathrm{at}}} \sum_{q_1=2}^{N_4^{(r)}} \sum_{q_2=q_1}^{N_4^{(r)}} \sum_{q_3=q_2}^{N_4^{(r)}} \sum_{\substack{(q_4,q_5,q_6) \in \\ \mathcal{M}_{N_4^{(\theta)}}(q_1,q_2,q_3)}} \frac{\tilde{c}_4^{(q_1\, q_2\, q_3\, q_4\, q_5\, q_6)}\, 64\,\pi^3}{(2\,q_4+1)\,(2\,q_5+1)\,(2\,q_6+1)} \times \\
& \times \sum_{q_7=-q_4}^{q_4} \sum_{q_8=-q_5}^{q_5} \sum_{q_9=-q_6}^{q_6} \left(\kappa^{(4i)}_{q_2 q_4 q_7 q_6 q_9} \left(\chi^{(4i)}_{q_3 q_5 q_8 q_6 q_9}\right)^* \nabla_{\mathbf{r}_k} \chi^{(4i)}_{q_1 q_4 q_7 q_5 q_8} \right. \\
& + \chi^{(4i)}_{q_1 q_4 q_7 q_5 q_8} \left(\chi^{(4i)}_{q_3 q_5 q_8 q_6 q_9}\right)^* \nabla_{\mathbf{r}_k} \kappa^{(4i)}_{q_2 q_4 q_7 q_6 q_9} \\
& \left. + \chi^{(4i)}_{q_1 q_4 q_7 q_5 q_8}\, \kappa^{(4i)}_{q_2 q_4 q_7 q_6 q_9}\, \nabla_{\mathbf{r}_k} \left(\chi^{(4i)}_{q_3 q_5 q_8 q_6 q_9}\right)^* \right).
\end{aligned}
\tag{6.148}
$$

We obtain for atom i

$$\nabla_{\mathbf{r}_i} \chi^{(4i)}_{q_1 q_4 q_7 q_5 q_8} = \sum_{\substack{j=1 \\ j \neq i \\ r_{ij} < r_4^{(c)}}}^{N_{\text{at}}} \nabla_{\mathbf{r}_{ij}} \left(\left(1 - \frac{r_{ij}}{r_4^{(c)}} \right)^{q_1} Y_{q_4 q_7}(\vartheta_{ij}, \varphi_{ij})\, Y_{q_5 q_8}(\vartheta_{ij}, \varphi_{ij}) \right) \tag{6.149}$$

and for atom j

$$\nabla_{\mathbf{r}_j} \chi^{(3i)}_{q_1 q_3 q_4} = -\nabla_{\mathbf{r}_{ij}} \left(\left(1 - \frac{r_{ij}}{r_4^{(c)}} \right)^{q_1} Y_{q_4 q_7}(\vartheta_{ij}, \varphi_{ij})\, Y_{q_5 q_8}(\vartheta_{ij}, \varphi_{ij}) \right) \tag{6.150}$$

with

$$\begin{aligned}
&\nabla_{\mathbf{r}_{ij}} \left(\left(1 - \frac{r_{ij}}{r_4^{(c)}} \right)^{q_1} Y_{q_4 q_7}(\vartheta_{ij}, \varphi_{ij})\, Y_{q_5 q_8}(\vartheta_{ij}, \varphi_{ij}) \right) \\
&= -\,\hat{\mathbf{r}}_{ij} \frac{q_1 \left(1 - \frac{r_{ij}}{r_4^{(c)}} \right)^{q_1 - 1}}{r_4^{(c)}} Y_{q_4 q_7}(\vartheta_{ij}, \varphi_{ij})\, Y_{q_5 q_8}(\vartheta_{ij}, \varphi_{ij}) \\
&\quad + \hat{\mathbf{e}}_{\vartheta_{ij}} \frac{\left(1 - \frac{r_{ij}}{r_4^{(c)}} \right)^{q_1}}{r_{ij}} \times \\
&\quad \times \left(\frac{\partial Y_{q_4 q_7}(\vartheta_{ij}, \phi_{ij})}{\partial \vartheta_{ij}} Y_{q_5 q_8}(\vartheta_{ij}, \varphi_{ij}) + Y_{q_4 q_7}(\vartheta_{ij}, \varphi_{ij}) \frac{\partial Y_{q_5 q_8}(\vartheta_{ij}, \phi_{ij})}{\partial \vartheta_{ij}} \right) \\
&\quad + \hat{\mathbf{e}}_{\phi_{ij}} \frac{\left(1 - \frac{r_{ij}}{r_4^{(c)}} \right)^{q_1}}{r_{ij}\, \sin(\vartheta_{ij})} \times \\
&\quad \times \left(\frac{\partial Y_{q_4 q_7}(\vartheta_{ij}, \phi_{ij})}{\partial \phi_{ij}} Y_{q_5 q_8}(\vartheta_{ij}, \varphi_{ij}) + Y_{q_4 q_7}(\vartheta_{ij}, \varphi_{ij}) \frac{\partial Y_{q_5 q_8}(\vartheta_{ij}, \phi_{ij})}{\partial \phi_{ij}} \right).
\end{aligned} \tag{6.151}$$

The gradient of $\kappa^{(4i)}_{q_2 q_4 q_7 q_6 q_9}$ is analogously calculated. The above method allows us to calculate the energy and forces of the four-body term like for a two-body term. Therefore, we expect that this method will scale linearly with the number of considered neighbors in contrast to the method using power lists, which scales cubically.

6.4 Summary

An intense femtosecond laser-pulse typically excites the electrons to a high electronic temperature T_e while the ions remain mostly cold. The high T_e changes dramatically the potential energy surface (PES), on which the ions move. The electron-phonon coupling (EPC) induces an energy transfer from the electrons to the ions over time, so that the electronic and ionic temperature become finally equal.

In this chapter, a novel method is presented to simulate simultaneously the excited PES and the EPC in such a way, that the energy, which is removed from the electronic subsystem by EPC, is transferred to the ions. The excited PES corresponds to the potential part of the internal energy of the ions and to the Helmholtz free energy of the electrons, since the electrons are modeled by a temperature T_e. This PES is responsible for the ionic forces that are related to interatomic bonding. However, the energy conservation can only be formulated with the internal energy. Thus in the case of the electrons, the Helmholtz free energy has to be transformed to the internal energy. An integral expression is presented, which is able to perform this transformation such that the potential part of the internal energy of the ions and the internal energy of the electrons as well as the ionic forces are still derived from the Helmholtz free energy of the electrons. Similar to the commonly used simulation setups, the time propagation of the electronic temperature is controlled by the total change of the internal energy of the electrons. This energy change is given by the energy absorption rate from an external laser field, the energy transfer rate to the ions due to the different electronic and ionic temperatures and, if considered, the local heat conduction related to different local electronic temperatures. The ballistic electrons can be taken into account by increasing the penetration depth of the laser. The ionic equations of motions are directly derived from the energy conservation. The novel method can be used in MD simulations, in which the excited PES is obtained from T_e-dependent DFT or from a T_e-dependent interatomic potential. Also a numerical implementation of the novel method within the Velocity Verlet algorithm is presented. In the novel method, the electronic specific heat must be derived from the excited PES using thermodynamic relations. Thus, if a T_e-dependent interatomic potential is used, it must provide a meaningful electronic specific heat.

The available T_e-dependent interatomic potentials provide, so far, only a rough or even a wrong description of the excited PES obtained from T_e-dependent DFT. Therefore in the second part of the chapter, a novel method is presented to construct highly accurate T_e-dependent interatomic potentials. These interatomic potentials consists of a sum of local interaction terms that are commonly used in the classical interatomic potentials. These terms are expanded into polynomials to be very flexible and to be able to find the global minimum in the parameter space during fitting. The degrees of the polynomials can be optimally adapted to describe the ab-initio data most accurately and efficiently. The reason for this is that the global minimum of the fit error can be determined for all existing reasonable interatomic potentials of the new type. Furthermore, an iterative procedure is developed that selects a small subset of interatomic potentials with optimally adjusted polynomial degrees from the complete list of reasonable interatomic potentials. The small number of candidates for the interatomic potential function in the subset can be manually analyzed to select the final interatomic potential.

The obtained interatomic potentials can be very efficiently evaluated. Using the addition theorem of the spherical harmonics, a calculation method was derived, which allows to evaluate the three- and four-body interaction terms of the interatomic potential as efficiently as a two-body potential.

References

1. N.D. Mermin, Phys. Rev. **137**, A1441 (1965). https://doi.org/10.1103/PhysRev.137.A1441. https://link.aps.org/doi/10.1103/PhysRev.137.A1441
2. R.M. Wentzcovitch, J.L. Martins, P.B. Allen, Phys. Rev. B **45**, 11372 (1992). https://doi.org/10.1103/PhysRevB.45.11372. https://link.aps.org/doi/10.1103/PhysRevB.45.11372
3. M. Weinert, J.W. Davenport, Phys. Rev. B **45**, 13709 (1992). https://doi.org/10.1103/PhysRevB.45.13709. https://link.aps.org/doi/10.1103/PhysRevB.45.13709
4. P.L. Silvestrelli, A. Alavi, M. Parrinello, D. Frenkel, Phys. Rev. Lett. **77**, 3149 (1996). https://doi.org/10.1103/PhysRevLett.77.3149. https://link.aps.org/doi/10.1103/PhysRevLett.77.3149
5. E.S. Zijlstra, A. Kalitsov, T. Zier, M.E. Garcia, Advanced Materials **25**(39), 5605 (2013). https://doi.org/10.1002/adma201302559. http://dx.doi.org/10.1002/adma201302559
6. E.S. Zijlstra, A. Kalitsov, T. Zier, M.E. Garcia, Phys. Rev. X **3**, 011005 (2013). https://doi.org/10.1103/PhysRevX.3.011005. http://link.aps.org/doi/10.1103/PhysRevX.3.011005
7. J.C. Tannehill, D.A. Anderson, R.H. Pletcher, *Computational Fluid Mechanics and Heat Transfer*, 2nd edn. (Taylor & Francis, Milton Park, 1997)
8. J. Hohlfeld, S.S. Wellershoff, J. Güdde, U. Conrad, V. Jähnke, E. Matthias, Chemical Physics **251**(1), 237 (2000). https://doi.org/10.1016/S0301-0104(99)00330-4. http://www.sciencedirect.com/science/article/pii/S0301010499003304
9. F.H. Stillinger, T.A. Weber, Phys. Rev. B **31**, 5262 (1985). https://doi.org/10.1103/PhysRevB.31.5262. http://link.aps.org/doi/10.1103/PhysRevB.31.5262
10. M.S. Daw, M.I. Baskes, Phys. Rev. Lett. **50**, 1285 (1983). https://doi.org/10.1103/PhysRevLett.50.1285. http://link.aps.org/doi/10.1103/PhysRevLett.50.1285
11. V.V. Zhakhovskii, N.A. Inogamov, Y.V. Petrov, S.I. Ashitkov, K. Nishihara, Appl. Surface Sci. **255**(24), 9592 (2009). https://doi.org/10.1016/j.apsusc.2009.04.082. http://www.sciencedirect.com/science/article/pii/S0169433209004346
12. J.M. Bowman, Accounts of Chemical Research **19**(7), 202 (1986). https://doi.org/10.1021/ar00127a002
13. B.J. Braams, J.M. Bowman, Int. Rev. Phys. Chem. **28**(4), 577 (2009). https://doi.org/10.1080/01442350903234923
14. R. Conte, C. Qu, J.M. Bowman, J. Chem. Theory Comput. **11**(4), 1631 (2015). (PMID: 26574372). https://doi.org/10.1021/acs.jctc.5b00091
15. K. Atkinson, W. Han, *Spherical Harmonics and Approximations on the Unit Sphere: An Introduction* (Springer, Berlin, 2012)
16. D.J. Griffiths, *Introduction to Electrodynamics*, 4th edn. (Pearson, 2013)
17. R. Biswas, D.R. Hamann, Phys. Rev. Lett. **55**, 2001 (1985). https://doi.org/10.1103/PhysRevLett.55.2001. http://link.aps.org/doi/10.1103/PhysRevLett.55.2001

Chapter 7
Study of Femtosecond-Laser Excited Si

Abstract In spite of a surge of new candidates, silicon (Si) is still the most relevant material for producing integrated circuits used in common electronic devices like computers. One reason for this is that Si can be produced as high-purity Si monocrystalline wafers that can be patterned at the nanoscale to form integrated circuits. Today, this patterning is usually done by photolithography, which requires several process steps. Femtosecond-laser patterning for a precise surface structuring is possible in just a single process step [1–6]. The reason behind this precise material processing lies in the ultrashort interaction time of the femtosecond laser-pulse with the matter, which drives it to an excited state far away from thermodynamic equilibrium. To simulate such a femtosecond-laser material processing, one has to take the bond changes due to the excited electrons into account and one has to simulate typically $10^8 - 10^9$ atoms [6]. This can only be done with a T_e-dependent interatomic potential [7–14]. Since the T_e-dependent interatomic potentials up to now available for Si provide only a rough or rather incorrect description of femtosecond laser-excited Si (see Sect. 5.4), a T_e-dependent interatomic potential $\Phi^{(Si)}(T_e)$ for Si was developed in this chapter. For the construction, the interatomic potential model and the fitting procedure described in great detail in Sect. 6.3 was applied based on extensive MD simulations of Si at various T_e's obtained from T_e-dependent DFT. The interatomic potential $\Phi^{(Si)}(T_e)$ describes sufficiently accurate the PES for a wide range of T_e's. Beside this, $\Phi^{(Si)}(T_e)$ can even describe electronic properties like the electronic specific heat at the required degree of accuracy. Additionally it is shown that such an accurate description of the excited PES cannot possibly be reached by just fitting the coefficients of commonly used reported interatomic potentials for Si to the extensive MD simulations. In the second part of the chapter, $\Phi^{(Si)}(T_e)$ was employed at MD simulations of femtosecond-laser excited Si taking the excited PES and the EPC into account using the methods described in Sect. 6.1. The femtosecond-laser induced Bragg peak decay was derived and an excellent agreement with available experimental data was found. The femtosecond-laser induced melting of a Si film was simulated and the influences of the excited PES and the EPC on the dynamics were analyzed. It was also studied how the additional consideration of the EPC in addition to the excited PES influences the thermal phonon antisqueezing and the non-thermal melting in bulk Si. $\Phi^{(Si)}(T_e)$ exhibits a melting temperature that is close to the DFT value, however, both deviate from the experimental value. Therefore, in the end of

© The Author(s), under exclusive license to Springer Nature Switzerland AG 2021

B. Bauerhenne, *Materials Interaction with Femtosecond Lasers*,
https://doi.org/10.1007/978-3-030-85135-4_7

the chapter, a modification of the potential parameters of $\Phi^{(Si)}(T_e)$ are presented, so that the experimental melting temperature is reproduced without changing other properties at higher T_e's.

7.1 T_e-Dependent Interatomic Potential for Si

7.1.1 Ab-Initio Reference Simulations Used for Fitting

The T_e-dependent interatomic potential for Si of Shokeen & Schelling provides only a rough description and the one of Darkins et al. an incorrect description of Si at increased T_e's, as laid out in Sects. 5.4.1 and 5.4.2, respectively. One of the reasons for this is that they only use, a small number of ideal bulk crystal structures for fitting, for which they only considered the cohesive energies. Therefore, we utilized a large ab-initio data set for fitting, which consists of cohesive energies and forces of several reference simulations obtained from T_e-dependent DFT. To take both bulk and surface effects into account, we performed the reference ab-initio simulations using the simulation cell with 320 Si atoms forming a thin film that we introduced in Sect. 4.4.7. To get the necessary sampling of the PES, we considered eleven T_e's, which are listed in Table 7.1 and performed the following sets of reference ab-initio simulations $\{s_k\}$ at each T_e:

s_1: We performed a MD simulation of the thin film at each T_e as described in Sect. 4.4.7. We consider these simulations to get insights in the atomic kinetics at this T_e. We initialized the atomic structure thermalized at an ionic temperature of $T_i = 316$ K (1 mHa) and used a time step of 2 fs. For fitting, we considered 500 times steps of the MD simulation, which corresponds to a total simulation time of 1 ps.

s_2: We took the thin film initiated at $T_i = 316$ K, compressed it step by step in z-direction and calculated the corresponding energy and forces for each obtained atomic structure at each studied T_e. We included this reference simulation to cover a compression of the thin film, since the laser excitation of solids can lead to high local positive pressures with high local densities.

s_3: We took the time-dependent coordinates obtained from the reference MD simulation s_1 at $T_e = 25262$ K (80 mHa) and used them to calculate the energies and forces at all $T_e < 25262$ K. We included this reference simulation to cover an expansion of the material at all studied T_e's, since the laser excitation of solids can also lead to strong local negative pressures with low local densities. Only at $T_e \geq 25262$ K, the thin film naturally expands due to bond softening in the simulation s_1 within the simulation time.

In total, we obtained around 10^6 ab-initio data-points for $T_e < 25262$ K and 5×10^5 for $T_e \geq 25262$ K.

7.1.2 Parameter Fitting of Classical Interatomic Potentials

At first, we fitted the parameters of several classical interatomic potentials for Si (see Sect. 5.1.1) to our reference simulations at each studied T_e. For this, we minimized our fit error $\mathbb{W}_{err}(T_e)$ from Eq. (5.95) and utilized our universal interatomic potential parameter fitting program described in detail in Sect. 5.5.

We considered the Modified Tersoff (MT) potential, since it was used as a basis to construct the two available T_e-dependent interatomic potentials for Si. In addition, we also considered the SW, T1, T2, T3, D, PM and MEAM potentials (see Sect. 5.1.1 for more details). The SW and T3 potential are widely used for Si with ground state electrons. The MT and PM potentials are modifications of the T3 potential and the D potential is a modification of the T1 potential. Since Si becomes metallic during melting (see Sect. 4.4.6) and metals are well described by embedded atom potentials [16], we considered the MEAM potential.

The resulting optimal parameters of the MT, SW, T1, T2, T3, D, PM and MEAM potentials are tabulated at all studied T_e's in Sect. A.6 of the appendix. We present the relative error in the cohesive energies (5.92) in Fig. 7.1 and the relative error in the atomic forces (5.91) in Fig. 7.2 of the reference MD simulation s_1 for the

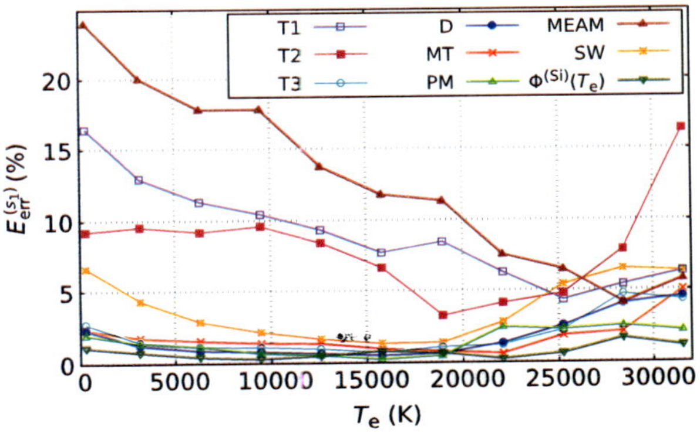

Fig. 7.1 The relative error in the cohesive energies of the reference MD simulation s_1 is shown for the different interatomic potentials as a function of T_e adapted from Fig. 8 of Ref. [15]. Lines are guides to the eye

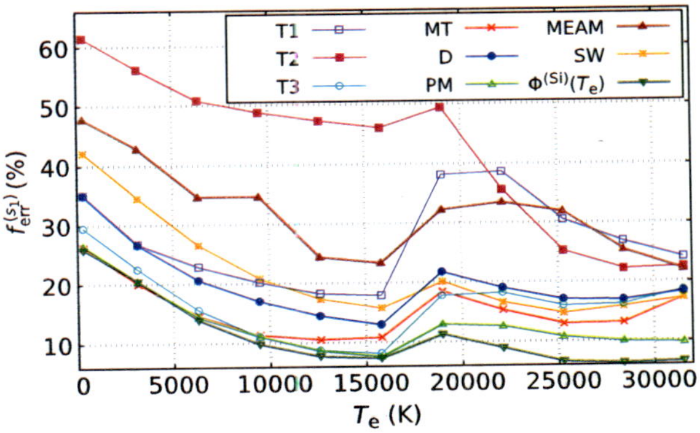

Fig. 7.2 The relative error in the atomic forces of the reference MD simulation s_1 is shown for the different interatomic potentials as a function of T_e adapted from Fig. 9 of Ref. [15]. Lines are guides to the eye

different interatomic potentials as a function of T_e. In both figures, we also include, for completeness, the corresponding errors of our polynomial interatomic potential $\Phi^{(Si)}(T_e)$, which construction and properties we present in the next section.

The relative errors in the atomic forces are much bigger than the relative errors in the cohesive energies. Our interatomic potential exhibits the smallest relative force and cohesive energy error at almost all T_e's compared to the other studied interatomic potentials. To analyze the obtained interatomic potentials, we repeated the calculations and MD simulations presented in Sect. 4.4 (except the MD simulations for the band gap) with them at constant T_e. To perform the MD simulations, we used the DFT initializations. Our results show, that none of the studied classical interatomic potentials can simultaneously describe all increased T_e-induced effects in Si, namely, the bond softening, the thermal phonon antisqueezing, the non-thermal melting, and the time-dependent root mean-square displacement in z-direction $\mathrm{RMSD}_z(t)$ of the thin-film, even after we reparametrized them using our reference MD simulations. As an example, in Fig. 7.3, we present the summary of the comparison between the

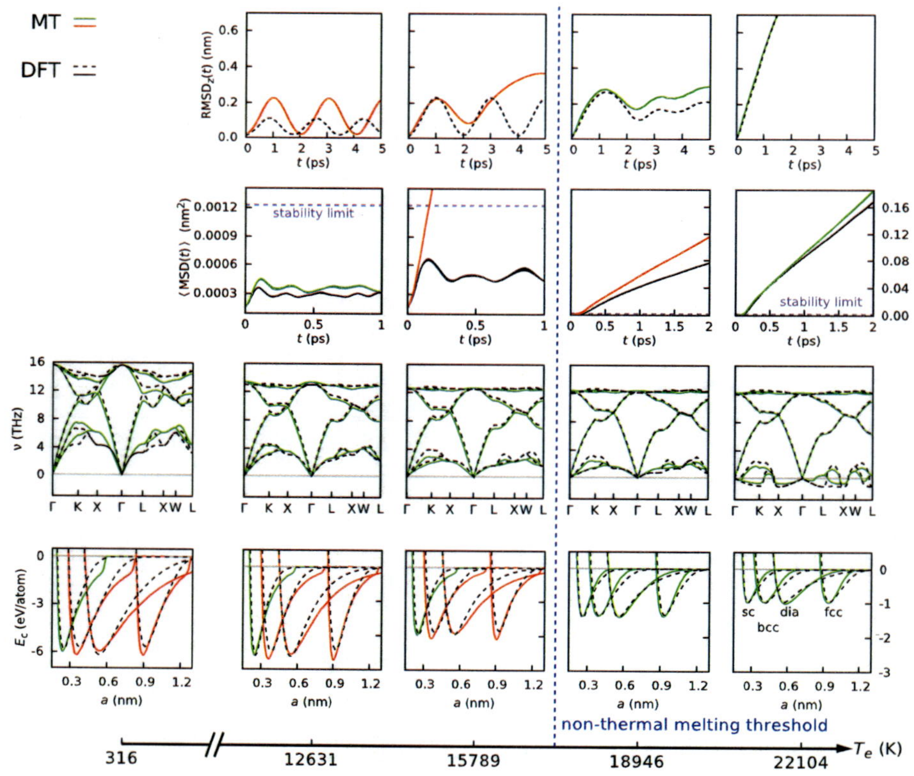

Fig. 7.3 Summary of the performance of the MT potential, reparametrized using our ab-initio reference simulations, in describing the effects occurring in Si induced by the increased T_e deformed PES. Description see Sect. 4.4.8

reparametrized MT potential and DFT results. The description of the DFT results is much better compared to the Shokeen & Schelling potential (see Fig. 5.11), for which the parameters of the MT potential were fitted to a few ab-initio cohesive energies. In Sect. A.7 of the appendix, we present the corresponding figures for the other studied interatomic potentials. In Ref. [15], we also reported these results, except for the MEAM potential.

As one can clearly see in Fig. 7.3 at $T_e = 15789$ K, the crystal structure already melts in bulk MD simulations with the MT potential in contrast to DFT, where thermal phonon antisqueezing occurs. Also the time-dependent root mean-square displacement in z-direction $\mathrm{RMSD}_z(t)$ of the thin-film is not accurately described below the non-thermal melting threshold.

The used classical interatomic potentials were developed to describe Si with ground state electrons. Thus, it is not surprising that these analytical functional forms do not describe accurately the PES with hot electrons, if their parameters are fitted to the ab-initio reference simulations at increased T_e. To get a proper description of the PES at increased T_e, one also has to adapt the functional form of the interatomic potential. This task would be quite difficult for the presented classical interatomic potentials. Beside this, we want to note that the fitting of the corresponding interatomic potential parameters is also quite difficult, since the fit error exhibits, in general, many local minima. These problems do not occur in our interatomic potential construction method described in Sect. 6.3.

7.1.3 Polynomial Interatomic Potential $\Phi^{(\mathrm{Si})}(T_e)$

We utilized our polynomial T_e-dependent interatomic potential model and our fitting procedure described in Sect. 6.3. At first, we derived the global minimum of the fit error $\mathbb{W}_{\mathrm{err}}(T_e)$ for all polynomial-degree combinations up to arbitrarily chosen upper degree limits.

For this, we considered the local interaction terms Φ_2, Φ_3, Φ_ρ, ignored the four-body term Φ_4, and set the upper degree limits $N_2^{(r,\max)} = 15$ for the two-body term and $N^{(\max)} = 9$ for all other terms. To determine the optimal cutoff radii, we checked all cutoff radii starting from 0.25 nm, which is slightly above the nearest neighbor distance of 0.234 nm in the equilibrium diamond-like structure (see Table 4.2), up to 0.90 nm with an increment of 0.01 nm. We obtained the global minimum of the fit error $\mathbb{W}_{\mathrm{err}}(T_e)$ for the resulting 165344 polynomial-degree combinations at $T_e = 316$ K and $T_e = 18946$ K (60 mHa), which is above the non-thermal melting threshold of 17052 K (54 mHa) [18]. We considered these two T_e's and averaged the fit error $\mathbb{W}_{\mathrm{err}}(T_e)$ over both T_e's, since the final polynomial-degree combination should work for all studied T_e's. In Fig. 7.4, we present the averaged fit error $\langle \mathbb{W}_{\mathrm{err}} \rangle$ for all polynomial-degree combinations up to the upper degree limits as a function of the number of coefficients N_C.

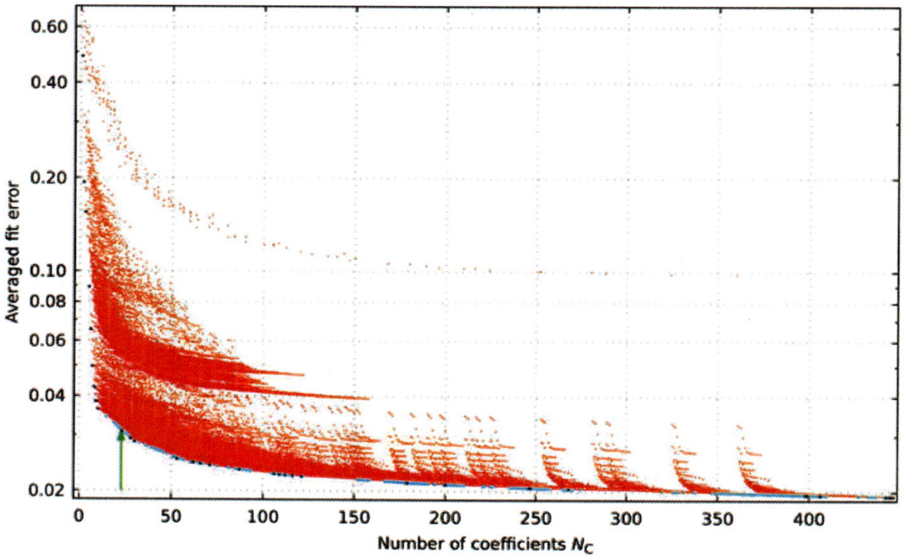

Fig. 7.4 Fit error $\langle \mathbb{W}_{\text{err}} \rangle$ averaged over $T_e = 316$ K and $T_e = 18946$ K is shown as a function of the number of coefficients N_C for all polynomial-degree combinations up to the upper degree limits adapted from Fig. 1 of [17]. Each red dot represents an interatomic potential which is, for the corresponding polynomial-degree combination, a global minimum in parameter space. The members of the subset $\mathcal{S}_{\text{best}}$ are highlighted in light blue, the series $\Phi^{(k)}$, see text, is highlighted in black, and the final choice, $\Phi^{(\text{Si})}(T_e)$, is highlighted in green and marked by an arrow

One can clearly see that $\langle \mathbb{W}_{\text{err}} \rangle$ decreases insignificantly for large values of N_C indicating that our chosen upper degree limits are high enough. In order to obtain the subset $\mathcal{S}_{\text{best}}$ that contains the optimally adjusted polynomial-degree combinations, we applied the iterative procedure described in detail in Sect. 6.3.3. The members of the subset $\mathcal{S}_{\text{best}}$ are highlighted in light blue and the series $\Phi^{(k)}$ from the iterative procedure is highlighted in black in Fig. 7.4. To select the final polynomial-degree combination, we analyzed the members of $\mathcal{S}_{\text{best}}$ with $N_C \leq 50$, because any further decrease of $\langle \mathbb{W}_{\text{err}} \rangle$ is quite low for $N_C > 50$ within the members of $\mathcal{S}_{\text{best}}$, as one can see in Fig. 7.4. For these polynomial-degree combinations of $\mathcal{S}_{\text{best}}$, we constructed the corresponding interatomic potential at each of the eleven studied T_e's by determining the related optimal coefficients and cutoff radii that minimize $\mathbb{W}_{\text{err}}(T_e)$. Using these interatomic potentials, we analyzed the description of the following physical properties at all studied T_e's:

(i) the phonon band structure of the diamond-like bulk crystal structure,
(ii) the cohesive energy curves of the diamond-like, fcc, bcc and sc bulk crystal structures
(iii) the time evolution of the atomic mean-square displacements (MSD) in the MD simulation of bulk Si.

Table 7.1 Fit error $\mathcal{W}_{\text{err}}(T_e)$ of the final polynomial-degree combination is tabulated at the studied T_e's for three cases: for the final interatomic potential $\Phi^{(\text{Si})}(T_e)$ with the chosen constant cutoff radii and polynomial approximated coefficients, for the chosen constant cutoff radii and T_e-optimized coefficients, and for T_e-optimized cutoff radii and coefficients. For the last case, also the optimal cutoff radii are tabulated

	$\Phi^{(\text{Si})}(T_e)$	$r_2^{(c)} = 0.63$ nm, $r_3^{(c)} = 0.42$ nm, $r_\rho^{(c)} = 0.48$ nm				
T_e (K)	$\mathcal{W}_{\text{err}}(T_e)$	$\mathcal{W}_{\text{err}}(T_e)$	$\mathcal{W}_{\text{err}}(T_e)$	$r_2^{(c)}$ (nm)	$r_3^{(c)}$ (nm)	$r_\rho^{(c)}$ (nm)
316	0.05693	0.05693	0.05661	0.650	0.421	0.468
3158	0.04000	0.03999	0.03937	0.642	0.419	0.461
6315	0.02415	0.02414	0.02341	0.634	0.417	0.456
9473	0.01494	0.01494	0.01443	0.632	0.418	0.454
12631	0.00952	0.00952	0.00931	0.629	0.418	0.458
15789	0.00628	0.00625	0.00622	0.626	0.420	0.468
18946	0.00594	0.00591	0.00591	0.626	0.419	0.479
22104	0.00358	0.00355	0.00350	0.628	0.411	0.493
25262	0.00220	0.00219	0.00208	0.636	0.402	0.510
28420	0.00211	0.00199	0.00186	0.639	0.396	0.527
31577	0.00221	0.00218	0.00194	0.526	0.403	0.769

Our analysis showed that the performance of the studied members of $\mathcal{S}_{\text{best}}$ varies considerably, so that not all members of $\mathcal{S}_{\text{best}}$ represent reliable interatomic potentials. We selected the final polynomial-degree combination with $N_C = 23$, since the corresponding interatomic potential best describes the above mentioned physical properties with a minimal possible N_C.

The resulting optimal cutoff radii vary insignificantly around $r_2^{(c)} = 0.63$ nm, $r_3^{(c)} = 0.42$ nm, $r_\rho^{(c)} = 0.48$ nm at the eleven studied T_e's and the fit error is not significantly influenced, if these values are used, as one can see in Table 7.1. Thus, we chose these values and kept them constant for all T_e's. Now only the coefficients depend on T_e. Up to now, we only inferred optimal values at the eleven studied T_e's. Using these optimal values as supporting points, we approximated the final coefficients by polynomials of degree 5 in T_e to obtain a functional dependence on T_e. This polynomial approximation of the coefficients has almost no influence on the fit error, as one can see in Table 7.1. The coefficients of the final interatomic potential $\Phi^{(\text{Si})}(T_e)$ are tabulated in Sect. A.8 of the appendix and have been also published in Ref. [17].

7.1.4 Physical Properties of Polynomial $\Phi^{(Si)}(T_e)$

For the final polynomial interatomic potential $\Phi^{(\mathrm{Si})}(T_\mathrm{e})$, we repeated the calculations and MD simulations presented in Sect. 4.4 (except the MD simulations for the band gap) at constant T_e. To perform the MD simulations, we used the DFT initializations. In Fig. 7.5, we present the summary of the comparison of $\Phi^{(\mathrm{Si})}(T_\mathrm{e})$ with DFT results.

As expected, $\Phi^{(\mathrm{Si})}(T_\mathrm{e})$ reproduces the physical properties (i)–(iii) that were checked for selecting the final polynomial-degree combination. Also the atomic root mean-square displacement in z-direction $\mathrm{RMSD}_z(t)$ of the thin film after laser excitation is well reproduced by $\Phi^{(\mathrm{Si})}(T_\mathrm{e})$ at all T_e's. For the ideal diamond-like structure as a function of T_e, we derived the electronic internal energy E_e from Eq. (5.77) and the electronic specific heat C_e from Eq. (5.78) using thermodynamic relations for $\Phi^{(\mathrm{Si})}(T_\mathrm{e})$. The resulting curves are presented in Figs. 7.6 and 7.7. $\Phi^{(\mathrm{Si})}(T_\mathrm{e})$ describes accurately both quantities, which indicates that the polynomial T_e-approximation of the coefficients works very well. Only for $T_\mathrm{e} > 26000$ K, C_e obtained from $\Phi^{(\mathrm{Si})}(T_\mathrm{e})$

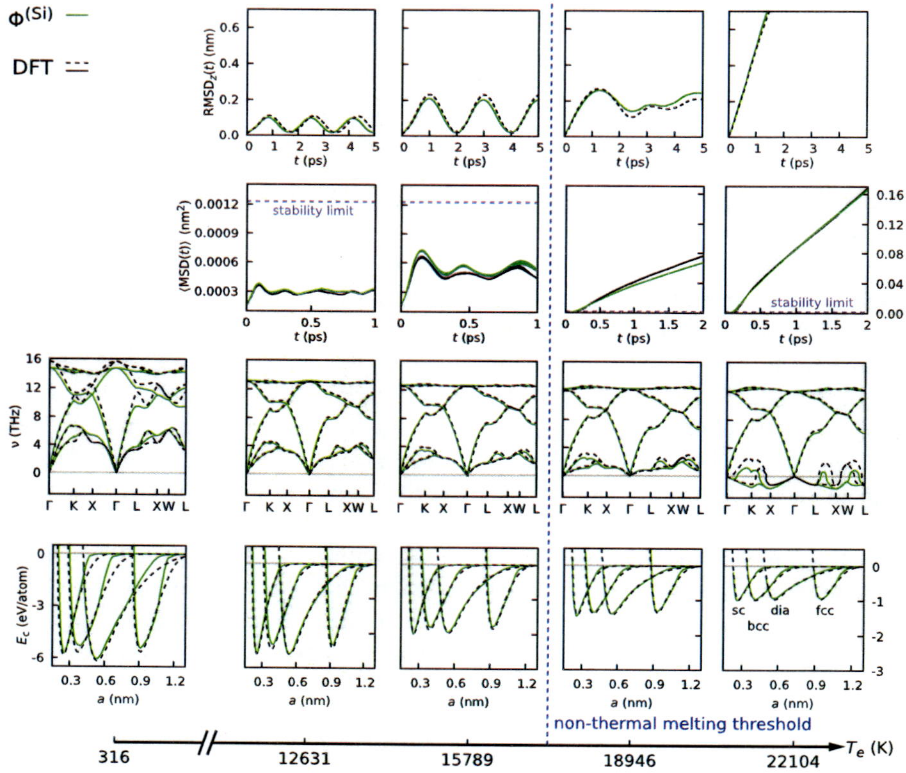

Fig. 7.5 Summary of the performance of $\Phi^{(\mathrm{Si})}(T_\mathrm{e})$ in describing the effects occurring in Si induced by the increased T_e deformed PES. Description see Sect. 4.4.8

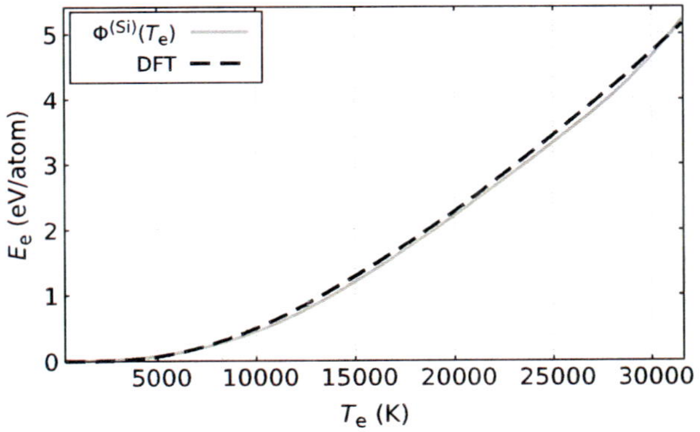

Fig. 7.6 Electronic internal energy E_e of the ideal diamond-like structure is shown as a function of T_e for DFT and $\Phi^{(Si)}(T_e)$ adapted from Figure S2 of Ref. [17]

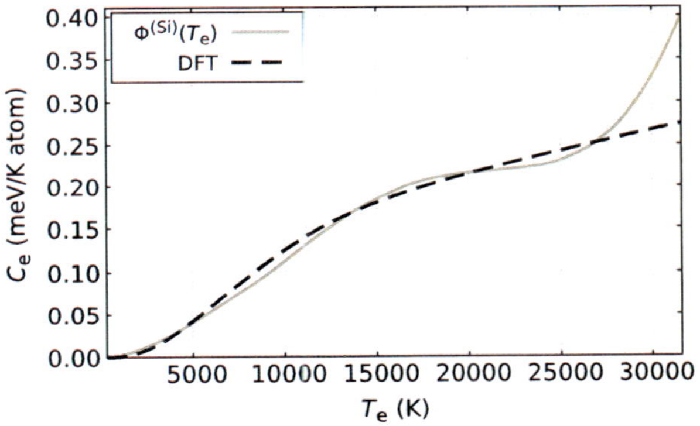

Fig. 7.7 Electronic specific heat C_e of the ideal diamond-like structure is shown as a function of T_e for DFT and $\Phi^{(Si)}(T_e)$ adapted from Fig. 2 of Ref. [17]

starts to deviate from the DFT results, but the behavior is still physically acceptable in a particular situation, because it is still increasing.

The relative error of $\Phi^{(Si)}(T_e)$ in the atomic forces of reference simulation s_1 decreases from 26% to 6% when T_e increases from 316 K to 31577 K, as one can see in Fig. 7.2. We conclude from this that the PES becomes less complex for higher T_e's. The relative error of $\Phi^{(Si)}(T_e)$ in the cohesive energies always lies below 1.7%.

We also checked that $\Phi^{(Si)}(T_e)$ describes forces in independent ab-initio MD simulations, which were not used for its development, with the same accuracy. For this, we initialized four additional samples at $T_i = 316$ K for the thin-film using the Andersen thermostat with CHIVES. Using these initializations, we performed MD simulations of laser excitation at various constant T_e's with CHIVES to get four additional runs beside the run, which was used for fitting as set s_1. We derived the relative force error (5.91) for all runs at various T_e's and list them in Table 7.2. Only run 1 was used for the interatomic potential fitting. One can see that the atomic forces in the independent runs are described at the same level of accuracy. This indicates that fitting additional independent ab-initio runs should not produce any improvement and hints to the transferability of $\Phi^{(Si)}(T_e)$, what is difficult to prove.

Table 7.2 Relative force error (5.91) in the thin-film MD simulation at constant T_e is listed for the different independent runs adapted from Table S5 of Ref. [17]. Only run 1 was used for fitting as set s_1

	Run 1	Run 2	Run 3	Run 4	Run 5
T_e (K)	f_{err} (%)	f_{err} (%)	f_{err} (%)	f_{err} (%)	f_{err} (%)
12631	7.7	7.6	7.4	7.5	7.8
15789	7.3	7.1	7.1	7.2	7.5
18946	11.2	11.1	11.0	10.7	11.8
22104	8.9	9.0	9.2	8.8	8.9
25262	6.6	6.7	6.7	6.7	6.7

$\Phi^{(\mathrm{Si})}(T_e)$ also reproduces properties that were not used in the fitting procedure. We analyzed the description of the independent elastic constants $\mathscr{C}_{11}, \mathscr{C}_{12}, \mathscr{C}_{44}$ of bulk diamond-like Si at various T_e's. To do this, we derived the elastic constants from the derivatives of the acoustic phonon branches at the Γ-Point, as described in Ref. [19]. In Table 7.3, we present the relative differences

$$\Delta\mathscr{C}_{11} = \frac{\left|\mathscr{C}_{11}^{(\Phi)} - \mathscr{C}_{11}^{(\mathrm{DFT})}\right|}{\left|\mathscr{C}_{11}^{(\mathrm{DFT})}\right|}, \quad \Delta\mathscr{C}_{12} = \frac{\left|\mathscr{C}_{12}^{(\Phi)} - \mathscr{C}_{12}^{(\mathrm{DFT})}\right|}{\left|\mathscr{C}_{12}^{(\mathrm{DFT})}\right|}, \quad \Delta\mathscr{C}_{44} = \frac{\left|\mathscr{C}_{44}^{(\Phi)} - \mathscr{C}_{44}^{(\mathrm{DFT})}\right|}{\left|\mathscr{C}_{44}^{(\mathrm{DFT})}\right|} \tag{7.1}$$

between the elastic constants derived from $\Phi^{(\mathrm{Si})}(T_e)$ and those derived from the T_e-dependent DFT calculations. In general, the values of the elastic constants show a reasonable agreement. We want to note that the derivatives of the acoustic phonon branches become negative for both, T_e-dependent DFT and $\Phi^{(\mathrm{Si})}(T_e)$ for electronic temperatures T_e's above 18946 K. Therefore, the elastic constants may not be properly defined any more for these T_e's.

Furthermore, we analyzed the description of the pair-correlation function and the bond-angle distribution between the neighboring atoms in bulk Si at increased T_e's. For this, we took the initial conditions of the ab-initio MD simulations of laser-induced non-thermal melting at 18946 K, 22104 K, and 25262 K described in Sect. 4.4.4 and of laser-induced thermal phonon antisqueezing at 12631 K and 15789 K described in Sect. 4.4.4, and repeated the simulations with $\Phi^{(\mathrm{Si})}(T_e)$. We also considered the additional MD simulations of laser-induced non-thermal melting at $T_e = 18315$ K of Ref. [20].

We present in Figs. 7.8, 7.10 and 7.12 typical pair-correlation functions obtained from the MD simulations 1 ps after the laser irradiation. In the same way, we present in Figs. 7.9, 7.11 and 7.13 typical bond-angle distributions, for which atoms were taken into account up to a distance of 0.41 nm. Our study indicates that at least for the simulated time of 1 ps, the pair-correlation function and bond-angle distribution are reliably described by $\Phi^{(\mathrm{Si})}(T_e)$. $T_e = 18315$ K is obtained from the polynomial T_e-approximation of the coefficients, since it lies between the T_e's 15789 K and 18946 K

Table 7.3 Relative errors $\Delta\mathscr{C}_{11}$, $\Delta\mathscr{C}_{12}$, $\Delta\mathscr{C}_{44}$ in the independent elastic constants $\mathscr{C}_{11}$, $\mathscr{C}_{12}$, $\mathscr{C}_{44}$ derived from $\Phi^{(\mathrm{Si})}(T_e)$ and from T_e-dependent DFT at the studied electronic temperatures T_e's are listed adapted from Table S6 of Ref. [17]

T_e (K)	$\Delta\mathscr{C}_{11}$ (%)	$\Delta\mathscr{C}_{12}$ (%)	$\Delta\mathscr{C}_{44}$ (%)
316	−7.6	−56.1	−17.5
3158	−1.3	−73.0	−18.9
6315	2.7	−104.1	−15.3
9473	6.8	−575.4	−10.4
12631	8.6	67.2	−6.6
15789	7.7	18.5	−4.0
18946	5.2	5.7	−3.3
22104	2.0	−3.6	35.4
25262	−1.2	−28.1	23.1
28420	−4.4	−149.4	28.1
31577	−7.6	609.2	27.4

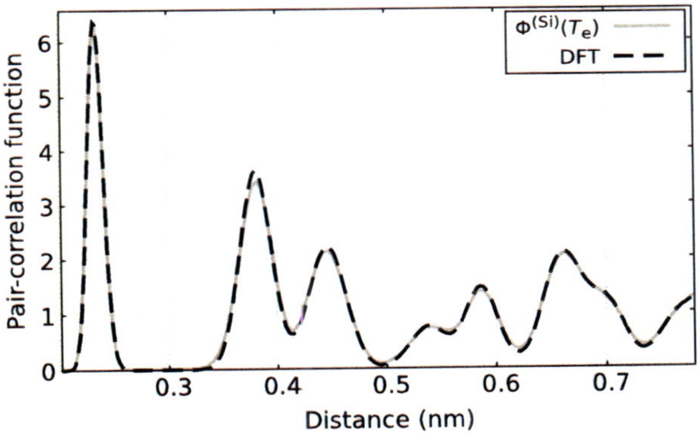

Fig. 7.8 Pair-correlation function 1 ps after the laser excitation at $T_e = 15789$ K adapted from Figure S5 of Ref. [17]

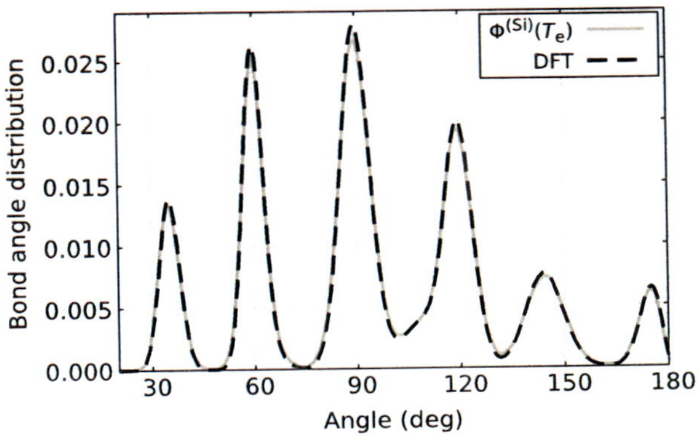

Fig. 7.9 Bond-angle distribution 1 ps after the laser excitation at $T_e = 15789$ K adapted from Figure S6 of Ref. [17]

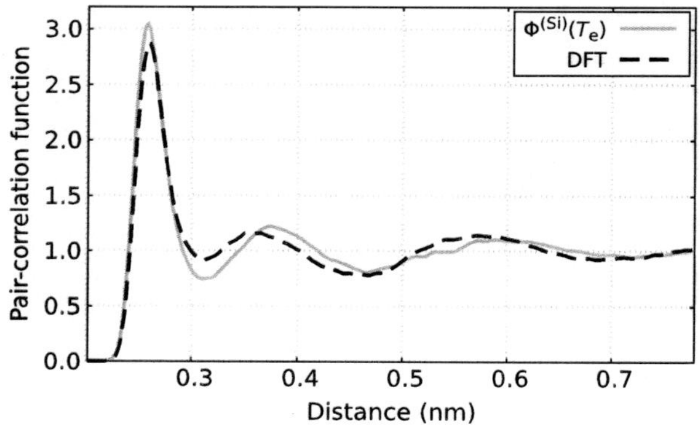

Fig. 7.10 Pair-correlation function 1 ps after the laser excitation at $T_e = 18315$ K adapted from Figure S7 of Ref. [17]

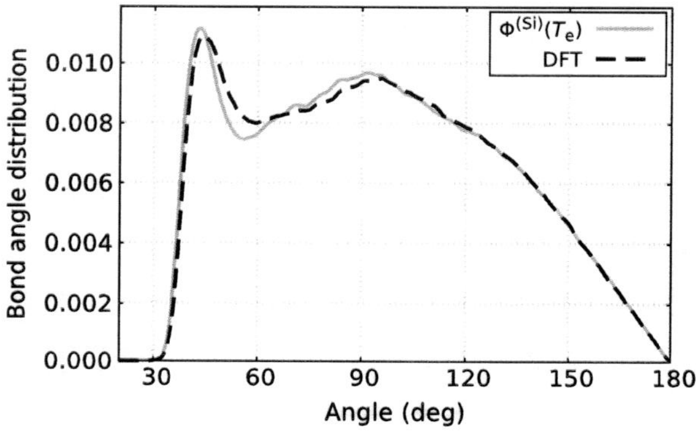

Fig. 7.11 Bond-angle distribution 1 ps after the laser excitation at $T_e = 18315$ K adapted from Figure S8 of Ref. [17]

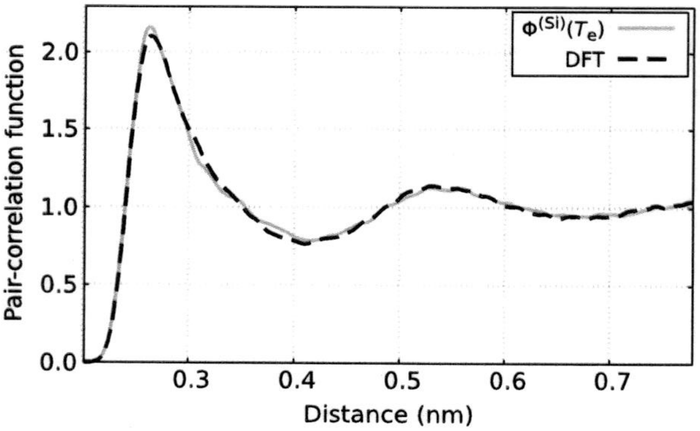

Fig. 7.12 Pair-correlation function 1 ps after the laser excitation at $T_e = 22104$ K adapted from Figure S9 of Ref. [17]

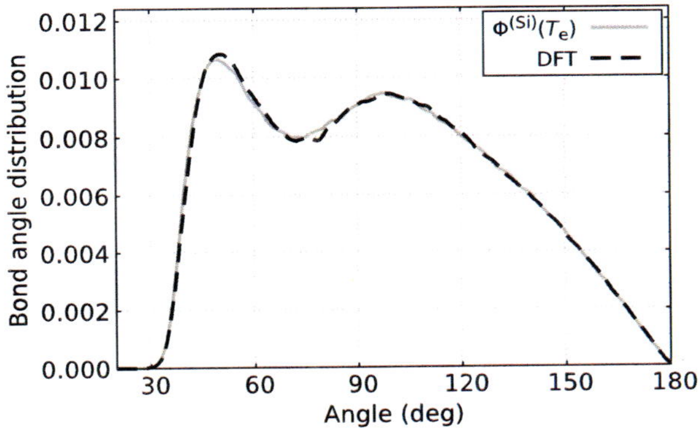

Fig. 7.13 Bond-angle distribution 1 ps after the laser excitation at $T_e = 22104$ K adapted from Figure S10 of Ref. [17]

that were used for fitting. The pair-correlation function and bond-angle distribution are well described at $T_e = 18315$ K, which indicates additionally that the polynomial T_e-approximation of the coefficients works well.

We also published the findings on $\Phi^{(Si)}(T_e)$, its outstanding properties and the development method in Ref. [17].

7.1.5 *Thermophysical Properties of Polynomial* $\Phi^{(Si)}(T_e)$

In a next step, we analyzed the thermophysical properties of $\Phi^{(Si)}(T_e)$. We set up a simulation cell that consists of $32 \times 16 \times 16$ conventional cells and contains $N_{at} = 65536$ Si atoms. We used periodic boundary conditions in all directions to get bulk Si. We set the ideal diamond-like structure and applied the Andersen thermostat (see Sect. 4.1.2), to thermalize the crystal structure at a given temperature $T_i = T_e$. We performed this thermalization at zero pressure in the whole simulation cell. We describe the calculation of the pressure in Sect. A.4 of the appendix. To reach the target pressure, we rescaled the simulation cell volume and the atomic coordinates in the MD simulation every picosecond. We considered 27 temperatures ranging from 50 K up to 1350 K with an increment of 50 K. At all of these temperatures, the crystal structure remains intact, whereas it melts in the thermalization at $T_i = T_e = 1400$ K within the simulated time. In addition, we melted the crystal structure by thermalizing it at 2500 K starting from the ideal diamond-like structure. Then we thermalized the molten structure at zero pressure at 30 temperatures ranging from 1050 K up to 2500 K with an increment of 50 K. At all of these temperatures, the atomic structures does not recrystallize, whereas it starts to crystallize at $T_i = T_e = 1000$ K within the simulated time. In Fig. 7.14, we present the energy as a function of temperature for the crystalline and the liquid phase. In Fig. 7.15, we present the lattice parameter as a function of temperature for the crystalline and the liquid phase. In both figures, the vertical line indicates the model melting temperature T_m at zero pressure, which we derive below.

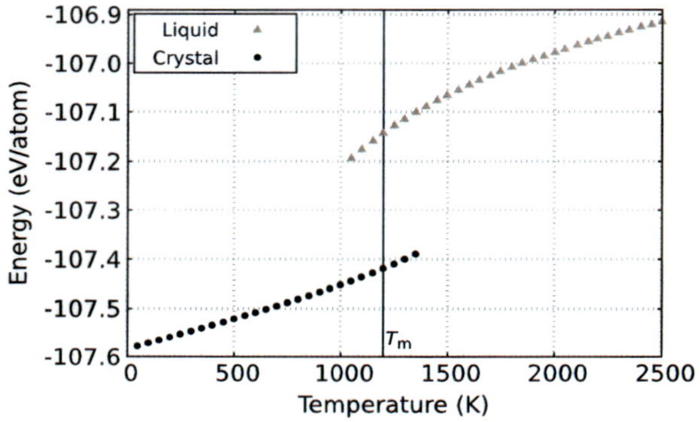

Fig. 7.14 Energies of the liquid and crystalline phases at zero pressure are shown as a function of time. The vertical line indicates the position of the melting temperature

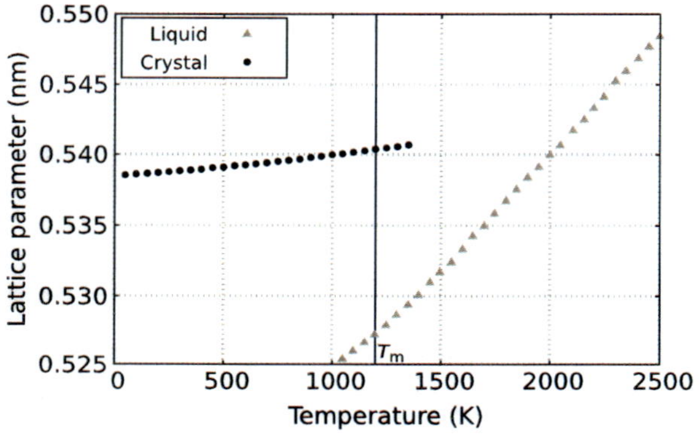

Fig. 7.15 Lattice parameters of the liquid and crystalline phases at zero pressure are shown as a function of time. The vertical line indicates the position of the melting temperature

We fitted the temperature dependence of the energy $E(T_\text{i})$ of the crystalline phase as a polynomial of degree two in the temperature T_i using the data points shown in Fig. 7.14. From this, we obtained the specific heat at constant pressure of the ions using the thermodynamic relation $C_{\text{i}p} = \left.\frac{\partial E}{\partial T_\text{i}}\right|_p$:

$$
\begin{aligned}
C_{\text{i}p}(T_\text{i}) &= 0.105489\,\frac{\text{meV}}{\text{atom K}} + 5.15644 \times 10^{-5}\,\frac{\text{meV}}{\text{atom K}^2} \times T_\text{i} \\
&= 362.341\,\frac{\text{J}}{\text{kg K}} + 0.177117\,\frac{\text{J}}{\text{kg K}^2} \times T_\text{i}.
\end{aligned}
$$

In order to change the unit, we used the Avogadro constant $N_\text{A} = 6.02214199 \times 10^{23}$ and the molar mass of Si $m_\text{Si} = 28.0855\,\frac{\text{g}}{\text{mol}}$ [21]. At $T_\text{e} = T_\text{i} = 300$ K, we got $C_{\text{i}p} = 415\,\frac{\text{J}}{\text{kg K}}$ and the literature value yields $C_{\text{i}p} = 702\,\frac{\text{J}}{\text{kg K}}$ [21]. The enthalpy of fusion E_fusion is defined as the energy difference between the liquid and crystalline phase at the melting temperature. We obtained $E_\text{fusion} = 0.276501\,\frac{\text{eV}}{\text{atom}} = 26.6782\frac{\text{kJ}}{\text{mol}}$ and the literature value yields $E_\text{fusion} = 50.208\,\frac{\text{kJ}}{\text{mol}}$ [22]. By comparing the lattice parameters of the liquid and crystal phase at the melting temperature, we got a volume shrinkage during melting of 7.1%, which agrees with the literature value of 8.37% [23]. We found the lattice parameter of $a = 0.538364$ nm for the ideal diamond-like structure in agreement with the DFT value of $a = 0.539872$ nm. For the diamond-like structure thermalized at $T_e = T_i = 300$ K, we determined an equilibrium lattice parameter of $a = 0.53837$ nm, which agrees with the literature value of 0.543072 nm [21]. At $T_e = T_i = 300$ K, we obtained a linear expansion coefficient of $2.25 \times 10^{-6}\,\frac{1}{\mathrm{K}}$, which agrees with the literature value of $2.49 \times 10^{-6}\,\frac{1}{\mathrm{K}}$ [24].

7.1.5.1 Melting Temperature

As depicted in Fig. 7.14, one cannot calculate the melting temperature just by heating up the crystal structure until it melts or just by cooling down the liquid until it crystallizes. If one heats up the crystal structure, one obtains an overheated crystal, since the melting does not occur due to missing crystal defects or liquid parts, at which the melting is initiated. On the other hand, if one cools down the liquid, one obtains an undercooled liquid, since the crystallization does not occur due to missing crystalline parts, at which the crystallization is initiated. Thus, to obtain reliably the melting temperature T_m, one has to simulate the coexistence of liquid and crystal parts [25]. For this, we used the already introduced bulk simulation cell with $N_{at} = 65536$ Si atoms. We fixed the coordinates of half of the atoms and melted the other part of the crystal structure by applying the Anderson thermostat at $T_i = 2500$ K. Then we allowed again that all atoms move. We applied the Anderson thermostat to all atoms at a temperature, which we assumed to be close to the melting temperature. During this thermalization, we rescaled the simulation cell volume and the atomic coordinates every picosecond to reach a given target pressure. We performed three of such thermalizations to get the pressures $p = -1$ GPa, 0 GPa, 1 GPa. In this way, we obtained an initialization of atomic coordinates and velocities at a given pressure and temperature, where half of the structure is molten and the other half is crystalline, so that there are two planar liquid-crystal interfaces. Then we performed a MD simulation at constant volume and energy for 500 ps. In Fig. 7.16, we show a snapshot of the atomic structure at the initialization and one after the MD simulation period of 500 ps for the initialization at zero pressure.

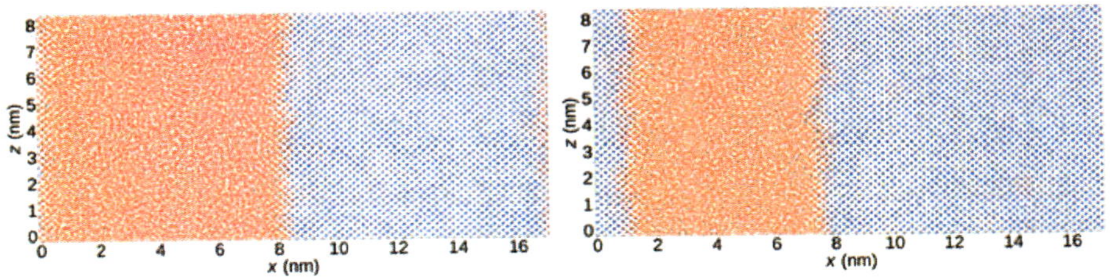

Fig. 7.16 Snapshots of the MD simulation with 65536 Si atoms at zero pressure are shown at the initialization $t = 0$ ps (left) and at $t = 500$ ps (right). The atoms are colored due to their CSP value (see below): blue corresponds to crystalline and red to molten environment

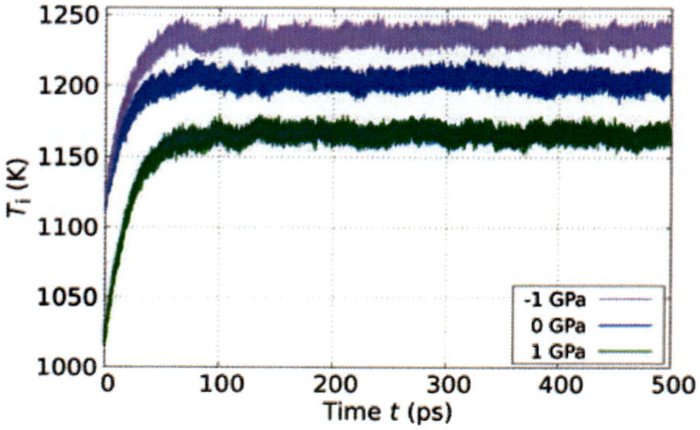

Fig. 7.17 Ionic temperatures occurring in the MD simulation of the bulk simulation cell with 65536 Si atoms are shown as a function of time for various constant pressures

In this MD simulation at constant energy (and volume), the ionic temperature T_i converges always to the melting temperature T_m. The reason for this is the following: If the temperature at the initialization is below the melting temperature, the atoms of the liquid at the interface start to crystallize, so that the size of the liquid part is reduced. The crystallization increases the temperature until the melting temperature is reached, since the heat of fusion is released from the crystallization. If the temperature at the initialization is above the melting temperature, the atoms of the crystal at the interface start to melt, so that the size of the crystal part is reduced. The melting deceases the temperature until the melting temperature is reached, since the heat of fusion is taken for the melting. If the temperature finally reaches the melting temperature, it remains constant, since the same amount of atoms melt and crystallize. There are only small fluctuations, which decrease with the size of the simulation cell. In Fig. 7.17, we present the ionic temperature T_i obtained from Eq. (4.16) of the MD simulations at constant energy as a function of time. One can clearly see, that T_i converges to the melting temperature at the given pressure and oscillates then around this value. We determined the melting temperature

$$T_m(p) = (1199 \pm 2)\,\mathrm{K} - (40 \pm 3)\,\frac{\mathrm{K}}{\mathrm{GPa}} \times p \tag{7.2}$$

near zero pressure p by performing a linear regression of the melting temperature values at the three studied pressures. This value agrees with $T_m(p) = (1300 \pm 50)\,\mathrm{K} - 58\,\frac{\mathrm{K}}{\mathrm{GPa}} \times p$ from LDA-DFT [26] and differs from the experimental value, $T_m(p) = (1687 \pm 5)\,\mathrm{K} - 58\,\frac{\mathrm{K}}{\mathrm{GPa}} \times p$ [27, 28].

We want to note, that the whole structure melts or crystallizes in the MD simulation at constant energy, if one initializes the temperature to far away from the melting temperature. In such a MD simulation, the temperature also remains constant in the end, since the material remains molten or crystalline. Therefore, one has to check additionally the atomic structure, if there is really a coexistence of a liquid-crystal interface, as presented in Fig. 7.16. To identify molten and crystalline structures more reliably, we utilized the crystal symmetry parameter (CSP) from Ref. [29]:

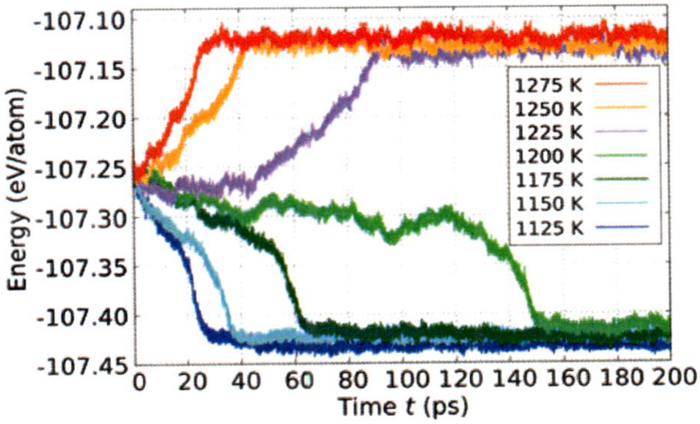

Fig. 7.18 The Helmholtz free energy in the MD simulation of the bulk simulation cell with 1024 Si atoms is shown as a function of time for various temperatures

Definition 7.1 *(Crystal symmetry parameter)* The crystal symmetry parameter of atom i is defined as

$$\mathrm{CSP}_i = 1 - \frac{\left(\sum_{\substack{j=1 \\ j \neq i \\ r_{ij} < r^{(c)}}}^{N_{at}} \mathbf{r}_{ij} \right)^2}{\left(\sum_{\substack{j=1 \\ j \neq i \\ r_{ij} < r^{(c)}}}^{N_{at}} 1 \right) \left(\sum_{\substack{j=1 \\ j \neq i \\ r_{ij} < r^{(c)}}}^{N_{at}} |\mathbf{r}_{ij}|^2 \right)}, \tag{7.3}$$

where $r^{(c)}$ is located between the first and second nearest neighbor distances for the ideal crystal structure.

The ideal diamond-like crystal structure keeps invariant, if one performs an inversion of all atomic coordinates. The reason for this is that, for any atom i and any neighbor j, there exists a neighbor k of atom i with $\mathbf{r}_{ij} = -\mathbf{r}_{ik}$. Due to this, the CSP value of any atom in the ideal diamond-like crystal structure is 1. If there are distortions of the crystal structure around an atom i, its CSP value decreases. Therefore, the CSP value allows to distinguish between crystalline and molten phases. For Si, we used $r^{(c)} = 0.3078$ nm, which is exactly located between the first and second nearest neighbor distance (see Table 4.2). From the MD simulations of liquid-crystal coexistence, we chose the criteria that CSP < 0.955 corresponds to molten and CSP ≥ 0.955 to crystalline environment. We want to note, that the value of the criteria is quite sensitive. However, in the MD simulations of liquid-crystal coexistence, one can clearly identify the molten and crystalline part from the atomic coordinates by the eye (see Fig. 7.16 for example). Thus, we can use these MD sim-

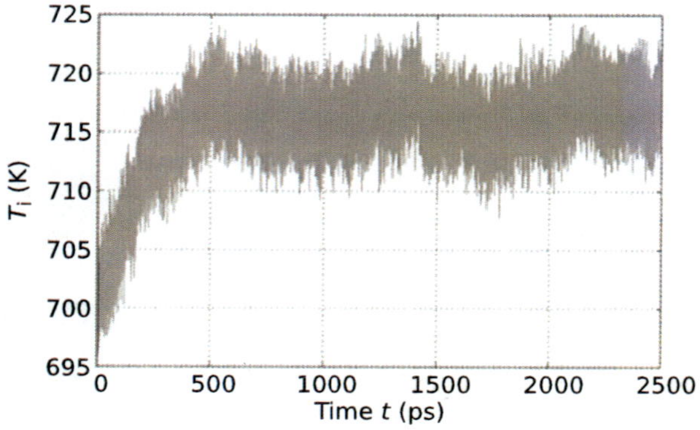

Fig. 7.19 Supercell containg 1024 Si atoms

ulations to determine accurately the criteria. To smooth the CSP values, we averaged the CSP value of every atom over its neighbors inside the sphere with radius $r^{(c)}$.

The larger the simulation cell, the smaller are the fluctuations of the temperature after the melting temperature is reached. On the other hand, if the simulation cell is to small, the fluctuations are so big that the liquid-crystal interface cannot be stabilized, so that the whole structure always melts or crystallizes. Nevertheless, we can use the MD simulations of a small simulation cell to derive an approximation of the melting temperature. For this, we set up a simulation cell that consists of $8 \times 4 \times 4$ conventional cells and contains $N_{at} = 1024$ Si atoms (see Fig. 7.19). We used periodic boundary conditions in all directions to get bulk Si. At first, we fixed the atomic coordinates of half of the atoms and applied the Anderson thermostat at $T_i = 2500$ K to the other atoms to melt their structure. In this way, we obtained a structure, where half is molten and the other half is in a crystalline state. Then we released the fixed atoms and applied the Andersen thermostat at a given temperature T_i on a long timescale. Now the whole structure melts or crystallizes, as one can see in Fig. 7.18, where the structural energy is shown as a function of time for several temperatures. If the structure melts entirely, the structural energy increases, and, if the structure crystallizes entirely, the structural energy decreases. The reason for this is that the energy of the crystal phase is smaller compared to the energy of the liquid phase, as one can see in Fig. 7.14. For temperatures significantly below the melting temperature, the structure always crystallizes and the structure always melts for temperatures significantly above the melting temperature. For temperatures close to the melting temperature, the structure melts or crystallizes depending on the actually used random numbers in the Anderson thermostat. Thus, one only obtains a rough approximation of the melting temperature from such simulations.

In Table 7.4, we finally list the obtained thermophysical properties of $\Phi^{(\mathrm{Si})}(T_e)$ together with experimental values from the literature. There are some deviations from the experimental values, which is not surprising, since the interatomic potential was fitted to T_e-dependent DFT calculations.

Table 7.4 Thermophysical properties of $\Phi^{(\mathrm{Si})}(T_e)$ together with experimental values

	$\Phi^{(\mathrm{Si})}(T_e)$	Experimental value	Refs.
Melting temperature T_m	(1199 ± 2) K	(1687 ± 5) K	[27]
Melting curve slope near $p = 0$	$(-40 \pm 3)\ \frac{\mathrm{K}}{\mathrm{GPa}}$	$-58\ \frac{\mathrm{K}}{\mathrm{GPa}}$	[28]
Lattice parameter at $T = 300$ K, $p = 0$	0.53837 nm	0.54305 nm	[21]
Linear expansion coefficient at $T = 300$ K	$2.25 \times 10^{-6}\ \frac{1}{\mathrm{K}}$	$2.49 \times 10^{-6}\ \frac{1}{\mathrm{K}}$	[24]
Volume shrinkage during melting	7.1%	8.37%	[23]
Enthalpy of fusion E_{fusion}	$26.6782\ \frac{\mathrm{kJ}}{\mathrm{mol}}$	$50.208\ \frac{\mathrm{kJ}}{\mathrm{mol}}$	[22]
Heat capacity C_{ip} at $T = 300$ K, $p = 0$	$415\ \frac{\mathrm{J}}{\mathrm{kg\,K}}$	$702\ \frac{\mathrm{J}}{\mathrm{kg\,K}}$	[21]

7.2 MD Simulations of Excited PES and EPC with Polynomial $\Phi^{(\mathrm{Si})}(T_e)$

For all presented MD simulations with $\Phi^{(\mathrm{Si})}(T_e)$, we used the lattice parameter of $a = 0.53837$ nm, which we obtained for the diamond-like structure thermalized at $T_e = T_i = 300$ K (see Table 7.4). Furthermore, we utilized

$$G_{\mathrm{ep}} = 1.8 \times 10^{17}\ \frac{\mathrm{W}}{\mathrm{K\,m^3}} = 2.19638 \times 10^{-8}\ \frac{\mathrm{eV}}{\mathrm{fs\,K\,atom}} \tag{7.4}$$

for the electron-phonon coupling constant of all phonon modes, which was calculated from LDA-DFT by Sadasivam et al. in 2017 with the help of Eq. (3.223) [30]. We took this value from Fig. 3c of Ref. [30]. Sadasivam et al. stated that G_{ep} depends on the electronic temperature T_e and on the ionic temperature T_i and differs for optical and acoustic phonon modes [30]. In addition, Sadasivam et al. presented calculations with their G_{ep} using a two and three temperature model for T_e's up to 3000 K [30]. Nevertheless, they only report one value for G_{ep}, which we took for our calculations. To transform G_{ep} in Eq. (7.4) to the units we are using in our calculations, we utilized the ab-initio equilibrium atomic density $\rho_{\mathrm{at}} = 50.8414\ \frac{\mathrm{atoms}}{\mathrm{nm}^3}$. In the presented calculations for Si, we utilize a global T_e, since we assume a fast equilibration of the T_e to a uniform value in the sample.

To gauge the influence of the excited PES with hot electrons and of the EPC on the ionic dynamics, we performed three different scenarios of MD simulations:

excited PES & EPC: We simulated the influences of the excited PES and the EPC by applying our method described in detail in Sect. 6.1.

only excited PES: We simulated only the influence of the excited PES by setting $G_{\mathrm{ep}} = 0$ in our method. This simulation procedure is equivalent to the T_e-dependent DFT MD simulations presented in Sect. 4.4.

only EPC: We simulated only the influence of the EPC by using $\Phi^{(\mathrm{Si})}(T_\mathrm{e})$ at $T_\mathrm{e} =$ 300 K equivalent to the MD simulations of Ivanov and Zhigilei described in Sect. 5.2.

To simulate only the EPC, we needed the electronic specific heat $C_\mathrm{e}(T_\mathrm{e})$ and the electronic internal energy $E_\mathrm{e}(T_\mathrm{e})$ as a function of T_e. We considered the ideal diamond-like structure of Si with the optimal lattice parameter $a = 0.539872$ nm and derived the Helmholtz free energy $F(T_\mathrm{e})$ from T_e-dependent DFT for various T_e's. We fitted $F(T_\mathrm{e})$ as a polynomial of degree 11 in T_e and derived $C_\mathrm{e}(T_\mathrm{e})$ from this polynomial using the thermodynamic relation (5.78):

$$C_\mathrm{e}(T_\mathrm{e}) = N_\mathrm{at} \sum_{k=1}^{10} a_{C_\mathrm{e}}^{(k)} \left(\frac{T_\mathrm{e}}{31577\,\mathrm{K}}\right)^k . \tag{7.5}$$

The coefficients $a_{C_\mathrm{e}}^{(k)}$ are tabulated in Table A.1 of the appendix.

Since we are using a global T_e, we obtain the total electronic internal energy just by

$$E_\mathrm{e}(T_\mathrm{e}) = \int_0^{T_\mathrm{e}} dT_\mathrm{e}'\, C_\mathrm{e}(T_\mathrm{e}'). \tag{7.6}$$

Now, the potential energy surface of ions and electrons is given by

$$E(\mathbf{r}_1, \ldots, \mathbf{r}_{N_\mathrm{at}}, T_\mathrm{e}) = \Phi^{(\mathrm{Si})}(\mathbf{r}_1, \ldots, \mathbf{r}_{N_\mathrm{at}}, T_\mathrm{e} = 300\,\mathrm{K}) + E_\mathrm{e}(T_\mathrm{e}). \tag{7.7}$$

For the time propagation of T_e, we insert $C_\mathrm{e}(T_\mathrm{e})$ from Eq. (7.5) in Eq. (6.29). The equation of motion for the ions is still given by Eq. (6.31).

7.2.1 Direct Comparison of the Bragg Peak Intensities with Experiments

7.2.1.1 Femtosecond-Laser Excitation Below the Damage Threshold

Harb et al. prepared a free standing polycrystalline Si film with a thickness of $d_\mathrm{film} =$ 50 nm [31]. They excited this film by an intense femtosecond laser-pulse with a central wavelength of $\lambda = 387$ nm and a FWHM-time width of $\tau = 150$ fs. The fluence of $I_{\mathrm{L_{tot}}} = 5.6\,\frac{\mathrm{mJ}}{\mathrm{cm}^2}$ was absorbed at the surface, which is below the damage threshold of $I_{\mathrm{L_{tot}}}^{(\mathrm{damage})} = 6.5\,\frac{\mathrm{mJ}}{\mathrm{cm}^2}$. They measured the time-dependent intensity of serveral Bragg peaks using ultrafast electron diffraction.

To compare directly with this measurement, we set up a simulation cell that consists of $11 \times 11 \times 93$ conventional cells and contains $N_\mathrm{at} = 90024$ Si atoms. We applied periodic boundary conditions in x- and y-direction and open boundary conditions in z-direction to get a 50 nm thick Si film. We used the Andersen thermostat (see Sect. 4.1.2) to initialize the atomic coordinates and velocities at $T_\mathrm{i} = 300$ K.

Then we performed MD simulations of the femtosecond-laser excitation using the three different previously described scenarios—excited PES & EPC, only excited PES, only EPC. We used a time step of $\Delta t = 1$ fs and simulated a Gaußian-shaped pulse with a FWHM-time width of $\tau = 150$ fs similar to the experiment. To obtain the energy $E_{\mathrm{L_{tot}}}$ absorbed from the laser from the measured absorbed fluence $I_{\mathrm{L_{tot}}}$ at the surface, we have to consider the optical properties of Si. Harb et al. excited the Si film with a femtosecond-laser having a central wavelength of $\lambda = 387$ nm, which corresponds to a photon energy of

$$E_{\mathrm{phot}} = \frac{2\pi\,\hbar\,\mathrm{c}}{\lambda} = 3.2\,\mathrm{eV}, \tag{7.8}$$

where $\hbar$ denotes the reduced Planck's constant and c the speed of light in vacuum. At this photon energy, we found $n = 6.062 + 0.630\,\mathrm{i}$ for the index of refraction of Si in literature [21]. Using this index of refraction, we obtained the absorption coefficient of Si of

$$\alpha_{\mathrm{abs}} = \frac{4\pi}{\lambda}\,\mathrm{Im}(n) = 0.0204569\,\frac{1}{\mathrm{nm}}, \tag{7.9}$$

where $\mathrm{Im}(n)$ denotes the imaginary part of the index of refraction n. Using the ab-initio equilibrium atomic density $\rho_{\mathrm{at}} = 50.8414\,\frac{\mathrm{atoms}}{\mathrm{nm}^3}$ and the experimental absorbed laser fluence $I_{\mathrm{L_{tot}}} = 5.6\,\frac{\mathrm{mJ}}{\mathrm{cm}^2} = 349.525\,\frac{\mathrm{eV}}{\mathrm{nm}^2}$ at the surface, we obtained for the from the laser absorbed energy $E_{\mathrm{L_{tot}}}$ in the 50 nm thick film (see Eq. (6.51))

$$\frac{E_{\mathrm{L_{tot}}}}{N_{\mathrm{at}}} = \left(1 - \mathrm{e}^{-\alpha_{\mathrm{abs}}\,d_{\mathrm{film}}}\right)\frac{I_{\mathrm{L_{tot}}}}{d_{\mathrm{film}}\,\rho_{\mathrm{at}}} \approx 0.1\,\frac{\mathrm{eV}}{\mathrm{atom}}, \tag{7.10}$$

which we utilized in our MD simulations.

From the atomic coordinates, we derived the time-dependent intensities of the experimental studied Bragg peaks (see Sect. 4.2). In order to be able to compare directly with the experiments, we took into account, that Harb et al. did not measure the intensity of a single Bragg peak. Since they utilized a polycrystalline Si film, they obtained rings in the diffraction image instead of spots, which would be present for a monocrystalline film. They averaged the intensities within a ring of a given radius and labeled the resulting value by a Bragg peak, which diffraction peak is located inside the ring. Therefore, to derive the intensity of such a measured Bragg peak (hkl), we have to average the intensity over every Bragg peak with scattering vector $\mathbf{q}$ fulfilling the condition $|\mathbf{q}| \in \left[|\mathbf{G}_{hkl}| - \Delta q, |\mathbf{G}_{hkl}| + \Delta q\right]$. We chose the broadening Δq in such a way that the calculated relative Bragg peak intensity matches as good as possible to the measured one. We show, as an example, the influence of the broadening Δq on the relative Bragg peak intensity for the (620) Bragg peak in Fig. 7.20, where we also include the corresponding experimental data points. Several Bragg peaks share the same absolute value $|\mathbf{q}|$ of the scattering vector. Due to the

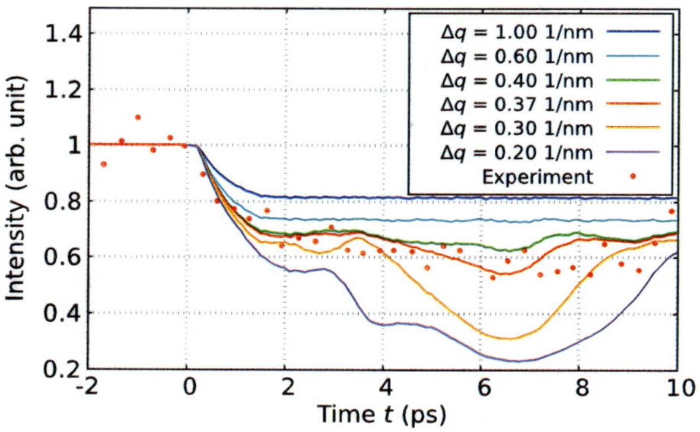

Fig. 7.20 Time-dependent relative intensity of the (620) Bragg peak is shown for several broadenings Δq. The points correspond to the measured (620) Bragg peak intensities taken from Fig. 4 of Ref. [31]

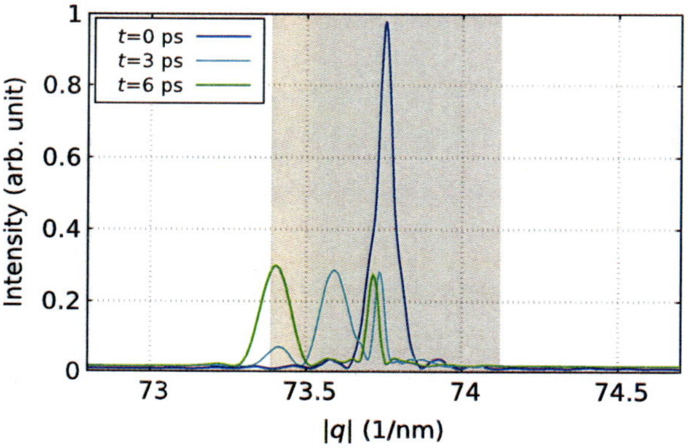

Fig. 7.21 Realtive scattering intensity is shown as a function of the absolute value $|\mathbf{q}|$ of the scattering vector for values close to $|\mathbf{G}_{620}| = 73.76\,\text{nm}^{-1}$. The gray area indicates the considered interval $\left[|\mathbf{G}_{620}| - \Delta q, |\mathbf{G}_{620}| + \Delta q\right]$ used to compare with the experiment

laser excitation, the thickness of the film oscillates and the absolute value $|\mathbf{q}|$ of the scattering vector changes for some of the Bragg peaks. In this way, $|\mathbf{q}|$ of some Bragg peaks moves out of the measured interval $\left[|\mathbf{G}_{hkl}| - \Delta q, |\mathbf{G}_{hkl}| + \Delta q\right]$, which dramatically influences the measured intensity. We show this, as an example, for the (620) Bragg peak in Fig. 7.21, where we present the $|\mathbf{q}|$-dependent relative intensity at selected times after the laser-excitation and where we indicate the used interval $\left[|\mathbf{G}_{620}| - \Delta q, |\mathbf{G}_{620}| + \Delta q\right]$ by a gray area.

In Fig. 7.22, we compare the measured relative intensities with our calculated ones for the six Bragg peaks that Harb et al. studied. We found an excellent agreement. In the next step, we analyzed the influences of the excited PES and EPC on the time behavior of the Bragg peak intensities.

By comparing the results from the three scenarios of MD simulations—excited PES & EPC, only EPC, only excited PES—we found out that the relative Bragg peak decay is dominated by the EPC whereas the influence of the excited PES is small. In Fig. 7.23, we present the calculated relative intensities of the Bragg peaks obtained from the MD simulation considering excited PES and EPC and from the MD simulation only considering the EPC. One can clearly see that the relative intensities obtained from the two different MD simulations are almost identical. In addition,

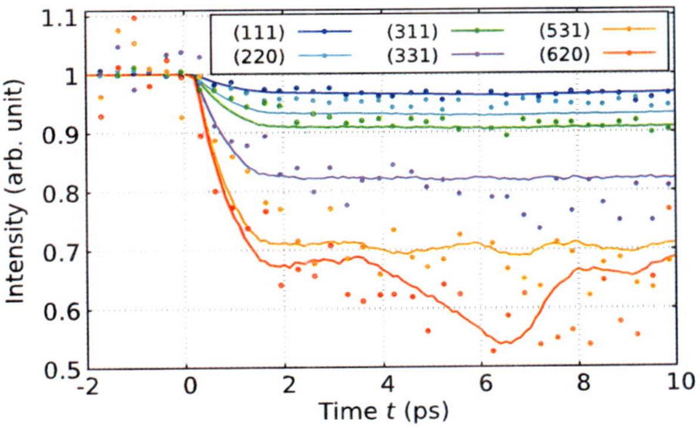

Fig. 7.22 Relative intensities of various Bragg peaks are shown as a function of time obtained from the experiment (points) and from our calculations (lines). The experimental values are taken from Fig. 4 of Ref. [31]

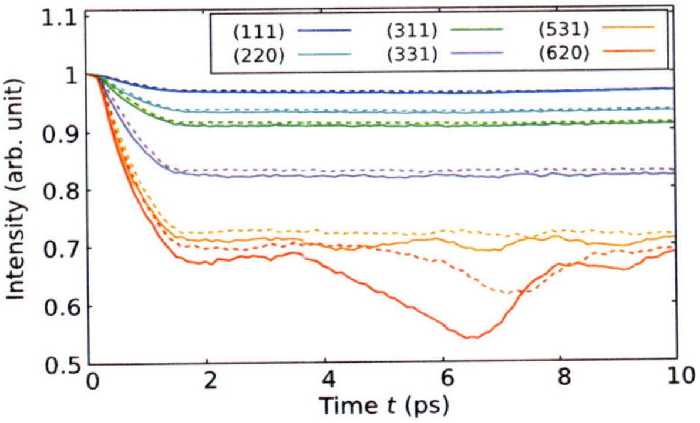

Fig. 7.23 Relative intensities of various Bragg peaks are shown as a function of time obtained from the MD simulation considering the excited PES & EPC (solid lines) and the MD simulation considering only the EPC (dashed lines)

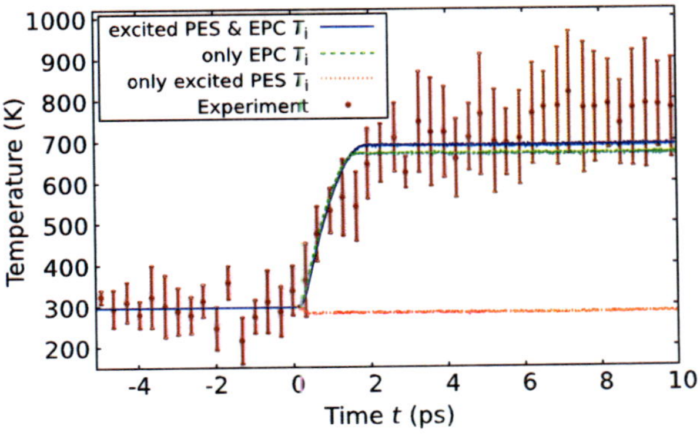

Fig. 7.24 Ionic temperature T_i is shown as a function of time obtained from the experiment (points with error bars) using Debye Waller theory and from our calculations (lines). The experimental values are taken from Fig. 5 of Ref. [31]

the relative intensities keep almost unaffected in the MD simulation which only considers the excited PES.

Harb et al. derived the time-dependent ionic temperature T_i of the Si film from the time-dependent Bragg peak intensities using Debye Waller theory. In Fig. 7.24, we show the ionic temperature T_i as a function of time obtained by Harb et al. and obtained from our three different MD simulations.

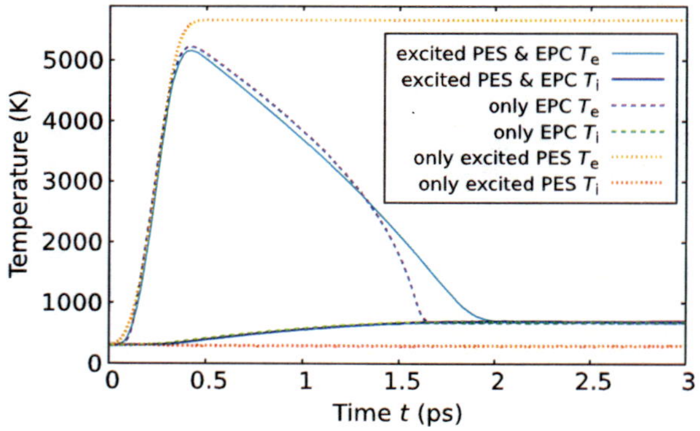

Fig. 7.25 Electronic and ionic temperatures are show as a function of time obtained from our calculations

Figure 7.24 clearly shows that the experimentally obtained T_i is well explained within the error bars by our MD simulation considering the excited PES and the EPC and by our MD simulation only considering the EPC. In Fig. 7.25, we present the electronic and ionic temperatures as a function of time obtained from our three different scenarios of MD simulations. T_e is firstly increased due to the laser excitation. Then, T_e is decreased and T_i is increased due to the EPC until both temperatures reach the same value.

7.2.1.2 Femtosecond-Laser Excitation Above the Damage Threshold

In a second measurement, Harb et al. prepared a free standing monocrystalline Si film with a thickness of $d_{\text{film}} = 30$ nm [32]. They again excited this film by an intense femtosecond laser-pulse with a central wavelength of $\lambda = 387$ nm and a FWHM-time width of $\tau = 150$ fs. The fluence of $I_{L_{\text{tot}}} = 65\,\frac{\text{mJ}}{\text{cm}^2}$ was absorbed at the surface, which is above the damage threshold. They again measured the time-dependent intensity of the (220) Bragg peak using ultrafast electron diffraction.

To compare directly with this measurement, we set up a simulation cell that consist of $11 \times 11 \times 56$ conventional cells and contains $N_{\text{at}} = 54208$ Si atoms. We applied periodic boundary conditions in x- and y-direction and applied open boundary conditions in z-direction ($[\bar{1}11]$ direction of the crystal structure) to get a 30 nm thick Si film. We applied the Andersen thermostat (see Sect. 4.1.2) to initialize the atomic coordinates and velocities at $T_i = 300$ K. Then we performed MD simulations of the femtosecond-laser excitation using the three different methods—excited PES & EPC, only excited PES, only EPC. We used a time step of $\Delta t = 1$ fs and simulated a Gaußian-shaped pulse with a FWHM-time width of $\tau = 150$ fs and set for the by the laser absorbed energy $E_{L_{\text{tot}}}$ (see Eq. (6.51)) in the 30 nm thick film

$$\frac{E_{L_{\text{tot}}}}{N_{\text{at}}} = \left(1 - e^{-\alpha_{\text{abs}}\, d_{\text{film}}}\right) \frac{I_{L_{\text{tot}}}}{d_{\text{film}}\, \rho_{\text{at}}} \approx 1.2\, \frac{\text{eV}}{\text{atom}} \tag{7.11}$$

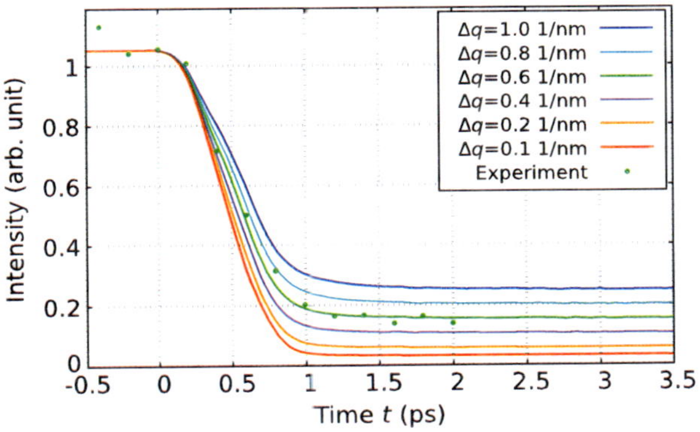

Fig. 7.26 Time-dependent relative intensity of the (220) Bragg peak is shown for several broadenings Δq. The points correspond to the measured (220) Bragg peak intensities taken from Fig. 3c of Ref. [32]

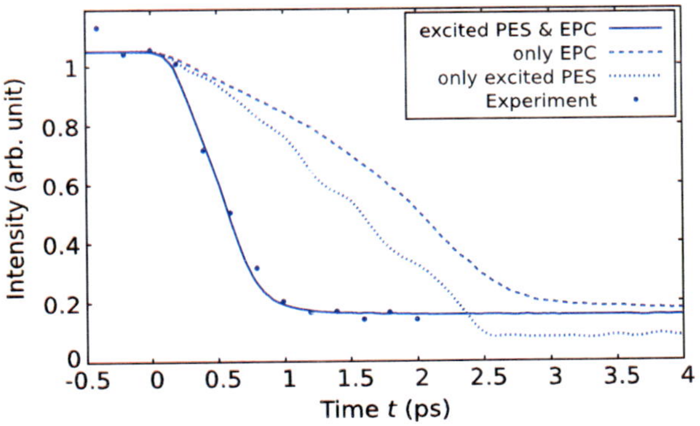

Fig. 7.27 Relative intensity of the (220) Bragg peak is shown as a function of time obtained from the experiment (points) and from our calculations (lines). The experimental values are taken from Fig. 3c of Ref. [32]

which we obtained from the previously derived absorption coefficient $\alpha_{\text{abs}} = 0.0204569\,\frac{1}{\text{nm}}$, the ab-initio equilibrium atomic density $\rho_{\text{at}} = 50.8414\,\frac{\text{atoms}}{\text{nm}^3}$, and the experimental absorbed laser fluence $I_{\text{L}_{\text{tot}}} = 65\,\frac{\text{mJ}}{\text{cm}^2} = 4056.98\,\frac{\text{eV}}{\text{nm}^2}$ at the surface.

We derived the time-dependent intensity of the (220) Bragg peak from the atomic coordinates by considering all Bragg peaks in the interval $\left[|\mathbf{G}_{220}| - \Delta q, |\mathbf{G}_{220}| + \Delta q\right]$, since Harb et al. measured the (220) Bragg peak again by averaging over the intensities inside a ring of the measured diffraction image. In Fig. 7.26, we present our obtained time-dependent intensities for various Δq for the MD simulation considering excited PES & EPC together with Harb's results. The value of Δq controls mainly the remaining intensity after the Bragg peak is decayed. The remaining intensity increases with increasing Δq, since one measures more of the background intensity for bigger Δq. We chose $\Delta q = 0.6\,\frac{1}{\text{nm}}$, since then the rest intensity of the experiment is well reproduced by our calculations. In Fig. 7.27, we present the calculated intensity decay of the (220) Bragg peak for our three different scenarios of MD simulations together with the experimental result. One can clearly see that the MD simulation considering excited PES & EPC describes excellently the experimental results, whereas the two other ones produce a to slow intensity decay.

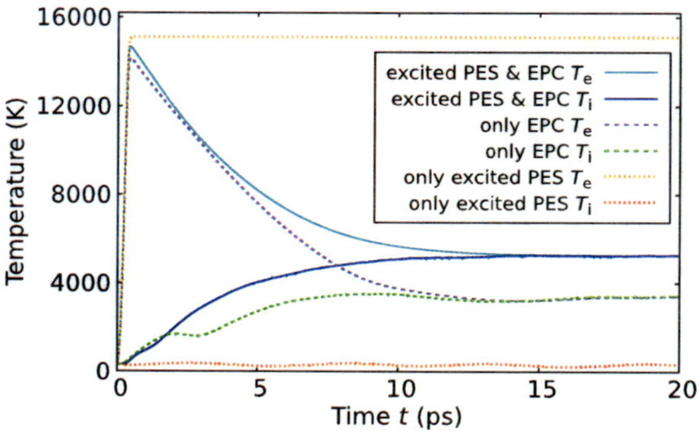

Fig. 7.28 Electronic and ionic temperatures are show as a function of time for the 30 nm Si film obtained from our calculations

In Fig. 7.28, we present the electronic and ionic temperatures as a function of time for our three different scenarios of MD simulations. If the EPC is taken into account, the final ionic temperature is higher, when the excited PES is also taken into account, since the bonds become weaker.

Harb et al. also prepared a polycrystalline Si film of $d_{\text{film}} = 50$ nm thickness. They excited also this film by an intense femtosecond laser-pulse with a central wavelength of $\lambda = 387$ nm and a FWHM-time width of $\tau = 150$ fs, so that the fluence of $I_{\text{L}_{\text{tot}}} = 65\,\frac{\text{mJ}}{\text{cm}^2}$ was absorbed at the surface. They measured the time-dependent intensity of serveral Bragg peaks using ultrafast electron diffraction.

In order to compare with this measurement, we utilized the previously generated simulation cell that consists of $11 \times 11 \times 93$ conventional cells and contains $N_{\text{at}} = 90024$ Si atoms. The atoms form a 50 nm thick Si film and were previously thermalized at $T_{\text{i}} = 300$ K. Again, we performed MD simulations of the femtosecond-laser excitation using the three different scenarios. We simulated a Gaußian-shaped pulse with a FWHM-time width of $\tau = 150$ fs and we set for the energy $E_{\text{L}_{\text{tot}}}$ (see Eq. (6.51)) that is absorbed from the laser in the 50 nm thick film

$$\frac{E_{\text{L}_{\text{tot}}}}{N_{\text{at}}} = \left(1 - \text{e}^{-\alpha_{\text{abs}}\, d_{\text{film}}}\right) \frac{I_{\text{L}_{\text{tot}}}}{d_{\text{film}}\, \rho_{\text{at}}} \approx 1.0\, \frac{\text{eV}}{\text{atom}}. \tag{7.12}$$

We derived the time-dependent intensities of the three measured Bragg peaks (111), (220), and (311) from the atomic coordinates and present the obtained results together with the experimental results in Fig. 7.29. One can clearly see that the experimental data points spread a lot indicating that the measurement may be not so accurate. The inaccuracy is further confirmed by the presence of points with a negative intensity. Harb et al. concluded from their measurement that the three Bragg peaks decay with a similar speed. This conclusion may be correct from the available data points of the measurement. However, our calculations show that the (111) Bragg peak decays significantly slower than the two other ones. This agrees with Debye Waller theory, since the (111) Bragg peak belongs to a wave vector with a significant smaller absolute value than the other Bragg peaks. The MD simulation considering

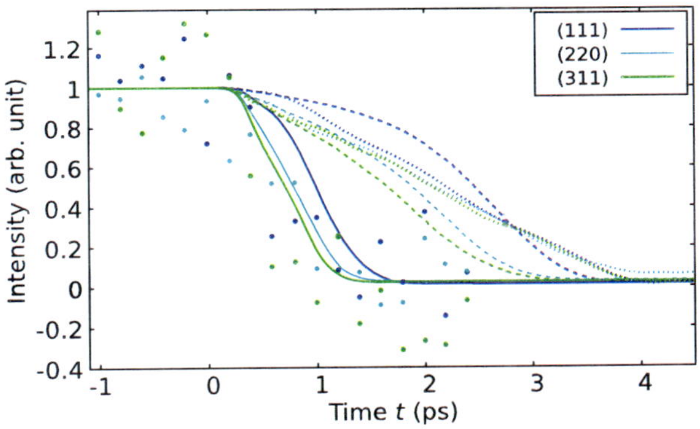

Fig. 7.29 Relative intensities of several Bragg peaks are shown as a function of time for the 50 nm Si film obtained from the experiment (points) and from the MD simulation considering excited PES & EPC (solid lines), from the MD simulation considering excited PES (dotted lines) and from the MD simulation considering EPC (dashed lines). The experimental values are taken from Fig. 3a of Ref. [32]

only the excited PES and the MD simulation considering only the EPC generate again a to slow Bragg peak decay compared to the experiment. The (111) Bragg peak also decays slower compared to the other ones for the MD simulation considering only the EPC. For the MD simulation considering only the excited PES, all Bragg peaks decay in a similar way.

7.2.2 *MD Simulations of a Femtosecond-Laser Excited Si Film*

In this section, we focus on the laser-induced melting of a Si film. We used the previously introduced simulation cell that contains $N_{at} = 90024$ Si atoms forming a 50 nm thick film and took our previously generated atomic coordinates and velocities at $T_i = 300$ K for the initialization.

We performed MD simulations considering an excitation by a Gaußian-shaped pulse with a FWHM-time width of $\tau = 10$ fs and several absorbed energies and revisited the three different simulation scenarios—excited PES & EPC, only excited PES, only EPC. We used again a time step of $\Delta t = 1$ fs. In order to identify molten and crystalline structures, we utilized the crystal symmetry parameter (CSP) introduced in Sect. 7.1.5. In addition, the crystal symmetry parameter allows us to determine the percentage of molten material.

We present in Fig. 7.30 the thickness of the film and in Fig. 7.31 the percentage of molten material as a function of time at several energies, that were absorbed from the laser, for the MD simulation with excited PES & EPC. At weak energies that are absorbed from the laser, the film only shows a periodical oscillation of the thickness

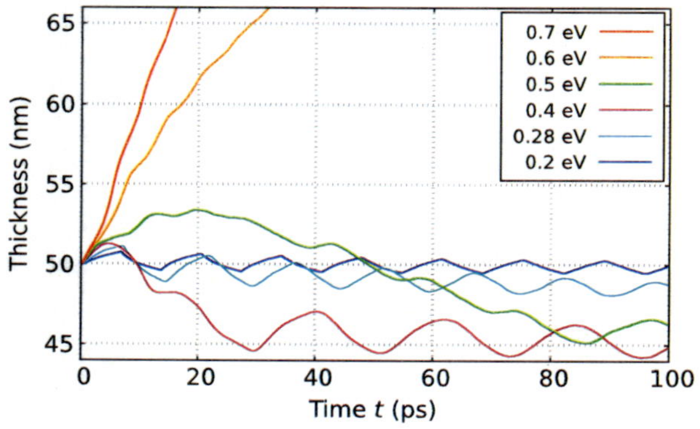

Fig. 7.30 Time-dependent thickness of the film is shown at several absorption energies for the MD simulations with excited PES & EPC. We derived the thickness as the difference between the averaged *z*-positions of the top and bottom crystal plane

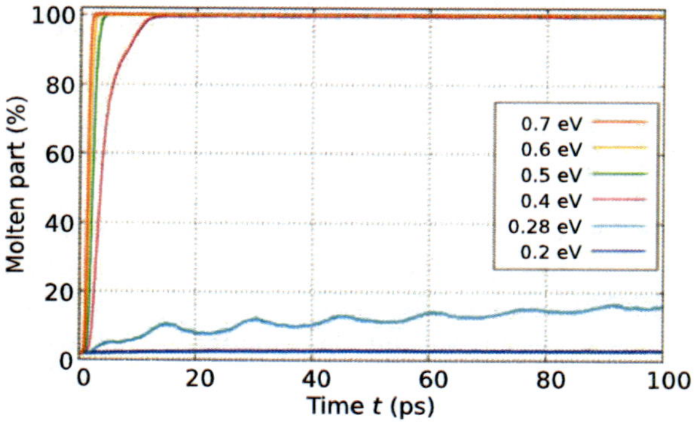

Fig. 7.31 Time-dependent fraction of molten material of the film is shown at several absorption energies for the MD simulations with excited PES & EPC

and no melting occurs. This behavior is present in all scenarios indicating that EPC and excited PES cause these oscillations alone and in combination. The oscillations become bigger, if the excited PES is also taken into account, since the atoms prefer larger interatomic distances at higher electronic temperatures.

If more energy is absorbed from the laser, the film starts to melt from the surface. In addition, a part of the crystal structure inside the film starts to melt and liquid regions are formed, which grow. In Fig. 7.32, we present snapshots of the MD simulation with excited PES & EPC, where an energy of $0.28 \frac{\text{eV}}{\text{atom}}$ was absorbed from the laser. Here, the film melts from the surface and one liquid region is formed in the middle of the film. At slightly higher energies that were absorbed from the laser, we observed the forming of two liquid regions.

One can only observe the melting starting from the surface, if the EPC is taken into account. The surface is a defect of the ideal crystal structure, so that the atoms are weaker bonded and can go over the barrier due to the increased velocities. If one only takes the excited PES into account, the film exclusively starts to melt inside at moderate excitation like for the DFT MD simulations (see Fig. 4.27). The reason for this is that the pressure increases in the center of the film, when the thickness shrinks again. In the region with higher pressure the crystal melts, since the melting temperature decreases at higher pressures (see Fig. 7.17). This excited PES induced

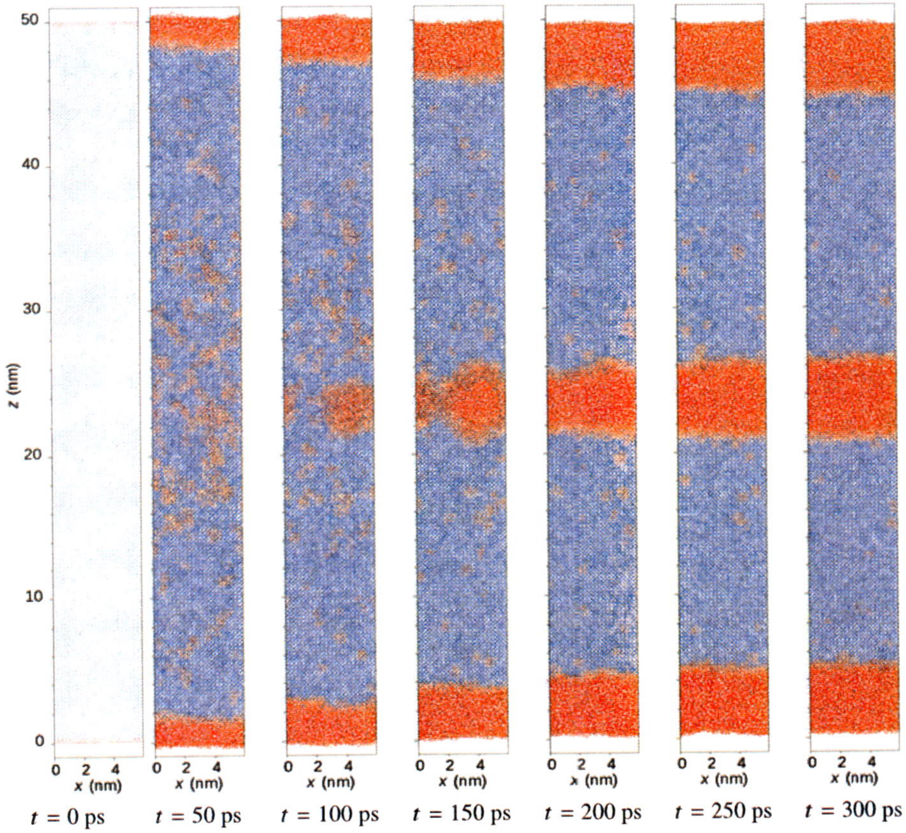

Fig. 7.32 Snapshots of the MD simulation considering excited PES & EPC of the 50 nm film containing 90024 Si atoms. The film was excited by a femtosecond laser-pulse with a FWHM time width of 10 fs, where the energy of $0.28\,\frac{\text{eV}}{\text{atom}}$ was absorbed. The atoms are colored related to their CSP value: blue corresponds to crystalline and red to molten environment

melting cannot be observed, if the EPC is taken into account, since the entire crystal structure immediately melts at this absorbed energy levels. On the other hand, one can also observe the melting inside the film, if only the EPC is taken into account. Here, small crystal defects inside the film induce the development of liquid nuclei, which then start to grow.

If more energy is absorbed from the laser, the entire film melts immediately. Due to this, the thickness is reduced, since the liquid phase prefers a smaller volume than the crystal phase, as one can see in Fig. 7.15. Beside this reduction on a long timescale, the thickness initially expands, since the atoms first prefer larger distances at increased T_e. The expansion causes the forming of voids and regions with low atomic densities inside the film, which vanish, when the film shrinks. This can be clearly seen in Fig. 7.33, where we present snapshots of the MD simulation with

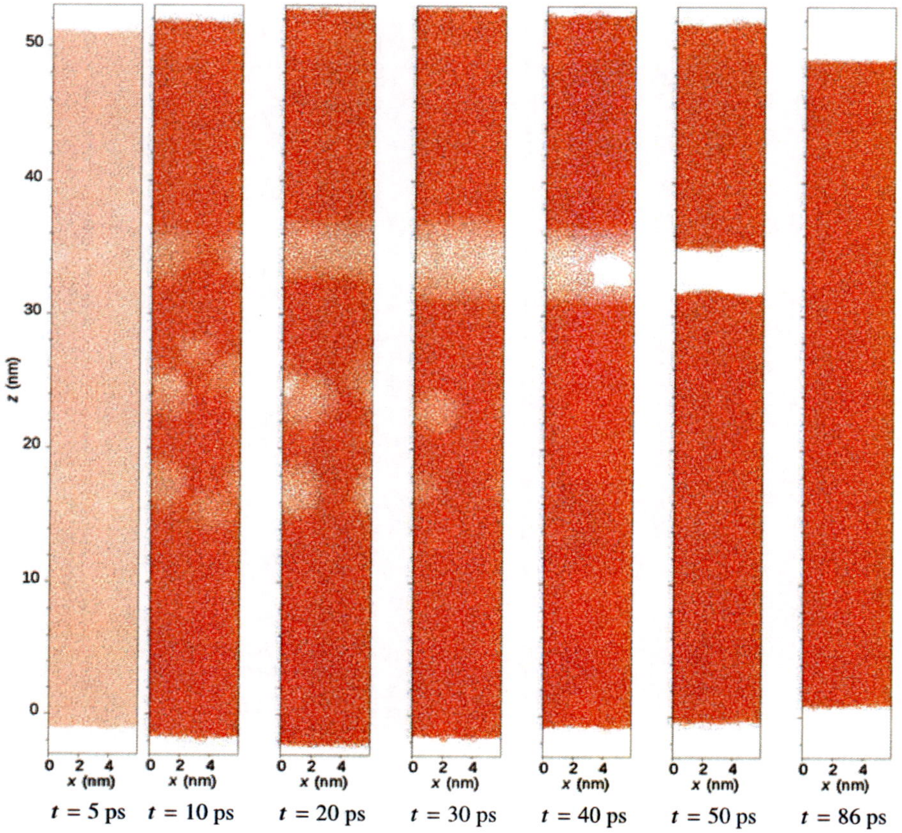

Fig. 7.33 Snapshots of the MD simulation considering excited PES & EPC of the 50 nm film containing 90024 Si atoms. The film was excited by a femtosecond laser-pulse with a FWHM time width of 10 fs, where the energy of $0.5\frac{\text{eV}}{\text{atom}}$ was absorbed. The atoms are colored related to their CSP value: blue corresponds to crystalline and red to molten environment

excited PES & EPC at an energy of $0.5\frac{\text{eV}}{\text{atom}}$ that was absorbed from the laser. The forming of voids and regions with low atomic densities is also present in the two other scenarios.

If more energy is absorbed, the entire film melts immediately and only expands. During the expansion, the upper layers of the film detach themselves and fly away alone. This can be clearly seen in Fig. 7.34, where we present snapshots of the MD simulation with excited PES & EPC at an energy of $1.0\frac{\text{eV}}{\text{atom}}$ that was absorbed from the laser. This so called ablation is also present, if one simulates only the EPC or only the excited PES.

For the excited PES only or the EPC only scenarios, one needs higher absorbed energies to reach the same melting speed compared to the MD simulation including both effects. This is not surprising, since both effects destabilize the crystal structure.

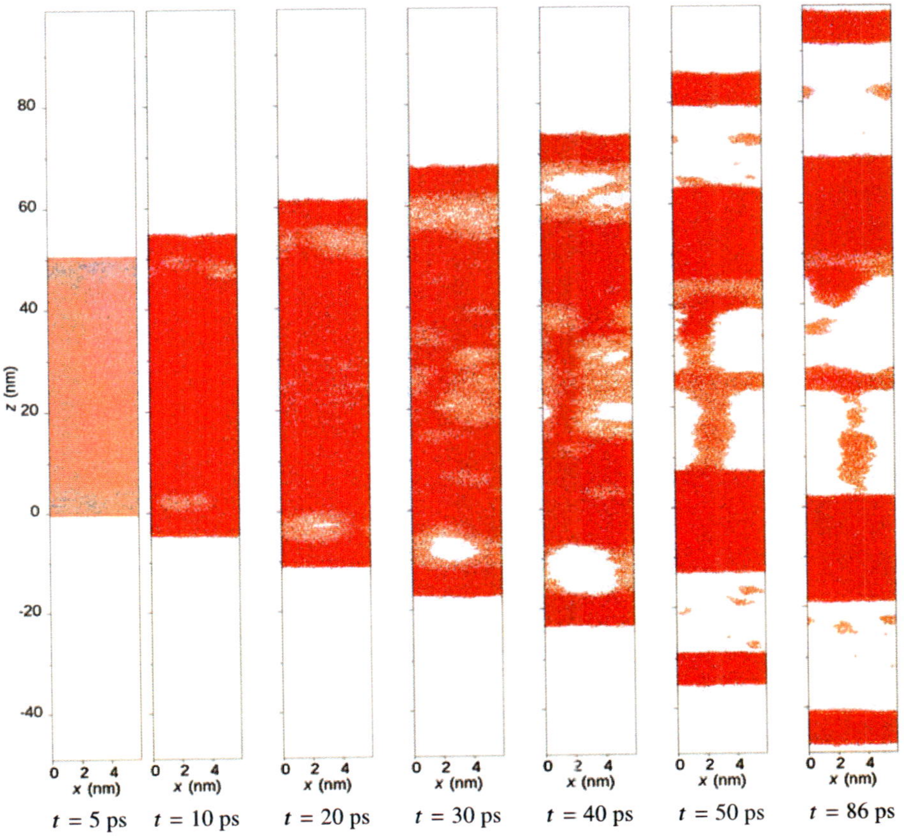

Fig. 7.34 Snapshots of the MD simulation considering excited PES & EPC of the 50 nm film containing 90024 Si atoms. The film was excited by a femtosecond laser-pulse with a FWHM time width of 10 fs, where the energy of $1.0\frac{\mathrm{eV}}{\mathrm{atom}}$ was absorbed. The atoms are colored related to their CSP value: blue corresponds to crystalline and red to molten environment

Moreover, the laser induced melting is mainly dominated by the EPC in contrast to the excited PES, which only plays a small role.

Due to the excited PES, the oscillation of the thickness at moderate excitation or the expansion at high excitation is somewhat bigger compared to the simulations with only EPC. This bigger expansion related to the excited PES can be observed, since the expansion causes a delay in the melting below the surface. In Fig. 7.35, we present the averaged z-positions of the crystal planes as a function of time for an energy of $1.0\,\frac{\mathrm{eV}}{\mathrm{atom}}$ that is absorbed from the laser for the MD simulation considering excited PES & EPC and the MD simulation considering only the EPC. The line width of the curves corresponds to the standard deviations of the atomic z-coordinates in each plane. A melting of the local crystal structure is indicated by a significant increase of the line width, especially in such a way, that the line overlaps with the neighboring

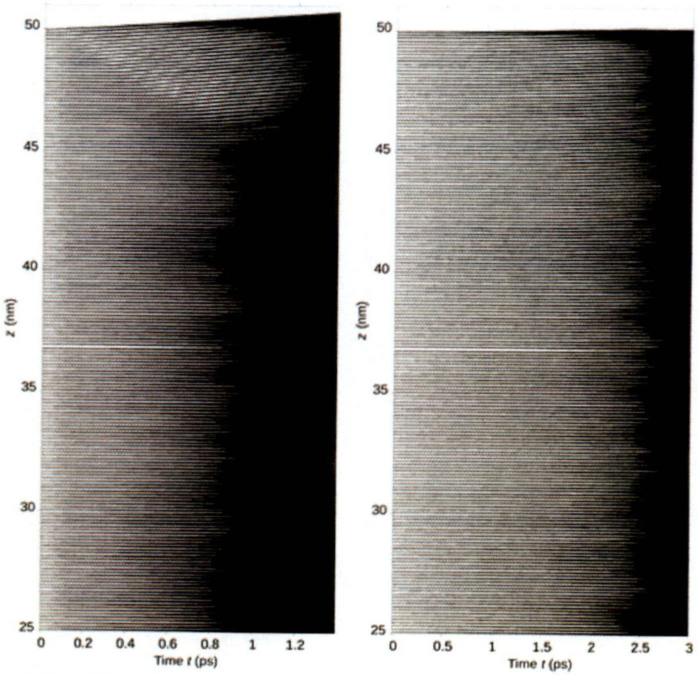

Fig. 7.35 Averaged z-positions of the crystal planes of the 50 nm film are shown as a function of time. The line width corresponds to the standard deviations of the atomic z-coordinates in each plane. A melting of the local crystal structure is indicated by a significant increase of the line width, especially in such a way, that an overlap with neighboring lines occurs. The film was excited by a femtosecond laser-pulse with a FWHM time width of 10 fs, where the energy of $1.0\,\frac{\text{eV}}{\text{atom}}$ was absorbed

lines. One can clearly see in Fig. 7.35 that, up to 5 nm below the surface, the laser-induced melting is delayed by approximately 300 fs compared to the center of the film. This effect cannot be seen in the MD simulation, which only includes the EPC, since the expansion at the surface is much smaller compared to the MD simulation with excited PES & EPC.

The delayed melting at the surface may be also experimentally observable by depth-resolved x-ray diffraction. The Bragg peaks measured at the surface will decay slower compared to ones that are measured more deeper.

7.2.3 MD Simulations of Femtosecond-Laser Excited Bulk Si

In the T_e-dependent DFT MD simulations of Si at increased T_e without EPC, one observes thermal phonon antisqueezing (see Sect. 4.4.4) and non-thermal melting

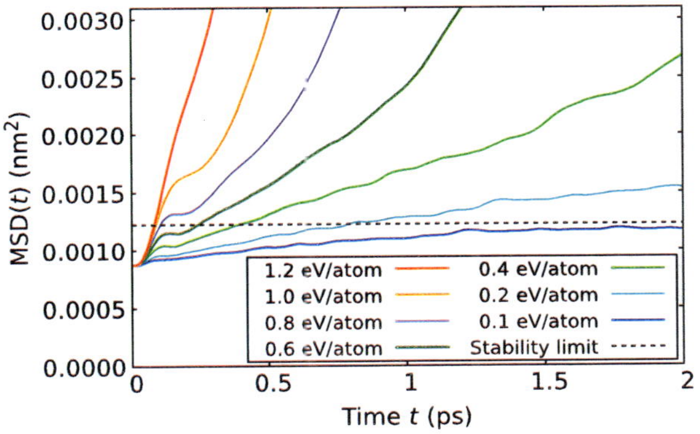

Fig. 7.36 Time-dependent mean-square displacement is shown at several absorption energies for the MD simulations with excited PES & EPC. The black dashed line indicates the Lindemann stability limit

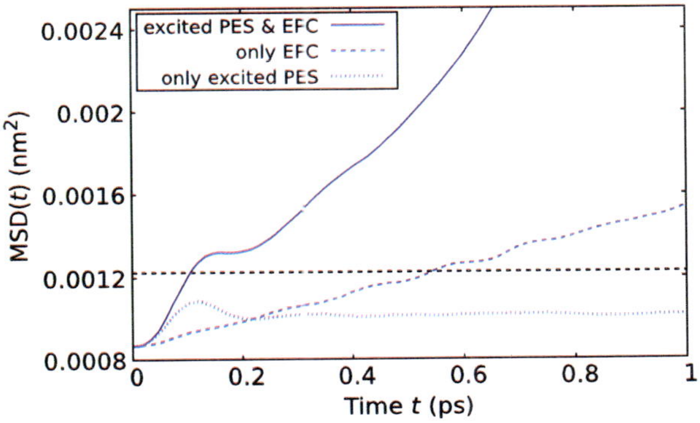

Fig. 7.37 Time-dependent mean-square displacement is shown at an absorbed energy of 0.8 $\frac{\text{eV}}{\text{atom}}$ for the three different scenarios of MD simulations. The black dashed line indicates the Lindemann stability limit

in three stages (see Sect. 4.4.5) driven by the excited PES. In order to analyze the influence of the EPC on both effects, we performed additional bulk MD simulations using our same three scenarios. We set up a simulation cell that consists of $32 \times 16 \times 16$ conventional cells and $N_{at} = 65536$ Si atoms. This simulation cell was also used to calculate the thermophysical properties in Sect. 7.1.5 We applied periodic boundary conditions in all directions to get bulk Si. By applying the Andersen thermostat (see Sect. 4.1.2), we initialized the atomic coordinates and velocities at $T_i = 300$ K. Using this initialization, we performed MD simulations of the laser excitation. We simulated a Gaußian-shaped pulse with a FWHM-time width of $\tau = 10$ fs and we considered several energies $E_{L_{tot}}$ that were absorbed from the laser.

At first, we studied the thermal phonon antisqueezing at moderate absorbed energies. In Fig. 7.36, we present the atomic mean-square displacement as a function of time for several energies $E_{L_{tot}}$ for the MD simulations with excited PES & EPC.

At low and moderate absorbed energies, the atomic mean-square displacement increases only with tiny oscillations. For energies $E_{L_{tot}}$ between 0.4 $\frac{\text{eV}}{\text{atom}}$ and 1.0 $\frac{\text{eV}}{\text{atom}}$, one can clearly see a kink in the curve of the mean-square displacement approximately 200 fs after the laser excitation. This kink is driven by the excited PES and is related to the thermal phonon antisqueezing. This can be seen in Fig. 7.37, in which

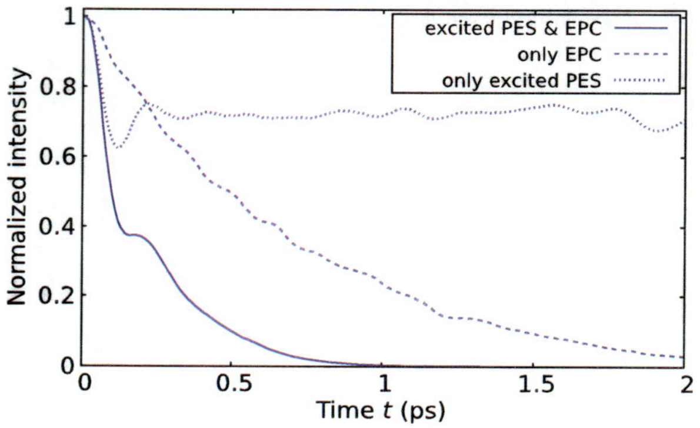

Fig. 7.38 Normalized intensity of the (400) Bragg peak is shown as a function of time at an absorbed energy of $0.8\,\frac{\text{eV}}{\text{atom}}$ for the three different scenarios of MD simulations

we present the mean-square displacement of the atoms as a function of time at an absorbed energy of $E_{\text{L}_{\text{tot}}} = 0.8\,\frac{\text{eV}}{\text{atom}}$ for the three different scenarios of MD simulations. If only the EPC is simulated, the atomic mean-square displacement increases only with some tiny oscillations. If only the excited PES is simulated, the atomic mean-square displacement first increases and then decreases significantly due to the antisqueezing. Then the atomic mean-square displacement oscillates weakly below the Lindemann stability limit indicating that the crystal structure remains intact. If both effects are simulated, the atomic mean-square displacement shows firstly a kink and then increases further. The kink is located at the same time as the decrease that is related to the antisqueezing in the MD simulation with only excited PES. Also in realizable experiments, one would still observe the thermal antisqueezing as a kink in the time-dependent intensities of the Bragg peaks. One can clearly see this in Fig. 7.38, where we present the time-dependent intensity of the (400) Bragg peak, as an example, for the three different scenarios. However, the effect of the thermal phonon squeezing is quite small and is overlaid by the EPC driven melting.

In a next step, we analyzed the non-thermal melting. In Fig. 7.39, we present the atomic mean-square displacement as a function of time for several higher absorbed energies $E_{\text{L}_{\text{tot}}}$ for the MD simulations with excited PES & EPC. In order to obtain a deeper insight, we show in Fig. 7.40 the time derivative of the atomic mean-square displacement as a function of time.

At low and moderate absorbed energies, the melting takes place only in two stages, as one can see in Fig. 7.40 (see also Sect. 4.4.5):

1. The time derivative of the mean-square displacement increases during the super diffusive stage.
2. The time derivative of the mean-square displacement remains constant during the normally diffusive stage. The electronic and electronic temperature are equal.

At very high absorbed energies, the melting takes place in four stages in the MD simulations with excited PES & EPC included:

1. The time derivative of the mean-square displacement increases during the super diffusive stage.

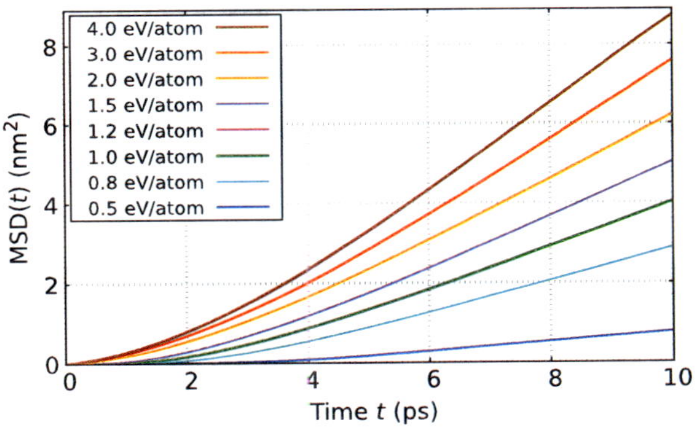

Fig. 7.39 Time-dependent mean-square displacement is shown at several absorption energies for the MD simulations with excited PES & EPC

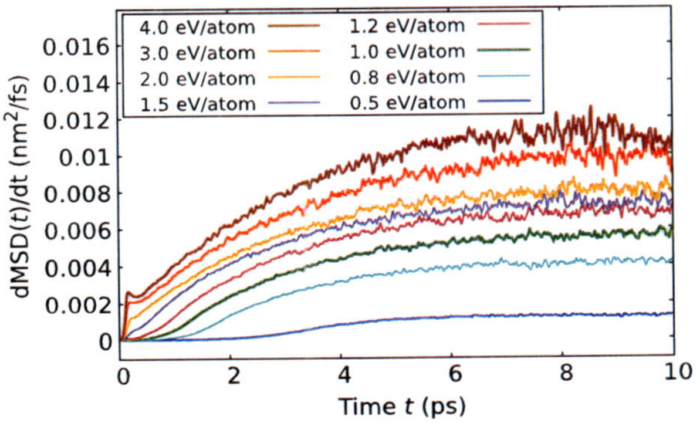

Fig. 7.40 Time derivative of the mean-square displacement is shown as a function of time at several absorption energies for the MD simulations with excited PES & EPC

2. The time derivative decreases shortly and weakly during the fractional diffusive stage.
3. The time derivative increases again.
4. The time derivative remains constant during the normally diffusive stage. The electronic and ionic temperature are equal.

One can see this behavior with four stages most clearly at the highest studied absorbed energy, namely 4.0 $\frac{\mathrm{eV}}{\mathrm{atom}}$. We present the time-dependent mean-square displacement in Fig. 7.41 and the time derivative of the mean-square displacement in Fig. 7.42 at an absorbed energy of 4.0 $\frac{\mathrm{eV}}{\mathrm{atom}}$ for the three different scenarios of MD simulations. If only the EPC is simulated, the melting occurs in two stages. If only the excited PES is simulated, the melting occurs in three stages as described in Sect. 4.4.5. If the excited PES and EPC is taken into account, one observes the four stages described above. Moreover, during the fractional diffusive stage, the decrease of the time derivative is quite small and cannot be seen in the time behavior of the mean-square displacement in Fig. 7.39 in contrast to the DFT MD simulations with only excited PES presented in Fig. 4.54. The three stages of melting induced by the excites PES are strongly overlaid by the EPC driven melting, which only consists of two stages. In Fig. 7.43, we present the time-dependent intensities of various Bragg

Fig. 7.41 Time-dependent mean-square displacement is shown at an absorbed energy of 4.0 $\frac{\text{eV}}{\text{atom}}$ for the three different scenarios of MD simulations

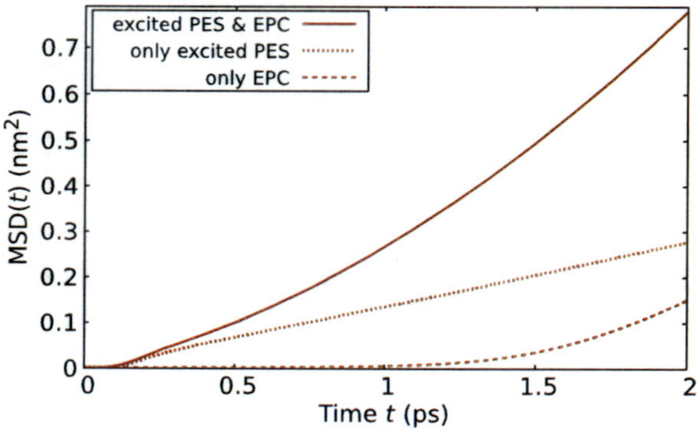

Fig. 7.42 Time derivative of the mean-square displacement is shown as a function of time at an absorbed energy of 4.0 $\frac{\text{eV}}{\text{atom}}$ for the three different scenarios of MD simulations

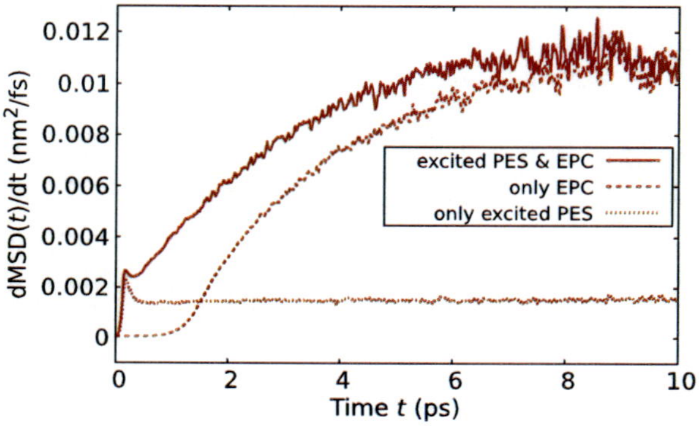

Fig. 7.43 Normalized intensities of various Bragg peak are shown as a function of time at an absorbed energy of 4.0 $\frac{\text{eV}}{\text{atom}}$ for the MD simulation with excited PES & EPC (solid lines), MD simulation only with EPC (dashed lines) and MD simulations only with excited PES (dotted lines)

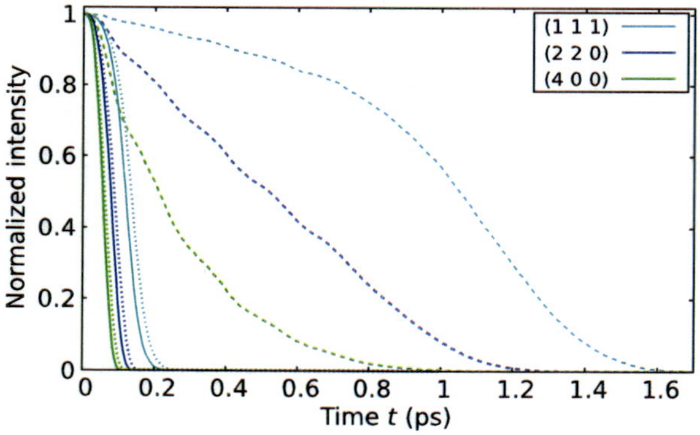

peaks at an absorbed energy of 4.0 $\frac{\text{eV}}{\text{atom}}$ for the three different scenarios of MD simulations. In all of the three different MD simulations, the Bragg peak intensities decay with significantly different times to zero. The behavior of the Bragg peak intensities is quite similar between the MD simulation considering the exited PES & EPC and the MD simulation considering only the excited PES, whereas the Bragg peak intensities decay much slower for the MD simulation considering only the EPC.

7.3 Correction of the Melting Temperature

$\Phi^{(\mathrm{Si})}(T_\mathrm{e})$ exhibits a melting temperature of $T_\mathrm{m} = 1199 \pm 2$ K, as we derived in Sect. 7.1.5. This value agrees with $T_\mathrm{m} = (1300 \pm 50)$ K from LDA-DFT [26], which is not surprising, since $\Phi^{(\mathrm{Si})}(T_\mathrm{e})$ was fitted to LDA-DFT reference simulations. However, the melting temperature of $\Phi^{(\mathrm{Si})}(T_\mathrm{e})$ is significantly below the experimental value of $T_\mathrm{m} = (1687 \pm 5)$ K [27]. If one performs MD simulations with $\Phi^{(\mathrm{Si})}(T_\mathrm{e})$ and wants to compare directly with experiments, $\Phi^{(\mathrm{Si})}(T_\mathrm{e})$ should reproduce the experimental melting temperature. Consequently, now we present modifications of the coefficients of $\Phi^{(\mathrm{Si})}(T_\mathrm{e})$, which brings its melting temperature to the experimental value. This is possible, since, firstly, $\Phi^{(\mathrm{Si})}(T_\mathrm{e})$ is constructed as a sum of the physical interpretable potential terms Φ_0, Φ_2, Φ_3, Φ_ρ and since, secondly, this potential terms are formulated in a simple polynomial functional form. We want to note that such a manipulation to adapt properties, which cannot be just fitted, would be very complicated or even impossible for the commonly used machine-learning interatomic potentials introduced in Sect. 5.1.3.

But how can the melting temperature be corrected?

The basic idea to adapt the melting temperature: The Si crystal, which forms the diamond-like structure, must be stabilized in the interatomic potential description. A stabilization of a structure means that the bonding energy of this structure should become higher within the interatomic potential description. Figure 7.44 presents the conventional cell of the diamond-like structure. For the dark green atoms, all nearest neighbors are included in the cell and, consequently, are shown. Indeed, in the diamond-like structure, each atom joins four nearest neighbors. The angle θ_{ijk} between any of these neighbors is always the same and obeys $\cos(\theta_{ijk}) = -\frac{1}{3}$. Consequently, this angle should be stabilized for the nearest neighbors. This can be easily done by adding the following correction term to the three-body potential:

$$\Phi_3^{(\mathrm{cor})}\left(r_{ij}, r_{ik}, \cos(\theta_{ijk})\right) = g(r_{ij})\, g(r_{ik}) \left(\cos(\theta_{ijk}) + \frac{1}{3}\right)^2, \tag{7.13}$$

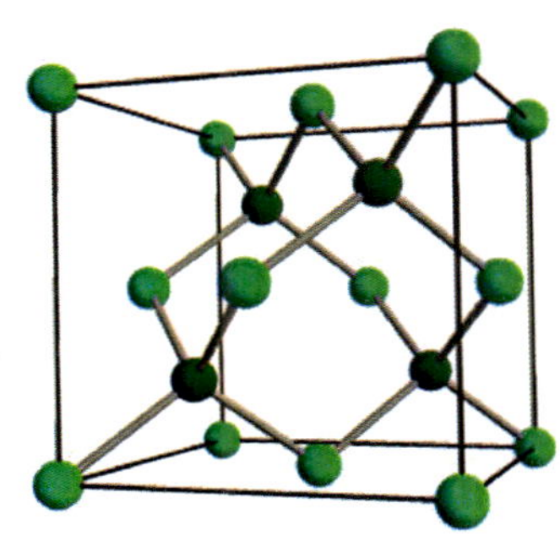

Fig. 7.44 Primitive cell of the diamond-like structure

with $g(r_{ij}) \geq 0$ and $g(r_{ik}) \geq 0$. By this construction, the preferred nearest neighbor angle of the diamond-like structure is stabilized by the parabola that exhibits its minimum at $-\frac{1}{3}$ for $\cos(\theta_{ijk})$. The distance function $g(r)$ should take care, that mainly the nearest neighbors are affected, which are located at a distance of 0.234 nm for Si derived ab-initio using CHIVES (see Table 4.4).

7.3.1 Correction of the 3-Body Potential Coefficients

At first, we modified only the coefficients of the three-body potential in such a way, that the melting temperature is increased, i.e., that the above mentioned term (7.13) is added to the potential. The polynomial Si potential $\Phi^{(\mathrm{Si})}(T_\mathrm{e})$ uses a three-body cutoff radius of $r_3^{(\mathrm{c})} = 0.42$ nm and the degrees $N_3^{(r)} = 3$ and $N_3^{(\theta)} = 2$ for the three-body potential (see Sect. A.8):

$$\Phi_3\big(r_{ij}, r_{ik}, \cos(\theta_{ijk})\big) = \sum_{q_1=2}^{3} \sum_{q_2=q_1}^{3} \sum_{q_3=0}^{2} c_3^{(q_1\, q_2\, q_3)} \left(1 - \frac{r_{ij}}{r_3^{(\mathrm{c})}}\right)^{q_1} \left(1 - \frac{r_{ik}}{r_3^{(\mathrm{c})}}\right)^{q_2} \times \times \Big(\cos(\theta_{ijk})\Big)^{q_3}. \quad (7.14)$$

To add a correction term like (7.13) to the three-body potential, we use the following construction, which just corresponds to a modification of three existing coefficients:

$$\begin{aligned}
\Phi_3^{(\mathrm{cor})}\big(r_{ij}, r_{ik}, \cos(\theta_{ijk})\big) &= \underbrace{\sqrt{\aleph_3}\left(1 - \frac{r_{ij}}{r_3^{(\mathrm{c})}}\right)^3}_{=g(r_{ij})} \underbrace{\sqrt{\aleph_3}\left(1 - \frac{r_{ik}}{r_3^{(\mathrm{c})}}\right)^3}_{=g(r_{ik})} \left(\cos(\theta_{ijk}) + \frac{1}{3}\right)^2 \\
&= \aleph_3 \left(1 - \frac{r_{ij}}{r_3^{(\mathrm{c})}}\right)^3 \left(1 - \frac{r_{ik}}{r_3^{(\mathrm{c})}}\right)^3 \Big(\cos(\theta_{ijk})\Big)^2 \\
&\quad + \frac{2}{3}\aleph_3 \left(1 - \frac{r_{ij}}{r_3^{(\mathrm{c})}}\right)^3 \left(1 - \frac{r_{ik}}{r_3^{(\mathrm{c})}}\right)^3 \cos(\theta_{ijk}) \\
&\quad + \frac{1}{9}\aleph_3 \left(1 - \frac{r_{ij}}{r_3^{(\mathrm{c})}}\right)^3 \left(1 - \frac{r_{ik}}{r_3^{(\mathrm{c})}}\right)^3 .
\end{aligned} \quad (7.15)$$

Here, the correction strength is controlled by $\aleph_3$ and $\aleph_3 = 0$ corresponds to the uncorrected original potential. We chose the term

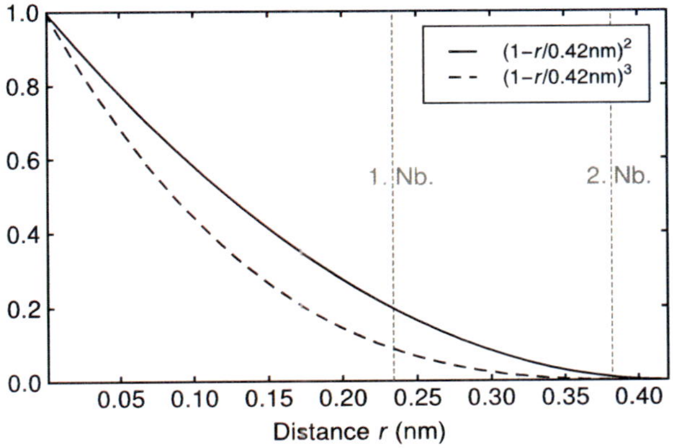

Fig. 7.45 Plot of the two different terms used in the potenial distance functions are shown. The two gray vertical lines indicate the position of the first and second nearest neighbors in Si

$$\left(1 - \frac{r}{r_3^{(c)}}\right)^3$$

for the distance function $g(r)$ instead of

$$\left(1 - \frac{r}{r_3^{(c)}}\right)^2,$$

since power of three converges faster to zero at reaching the cutoff-radius of $r_3^{(c)} = 0.42$ nm and, consequently, the correction is more dominated at the nearest neighbor distance of 0.234 nm (see Fig. 7.45). To add the above mentioned term to the potential corresponds to add $\frac{1}{9}\aleph_3$ to $c_3^{(3\,3\,0)}$, $\frac{2}{3}\aleph_3$ to $c_3^{(3\,3\,1)}$, and $\aleph_3$ to $c_3^{(3\,3\,2)}$. Since the potential correction should only take place at low electronic temperatures T_e around the experimental melting temperature $T_m = 1687$ K, we did the following: The interatomic potential coefficients are obtained from a polynomial approximation of the fitted ideal coefficient values at the eleven electronic temperatures of 316 K (1 mHa), 3158 K (10 mHa), 6315 K (20 mHa), ..., 31578 K (100 mHa). Consequently, we added the corresponding correction value to the ideal coefficient values for $c_3^{(3\,3\,0)}$, $c_3^{(3\,3\,1)}$, and $c_3^{(3\,3\,2)}$. Furthermore, we added the correction value at 316 K and added half of it at 3158 K, since the correction should only take place at low electronic temperatures T_e and should smoothly vanish above the experimental melting temperature of $T_m = 1687$ K. Finally, the corrected polynomial is approximated from these at two low T_e's shifted ideal coefficient values. To demonstrate this procedure, we show the original and corrected ideal coefficient values together with the corresponding original and corrected polynomial in Fig. 7.47 for $\aleph_3 = 12$ eV, which leads to the experimental melting temperature. Indeed, the corrected polynonial significantly differs from the original only at electronic temperatures below 7000 K.

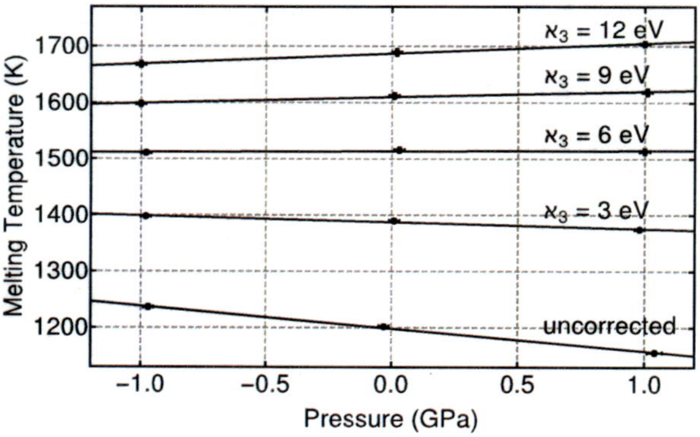

Fig. 7.46 The melting temperatures T_{m} of several corrected and the uncorrected interatomic potential are shown for different pressures p together with the linear regression line. The error bars indicate the standard deviation of the ionic temperature and of the pressure in the liquid-crystal coexistence MD simulations

Table 7.5 Melting temperature $T_{\mathrm{m}}\big|_{p=0}$ and slope $\frac{dT_{\mathrm{m}}}{dp}\big|_{p=0}$ in the T_{m} *versus* p diagram near zero pressure are listed for different $\aleph_3$-corrections of the interatomic potential coefficients. The presented errors have their origin in the standard deviation of the ionic temperature in the liquid-crystal coexistence MD simulations, since the error of the linear regression is much smaller

$\aleph_3$ (eV)	$T_{\mathrm{m}}\big\|_{p=0}$ (K)	$\frac{dT_{\mathrm{m}}}{dp}\big\|_{p=0}$ $\left(\frac{\mathrm{K}}{\mathrm{GPa}}\right)$
0.0	1199 ± 2	-40 ± 3
3.0	1388 ± 2	-12 ± 3
6.0	1514 ± 3	2 ± 3
9.0	1610 ± 3	11 ± 4
12.0	1687 ± 3	18 ± 4

We performed the above described correction of the interatomic potential for several values of $\aleph_3$. We calculated the melting temperature T_{m} for each corrected interatomic potential at the pressures p of -1 GPa, 0 GPa, and 1 GPa by performing large scale liquid-crystal coexistence MD simulations using the simulation cell with 65536 atoms as described in Sect. 7.1.5.1. The melting temperature T_{m} *versus* pressure slope $\frac{dT_{\mathrm{m}}}{dp}\big|_{p=0}$ and melting temperature $T_{\mathrm{m}}\big|_{p=0}$ at zero pressure was determined from the obtained three T_{m} values by a linear regression. In Fig. 7.46, the derived pressure-depend melting temperatures are shown together with the linear regression for different $\aleph_3$ values. The errorbars indicate the standard deviation coming from the time average in the MD simulation. In Table 7.5, we present the results of the linear regression.

The melting temperature of the interatomic potential increases with increasing $\aleph_3$ correction, as expected, and the experimental value is reached at $\aleph_3 = 12$ eV. The manipulation for getting the experimental melting temperature introduces additional changes to the potential, which are shown in Figs. 7.48, 7.49, 7.50 and 7.51. Here,

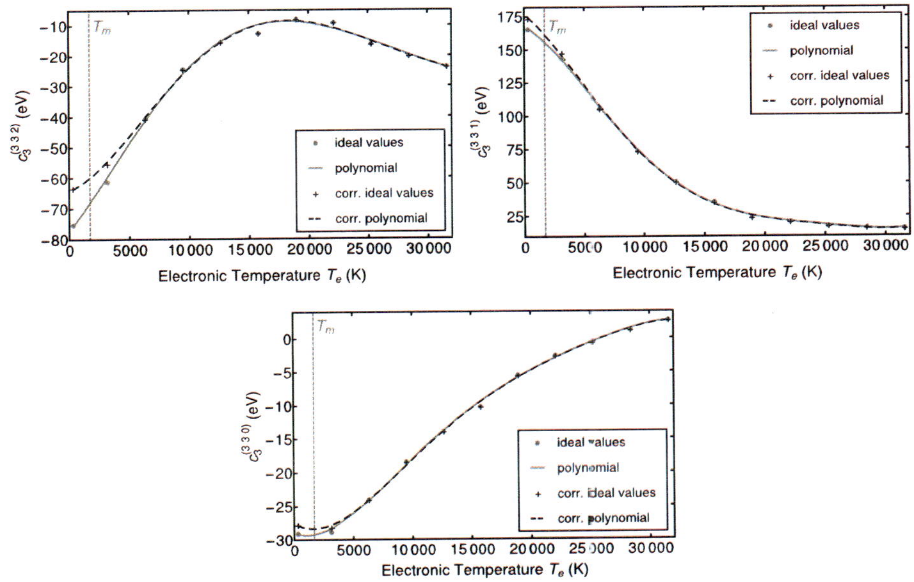

Fig. 7.47 The polynomial approximation of the potential coefficients $c_3^{(330)}$, $c_3^{(331)}$, and $c_3^{(332)}$ is shown together with the ideal values before (gray) and after (black) the correction of the ideal values at $T_e = 316$ K and $T_e = 3158$ K. $\aleph_3 = 12$ eV is shown, which yields the experimental melting temperature $T_m = 1687$ K highlighted by a gray vertical line

we present the lowest $T_e = 316$ K, since the biggest modification of the interatomic potential occurs at the lowest T_e by construction (see Fig. 7.47). The potential modification slightly increases the phonon frequencies, but it does not significantly change the electronic specific heat C_e and the absorbed electronic energy of the diamond-like structure. The cohesive energy of the diamond-like structure and of the fcc, bcc, and sc structure is moderately decreased.

The added correction term (7.15) does not influence the diamond-like structure, since this term is zero for the nearest neighbor angle within the diamond-like structure by construction. Nevertheless, the cohesive energy of the diamond-like structure moderately decreases. The reason for this decrease is the used polynomial approximation of the shifted ideal coefficient values to get a smooth T_e-depend interatomic potential. The ideal coefficient values are shifted in such a way, that the correction term (7.15) is added. But the approximated corrected polynomial is not exactly equal to the fitted shifted ideal coefficient value at $T_e = 316$ K, especially for $c_3^{(331)}$, as it can be seen in Fig. 7.47. Consequently, the interatomic potential, which uses the T_e-depend polynomial approximation for the coefficients, exhibits a change in the cohesive energy of the diamond-like structure.

The relative error in the atomic forces and structural free cohesive energies of reference simulation s_1 is listed for the melting temperature corrected and uncorrected interatomic potential in Table 7.6 at different T_e's. The melting temperature correction

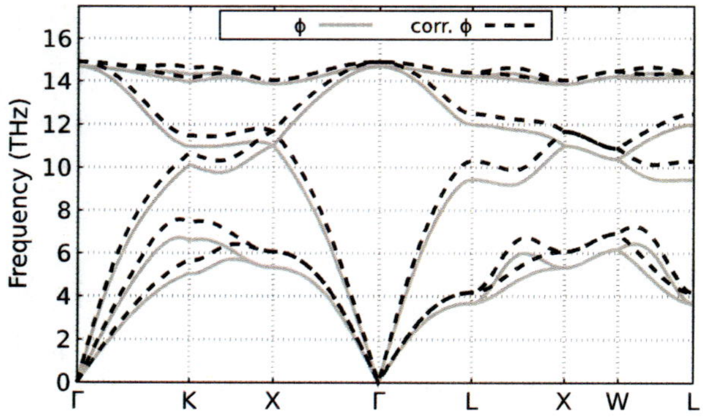

Fig. 7.48 The phonon bandstructure of the diamond structure is shown for the $\aleph_3 = 12$ eV corrected (black dashed) and uncorrected (gray solid) interatomic potential Φ at $T_e = 316$ K

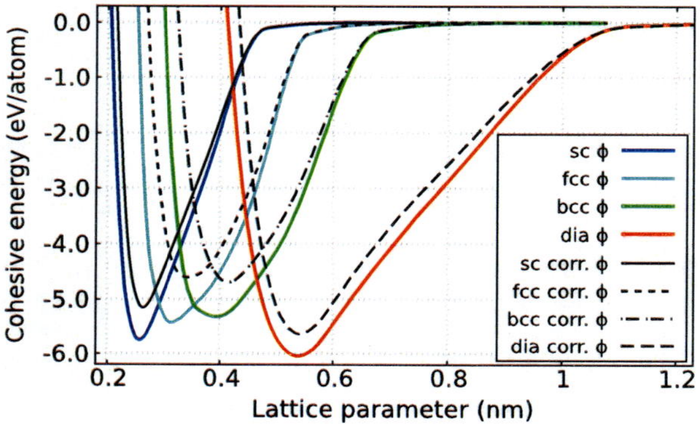

Fig. 7.49 The cohesive energies of different structures are shown for the $\aleph_3 = 12$ eV corrected (black dashed) and uncorrected (colored solid) interatomic potential Φ at $T_e = 316$ K

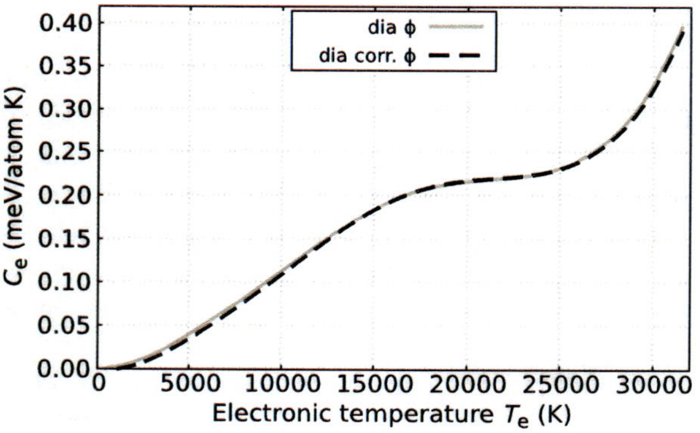

Fig. 7.50 The electronic specific heat C_e per atom of the diamond-like structure is shown for the $\aleph_3 = 12$ eV corrected (black dashed) and uncorrected (gray solid) interatomic potential Φ at $T_e = 316$ K

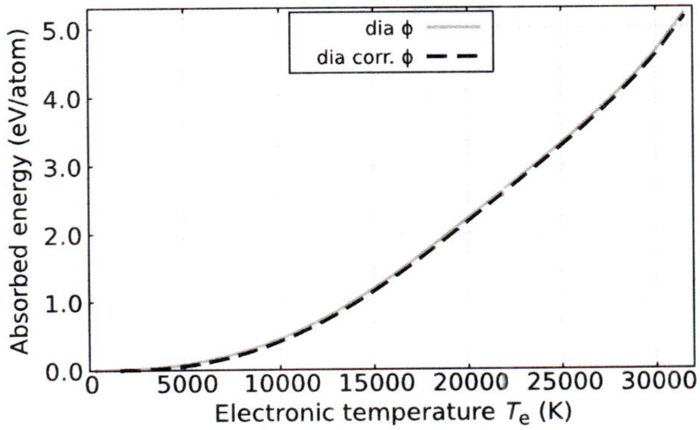

Fig. 7.51 The absorbed electronic energy per atom of the diamond-like structure is shown for the $\aleph_3 = 12$ eV corrected (black dashed) and uncorrected (gray solid) interatomic potential Φ at $T_e = 316$ K

Table 7.6 Relative error of the fitted atomic configurations in the atomic forces $f_{err}^{(s_1)}$ and in the structural free cohesive energies $E_{err}^{(s_1)}$ are shown for the uncorrected and the $\aleph_3 = 12$ eV corrected interatomic potential

	Uncorrected		$\aleph_3 = 12$ eV	
T_e (K)	$f_{err}^{(s_1)}$ (%)	$E_{err}^{(s_1)}$ (%)	$f_{err}^{(s_1)}$ (%)	$E_{err}^{(s_1)}$ (%)
316	25.8	1.1	43.1	2.1
3158	20.5	0.7	24.8	1.2
6315	13.9	0.4	14.2	0.5
9473	9.8	0.3	9.8	0.3
12631	7.7	0.5	7.8	0.4
15789	7.3	0.8	7.4	0.8
18946	11.2	0.6	11.2	0.7
22104	8.9	0.2	9.0	0.3
25262	6.6	0.6	6.6	0.6
28420	6.3	1.6	6.3	1.2
31577	6.6	1.1	6.6	1.0

leads to an error increase, which is the biggest at the lowest $T_e = 316$ K and vanishes almost for $T_e > 3158$ K.

In Summary, the melting temperature modification of the potential introduce some moderate changes. Nevertheless, with increasing $\aleph_3$ also the slope $\left.\frac{dT_m}{dp}\right|_{p=0}$ in the T_m *versus* p diagram increases. The slope rises from −40 K/GPa at $\aleph_3 = 0$ up to 18 K/GPa at $\aleph_3 = 12$ eV (see Table 7.5). The slope was experimentally measured to be −58 K/GPa [28]. Consequently, the slope 18 K/GPa of the melting temperature corrected interatomic potential cannot be accepted, since it is even positive compared to the experimental value.

But how the coefficient correction can be done to reach, on the one hand, the experimental melting temperature and to obtain, on the other hand, a negative slope?

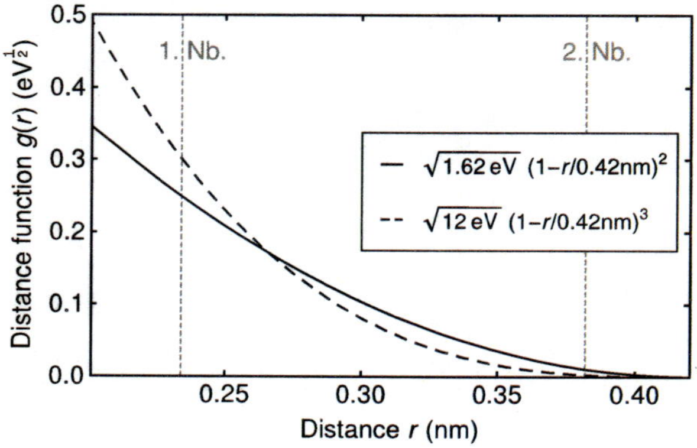

Fig. 7.52 Plot of the two different distance functions used in the potential correction term to reach the experimental melting temperature is shown. The two gray vertical lines indicate the position of the first and second nearest neighbors in Si

One way could be to use a different distance function $g(r)$ for the melting temperature correction. If the term

$$\left(1 - \frac{r}{r_3^{(c)}}\right)^2$$

is chosen for the distance function $g(r)$ instead of

$$\left(1 - \frac{r}{r_3^{(c)}}\right)^3,$$

the melting temperature of the interatomic potential can rise up to the experimental value. The melting temperature $T_m = 1685 \pm 3$ K is reached for $\aleph_3 = 1.62$ eV and the slope yields 20 ± 4 K/GPa. But this slope value is even further away from the experimental value of -58 K/GPa. In Fig. 7.52, the two mentioned distance functions $g(r)$ are plotted, which both yield the same melting temperature close to the experimental value but slightly different slopes.

To get a deeper insight in the slope behavior of an interatomic potential, we constructed and studied a series of test potentials, which all exhibit the experimental melting temperature, but different slopes. We report the results in the next subsection.

7.3.2 Melting Temperature and Slope Study on Test Potentials

The starting point is the widely used Stillinger & Weber potential [33], since, on the one hand, it exhibits the experimental melting temperature and a significantly negative slope and, on the other hand, the polynomial Si potential is a generalization of it. The Stillinger & Weber potential just consist of a two-body potential $\Phi_2^{(SW)}$ and a three-body potential $\Phi_3^{(SW)}$ (see Sect. 5.1.1):

$$\Phi^{(\mathrm{SW})} = \sum_{i=1}^{N_{\mathrm{at}}} \sum_{\substack{j=1 \\ j>i \\ r_{ij}<r_2^{(c)}}}^{N_{\mathrm{at}}} \Phi_2^{(\mathrm{SW})}(r_{ij}) + \sum_{i=1}^{N_{\mathrm{at}}} \sum_{\substack{j=1 \\ j\neq i \\ r_{ij}<r_3^{(c)}}}^{N_{\mathrm{at}}} \sum_{\substack{k=1 \\ k\neq i, k\neq j \\ r_{ik}<r_3^{(c)}}}^{N_{\mathrm{a}}} \Phi_3^{(\mathrm{SW})}\big(r_{ij}, r_{ik}, \cos(\theta_{ijk})\big), \tag{7.16}$$

whereas the three-body potential is constructed like the melting temperature correction term (7.13)

$$\Phi_3^{(\mathrm{SW})}\big(r_{ij}, r_{ik}, \cos(\theta_{ijk})\big) = g^{(\mathrm{SW})}(r_{ij})\, g^{(\mathrm{SW})}(r_{ik})\left(\cos(\theta_{ijk}) + \frac{1}{3}\right)^2 \tag{7.17}$$

using the distance function $g^{(\mathrm{SW})}(r)$. The Stillinger & Weber potential $\Phi^{(\mathrm{SW})}$ uses the two-body potential

$$\Phi_2^{(\mathrm{SW})}(r_{ij}) = \exp\left(\frac{\sigma}{r - r^{(c)}}\right)\left(A_0\, r_{ij}^{-p} - B_0\, r_{ij}^{-q}\right) \tag{7.18}$$

and the distance function

$$g^{(\mathrm{SW})}(r) = \sqrt{\frac{\lambda_0}{2}}\, \exp\left(\frac{\sigma}{r - r^{(c)}}\right)^{\gamma}. \tag{7.19}$$

The parameters $\sigma = 0.20951$ nm, $\lambda_0 = 45.532305023389895$ eV, $\gamma = 1.2$ are used and both, two- and three-body potential use the cutoff radius $r^{(c)} = 0.377118$ nm. To verify our results, we determined the melting temperature $T_{\mathrm{m}}\big|_{p=0}$ and the slope $\frac{dT_{\mathrm{m}}}{dp}\big|_{p=0}$ at zero pressure of the Stillinger & Weber potential by performing the liquid-crystal coexistence MD simulations described in Sect. 7.1.5. Indeed, our obtained melting temperature of 1688 ± 3 K is close to the literature value of 1691 ± 20 K [34] and also the obtained slope -49 ± 3 K/GPa is close to the literature value of -50 ± 10 K/GPa [35].

To define a simple polynomial test potential $\Phi^{(\mathrm{pol})}$ following the construction of the polynomial Si potential, we construct it as a sum of a two-body and a three-body potential similar to the Stillinger & Weber potential. We set the two-body potential as

$$\Phi_2^{(\mathrm{pol})}(r_{ij}) \;=\; -\frac{3\,\aleph_2 \left(r_2^{(c)}\right)^2}{\left(r_2^{(c)} - r^{(\min)}\right)^2}\left(1 - \frac{r_{ij}}{r_2^{(c)}}\right)^2 + \frac{2\,\aleph_2 \left(r_2^{(c)}\right)^3}{\left(r_2^{(c)} - r^{(\min)}\right)^3}\left(1 - \frac{r_{ij}}{r_2^{(c)}}\right)^3 \tag{7.20}$$

and we construct the three-body potential like Eq. (7.17) using the simple distance function (Fig. 7.53)

$$g^{(\mathrm{pol})}(r) = \sqrt{\aleph_3}\left(1 - \frac{r}{r_3^{(c)}}\right)^2. \tag{7.21}$$

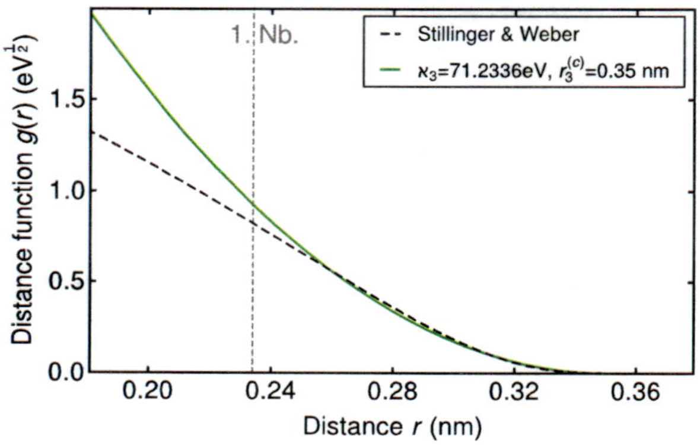

Fig. 7.53 Distance function $g(r)$ of the Stillinger & Weber and the test potential with $\Phi_2^{(\mathrm{SW})} \approx \Phi_2^{(\mathrm{pol})}$ and same T_m is shown. The gray vertical line indicates the distance of the nearest neighbors in Si

Now $\Phi_2^{(\mathrm{pol})}$ exhibits one single minimum, which is reached at $r^{(\mathrm{min})}$ and has got the value of $-\aleph_2$. The position of the minimum is set to $r^{(\mathrm{min})} = 0.234$ nm in the following, since at this distance the nearest neighbors are located in Si.

To initialize the study, we chose the strength $\aleph_2 = 2.18$ eV and the cutoff-radius $r_2^{(\mathrm{c})} = 0.35$ nm to get a similar course of the polynomial two-body potential $\Phi_2^{(\mathrm{pol})}$ compared with the Stillinger & Weber two-body potential $\Phi_2^{(\mathrm{SW})}$ for distances bigger than the first neighbor distance 0.234 nm (see Fig. 7.54). The strength $\aleph_3$ of the corresponding three-body potential $\Phi_3^{(\mathrm{pol})}$ was optimized to get the same melting temperature at zero pressure like the Stillinger & Weber potential. To do so, several small cell liquid-crystal coexistence MD simulations were initially performed to get a prediction of the corresponding $\aleph_3$ value, then large scale liquid-crystal coexistence MD simulations were performed for a few $\aleph_3$ values to get the searched value. This procedure was repeated for several strengths $\aleph_2$ of the polynomial two-body potential to get the corresponding $\aleph_3$ values for reaching the same melting temperature at zero pressure. Also the cutoff-radius of the polynomial two-body potential was reduced to $r_2^{(\mathrm{c})} = 0.33$ nm at $\aleph_2 = 2.18$ eV and the corresponding $\aleph_3$ was determined. The obtained results are listed in Table 7.7. In addition, the cutoff-radius of the polynomial two-body potential was increased to $r_2^{(\mathrm{c})} = 0.37$ nm at $\aleph_2 = 2.18$ eV. But this setting leads to a crystallization in the hexagonal closed-packed (hcp) structure instead of the diamond-like structure. Consequently, this parameter combination is skipped in Table 7.7.

The strength $\aleph_2$ of the two-body potential is not relevant for the melting temperature $T_\mathrm{m}\big|_{p=0}$ at zero pressure, as it can be seen in Table 7.7. Rather, the relationship of $\aleph_2$ and $\aleph_3$ defines $T_\mathrm{m}\big|_{p=0}$.

Figure 7.55 shows $\sqrt{\aleph_3}$ as a function of $\aleph_2$ for the test potentials with $r_2^{(\mathrm{c})} = r_3^{(\mathrm{c})} = 0.35$ nm and $T_\mathrm{m}\big|_{p=0} \approx 1687$ K. All these values lie on a straight line, which can be obtained from a linear regression. To reach the same melting temperature, $\aleph_3$ must increase quadratically for increasing $\aleph_2$ and $\aleph_3$ must decrease quadratically for decreasing $\aleph_2$. Of course, this dependence is only valid in a certain interval for $\aleph_2$,

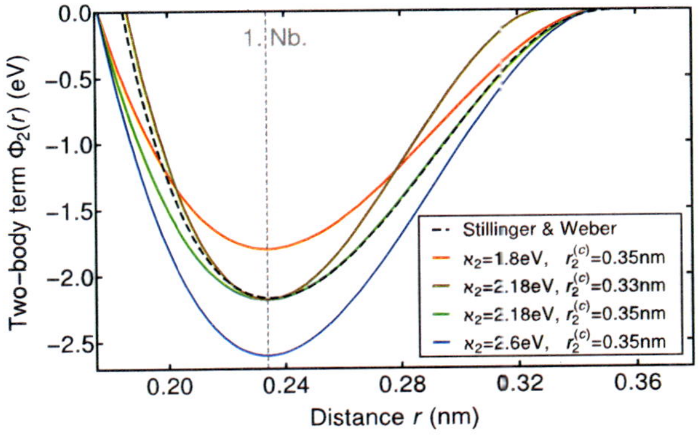

Fig. 7.54 The two-body potential $\Phi_2^{(SW)}$ of Stillinger & Weber is shown together with the two-body potentials $\Phi_2^{(pol)}$ of selected test potentials. The gray vertical line indicates the distance of the nearest neighbors in Si

Table 7.7 Melting temperature $T_m|_{p=0}$ and slope $\frac{dT_m}{dp}|_{p=0}$ at zero pressure for the different test and the Stillinger & Weber potential

$\aleph_2$ (eV)	$\aleph_3$ (eV)	$r_2^{(c)}$ (nm)	$r_3^{(c)}$ (nm)	$T_m\|_{p=0}$ (K)	$\frac{dT_m}{dp}\|_{p=0}$ $\left(\frac{K}{GPa}\right)$
Stillinger & Weber		0.377118	0.377118	1688 ± 3	−49 ± 3
1.8	61.999876	0.35	0.35	1689 ± 3	−29 ± 4
2.0	66.7489	0.35	0.35	1688 ± 3	−43 ± 3
2.18	71.2336	0.35	0.35	1689 ± 3	−56 ± 4
2.2	71.723961	0.35	0.35	1687 ± 3	−58 ± 4
2.4	76.9831	0.35	0.35	1689 ± 3	−68 ± 3
2.6	82.337476	0.35	0.35	1689 ± 3	−89 ± 3
2.18	45.104656	0.33	0.35	1688 ± 3	−80 ± 6

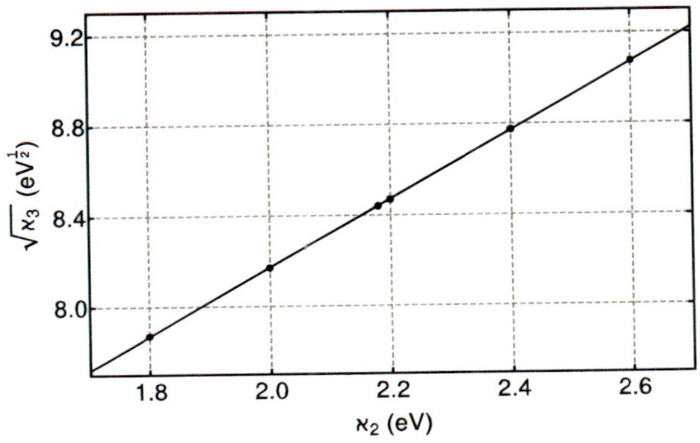

Fig. 7.55 $\sqrt{\aleph_3}$ is shown as a function of $\aleph_2$ for the test potentials with $r_2^{(c)} = r_3^{(c)} = 0.35$ nm and $T_m \approx 1687$ K (see Table 7.7). Also a linear regression line is inserted

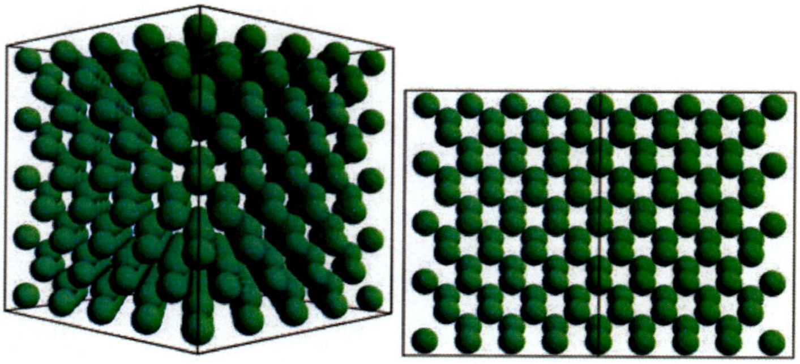

Fig. 7.56 Three-dimensional and projected view of the diamond-like structure are shown. The spheres touch each other

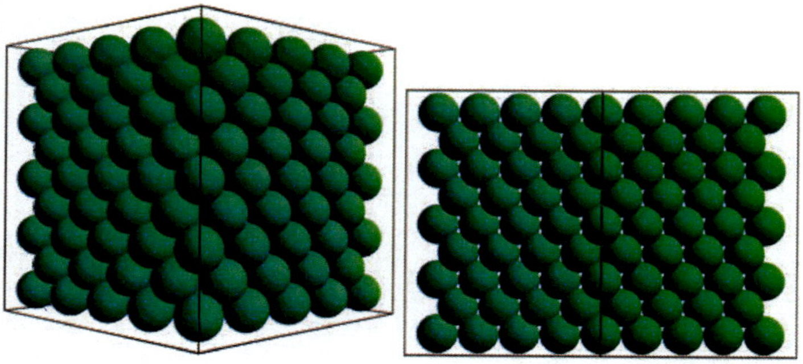

Fig. 7.57 Three-dimensional and projected view of the fcc structure are shown. The spheres touch each other

since the nearest neighbor distance is not stabilized any more for $\aleph_2 \to 0$ and the crystal will melt easily at very low temperatures.

Indeed, the two-body potential takes care, that the nearest neighbors prefer the distance, which corresponds to its minimum. If there is no three-body potential, each atom prefers as much neighbors as possible at this distance, which is fulfilled for a closed-packed structure like fcc or hcp. But the present three-body potential forces the nearest neighbors to exhibit always an angle θ obeying $\cos(\theta) = -\frac{1}{3}$. This is only possible for four nearest neighbors building a tetragonal structure. Hence, the diamond-like structure is formed, which is called open, since there exists free space between the atoms, to which the atoms could move during melting. To demonstrate this, the diamond-like structure is compared with the fcc structure in Figs. 7.56 and 7.57. In addition, if the cutoff radius of the two-body potential is bigger than that of the three-body potential, the atoms still crystallize in the hcp structure like it would be without three-body potential.

If the pressure p is increased, the atoms are located closer together. Now, the nearest neighbors are to close and the two-body potential associated forces want to move them further away. Consequently, the nearest neighbors are less prevented to move into the free space and the crystal will melt more easily, i.e., T_m decreases with increasing pressure. On the other hand, if the pressure is decreased, the atoms will be located further away from each other. Then, the nearest neighbors are to far away and the two-body potential wants to move them closer together. Consequently, the nearest neighbors are more prevented to move into the free space and the crystal will melt more hardly, i.e., T_m increases with decreasing pressure. This explains the negative slope.

Moreover, if the increase besides the minimum of the two-body potential becomes stronger, like for increasing $\aleph_2$ or decreasing $r_2^{(\mathrm{c})}$ (see Table 7.7), the slope will become more negative, since the pressure-conditioned displacement of the nearest neighbors from their equilibrium distance will cause stronger forces on them. Hence, a function with a minimum at the nearest neighbor distance and a strong increase beside this minimum should be added to the two-body potential of the polynomial Si potential for getting a negative slope.

7.3.3 Correction of the 2-Body and 3-Body Potential Coefficients

We need to manipulate the two- and three-body potentials of the polynomial Si potential $\Phi^{(\mathrm{Si})}(T_\mathrm{e})$ to control the melting temperature $T_\mathrm{m}\big|_{p=0}$ and the slope $\frac{dT_\mathrm{m}}{dp}\big|_{p=0}$ at zero pressure while keeping the previous results in mind. We performed the manipulation of the three-body potential in the same way as described in Sect. 7.3.1 using $\aleph_3$. The two-body potential of $\Phi^{(\mathrm{Si})}(T_\mathrm{e})$ uses the cutoff-radius of $r_2^{(\mathrm{c})} = 0.63$ nm and the degree $N_2^{(r)} = 10$ (see Sect. A.8):

$$\Phi_2(r_{ij}) = \sum_{q=2}^{10} c_2^{(q)} \left(1 - \frac{r_{ij}}{r_2^{(\mathrm{c})}}\right)^q . \tag{7.22}$$

To get a negative slope, the coefficients $\left\{c_2^{(q)}\right\}$ of the two-body potential should be modified in such a way that a correction two-body term $\Phi_2^{(\mathrm{cor})}$ is added, which exhibits a single minimum at the nearest neighbor distance 0.234 nm of the diamond-like structure of Si and which increases strongly beside the minimum. For this, we used a linear combination of the three highest powers of the term

$$\left(1 - \frac{r_{ij}}{r_2^{(\mathrm{c})}}\right)$$

for $\Phi_2^{(\mathrm{cor})}$. We chose the three highest powers for the manipulation, since lower powers cause a weaker increase beside the minimum of $\Phi_2^{(\mathrm{cor})}$. We found the coefficients of $\Phi_2^{(\mathrm{cor})}$ by the following conditions: $\Phi_2^{(\mathrm{cor})}$ exhibits a minimum at $r^{(\mathrm{min})} = 0.234$ nm

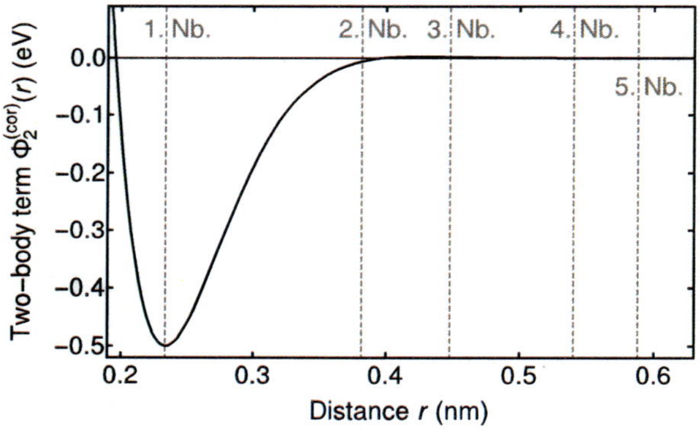

Fig. 7.58 The correction two-body term $\Phi_2^{(cor)}$ is shown for $\aleph_2 = 0.5$ eV. The vertical lines indicate the positions of the neighbors in the diamond-like structure of Si

and sets to $-\aleph_2$ at this minimum and sets to zero at $r^{(1)} = 0.4$ nm. The last constraint takes care that $\Phi_2^{(cor)}$ stays approximately at 0 for distances between $r^{(1)}$ and the cutoff radius $r_2^{(c)} = 0.63$ nm. $r^{(1)}$ should be as small as possible, since the increase beside the minimum becomes bigger for decreasing $r^{(1)}$ allowing a stronger slope $\frac{dT_m}{dp}\big|_{p=0}$ control. But, if $r^{(1)}$ is set smaller than 0.4 nm, $\Phi_2^{(cor)}$ becomes significantly positive for distances bigger than $r^{(1)}$, which should be avoided. The correction two-body term $\Phi_2^{(cor)}$, which fulfills the above mentioned conditions, is shown in Fig. 7.58 and is defined as

$$\begin{aligned}\Phi_2^{(cor)}(r_{ij}) &= 647.3458562015724\,\aleph_2\left(1-\frac{r_{ij}}{r_2^{(c)}}\right)^8\\ &\quad -2712.579093531517\,\aleph_2\left(1-\frac{r_{ij}}{r_2^{(c)}}\right)^9\\ &\quad +2573.178456073968\,\aleph_2\left(1-\frac{r_{ij}}{r_2^{(c)}}\right)^{10}. \qquad (7.23)\end{aligned}$$

Indeed, adding the above mentioned correction two-body term $\Phi_2^{(cor)}$ to Φ_2 corresponds just to modify the coefficients $c_8^{(2)}$, $c_9^{(2)}$ and $c_{10}^{(2)}$ of Φ_2. More detailed, $647.3458562015724\,\aleph_2$ is add to $c_2^{(8)}$, $-2712.579093531517\,\aleph_2$ to $c_2^{(9)}$ and $2573.178456073968\,\aleph_2$ to $c_2^{(10)}$.

Similar to the $\aleph_3$ manipulation, we add the corresponding correction value to the ideal coefficient values at $T_e = 316$ K and half of it at $T_e = 3158$ K before the polynomial is approximated from the ideal coefficient values to get the smooth T_e-dependence of the potential coefficients. Figure 7.59 shows the original and corrected ideal coefficient values together with the corresponding fitted polynomials for the modified coefficients at $\aleph_2 = 0.5$ eV and $\aleph_3 = 20.3$ eV.

$\aleph_2$ was gradually increased and, for each $\aleph_2$, the corresponding $\aleph_3$ was determined to get the experimental melting temperature of $T_m = 1687$ K at zero pressure by

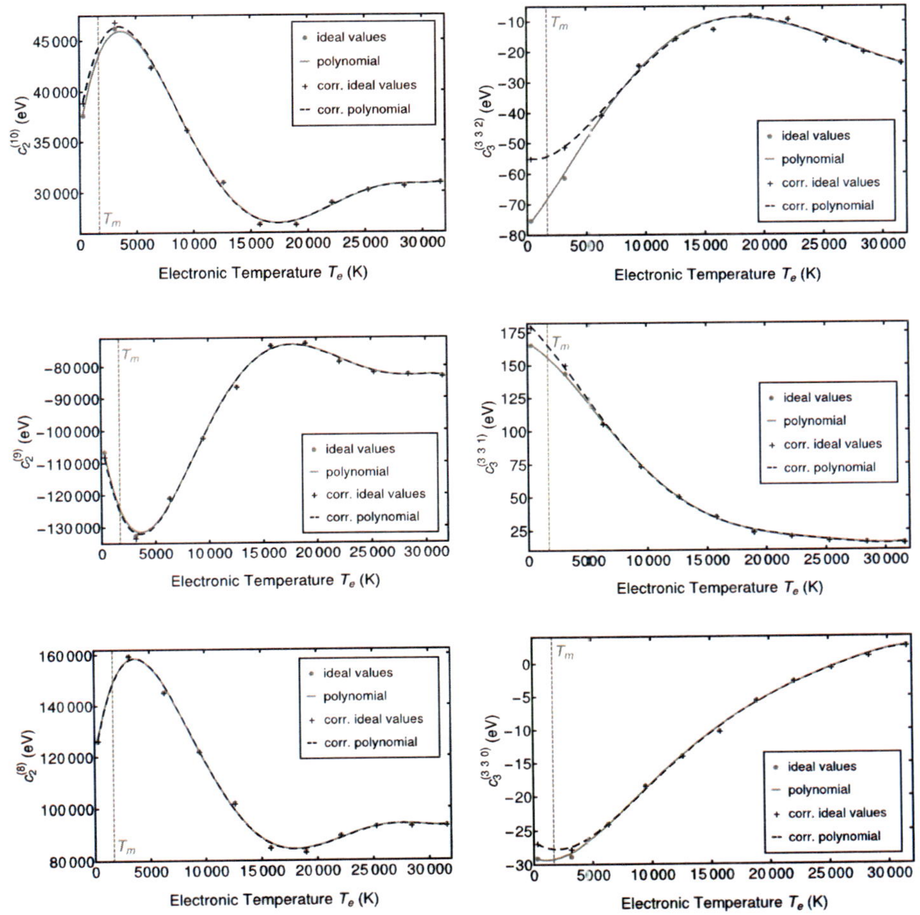

Fig. 7.59 The polynomial approximation of the coefficients $c_8^{(2)}$, $c_9^{(2)}$, $c_{10}^{(2)}$, $c_{3\,3\,0}^{(3)}$, $c_{3\,3\,1}^{(3)}$, and $c_{3\,3\,2}^{(3)}$ is shown together with the ideal values before (gray) and after (black) the correction of the ideal values at $T_e = 316$ K and $T_e = 3158$ K. $\aleph_2 = 0.5$ eV and $\aleph_3 = 20.3$ eV are shown, which yields the experimental melting temperature $T_m = 1687$ K highlighted by a gray vertical line

using the procedure described in Sect. 7.3.2. Table 7.8 lists the obtained $\aleph_2$, $\aleph_3$ pairs together with the corresponding melting temperature and slope at zero pressure.

Indeed, the slope decreases with increasing $\aleph_2$ and it becomes negative at $\aleph_2 = 0.3$ eV. The phonon frequencies increase with increasing $\aleph_2$ value, since the bonding becomes stronger. At $\aleph_2 = 0.5$ eV, the corresponding phonon bandstructure of the diamond-like structure at $T_e = 316$ K looks similar to the one of the famous Stillinger & Weber potential, as it can be seen in Fig. 7.60. Especially the acoustic phonon branches are in an excellent agreement. Consequently, $\aleph_2 = 0.5$ eV is selected for the final corrected interatomic potential. We tabulate the corresponding modified coefficients in Sect. A.8.1.

Table 7.8 Melting temperature $T_\mathrm{m}\big|_{p=0}$ and slope $\frac{dT_\mathrm{m}}{dp}\big|_{p=0}$ at zero pressure for the different coefficient corrections to the interatomic potential

$\aleph_2$ (eV)	$\aleph_3$ (eV)	$T_\mathrm{m}\big\|_{p=0}$ (K)	$\frac{dT_\mathrm{m}}{dp}\big\|_{p=0}$ $\left(\frac{\mathrm{K}}{\mathrm{GPa}}\right)$
0.0	12.0	1687 ± 3	18 ± 4
0.1	13.1	1686 ± 3	12 ± 4
0.2	14.6	1687 ± 3	3 ± 4
0.3	16.4	1689 ± 3	-1 ± 4
0.4	18.2	1685 ± 3	-7 ± 4
0.5	20.3	1689 ± 3	-11 ± 4
0.6	22.4	1688 ± 3	-14 ± 4

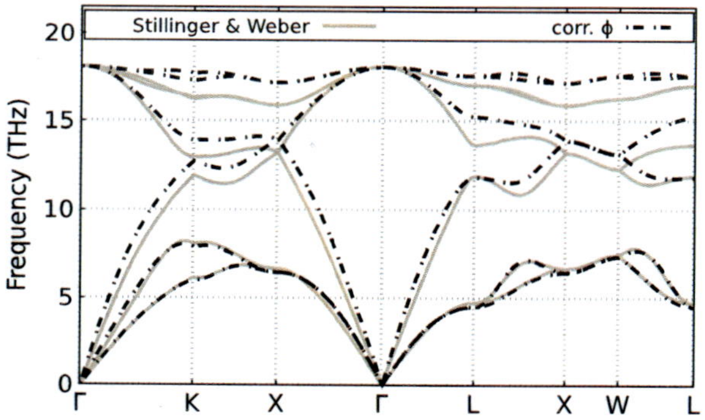

Fig. 7.60 The phonon bandstructure of the diamond-like structure is shown for the $\aleph_2 = 0.5$ eV, $\aleph_3 = 20.3$ eV corrected Φ at $T_\mathrm{e} = 316$ K (black dot-dashed) and the Stillinger & Weber potential (gray solid)

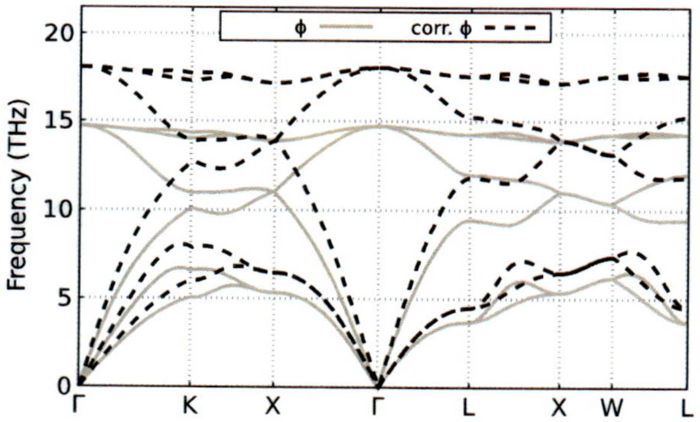

Fig. 7.61 The phonon bandstructure of the diamond-like structure is shown for the $\aleph_2 = 0.5$ eV, $\aleph_3 = 20.3$ eV corrected (black dashed) and uncorrected (gray solid) interatomic potential Φ at $T_\mathrm{e} = 316$ K

The correction of the melting temperature and the slope introduces greater changes to the interatomic potential at low T_e than just the melting temperature correction in Sect. 7.3.1.

Figure 7.61 shows the comparison of the phonon bandstructure, Fig. 7.62 the comparison of the cohesive energies of several bulk crystal structures, Fig. 7.64 the comparison of the absorbed electronic energy, and Fig. 7.63 the comparison of the electronic specific heat between the corrected and uncorrected interatomic potential at

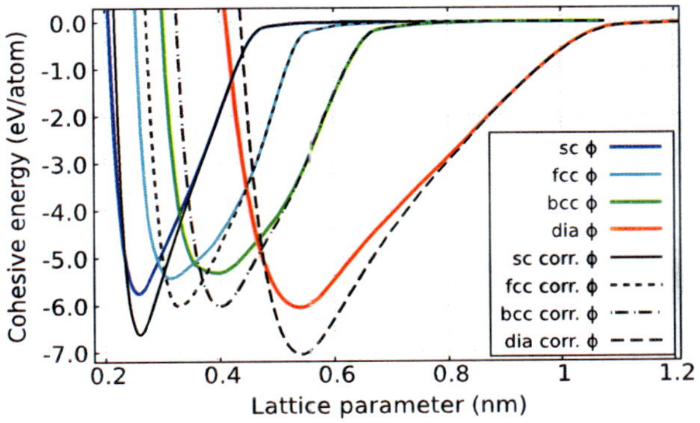

Fig. 7.62 The cohesive energies of different structures are shown for the $\aleph_2 = 0.5$ eV, $\aleph_3 = 20.3$ eV corrected (black dashed) and uncorrected (colored solid) interatomic potential Φ at $T_e = 316$ K

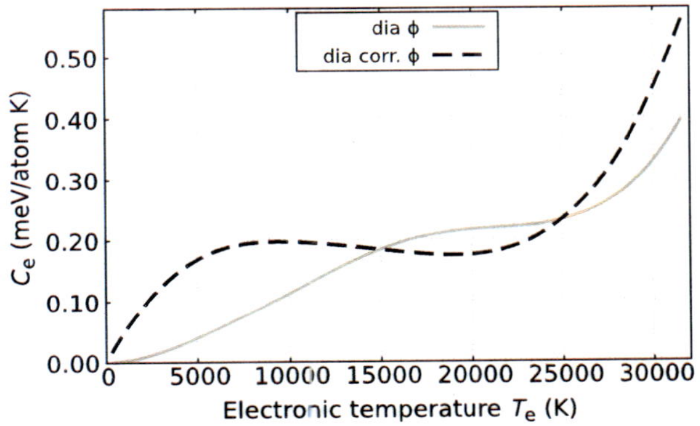

Fig. 7.63 The electronic specific heat C_e of the diamond-like structure is shown for the $\aleph_2 = 0.5$ eV, $\aleph_3 = 20.3$ eV corrected (black dashed) and uncorrected (gray solid) interatomic potential Φ at $T_e = 316$ K

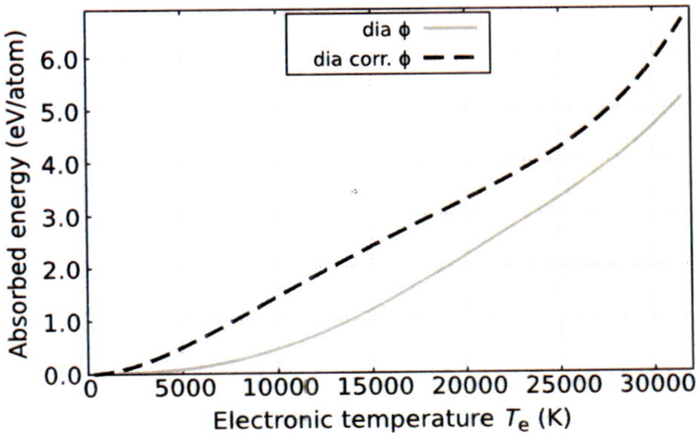

Fig. 7.64 The absorbed electronic energy E_{abs} of the diamond-like structure is shown for the $\aleph_2 = 0.5$ eV, $\aleph_3 = 20.3$ eV corrected (black dashed) and uncorrected (gray solid) interatomic potential Φ at $T_e = 316$ K

Table 7.9 Relative error in the atomic forces $f_{\text{err}}^{(s_1)}$ and in the structural free cohesive energies $E_{\text{err}}^{(s_1)}$ of reference simulation s_1 are shown for the uncorrected and the $\aleph = 0.5$ eV, $\aleph_3 = 20.3$ eV corrected interatomic potential

			$\aleph_2 = 0.5$ eV	
	Uncorrected		$\aleph_3 = 20.3$ eV	
T_e (K)	$f_{\text{err}}^{(s_1)}$ (%)	$E_{\text{err}}^{(s_1)}$ (%)	$f_{\text{err}}^{(s_1)}$ (%)	$E_{\text{err}}^{(s_1)}$ (%)
316	25.8	1.1	65.7	14.6
3158	20.5	0.7	32.0	6.9
6315	13.9	0.4	14.7	1.7
9473	9.8	0.3	10.0	0.9
12631	7.7	0.5	8.2	1.6
15789	7.3	0.8	7.5	1.1
18946	11.2	0.6	11.2	0.2
22104	8.9	0.2	9.0	1.0
25262	6.6	0.6	6.6	0.6
28420	6.3	1.6	6.4	4.3
31577	6.6	1.1	6.6	2.9

$T_e = 316$ K. The phonon frequencies and the absolute value of the cohesive energies increase, since the corrected potential contains a stronger two-body term. The T_e-dependence of the electronic specific heat looks slightly different compared to the uncorrected one. But it can be still physically accepted, since it is positive. The absorbed electronic energy of the corrected interatomic potential is $\sim$1.0 eV higher than for the uncorrected one for $T_e > 10000$ K, but the functional shape is the same.

In addition, we calculated the relative error in the atomic forces and the structural free cohesive energies of reference simulation s_1 for the corrected interatomic potential and compared it with the uncorrected one in Table 7.9. The correction does not produce any significant changes in the relative errors at high electronic temperatures $T_e \geq 9473$ K. Indeed, the corrected polynomial does not differ from the uncorrected polynomial for the smooth T_e-approximation of the coefficients at high T_e's, as it can be seen in Fig. 7.59. Thus, the physical properties at high T_e are not influenced by the melting temperature correction.

7.4 Summary

A T_e-dependent interatomic potential $\Phi^{(\text{Si})}(T_e)$ was developed for Si. For this, the polynomial interatomic potential model and the fitting procedure described in Sect. 6.3 was successfully utilized. An extensive set of atomic forces and structural cohesive energies obtained from T_e-dependent DFT reference simulations were fitted. The finally obtained interatomic potential $\Phi^{(\text{Si})}(T_e)$ describes accurately, within the scope of use, the femtosecond laser-induced effects in Si, namely the bond softening, thermal phonon antisqueezing, non-thermal melting, and ablation. Furthermore, the laser-driven atomic mean-square displacements perpendicular to the surface of a thin-film are properly described with respect to DFT. This indicates that the description by means of $\Phi^{(\mathrm{Si})}(T_\mathrm{e})$ does not physically break down at the symmetry breaking surface and the effects associated. In addition, electronic properties like the electronic absorbed energy or the electronic specific heat are well described by $\Phi^{(\mathrm{Si})}(T_\mathrm{e})$. The explanation that $\Phi^{(\mathrm{Si})}(T_\mathrm{e})$ describes accurately the exited PES is, on the one hand, the large data set used for fitting, and, on the other hand, the adaption of its functional form. The used data set contains many different atomic configurations and includes also surfaces. The well understood polynomial functional form of $\Phi^{(\mathrm{Si})}(T_\mathrm{e})$ allows to adapt it optimally to describe the ab-initio data most accurately and efficiently. Such a adaption of the functional form is crucial, since a reparametrization of widely used interatomic potentials for ground state Si does not produce such an accurate description of the excited PES like $\Phi^{(\mathrm{Si})}(T_\mathrm{e})$ can do it.

$\Phi^{(\mathrm{Si})}(T_\mathrm{e})$ was utilized to simulate the femtosecond-laser excitation of a thin Si film including the effects of the excited PES and the EPC. The time-dependent intensity of several Bragg peaks were derived and compared with available experiments, in which the time-dependent intensities of several Bragg peaks was measured using ultrafast electron diffraction. An excellent agreement was found, which confirms the validity of the theoretical description of the femtosecond-laser excitation using the excited PES and the EPC. Moreover, if only the excited PES or only the EPC is taken into account, the obtained results disagree with the experiments at high laser intensities. This indicates that both effects are important and must be taken into account for a valid description.

The influence of the EPC on the evolution of laser excited Si dominates compared to the influence of the excited PES. One evidence of the influence of the excited PES can be seen in the laser excitation of a thin film: Due to the excited PES, the film experiences a delay of 300 fs in the onset of melting at the surface in comparison to the central volume. The reason for this is the expansion of the film at the surface, since the resulting lower pressures stabilize the crystal structure. Furthermore, it was found that one can still identify the thermal phonon antisqueezing in the simulation, which is driven by the excited PES, in scenarios with both, excited PES and EPC present. At high intensities, Si melts non-thermally in two distinguishable stages and, at very high intensities, in four stages, if the most complex scenario with excited PES and EPC is taken into account. If only the excited PES is taken into account, Si melts non-thermally in three stages.

The polynomial interatomic potential $\Phi^{(\mathrm{Si})}(T_\mathrm{e})$ for Si was developed from LDA-DFT atomic forces and cohesive energies. Therefore, its melting temperature T_m reproduces the LDA-DFT value, which is significantly below the experimental value. Furthermore, $\Phi^{(\mathrm{Si})}(T_\mathrm{e})$ exhibits a negative slope $\frac{dT_\mathrm{m}}{dp}\big|_{p=0}$ at zero pressure p in the T_m *versus* p diagram similar to the experiment. The physical interpretation for a negative slope rests on the fact that the crystal structure of Si is stabilized at low pressures, or in other words, Si melts slower at lower pressures.

Since $\Phi^{(Si)}(T_e)$ consists of a sum of the physically interpretable terms Φ_ρ, Φ_2, Φ_3, which are formulated as polynomials, these terms can be adjusted for in an attempt to push T_m up to the experimental value. By the modification of some coefficients of the two- and three-body potential term at low electronic temperatures T_e, the melting temperature can be set to the experimental value while maintaining the necessary negative slope. This tweak of the interatomic potential at low T_e's does not produce any significant changes in the description of the laser excited PES at higher T_e's. In addition, the electronic spezific heat C_e as a function of T_e shows still a physical behavior, which indicates that the tailored approximation works well.

References

1. X. Liu, D. Du, G. Mourou, IEEE J. Quantum Electron. **33**(10), 1706 (1997). https://doi.org/10.1109/3.631270
2. F. Courvoisier, P.A. Lacourt, M. Jacquot, M.K. Bhuyan, F. L., J.M. Dudley, Opt. Lett. **34**(20), 3163 (2009). https://doi.org/10.1364/OL.34.003163. http://ol.osa.org/abstract.cfm?URI=ol-34-20-3163
3. X.C. Wang, H.Y. Zheng, C.W. Tan, F. Wang, H.Y. Yu, K.L. Pey, Opt. Express **18**(18), 19379 (2010). https://doi.org/10.1364/OE.18.019379. http://www.opticsexpress.org/abstract.cfm?URI=oe-18-18-19379
4. S. Höhm, M. Rohloff, A. Rosenfeld, J. Krüger, J. Bonse, Appl. Phys. A **110**(3), 553 (2013). https://doi.org/10.1007/s00339-012-7184-z
5. S. Höhm, M. Herzlieb, A. Rosenfeld, J. Krüger, J. Bonse, Opt. Express **23**(1), 61 (2015). https://doi.org/10.1364/OE.23.000061. http://www.opticsexpress.org/abstract.cfm?URI=oe-23-1-61
6. D.S. Ivanov, V.P. Lipp, A. Blumenstein, F. Kleinwort, V.P. Veiko, E. Yakovlev, V. Roddatis, M.E. Garcia, B. Rethfeld, J. Ihlemann, P. Simon, Phys. Rev. Appl. **4**, 064006 (2015). https://doi.org/10.1103/PhysRevApplied.4.064006. https://link.aps.org/doi/10.1103/PhysRevApplied.4.064006
7. S. Khakshouri, D. Alfè, D.M. Duffy, Phys. Rev. B **78**, 224304 (2008). https://doi.org/10.1103/PhysRevB.78.224304
8. L. Shokeen, P.K. Schelling, Appl. Phys. Lett. **97**(15), 151907 (2010). https://doi.org/10.1063/1.3499296
9. L. Shokeen, P.K. Schelling, J. Appl. Phys. **109**(7), 073503 (2011). https://doi.org/10.1063/1.3554410
10. L. Shokeen, P.K. Schelling, Comput. Mater. Sci. **67**, 316 (2013). https://doi.org/10.1016/j.commatsci.2012.07.042. http://www.sciencedirect.com/science/article/pii/S0927025612004910
11. J.A. Moriarty, R.Q. Hood, L.H. Yang, Phys. Rev. Lett. **108**, 036401 (2012). https://doi.org/10.1103/PhysRevLett.108.036401. http://link.aps.org/doi/10.1103/PhysRevLett.108.036401
12. G.E. Norman, S.V. Starikov, V.V. Stegailov, J. Exp. Theor. Phys. **114**(5), 792 (2012). https://doi.org/10.1134/S1063776112040115
13. S.T. Murphy, S.L. Daraszewicz, Y. Giret, M. Watkins, A.L. Shluger, K. Tanimura, D.M. Duffy, Phys. Rev. B **92**, 134110 (2015). https://doi.org/10.1103/PhysRevB.92.134110
14. R. Darkins, P.W. Ma, S.T. Murphy, D.M. Duffy, Phys. Rev. B **98**, 024304 (2018). https://doi.org/10.1103/PhysRevB.98.024304. https://link.aps.org/doi/10.1103/PhysRevB.98.024304
15. B. Bauerhenne, M.E. Garcia, The European Physical Journal Special Topics **227**(14), 1615 (2019). https://doi.org/10.1140/epjst/e2019-800181-3
16. V.V. Zhakhovskii, N.A. Inogamov, Y.V. Petrov, S.I. Ashitkov, K. Nishihara, Appl. Surf. Sci. **255**(24), 9592 (2009). https://doi.org/10.1016/j.apsusc.2009.04.082. http://www.sciencedirect.com/science/article/pii/S0169433209004346

17. B. Bauerhenne, V.P. Lipp, T. Zier, E.S. Zijlstra, M.E. Garcia, Phys. Rev. Lett. **124**, 085501 (2020). https://doi.org/10.1103/PhysRevLett.124.085501
18. T. Zier, E.S. Zijlstra, M.E. Garcia, Appl. Phys. A **117**(1), 1 (2014). https://doi.org/10.1007/s00339-014-8316-4
19. R. Gross, A. Marx, *Festkörperphysik* (Oldenburg Wissenschaftsverlag GmbH, München, 2012)
20. E.S. Zijlstra, A. Kalitsov, T. Zier, M.E. Garcia, Adv. Mater. **25**(39), 5605 (2013). https://doi.org/10.1002/adma201302559
21. D.R. Lide, *Handbook of Chemistry and Physics*, 84th edn. (CRC Press, 2004)
22. W.C. Malcolm, *NIST-JANAF Thermochemical Tables*, 4th edn. (American Chemical Society, 1998)
23. V.M. Glazov, O.D. Shchelikov, High Temp. **38**, 405 (2000). https://doi.org/10.1007/BF02756000
24. Z. Lin, L.V. Zhigilei, V. Celli, Phys. Rev. B **77**, 075133 (2008). https://doi.org/10.1103/PhysRevB.77.075133
25. P. Keblinski, M.Z. Bazant, R.K. Dash, M.M. Treacy, Phys. Rev. B **66**, 064104 (2002). https://doi.org/10.1103/PhysRevB.66.064104
26. D. Alfè, M.J. Gillan, Phys. Rev. B **68**, 205212 (2003). https://doi.org/10.1103/PhysRevB.68.205212. https://link.aps.org/doi/10.1103/PhysRevB.68.205212
27. K. Yamaguchi, K. Itagaki, J. Therm. Anal. Calorim. **69**(3), 1059 (2002). https://doi.org/10.1023/A:1020609517891
28. A. Jayaraman, W. Klement, G.C. Kennedy, Phys. Rev. **130**, 540 (1963). https://doi.org/10.1103/PhysRev.130.540
29. V.P. Lipp, B. Rethfeld, M.E. Garcia, D.S. Ivanov, Phys. Rev. B **90**, 245306 (2014). https://doi.org/10.1103/PhysRevB.90.245306
30. S. Sadasivam, M.K.Y. Chan, P. Darancet, Phys. Rev. Lett. **119**, 136602 (2017). https://doi.org/10.1103/PhysRevLett.119.136602. https://link.aps.org/doi/10.1103/PhysRevLett.119.136602
31. M. Harb, R. Ernstorfer, T. Dartigalongue, C.T. Hebeisen, R.E. Jordan, R.J.D. Miller, J. Phys. Chem. B **110**(50), 25308 (2006). (PMID: 17165976). https://doi.org/10.1021/jp064649n
32. M. Harb, R. Ernstorfer, C.T. Hebeisen, G. Sciaini, W. Peng, T. Dartigalongue, M.A. Eriksson, M.G. Lagally, S.G. Kruglik, R.J.D. Miller, Phys. Rev. Lett. **100**, 155504 (2008). https://doi.org/10.1103/PhysRevLett.100.155504
33. F.H. Stillinger, T.A. Weber, Phys. Rev. B **31**, 5262 (1985). https://doi.org/10.1103/PhysRevB.31.5262
34. J.Q. Broughton, X.P. Li, Phys. Rev. B **35**, 9120 (1987). https://doi.org/10.1103/PhysRevB.35.9120. https://link.aps.org/doi/10.1103/PhysRevB.35.9120
35. S. Yoo, X.C. Zeng, J.R. Morris, J. Chem. Phys. **120**(3), 1654 (2004). https://doi.org/10.1063/1.1633754

Chapter 8
Study of Femtosecond-Laser Excited Sb

Abstract One of the most interesting effects induced by a femtosecond-laser excitation is an ultrafast phase transition. In the previous chapters, the non-thermal melting of Si was already studied, which is a laser-induced ultrafast solid to liquid transition. Even more interesting is a laser-induced ultrafast solid to solid transition. For such a study, antimony (Sb) is a promising candidate, since it crystallize in the A7 structure, similar to bismuth and arsenic of the group in the periodic table. The A7 structure is stabilized by a so called Peierls distortion. The atoms form planes with hexagonal atom arrangement, where the planes are stacked on top of each other using two alternating distances (see Sect. 4.5.1). The difference between these two plane distances is expressed by the Peierls parameter. An intense femtosecond laser-pulse potentially excites an oscillation of the planes against each other in Sb (see Sect. 4.5.3), which correspond to an oscillation of the Peierls parameter caused by the excitation of the so called A_{1g} phonon. This laser excitation of the A_{1g} phonon has been confirmed experimentally in Sb [1–4] and bismuth [5, 6]. Arsenic transforms from the A7 structure into the distortion free simple cubic (sc) structure under the influence of pressure [7]. If arsenic is excited with a femtosecond laser-pulse, the necessary pressure for this transition can be reduced [8, 9]. At high pressures, it was experimentally found that Sb transforms from A7 to sc [10]. Furthermore, if the temperature is increased, this transformation is induced at lower pressures. In 2009, Fausti et al. measured in Sb a laser-induced reversible transition from the A7 to a structure with lower symmetry by using picosecond Raman scattering [11]. In the semimetal bismuth, Murray et al. [12] predicted 2005 that the presence of excited carriers can remove the Peierls distortion by means of DFT calculations. Fritz et al. [13] confirmed this prediction in 2007 by measuring the equilibrium position of the Peierls parameter using x-ray diffraction after a femtosecond laser excitation. In 2018, Teitelbaum et al. [14] observed the laser-induced phase transition of bismuth from the A7 structure to a structure with higher symmetry using dual echelon femtosecond single-shot spectroscopy [15]. Therefore, it is reasonable to analyze, if an A7 to sc transition can be induced in Sb by the femtosecond laser-pulse alone. The A_{1g} phonon mode and the Peierls parameter play a key role in the A7 to sc transition and will aid as the signature in the simulation. The laser-excitation of the A_{1g} phonon in Sb is related to the laser-induced changes in the PES. The MD simulations with T_e-dependent DFT showed finite size effects at the relevant laser intensities

© The Author(s), under exclusive license to Springer Nature Switzerland AG 2021

B. Bauerhenne, *Materials Interaction with Femtosecond Lasers*,
https://doi.org/10.1007/978-3-030-85135-4_8

and T_e's (see Sect. 4.5.5), even at the larger simulation cells that could be afforded. Thus, in order to simulate the femtosecond laser excitation of Sb with sufficient large number of atoms, which is indispensable for comparison with experiments, a T_e-dependent interatomic potential is needed. In this chapter, the interatomic potential model and the fitting procedure described in Sect. 6.3 is applied to develop a T_e-dependent interatomic potential $\Phi^{(\mathrm{Sb})}(T_e)$ for Sb. An extensive set of reference simulations were used obtained from T_e-dependent DFT for fitting. It was confirmed that $\Phi^{(\mathrm{Sb})}(T_e)$ describes accurately the PES for a wide range of T_e's and that the ab-initio electronic specific heat is well reproduced. In order to compare directly with experiments, in which the time-dependent intensity of a Bragg peak after a femtosecond laser-excitation was measured using ultrafast x-ray diffraction, additionally the optical properties of Sb were determined using the DFT code WIEN2K [16]. In Sb, the absorption coefficient and the reflectivity depends strongly on the Peierls parameter, a dependency not systematically studied so far. The time-dependent intensity of the experimentally studied Bragg peak was derived using $\Phi^{(\mathrm{Sb})}(T_e)$ and the simulation method including excited PES and EPC (see Sects. 6.1 and 6.2). An excellent agreement with the experimentally measured decay was found, which confirms the validity of the approach. Finally, the femtosecond laser excitation of a Sb film was studied using $\Phi^{(\mathrm{Sb})}(T_e)$ and it was analyzed, if a A7 to sc transition can be induced by the laser.

8.1 T_e-dependent Interatomic Potential for Sb

8.1.1 Ab-initio Reference Simulations Used for Fitting

Similar to our polynomial interatomic potential for Si (see Sect. 7.1.1), we utilized a large ab-initio data set for fitting, which consists of cohesive energies and forces of several reference simulations obtained from T_e-dependent DFT.

In order to take the oscillation of the A_{1g} phonon mode into account, we performed ab-initio simulations using the bulk simulation cell with 432 Sb atoms that we introduced in Sect. 4.5.7. In order to take surface effects into account, we performed also ab-initio simulations using the two simulation cells with 384 Sb atoms forming a thin film in z- and y-direction that we introduced in Sect. 4.5.8. We considered two different films, since the z-direction is not equivalent to the y- and z-direction. In order to get the necessary sampling of the PES, we considered eleven T_e's, which are listed in Tab. 8.1 and performed the following sets of reference ab-initio simulations $\{s_k\}$ at each T_e:

s_1: MD simulation of the bulk cell for $T_e < 12000$ K to consider the movement of the A_{1g} phonon mode at the considered T_e. Since the crystal structure melts ultrafast for $T_e \geq 12000$ K, we did not consider these high T_e's. We initialized the atomic structure thermalized at an ionic temperature of $T_i = 300$ K and used a time step of 5 fs. For fitting, we considered 500 times steps of the MD simulation, which corresponds to a total simulation time of 2.5 ps.

s_2: Time-dependent coordinates obtained from the reference MD simulation of scenario s_1 at $T_e = 9000$ K were used to calculate the energies and forces at all $T_e < 9000$ K. We included this reference simulation to cover large A_{1g} phonon oscillation amplitudes at low T_e's. At $T_e = 9000$ K, the A_{1g} phonon mode shows a big oscillation amplitude and the crystal structure keeps still intact. At lower T_e's, the amplitude is much smaller.

s_3: MD simulation of the thin film in z-direction ([111] direction of the crystal structure) at each T_e. We consider these simulations to get insights in the atomic kinetics at this T_e. We initialized the atomic structure thermalized at an ionic temperature of $T_i = 300$ K and used a time step of 5 fs. For fitting, we considered 500 times steps of the MD simulation, which corresponds to a total simulation time of 2.5 ps.

s_4: Time-dependent coordinates obtained from the reference MD simulation s_3 at $T_i = 18000$ K were used to calculate the energies and forces at all $T_e < 18000$ K. This reference simulation covers an expansion of the material at all studied T_e's, since the laser excitation of solids can also lead to strong local negative pressures with low local densities. Only at $T_e \geq 18000$ K, the thin film naturally expands due to bond softening in the simulation s_3 within the simulation time.

s_5: MD simulation of the thin film in y-direction ($[\overline{1}\overline{1}2]$ direction of the crystal structure) at each T_e similar to the reference simulation s_3, since the z direction is not equivalent to the x- and y-direction.

s_6: Time-dependent coordinates obtained from the reference MD simulation s_5 at $T_i = 18000$ K were used to calculate the energies and forces at all $T_e < 18000$ K analogously to reference simulation s_4.

s_7: The thin film initiated at $T_i = 300$ K was compressed step by step in z-direction and the corresponding energy and forces at each studied T_e were calculated. We included this reference simulation to cover a compression of the thin film, since the laser excitation of solids can lead to high local positive pressures with high local densities.

s_8: Accordingly to scenario s_7 but compression in y- instead of z-direction.

In total, we obtained around 3.7×10^6 ab-initio data-points for $T_e < 9000$ K, 3.0×10^6 for $T_e = 9000$ K, 2.4×10^6 for $9000\text{ K} < T_e < 18000$ K, and 1.2×10^6 for $T_e \geq 18000$ K.

8.1.2 Optimization of the Functional Form of the Polynomial Potential

We utilized our polynomial T_e-dependent interatomic potential model and our fitting procedure described in Sect. 6.3. At first, we derived the global minimum of the fit error $\mathbb{W}_{err}(T_e)$ for all polynomial-degree combinations up to arbitrarily chosen upper degree limits.

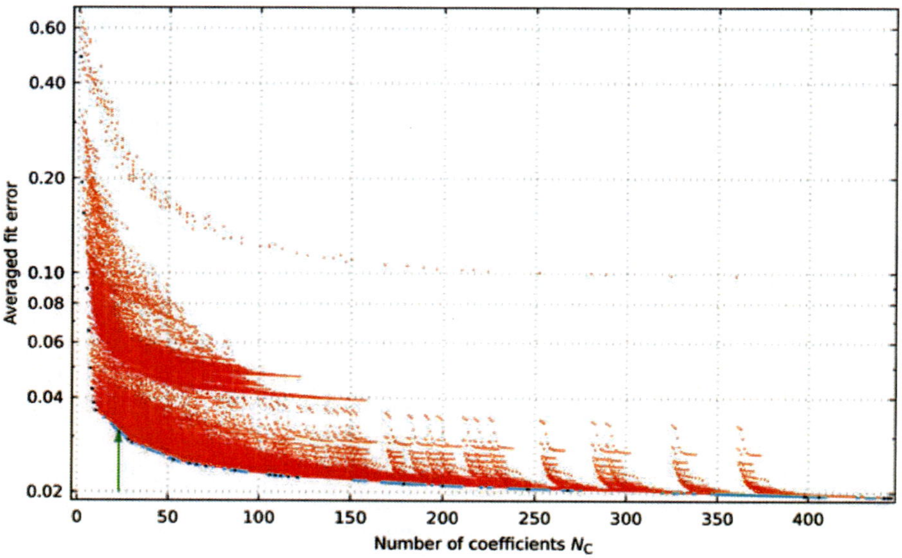

Fig. 8.1 Fit error $\langle \mathbb{W}_{err} \rangle$ averaged over $T_e = 300$ K and $T_e = 18000$ K as a function of the number of coefficients N_C for all polynomial-degree combinations up to the upper degree limits. Each blue dot represents an interatomic potential which is, for the corresponding polynomial-degree combination, a global minimum in parameter space. The members of the subset $\mathcal{S}_{best}$ are highlighted in magenta, the series $\Phi^{(k)}$ is highlighted in black, and the final choice, $\Phi^{(Sb)}(T_e)$, is highlighted in green and marked by an arrow

For this, we considered the local interaction terms Φ_2, Φ_3, Φ_ρ, ignored the four-body term Φ_4, and set the upper degree limits $N_2^{(r,\max)} = 15$ for the two-body term and $N^{(\max)} = 9$ for all other terms. To determine the optimal cutoff radii, we checked all cutoff radii starting from 0.30 nm, which is slightly above the nearest neighbor distance of 0.2907 nm in the equilibrium structure (see Table 4.4), up to 0.90 nm with an increment of 0.01 nm. We obtained the global minimum of the fit error $\mathbb{W}_{err}(T_e)$ for the resulting 165344 polynomial-degree combinations at $T_e = 300$ K and $T_e = 18000$ K, which is above the non-thermal melting threshold of around 12000 K (see Sect. 4.5.7). We considered these two T_e's and averaged the fit error $\mathbb{W}_{err}(T_e)$ over both T_e's, since the final polynomial-degree combination should work for all studied T_e's. In Fig. 8.1, we present the averaged fit error $\langle \mathbb{W}_{err} \rangle$ for all polynomial-degree combinations up to the upper degree limits as a function of the number of coefficients N_C.

Figure 8.1 looks quite similar to Fig. 7.4 for Si, where we used the same upper degree limits.

Our chosen upper degree limits are high enough, since $\langle \mathbb{W}_{err} \rangle$ decreases insignificantly for large values of N_C (see Fig. 8.1). In order to obtain the subset $\mathcal{S}_{best}$ that contains the optimally adjusted polynomial-degree combinations, we applied the iterative procedure described in detail in Sect. 6.3.3. The members of the subset $\mathcal{S}_{best}$

are highlighted in light blue and the series $\Phi^{(k)}$ from the iterative procedure is highlighted in black in Fig. 8.1. To select the final polynomial-degree combination, we analyzed the members of $\mathcal{S}_{\text{best}}$ with $N_C < 60$, because any further decrease of $\langle \mathbb{W}_{\text{err}} \rangle$ is quite low for $N_C \geq 60$ within the members of $\mathcal{S}_{\text{best}}$, as one can see in Fig. 8.1. For these polynomial-degree combinations of $\mathcal{S}_{\text{best}}$, we constructed the corresponding interatomic potential at each of the eleven studied T_e's by determining the related optimal coefficients and cutoff radii that minimize $\mathbb{W}_{\text{err}}(T_e)$. Using these interatomic potentials, we analyzed the description of the following physical properties at all studied T_e's:

(i) the phonon band structure of the A7 structure,
(ii) the cohesive energy curves of the diamond-like, fcc, bcc and sc bulk crystal structures,
(iii) the time evolution of the atomic root mean-square displacements in z- and y-direction of the MD simulation of the thin film in z- and y-direction.

Similar to Si, we found that the performance of the studied members of $\mathcal{S}_{\text{best}}$ varies considerably, so that not all members of $\mathcal{S}_{\text{best}}$ represent reliable interatomic potentials. We selected the final polynomial-degree combination with $N_C = 38$, since the corresponding interatomic potential best describes the above mentioned physical properties with a minimal possible N_C. The selected polynomial-degree combination for the interatomic potential for Si has only $N_C = 23$ coefficients (see Sect. 7.1.3), which is not surprising, since the A7 structure of Sb exhibiting a Peierls distortion is much more complicated to describe than the diamond-like structure of Si.

The resulting optimal cutoff radii vary at the eleven studied T_e's. For the final interatomic potential, the cutoff radii should be constant for all T_e, so that one gets a meaningful electronic specific heat, which is the second derivative of the interatomic potential with respect to T_e. The biggest fit error is found at the lowest T_e, namely $T_e = 300$ K. Thus, we set for all T_e's the cutoff radii at $r_2^{(c)} = 0.71$ nm, $r_3^{(c)} = 0.44$ nm, $r_\rho^{(c)} = 0.56$ nm close to the optimal vales at $T_e = 300$ K. With this, only the coefficients depend on T_e. For these coefficients we know, so far, only the optimal values at the eleven studied T_e's. Using these optimal values as supporting points, we approximated the complete set of coefficients by fitting of polynomials of degree 5 in T_e to obtain a continuous dependence on T_e. This polynomial approximation of the coefficients has almost no influence on the fit error, as one can see in Table 8.1. The coefficients of the final interatomic potential $\Phi^{(\text{Si})}(T_e)$ are tabulated in Sect. A.9 of the appendix.

8.1.3 Physical Properties of Polynomial $\Phi^{(Sb)}(T_e)$

For the final polynomial interatomic potential $\Phi^{(\text{Sb})}(T_e)$, we repeated the calculations and MD simulations presented in Sect. 4.5 at constant T_e. To perform the MD simulations, we used the DFT initializations. In Fig. 8.4, we present the summary of the comparison of $\Phi^{(\text{Sb})}(T_e)$ with DFT results.

Table 8.1 Fit error $\mathbb{W}_{err}(T_e)$ of the final polynomial-degree combination is tabulated at the studied T_e's for three cases: for the final interatomic potential $\Phi^{(Sb)}(T_e)$ with the chosen constant cutoff radii and polynomial approximated coefficients, for the chosen constant cutoff radii and T_e-optimized coefficients, and for T_e-optimized cutoff radii and coefficients. For the last case, also the optimal cutoff radii are tabulated

T_e (K)		$r_2^{(c)} = 0.71$ nm				
		$r_3^{(c)} = 0.44$ nm				
	$\Phi^{(Sb)}(T_e)$	$r_\rho^{(c)} = 0.56$ nm				
	$\mathbb{W}_{err}(T_e)$	$\mathbb{W}_{err}(T_e)$	$\mathbb{W}_{err}(T_e)$	$r_2^{(c)}$ (nm)	$r_3^{(c)}$ (nm)	$r_\rho^{(c)}$ (nm)
300	0.04343	0.04339	0.04335	0.713	0.440	0.558
3000	0.02943	0.02903	0.02892	0.714	0.452	0.576
6000	0.01421	0.01399	0.01362	0.775	0.469	0.561
9000	0.00646	0.00633	0.00571	0.736	0.534	0.558
12000	0.00561	0.00544	0.00431	0.847	0.550	0.521
15000	0.00349	0.00325	0.00235	0.759	0.551	0.525
18000	0.00212	0.00209	0.00146	0.760	0.541	0.601
21000	0.00194	0.00184	0.00155	0.852	0.508	0.636
24000	0.00219	0.00216	0.00191	0.747	0.455	0.999
27000	0.00272	0.00259	0.00234	0.744	0.457	0.999
30000	0.00300	0.00300	0.00274	0.679	0.457	0.759

As expected, $\Phi^{(Sb)}(T_e)$ reproduces the physical properties (i)–(iii) that were checked for selecting the final polynomial-degree combination. The derivatives of the phonon branches at the Γ-point are correct indicating that the elastic constants are well described. Also the movement of the A_{1g} phonon mode in bulk Sb is reproduced for $T_e < 12000$ K. In addition, the in plane atomic mean-square displacements MSD_{xy} in bulk Sb are accurately described by $\Phi^{(Sb)}(T_e)$ at all T_e's. For the ideal A7 structure as a function of T_e, we derived the electronic internal energy E_e from Eq. (5.77) and the electronic specific heat C_e from Eq. (5.78) using thermodynamic relations for $\Phi^{(Sb)}(T_e)$. We present the resulting curves in Figs. 8.2 and 8.3, respectively. $\Phi^{(Sb)}(T_e)$ describes reasonably both quantities, which indicates that the polynomial T_e-approximation of the coefficients works very well.

We also checked that $\Phi^{(Sb)}(T_e)$ describes forces in independent ab-initio MD simulations, which were not used for its development, with the same accuracy. For this, we generated a thin-film in x-direction ([$\bar{1}10$] direction of the crystal structure). We utilized again the supercell containing 384 atoms. But now we increased the simulation volume four times in x-direction instead of increasing it in the y- or z-direction as we did it in the reference simulations used for fitting. In this way, we obtained a film of 1.7 nm thickness. We initialized the atomic coordinates and velocities at $T_i = 300$ K using the Andersen thermostat with CHIVES. Using this initialization, we performed MD simulations of laser excitation at various constant T_e's with CHIVES. We derived the relative force error (5.91) and the relative cohesive

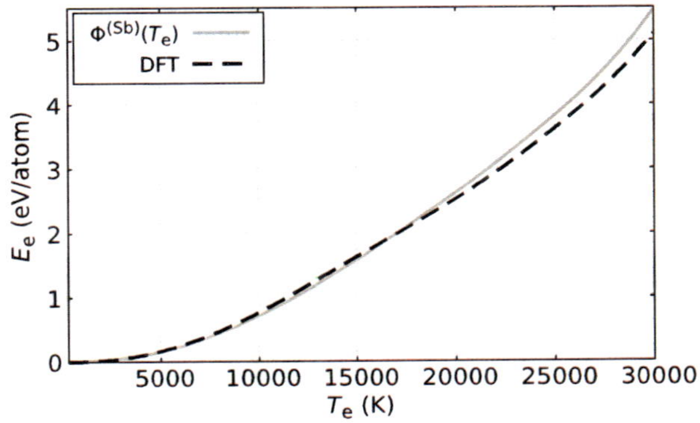

Fig. 8.2 Electronic internal energy E_e of the ideal A7 structure as a function of T_e for DFT and $\Phi^{(Sb)}(T_e)$

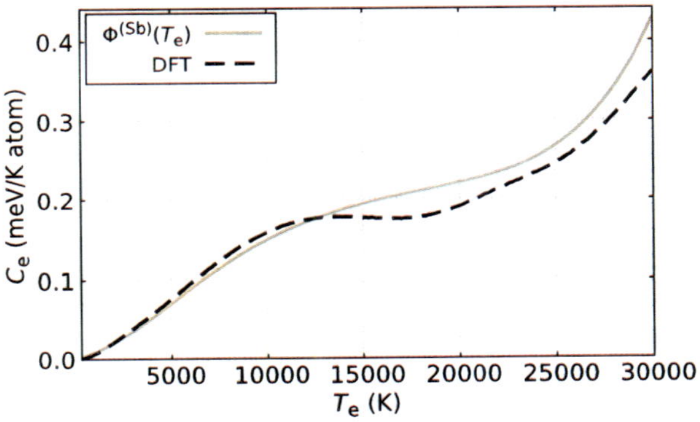

Fig. 8.3 Electronic specific heat C_e of the ideal A7 structure as a function of T_e for DFT and $\Phi^{(Sb)}(T_e)$

energy error (5.92) for the three films at various T_e's and list them in Table 8.2. Only the films with vacuum in y- and z-direction were used for the interatomic potential fitting as reference simulations s_5 and s_3, respectively. One can see that the atomic forces and cohesive energies of the film with vacuum in x-direction are described at the same level of accuracy as for the film with vacuum in y-direction. The errors for the film with vacuum in z-direction are slightly smaller. The reason for this is that the films in x- and y-direction are thinner and expand more within the simulation time of 2.5 ps. In addition, the x- and y-direction are equivalent to each other, but differ from the z-direction. This indicates that fitting additional independent ab-initio runs should not produce any improvement and hints to the transferability of $\Phi^{(Sb)}(T_e)$.

In the next step, we derived the melting temperature of $\Phi^{(Sb)}(T_e)$ at zero pressure. We set up a simulation cell that consists of $50 \times 40 \times 9$ minimal cells (see Sect. 4.5.1) and contains $N_{at} = 75600$ Sb atoms. We used periodic boundary conditions in all directions to get bulk Sb.

For deriving the melting temperature T_m, we simulated the coexistence of liquid and crystal parts (see Sect. 7.1.5). For this, we fixed the coordinates of half of the atoms and melted the other part of the crystal structure by applying the Anderson thermostat (see Sect. 4.1.2) at $T_i = 2500$ K. Then we allowed again that all atoms move. We applied the Anderson thermostat to all atoms at a temperature, which

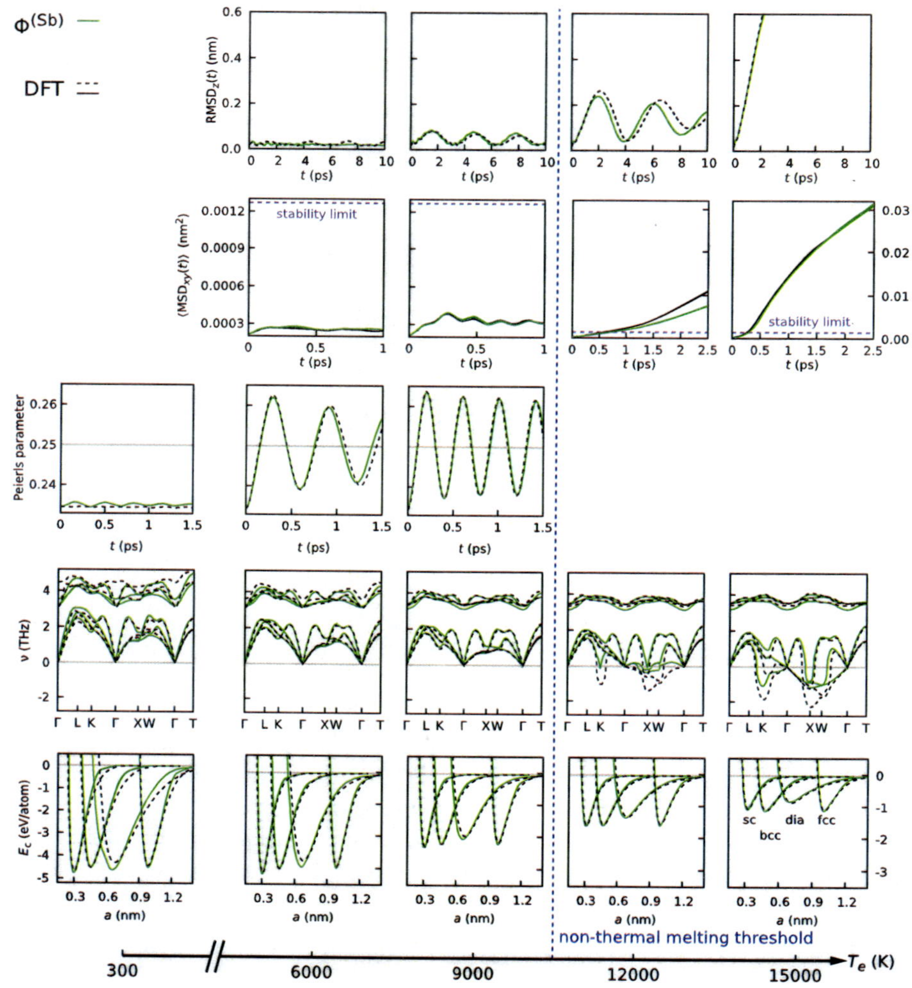

Fig. 8.4 Summary of the performance of $\Phi^{(\mathrm{Sb})}(T_e)$ in describing the effects occurring in Sb induced by the increased T_e deformed PES. Description see Sect. 4.5.9

we assumed to be close to the melting temperature. During this thermalization, we rescaled the simulation cell volume and the atomic coordinates every 2.5 ps to reach again zero pressure. In Sect. A.4 of the appendix, we describe the calculation of the pressure. Then we performed a MD simulation (at constant energy) for 2500 ps using a time step of 5 fs. In Fig. 8.6, we show a snapshot of the atomic structure at the initialization and one after the MD simulation period of 2500 ps. We want to note that we cannot use the CSP value from Eq. (7.3) to identify molten or crystalline parts, since the A7 structure of Sb is not inversion invariant in contrast to the diamond-like structure of Si. In Fig. 8.5, we present the ionic temperature T_i obtained from

Table 8.2 The relative force error (5.91) and relative cohesive energy error (5.92) in the MD simulation at constant T_e are listed for the three different types of films. Only the films with vacuum in y- and z-direction were used for fitting as set s_5 and s_3, respectively

T_e (K)	Vacuum x		Vacuum y		Vacuum z	
	f_{err} (%)	E_{err} (%)	f_{err} (%)	E_{err} (%)	f_{err} (%)	E_{err} (%)
9000	12.1	0.06	12.7	0.08	9.4	0.38
12000	11.8	0.23	11.8	0.27	8.5	0.23
15000	8.9	0.77	8.1	0.62	7.7	0.75
18000	7.2	0.44	6.9	0.46	6.0	0.60
21000	10.2	1.39	9.6	1.23	5.8	1.00

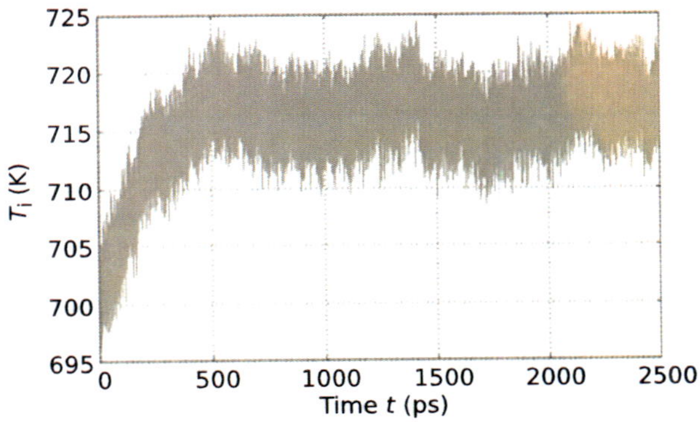

Fig. 8.5 The ionic temperature occurring in the MD simulation of the bulk simulation cell with 75500 Sb atoms is shown as a function of time

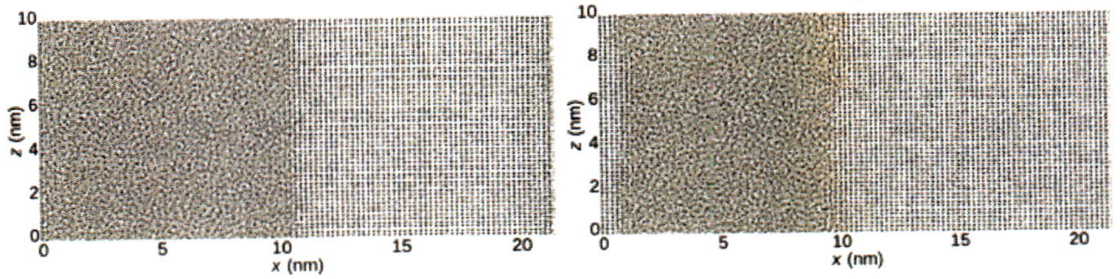

Fig. 8.6 The snapshots of the MD simulation with 75600 Sb atoms at zero pressure are shown at the initialization $t = 0$ ps (left) and at $t = 2500$ ps (right)

Eq. (4.16) of the MD simulation. One can clearly see, that T_i converges to the melting temperature and oscillates than around this value.

Finally, we determined the melting temperature

$$T_m = (717 \pm 2)\,\text{K} \tag{8.1}$$

by averaging T_i over the last 1000 ps. This value differs from the experimental value of $T_m = 903.78$ K [17], which is again not surprising, since $\Phi^{(Sb)}(T_e)$ was fitted to LDA-DFT reference simulations.

8.2 Optical Properties of Sb as a Function of the Peierls Parameter

In Sect. 8.3.1, we compare our $\Phi^{(Sb)}(T_e)$ calculations directly with experiments, in which a thin-film was firstly excited by a femtosecond laser-pulse and then probed by ultrafast x-ray diffraction. The laser excitation induces a strong oscillation of the Peierls parameter, which changes significantly the optical properties of Sb during the absorption of the laser pulse. So far, the optical properties of Sb are experimentally and theoretically only known at ambient conditions. Therefore, we hat to derive ab-initio the optical properties of Sb as a function of the Peierls parameter with the help of the DFT code WIEN2K [16] in the version 19.1. In these calculations, all electrons are taken explicitly into account. For this, the space is separated into two parts. Spherical volumes with radius r_{MT} are considered around the nuclei, in which spherically symmetric basis functions are used. Outside of the spheres, plane waves are used. This setup allows a very precise treatment of the electrons within DFT.

To study Sb, we used a simulation cell containing two Sb atoms and applied periodic boundary conditions. We used a fine mesh of **k**-points, namely $100 \times 100 \times 100$, which contains 85850 irreducible **k**-points. We set $r_{MT} = 0.12171$ nm (2.4 bohr) and set a Fermi distribution of the electrons with $T_e = 316$ K (1 mHa). By relaxing the atomic coordinates, we obtained an equilibrium Peierls parameter of $z = 0.23449091$ for the ideal A7 structure, i.e., $T_i = 0$ K. A change of the Peierls parameter modifies significantly the electronic properties. This can be clearly see in Figs. 8.7 and 8.8, where the electronic band structure and the electronic density of states are shown for $z = 0.23449091$ and $z = 0.2495$, where, at the latter, the Peierls distortion is almost removed. If the Peierls distortion becomes smaller, i.e. z moves closer to the value 0.25, the density of states at the Fermi energy increases, so that the semimetal Sb becomes more metallic.

In the next step, we derived the reflectivity $\mathfrak{R}$ and the absorption coefficient α_{abs} as a function of the wave length λ for the Peierls parameters $z = 0.23$, 0.23449091, 0.235, 0.24, 0.245, 0.2495, and 0.25. For this, we calculated the momentum matrix elements and integrated them over the first Brillouin zone to obtain the imaginary part of the dielectric function within the random-phase approximation (RPA) [18, 19]. Then we derived the real part of the dielectric function using the Kramers-Kronig relation. Finally, we obtained the reflectivity and the absorption coefficient from the dielectric function.

To obtain reliable results for the optical constants, a fine **k**-mesh with many **k**-points is needed. In Fig. 8.9, we present the reflectivity $\mathfrak{R}$ as a function of the wavelength λ at $z = 0.23449091$ for different meshes of **k**-points. The $100 \times 100 \times$

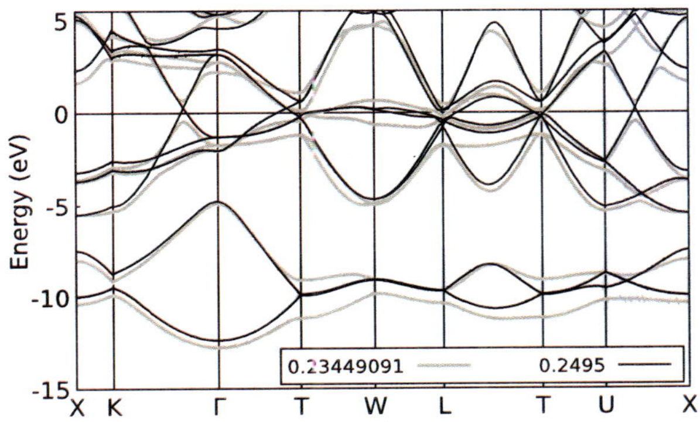

Fig. 8.7 Electronic band structure for two different Peierls parameters. The zero energy corresponds to the Fermi energy

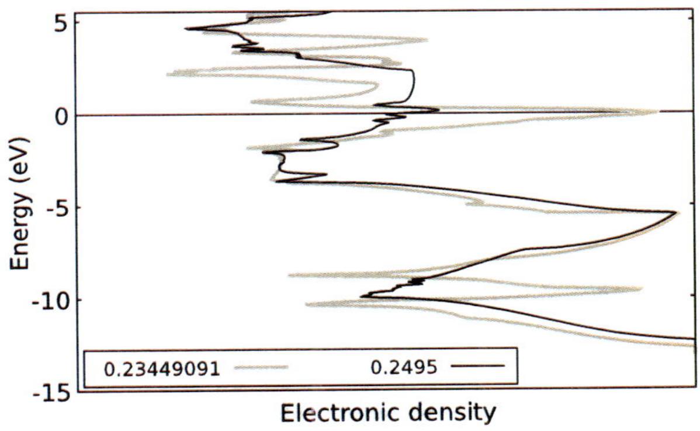

Fig. 8.8 Electronic density of states for two different Peierls parameters. The zero energy corresponds to the Fermi energy

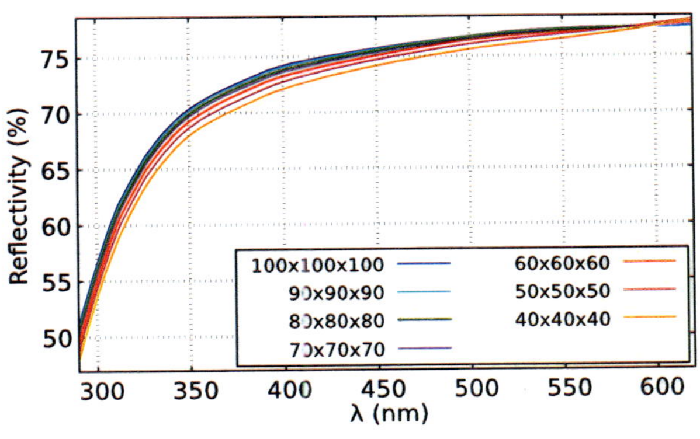

Fig. 8.9 The reflectivity is shown as a function of the wave length for different meshs of **k**-points at a Peierls parameter of z = 0.23449091

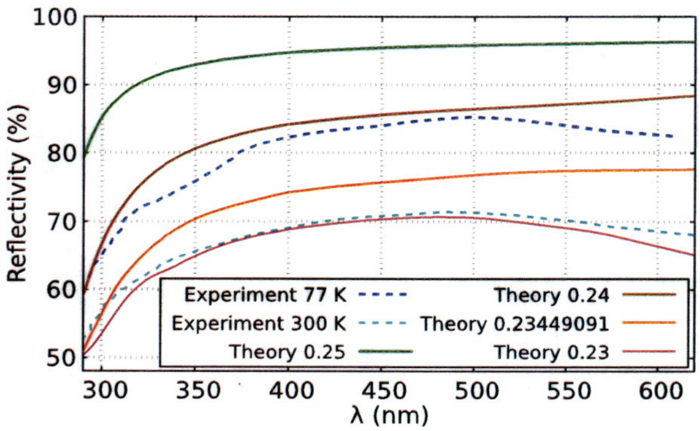

Fig. 8.10 The reflectivity is shown as a function of the wave length obtained from experiments (dashed lines) and obtained from our calculations for different Peierls parameters (solid lines). The experimental data was taken from FIG. 5 of Ref. [20]

100 mesh of **k**-points was finally used, since the reflectivity $\mathfrak{R}$ only slightly changes compared to the $90 \times 90 \times 90$ mesh of **k**-points. In Fig. 8.10, we show the calculated reflectivity $\mathfrak{R}$ as a function of the wavelength λ for several Peierls parameter $\mathbb{z}$ together with experimental results obtained at the temperatures 77 K and 300 K. The calculated reflectivity curve at equilibrium Peierls parameter $\mathbb{z} = 0.23449091$ lies between the two experimental curves at different temperatures. The reflectivity decreases within the experiments at increasing temperature. This indicates, that the thermal displacements of the atoms reduce the reflectivity. Moreover, the reflectivity increases in our calculations, if the Peierls distortion is decreased.

In Sect. 8.3.1, we compare our $\Phi^{(\mathrm{Sb})}(T_\mathrm{e})$ calculations with experiments that use a pump laser with a central wavelength of $\lambda = 534$ nm. Therefore, we considered the reflectivity and the absorption coefficient at $\lambda = 534$ nm for the different studied Peierls parameters $\mathbb{z}$. We fitted both quantities as a function of the Peierls parameter $\mathbb{z}$ to a polynomial of degree 4:

$$\mathfrak{R}(\mathbb{z}) = \sum_{k=0}^{4} a_{\mathfrak{R}}^{(k)}\, \mathbb{z}^k, \tag{8.2}$$

$$\alpha_{\mathrm{abs}}(\mathbb{z}) = \sum_{k=0}^{4} a_{\alpha_{\mathrm{abs}}}^{(k)}\, \mathbb{z}^k. \tag{8.3}$$

In Fig. 8.11, we present the reflectivity and, in Fig. 8.12, the absorption coefficient as a function of the Peierls parameter at $\lambda = 534$ nm. Table A.47 of the appendix contains the obtained parametrization for the reflectivity and Table A.48 of the appendix contains the obtained parametrization for the absorption coefficient.

We derived the optical properties for the ideal A7 structure and found that the reflectivity and the absorption coefficient are different for an oscillating electrical field in the z-direction compared to the x- and y-directions, if the hexagonal planes are located in the x, y-plane. The reflectivity measurements of Ref. [20] were performed on polycrystalline Sb samples. Thus, we averaged the calculated reflectivity over the z- and the x-, y-directions in Figs. 8.9 and 8.10. The experiments in Sect. 8.3.1

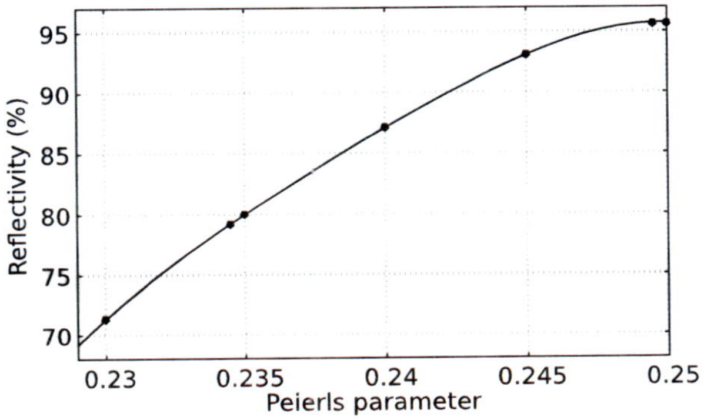

Fig. 8.11 The reflectivity is shown as a function of the Peierls Parameter for $\lambda = 534$ nm. The line corresponds to the polynomial fit to the data points

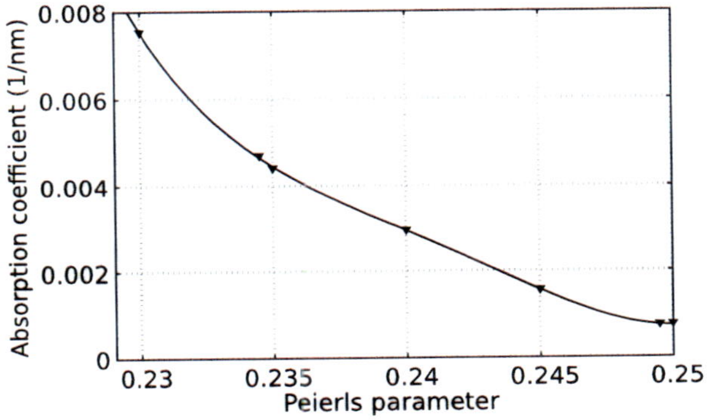

Fig. 8.12 The absorption coefficient is shown as a function of the Peierls Parameter for $\lambda = 534$ nm. The line corresponds to the polynomial fit to the data points

were performed with almost monocrystalline Sb films, where the electrical field of the pump laser oscillates within the x-, and y-direction. Therefore, we considered in Figs. 8.11 and 8.12 only the quantities in x-, y-direction. Moreover, we only considered $T_e = 316$ K, since, for increased T_e's, the Drude term used to derive the intraband transitions within WIEN2K must be extended [21].

The change of the reflectivity due to the laser-induced oscillation of the Peierls parameter / A_{1g} phonon mode was already experimentally observed within pump probe experiments in Sb [1–3].

8.3 MD Simulations of Excited PES and EPC with Polynomial $\Phi^{(Sb)}(T_e)$

For all presented MD simulations with $\Phi^{(Sb)}(T_e)$, we used the lattice parameters of $a = 0.43007$ nm and $c = 1.12221$ nm, which were experimentally found [22]. Professor Dr. Felipe Valencia Hernandez from the National university of Colombia calculated the electron-phonon coupling constant of all phonon modes in Sb for us. He used LDA-DFT and derived the electron-phonon coupling constant G_{ep} for

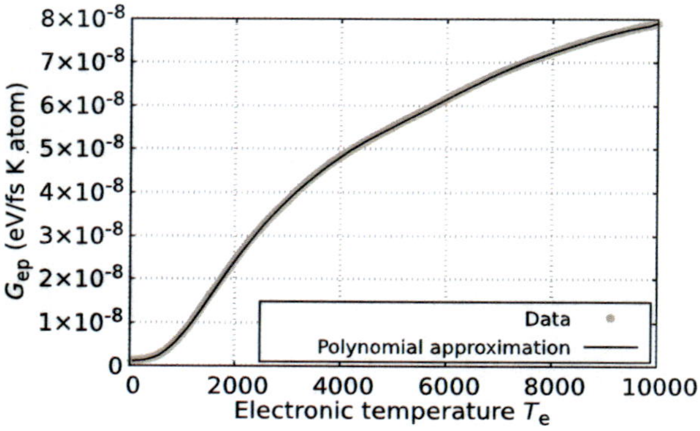

Fig. 8.13 The electron-phonon coupling constant G_{ep} of all phonon modes is shown as a function of T_{e}

different ionic T_{i} and electronic temperatures T_{e} with the help of Eq. (3.231). For this, he used the freely available DFT code QUANTUM ESPRESSO [23, 24]. The findings show that G_{ep} increases significantly with increasing T_{e}, whereas T_{i} has no significant influence on G_{ep}. Thus, we fitted his G_{ep} as a function of T_{e} for $T_{\text{i}} = 300$ K to a polynomial of degree 11:

$$G_{\text{ep}}(T_{\text{e}}) = \begin{cases} \sum_{k=0}^{11} a_{G_{\text{ep}}}^{(k)} \left(\frac{T_{\text{e}}}{10000\,\text{K}}\right)^k & T_{\text{e}} \leq 10000\,\text{K} \\ 7.907519999267578 \times 10^{-8} \,\frac{\text{eV}}{\text{fs K atom}} & T_{\text{e}} > 10000\,\text{K}. \end{cases} \tag{8.4}$$

We assumed that G_{ep} remains constant for $T_{\text{e}} > 10000$ K, since Professor Felipe Valencia Hernandez derived G_{ep} for T_{e}'s up to 10000 K and G_{ep} goes almost to saturation at $T_{\text{e}} = 10000$ K. In Table A.49 of the appendix, we list the coefficients of the polynomial approximation of G_{ep}. In Fig. 8.13, we present G_{ep} as a function of T_{e}. We show the data points obtained by Professor Felipe Valencia Hernandez and our polynomial approximation, which reproduces perfectly the ab-initio data points.

In order to gauge the influence of the excited PES with hot electrons and of the EPC on the atomic dynamics, we performed again the three different scenarios of MD simulations—excited PES & EPC, only excited PES, and only EPC—described in detail in Sect. 7.2. In order to simulate only the EPC, we needed the electronic specific heat $C_{\text{e}}(T_{\text{e}})$ as a function of T_{e}. For this, we considered the ideal A7 structure of Sb with the experimentally determined literature lattice parameters and Peierls parameter $z = 0.234$. For this structure we derived the Helmholtz free energy $F(T_{\text{e}})$ from T_{e}-dependent DFT for various T_{e}'s. We derived $C_{\text{e}}(T_{\text{e}})$ numerically from $F(T_{\text{e}})$ using again the thermodynamic relation (5.78). The obtained data points are shown in Fig. 8.3. We fitted $C_{\text{e}}(T_{\text{e}})$ as a polynomial of degree 14 in T_{e}:

$$C_{\text{e}}(T_{\text{e}}) = N_{\text{at}} \sum_{k=0}^{14} a_{C_{\text{e}}}^{(k)} \left(\frac{T_{\text{e}}}{30000\,\text{K}}\right)^k . \tag{8.5}$$

We list the coefficients $a_{C_e}^{(k)}$ in Table A.46 of the appendix. From $C_e(T_e)$, we derived the electronic internal energy $E_e(T_e)$ just by Eq. (7.6).

8.3.1 *Direct Comparison of the Bragg Peak Intensities with Experiments*

We have a collaboration with Dwayne Miller's group in the Max Planck Institute for the Structure and Dynamics of Matter located in Hamburg. Dr. Sascha Epp, a member, prepared three almost monocrystalline Sb films with a thickness of $d_{\text{film}} = 30$ nm and put each of them on a 4000 nm thick prolene substrate. He excited the films by one intense femtosecond laser-pulse with a central wavelength of $\lambda = 534$ nm, a FWHM-time width of $\tau = 100$ fs and a certain fluence. Then the time-dependent intensity of the (200) Bragg peak using ultrafast X-ray diffraction was measured as a function of the pump-probe delay. The experimental setup was similar to the already published measurements in bismuth [25]. The spotsize of the optical pump pulses yielded always 130 μm FWHM on the Sb surface, whereas the spotsize of the X-ray probe pulses yielded 5 μm FWHM.

To compare our simulations directly with these experiments, we set up a simulation cell that consists of $21 \times 12 \times 27$ minimal cells (see Sect. 4.5.1) and contains $N_{\text{at}} = 81648$ Sb atoms. We applied periodic boundary conditions in x- and y-direction and open boundary conditions in z-direction ([111] direction of the crystal structure) to adapt to a 30 nm thick Sb film. The top surface area A of the thin film within this simulation cell yields $A = 80.73\,\text{nm}^2$. To consider the prolene substrate, we put a barrier at the bottom of the film, so that the atoms cannot move downwards. Beside this, we did not make further adjustments related to the prolene. We applied the Andersen thermostat (see Sect. 4.1.2) to initialize the atomic coordinates and velocities at $T_i = 300$ K.

Then we performed MD simulations of the femtosecond-laser excitation using the three different previously described methods—excited PES & EPC, only excited PES, only EPC. We used a time step of $\Delta t = 1$ fs and simulated a Gaußian-shaped pulse with a FWHM-time width of $\tau = 100$ fs similar to the experiment. We can neglect a heat flow in x- and y-direction, since the pump pulse has a much bigger spotsize than the probe pulse, so that the fluence of the pump pulse can be assumed to be spatial constant within the spot size of the probe pulse. Thus, we assumed an uniform T_e and T_i in x- and y-direction. At first, we also neglected the heat flow in z-direction and assumed a global T_e and T_i within the simulation cell. The experiment was carried out at three different total incident laser fluences $I_{\text{inc}_{\text{tot}}}$, namely $(8000 \pm 1000)\,\frac{\text{eV}}{\text{nm}^2}$, $(2300 \pm 500)\,\frac{\text{eV}}{\text{nm}^2}$, and $(900 \pm 500)\,\frac{\text{eV}}{\text{nm}^2}$. We denote the incident laser intensity at time t by $I_{\text{inc}}(t)$, so that we have

$$I_{\text{inc}_{\text{tot}}} = \int_{-\infty}^{\infty} dt\, I_{\text{inc}}(t). \tag{8.6}$$

Since the experiment used a pump pulse with a Gaussian-shaped time profile with a FWHM time width of $\tau = 100$ fs, we set

$$I_{\text{inc}}(t) = \frac{I_{\text{inc}_{\text{tot}}}}{\tau}\sqrt{\frac{\log(16)}{\pi}}\,\exp\left(-\frac{(t-2\,\tau)^2}{\tau^2}\,\log(16)\right). \tag{8.7}$$

Thus, we obtain for the incident laser fluence, that is irradiated onto the film during the time step t_ℓ:

$$\int_{t_\ell}^{t_{\ell+1}} dt\; I_{\text{inc}}(t) = \frac{I_{\text{inc}_{\text{tot}}}}{2}\left(\operatorname{erf}\left(\frac{t_{\ell+1}-2\,\tau}{\tau}\sqrt{\log(16)}\right) - \operatorname{erf}\left(\frac{t_\ell-2\,\tau}{\tau}\sqrt{\log(16)}\right)\right). \tag{8.8}$$

The maximal intensity is irradiated at $t = 2\tau$. We started the MD simulations at $t = 0$ and not earlier, since 99.99975% of the incident laser intensity is absorbed during $t = 0$ and $t = 4\,\tau$. To obtain the energy $\Delta E_{\text{L}_{\text{abs}}}(t_\ell)$ that is absorbed from the laser at time step t_ℓ within the simulation cell, we used the reflectivity $\mathfrak{R}\big(\mathbb{z}(t)\big)$ and the absorption coefficient $\alpha_{\text{abs}}\big(\mathbb{z}(t)\big)$, that we derived as a function of the Peierls parameter $\mathbb{z}$ in Sect. 8.2:

$$\Delta E_{\text{L}_{\text{abs}}}(t_\ell) = \Big(1-\mathfrak{R}\big(\mathbb{z}(t_\ell)\big)\Big)\left(1-\mathrm{e}^{-\alpha_{\text{abs}}\left(\mathbb{z}(t_\ell)\right)d_{\text{film}}}\right)\int_{t_\ell}^{t_{\ell+1}} dt\; I_{\text{inc}}(t)\,A. \tag{8.9}$$

$A = 80.73\,\text{nm}^2$ denotes the top surface area of the thin film within the simulation cell. For the reflectivity $\mathfrak{R}$, we derived the Peierls parameter $\mathbb{z}(t)$ from the z-coordinates of the three top hexagonal planes at time t (see Sect. 4.5.1). For the absorption coefficient α_{abs}, we derived the Peierls parameter $\mathbb{z}(t)$ just by averaging over all local Peierls parameters within the film at time t.

From the atomic coordinates, we derived the time-dependent intensities of the (200) Bragg peak (see Sect. 4.2). In order to be able to compare directly with the experiments, we took into account, that within the experiment they did not measure the intensity of the single (200) Bragg peak but of rings with radial width Δq in the diffraction image, since the sample is not entirely monocrystalline and shows some polycrystalline nature. Therefore, to derive the intensity in the simulation, we have to average the intensity over every Bragg peak with scattering vector $\mathbf{q}$ fulfilling the condition $|\mathbf{q}| \in \big[|\mathbf{G}_{200}| - \Delta q, |\mathbf{G}_{200}| + \Delta q\big]$. At the two highest incident laser fluences, the Sb film melts and the (200) Bragg peak decays entirely. The broadening Δq controls, how big the rest intensity is after the decay, as one can see in Fig. 8.14. The bigger Δq is, the more is taken into account from the background scattering intensity, as one can see in Fig. 8.15. We chose the broadening $\Delta q = 0.06\,\frac{1}{\text{nm}}$, since then the rest intensity matches best to the measured one at the highest incident laser fluence in the experiment (see Fig. 8.14).

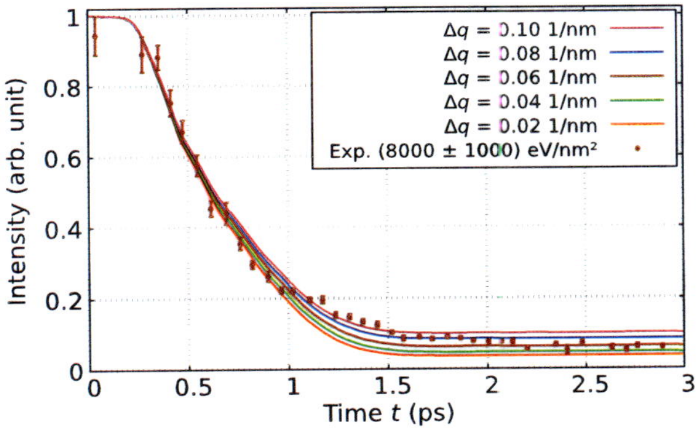

Fig. 8.14 Time-dependent normalized intensity of the (200) Bragg peak is shown for several broadenings Δq for a total incident laser fluence of $I_{\text{inc}_{\text{tot}}} = 4200\ \frac{\text{eV}}{\text{nm}^2}$ in the MD simulation considering the excited PES & EPC. Also the corresponding experimental data is shown

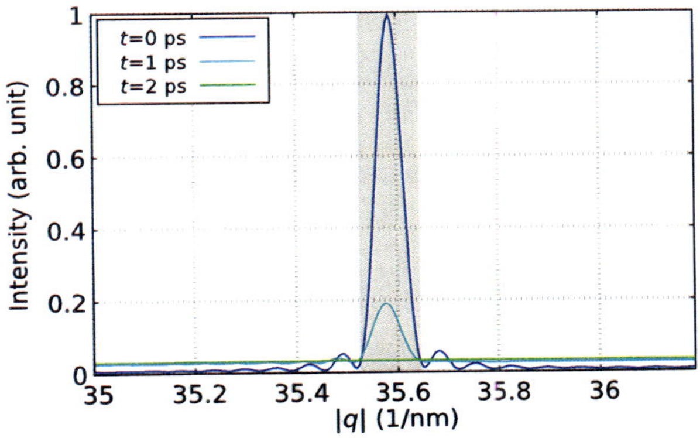

Fig. 8.15 Realtive scattering intensity is shown as a function of the absolute value $|\mathbf{q}|$ of the scattering vector for values close to $|\mathbf{G}_{200}| = 35.58\ \text{nm}^{-1}$ at different times t. The gray area indicates the considered interval $\left[|\mathbf{G}_{200}| - \Delta q, |\mathbf{G}_{200}| + \Delta q\right]$ used to compare with the experiments. In the MD simulation, a total incident laser fluence of $I_{\text{inc}_{\text{tot}}} = 4200\ \frac{\text{eV}}{\text{nm}^2}$ was used and the excited PES & EPC was taken into account

At first, we performed MD simulations with excited PES & EPC for various total incident laser fluences, derived the time-dependent intensity of the (200) Bragg peak and compared it with the three measurements. In this way, we found, for every measurement, the corresponding total incident laser fluence, so that the calculated intensity behavior of the (200) Bragg peak matches best the measured one. We present in Figs. 8.16, 8.18, and 8.19 the measured and calculated intensities for the three measurements. The time-dependent intensities of the (200) Bragg peak obtained from the MD simulations with excited PES & EPC reproduce the measured ones at all studied incident fluences. At the experimental fluence of $(2300 \pm 500)\ \frac{\text{eV}}{\text{nm}^2}$, which is presented in Fig. 8.18, the calculated Bragg peak intensity is significantly bigger than

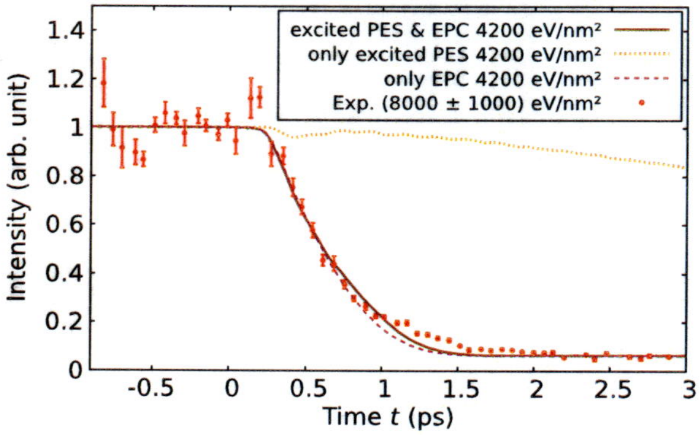

Fig. 8.16 The normalized intensity of the (200) Bragg peak is shown as a function of time obtained from the experiment (points) and from our three different scenarios of MD simulations

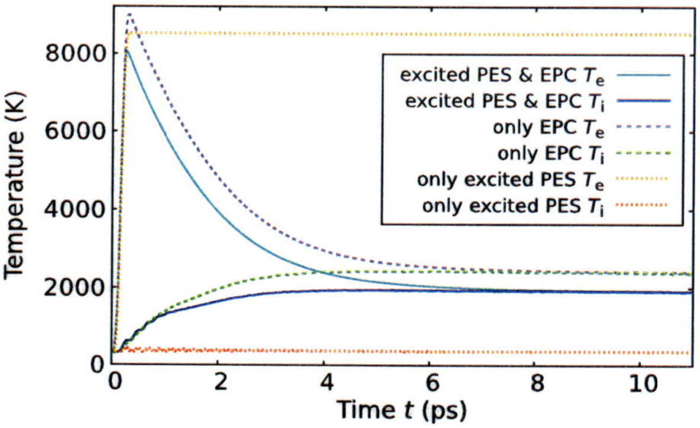

Fig. 8.17 Electronic and ionic temperatures are show as a function of time at a total incident laser fluence of $I_{inc_{tot}} = 4200\,\frac{eV}{nm^2}$ obtained from our three different scenarios of MD simulations

the experimental one for times between $t = 0.1$ ps and $t = 0.4$ ps. In the two other experiments, the data points for $t < 0.4$ ps show strong fluctuations and several of them exhibit large error bars. Thus, the precision of the measurements at times below 0.4 ps may be not so high and the errorbars may be underestimated, so that the above mentioned discrepancy is just related to the error in the measurements. For the two lowest fluences, the total incident fluence used in theory matches the experimental one within the fluence error of the experiment. However, at the highest measured fluence, the total incident fluence used in theory yields only 53% of the experimental one. The reason for this discrepancy may be the error in the reflectivity and absorption coefficient. We derived both for an electronic temperature of $T_e = 300$ K. But at the highest incident fluence, the electronic temperature reaches 8000 K during the pump laser pulse, as one can see in Fig. 8.17. At this high T_e's, the influence of an increased T_e may be significant on the optical properties. We are convinced that the reflectivity increase and the absorption coefficient decreases with increasing T_e, since Sb becomes then more metallic. However, we cannot take increased T_e's into account in our calculations with WIEN2K as mentioned in Sect. 8.2.

In the next step, we repeated our MD simulations using only EPC and using only the excited PES to study the influences of the EPC and the excited PES. We present

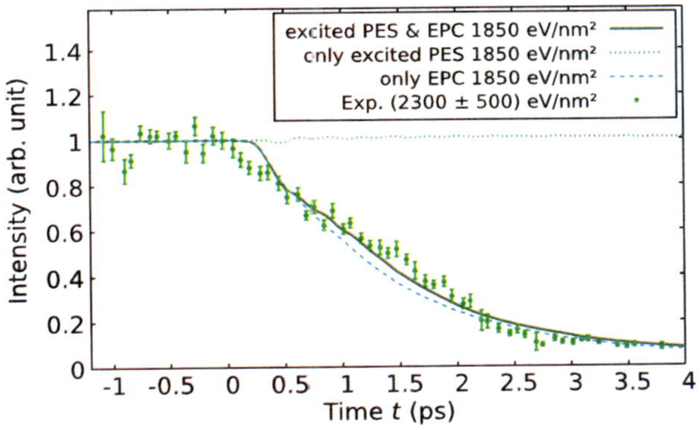

Fig. 8.18 The normalized intensity of the (200) Bragg peak is shown as a function of time obtained from the experiment (points) and from our three different scenarios of MD simulations

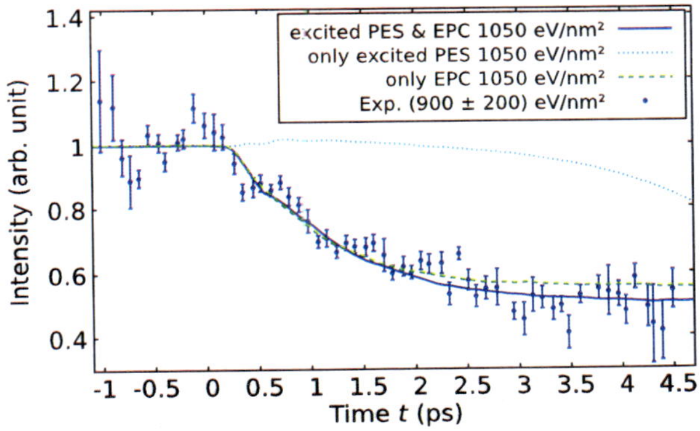

Fig. 8.19 The normalized intensity of the (200) Bragg peak is shown as a function of time obtained from the experiment (points) and from our three different scenario of MD simulations

the resulting curves also in Figs. 8.16, 8.17, 8.18, and 8.19. At all studied incident fluences, one can clearly see that the Bragg peak behavior is mostly determined by the EPC and that the effect of the excited PES is quite small. The curve obtained from the MD simulation considering the excited PES & EPC is almost identical to the curve obtained from the MD simulation considering only the EPC. In the MD simulations considering only the excited PES, the thin film does not melt at the three presented incident fluences. Thus, the Bragg peak intensity remains always almost constant. If, in addition to the EPC, the excited PES is also taken into account, the ionic and electronic temperatures become smaller, as one can see in Fig. 8.17. The reason for this is that less energy is absorbed from the laser, if the excited PES is taken into account, since the Peierls parameter moves closer to 0.25 due to the excited PES during the pump pulse, which causes a bigger reflectivity and a smaller absorption coefficient (see Sect. 8.2).

Our MD simulations using excited PES & EPC are able to reproduce the experimental (200) Bragg peak intensities. The used total incident laser fluence is in excellent agreement with the experimental one for the two lowest fluences. This verifies our MD simulation setup. At the highest measured fluence, we can reproduce the experimental Bragg peak decay only with a lower incident fluence compared to the experiment. The reason for this error may be the dependence of the reflectivity and the absorption coefficient on the electronic temperature, which we neglected. We took the influence of the Peierls parameter on the reflectivity and on the absorption coefficient into account. If we did not take this into account, the used incident fluence would disagree even for the two measurements with low fluences. Thus, the dependence of the reflectivity and the absorption coefficient on the Peierls parameter is quite important.

8.3.1.1 Local Electronic and Ionic Temperatures

The spotsize of the pump pulse is about 26 times bigger than the spot size of the X-ray probe pulse. Since we only simulate a part of the probed volume to derive the Bragg peak behavior, we can for the scope of this simulation neglect the heat flow in x- and y-direction, so that we can assume an uniform T_e and T_i in these directions. In the next step, we simulated a spatially dependent variable T_e and T_i in z-direction ([111] direction of the crystal structure) to model the penetration of the pump pulse into the film. The electrons can transport heat in z-direction. In order to model this, knowledge of the electronic heat conductivity K_e (see Sect. 6.2) is required. We obtained K_e from the Law of Wiedemann and Franz [26], which was empirically found for metals:

Theorem 8.1 (Law of Wiedemann and Franz) *For a metal, the electronic heat conductivity K_e is given by*

$$K_e = \mathbb{L}\, \sigma_{el}\, T_e, \tag{8.10}$$

where $\mathbb{L} = 2.44 \times 10^{-8}\, \frac{W\Omega}{K^2}$ is the Lorenz constant, σ_{el} is the electrical conductivity, and T_e is the electronic temperature.

We took the literature value $\sigma_\mathrm{el} = 2.5 \times 10^6\, \frac{1}{\Omega\,\mathrm{m}}$ for the electrical conductivity of Sb [17] and obtained

$$K_\mathrm{e}(T_\mathrm{e}) = 0.061\, \frac{\mathrm{W}}{\mathrm{K}^2\,\mathrm{m}} \times T_\mathrm{e} = 3.8073207792338285 \times 10^{-7}\, \frac{\mathrm{eV}}{\mathrm{fs}\,\mathrm{K}^2\,\mathrm{nm}} \times T_\mathrm{e}, \tag{8.11}$$

which we used in our calculations. We used again the simulation cell containing the thin film with 81648 Sb atoms and set the barrier at the bottom of the film to consider prolene, so that the atoms cannot move downwards. We also considered the Peierls parameter dependent reflectivity and absorption coefficient derived in Sect. 8.2. We assumed no electronic heat flow to prolene or out of the simulation cell in x- and y-direction. In z-direction, we divided the simulation cell into sub cells of length 0.724 nm. Each sub cell has got a local electronic and ionic temperature. We used a time step of $\Delta t = 1$ fs for moving the ions and a time step $\Delta t_\mathrm{D} = 0.1$ fs to

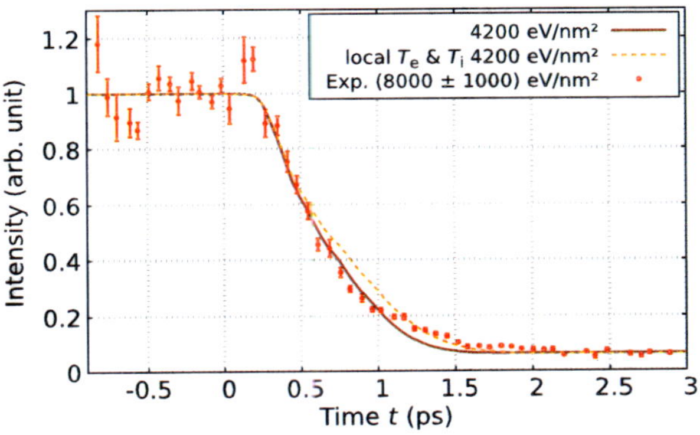

Fig. 8.20 The normalized intensity of the (200) Bragg peak is shown as a function of time obtained from the experiment (points) and from our calculation considering the excited PES & EPC with global temperatures (solid line) and with local temperatures (dashed line)

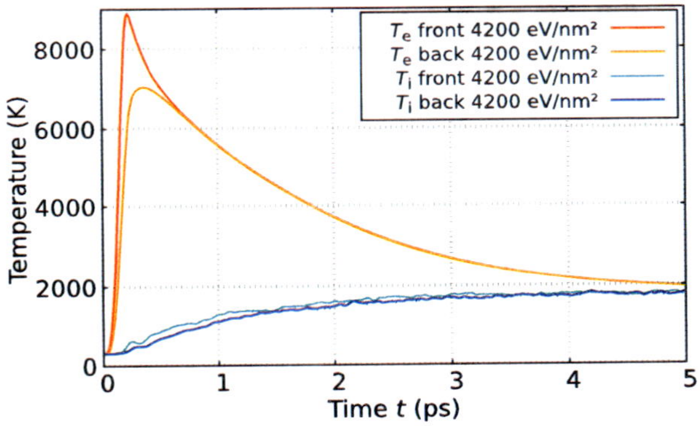

Fig. 8.21 Electronic and ionic temperature at the front and at the back of the film are show as a function of time at a total incident laser fluence of $I_{\text{inc}_{\text{tot}}} = 4200\ \frac{\text{eV}}{\text{nm}^2}$ obtained from our calculation considering the excited PES & EPC

solve the heat diffusion equations for the electrons (see Sect. 6.2). We performed the MD simulations considering the excited PES & EPC with local temperatures. We considered the total incident laser fluences, which produced the best agreement of the time dependent (200) Bragg peak intensity with the experiment within the MD simulations using global temperatures. We derived again the time dependent (200) Bragg peak intensity from the atomic coordinates.

In Figs. 8.20, 8.22, and 8.23, we present the obtained Bragg peak intensities together with the corresponding experimental results and the corresponding results from calculations with global temperatures. One can clearly see that the Bragg peak intensities obtained with the help of local temperatures differ only slightly from the ones obtained from global temperatures. The electronic heat conductivity is quite large in Sb, so that the local electronic temperatures become equal almost 500 fs after the maximum intensity of the pump pulse at $t = 200$ fs, as one can see in Fig. 8.21. Beside this, the ionic temperatures of the front and back of the film differ only slightly after the laser excitation. Thus, the assumption of always global temperatures is a quite good approximation.

In addition, we analyzed the influence of an electronic heat flow at the bottom of the film into prolene. For this, we assumed a constant temperature of 300 K of

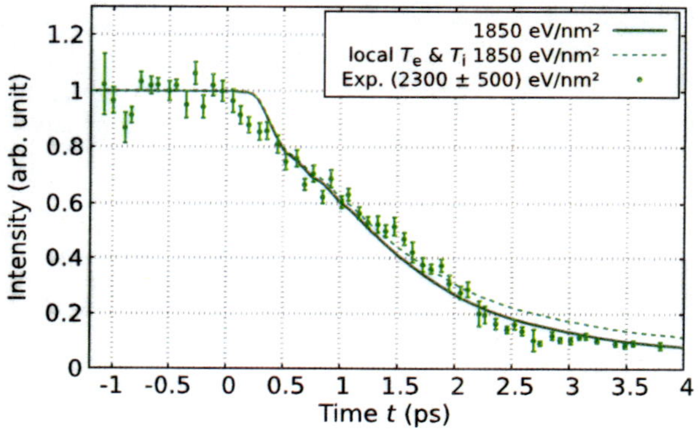

Fig. 8.22 The normalized intensity of the (200) Bragg peak is shown as a function of time obtained from the experiment (points) and from our calculation considering the excited PES & EPC with global temperatures (solid line) and with local temperatures (dashed line)

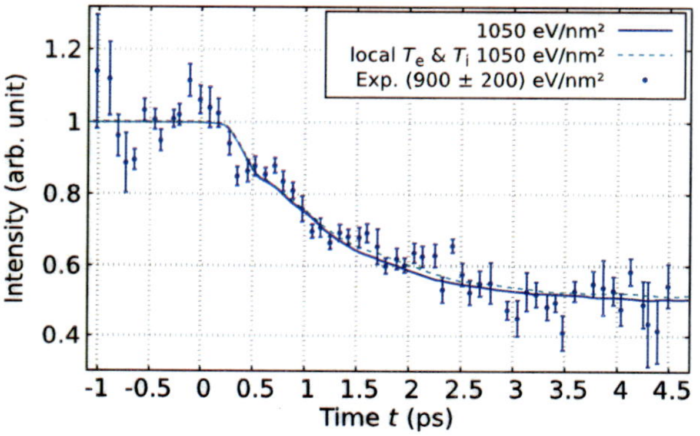

Fig. 8.23 The normalized intensity of the (200) Bragg peak is shown as a function of time obtained from the experiment (points) and from our calculation considering the excited PES & EPC with global temperatures (solid line) and with local temperatures (dashed line)

the prolene and that the electronic heat conductivity into prolene is similar to the electronic heat conductivity within Sb. Thus, our calculations consider the effect of the maximal possible heat flow, since, commonly, the electronic heat conductivity between different materials is worse compared to the conductivity within the material. If we include such a heat flow into prolene, the obtained time-dependent (200) Bragg peak intensities do not match any more to the experimental ones. Also an adjustment of the laser intensity does not produce an agreement. As an example, we show in Fig. 8.24 the obtained Bragg peak intensities together with the corresponding experimental results for the highest experimentally studied laser intensity. Figure 8.25 shows the related electronic and ionic temperatures at the front and the back of the Sb film. Due to the cooling, the ionic temperature at the back stays at room temperature, so that the bottom of the film does not melt. We conclude from the mismatch with respect to the experiment that a heat flow into prolene can be neglected.

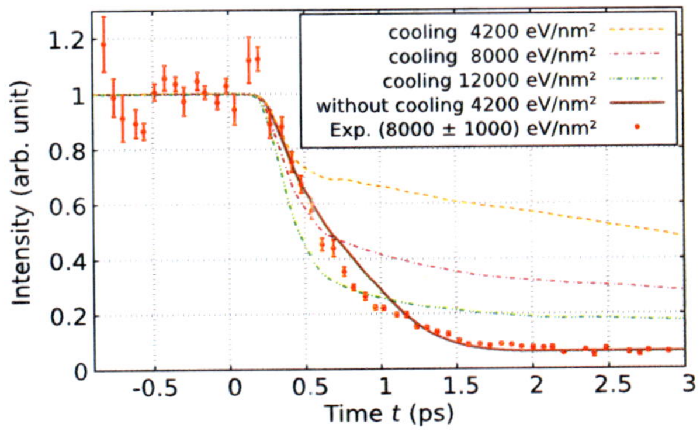

Fig. 8.24 The normalized intensity of the (200) Bragg peak is shown as a function of time obtained from the experiment (points) and from our calculation considering the excited PES & EPC using local temperatures without (solid line) and with heat flow into prolene (dashed lines)

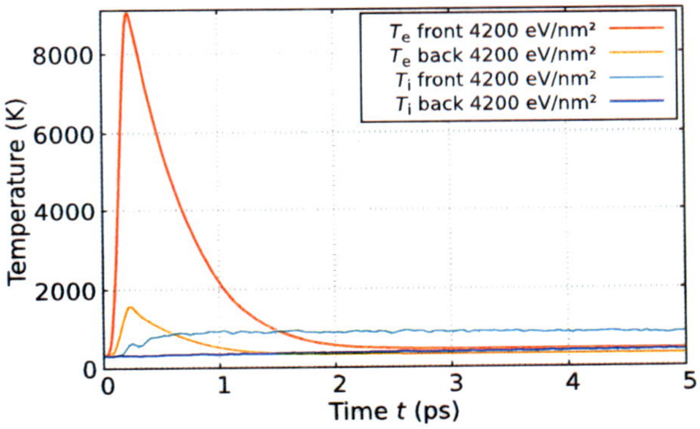

Fig. 8.25 Electronic and ionic temperature at the front and at the back of the film are show as a function of time at a total incident laser fluence of $I_{inc_{tot}} = 4200\ \frac{eV}{nm^2}$ obtained from our calculation considering the excited PES & EPC. Here, the heat flow into prolene is considered

8.3.2 Laser-Induced A7 to Sc Transition

If the Peierls parameter z equals 0.25, the Peierls distortion is gone and the first step towards the sc structure is done (see Sect. 4.5.1). Our T_e-dependent DFT calculations in Sect. 4.5.3 indicated that a femtosecond laser-pulse excites an oscillation of the A_{1g} phonon mode, which correspond to an oscillation of the Peierls parameter z. At moderate energies that are absorbed from the laser, the crystal structure remains intact and the oscillation of the Peierls parameter reaches and might exceed 0.25. However, our corresponding T_e-dependent DFT MD simulations in Sect. 4.5.5 show finite size effects at the relevant electronic temperatures and related absorbed energies. The used simulation cell containing 864 Sb atoms, which is almost the maximum number of atoms we can treat with DFT, is to small to simulate the laser-induced A7 to sc transition.

Thus, we revisited the laser-induced A7 to sc transition using $\Phi^{(Sb)}(T_e)$, which allows to treat the necessary large number of atoms. Furthermore, we took the EPC and surface effects into account, which both were neglected in the T_e-dependent DFT MD simulations in Sect. 4.5.5. We set up a simulation cell that consists of $117 \times 68 \times 45$ minimal cells (see Sect. 4.5.1) and contains $N_{at} = 4296240$ Sb atoms.

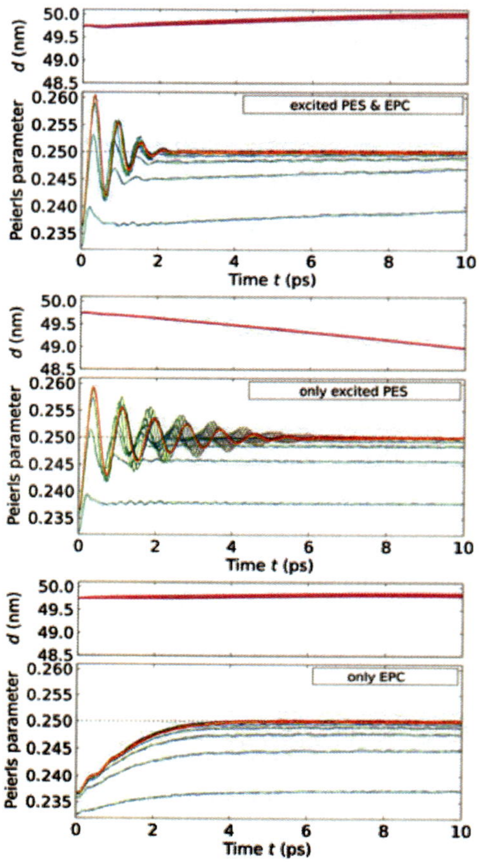

Fig. 8.26 The film thickness d and all individual local Peierls parameters are shown as a function of time t at an absorbed energy of $0.15\ \frac{\text{eV}}{\text{atom}}$ obtained from our three different scenarios of MD simulations. The color of the Peierls parameters is organized as followed: green $\hat{=}$ top, red $\hat{=}$ center, and blue $\hat{=}$ bottom. The thickness of the d curve corresponds to the standard deviation of the z-coordinates of the atoms in the top and bottom plane

We applied periodic boundary conditions in x- and y-direction and open boundary conditions in z-direction ($[\bar{1}11]$ direction of the crystal structure) to get a 50 nm thick Sb film, which is freestanding. We used the Andersen thermostat (see Sect. 4.1.2) to initialize the atomic coordinates and velocities at an ionic temperature of $T_{\text{i}} = 300$ K. Then we performed MD simulations of the femtosecond laser-excitation using the three different MD simulation scenarios—excited PES & EPC, only excited PES, only EPC. We used a time step of $\Delta t = 1$ fs and simulated a Gaußian-shaped pulse in the time domain with a FWHM-time width of $\tau = 10$ fs. We used a global T_{e} and T_{i} within the simulation cell, since both temperatures become very quickly spatial uniform after the laser excitation, as we found out in the simulations described before. We studied the physics for several energies absorbed from the laser.

The relative z-position of three neighboring hexagonal planes determine one local Peierls parameter (see Sect. 4.5.1). This allows us to derive local Peierls parameters as a function of the z-coordinate from the atomic coordinates. In Fig. 8.26, we present the film thickness d and all individual local Peierls parameter as a function of the

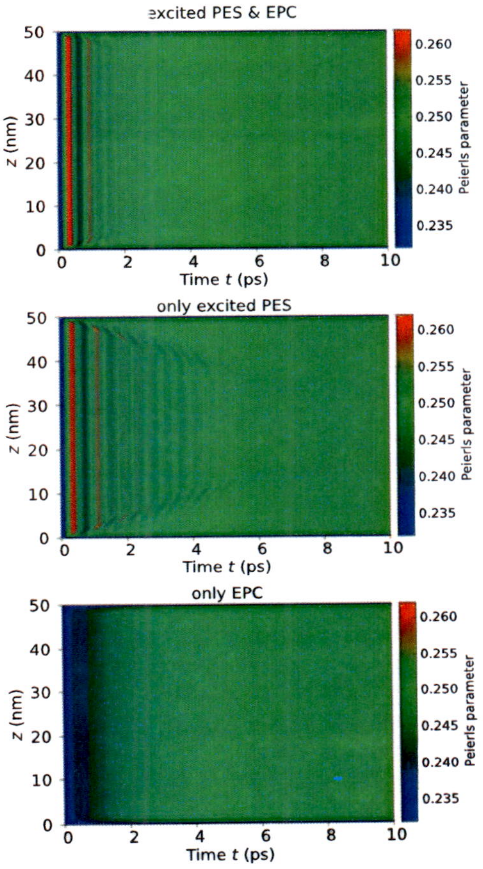

Fig. 8.27 The local Peierls parameters are shown as a function of time t and z-coordinate at an absorbed energy of $0.15\ \frac{\mathrm{eV}}{\mathrm{atom}}$ for the three different scenarios of MD simulations

time t at an absorbed energy of $0.15\ \frac{\mathrm{eV}}{\mathrm{atom}}$ for the three different scenarios of MD simulations. In addition in Fig. 8.27, we present the local Peierls parameters as a function of time t and the z-coordinate.

If the excited PES & EPC is simulated, the film slightly expands within the simulation time of 10 ps. The local Peierls parameters inside the film show first a strong oscillation, which amplitude exceeds $\mathbb{z} = 0.25$ several times. This oscillation is entirely damped after 2.5 ps and then the local Peierls parameters stay at $\mathbb{z} = 0.25$. The local Peierls parameters close to the surface show first a weak oscillation, which amplitude does not reach $\mathbb{z} = 0.25$ and is strongly damped. Then the local Peierls parameters weakly increase, since the film starts to melt slowly from the surface. Such an increase is not present in the two other scenarios, where no melting occurs. If only the excited PES is simulated and the EPC is ignored, the thickness of the film shrinks within the simulation time. The initial oscillation of the local Peierls parameters inside the film are less damped compared to the MD simulation considering the excited PES & EPC. If only the EPC is simulated and the excited PES is ignored, the film thickness

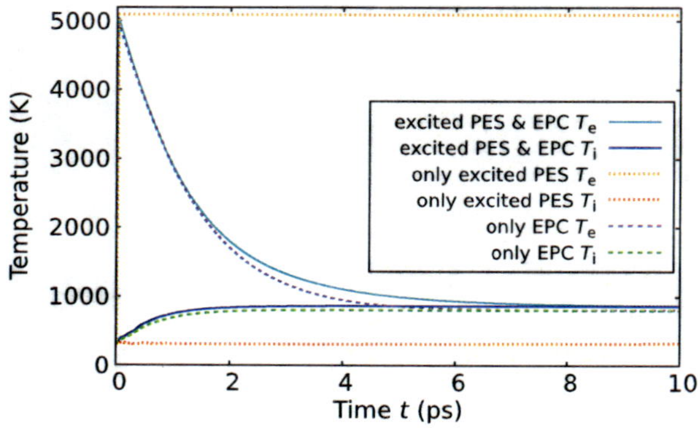

Fig. 8.28 Electronic and ionic temperatures are shown as a function of time at an absorbed energy of 0.15 $\frac{\text{eV}}{\text{atom}}$ obtained from our three different scenarios of MD simulations

remains almost constant and the local Peierls parameters do not show any oscillations. The local Peierls parameters inside the film increase monotonously to $z = 0.25$, which is reached at $t = 3$ ps, and remain constant. The local Peierls parameters close to the surface increase and stay at values significantly below $z = 0.25$.

The femtosecond laser-pulse increases significantly the electronic temperature T_e within tens of femtoseconds, as one can see in Fig. 8.28, where the electronic and ionic temperatures are shown as a function of time at an absorbed energy of 0.15 $\frac{\text{eV}}{\text{atom}}$ for the three different scenarios of MD simulations. At increasing T_e, the PES as a function of the Peierls parameter z changes significantly, as one can see in Fig. 4.38, which we derived using T_e-dependent DFT for the ideal A7 structure. At $T_e = 300$ K, the PES exhibits two minima that are separated by a barrier located at $z = 0.25$. We want to note that the two minima are equivalent, since the Peierls parameter z is equivalent to the Peierls parameter $0.5 - z$ (see Sect. 4.5.1). If T_e is increased, the barrier at $z = 0.25$ decreases and the two minima move closer to $z = 0.25$ and, finally, $z = 0.25$ is the only minimum. As a consequence of this behavior after the increase of T_e, the actual position of the Peierls parameter on the changed PES is far away from the minimum. Thus, the Peierls parameter starts to oscillate with a large amplitude, if the excited PES is taken into account. The local Peierls parameters close to the surface behave a little different due to surface effects.

An ionic temperature $T_i > 0$ corresponds to displacements of the atoms from their equilibrium positions inside the hexagonal planes and, likewise, to velocities attached to the atoms. Displacements of the atoms in the hexagonal planes exert a significant influence on the PES as a function of the Peierls parameter z. In Sect. 4.5.5, we analyzed the influence of the T_i on the PES as a function of the Peierls parameter z within the harmonic approximation using T_e-dependent DFT. We found that the minimum of the PES moves closer to $z = 0.25$, if T_i is increased from 0 K to 300 K (see Fig. 4.46).

In order to study the influence of the T_i-induced displacements of the atoms on the PES beyond the harmonic approximation, we set up a simulation cell that consists of $117 \times 68 \times 1$ minimal cells (see Sect. 4.5.1) and contains $N_{at} = 95472$ Sb atoms. We applied periodic boundary conditions in all three directions to get bulk Sb. We used the Anderson thermostat (see Sect. 4.1.2) to thermalize the atomic coordinates and velocities at several T_i's below the melting temperature of $\Phi^{(\mathrm{Sb})}(T_e)$ of $T_m = (717 \pm 2)$ K, which differs from the experimental value of $T_m = 903.78$ K [17] (see Sect. 8.1.3). Then we took the atomic coordinates, set the velocities to zero and moved step wise the hexagonal planes in z-direction against each other to vary the Peierls parameter $\mathbb{z}$. For each obtained atomic structure, we calculated the Helmholtz free energy from $\Phi^{(\mathrm{Sb})}(T_e)$. In this way, we obtained the PES as a function of the Peierls parameter $\mathbb{z}$ at atomic displacements compatible to a given T_i, as shown in Fig. 8.30. We always set $T_e = 300$ K. If T_i is increased, the barrier at $\mathbb{z} = 0.25$ decreases and the two minima beside $\mathbb{z} = 0.25$ move closer to $\mathbb{z} = 0.25$. At $T_i \approx 600$ K, the barrier vanishes and $\mathbb{z} = 0.25$ is the single minimum.

If the EPC is taken into account, the ionic temperature T_i increases significantly (see Fig. 8.28) and, thus, the Peierls parameter increases to $\mathbb{z} = 0.25$, since the minima of the PES move to $\mathbb{z} = 0.25$ at increasing T_i. The increase of T_i and the T_i-dependence of the PES do not induce an oscillation of the Peierls parameter, since the increase of T_i and the related deformation of the PES occur slowly. An oscillation of the Peierls parameter corresponds to the excitation of the A_{1g} phonon mode. Due to the phonon-phonon coupling, the energy of the A_{1g} phonon mode is transferred to the other phonon modes, so that the oscillation is damped. This damping is stronger, if the EPC is taken into account. The reason for this is that T_e decreases significantly due to the EPC (see Fig. 8.28) and the T_e-induced distortion of the PES becomes weaker. In order to accurately describe the phonon-phonon coupling, one needs a large simulation cell. Especially, the hexagonal planes must be big enough in x- and y-direction. Due to the increase of T_e, the hexagonal planes move at first coherently against each other, which correspond to an oscillation of the Peierls parameter or the A_{1g} phonon mode. Then the atoms within the hexagonal planes start to move incoherently against each other and the coherent movement of the hexagonal planes is damped and vanishes finally, since the energy of the A_{1g} phonon mode is transferred to the other phonon modes due to the phonon-phonon coupling. If the simulation cell or the hexagonal planes are to small, one observes finite size effects, as one can clearly see in Fig. 4.47, where we showed the oscillation of the Peierls parameter at increased T_e's obtained from T_e-dependent DFT MD simulations using 864 Sb atoms.

Our MD simulations with $\Phi^{(\mathrm{Sb})}(T_e)$ show that, after a moderate femtosecond laser-excitation, the Peierls parameter inside the film finally stays at $\mathbb{z} = 0.25$ for all three scenarios of MD simulations. If the excited PES is taken into account, the Peierls parameter shows an oscillation at the beginning. The behavior of the Peierls parameter can be experimentally determined, if one measures the time-dependent intensity of a Bragg peak that is sensitive to the Peierls parameter $\mathbb{z}$, like the (221) Bragg peak. Especially, the (221) Bragg peak vanishes entirely for $\mathbb{z} = 0.25$. In Fig. 8.29, we present the normalized intensity of the (221) Bragg peak as a function of time at an absorbed energy of $0.15\,\frac{\mathrm{eV}}{\mathrm{atom}}$ for all three different scenarios of MD simulations. One can clearly see that the intensity oscillates significantly, if the excited PES is taken into account. If only the EPC is taken into account, the intensity decays to zero with

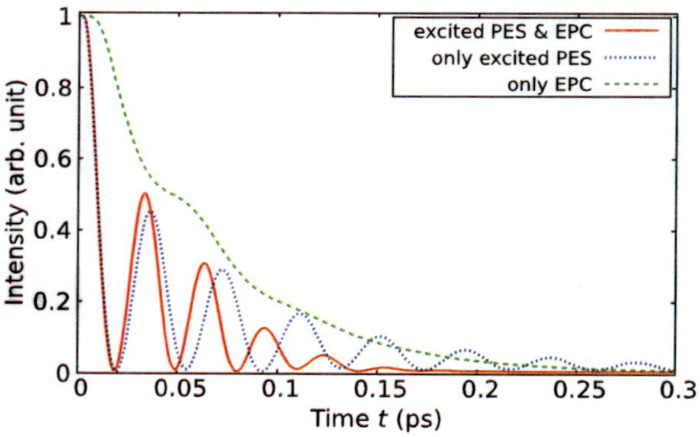

Fig. 8.29 The normalized intensity of the (221) Bragg peak is shown as a function of time at an absorbed energy of 0.15 $\frac{\text{eV}}{\text{atom}}$ obtained from our three different scenarios of MD simulations

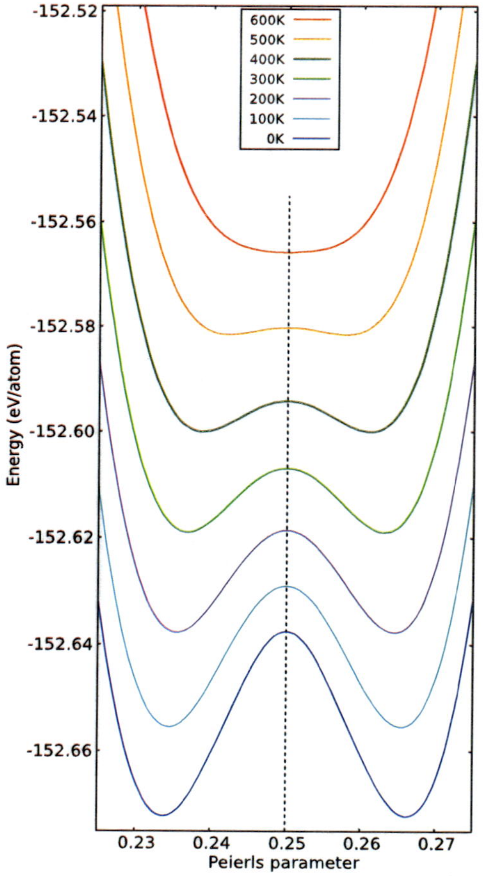

Fig. 8.30 The PES is shown as a function of the Peierls parameter at various T_i's. T_e is always set to 300 K. The atoms of the hexagonal planes are displaced from their equilibrium positions compatible to T_i

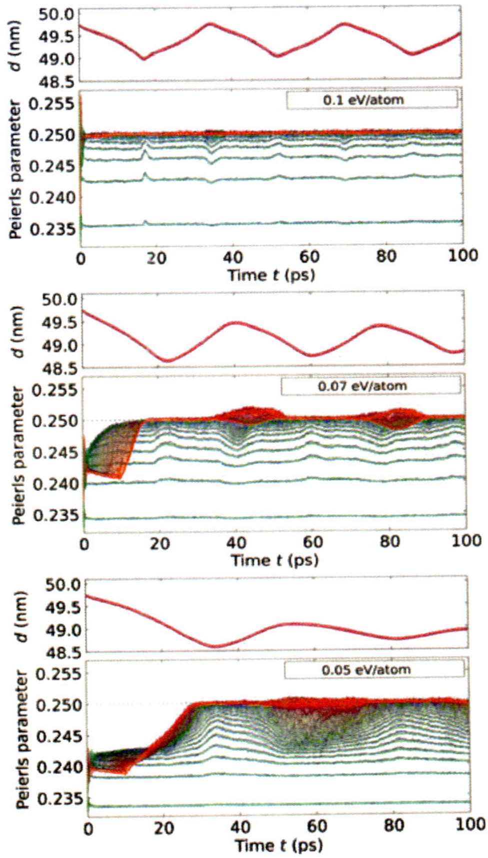

Fig. 8.31 The film thickness d and all individual local Peierls parameters are shown as a function of time t at various moderate absorbed energies for the MD simulation considering the excited PES & EPC. The color of the Peierls parameters is organized as followed: green $\hat{=}$ top, red $\hat{=}$ center, and blue $\hat{=}$ bottom. The thickness of the d curve corresponds to the standard deviation of the z-coordinates of the atoms in the top and bottom plane

tiny oscillations. Thus, the measurement of the time-dependent (221) Bragg peak intensity can show that the excited PES induced oscillation of the Peierls parameter is present. Another possibility is the measurement of the time-dependent reflectivity changes, since the Peierls parameter influences the optical properties (see Sect. 8.2).

At an absorbed energy of 0.15 $\frac{\text{eV}}{\text{atom}}$ for the MD simulation considering excited PES & EPC, the crystal structure inside the film still keeps intact within the simulation time. Especially, the displacements of the atoms out of the hexagonal planes are significantly smaller than the distances between the planes. The ionic temperature reaches $T_i = 865$ K, which is above the melting temperature of $\Phi^{(Sb)}(T_e)$ of $T_m = (717 \pm 2)$ K (see Sect. 8.1.3). The crystal structure starts to melt slowly from the surface, since the surface atoms are weaker bonded. If less energy is absorbed, no melting from the surface is initiated and the local Peierls parameters inside the film still move to $z = 0.25$ and stay there for a given time. In Fig. 8.31, we show the film thickness d and all individual local Peierls parameters as a function of time and, in

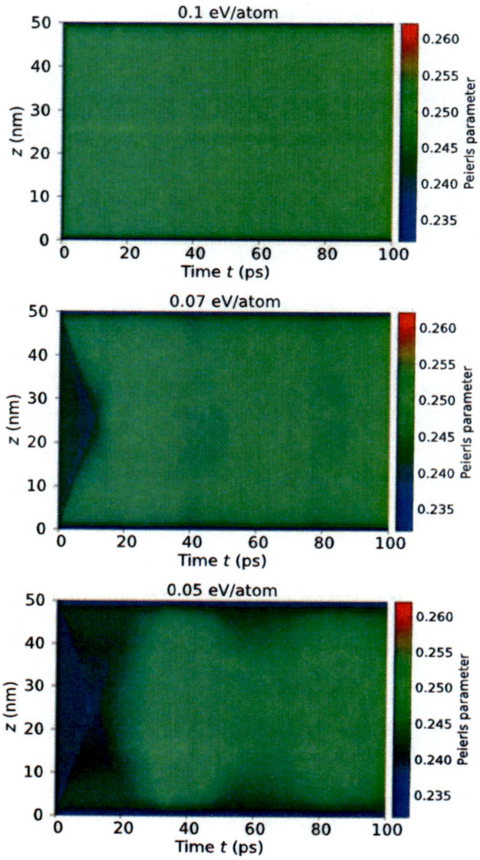

Fig. 8.32 The local Peierls parameters are shown as a function of time t and z-coordinate at various moderate absorbed energies for the MD simulation considering the excited PES & EPC

Fig. 8.32, we show the local Peierls parameter as a function of time t and z-coordinate at various moderate absorbed energies for the MD simulation considering the excited PES & EPC.

At an absorbed energy of 0.1 $\frac{\text{eV}}{\text{atom}}$, the local Peierls parameters inside the film show first a damped oscillation and stay then at $\mathbb{z} = 0.25$. No melting is induced by the laser within the simulation time of 100 ps, since the final ionic temperature yields 705 K, which is below the melting temperature of $\Phi^{(\text{Sb})}(T_e)$ of $T_m = (717 \pm 2)$ K (see Sect. 8.1.3). Here the main reason that the local Peierls parameter stay at $\mathbb{z} = 0.25$ is the increased T_i. For lower absorbed energies the pressure increase due to the contraction of the film is additionally necessary to move the local Peierls parameters inside the film finally to $\mathbb{z} = 0.25$.

At an absorbed energy of 0.07 $\frac{\text{eV}}{\text{atom}}$, the local Peierls parameters inside the film show fist a damped oscillation, which does not reach $\mathbb{z} = 0.25$. This oscillation is induced by the increased T_e. Beside this, the film initially contracts and one pressure wave propagates from the top and one from the bottom surface to the center with a speed of 2580 $\frac{\mathrm{m}}{\mathrm{s}}$ (see Fig. 8.32), which corresponds to the speed of sound for Sb obtained from $\Phi^{(\mathrm{Sb})}(T_e)$. These pressure waves induce a shift of the local Peierls parameters to $\mathbb{z} = 0.25$. First the local Peierls parameters close to the surface start to move. Then the local Peierls parameters in the inner regions start successively to move. After 14 ps, all local Peierls parameters, except for those close to the surface, reached $\mathbb{z} = 0.25$ and stayed at this value. The final ionic temperature yields $T_i = 585\,\mathrm{K}$, so that the PES as a function of $\mathbb{z}$ exhibits still two minima close to $\mathbb{z} = 0.25$ (see Fig. 8.30). Thus, the pressure waves induce the final movement to $\mathbb{z} = 0.25$, when the film thickness initially contracts. Then the film expands and contracts periodically. When the film expands, the local Peierls parameters inside the film move slightly away from $\mathbb{z} = 0.25$. When the film contracts, the local Peierls parameters inside the film move back to $\mathbb{z} = 0.25$ and stay at this value until the next film expansion. The local Peierls parameters move to $\mathbb{z} = 0.25$ and stay at this value only if the pressure is increased due to the contraction of the film. Here, the final ionic temperature is not high enough to induce a movement to $\mathbb{z} = 0.25$.

At an absorbed energy of 0.05 $\frac{\mathrm{eV}}{\mathrm{atom}}$, the final ionic temperature just yields 488 K. The pressure waves are weaker and the initial contraction of the film takes more time. In this situation one can see the two pressure waves until they reach the opposite surface in the bottom figure of Fig. 8.32. Again, following these pressure waves, the local Peierls parameters start to move to $\mathbb{z} = 0.25$. After 30 ps, the local Peierls parameters in the center finally move to $\mathbb{z} = 0.25$, when the film reaches approximately the minimal thickness during the initial contraction. Then the film thickness increases and decreases periodically with a smaller amplitude compared to the situation at absorbed energy of 0.07 $\frac{\mathrm{eV}}{\mathrm{atom}}$. Thus, the local Peierls parameters in the center move slightly away from $\mathbb{z} = 0.25$, when the film is expanded again. But they do not move as far away as for the absorbed energy of 0.07 $\frac{\mathrm{eV}}{\mathrm{atom}}$ due to the reduced amplitude of the film thickness oscillation. For lower absorbed energies the local Peierls parameters do not reach $\mathbb{z} = 0.25$ any more, which is presented in Fig. 8.33 and Fig. 8.34. At an absorbed energy of 0.04 $\frac{\mathrm{eV}}{\mathrm{atom}}$, the final ionic temperature yields only 450 K. Consequently, the T_i-related equilibrium position of the local Peierls parameters is slightly increased, but it is significantly below $\mathbb{z} = 0.25$. The increased T_e at the beginning only induces a weak oscillation of the local Peierls parameters that does not reach $\mathbb{z} = 0.25$. Thus, after the laser excitation, the local Peierls parameters increase but stay below $\mathbb{z} = 0.25$. One can clearly observe the influence of the pressure on the local Peierls parameters inside the film. When the film thickness decreases, the pressure increases and the local Peierls parameters inside the film move closer to $\mathbb{z} = 0.25$. The local Peierls parameters directly in the center move the closet towards $\mathbb{z} = 0.25$, since this is the region of highest pressure. When the film thickness increases, the pressure decreases and the local Peierls parameters inside the film decrease again. In this way, the oscillation of the film thickness induces an oscillation of the Peierls parameters inside the film with a time period of 80 ps. At an absorbed energy of 0.03 $\frac{\mathrm{eV}}{\mathrm{atom}}$, the final ionic temperature just yields 414 K and the pressure fluctuations are smaller. Thus, the local Peierls Parameters inside the film only weakly increase at the beginning and show then tiny oscillations.

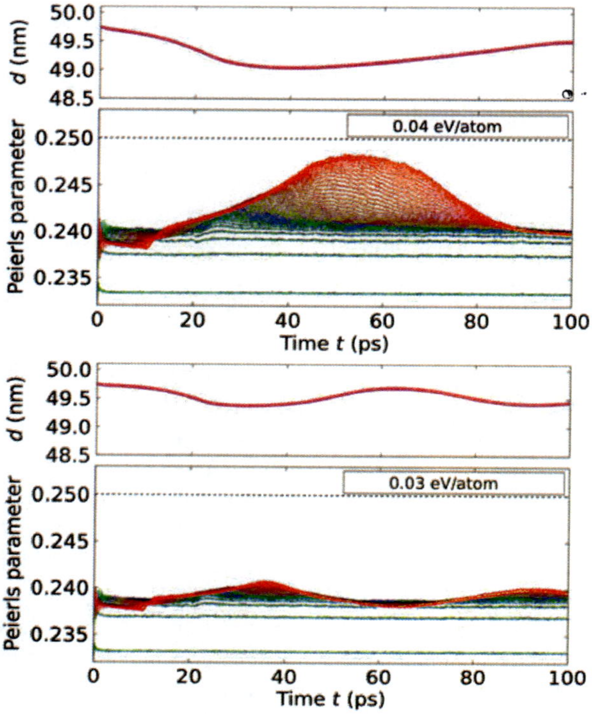

Fig. 8.33 The film thickness d and all individual local Peierls parameters are shown as a function of time t at various small absorbed energies for the MD simulation considering the excited PES & EPC. The color of the Peierls parameters is organized as followed: green $\hat{=}$ top, red $\hat{=}$ center, and blue $\hat{=}$ bottom. The thickness of the d curve corresponds to the standard deviation of the z-coordinates of the atoms in the top and bottom plane

Our calculations indicate that a moderate femtosecond laser excitation effectively removes the Peierls distortion. This is the first step towards the sc structure. The sc structure is reached, if the hexagonal lattice parameter ratio $\frac{c}{a}$ decreases from initially $\frac{c}{a} = 2.609$ to $\frac{c}{a} = \sqrt{6} \approx 2.449$ (see Sect. 4.5.1). This would be the case, if the film contracts to $d = 46.7$ nm. However, after the laser excitation, the film contracts minimally to a thickness of $d = 48.6$ nm corresponding to $\frac{c}{a} = 2.549$. Thus, the new state with removed Peierls distortion corresponds still to a sc structure that is distorted in the direction of the body diagonal. Beside this, the atoms are significantly displaced from the equilibrium positions due to the increased T_{i}. The main reason for this transformation is the increase of the ionic temperature and the increase of the pressure, where the latter is caused by the contraction of the film.

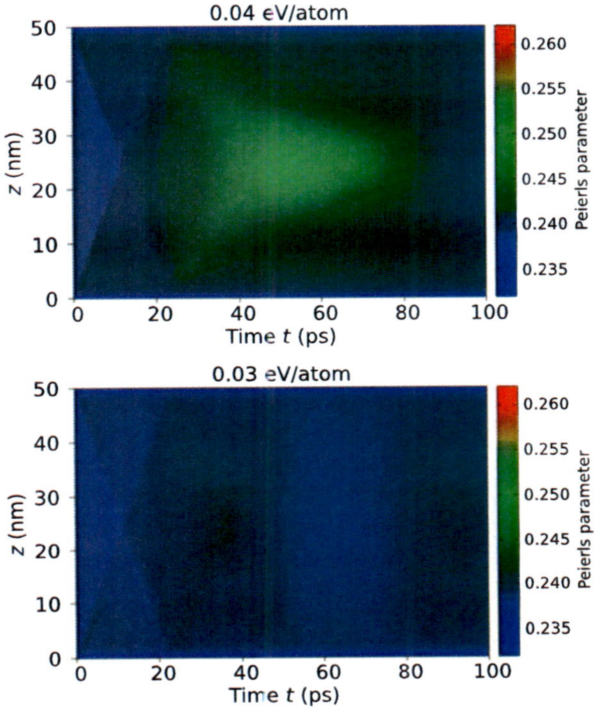

Fig. 8.34 The local Peierls parameters are shown as a function of time t and z-coordinate at various small absorbed energies for the MD simulation considering the excited PES & EPC

8.4 Summary

A T_e-dependent interatomic potential $\Phi^{(Sb)}(T_e)$ was developed for Sb using the polynomial interatomic potential model and the fitting procedure described in Sect. 6.3. An extensive set of interatomic forces and structural cohesive energies was fitted obtained from many T_e-dependent DFT reference simulations. The reference simulations were performed in thin-film geometry and in bulk. The finally obtained interatomic potential $\Phi^{(Sb)}(T_e)$ describes accurately the femtosecond laser-induced effects in Sb, namely, bond softening and hardening, excitation of the A_{1g} phonon mode, thermal phonon antisqueezing, non-thermal melting, and ablation effects. The laser-driven atomic mean-square displacements perpendicular to the surface of a thin-film are properly described. The error of $\Phi^{(Sb)}(T_e)$ in describing ab-initio forces and cohesive energies is reasonably low. $\Phi^{(Sb)}(T_e)$ reproduces forces and energies in independent ab-initio MD simulations, which were not used for its development, with the same accuracy. In addition, the electronic absorbed energy and the electronically specific heat are well described by $\Phi^{(Sb)}(T_e)$.

In order to compare directly with experiments, where the time-dependent intensity of the (200) Bragg peak was measured after a femtosecond laser-pulse excitation using ultrafast x-ray diffraction, the optical properties of Sb were calculated. The Peierls parameter and the associated physics influence significantly the optical properties. Thus, the reflectivity and the absorption coefficient were derived as a function of the Peierls parameter using the DFT code WIEN2K. $\Phi^{(Sb)}(T_e)$ was utilized to simulate the femtosecond laser-excitation of a 30 nm thick Sb film, which was considered in the experiments, including the effects of the excited PES and the EPC. The calculated optical properties were used, MD simulations were performed with $\Phi^{(Sb)}(T_e)$ and the time-dependent intensity of the (200) Bragg peak was derived. It was possible to reproduce the measured intensity curves at the same laser intensities as used in the experiments. This confirms the validity of $\Phi^{(Sb)}(T_e)$ and the theoretical description of the femtosecond-laser excitation using the excited PES and the EPC. However, only at very high laser intensities, the laser intensity had to be reduced in the calculation to reproduce the measured intensity decay of the (200) Bragg peak. The reason for this discrepancy is that the optical properties of Sb were derived for the electrons in the ground state. Unfortunately, the optical properties cannot be derived for excited electrons from WIEN2K, since, up to now, WIEN2K describes only inaccurately hot electrons. Furthermore, the electronic heat flow within the film was additionally simulated using local electronic and ionic temperatures. It was found out that the local electronic temperatures become equal within 500 fs after the laser excitation and that the intensity behavior of the (200) Bragg peak is the same as for the simulation using a uniform T_e and T_i from the beginning.

Finally, the femtosecond-laser excitation of a 50-nm thick film was studied using $\Phi^{(Sb)}(T_e)$. It was found out that a moderate, but very short laser excitation can remove the Peierls distortion without inducing a melting of the crystal structure. The reason for this is the increased ionic temperature and the laser induced pressure waves. The finally obtained crystal structure corresponds to a sc structure that is distorted in direction of the body diagonal. Thus, the femtosecond laser excitation induces a A7 to a sc-like structure transition. At lower laser intensities, the femtosecond laser-pulse no longer removes the Peierls distortion completely, but rather induces a periodic oscillation in the crystal reflected in an oscillation of the Peierls parameter.

References

1. K. Ishioka, M. Kitajima, O.V. Misochko, J. Appl. Phys. **103**(12), 123505 (2008) https://doi.org/10.1063/1.2940130
2. M. Hase, K. Ushida, M. Kitajima, J. Phys. Soc. Jpn. **84**(2), 024708 (2015). https://doi.org/10.7566/JPSJ.84.024708
3. O.V. Misochko, J. Exp. Theor. Phys. **123**(2), 292 (2016) https://doi.org/10.1134/S1063776116070116
4. B.N. Mironov, V.O. Kompanets, S.A. Aseev, A.A. Ischenko, I.V. Kochikov, O.V. Misochko, S.V. Chekalin, E.A. Ryabov, J. Exp. Theor. Phys. **124**(3), 422 (2017). https://doi.org/10.1134/S1063776117020145
5. S.L. Johnson, P. Beaud, E. Möhr-Vorobeva, A. Caviezel, G. Ingold, C.J. Milne, Phys. Rev. B **87**, 054301 (2013). https://doi.org/10.1103/PhysRevB.87.054301
6. A.A. Melnikov, O.V. Misochko, S.V. Chekalin, J. Appl. Phys. **114**(3), 033502 (2013). https://doi.org/10.1063/1.4813141
7. T.N. Kolobyanina, S.S. Kabalkina, L.F. Vereshchagin, L.V. Fedina, Zh. Eksp. Teor. Fiz. **55**, 164 (1968). http://www.jetp.ac.ru/cgi-bin/dn/e_031_02_0259.pdf

8. E.S. Zijlstra, N. Huntemann, M.E. Garcia, New J. Phys. **10**(3), 033010 (2008). http://stacks.iop.org/1367-2630/10/i=3/a=033010
9. N. Huntemann, E.S. Zijlstra, M.E. Garcia, Appl. Phys. A **96**(1), 19 (2009). https://doi.org/10.1007/s00339-009-5175-5
10. D.A. Young, *Phase Diagrams of the Elements* (University of California Press, 1991)
11. D. Fausti, O.V. Misochko, P.H.M. van Loosdrecht, Phys. Rev. B **80**, 161207 (2009). https://doi.org/10.1007/s00339-009-5175-5
12. E.D. Murray, D.M. Fritz, J.K. Wahlstrand, S. Fahy, D.A. Reis, Phys. Rev. B **72**, 060301 (2005). https://doi.org/10.1103/PhysRevB.72.060301
13. D.M. Fritz, D.A. Reis, B. Adams, R.A. Akre, J. Arthur, C. Blome, P.H. Bucksbaum, A.L. Cavalieri, S. Engemann, S. Fahy, R.W. Falcone, P.H. Fuoss, K.J. Gaffney, M.J. George, J. Hajdu, M.P. Hertlein, P.B. Hillyard, M. Horn-von Hoegen, M. Kammler, J. Kaspar, R. Kienberger, P. Krejcik, S.H. Lee, A.M. Lindenberg, B. McFarland, D. Meyer, T. Montagne, É.D. Murray, A.J. Nelson, M. Nicoul, R. Pahl, J. Rudati, H. Schlarb, D.P. Siddons, K. Sokolowski-Tinten, T. Tschentscher, D. von der Linde, J.B. Hastings, Science **315**(5812), 633 (2007). https://doi.org/10.1126/science.1135009, https://science.sciencemag.org/content/315/5812/633
14. S.W. Teitelbaum, T. Shin, J.W. Wolfson, Y.H. Cheng, I.J. Porter, M. Kandyla, K.A. Nelson, Phys. Rev. X **8**, 031081 (2018). https://doi.org/10.1103/PhysRevX.8.031081
15. T. Shin, J.W. Wolfson, S.W. Teitelbaum, M. Kandyla, K.A. Nelson, Rev. Sci. Instrum. **85**(8), 083115 (2014). https://doi.org/10.1063/1.4893641
16. P. Blaha, K. Schwarz, G.K.H. Madsen, D. Kvasnicka, J. Luitz, R. Laskowski, F. Tranand, L.D. Marks, *WIEN2k, An Augmented Plane Wave + Local Orbitals Program for Calculating Crystal Properties* (Karlheinz Schwarz, Techn. Universität Wien, Austria, ISBN 3-9501031-1-2, 2018)
17. D.R. Lide, *Handbook of Chemistry and Physics*, 84th edn. (CRC press, 2004)
18. R. Abt, C. Ambrosch-Draxl, P. Knoll, Physica B: Condens. Matter **194–196**, 1451 (1994). https://doi.org/10.1016/0921-4526(94)91225-4
19. C. Ambrosch-Draxl, J.O. Sofo, Comput. Phys. Commun. **175**(1), 1 (2006). https://doi.org/10.1016/j.cpc.2006.03.005, http://www.sciencedirect.com/science/article/pii/S0010465506001299
20. M. Cardona, D.L. Greenaway, Phys. Rev. **133**, A1685 (1964). https://doi.org/10.1103/PhysRev.133.A1685
21. A. Blumenstein, E.S. Zijlstra, D.S. Ivanov, S.T. Weber, T. Zier, F. Kleinwort, B. Rethfeld, J. Ihlemann, P. Simon, M.E. Garcia, (2018), arXiv:1802.00101
22. Y. Liu, R.E. Allen, Phys. Rev. B **52**, 1566 (1995). https://doi.org/10.1103/PhysRevB.52.1566
23. P. Giannozzi, S. Baroni, N. Bonini, M. Calandra, R. Car, C. Cavazzoni, D. Ceresoli, G.L. Chiarotti, M. Cococcioni, I. Dabo, A. Dal Corso, S.d. Gironcoli, S. Fabris, G. Fratesi, R. Gebauer, U. Gerstmann, C. Gougoussis, A. Kokalj, M. Lazzeri, L. Martin-Samos, N. Marzari, F. Mauri, R. Mazzarello, S. Paolini, A. Pasquarello, L. Paulatto, C. Sbraccia, S. Scandolo, G. Sclauzero, A.P. Seitsonen, A. Smogunov, P. Umari, R.M. Wentzcovitch, J. Phys.: Condens. Matter **21**(39), 395502 (2009). https://doi.org/10.1088/0953-8984/21/39/395502
24. P. Giannozzi, O. Andreussi, T. Brumme, O. Bunau, M.B. Nardelli, M. Calandra, R. Car, C. Cavazzoni, D. Ceresoli, M. Cococcioni, N. Colonna, I. Carnimeo, A.D. Corso, S.d. Gironcoli, P. Delugas, R.A.J. DiStasio, A. Ferretti, A. Floris, G. Fratesi, G. Fugallo, R. Gebauer, U. Gerstmann, F. Giustino, T. Gorni, J. Jia, M. Kawamura, H.Y. Ko, A. Kokalj, E. Küçükbenli, M. Lazzeri, M. Marsili, N. Marzari, F. Mauri, N.L. Nguyen, H.V. Nguyen, A. Otero-de-la Roza, L. Paulatto, S. Poncé, D. Rocca, R. Sabatini, B. Santra, M. Schlipf, A.P. Seitsonen, A. Smogunov, I. Timrov, T. Thonhauser, P. Umari, N. Vast, X. Wu, S. Baroni, J. Phys.: Condens. Matter **29**(46), 465901 (2017). https://doi.org/10.1088/1361-648x/aa8f79
25. S.W. Epp, M. Hada, Y. Zhong, Y. Kumagai, K. Motomura, S. Mizote, T. Ono, S. Owada, D. Axford, S. Bakhtiarzadeh, H. Fukuzawa, Y. Hayashi, T. Katayama, A. Marx, H.M. Müller-Werkmeister, R.L. Owen, D.A. Sherrell, K. Tono, K. Ueda, F. Westermeier, R.J.D. Miller, Struct. Dyn. **4**(5), 054308 (2017). https://doi.org/10.1063/1.4999701
26. N.W. Ashcroft, N.D. Mermin, *Solid State Physics* (Saunders College Publishing, 1976)

Chapter 9
Summary and Outlook

9.1 Overview

In this book two fundamental advancements in the ab-initio theoretical description of the ultrafast dynamics of femtosecond laser-pulse excited condensed matter could be achieved:

1. The development of a theory that leads to a molecular dynamics (MD) simulation method that, on the one hand, fulfills by design one of the most basic foundations in physics, that is energy conservation. And on the other hand, it considers the electronically driven substantial changes in interatomic bonding as well as the dynamics of the energy transfer by the electron-phonon collisions induced by the laser-excited electron hole pairs. In other words, the MD simulation method takes the changes in the potential energy surface (PES) and the electron-phonon coupling (EPC) dynamics due to excited electrons into account.
2. Realization of ultra-large scale atomistic simulations of femtosecond laser-pulses exited matter based on a novel method to construct computationally efficient and highly accurate T_e-dependent interatomic potentials from T_e-dependent DFT reference simulations. This class of interatomic potentials includes the influences of the hot electrons on interatomic bonding and describes electronic properties like the electronic specific heat or the energy that is absorbed by the electrons.

The interatomic potential construction and MD simulation method were applied to two different materials, namely silicon (Si) and antimony (Sb), for simulating the response to a femtosecond laser-excitation. The semiconductor Si was chosen, since it is the most important material in common electronics, and Sb as chosen, since it exhibits a crystal structure that allows to study a laser-induced solid-to-solid phase transition. For both materials, the results obtained from the MD simulations using the previously described approaches were compared with available experimental data and an excellent agreement was found, which validates the approaches.

In addition, an extension of the Code for Highly excIted Valence Electron Systems (CHIVES) was performed, which was developed in the group of Prof. Dr. Martin E.

© The Author(s), under exclusive license to Springer Nature Switzerland AG 2021

B. Bauerhenne, *Materials Interaction with Femtosecond Lasers*,
https://doi.org/10.1007/978-3-030-85135-4_9

Garcia to simulate the effects of hot electrons on matter using a very efficient implementation of T_e-dependent density functional theory (DFT). The calculation of local electric dipole moments was implemented, which allows to study the electromagnetic radiation from atomic oscillations caused by a femtosecond laser-excitation.

9.1.1 THz Emission from Coherent Phonon Oscillations

Using this extension of CHIVES, we found that the laser-induced coherent phonon oscillations in the (5, 0) zigzag boron nitride nanotube (BNNT) emit THz radiation. In the (5, 0) zigzag BNNT, a femtosecond laser-pulse excites three coherent phonon modes—the radial breathing mode, the radial bucking mode, and the longitudinal bond stretching mode. The corresponding atomic oscillations of each phonon mode emit THz radiation, since the boron-nitride bonds exhibit a permanent electric dipole moment due to the different electronegativities of the boron and nitrogen atoms. The longitudinal bond stretching mode emits most efficiently, since, in this optical mode, several boron and nitrogen atoms oscillate directly against each other.

9.1.2 Universal Behavior of the Indirect Electronic Band Gap in Laser-Excited Si

An intense femtosecond laser-excitation of Si induces an ultrafast melting of the crystal structure and a metallization. By performing MD simulations of bulk Si with increased T_e using CHIVES, we found that the indirect electronic band gap of Si decreases to zero as a universal function of the atomic mean square displacements almost independent of T_e during non-thermal melting. Moreover, this dependency is linear for a wide range of atomic mean square displacements.

9.1.3 Theory Allowing MD Simulations Considering Excited Potential Energy Surface and Electron-Phonon Coupling

The femtosecond laser-pulse excites the electrons to a high electronic temperature T_e, while the ions remain at their initially temperature T_i. The EPC is responsible for an energy transfer between the hot electrons and the cold ions so that the electronic and ionic temperature finally become equal. In an appropriate MD simulation setup, the amount of energy that is taken from the hot electrons due to the EPC should be transferred to the ions. For this, the MD simulation must allow for energy conservation. However, the PES that defines the forces on the ions for electrons at T_e

corresponds to the Helmholtz free energy of the electrons. But energy conservation can be only formulated in terms of the internal energy. To solve this problem, we transform the Helmholtz free energy to the internal energy of the electrons using an integral transformation of the PES. From the final integral expression, we obtain directly the equations of motion for the ions using the energy conservation. Especially, the ionic equation of motions contain, on the one hand, the correct interatomic forces from the PES and, on the other hand, the effects of the EPC on the ions. To define the time evolution of T_e, we used the corresponding differential equation for T_e from the two-temperature model (TTM). In our approach, the PES can be calculated from T_e-dependent density functional theory (DFT) or from a T_e-dependent interatomic potential. If the latter is used, ultra large-scale atomistic simulations are possible. For such simulations, we extended our theory to local electronic and ionic temperatures. The simulation cell is divided into small sub cells, each one with its own electronic and ionic temperature. Now the energy conservation with our integral expression is considered locally within each sub cell. This allows to simulate additionally the spatial distribution of the laser radiation and the electronic heat flow within the material.

9.1.4 Construction of Efficient and Highly Accurate T_e-Dependent Interatomic Potentials

A T_e-dependent interatomic potential is a function of the atomic coordinates and the electronic temperature T_e that can describe the PES before, during and after laser excitation. In order to reach an accurate description of the PES, we fit our interatomic potentials to an extensive set of forces and structural cohesive energies obtained from ab-initio reference simulations. We construct our interatomic potentials as a sum of local interaction terms that are physically motivated and commonly used in classical interatomic potentials. To be as flexible as possible and to allow the optimization procedure to find the global minimum in the parameter space, we expand the local interaction terms into polynomials, which can reproduce any physically reasonable function. Moreover, we can optimally adjust the analytical functional form of our interatomic potentials in such a way that the fitted ab-initio data is most accurately and efficiently described. For this, we have to adjust the polynomial degrees of the different local interaction terms. The corresponding polynomial expansions converge to the ab-initio determined PES relatively quickly, so that high degrees can be excluded, or equivalently, the inclusion of high degrees does not bring significant improvement in accuracy. Thus, we initially choose arbitrary upper degree limits and calculate the minimal fit error for all polynomial-degree combinations/interatomic potentials up to these limits. Using the results of these calculations, we can check whether the initially chosen upper degree limits were high enough. If not, we can eventually increase them. In this way, we considered the complete set of all "reasonable" interatomic potentials of our functional form. In order to select the final interatomic potential showing physical behavior and having a minimal fit error as well as reasonably low computational cost, we first extract a relatively small subset of reliable interatomic potentials that exhibit a low fit error combined with a relatively small number of coefficients. By analyzing the physical properties of the interatomic potentials within the small subset we select the final interatomic potential.

Our interatomic potentials can be very efficiently evaluated using the well-known addition theorem of the spherical harmonics, which allows to evaluate the three- and four-body interaction terms occurring in the interatomic potential as efficiently as for a two-body potential.

9.1.5 T_e-Dependent Interatomic Potential $\Phi^{(Si)}(T_e)$ for Si

Using the construction method described above, we developed a T_e-dependent interatomic potential $\Phi^{(Si)}(T_e)$ for Si. We fitted an extensive set of T_e-dependent DFT reference simulations in thin film geometry. $\Phi^{(Si)}(T_e)$ describes accurately the femtosecond-laser induced effects in Si, namely, bond softening, thermal phonon antisqueezing, non-thermal melting, and ablation. $\Phi^{(Si)}(T_e)$ also describes well the electronically absorbed energy and the electronic specific heat.

We analyzed the two available T_e-dependent interatomic potentials for Si from the literature and found that their description of the femtosecond-laser induced effects in Si is very incomplete or rather wrong. We also fitted the parameters of several widely used classical interatomic potentials developed for Si with electrons in the ground state to our ab-initio reference simulations. For this, we used our universal potential parameter fitting program, which allows to fit the parameters of an interatomic potential with almost arbitrary functional form to ab-initio forces and structural cohesive energies. We found out, that the reparametrization of classical interatomic potentials developed for ground state Si produces a better description of the femtosecond-laser induced effects in Si compared to the two available T_e-dependent interatomic potentials, which were developed from less reference data. However, the reparametrized T_e-dependent interatomic potentials cannot describe all relevant femtosecond-laser induced effects in Si as accurately as $\Phi^{(Si)}(T_e)$.

9.1.6 Correction of the Melting Temperature of $\Phi^{(Si)}(T_e)$ to the Experimental Value

By design, $\Phi^{(Si)}(T_e)$ reproduces the LDA-DFT melting temperature of Si, which is significantly below the experimental melting temperature. However, the parameters of an interatomic potential cannot be directly fit to reproduce a specific value of the melting temperature. Thanks to the construction of $\Phi^{(Si)}(T_e)$ as a sum of physically interpretable interaction terms and their easy polynomial representation, we can adjust the parameters of $\Phi^{(\mathrm{Si})}(T_\mathrm{e})$ in such a way that the experimental melting temperature is reproduced without causing any significant changes of the PES at increased T_e's, so that the excited PES is still well described.

9.1.7 MD Simulations of Femtosecond Laser-Pulse Excited Si

Using the original $\Phi^{(\mathrm{Si})}(T_\mathrm{e})$ without corrections and our MD simulation method, we performed MD simulations of femtosecond laser-pulse excited Si including the effects of the excited PES and the EPC. We simulated the laser-excitation of a thin Si film and derived the time-dependent intensities of several experimentally accessible Bragg peaks. We compared our calculated Bragg peak intensities with measured Bragg peak intensities obtained from ultrafast electron diffraction. We found an excellent agreement. Moreover, we found out that the effects of the EPC dominate over the non-thermal effects attributable to the excited PES. At low intensities, the measured time-dependent Bragg peak intensities are well described, if only the EPC combined with the ground state PES is taken into account. At high intensities, the EPC and the excited PES must be taken into account to describe the experimental Bragg peak decay. We also found that the excited PES is such that a laser-excited Si film experiences a delay of 300 fs in the onset of melting at the surface in comparison to the central volume. The reason for this is a non-thermal expansion of the film at the surface, which stabilizes, by lower pressures, the crystal structure for some time. On the other hand, the EPC initiates a melting of the film starting at the surface on a picosecond timescale. We also performed several MD simulations of femtosecond laser-pulse excited bulk Si. We found out that the thermal phonon antisqueezing, which is driven by the excited PES, is still recognizable, if the EPC is taken into account in the scenario. In addition, we found out that Si melts non-thermally in two stages at high and in four stages at very high intensities in comparison of three stages, if both effects, EPC and PES, are taken into account.

9.1.8 T_e-Dependent Interatomic Potential $\Phi^{(Sb)}(T_e)$ for Sb

We applied our method to antimony and developed a T_e-dependent interatomic potential $\Phi^{(\mathrm{Sb})}(T_\mathrm{e})$. We fitted an extensive set of T_e-dependent DFT reference simulations in bulk and thin film geometry. $\Phi^{(\mathrm{Sb})}(T_\mathrm{e})$ describes accurately the femtosecond-laser induced effects in Sb, namely, bond softening and hardening, the excitation of the A_{1g} phonon mode, thermal phonon antisqueezing, non-thermal melting, and ablation. Like in the case of Si, $\Phi^{(\mathrm{Sb})}(T_\mathrm{e})$ also describes well the electronically absorbed energy and the electronic specific heat.

9.1.9 MD Simulations of Femtosecond Laser-Pulse Excited Sb

Using $\Phi^{(\mathrm{Sb})}(T_e)$ and our MD simulation method, we performed MD simulations of a femtosecond laser-pulse excited Sb film. In order to compare directly with ultrafast x-ray diffraction experiments, in which the time-dependent intensity of the (200) Bragg peak was measured after a femtosecond laser-pulse excitation, we calculated the reflectivity and the absorption coefficient as functions of the Peierls parameter in an ab-initio approach. We calculated the time-dependent intensity behavior of the (200) Bragg peak for several laser intensities and compared it with the experimental results. We were able to reproduce the measured time-dependent behavior at almost all experimental studied laser intensities. In addition, we found that a moderate femtosecond laser-pulse excitation can induce an A7 to sc-like transition in Sb. The crystal structure does not melt and the finally reached structure corresponds to a sc structure that is still distorted in direction of the body diagonal as a remnant of the A7 structure.

9.2 Future Perspectives

Using our MD simulation setup, we can study the influences of non-thermal effects during ablation or laser-controlled formation of nanostructures. For example, using $\Phi^{(\mathrm{Si})}(T_e)$, we can extend our study to the influence of the non-thermal effects on the formation of laser induced periodic surface structures (LIPSS) in Si [1], which was not possible so far. We are not restricted to the materials Si and Sb, since we can develop T_e-dependent interatomic potentials using our construction procedure for any element of the periodic table, in principle. Moreover, we will extend our interatomic potential model to systems with two different types of atoms. This will allow us to develop T_e-dependent interatomic potentials for binary systems like SiC or SiO_2 and study the influences of a laser excitation in such systems. We can also extend the interatomic potential model to systems with more than two types of atoms. This will allow the simulation of biological systems with higher accuracy than hitherto available. For example, protein folding in solution could be simulated under ambient conditions or after a laser-excitation. In principle, our T_e-dependent interatomic potentials are developed for a Fermi distribution with temperature T_e of the electrons. We will analyze how to develop interatomic potentials for other distributions of the electrons to describe early stages of the laser-excitation before the excited electrons are thermalized to a Fermi distribution with a certain T_e. In addition, we will study the possibility to directly derive the electron-phonon coupling constant G_{ep} from the T_e-dependent interatomic potential on the fly during the MD simulations. We will analyze, which modifications of the interatomic potential are necessary and/or which additional information is needed for this. In addition, we will analyze, if our polynomial construction and our automatized optimization procedure can be modified for a usage in other machine learning approaches, where neural networks are commonly applied, since one benefit of our polynomial construction is the possibility of a perfect adaption.

Reference

1. J. Reif, O. Varlamova, S. Varlamov, M. Bestehorn, Appl. Phys. A **104**(3), 969 (2011). https://doi.org/10.1007/s00339-011-6472-3

Appendix A
Additional Information and Tables

A.1 Review of Vector Calculus

Between any two vectors $\mathbf{a}, \mathbf{b} \in \mathbb{R}^3$, one defines the cross product by

$$\mathbf{a} \times \mathbf{b} = \begin{bmatrix} a_y\, b_z - a_z\, b_y \\ a_z\, b_x - a_x\, b_z \\ a_x\, b_y - a_y\, b_x \end{bmatrix}. \tag{A.1}$$

One has for any three vectors $\mathbf{a}, \mathbf{b}, \mathbf{c} \in \mathbb{R}^3$

$$\mathbf{a}^{\mathrm{t}} \cdot \left(\mathbf{b} \times \mathbf{c}\right) = \mathbf{b}^{\mathrm{t}} \cdot \left(\mathbf{c} \times \mathbf{a}\right) = \mathbf{c}^{\mathrm{t}} \cdot \left(\mathbf{a} \times \mathbf{b}\right) \tag{A.2}$$

and

$$\mathbf{a} \times \left(\mathbf{b} \times \mathbf{c}\right) = \mathbf{b}\left(\mathbf{a}^{\mathrm{t}} \cdot \mathbf{c}\right) - \mathbf{c}\left(\mathbf{a}^{\mathrm{t}} \cdot \mathbf{b}\right). \tag{A.3}$$

For any differentiable vector field $\mathbf{E}(\mathbf{r}) \in \mathbb{R}^3$, one defines the divergence of $\mathbf{E}$ by

$$\nabla \cdot \mathbf{E} = \frac{\partial E_x}{\partial x} + \frac{\partial E_y}{\partial y} + \frac{\partial E_z}{\partial z} \tag{A.4}$$

and the rotation of $\mathbf{E}$ by

$$\nabla \times \mathbf{E} = \begin{bmatrix} \frac{\partial E_z}{\partial y} - \frac{\partial E_y}{\partial z} \\ \frac{\partial E_x}{\partial z} - \frac{\partial E_z}{\partial x} \\ \frac{\partial E_y}{\partial x} - \frac{\partial E_x}{\partial y} \end{bmatrix}. \tag{A.5}$$

For any differentiable scalar field $f(\mathbf{r}) \in \mathbb{R}$, one defines the gradient of f by

© The Editor(s) (if applicable) and The Author(s), under exclusive license to Springer Nature Switzerland AG 2021
B. Bauerhenne, *Materials Interaction with Femtosecond Lasers*,
https://doi.org/10.1007/978-3-030-85135-4

$$\nabla f = \begin{bmatrix} \frac{\partial f}{\partial x} \\ \frac{\partial f}{\partial y} \\ \frac{\partial f}{\partial z} \end{bmatrix}. \tag{A.6}$$

and the Laplace operator of f by

$$\nabla^2 f = \nabla \cdot (\nabla f) = \frac{\partial^2 f}{\partial x^2} + \frac{\partial^2 f}{\partial y^2} + \frac{\partial^2 f}{\partial z^2}. \tag{A.7}$$

The Laplace operator is also defined for a differentiable vector field $\mathbf{E}(\mathbf{r}) \in \mathbb{R}^3$ by

$$\nabla^2 \mathbf{E} = \begin{bmatrix} \nabla^2 E_x \\ \nabla^2 E_y \\ \nabla^2 E_z \end{bmatrix}. \tag{A.8}$$

For any differentiable vector field $\mathbf{E}(\mathbf{r}) \in \mathbb{R}^3$ and Volume $\mathcal{V}$, the **divergence theorem** is valid:

$$\int_{\mathbb{S}(\mathcal{V})} d\mathbf{a}^{\mathrm{t}} \cdot \mathbf{E}(\mathbf{r}, t) = \int_{\mathcal{V}} d^3 r \left(\nabla \cdot \mathbf{E}(\mathbf{r}, t)\right). \tag{A.9}$$

Here, $\mathbb{S}(\mathcal{V})$ denotes the surface of $\mathcal{V}$ and $d\mathbf{a}$ is the normal vector that is orthogonal to the surface and points out of the volume $\mathcal{V}$.

A.2 Method of Least Squares and Givens Rotations

The fit error can be written as

$$\mathbb{W}_{\mathrm{err}}(T_{\mathrm{e}}) = \left| \sum_{k=1}^{N_{\mathrm{C}}} \alpha_k \, \mathbf{v}_k - \mathbf{V} \right|^2.$$

The optimal coefficients $\mathbf{a} = \left[\alpha_1, \ldots, \alpha_{N_{\mathrm{C}}}\right]^{\mathrm{t}}$ that minimize the fit error can be easily found by using the following theorem [1]:

Mathematical Theorem A.1 (Method of least squares) *Let the vector* $\mathbf{V} \in \mathbb{R}^{N_d}$ *and the* N_C *vectors* $\mathbf{v}_k \in \mathbb{R}^{N_d}$, $k = 1, \ldots, N_C$ *be given. The best parameters* $\mathbf{a} = \left[\alpha_1, \ldots, \alpha_{N_C}\right]^t$ *to approximate the vector* $\mathbf{V}$ *by a linear combination of the vectors* $\mathbf{v}_k$, $k = 1, \ldots, N_C$ *using the approximation error*

$$\left|\sum_{k=1}^{N_C} \alpha_k \, \mathbf{v}_k - \mathbf{V}\right|^2$$

is given by the solution of the linear equation

$$\mathbf{Q} \cdot \mathbf{a} = \mathbf{b}, \tag{A.10}$$

where the Matrix $\mathbf{Q} \in \mathbb{R}^{N_C \times N_C}$ *is defined by*

$$Q_{ij} = \mathbf{v}_i^t \cdot \mathbf{v}_j \tag{A.11}$$

and the vector $\mathbf{b} \in \mathbb{R}^{N_C}$ *is defined by*

$$b_i = \mathbf{V}^t \cdot \mathbf{v}_i. \tag{A.12}$$

If the matrix $\mathbf{Q}$ *can be inverted, the solution of Eq. (A.10) corresponds to the global minimum.*

Proof At first we define the function $f : \mathbb{R}^n \rightarrow \mathbb{R}$ by

$$f(\alpha_1, \ldots, \alpha_n) := \left|\sum_{k=1}^{N_C} \alpha_k \, \mathbf{v}_k - \mathbf{V}\right|^2 = \sum_{i=1}^{N_d} \left(\sum_{k=1}^{N_C} \alpha_k \, v_{ki} - V_i\right)^2.$$

Here, v_{ki} denotes the i-th component of the vector $\mathbf{v}_k$. $f(\alpha_1, \ldots, \alpha_{N_C})$ corresponds to the approximation error related to the parameters $(\alpha_1, \ldots, \alpha_{N_C})$. The approximation error is extreme, if

$$\frac{\partial f}{\partial \alpha_j} = 0, \qquad j = 1, \ldots, N_C.$$

By inserting $f(\alpha_1, \ldots, \alpha_{N_C})$, we obtain further

$$\begin{aligned}
& 2 \sum_{i=1}^{N_d} \left(\sum_{k=1}^{N_C} \alpha_k \, v_{ki} - V_i\right) v_{ji} = 0, && j = 1, \ldots, N_C, \\
\Leftrightarrow \quad & \sum_{k=1}^{N_C} \alpha_k \sum_{i=1}^{N_d} v_{ki} \, v_{ji} = \sum_{i=1}^{N_d} V_i \, v_{ji}, && j = 1, \ldots, N_C, \\
\Leftrightarrow \quad & \sum_{k=1}^{N_C} \alpha_k \, \mathbf{v}_k^t \cdot \mathbf{v}_j = \mathbf{V}^t \cdot \mathbf{v}_j, && j = 1, \ldots, N_C.
\end{aligned}$$

Using the definition of $\mathbf{Q}$, $\mathbf{a}$, and $\mathbf{b}$, we get

$$\sum_{k=1}^{N_C} Q_{kj}\, a_k = b_j, \qquad j = 1, \ldots, N_C.$$

This is nothing else than the component-wise notation of (A.10). If the matrix $\mathbf{Q}$ can be inverted, there exist only one solution of Eq. (A.10), which must be a global minimum. If this solution would be a maximum, this solution would be the global maximum, since there are no further extremes. But the function f cannot be bounded, because the parameters $\mathbf{a}$ can be chosen arbitrarily badly. Thus, the solution of Eq. (A.10) corresponds to the global minimum, if the matrix $\mathbf{Q}$ can be inverted. ∎

Since a large number of forces and cohesive energies should be fit for an accurate interatomic potential, N_d is quite large in the range of millions. Due to this, the solution of Eq. (A.10) will contain large numerical errors, so that one has to use a different numerically much more stable method [1]. For this, we write the fit error as

$$\mathbb{W}_{\text{err}}(T_e) = \left|\mathbf{M} \cdot \mathbf{a} - \mathbf{V}\right|^2$$

using the matrix

$$\mathbf{M} = \left[\mathbf{v}_1, \ldots, \mathbf{v}_{N_C}\right] \in \mathbb{R}^{N_d \times N_C},$$

which column vectors correspond to the vectors $\mathbf{v}_k$. Now we transform the vector $\mathbf{M} \cdot \mathbf{a} - \mathbf{V}$ without changing its norm with the help of a series of rotations to determine the optimal coefficients.

Definition A.1 *(Rotation matrix)* The matrix $\mathbf{R}^{(k\,\ell\,\theta)} \in \mathbb{R}^{N_d \times N_d}$ defined by

$$\mathbf{R}^{(k\,\ell\,\theta)} = \begin{bmatrix} 1 & & & & & & & & & \\ & \ddots & & & & & & & & \\ & & 1 & & & & & & & \\ & & & \cos(\theta) & & & -\sin(\theta) & & & \\ & & & & 1 & & & & & \\ & & & & & \ddots & & & & \\ & & & & & & 1 & & & \\ & & & \sin(\theta) & & & \cos(\theta) & & & \\ & & & & & & & 1 & & \\ & & & & & & & & \ddots & \\ & & & & & & & & & 1 \end{bmatrix} \begin{matrix} \\ \\ \\ \leftarrow k \\ \\ \\ \\ \leftarrow \ell \\ \\ \\ \\ \end{matrix} \qquad \text{(A.13)}$$

with $1 \leq k < \ell \leq N_d$ and $\theta \in \mathbb{R}$ is called rotation matrix. We can imagine $\mathbf{R}^{(k\,\ell\,\theta)} \in \mathbb{R}^{N_d \times N_d}$ as the linear transformation in $\mathbb{R}^{N_d}$ that rotates by the angle θ in the two-dimensional plane spanned by the k-th and ℓ-th coordinate direction.

By definition, $\mathbf{R}^{(k\,\ell\,\theta)}$ is orthogonal and fulfills ($\mathbb{1}$ denotes the $N_\mathrm{d} \times N_\mathrm{d}$ unity matrix)

$$\left(\mathbf{R}^{(k\,\ell\,\theta)}\right)^\mathrm{t} \cdot \mathbf{R}^{(k\,\ell\,\theta)} = \mathbb{1} \quad \text{and} \quad \left(\mathbf{R}^{(k\,\ell\,\theta)}\right)^{-1} = \left(\mathbf{R}^{(k\,\ell\,\theta)}\right)^\mathrm{t}. \tag{A.14}$$

Mathematical Theorem A.2 (Transformation by a Givens rotation) *Let* $\mathbf{M} \in \mathbb{R}^{N_d \times N_C}$ *be an arbitrary matrix with* $M_{\ell k} \neq 0$, $1 \leq k < \ell \leq N_d$. *The rotation matrix* $\mathbf{G}^{(\ell\,k)} := \mathbf{R}^{(k\,\ell\,\theta)} \in \mathbb{R}^{N_d \times N_d}$ *with*

$$\cos(\theta) = \frac{M_{kk}}{\sqrt{M_{kk}^2 + M_{\ell k}^2}}, \qquad \sin(\theta) = -\frac{M_{\ell k}}{\sqrt{M_{kk}^2 + M_{\ell k}^2}} \tag{A.15}$$

is called Givens rotation. For the transformed matrix $\mathbf{M}' = \mathbf{R}^{(k\,\ell\,\theta)} \cdot \mathbf{M}$, *the element* $M'_{\ell k}$ *is zero. The column vectors of the transformed matrix* $\mathbf{M}'$ *exhibit the same norm than the column vectors of the matrix* $\mathbf{M}$.

Proof We have for all $j = 1, \ldots, N_\mathrm{C}$:

$$\begin{aligned}
M'_{kj} &= M_{kj}\,\cos(\theta) - M_{\ell j}\,\sin(\theta),\\
M'_{\ell j} &= M_{kj}\,\sin(\theta) + M_{\ell j}\,\cos(\theta),\\
M'_{ij} &= M_{ij}, \qquad\qquad i \neq k, i \neq \ell.
\end{aligned}$$

Due to this, we obtain for any $j = 1, \ldots, N_\mathrm{C}$:

$$\begin{aligned}
&\sum_{i=1}^{N_\mathrm{d}} \left(M'_{ij}\right)^2\\
&= \overbrace{\Big(M_{kj}\,\cos(\theta) - M_{\ell j}\,\sin(\theta)\Big)^2}^{=(M'_{kj})^2} + \overbrace{\Big(M_{kj}\,\sin(\theta) + M_{\ell j}\,\cos(\theta)\Big)^2}^{=(M'_{\ell j})^2} + \sum_{\substack{i=1\\ i\neq k, i\neq \ell}}^{N_\mathrm{d}} \overbrace{M_{ij}^2}^{=\left(M'_{ij}\right)^2}\\
&= M_{kj}^2\,\cos^2(\theta) - 2\,M_{kj}\,M_{\ell j}\,\sin(\theta)\,\cos(\theta) + M_{\ell j}^2\,\sin^2(\theta)\\
&\quad + M_{kj}^2\,\sin^2(\theta) + 2\,M_{kj}\,M_{\ell j}\,\sin(\theta)\,\cos(\theta) + M_{\ell j}^2\,\cos^2(\theta) + \sum_{\substack{i=1\\ i\neq k, i\neq \ell}}^{N_\mathrm{d}} M_{ij}^2
\end{aligned}$$

$$
\begin{aligned}
&= M_{kj}^2 + M_{\ell j}^2 + \sum_{\substack{i=1 \\ i \neq k, i \neq \ell}}^{N_\mathrm{d}} M_{ij}^2 \\
&= \sum_{i=1}^{N_\mathrm{d}} M_{ij}^2 .
\end{aligned}
$$

Thus, the norm of the column vectors is preserved after the transformation with the Givens rotation. We want to note that any rotation preserves the norm. The definition of the Givens rotation is meaningful, since we have

$$
\underbrace{\frac{M_{kk}^2}{M_{kk}^2 + M_{\ell k}^2}}_{=\cos^2(\theta)} + \underbrace{\frac{M_{\ell k}^2}{M_{kk}^2 + M_{\ell k}^2}}_{=\sin^2(\theta)} = 1.
$$

We obtain further

$$
M'_{\ell k} = M_{kk} \sin(\theta) + M_{\ell k} \cos(\theta) = M_{kk} \frac{-M_{\ell k}}{\sqrt{M_{kk}^2 + M_{\ell k}^2}} + M_{\ell k} \frac{M_{kk}}{\sqrt{M_{kk}^2 + M_{\ell k}^2}} = 0.
$$

■

If a Givens rotation $\mathbf{G}^{(\ell\, k)}$ is applied to an arbitrary matrix $\mathbf{M}$, it replaces each of the k-th and ℓ-th rows of $\mathbf{M}$ by a specific linear combination of both rows and sets the element $M_{\ell k}$ to zero. Now we consider the matrix $\mathbf{M}$ occurring in Eq. (6.108) and we assume $N_\mathrm{d} \geq N_\mathrm{C}$, since a large amount of ab-initio data is fitted. We can set all elements below the diagonal of $\mathbf{M}$ to zero by applying successively the following Givens rotations:

$$
\mathbf{G}^{(\mathrm{all})} = \mathbf{G}^{(N_\mathrm{d}\, N_\mathrm{C})} \cdot \ldots \cdot \mathbf{G}^{(4\,3)} \cdot \mathbf{G}^{(N_\mathrm{d}\, 2)} \cdot \ldots \cdot \mathbf{G}^{(4\,2)} \cdot \mathbf{G}^{(3\,2)} \cdot \mathbf{G}^{(N_\mathrm{d}\, 1)} \cdot \ldots \cdot \mathbf{G}^{(3\,1)} \cdot \mathbf{G}^{(2\,1)}. \tag{A.16}
$$

This is, indeed, possible due to the following: Before the Givens rotation $\mathbf{G}^{(\ell\, k)}$ is applied, the matrix $\mathbf{M}'$, which was generated by applying the previous Givens rotations on $\mathbf{M}$, looks like

$$
\mathbf{M}' = \begin{bmatrix}
M'_{11} & \cdots & \cdots & \cdots & \cdots & \cdots & \cdots & \cdots & * \\
0 & \ddots & & & & & & & \vdots \\
\vdots & \ddots & M'_{k-1k-1} & & & & & & \vdots \\
\vdots & & 0 & M'_{kk} & & & & & \vdots \\
\vdots & & \vdots & 0 & M'_{k+1k+1} & & & & \vdots \\
\vdots & & \vdots & \vdots & \vdots & \ddots & & & \vdots \\
\vdots & & \vdots & 0 & \vdots & & \ddots & & \vdots \\
\vdots & & \vdots & M'_{\ell k} & \vdots & & M'_{\ell\ell} & & \vdots \\
\vdots & & \vdots & \vdots & \vdots & & & \ddots & \vdots \\
\vdots & & \vdots & \vdots & \vdots & & & & M'_{N_\mathrm{C}N_\mathrm{C}} \\
\vdots & & \vdots & \vdots & \vdots & & & & \vdots \\
\vdots & & \vdots & \vdots & \vdots & & & & \vdots \\
0 & \cdots & 0 & * & * & \cdots & \cdots & \cdots & *
\end{bmatrix}
\begin{matrix} \\ \\ \\ \leftarrow k \\ \\ \\ \\ \leftarrow \ell \\ \\ \\ \\ \\ \\ \end{matrix}
\tag{A.17}
$$

and $*$ stands for any real number. There are zeros below the diagonal in the first $k-1$ columns and, in the k-th column, there are zeros below the diagonal element M'_{kk} until the element $M'_{\ell k}$. The Givens rotation $\mathbf{G}^{(\ell\,k)}$ sets the element $M'_{\ell k}$ to zero. For this, $\mathbf{G}^{(\ell\,k)}$ replaces only the k-th and ℓ-th rows by a linear combination of both rows. The elements of the k-th and ℓ-th rows with a column index smaller than k are already zero due to the previous Givens rotations. Thus, these elements keep zero after $\mathbf{G}^{(\ell\,k)}$ is applied, since a linear combination of zeros keep zero. Finally, we obtain

$$
\mathbf{G}^{(\mathrm{all})} \cdot \mathbf{M} = \begin{bmatrix}
M''_{11} & \cdots & * \\
0 & \ddots & \vdots \\
\vdots & \ddots & M''_{N_\mathrm{C}N_\mathrm{C}} \\
\vdots & & 0 \\
\vdots & & \vdots \\
0 & \cdots & 0
\end{bmatrix} . \tag{A.18}
$$

A.3 Implementation of the e^--Phonon Coupling in Velocity Verlet

Here, we summarize the calculation procedure described in Sect. 6.1.1:

$$
\begin{aligned}
\mathcal{R}(t_{\ell+1}) &\overset{(6.34)}{=} \mathcal{R}(t_\ell) + \Delta t\ \mathcal{V}(t_\ell) + \frac{\Delta t^2}{2m}\ \mathcal{F}_{\text{tot}}(t_\ell),\\
\Delta E_{\text{L}_{\text{abs}}}(t_\ell) &\overset{(6.36)}{=} E_{\text{L}_{\text{abs}}}(t_{\ell+1}) - E_{\text{L}_{\text{abs}}}(t_\ell),\\
\Delta E_{\text{ep}}(t_\ell) &\overset{(6.37)}{=} -\sum_{k=1}^{N_{\mathcal{M}}} |\mathcal{M}_k|\, G_{\text{ep}_{\mathcal{M}_k}}(t_\ell)\left(T_{\text{e}}(t_\ell) - T_{\text{i}_{\mathcal{M}_k}}(t_\ell)\right)\Delta t,\\
\Delta T_{\text{e}}(t_\ell) &\overset{(6.39)}{=} \frac{\Delta E_{\text{ep}}(t_\ell) + \Delta E_{\text{L}_{\text{abs}}}(t_\ell)}{C_{\text{e}}(t_\ell)},\\
T_{\text{e}}(t_{\ell+1}) &\overset{(6.40)}{=} T_{\text{e}}(t_\ell) + \Delta T_{\text{e}}(t_\ell),\\
S_{\text{e}}(t_{\ell+1}) &\overset{(6.4)}{=} -\frac{\partial\Phi\Big(T_{\text{e}}(t_{\ell+1}), \mathcal{R}(t_{\ell+1})\Big)}{\partial T_{\text{e}}},\\
C_{\text{e}}(t_{\ell+1}) &\overset{(5.78)}{=} -\,T_{\text{e}}(t_{\ell+1})\,\frac{\partial^2\Phi\Big(T_{\text{e}}(t_{\ell+1}), \mathcal{R}(t_{\ell+1})\Big)}{\partial T_{\text{e}}^2},\\
\mathcal{F}(t_{\ell+1}) &\overset{(6.3)}{=} \begin{bmatrix} -\nabla_{\mathbf{r}_1}\Phi\Big(T_{\text{e}}(t_{\ell+1}), \mathcal{R}(t_{\ell+1})\Big)\\ \vdots \\ -\nabla_{\mathbf{r}_{N_{\text{at}}}}\Phi\Big(T_{\text{e}}(t_{\ell+1}), \mathcal{R}(t_{\ell+1})\Big)\end{bmatrix},\\
G_{\text{ep}_{\mathcal{M}_k}}(t_{\ell+1}) &\overset{(6.42)}{=} G_{\text{ep}_{\mathcal{M}_k}}\left(T_{\text{e}}(t_{\ell+1}), \mathcal{R}(t_{\ell+1}), \mathcal{V}(t_\ell)\right),\\
\mathcal{W}(t_{\ell+1}) &\overset{(6.43)}{=} \mathcal{V}(t_\ell) + \frac{\Delta t}{2m}\Big(\mathcal{F}_{\text{tot}}(t_\ell) + \mathcal{F}(t_{\ell+1})\Big),\\
\mathcal{H}_{\mathcal{M}_k}(t_{\ell+1}) &\overset{(6.45)}{=} \frac{m}{2}\left(\mathbf{P}_{\mathcal{M}_k}\cdot\mathcal{W}(t_{\ell+1})\right)^{\text{t}}\cdot\mathbf{P}_{\mathcal{M}_k}\cdot\mathcal{W}(t_{\ell+1}),\\
\xi_{\mathcal{M}_k}(t_{\ell+1}) &\overset{(6.48)}{=} \frac{4\,\mathcal{H}_{\mathcal{M}_k}(t_{\ell+1})}{|\mathcal{M}_k|\,G_{\text{ep}_{\mathcal{M}_k}}(t_{\ell+1})\,T_{\text{e}}(t_{\ell+1})\,\Delta t^2} + \frac{2}{\Delta t}\\
&\quad - \sqrt{\begin{aligned}&\left(\frac{4\,\mathcal{H}_{\mathcal{M}_k}(t_{\ell+1})}{|\mathcal{M}_k|\,G_{\text{ep}_{\mathcal{M}_k}}(t_{\ell+1})\,T_{\text{e}}(t_{\ell+1})\,\Delta t^2} + \frac{2}{\Delta t}\right)^2\\ &+ \frac{8\,\mathcal{H}_{\mathcal{M}_k}(t_{\ell+1})}{k_{\text{B}}\,|\mathcal{M}_k|\,G_{\text{ep}_{\mathcal{M}_k}}\,T_{\text{e}}(t_{\ell+1})\,\Delta t^2} - \frac{4}{\Delta t^2}\end{aligned}},\\
\mathcal{V}(t_{\ell+1}) &\overset{(6.44)}{=} \sum_{k=1}^{N_{\mathcal{M}}} \frac{1}{1 - \frac{\Delta t}{2}\,\xi_{\mathcal{M}_k}(t_{\ell+1})}\,\mathbf{P}_{\mathcal{M}_k}\cdot\mathcal{W}(t_{\ell+1}).
\end{aligned}
$$

$$
\begin{aligned}
\mathcal{F}_{\text{tot}}(t_{\ell+1}) &\overset{(6.15)}{=} \mathcal{F}(t_{\ell+1}) + \sum_{k=1}^{N_{\mathcal{M}}} \xi_{\mathcal{M}_k}(t_{\ell+1})\, m\, \mathbf{P}_{\mathcal{M}_k} \cdot \mathcal{V}(t_{\ell+1}), \\
E_{\text{kin}_{\mathcal{M}_k}}(t_{\ell+1}) &\overset{(6.33)}{=} \frac{m}{2}\, \mathcal{V}(t_{\ell+1})^{\text{t}} \cdot \mathbf{P}_{\mathcal{M}_k} \cdot \mathcal{V}(t_{\ell+1}), \\
T_{\text{i}_{\mathcal{M}_k}}(t_{\ell+1}) &\overset{(6.17)}{=} \frac{2\, E_{\text{kin}_{\mathcal{M}_k}}(t_{\ell+1})}{|\mathcal{M}_k|\, k_{\text{B}}}, \\
\Delta E_{\text{L}_{\text{abs}}}(t_{\ell+1}) &\overset{(6.36)}{=} E_{\text{L}_{\text{abs}}}(t_{\ell+2}) - E_{\text{L}_{\text{abs}}}(t_{\ell+1}), \\
\Delta E_{\text{ep}}(t_{\ell+1}) &\overset{(6.37)}{=} -\sum_{k=1}^{N_{\mathcal{M}}} |\mathcal{M}_k|\, G_{\text{ep}_{\mathcal{M}_k}}(t_{\ell+1}) \Big(T_{\text{e}}(t_{\ell+1}) - T_{\text{i}_{\mathcal{M}_k}}(t_{\ell+1}) \Big)\, \Delta t, \\
\Delta T_{\text{e}}(t_{\ell+1}) &\overset{(6.39)}{=} \frac{\Delta E_{\text{ep}}(t_{\ell+1}) + \Delta E_{\text{L}_{\text{abs}}}(t_{\ell+1})}{C_{\text{e}}(t_{\ell+1})}, \\
I(t_{\ell+1}) &\overset{(6.50)}{=} I(t_\ell) + \frac{1}{2}\Big(\big(S_{\text{e}}(t_\ell) + C_{\text{e}}(t_\ell)\big)\, \Delta T_{\text{e}}(t_\ell) + \big(S_{\text{e}}(t_{\ell+1}) + C_{\text{e}}(t_{\ell+1})\big)\, \Delta T_{\text{e}}(t_{\ell+1}) \Big), \\
E(t_{\ell+1}) &= \Phi\Big(T_{\text{e}}(t_{\ell+1}), \mathcal{R}(t_{\ell+1}) \Big) + I(t + \Delta t).
\end{aligned}
$$

If the above algorithm is implemented, every quantity only needs to be stored at the actual time step $t_{\ell+1}$ and the previous time step t_ℓ.

A.4 Calculation of the Pressure in a MD Simulation

The macroscopic pressure p of a set of N_{at} interacting atoms contained in a simulation cell of volume $\mathcal{V}$ treated with periodic boundary conditions is quite difficult [2]. In our case, the calculation is simpler, since the interaction is described by a T_{e}-dependent interatomic potential $\Phi(\mathbf{r}_1, \ldots, \mathbf{r}_{N_{\text{at}}}, T_{\text{e}})$, which corresponds to the internal energy for the nuclei at coordinates $\mathbf{r}_1, \ldots, \mathbf{r}_{N_{\text{at}}}$. Thus, the total energy of the nuclei is given by

$$
E = \underbrace{\sum_{i=1}^{N_{\text{at}}} \frac{m_i}{2}\, \mathbf{v}_i^2}_{=E_{\text{kin}}} + \Phi(\mathbf{r}_1, \ldots, \mathbf{r}_{N_{\text{at}}}, T_{\text{e}}), \tag{A.19}
$$

where m_i denotes the energy and $\mathbf{v}_i$ the velocity of the i-th nuclei. Following the first law of thermodynamics from Eq. (2.166), one obtains for the pressure

$$
p = -\frac{\partial E}{\partial \mathcal{V}}. \tag{A.20}
$$

In order to calculate the derivative with respect to the volume $\mathcal{V}$, we introduce a temporary factor κ, with which we scale uniformly the space in each room direction, so that we obtain new scaled variables with a prime:

$$\mathbf{r}'_i = \kappa\,\mathbf{r}_i, \qquad \mathbf{v}'_i = \kappa\,\mathbf{v}_i. \tag{A.21}$$

Using the scaled variables, we obtain for the pressure

$$p = -\left.\frac{\partial E}{\partial \kappa}\right|_{\kappa=1} \left.\frac{\partial \kappa}{\partial \mathcal{V}}\right|_{\kappa=1}. \tag{A.22}$$

Since we obtain for the scaled volume

$$\mathcal{V}' = \kappa^3\,\mathcal{V}, \tag{A.23}$$

we get

$$\kappa = \mathcal{V}^{-\frac{1}{3}}\,(\mathcal{V}')^{\frac{1}{3}}, \tag{A.24}$$

from which we obtain directly

$$\left.\frac{\partial \kappa}{\partial \mathcal{V}}\right|_{\kappa=1} = -\frac{1}{3}\mathcal{V}^{-\frac{4}{3}}\,\left.(\mathcal{V}')^{\frac{1}{3}}\right|_{\mathcal{V}'=\mathcal{V}} = -\frac{1}{3\,\mathcal{V}}, \tag{A.25}$$

since $\kappa = 1$ results in $\mathcal{V}' = \mathcal{V}$. Furthermore, we have

$$\left.\frac{\partial E_{\text{kin}}}{\partial \kappa}\right|_{\kappa=1} = \left.\frac{\partial\left(\sum_{i=1}^{N_{\text{at}}} \frac{m_i}{2}\,\kappa^2\,\mathbf{v}_i^2\right)}{\partial \kappa}\right|_{\kappa=1} = 2\sum_{i=1}^{N_{\text{at}}} \frac{m_i}{2}\,\kappa\,\left.\mathbf{v}_i^2\right|_{\kappa=1} = 2\,E_{\text{kin}}. \tag{A.26}$$

Thus, we get for the pressure

$$p = \frac{2\,E_{\text{kin}}}{3\,\mathcal{V}} + \frac{1}{3\,\mathcal{V}}\,\left.\frac{\partial \Phi(\kappa\,\mathbf{r}_1,\ldots,\kappa\,\mathbf{r}_{N_{\text{at}}},T_{\text{e}})}{\partial \kappa}\right|_{\kappa=1}. \tag{A.27}$$

The calculation of $\frac{\partial \Phi}{\partial \kappa}\big|_{\kappa=1}$ is quite simple, since our polynomial interatomic potential consists of the sum over the local interaction terms Φ_0, Φ_2, Φ_3, Φ_4, Φ_ρ (see Eq. 6.99). Φ_0 do not contribute, since it only depends on T_{e}, which is not influenced by a space scaling. If the space is uniformly scaled with κ, bond angles remain unchanged and the interatomic distances scale just by

$$r'_{ij} = \kappa\,r_{ij}. \tag{A.28}$$

Therefore, we obtain

$$\left.\frac{\partial \Phi_2(\kappa\,r_{ij},T_{\text{e}})}{\partial \kappa}\right|_{\kappa=1} = \left.\frac{\partial \Phi_2(\kappa\,r_{ij},T_{\text{e}})}{\partial(\kappa\,r_{ij})}\,\frac{\partial(\kappa\,r_{ij})}{\partial \kappa}\right|_{\kappa=1} = \frac{\partial \Phi_2(r_{ij},T_{\text{e}})}{\partial r_{ij}}\,r_{ij} \tag{A.29}$$

and

$$\left.\frac{\partial \Phi_3\Big(\kappa\, r_{ij}, \kappa\, r_{ik}, \cos(\theta_{ijk}), T_\mathrm{e}\Big)}{\partial \kappa}\right|_{\kappa=1}$$
$$= \frac{\partial \Phi_3\Big(r_{ij}, r_{ik}, \cos(\theta_{ijk}), T_\mathrm{e}\Big)}{\partial r_{ij}}\, r_{ij} + \frac{\partial \Phi_3\Big(r_{ij}, r_{ik}, \cos(\theta_{ijk}), T_\mathrm{e}\Big)}{\partial r_{ik}} \tag{A.30}$$

and

$$\left.\frac{\partial \Phi_4\Big(\kappa\, r_{ij}, \kappa\, r_{ik}, \kappa\, r_{i\ell}, \cos(\theta_{ijk}), \cos(\theta_{ij\ell}), \cos(\theta_{ik\ell}), T_\mathrm{e}\Big)}{\partial \kappa}\right|_{\kappa=1}$$
$$= \frac{\partial \Phi_4\Big(r_{ij}, r_{ik}, r_{i\ell}, \cos(\theta_{ijk}), \cos(\theta_{ij\ell}), \cos(\theta_{ik\ell}), T_\mathrm{e}\Big)}{\partial r_{ij}}\, r_{ij}$$
$$+ \frac{\partial \Phi_4\Big(r_{ij}, r_{ik}, r_{i\ell}, \cos(\theta_{ijk}), \cos(\theta_{ij\ell}), \cos(\theta_{ik\ell}), T_\mathrm{e}\Big)}{\partial r_{ik}}\, r_{ik}$$
$$+ \frac{\partial \Phi_4\Big(r_{ij}, r_{ik}, r_{i\ell}, \cos(\theta_{ijk}), \cos(\theta_{ij\ell}), \cos(\theta_{ik\ell}), T_\mathrm{e}\Big)}{\partial r_{i\ell}}\, r_{i\ell}. \tag{A.31}$$

For the embedded atom potential term Φ_ρ, we obtain

$$\left.\frac{\partial \Phi_\rho\left(\rho_i^{(2)}, \rho_i^{(3)}, \ldots, \rho_i^{\left(N_\rho^{(r)}\right)}, T_\mathrm{e}\right)}{\partial \kappa}\right|_{\kappa=1} = \sum_{q_1=2}^{N_\rho^{(r)}} \frac{\partial \Phi_\rho}{\partial \rho_i^{(q_1)}} \left.\frac{\partial \rho_i^{(q_1)}}{\partial \kappa}\right|_{\kappa=1} \tag{A.32}$$

with

$$\left.\frac{\partial \rho_i^{(q_1)}}{\partial \kappa}\right|_{\kappa=1} = \left.\frac{\partial}{\partial \kappa}\left(\sum_{\substack{j=1 \\ j \neq i \\ r_{ij} < r_\rho^{(c)}}}^{N_\mathrm{at}} \left(1 - \frac{\kappa\, r_{ij}}{r_\rho^{(c)}}\right)^{q_1} \right)\right|_{\kappa=1} = \sum_{\substack{j=1 \\ j \neq i \\ r_{ij} < r_\rho^{(c)}}}^{N_\mathrm{at}} -\frac{q_1\, r_{ij}}{r_\rho^{(c)}} \left(1 - \frac{\kappa\, r_{ij}}{r_\rho^{(c)}}\right)^{q_1 - 1}. \tag{A.33}$$

Table A.1 Parametrization of the electronic specific heat $C_e(T_e)$ using Eq. (7.5). The unit of $a_{C_e}^{(k)}$ is $\frac{\text{eV}}{\text{K atom}}$

k	$a_{C_e}^{(k)}$	k	$a_{C_e}^{(k)}$	k	$a_{C_e}^{(k)}$
1	9.990955456836453E-6	2	−6.188280768791413E-4	3	0.040068462158504514
4	−0.26312331638621433	5	0.8576043019886007	6	−1.679202915977002
7	2.069435128552068	8	−1.5768394499029128	9	0.6799728970274491
10	−0.12703240946979374				

A.5 Electronic Specific Heat of Si

See Table A.1.

A.6 Adapted Parameters of Classical Potentials to Describe FS-Laser Excited Si

On the following pages, the fitted parameters of the studied interatomic potential models MT, T1, T2, T3, D, PM, SW, and MEAM are tabulated at all fitted T_e's. The unit of T_e is mHa. Furthermore, the units of all parameters are selected, so that the unit of length is Å and the unit of energy is eV. Except for MEAM, the parameters are also reported in Ref. [3] (Tables A.2, A.3, A.4, A.5, A.6, A.7, A.8, A.9, A.10, A.11, A.12, A.13, A.14, A.15, A.16, A.17, A.18, A.19, A.20, A.21, A.22, A.23, A.24, A.25, A.26, A.27, A.28, A.29, A.30, A.31, A.32, A.33, A.34 and A.35).

Table A.2 MT-coefficients r_c, r_1 and A

T_e	r_c	r_1	A
1	6.0215334842249550	5.4866188724220484	12027.619710407400
10	6.0010921757988642	5.4810948025383341	12027.625198810960
20	5.8718043582044599	5.5685691819183036	11772.057681172701
30	5.9352824664584443	5.5872286197855150	9438.8883415965174
40	5.9279345590339600	5.5832596476252592	9438.8936457684449
50	5.8962829950819478	5.5410928562426829	15681.421186892900
60	5.5265678273205463	−4.4804948676901022	53694.901278706740
70	5.4143540053357491	−4.9085710164745322	53694.312778375337
80	5.2562541640216338	−6.0895967978950933	59570.060667365557
90	4.9814521546701274	−8.5694001638198216	59162.910495947261
100	5.7089787127027503	−19.351780595100891	75403.818727448874

Table A.3 MT-coefficients B, λ_1 and λ_2

T_e	B	λ_1	λ_2
1	23.154902226486431	4.3734426307812422	0.61719886444453387
10	24.010574689984718	4.3553643151622436	0.72897264447240884
20	23.666142739518179	4.3429503889174583	0.81623580606404877
30	21.870366885269050	4.2357871017824049	0.89562773008561425
40	19.491348564530419	4.2635190181974636	0.92531291181535802
50	14.343222720504979	4.5645120703398998	0.89839900955361252
60	5.8634868795772324	4.4559118851706394	−0.12888136659322971
70	3.3278076195803621	4.3733567418116239	−0.33155505979351052
80	1.1615927908957040	4.1896153946833294	−0.73448689136299650
90	0.49447553144266648	3.6984099179269418	−1.2853258703283510
100	2.0729875460056002	3.0397647213638241	−1.3936705144168620

Table A.4 MT-coefficients η, δ and α

T_e	η	δ	α
1	1.0003754531241660	0.81329675308803273	1.7049398715818751
10	0.96164282456512162	0.83369484002540717	1.7126459483907910
20	0.96153418242992816	0.77953524472464908	1.7749635282731060
30	1.1026786876146890	0.57846992133316022	1.8051214104240150
40	1.0243234039771500	0.61673674268914536	1.9049795464236750
50	0.84640008869463490	0.75679777310293339	1.9391514844562161
60	0.74949716572346914	10.387486138384830	0.45834636919218569
70	0.70901950141208847	12.733917230092629	0.40512631122858861
80	0.81779768443478251	42.478881837402042	0.24708481781064839
90	0.97979633919550768	658.55121458020074	6.9325035351605821E-002
100	0.47514342661900533	26.833793155829191	0.67875258892152668

Table A.5 MT-coefficients β, h and c_1

T_e	β	h	c_1
1	1.0000000000000000	−0.30460801841760249	0.19919305305847379
10	1.0000000000000000	−0.31363662036502737	0.12645735161018759
20	1.0000000000000000	−0.31075318974877919	0.11372042142583340
30	1.0000000000000000	−0.31467592286537482	0.12560982273355470
40	1.0000000000000000	−0.32871675653820498	0.13581120271766139
50	1.0000000000000000	−0.36051903812922281	9.0266709645294443E-002
60	1.0000000000000000	−0.33839492895988671	7.5673024297687689E-003
70	1.0000000000000000	−0.33618826393860041	5.2228760963612614E-003
80	1.0000000000000000	−0.32532900330980269	−3.1707884151181399E-003
90	1.0000000000000000	−0.28891868784695218	−1.4642682428284591E-003
100	1.0000000000000000	−1.6964571112822320	−0.40054976491598743

Table A.6 MT-coefficients c_2, c_3 and c_4

T_e	c_2	c_3	c_4
1	574648.16107784235	1000124.8521391940	0.98409085589196454
10	574648.16155075689	1000124.8518674650	0.70665465165783636
20	640685.17111876712	999765.49062146957	0.53290692448487076
30	848582.39810465579	999685.10361938213	0.42698229328616122
40	848582.39731804584	999685.10428709525	0.31681714735704269
50	848956.38564381737	999367.64481257193	2.4227924609123500E-002
60	136314.35679546671	999444.40349505353	0.14006435218757610
70	136314.34774717811	999444.40472912835	0.16378273246083291
80	108657.14212005370	999444.40496367903	0.20899085079243909
90	62883.186292991828	999444.39873531740	0.57091600563903400
100	190702.87202421451	999444.40527865535	44.576654131789951

Table A.7 MT-coefficient c_5

T_e	c_5
1	14.467794477750269
10	15.127799772674271
20	15.265316840640381
30	14.495915350248350
40	14.012929353360510
50	16.598976522703861
60	47.752975337406383
70	23.935048797742219
80	17.944771439077229
90	15.141430458015320
100	3.3512282953793551

Table A.8 T1-coefficients r_c, r_1 and A

T_e	r_c	r_1	A
1	5.9855276270387936	4.7223515603645927	5308.0326985835427
10	6.0575655914696052	4.7087197906720446	39976.992819888073
20	5.8670366349824477	4.7417484707999256	15069.040207472681
30	5.6442199658052692	4.8130356268418311	4915.5915871848501
40	5.5819425235245239	4.8823785476034827	3822.8467045318730
50	5.9548396343057819	4.1932742539449261	3029.0787148995760
60	5.8464876551616909	4.2874476917068014	3801.7761906622641
70	5.7614875371452117	4.6231354896217498	3780.5714572391639
80	5.8592430087516068	4.6501923974449282	5638.9389736932353
90	6.0268334659446570	4.6315001897045853	4737.8285160288979
100	6.0618590152991407	4.6600379963795726	1988.8736819707090

Table A.9 T1-coefficients B_0, λ_1 and λ_2

T_e	B_0	λ_1	λ_2
1	45.338445738427403	3.7875180577253111	1.0294170818834310
10	29.901904231457920	4.9161223603894486	0.93667692451226359
20	35.345932833824790	4.3923675689934711	1.0731920764040559
30	52.242788743512932	3.7514873108290838	1.2855348608362220
40	54.569860902291367	3.6290363420168150	1.3732447098107550
50	55.479122336178527	3.5228969511252322	1.4511601361479789
60	29.307596972731201	3.7590632876084080	1.3096361233079230
70	13.183412315812131	3.8479911984588520	1.1073115766174251
80	4.8796237572762919	4.1165391367122606	0.85649163865355948
90	2.1592624858574099	4.0386612872770202	0.65335163526990425
100	1.5772083523409699	3.5830496926071032	0.61199915242052028

Table A.10 T1-coefficients b, n and c

T_e	b	n	c
1	182.80853175670850	1.5155916263331290	9.0146710638939927E-003
10	141.49619861301289	1.6656388370107400	1.0086003767335251E-002
20	187.92250758752499	1.5400621050058401	8.4354917973460120E-003
30	286.96506501368299	1.3781438542099300	6.6631294783616934E-003
40	380.11055106607068	1.3354420459072240	5.4074432225267180E-003
50	406.46070514627979	1.3040034405836129	5.1720846911523016E-003
60	520.85964177426763	1.3501539685884389	3.0762125346896559E-003
70	618.59662752597058	1.4929219477416580	2.6533488089959482E-003
80	735.82834583016142	1.9616639900119051	1.7305220629076170E-003
90	1260.1580452623120	2.6293300968911102	6.2073117653685568E-004
100	2580.6519422723918	3.0397537452846608	1.8419214898486591E-004

Table A.11 T1-coefficient d

T_e	d
1	7.9141704489364280
10	7.4240285250181888
20	7.9504982803163546
30	8.6916496311690814
40	9.2261242504614849
50	9.3426738775788145
60	10.144346065837031
70	10.681489036477920
80	11.242326755786060
90	12.557369592825880
100	14.229814417392030

Table A.12 T2-coefficients r_c, $r1$ and A

T_e	r_c	r_1	A
1	5.7341490608174581	−20.243393510774311	14822730.762776989
10	5.6624144494710071	−21.128545282352452	13072739.176011140
20	5.6828591578987258	−35.531329717214788	230737377.56472000
30	5.6495258202805108	−71.860407018066610	230737377.56397760
40	5.7200226775922198	−80.844651217161399	731454565.16499913
50	5.7506824441691968	−90.764629874613178	731454565.16513610
60	5.5176242175908650	−129.21870545658379	731454565.16703320
70	5.2733017138023941	6.9901493428402711E-002	117215.59645538770
80	5.1040591022923110	−7.8218958601209101	108057.89857997101
90	4.9414430320327121	−27.577471980058949	136483.98668845539
100	4.8614229454611451	−13.692263318212840	25106.645292160531

Table A.13 T2-coefficients B, λ_1 and λ_2

T_e	B	λ_1	λ_2
1	708.42827923402763	6.5096780133809347	7.9028100674469193E-002
10	499.20048060410448	6.3904621794618208	0.14105524079844520
20	140.80848900253031	7.3878888889646372	0.13434672809822590
30	469.70441973735620	6.7798039327938442	0.16709400668773319
40	317.12076524969882	7.2682386319872458	9.5993358806893136E-002
50	129.17748776865821	7.1949621046128387	−0.10512920107369431
60	59.648578014606699	6.8277889227076152	−0.47543966967949758
70	0.11744233326973150	5.5114964995240907	−0.45595795457970212
80	0.19360840560390000	4.6406839482643738	−0.76773442312205187
90	0.30421106257211572	3.7991637113194092	−1.1480510331139340
100	2.3688601683989018E-002	3.4309191835769721	−1.5690543807036961

Table A.14 T2-coefficients a and n

T_e	a	n
1	932.69603610101785	1.7643952175449520
10	377.03046139570313	1.7768726252270650
20	20.551165092833951	8.3617223079292220
30	77.793021173807404	7.4088014810600624
40	76.182503218574283	7.6251013008664028
50	65.388174540843508	7.3362829835075756
60	130.96372653720641	9.5057016317331904
70	0.36002872532145658	6.8730356907956391
80	2.2201438901164821	7.5275322306638683
90	20.665929132040329	8.2582952471179336
100	20.232950402651099	5.9927936603144802

Table A.15 T3-coefficients r_c, r_1 and A

T_e	r_c	r_1	A
1	5.2610000565100519	−3.4297688828791011	28803.886217027179
10	5.1033921081029252	−4.0331193770748159	28803.904702603930
20	4.9621941872759310	−4.7343565575679403	28803.880047042290
30	4.8695102661152152	−4.8207267709441997	28803.863901098481
40	4.8156296524933397	−4.4008748487887797	28803.750028385290
50	4.7806064184929022	−3.7872957251398840	28803.747257208110
60	4.6572523810924400	−3.1151837510928040	28803.755899509400
70	4.6391614004370050	−2.9649076063696742	28803.809303854508
80	4.6526186690524174	−13.885005266900720	128965.36309800160
90	4.5912511894177106	−45.191762094072622	313041.45155155333
100	5.1898058465341936	−331.88427807662401	1044078.5688747690

Table A.16 T3-coefficients B, λ_1 and λ_2

T_e	B	λ_1	λ_2
1	19.856699783927201	4.3257072668179690	0.22866300225342931
10	22.772516879314932	4.2273787405844114	0.24408983924300651
20	24.916709061758599	4.1227252297743826	0.25626879809787428
30	22.427434645585851	4.1023008621930916	0.26180535836725510
40	16.647433845304931	4.1517765355377856	0.24903149253696380
50	10.462254455375390	4.2338211782589834	0.20383172977061639
60	5.1292851689832659	4.3160958270278691	6.8131574850418139E-002
70	2.2878749437268029	4.3457459786427286	−0.10777650015298390
80	3.8336076305902349	4.2428827054940310	−0.41703370930705208
90	6.8501810079042498	3.6835512312469709	−0.81284682621747417
100	1158.4646428498399	2.5358549323782542	−0.11834057808405921

Table A.17 T3-coefficients β, n and c

T_e	β	n	c
1	0.42628738436504221	3.5975510640357440	9993.9449105130680
10	0.69075634323442958	3.3212079403822301	9993.9295596894117
20	0.91206193428473958	3.0171299239582621	9993.9274572095292
30	0.98679428656315027	2.7936031602928639	9993.9277419497521
40	0.92101111239520095	2.6238484989476110	9993.9353362800684
50	0.77366621180482742	2.4998977582529118	9993.9365487458017
60	0.63634274434461702	2.5782526789968898	9993.9420069287553
70	0.27498328627853719	2.4523191907758561	9993.9883341932255
80	8.9130686497997103E-003	2.4989390739165001	21917.986614526009
90	5.3455311172955176E-003	3.3612688746165080	25848.588843555051
100	269.32695015283122	4.9410535145373080	34460.506316307939

Table A.18 T3-coefficients d, h and λ_3

T_e	d	h	λ_3
1	55.564750472793300	−0.25594747471834528	0.64261330954843177
10	60.035744558914878	−0.26774727161086492	0.67418921867850479
20	60.497555717800367	−0.27430628472151503	0.69047400263271719
30	60.438501091389817	−0.29661529205397391	0.70456692994470149
40	60.046861229544326	−0.32936049922897259	0.71633417513193820
50	58.604791884637798	−0.36483933308920308	0.72314481361945537
60	56.369200996750394	−0.42445172494198058	0.75979977892493777
70	42.733523522324859	−0.41511384591618422	0.73323044855949793
80	15.857611155704809	−0.37920316923618802	0.69208243358982313
90	8.0394868296385535	−0.33276899081435979	0.73735793524451165
100	64.228214069348880	−0.22436680161527270	0.83797994563349898

Table A.19 T3-coefficient α

T_e	α
1	0.0000000000000000
10	0.0000000000000000
20	0.0000000000000000
30	0.0000000000000000
40	0.0000000000000000
50	0.0000000000000000
60	0.0000000000000000
70	0.0000000000000000
80	0.0000000000000000
90	0.0000000000000000
100	0.0000000000000000

Table A.20 D-coefficients r_c, r_1 and A

T_e	r_c	r_1	A
1	6.4203801838901811	5.0659557963169348	18564.003405009971
10	6.3839033357587294	5.0592818712858838	45907.089130213179
20	6.2619528769606072	5.0485512233762480	31033.519646578268
30	6.1565365667734504	5.0474259090785329	18564.001658632620
40	6.1180183646672122	5.0607114097065216	18564.000295257822
50	5.8875195312824804	5.8875195301551297	25049.879816849570
60	5.9090890012947579	5.5982964991796482	25613.868195797830
70	5.8875195312824804	5.8875195301551297	35515.855910369697
80	5.8875195312824804	5.8875195301551297	34829.721560303893
90	5.8875195312824804	5.8875195301551297	11106.405962195440
100	6.7692923754077103	352.51933845699659	1924.2238361709619

Table A.21 D-coefficients B_0, λ_1 and λ_2

T_e	B_0	λ_1	λ_2
1	36.992836980975120	4.5159964871328624	0.74855618803570767
10	32.382608410392329	4.9930141311623863	0.78694830976732966
20	33.100028349794840	4.7935284906662723	0.88675199068404809
30	34.383734316059098	4.5353734393947134	0.98946430847513733
40	30.047745848871141	4.5678380012537909	1.0192005561620530
50	22.118022910439429	4.7704077053660470	0.97116977158225881
60	15.082583850089129	4.8358727206144616	0.86460336672186899
70	9.8090677773406192	5.0431980596285761	0.64842165473098001
80	4.8150516624464013	5.0539697857112200	0.36290552216724548
90	3.5840240735887159	4.4896260410220759	3.9898385746732733E-003
100	8.6944756604694309	3.5811467658539100	7.1019866028728656E-003

Table A.22 D-coefficients b, n and c

T_e	b	n	c
1	1.7584013883529270	1.9664357380860129	0.30515951311249417
10	1.8705939966490399	2.0554489466583199	0.30210990925620168
20	2.0050188792103900	1.9490071661339481	0.26741692841660608
30	2.1120490985825269	1.8639663493045631	0.22382508777566329
40	2.0238731319458410	1.9246387955416839	0.19097654501803030
50	1.7230362204674650	2.1444735508865920	0.16455049641745301
60	1.3813748150980869	2.5020062980089528	0.13638523919701551
70	0.82276223631782941	3.5036998601936720	0.14267915676514831
80	0.59276640309207840	7.0951204175016551	0.12993474263802940
90	0.36784574844625478	695.33162226710249	0.13161829162504360
100	0.26882337048562238	344.43687990689853	0.21027822776963059

Table A.23 D-coefficients d and η

T_e	d	η
1	5.4322549716868913	0.35254365667862991
10	5.5160019296465714	0.35781287920169991
20	5.7786166866384177	0.34935089342404391
30	5.9024036085334188	0.33379893646875408
40	5.8933237951828943	0.30914163187867538
50	5.8996532416891370	0.27110226759979339
60	5.5461822450063378	0.24395554697695909
70	5.5626275702115482	0.17950340928600561
80	5.8672802739548482	0.14449060878216971
90	6.3969484604391234	0.11428091361914060
100	7.1138164812369604	0.10864095179255560

Table A.24 PM-coefficients $r^{(c)}$, $r^{(1)}$ and A

T_e	$r^{(c)}$	$r^{(1)}$	A
1	6.0190885417270668	−0.31682087877975362	53733.190270989682
10	5.9653803401015377	−0.90703980100896520	53733.190360783483
20	5.8177859861727459	−0.72611582765579286	53733.190858991373
30	6.4159678850007804	−8.1524141830315244	53733.177338276473
40	6.5025710905101706	−9.5425530387044866	53733.142427415143
50	6.6100372250111858	−10.491538708914121	53732.425373451137
60	7.0682266592509020	−13.558644549472890	53694.629792748339
70	9.3935732152359925	−18.229582913199771	53737.212700727127
80	5.7468268721970697	−3.8897651368247659	59570.429976034553
90	5.7979826813764346	−3.4338785519838870	59570.447558072483
100	5.7848380569285709	−3.2766022263902528	59570.449152557681

Table A.25 PM-coefficients B, λ_1 and λ_2

T_e	B	λ_1	λ_2
1	12.047402159506669	5.1035298350804483	0.22887569370408201
10	11.257803885927380	5.0107497929473528	0.25569999823927769
20	9.0411800663482929	5.0261808706325866	0.29113332513336421
30	50.155142643654457	4.0512572595198488	0.34995322503961979
40	56.976312352261289	3.9274470094825942	0.43072665530453969
50	52.669402486983167	3.8774320979662580	0.50591445420030356
60	87.302507604291094	3.7122877135868459	0.77183625890761720
70	385.44031226763872	3.8123427430816812	1.7091789985452110
80	1609.8526030635551	4.6972516898914440	−6.5641855039341528E-004
90	1608.9538654916710	4.7942860513102481	−6.8244920026389738E-004
100	1608.2015558179630	4.8252805176750480	−8.5537198446291489E-004

Table A.26 PM-coefficients η, δ and α

T_e	η	δ	α
1	0.94017604963116319	2.7830684247948270	1.1098672275365871
10	1.0262453096573920	3.8085539956956271	1.0133766722496440
20	1.0724540134693590	4.4318384495770804	0.96095978362675094
30	1.1328720289319809	43.840591399562598	0.91654490144834688
40	1.2065960083098450	89.184580867530585	0.87506997844421219
50	1.3523493100771979	234.73929099207109	0.78205147005713738
60	1.5754092179801080	1212.5891748737829	0.79449246625009418
70	1.2749653762584769	281.72870172975098	1.1870830032379569
80	0.30755154073593127	5.2854812758122444E-003	0.69478487423254409
90	0.32254260815822539	3.2655670261941419E-003	1.0401416094040170
100	0.26120298574164330	2.6408137841257570E-003	1.3656528636089900

Table A.27 PM-coefficients β, h and c_1

T_e	β	h	c_1
1	1.0000000000000000	−0.29484836846101031	0.10346288790228060
10	1.0000000000000000	−0.29780540477691042	7.5740831041105169E-002
20	1.0000000000000000	−0.29471708882076192	5.9350030535158992E-002
30	1.0000000000000000	−0.28535434566553503	5.2142839998631577E-002
40	1.0000000000000000	−0.29738269115441962	4.2646701259201648E-002
50	1.0000000000000000	−0.31047103172681362	3.0689419860168311E-002
60	1.0000000000000000	−0.31420421681622651	1.8506195689096439E-002
70	1.0000000000000000	−0.26683463800809482	4.1081083180859429E-004
80	1.0000000000000000	−0.30230443329122292	−3.3045834436263118E-004
90	1.0000000000000000	−0.25589152745592791	8.7073946087091233E-004
100	1.0000000000000000	−0.10122844988940830	−1.3393015135626139E-004

Table A.28 PM-coefficients c_2, c_3 and c_4

T_e	c_2	c_3	c_4
1	136299.39930543071	999446.44354064064	1.4151086665792421
10	136299.39949404000	999446.44351491902	1.0394032256926660
20	136299.39964766131	999446.44349396927	0.79637308440859711
30	136299.41371349950	999446.44157574687	0.64522973128405769
40	136299.44648955661	999446.43710592831	0.59626914154361443
50	136299.77871508399	999446.39179868519	0.52324895144533035
60	136314.43786995061	999444.39243729913	0.45021280211248887
70	136301.43523338990	999446.16585996887	−0.83811005283842899
80	108657.18374292550	999444.40043889266	0.13671872413292391
90	108657.18395630650	999444.40041569283	0.29262255610813398
100	108657.18416816140	999444.40039266297	−0.91522133961813845

Table A.29 PM-coefficients c_5 and c_0

T_e	c_5	c_0
1	12.764395572044100	−0.64970250186599454
10	13.280268460167800	−0.82276612017039241
20	11.383205611273940	−0.76277890189078001
30	11.573329147595251	−5.8355712117033640
40	10.515440108742000	−6.9164224888318104
50	8.9279938321850292	−7.0205733808291102
60	9.5108380313904917	−7.4765866376547452
70	2095.0445607036581	−2.3569439721278860
80	20.965482857974859	1609.2785359847960
90	18.325169745492879	1610 1795005590980
100	2.5788332798092028	1610.9350602511890

Table A.30 SW-coefficients r_c, σ and A_0

T_e	r_c	σ	A_0
1	6.7255622062514302	7.2485315485046993	3380.5883833971279
10	6.6584748438371202	7.5842314323706717	3998.9313818513720
20	6.5885074215025456	7.7496538488780224	4014.7910319479129
30	6.4148495399437877	7.0006999864583674	3823.0055572701799
40	6.5779750900265670	6.8353551825466843	3356.2316566238992
50	6.8720999264068006	5.2144107188371880	6070.3463981768227
60	6.5681878517691539	0.74042229346987709	40348.507344862621
70	6.5206872698418668	0.41390122460908230	40264.982536051823
80	6.8715243331143512	0.20643936119872630	40178.891260546858
90	7.1524299077607294	0.19330351337144749	40112.340125497511
100	7.3684829730554391	0.28895282051876858	40066.394895084290

Table A.31 SW-coefficients B_0, p and q

T_e	B_0	p	q
1	2436.3601606404559	4.4476200464329150	3.8987583306203550
10	2869.0372512460140	4.5747698789498878	4.0309370085014056
20	2741.5454058222622	4.7455178815178662	4.1377557807399983
30	2608.6363267624420	4.9119270685423464	4.3217990289223600
40	2062.3552036899341	5.3008276355522446	4.5762996721885987
50	4914.2226762909131	5.6629551162085834	5.3624540737984168
60	39577.271593278158	6.4018289464424800	6.3760175797466907
70	39660.796269239559	6.4035428640289016	6.3843879614580006
80	39746.887399990963	6.2543712578733528	6.2416362316713760
90	39813.438501232798	5.9792007074680811	5.9711869471689329
100	39859.383633598307	5.6122288303858170	5.6072530630608002

Table A.32 SW-coefficients λ_0 and γ

T_e	λ_0	γ
1	2638.9409944925169	2.6075120503842677
10	3719.7448417365999	2.5631513161149946
20	4450.7182891328666	2.5411708589660758
30	3030.7834431530559	2.6285669047163607
40	6609.2777490318094	3.0911217861580367
50	25684.965562739599	5.0041491678997865
60	22554.137194062911	33.203030414744440
70	22554.311698466219	59.771783702310053
80	22554.803416324019	132.53061121283127
90	22554.915953547719	152.70559990631389
100	22554.985006254250	107.74779523429970

Table A.33 MEAM-coefficients $r^{(c)}$, $r^{(1)}$ and r_g

T_e	$r^{(c)}$	$r^{(1)}$	r_g
1	5.0000000000000000	4.0019803122122664	2.3475145390896031
10	5.0000000000000000	4.0342686406596071	2.3547196469611378
20	5.0000000000000000	4.2127066833140256	2.3674978461515850
30	5.0000000000000000	4.2127067852467013	2.3674978529382091
40	5.0000000000000000	4.3346290532138800	2.3979427319489668
50	5.0000000000000000	4.4328838475575516	2.4287503706587090
60	5.0000000000000000	4.5843813924016592	2.4660879277889252
70	5.0000000000000000	3.3893847589097779	2.4349081612112031
80	5.0000000000000000	4.0707087468428762	2.6353224283095429
90	5.0000000000000000	4.1610995152748584	2.7549087900050808
100	5.0000000000000000	4.1772209563337208	2.9592784693165122

Table A.34 MEAM-coefficients E_0, B and m_1

T_e	E_0	B	m_1
1	4.6568526230643590	0.53438847223489960	4.4612477173719496
10	4.1485974954686231	0.49298678784738909	4.2474557219507636
20	3.4641795016274690	0.42550371950169080	4.0101000694113358
30	3.4641794660503451	0.42550371548893479	4.0101000403356624
40	2.2344780413699281	0.28671854809861452	3.2041404046529278
50	1.7177631750007620	0.22280635478948610	2.6022743139042839
60	1.2467533375963440	0.16853768200608971	1.7963970523859720
70	0.80144972956889649	7.9541980047379238E-002	−2.3676591505582998
80	0.53250554857788057	9.1569308583615366E-002	27.844369417344812
90	0.30507765362515982	5.3417203458667739E-002	36.251441740039262
100	0.14396409852595521	2.3494829356315378E-002	37.694535980974678

Table A.35 MEAM-coefficients α, a_1 and a_2

T_e	α	a_1	a_2
1	3.1060659527036738	0.36216136599048648	0.40271150079888079
10	3.0673296983961431	0.38393627109591771	0.40442274017053270
20	2.9953233209601819	0.41956740610789550	0.41385705045813348
30	2.9953232778768339	0.41956740648357121	0.41385705760928498
40	2.8066150131877592	0.46129058403861289	0.40804363084856032
50	2.6649649750441111	0.46337588287794768	0.38975636617184328
60	2.5393852836062241	0.45831846671714072	−0.35029716639936193
70	1.0506570979976351	0.55538805402524027	0.43207923021303601
80	12.338918785521191	−0.23706099980113390	−0.22167740165584651
90	15.237216979331430	−0.27954612673113810	−0.23214930788964450
100	15.818623015267780	−0.16988855877049699	−0.17050334747935050

A.7 Performance of Reparametrized Classical Potentials for Si

Here we present the performance of the the reparametrized classical potentials T1, T2, T3, D, PM, SW and MEAM. The performance of the MT potential was already presented in Fig. 7.3. Black dashed or solid lines correspond to ab-initio results. Green or red lines correspond to the interatomic potential. The red color indicates a significant deviation from ab-initio, whereas the green color indicates an acceptable description. We want to note that sometimes the choice of red or green color is not so clear (Figs. A.1, A.2, A.3, A.4, A.5, A.6 and A.7).

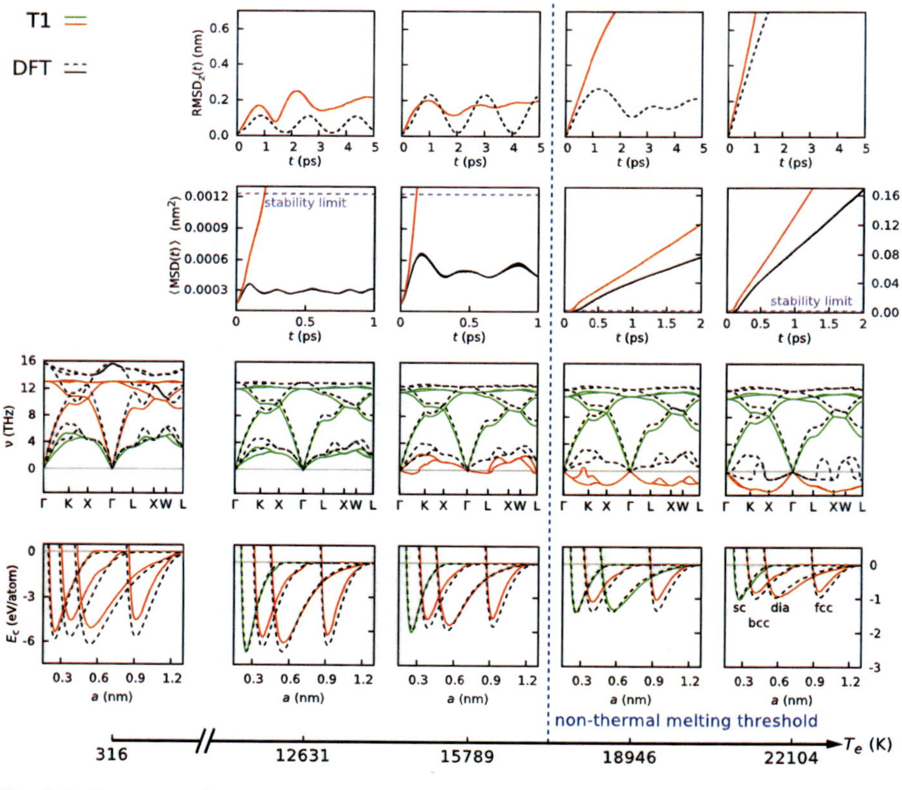

Fig. A.1 Summary of the performance of the T1 potential, reparametrized using our ab-initio reference simulations, in describing the effects occurring in Si induced by the increased T_e deformed PES. Description see Sect. 4.4.8

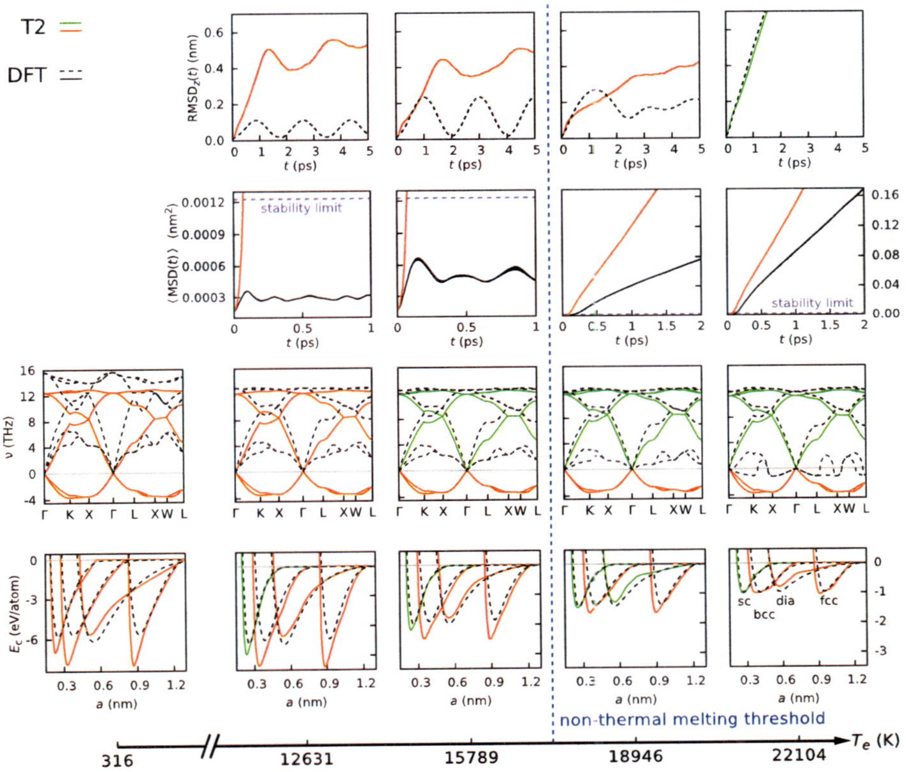

Fig. A.2 Summary of the performance of the T2 potential, reparametrized using our ab-initio reference simulations, in describing the effects occurring in Si induced by the increased T_e deformed PES. Description see Sect. 4.4.8

A.8 Coefficients of the Polynomial Interatomic Potential $\Phi^{(\mathrm{Si})}(T_e)$ for Si

Our final interatomic potential $\Phi^{(\mathrm{Si})}(T_e)$ for Si has the polynomial degrees

$$N_2^{(r)} = 10, \qquad N_3^{(r)} = 3, \qquad N_3^{(\theta)} = 3 \qquad N_\rho^{(r)} = 2, \qquad N_\rho^{(\rho)} = 2. \tag{A.34}$$

It needs in total 23 coefficients. The two-body term Φ_2 has 9, the three-body term Φ_3 has 12, and the embedding function Φ_ρ has 2 coefficients. Furthermore, we use the constant cutoff radii

$$r_2^{(c)} = 0.63 \text{ nm}, \qquad r_3^{(c)} = 0.42 \text{ nm}, \qquad r_\rho^{(c)} = 0.48 \text{ nm}. \tag{A.35}$$

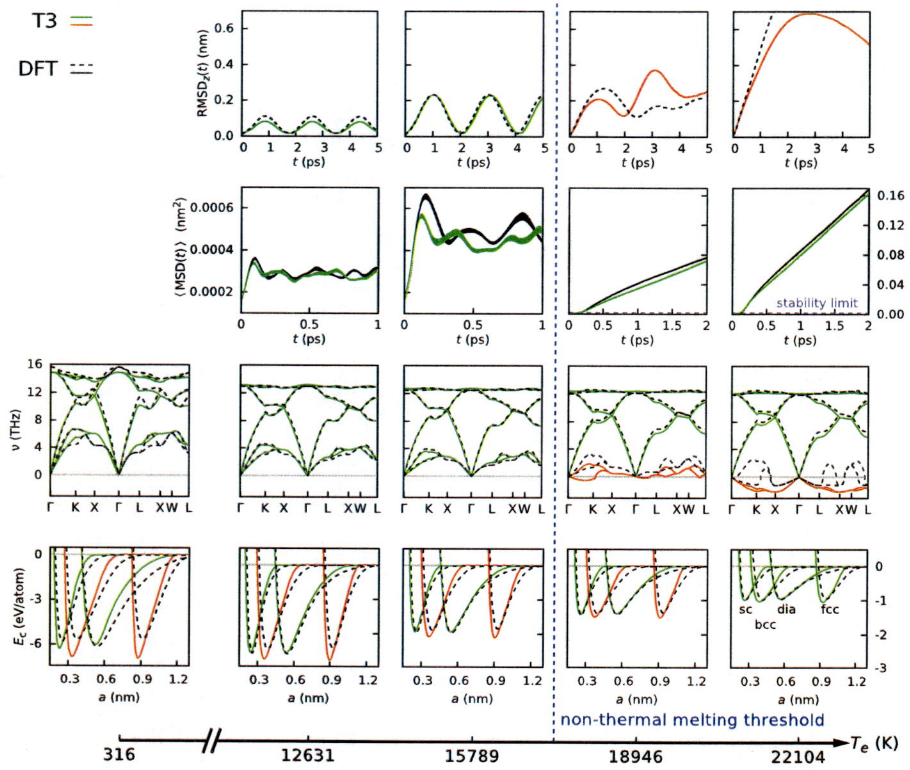

Fig. A.3 Summary of the performance of the T3 potential, reparametrized using our ab-initio reference simulations, in describing the effects occurring in Si induced by the increased T_e deformed PES. Description see Sect. 4.4.8

In order to describe the parametrization of the term $\Phi_0(T_e)$, let us consider an isolated atom. It contains discrete energy levels of the electrons. The electronic occupation of these energy levels is given by a Fermi distribution with T_e. Since there are big gaps between the discrete energy levels in the atom, the internal energy E_0 and the entropy S_0 of the atom do not change with increasing T_e at low T_e's. Consequently, the Helmholtz free energy

$$\Phi_0(T_e) = E_0 - T_e\, S_0 \tag{A.36}$$

of the isolated atom is a linear function of T_e at low T_e's. Starting from $\sim$ 4500 K, the internal energy E_0 and the entropy S_0 start to depend on T_e. Above this temperature, the Helmholtz free energy $\Phi_0(T_e)$ behaves in a non-linear fashion. Hence, we fitted the Helmholtz free energy $\Phi_0(T_e)$ of an isolated atom to a polynomial of degree 1 at $T_e \leq 4500$ K and to a polynomial of degree 13 at $T_e > 4500$ K:

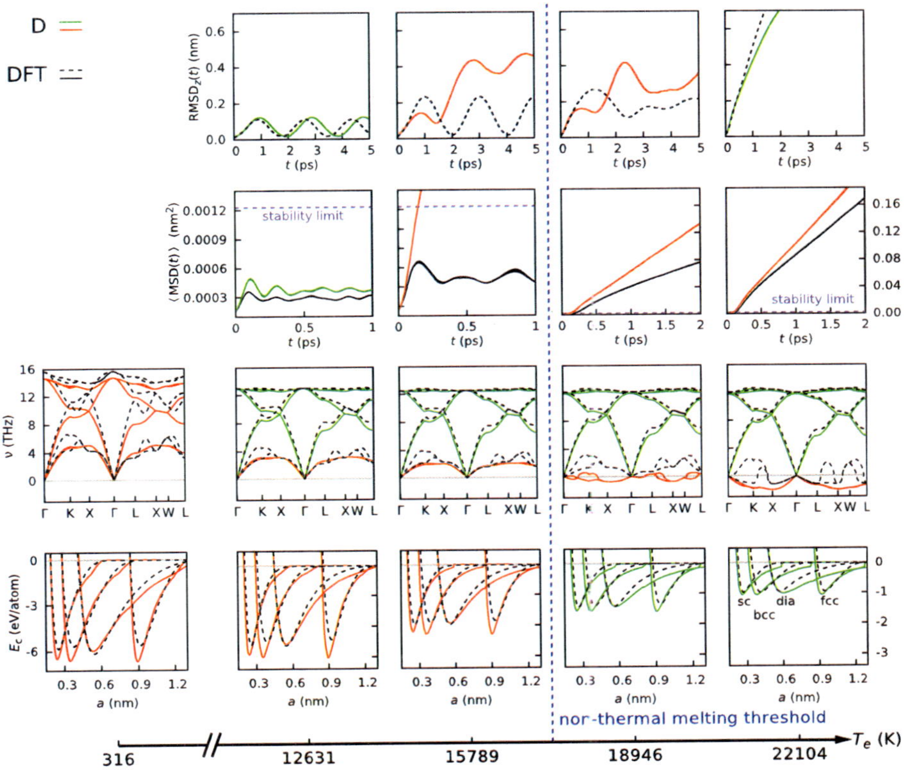

Fig. A.4 Summary of the performance of the D potential, reparametrized using our ab-initio reference simulations, in describing the effects occurring in Si induced by the increased T_e deformed PES. Description see Sect. 4.4.8

$$\Phi_0(T_e) = \begin{cases} a_0^{(0)} + a_0^{(1)} \left(\frac{T_e - 4500\,\mathrm{K}}{31577\,\mathrm{K}}\right), & T_e \leq 4500\,\mathrm{K}; \\ a_0^{(0)} + a_0^{(1)} \left(\frac{T_e - 4500\,\mathrm{K}}{31577\,\mathrm{K}}\right) + \sum_{k=4}^{13} a_0^{(k)} \left(\frac{T_e - 4500\,\mathrm{K}}{31577\,\mathrm{K}}\right)^k, & T_e > 4500\,\mathrm{K}. \end{cases} \tag{A.37}$$

By construction of $\Phi_0(T_e)$, its first, second, and third derivative are continuous functions of T_e. We present the electronic absorbed energy as a function of T_e in Fig. A.8 and the electronic specific heat as a function of T_e in Fig. A.9. One can clearly see that the interatomic potential reproduces the DFT values.

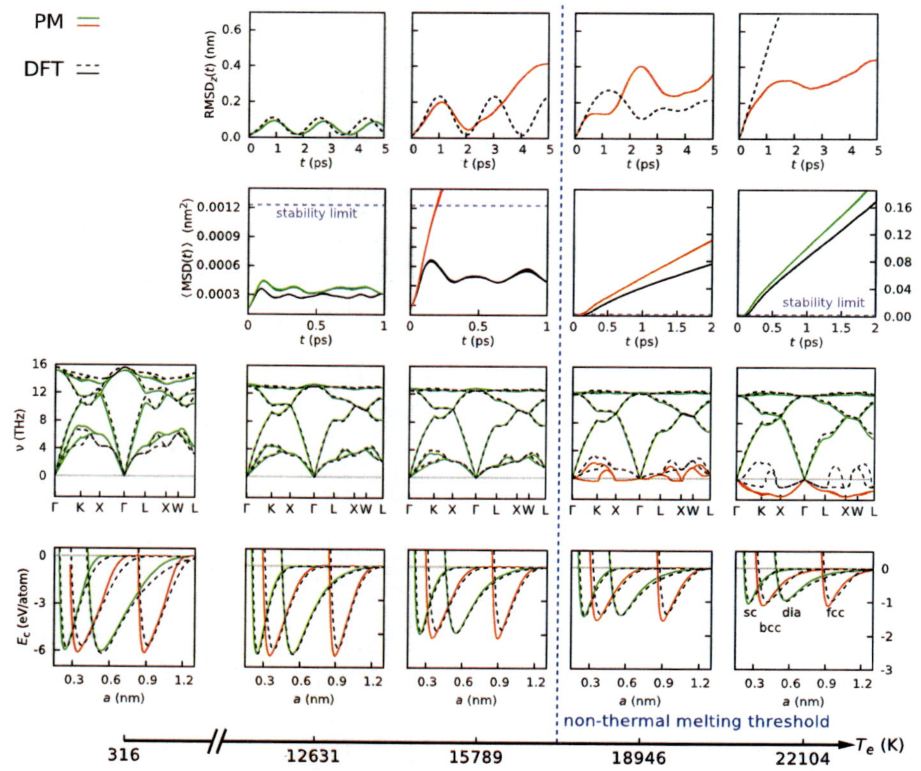

Fig. A.5 Summary of the performance of the PM potential, reparametrized using our ab-initio reference simulations, in describing the effects occurring in Si induced by the increased T_e deformed PES. Description see Sect. 4.4.8

We fitted the T_e-dependence of the remaining coefficients to a polynomial of degree 5. Consequently, any coefficient c of the interatomic potential depends on T_e as

$$c = \sum_{k=0}^{5} a^{(k)} \left(\frac{T_e}{31577\,\mathrm{K}} \right)^k . \tag{A.38}$$

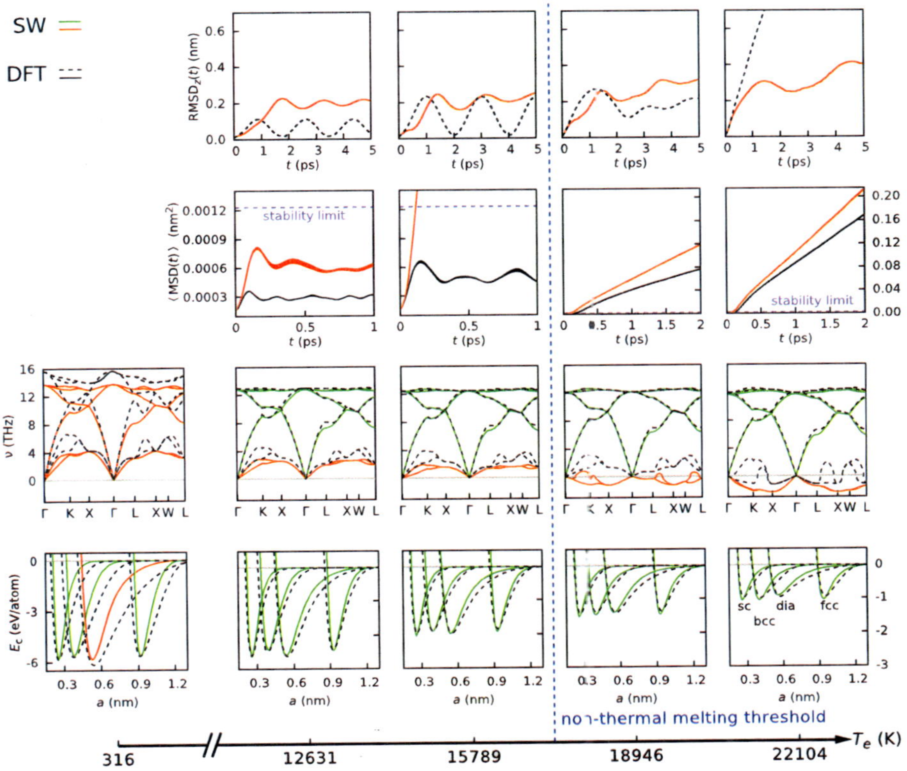

Fig. A.6 Summary of the performance of the SW potential, reparametrized using our ab-initio reference simulations, in describing the effects occurring in Si induced by the increased T_e deformed PES. Description see Sect. 4.4.8

The parametrization of $\Phi^{(Si)}(T_e)$ is tabulated in Tables A.36, A.37, A.38 and A.39. The unit of the coefficients is eV and the unit of T_e is K. The fitted range of the polynomial expansion yields 316 K $\leq T_e \leq$ 31577 K. The parametrization of $\Phi^{(Si)}(T_e)$ is also reported in Ref. [4].

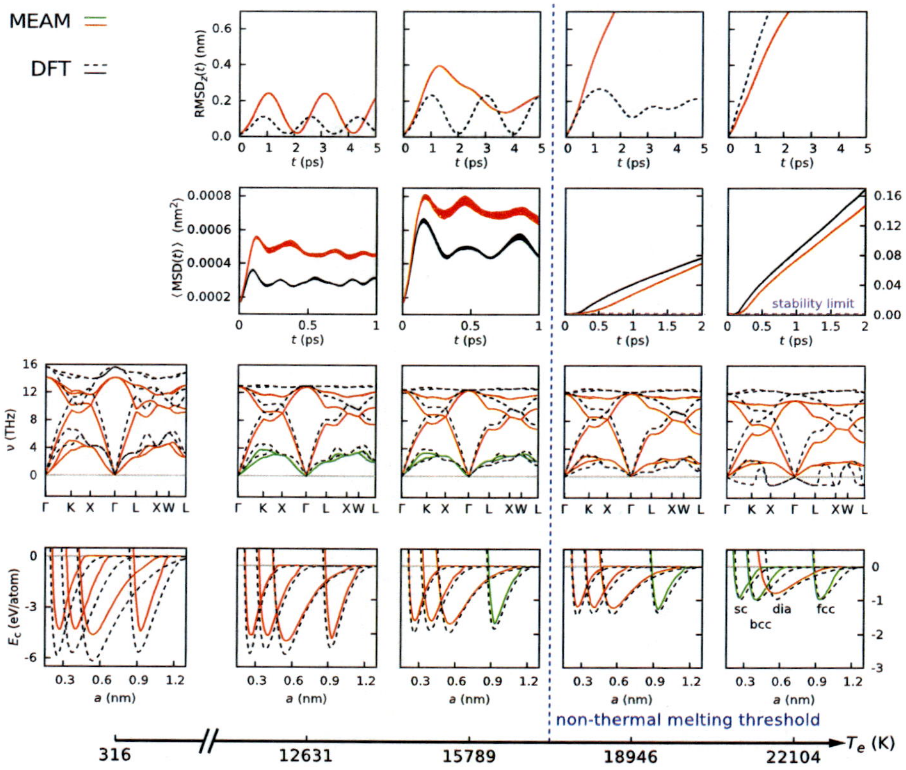

Fig. A.7 Summary of the performance of the MEAM potential, reparametrized using our ab-initio reference simulations, in describing the effects occurring in Si induced by the increased T_e deformed PES. Description see Sect. 4.4.8

Fig. A.8 Energy absorbed by the electrons E_e is shown as a function of T_e for an isolated Si atom adapted from Figure S3 of Ref. [4]

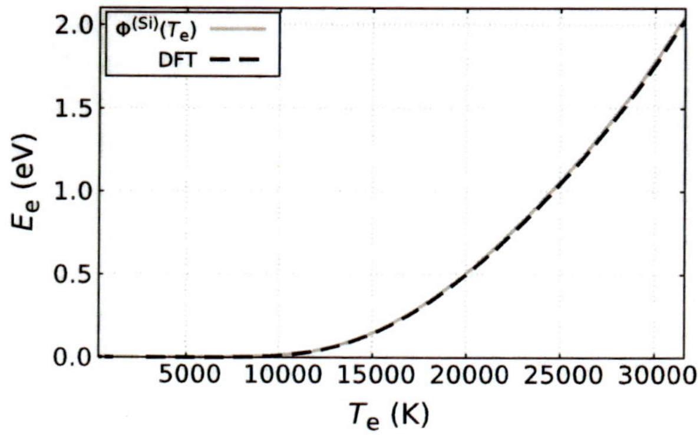

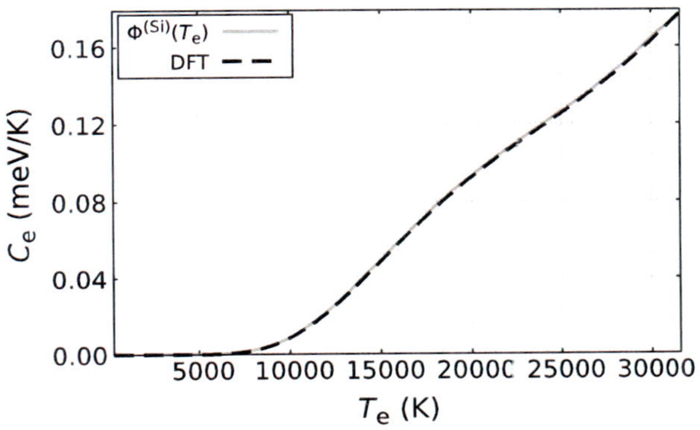

Fig. A.9 Specific heat of the electrons C_e is shown as a function of T_e for an isolated Si atom adapted from Figure S4 of Ref. [4]

Table A.36 Parametrization of the Helmholtz free energy of an isolated atom $\Phi_0(T_e)$, see Eq. (A.37)

k	$a_0^{(k)}$	k	$a_0^{(k)}$	k	$a_0^{(k)}$
0	−102.905307363449	1	−10.3921123486934	4	−1.06389593948069
5	12.4244554652269	6	−183.607940058980	7	816.889082583121
8	−1946.46133504029	9	2899.84763931852	10	−2817.24323376719
11	1750.31164692968	12	−635.201199559016	13	102.801048351004

Table A.37 Parametrization of $c_\rho^{(q_1\,q_2)} = \sum_{k=0}^{5} a_\rho^{(k\,q_1\,q_2)} \left(\frac{T_e}{31577\,\mathrm{K}}\right)^k$ for the embedding function Φ_ρ, see Eq. (6.105)

k	q_1	q_2	$a_\rho^{(k\,q_1\,q_2)}$	k	q_1	q_2	$a_\rho^{(k\,q_1\,q_2)}$
0	2	1	−14.1226298367901	0	2	2	15.1973595898918
1	2	1	42.8178754251152	1	2	2	−76.2426696199033
2	2	1	−46.9725273143709	2	2	2	180.466580679766
3	2	1	−7.49328100982082	3	2	2	−227.822610054901
4	2	1	50.9341743266501	4	2	2	143.162975148651
5	2	1	−26.0690457870547	5	2	2	−34.7706099024473

Table A.38 Parametrization of $c_2^{(q)} = \sum_{k=0}^{5} a_2^{(kq)} \left(\frac{T_e}{31577\,\mathrm{K}}\right)^k$ for the two-body term Φ_2, see Eq. (6.100)

k	q	$a_2^{(kq)}$	k	q	$a_2^{(kq)}$	k	q	$a_2^{(kq)}$
0	2	−0.652029301662623	0	3	−19.0589296869611	0	4	546.769257376815
1	2	9.26314534875668	1	3	−400.089680112155	1	4	6319.96355713893
2	2	−50.0186051618631	2	3	2167.05447438595	2	4	−35385.6550126843
3	2	115.557349548295	3	3	−4520.34437699249	3	4	71794.8635849514
4	2	−114.529669723874	4	3	4095.42902709553	4	4	−62858.4678911266
5	2	40.7054174379305	5	3	−1352.35365149216	5	4	20099.4017864489
0	5	−5494.99348657106	0	6	27386.1918081797	0	7	−75861.4334859187
1	5	−48814.0360404613	1	6	210387.578621541	1	7	−529386.428965916
2	5	283940.662885404	2	6	−1261022.76787976	2	7	3245873.44871437
3	5	−573728.968470894	3	6	2555763.80515297	3	7	−6616167.17139263
4	5	496377.262929330	4	6	−2204123.17465255	4	7	5709449.19416504
5	5	−156654.734823045	5	6	692217.214417601	5	7	−1791097.75814672
0	8	119482.732077564	0	9	−101003.484900608	0	10	35858.0271585324
1	8	775653.843416637	1	9	−614687.832893782	1	10	203955.905899671
2	8	−4831404.94060070	2	9	3865581.63043600	2	10	−1288401.57956179
3	8	9911649.99101982	3	9	−7978309.71558179	3	10	2672708.20259957
4	8	−8574228.73882666	4	9	6923565.00773351	4	10	−2326924.03159739
5	8	2692171.33820813	5	9	−2178041.48087368	5	10	733755.194711132

Table A.39 Parametrization of $c_3^{(q_1\,q_2\,q_3)} = \sum_{k=0}^{5} a_3^{(k\,q_1\,q_2\,q_3)} \left(\frac{T_e}{31577\,\mathrm{K}}\right)^k$ for the three-body term Φ_3, see Eq. (6.101)

k	q_1	q_2	q_3	$a_3^{(k\,q_1\,q_2\,q_3)}$	k	q_1	q_2	q_3	$a_3^{(k\,q_1\,q_2\,q_3)}$
0	2	2	0	−0.484742122936188	0	2	3	0	10.9087150193338
1	2	2	0	−9.06449732747187	1	2	3	0	32.6476593724955
2	2	2	0	41.9525344564407	2	2	3	0	−234.821794328009
3	2	2	0	−73.0549723173192	3	2	3	0	445.128565939339
4	2	2	0	62.1731314312558	4	2	3	0	−384.590065348177
5	2	2	0	−20.9735149728352	5	2	3	0	128.088250726806
0	3	3	0	−29.0228451088965	0	2	2	1	22.1886312645380
1	3	3	0	−20.4131309827352	1	2	2	1	−35.1166767655775
2	3	3	0	338.473408313579	2	2	2	1	−105.001706438450
3	3	3	0	−686.718447255702	3	2	2	1	325.429922575188
4	3	3	0	593.638243677844	4	2	2	1	−302.151293271689
5	3	3	0	−193.368789264996	5	2	2	1	96.1147759289952
0	2	3	1	−120.252372821342	0	3	3	1	168.069490773212
1	2	3	1	159.047356694074	1	3	3	1	−161.910998601428
2	2	3	1	749.046048760271	2	3	3	1	−1315.34984931164
3	2	3	1	−2216.09205938898	3	3	3	1	3702.73203746780
4	2	3	1	2126.62367753068	4	3	3	1	−3605.91631090411
5	2	3	1	−707.455756431839	5	3	3	1	1228.41451611926

(continued)

Table A.39 (continued)

k	q_1	q_2	q_3	$a_3^{(k\,q_1\,q_2\,q_3)}$	k	q_1	q_2	q_3	$a_3^{(k\,q_1\,q_2\,q_3)}$
0	2	2	2	−19.4315027093624	0	2	3	2	96.3144087590581
1	2	2	2	42.1788604704338	1	2	3	2	−173.588114464854
2	2	2	2	9.67634296126227	2	2	3	2	−157.818542202707
3	2	2	2	−141.700074704502	3	2	3	2	751.291373937404
4	2	2	2	157.523635751903	4	2	3	2	−702.941624093910
5	2	2	2	−54.5952459846017	5	2	3	2	212.581414043121
0	3	3	2	−77.3561827130667	0	2	2	3	−65.4726886857960
1	3	3	2	156.577143486935	1	2	2	3	102.053920198204
2	3	3	2	314.860740984096	2	2	2	3	425.612227408026
3	3	3	2	−1179.99411420945	3	2	2	3	−1403.05615543312
4	3	3	2	1097.63560691297	4	2	2	3	1437.50728878185
5	3	3	2	−335.520618998547	5	2	2	3	−504.993626721485
0	2	3	3	336.447897505456	0	3	3	3	−420.786323757512
1	2	3	3	−507.178567927645	1	3	3	3	602.981897592014
2	2	3	3	−2093.07884553079	2	3	3	3	2804.76991700882
3	2	3	3	6834.03239451122	3	3	3	3	−8941.73384478093
4	2	3	3	−6988.05320202182	4	3	3	3	9095.82514910275
5	2	3	3	2453.90944239108	5	3	3	3	−3187.19934480896

A.8.1 Modified Coefficients for the T_m-Corrected Interatomic Potential

Now we tabulate the modified coefficients of the two-body term Φ_2 and the three-body term Φ_3 to correct the melting temperature T_{m} (Tables A.40 and A.41).

A.9 Coefficients of the Polynomial Interatomic Potential $\Phi^{(\mathrm{Sb})}(T_{\mathrm{e}})$ for Sb

Our final interatomic potential $\Phi^{(\mathrm{Sb})}(T_{\mathrm{e}})$ for Sb has the polynomial degrees

$$N_2^{(r)} = 11, \qquad N_3^{(r)} = 4, \qquad N_3^{(\theta)} = 3 \qquad N_\rho^{(r)} = 3, \qquad N_\rho^{(\rho)} = 2. \tag{A.39}$$

It needs in total 38 coefficients. The two-body term Φ_2 has 10, the three-body term Φ_3 has 24, and the embedding function Φ_ρ has 4 coefficients. Furthermore, we use the constant cutoff radii

$$r_2^{(\mathrm{c})} = 0.71\ \mathrm{nm}, \qquad r_3^{(\mathrm{c})} = 0.44\ \mathrm{nm}, \qquad r_\rho^{(\mathrm{c})} = 0.56\ \mathrm{nm}. \tag{A.40}$$

Analogously to the interatomic potential for Si, we fitted the Helmholtz free energy $\Phi_0(T_{\mathrm{e}})$ of an isolated atom to a polynomial of degree 1 at $T_{\mathrm{e}} \leq 4900$ K and to a polynomial of degree 13 at $T_{\mathrm{e}} > 4900$ K:

$$\Phi_0(T_{\mathrm{e}}) = \begin{cases} a_0^{(0)} + a_0^{(1)} \left(\frac{T_{\mathrm{e}}-4900\,\mathrm{K}}{30000\,\mathrm{K}}\right), & T_{\mathrm{e}} \leq 4900\,\mathrm{K}; \\ a_0^{(0)} + a_0^{(1)} \left(\frac{T_{\mathrm{e}}-4900\,\mathrm{K}}{30000\,\mathrm{K}}\right) + \sum\limits_{k=4}^{13} a_0^{(k)} \left(\frac{T_{\mathrm{e}}-4900\,\mathrm{K}}{30000\,\mathrm{K}}\right)^k, & T_{\mathrm{e}} > 4900\,\mathrm{K}. \end{cases} \tag{A.41}$$

We present the electronic absorbed energy as a function of T_{e} in Fig. A.10 and the electronic specific heat as a function of T_{e} in Fig. A.11. One can clearly see that the interatomic potential reproduces the DFT values.

We fitted the T_{e}-dependence of the remaining coefficients to a polynomial of degree 5. Consequently, any coefficient c of the interatomic potential depends on T_{e} as

$$c = \sum_{k=0}^{5} a^{(k)} \left(\frac{T_{\mathrm{e}}}{30000\,\mathrm{K}}\right)^k. \tag{A.42}$$

The parametrization of $\Phi^{(\mathrm{Sb})}(T_{\mathrm{e}})$ is tabulated in Tables A.42, A.43, A.44 and A.45. The unit of the coefficients is eV and the unit of T_{e} is K. The fitted range of the polynomial expansion yields $300\ \mathrm{K} \leq T_{\mathrm{e}} \leq 30000$ K.

Table A.40 Parametrization of the modified coefficients $c_2^{(q)} = \sum_{k=0}^{5} a_2^{(k\,q)} \left(\frac{T_e}{31577\,\mathrm{K}}\right)^k$ for the two-body term Φ_2, see Eq. (6.100)

k	q	$a_2^{(k\,q)}$	k	q	$a_2^{(k\,q)}$	k	q	$a_2^{(k\,q)}$
0	8	119842.7335425191	0	9	−102512.0022381539	0	10	37289.02112395763
1	8	772629.7133831963	1	9	−602015.7939424901	1	10	191935.0886259170
2	8	−4822352.035766482	2	9	3827647.162865941	2	10	−1252416.581031599
3	8	9899400.748726800	3	9	−7926981.606755933	3	10	2624017.867952236
4	8	−8566713.640416473	4	9	6892074.424087163	4	10	−2297051.762950559
5	8	2690518.369643806	5	9	−2171115.032010596	5	10	727184.6991137798

Table A.41 Parametrization of the modified coefficients $c_3^{(q_1\,q_2\,q_3)} = \sum_{k=0}^{5} a_3^{(k\,q_1\,q_2\,q_3)} \left(\frac{T_e}{31577\,\mathrm{K}}\right)^k$ for the three-body term Φ_3, see Eq. (6.101)

k	q_1	q_2	q_3	$a_3^{(k\,q_1\,q_2\,q_3)}$	k	q_1	q_2	q_3	$a_3^{(k\,q_1\,q_2\,q_3)}$
0	3	3	0	−26.51412956990250	0	3	3	1	183.1217840071796
1	3	3	0	−41.48716191137632	1	3	3	1	−288.3551841733145
2	3	3	0	401.5597155688950	2	3	3	1	−936.8320057795876
3	3	3	0	−772.0788345667706	3	3	3	1	3190.569713601079
4	3	3	0	646.0081532301424	4	3	3	1	−3291.696853590043
5	3	3	0	−204.8877088427139	5	3	3	1	1159.300998652859
0	3	3	2	−54.77774286211852					
1	3	3	2	−33.08913487085887					
2	3	3	2	882.6375062820463					
3	3	3	2	−1948.237600009282					
4	3	3	2	1568.964792883849					
5	3	3	2	−439.1908951980764					

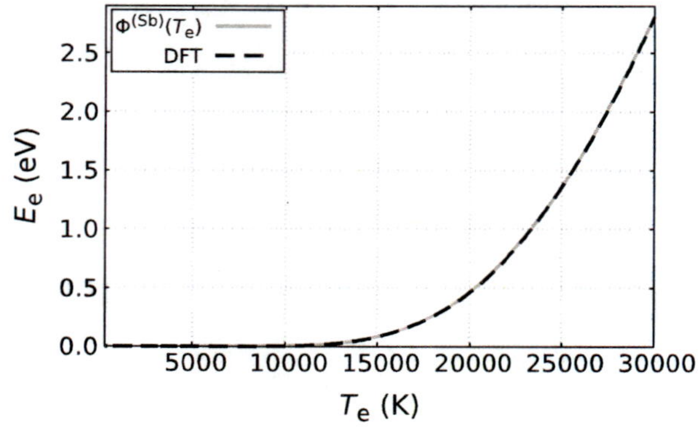

Fig. A.10 Energy absorbed by the electrons E_e is shown as a function of T_e for an isolated Sb atom

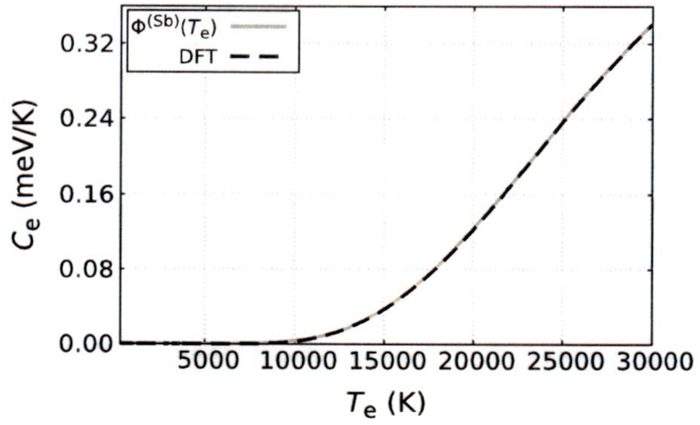

Fig. A.11 Specific heat of the electrons C_e is shown as a function of T_e for an isolated Sb atom

Table A.42 Parametrization of the Helmholtz free energy of an isolated atom $\Phi_0(T_e)$, see Eq. (A.41)

k	$a_0^{(k)}$	k	$a_0^{(k)}$	k	$a_0^{(k)}$
0	−149.5595703655766	1	−10.75140477862161	4	−0.1895026540743273
5	1.986986331162235	6	−22.41830172046881	7	14.72127712319645
8	122.9794157211983	9	−373.1907105264324	10	519.8291103122293
11	−408.4534308753481	12	175.4307182770214	13	−32.20413349991487

Table A.43 Parametrization of $c_\rho^{(q_1\,q_2)} = \sum_{k=0}^{5} a_\rho^{(k\,q_1\,q_2)} \left(\frac{T_e}{30000\,\mathrm{K}}\right)^k$ for the embedding function Φ_ρ, see Eq. (6.105)

k	q_1	q_2	$a_\rho^{(k\,q_1\,q_2)}$	k	q_1	q_2	$a_\rho^{(k\,q_1\,q_2)}$
0	2	1	−0.6940784544756305	0	2	2	13.04334650281262
1	2	1	63.95004404021758	1	2	2	−65.52313228985264
2	2	1	−363.0476414404053	2	2	2	94.83011134817697
3	2	1	771.8760651681542	3	2	2	23.22590585783060
4	2	1	−717.9698926040879	4	2	2	−139.7992773969778
5	2	1	245.6969379054431	5	2	2	75.09374450272083
0	3	1	−20.56349411764695	0	3	2	33.17112873652015
1	3	1	−32.79371522388591	1	3	2	85.32121518663481
2	3	1	544.5897813040112	2	3	2	−1048.284865077498
3	3	1	−1371.807630796040	3	3	2	2546.300710018568
4	3	1	1365.188190273409	4	3	2	−2484.289890653045
5	3	1	−483.6536758533985	5	3	2	868.7492617450540

Table A.44 Parametrization of $c_2^{(q)} = \sum_{k=0}^{5} a_2^{(k\,q)} \left(\frac{T_e}{30000\,\mathrm{K}}\right)^k$ for the two-body term Φ_2, see Eq. (6.100)

k	q	$a_2^{(k\,q)}$	k	q	$a_2^{(k\,q)}$	k	q	$a_2^{(k\,q)}$
0	2	−8.636232409445467	0	3	235.8653388866700	0	4	−3908.513785120673
1	2	6.599535085642408	1	3	−475.6803504615891	1	4	7662.724068694798
2	2	139.7737392464568	2	3	−1768.858138316761	2	4	29795.58076034468
3	2	−450.7520648409792	3	3	7269.526751400568	3	4	−118285.2210509798
4	2	501.1698835732757	4	3	−8554.975829435855	4	4	137092.5992154802
5	2	−189.1530631727320	5	3	3316.868722109235	5	4	−52719.52065292928
0	5	36270.64118170563	0	6	−196369.6673124892	0	7	644386.9810887519
1	5	−58465.06255414510	1	6	261759.4586325263	1	7	−745539.2051933097
2	5	−364575.8658408646	2	6	2375556.150462489	2	7	−8695413.657471943
3	5	1296266.205864705	3	6	−7967617.355495239	3	7	28376072.97962829
4	5	−1459908.275813274	4	6	8832888.034196274	4	7	−31226081.17024808
5	5	553778.2334917767	5	6	−3324615.761720433	5	7	11708188.89793409
0	8	−1303762.294794598	0	9	1594614.647571070	0	10	−1085193.674956167
1	8	1372968.561266020	1	9	−1581878.851861374	1	10	1032066.212524234
2	8	18810102.74542142	2	9	−24004849.19989146	2	10	16775670.00975985
3	8	−60644250.61551363	3	9	77068884.25131882	3	10	−53842222.50179398
4	8	66544105.81460975	4	9	−84536181.50433896	4	10	59111933.23930446
5	8	−24909482.20003287	5	9	31631869.32716605	5	10	−22123955.62084943
0	11	316873.3617459274						
1	11	−289381.4914016659						
2	11	−4967551.121869861						
3	11	15959903.16678988						
4	11	−17544597.69079288						
5	11	6569367.609977663						

Table A.45 Parametrization of $c_3^{(q_1\,q_2\,q_3)} = \sum_{k=0}^{5} a_3^{(k\,q_1\,q_2\,q_3)} \left(\frac{T_e}{30000\,\mathrm{K}}\right)^k$ for the three-body term Φ_3, see Eq. (6.101)

k	q_1	q_2	q_3	$a_3^{(k\,q_1\,q_2\,q_3)}$	k	q_1	q_2	q_3	$a_3^{(k\,q_1\,q_2\,q_3)}$
0	2	2	0	−43.56722574434935	0	2	3	0	451.2621956005495
1	2	2	0	26.26513336555701	1	2	3	0	−1059.064330790318
2	2	2	0	811.0670595280522	2	2	3	0	−4107.531676198178
3	2	2	0	−2465.815431356520	3	2	3	0	16466.09913153586
4	2	2	0	2692.685674061864	4	2	3	0	−19479.68398794441
5	2	2	0	−1014.014958470357	5	2	3	0	7631.235451543319
0	3	3	0	−1456.525647081006	0	2	4	0	−601.4431049904534
1	3	3	0	5230.719482343495	1	2	4	0	2469.879156790622
2	3	3	0	2202.761112280622	2	2	4	0	−230.8736311860961
3	3	3	0	−28597.33581690267	3	2	4	0	−9889.728140215620
4	3	3	0	39254.06686113183	4	2	4	0	14813.88035250671
5	3	3	0	−16319.25657937852	5	2	4	0	−6375.457854799374
0	3	4	0	4221.820654904389	0	4	4	0	−3106.422164401547
1	3	4	0	−20275.47840092985	1	4	4	0	18091.53148683341
2	3	4	0	23953.41955403770	2	4	4	0	−37141.70377659073
3	3	4	0	16654.17659384057	3	4	4	0	30323.53149595635
4	3	4	0	−51580.12940907154	4	4	4	0	−1587.044369538699
5	3	4	0	25968.79936469788	5	4	4	0	−5730.126877752481

(continued)

Table A.45 (continued)

k	q_1	q_2	q_3	$a_3^{(k\ q_1\ q_2\ q_3)}$	k	q_1	q_2	q_3	$a_3^{(k\ q_1\ q_2\ q_3)}$
0	2	2	1	−100.5004628448294	0	2	3	1	665.0452466965809
1	2	2	1	−302.9688055096382	1	2	3	1	3384.734386286354
2	2	2	1	5236.111732768285	2	2	3	1	−45470.25053395272
3	2	2	1	−14846.16642476725	3	2	3	1	125606.0420194058
4	2	2	1	15951.55154574616	4	2	3	1	−133572.1364700016
5	2	2	1	−5953.664132425367	5	2	3	1	49592.78895656631
0	3	3	1	−365.7340198537083	0	2	4	1	−357.7639861928289
1	3	3	1	−8251.206714060945	1	2	4	1	−4462.846423724260
2	3	3	1	77039.32681745353	2	2	4	1	47664.03223757816
3	3	3	1	−201668.0704446783	3	2	4	1	−128022.7961894182
4	3	3	1	209181.9488790141	4	2	4	1	134444.0623741686
5	3	3	1	76523.62748560121	5	2	4	1	−49557.47906700610
0	3	4	1	−2809.927843269936	0	4	4	1	4934.205364376948
1	3	4	1	18500.45710481897	1	4	4	1	−7708.808549183865
2	3	4	1	−88337.22785907785	2	4	4	1	−50187.24660150354
3	3	4	1	194019.0185089058	3	4	4	1	185757.0914024885
4	3	4	1	−181796.0965800533	4	4	4	1	−221571.4088202129
5	3	4	1	62004.83632346498	5	4	4	1	87792.77321886060

(continued)

Table A.45 (continued)

k	q_1	q_2	q_3	$a_3^{(k\,q_1\,q_2\,q_3)}$	k	q_1	q_2	q_3	$a_3^{(k\,q_1\,q_2\,q_3)}$
0	2	2	2	83.14489657207572	0	2	3	2	−446.0763537291387
1	2	2	2	266.8716668144791	1	2	3	2	−1866.987870015521
2	2	2	2	−4001.448576326827	2	2	3	2	28797.27242669591
3	2	2	2	10796.84055704430	3	2	3	2	−79633.61004395256
4	2	2	2	−11283.53985758435	4	2	3	2	83629.67742571829
5	2	2	2	4139.934790966999	5	2	3	2	−30657.86665296226
0	3	3	2	646.3904424118170	0	2	4	2	−14.89814649967197
1	3	3	2	3863.394089905582	1	2	4	2	1291.356966496785
2	3	3	2	−58101.17805221484	2	2	4	2	−19805.91422485548
3	3	3	2	162369.1138062980	3	2	4	2	57959.89956745392
4	3	3	2	−167845.3190512352	4	2	4	2	−61733.39999632576
5	3	3	2	60298.38245984744	5	2	4	2	22670.09580873388
0	3	4	2	2184.287032643392	0	4	4	2	−5284.367220680520
1	3	4	2	−11331.04540536191	1	4	4	2	14898.27394724923
2	3	4	2	86944.58851600113	2	4	4	2	−25281.59259903224
3	3	4	2	−238579.9598465826	3	4	4	2	41056.76497880089
4	3	4	2	236866.4498545296	4	4	4	2	−19946.99685704659
5	3	4	2	−81182.85294733283	5	4	4	2	−45.70932333963542

(continued)

Table A.45 (continued)

k	q_1	q_2	q_3	$a_3^{(k\,q_1\,q_2\,q_3)}$	k	q_1	q_2	q_3	$a_3^{(k\,q_1\,q_2\,q_3)}$
0	2	2	3	−35.73492793184669	0	2	3	3	1891.192488689694
1	2	2	3	1347.762428437422	1	2	3	3	−16878.19954720905
2	2	2	3	−8291.696381722345	2	2	3	3	72834.56663416295
3	2	2	3	18951.50724867393	3	2	3	3	−146856.8693140604
4	2	2	3	−18482.45573815029	4	2	3	3	133950.6057943326
5	2	2	3	6512.183264669749	5	2	3	3	−45228.96335331312
0	3	3	3	−10269.91148234739	0	2	4	3	−4608.349981641225
1	3	3	3	55470.73283920626	1	2	4	3	26226.22545043905
2	3	3	3	−136557.7955312003	2	2	4	3	−77463.72831826756
3	3	3	3	171486.8497751762	3	2	4	3	122931.7143061951
4	3	3	3	−97631.79888949100	4	2	4	3	−94590.38277108674
5	3	3	3	19095.43166730884	5	2	4	3	28011.03320976359
0	3	4	3	42385.01951288608	0	4	4	3	−41403.60440679067
1	3	4	3	−179247.5212736523	1	4	4	3	148289.2328877183
2	3	4	3	219534.8748500119	2	4	4	3	−15248.74753666111
3	3	4	3	116178.7831488915	3	4	4	3	−604780.5203022564
4	3	4	3	−420560.3535236654	4	4	4	3	910906.1076217494
5	3	4	3	215950.4218521583	5	4	4	3	−392251.6785638268

Table A.46 Parametrization of the electronic specific heat $C_e(T_e)$ using Eq. (8.5). The unit of $a_{C_e}^{(k)}$ is $\frac{\text{eV}}{\text{K atom}}$

k	$a_{C_e}^{(k)}$	k	$a_{C_e}^{(k)}$	k	$a_{C_e}^{(k)}$
0	5.796188877858925E-7	1	−1.3406454313227224E-4	2	0.011380113652383794
3	−0.10502307663263998	4	0.4574830096277418	5	0.4392825678210371
6	−16.07123735599759	7	90.81303822923645	8	−285.0919458280668
9	574.4851206439896	10	−771.7423048691147	11	688.6220099565475
12	−392.306879234134	13	129.21793071539435	14	−18.728363417323045

Table A.47 Parametrization of the reflectivity $\mathfrak{R}(z)$ using Eq. (8.2) for a wavelength of $\lambda = 534$ nm

k	$a_{\mathfrak{R}}^{(k)}$	k	$a_{\mathfrak{R}}^{(k)}$	k	$a_{\mathfrak{R}}^{(k)}$
0	−5869.329652632475	1	98451.9557493141	2	−619519.2574155919
3	1733275.2442907325	4	−1818924.9154050334		

Table A.48 Parametrization of the absorption coefficient $\alpha_{\text{abs}}(z)$ using Eq. (8.3) for a wavelength of $\lambda = 534$ nm. The unit of $a_{\alpha_{\text{abs}}}^{(k)}$ is $\frac{1}{\text{nm}}$

k	$a_{\alpha_{\text{abs}}}^{(k)}$	k	$a_{\alpha_{\text{abs}}}^{(k)}$	k	$a_{\alpha_{\text{abs}}}^{(k)}$
0	406.9512436659842	1	−6743.246751969342	2	41903.54984013184
3	−115732.00436903365	4	119859.68735985583		

A.10 Electronic Specific Heat of Sb

See Table A.46.

A.11 Optical Properties of Sb as a Function of the Peierls Parameter

See Tables A.47 and A.48.

A.12 Electron-Phonon Coupling Constant of Sb

See Table A.49.

Table A.49 Parametrization of the electron-phonon coupling constant $G_{ep}(T_e)$ to all phonon modes using Eq. (8.4). The unit of $a^{(k)}_{G_{ep}}$ is $\frac{eV}{fs\ K\ atom}$

k	$a^{(k)}_{G_{ep}}$	k	$a^{(k)}_{G_{ep}}$	k	$a^{(k)}_{G_{ep}}$
0	1.0583551816882315E-9	1	1.9958090566831475E-8	2	−1.0282388146176964E-6
3	3.0937319055901753E-5	4	−2.479278693221162E-4	5	1.0629369362957026E-3
6	−2.834689393441191E-3	7	4.926458956141932E-3	8	−5.585915659693175E-3
9	3.9830048381366764E-3	10	−1.6204165265759753E-3	11	2.866976969711105E-3

Table A.50 Gaussian basis set exponents in a_0^{-2} implemented in CHIVES for the elements studied in this book. The angular momenta used for each exponent are denoted in brackets

B	N	Si	Sb
2.72048834871290880 (*sp*)	7.12204011794196660 (*sp*)	1.17769 (*sp*)	1.73466 (*s*)
0.76193178360145564 (*spd*)	2.45549630047134480 (*sp*)	0.40348 (*spd*)	1.17728 (*sp*)
0.24651177349075326 (*sp*)	0.83772538808000141 (*spd*)	0.12989 (*sp*)	0.24788 (*spd*)
	0.30000368689110074 (*sp*)		0.06955 (*sp*)
	0.10580016930105678 (*sp*)		

A.13 Gaussian Basis Sets Used in CHIVES

See Table A.50.

References

1. H.R. Schwarz, N. Koeckler, *Numerische Mathematik*, 6th edn. (Teubner, 2006)
2. A.P. Thompson, S.J. Plimpton, W. Mattson, J. Chem. Phys. **131**(15), 154107 (2009). https://doi.org/10.1063/1.3245303
3. B. Bauerhenne, M.E. Garcia, Eur. Phys. J. Spec. Top. **227**(14), 1615 (2019). https://doi.org/10.1140/epjst/e2019-800181-3
4. B. Bauerhenne, V.P. Lipp, T. Zier, E.S. Zijlstra, M.E. Garcia, Phys. Rev. Lett. **124**, 085501 (2020). https://doi.org/10.1103/PhysRevLett.124.085501. https://link.aps.org/doi/10.1103/PhysRevLett.124.085501

Index

A

B

© The Editor(s) (if applicable) and The Author(s), under exclusive license to Springer Nature Switzerland AG 2021
B. Bauerhenne, *Materials Interaction with Femtosecond Lasers*,
https://doi.org/10.1007/978-3-030-85135-4

F

G

M

N

O

P

Q

R

S

T